Handbook of Mixed Membership Models and Their Applications

Chapman & Hall/CRC
Handbooks of Modern Statistical Methods

Series Editor

Garrett Fitzmaurice

Department of Biostatistics
Harvard School of Public Health
Boston, MA, U.S.A.

Aims and Scope

The objective of the series is to provide high-quality volumes covering the state-of-the-art in the theory and applications of statistical methodology. The books in the series are thoroughly edited and present comprehensive, coherent, and unified summaries of specific methodological topics from statistics. The chapters are written by the leading researchers in the field, and present a good balance of theory and application through a synthesis of the key methodological developments and examples and case studies using real data.

The scope of the series is wide, covering topics of statistical methodology that are well developed and find application in a range of scientific disciplines. The volumes are primarily of interest to researchers and graduate students from statistics and biostatistics, but also appeal to scientists from fields where the methodology is applied to real problems, including medical research, epidemiology and public health, engineering, biological science, environmental science, and the social sciences.

Published Titles

Handbook of Mixed Membership Models and Their Applications
Edited by Edoardo M. Airoldi, David M. Blei,
Elena A. Erosheva, and Stephen E. Fienberg

Handbook of Markov Chain Monte Carlo
Edited by Steve Brooks, Andrew Gelman,
Galin L. Jones, and Xiao-Li Meng

Longitudinal Data Analysis
Edited by Garrett Fitzmaurice, Marie Davidian,
Geert Verbeke, and Geert Molenberghs

Handbook of Spatial Statistics
Edited by Alan E. Gelfand, Peter J. Diggle,
Montserrat Fuentes, and Peter Guttorp

Handbook of Survival Analysis
Edited by John P. Klein, Hans C. van Houwelingen,
Joseph G. Ibrahim, and Thomas H. Scheike

Handbook of Missing Data Methodology
Edited by Geert Molenberghs, Garrett Fitzmaurice,
Michael G. Kenward, Anastasios Tsiatis, and Geert Verbeke

Chapman & Hall/CRC

Handbooks of Modern Statistical Methods

Handbook of Mixed Membership Models and Their Applications

Edited by

Edoardo M. Airoldi
Harvard University
Cambridge, Massachusetts, USA

David M. Blei
Columbia University
New York, New York, USA

Elena A. Erosheva
University of Washington
Seattle, Washington, USA

Stephen E. Fienberg
Carnegie Mellon University
Pittsburgh, Pennsylvania, USA

CRC Press
Taylor & Francis Group
Boca Raton London New York

CRC Press is an imprint of the
Taylor & Francis Group, an **informa** business
A CHAPMAN & HALL BOOK

CRC Press
Taylor & Francis Group
6000 Broken Sound Parkway NW, Suite 300
Boca Raton, FL 33487-2742

First issued in paperback 2021

© 2015 by Taylor & Francis Group, LLC
CRC Press is an imprint of Taylor & Francis Group, an Informa business

No claim to original U.S. Government works

ISBN-13: 978-1-4665-0408-0 (hbk)
ISBN-13: 978-0-367-33084-2 (pbk)

Publisher's Note

The publisher has gone to great lengths to ensure the quality of this reprint but points out that some imperfections in the original copies may be apparent.

Visit the Taylor & Francis Web site at
http://www.taylorandfrancis.com

and the CRC Press Web site at
http://www.crcpress.com

Contents

Preface

This volume is, in a sense, the culmination of over 20 years of statistical work and over 15 years of personal interactions. One of us, Fienberg, was exposed to the ideas of the Grade of Membership (GoM) model in a workshop on disability forecasting (National Research Council, 1994). The GoM work was done in the context of the National Long Term Care Survey, and the modeling ideas were likelihood-based and quite opaque to Fienberg at the time. Several years later Fienberg introduced Erosheva to the literature on the GoM model. A Bayesian approach to the GoM model, in the context of other latent structure models, became the main focus of Erosheva's dissertation research at Carnegie Mellon University (Erosheva, 2002). Independently, a third editor of this volume, Blei, began his Ph.D. research with his advisor Michael Jordan at the University of California at Berkeley on topic modeling, developing a method referred to as latent Dirichlet allocation (Blei et al., 2003), which resembles a Bayesian GoM. Around that time, conversations with Tom Minka prompted Erosheva to look deeper into that resemblance. Shortly afterwards, inspired by Matthew Stephens's pointer to the admixture model from genetics—another model that resembles a Bayesian GoM—Erosheva formulated the more general mixed membership framework, encompassing all three approaches, and provided ways to construct mixed membership models for other data structures (Erosheva, 2003). Blei and Erosheva met only once, for coffee at the 2003 Joint Statistical Meetings in San Francisco.

After completing his dissertation, Blei came to Carnegie Mellon as a postdoctoral fellow and began to collaborate on network modeling with Fienberg and Airoldi, and another colleague in the Machine Learning Department. This work culminated in the mixed membership stochastic block-model (Airoldi et al., 2008), and was also at the heart of Airoldi's Ph.D. thesis (Airoldi, 2006). In light of these and many subsequent interactions and joint work, it was natural for the four of us to collaborate on the present volume.

We have many people to thank for the completion of the present volume. First, we thank the authors of the chapters, many of whom are our friends, students, collaborators, and professional colleagues. They have contributed first-rate research to the volume, and the following pages are evidence of their great efforts and intellect. Second, we thank John Kimmel of Chapman & Hall, who gave us constant encouragement. Finally, and most of all, we thank Kira Bokalders. Kira is the real editor of this collection. She organized the effort, converted documents from varying formats, copy-edited every contribution, constructed the indexes, and guided us through by preparing the final camera-ready manuscript. Editors are listed in lexicographical order of their last names.

Edoardo M. Airoldi, Cambridge, MA
David M. Blei, New York City, NY
Elena A. Erosheva, Seattle, WA
Stephen E. Fienberg, Pittsburgh, PA

July 4, 2014.

References

Airoldi, E. M., Blei, D. M., Fienberg, S. E., and Xing, E. P. (2008). Mixed membership stochastic blockmodels. *Journal of Machine Learning Research* 9: 1981–2014.

Airoldi, E. M. (2006). Bayesian Mixed Membership Models of Complex and Evolving Networks. Ph.D. thesis, School of Computer Science, Carnegie Mellon University, Pittsburgh, Pennsylvania, USA.

Blei, D. M., Ng, A. Y., and Jordan, M. I. (2003). Latent Dirichlet allocation. *Journal of Machine Learning Research* 3: 993–1022.

Committee on National Statistics, National Research Council (1994). *Trends in Disability at Older Ages: Summary of a Workshop*. Washington, DC: The National Academies Press.

Erosheva, E. A. (2002). Grade of Membership and Latent Structure Models with Application to Disability Survey Data. Ph.D. thesis, Department of Statistics, Carnegie Mellon University, Pittsburgh, Pennsylvania, USA.

Erosheva, E. A. (2003). Bayesian estimation of the Grade of Membership model. In Bernardo, J. M., Bayarri, M. J., Berger, J. O., Dawid, A. P., Heckerman, D., Smith, A. F. M., and West, M. (eds), *Bayesian Statistics 7*. New York, NY: Oxford University Press, 501–510.

Editors

Edoardo M. Airoldi is Associate Professor of Statistics at Harvard University, where he leads the Harvard Laboratory for Applied Statistical Methodology. He holds a Ph.D. in Computer Science and an M.Sc. in Statistics from Carnegie Mellon University, and a B.Sc. in Mathematical Statistics and Economics from Bocconi University. His current research focuses on statistical theory and methods for designing and analyzing experiments in the presence of network interference, and on inferential issues that arise in models of network data. He works on applications in cellular biology and proteomics, and in social media analytics and marketing. Airoldi is the recipient of several research awards including the ONR Young Investigator Award, the NSF CAREER Award, and the Alfred P. Sloan Research Fellowship, and has received several outstanding paper awards including the Thomas R. Ten Have Award for his work on causal inference, and the John Van Ryzin Award for his work in biology. He has recently advised the Obama 4 America 2012 campaign on their social media efforts, and serves as a technical advisor at Nanigans.

David M. Blei is a Professor of Statistics and Computer Science at Columbia University. His research is in statistical machine learning involving probabilistic topic models, Bayesian nonparametric methods, and approximate posterior inference. He works on a variety of applications, including text, images, music, social networks, user behavior, and scientific data. David has received several awards for his research, including the Presidential Early Career Award for Scientists and Engineers (2011), Blavatnik Faculty Award (2013), and ACM-Infosys Foundation Award (2013).

Elena A. Erosheva is Associate Professor of Statistics and Social Work at the University of Washington and a core member of the Center for Statistics and the Social Sciences. Her research focuses on the development and application of modern statistical methods for diverse and heterogeneous data in the social, medical, and health sciences. Mixed membership models, the focus of her 2002 Ph.D. dissertation in Statistics from Carnegie Mellon University, are one way to provide structured explanations of such data. Erosheva's research draws on substantive disciplinary knowledge about the phenomenon under study to develop statistical models that are able to properly account for observed heterogeneity and are more in tune with overall scientific objectives. Her work includes statistical methodology development for multivariate and longitudinal data analysis, text analysis, survey methodology, and analysis of survey data. She is a recipient of the 2013 Mitchell Prize from the International Society of Bayesian Analysis for an outstanding paper that describes how a Bayesian analysis has solved an important applied problem. Erosheva is currently an Associate Editor of the *Journal of the American Statistical Association* and of the *Annals of Applied Statistics*, and a member of the Editorial Board of the *Journal of Educational and Behavioral Statistics*.

Stephen E. Fienberg is Maurice Falk University Professor of Statistics and Social Science at Carnegie Mellon University and co-director of the Living Analytics Research Centre (LARC), with appointments in the Department of Statistics, the Machine Learning Department, Heinz College, the Center for Human Rights Science, and Cylab. A Carnegie Mellon faculty member since 1980, he has also served as head of the Department of Statistics and Dean of the (Dietrich) College of Humanities and Social Sciences. He is currently Editor-in-Chief of the *Annals of Applied Statistics* and of the online *Journal of Privacy and Confidentiality*, and editor of *Annual Reviews of Statistics* *and its Application.* His research includes the development of statistical methods, especially tools for categorical data analysis and the analysis of network data, and his work on mixed membership spans these topics as well as text analysis. His research on confidentiality and privacy protection addresses issues that are especially relevant in applications of data science as well as in the context of large-scale sample surveys and censuses. Fienberg is the author or editor of over 20 books and 500 papers and related publications, and his two books on categorical data analysis are Citation Classics. He is a member of the U. S. National Academy of Sciences and a fellow of the Royal Society of Canada, the American Academy of Arts and Sciences, and the American Academy of Political and Social Science.

Contributors

Edoardo M. Airoldi
Harvard University
Cambridge, Massachusetts, USA

Christophe Ambroise
Université d'Évry-val-d'Essonne
Évry, France

Ramnath Balasubramanyan
Carnegie Mellon University
Pittsburgh, Pennsylvania, USA

Arindam Banerjee
University of Minnesota
Minneapolis, Minnesota, USA

Etienne Birmelé
Universite Paris Descartes
Paris, France

Jonathan M. Bischof
Harvard University
Cambridge, Massachusetts, USA

David M. Blei
Columbia University
New York, New York, USA

Jordan Boyd-Graber
University of Colorado
Boulder, Colorado, USA

Marcia C. Castro
Harvard School of Public Health
Boston, Massachusetts, USA

Yoon-Sik Cho
University of Southern California
Marina del Rey, California, USA

William Cohen
Carnegie Mellon University
Pittsburgh, Pennsylvania, USA

Elena A. Erosheva
University of Washington
Seattle, Washington, USA

Stephen E. Fienberg
Carnegie Mellon University
Pittsburgh, Pennsylvania, USA

Emily B. Fox
University of Washington
Seattle, Washington, USA

Aram Galstyan
University of Southern California
Marina del Rey, California, USA

April Galyardt
University of Georgia
Athens, Georgia, USA

Zoubin Ghahramani
University of Cambridge
Cambridge, United Kingdom

Isobel Claire Gormley
University College Dublin
Dublin, Ireland

Justin H. Gross
University of North Carolina at Chapel Hill
Chapel Hill, North Carolina, USA

Jonathan Gruhl
University of Washington
Seattle, Washington, USA

Daniel Heinz
Loyola University of Maryland
Baltimore, Maryland, USA

Katherine A. Heller
Duke University
Durham, North Carolina, USA

Qirong Ho
Carnegie Mellon University
Pittsburgh, Pennsylvania, USA

Michael I. Jordan
University of California
Berkeley, California, USA

Cyrille Joutard
Université Montpellier
Montpellier, France

Brian W. Junker
Carnegie Mellon University
Pittsburgh, Pennsylvania, USA

Pierre Latouche
Université Paris
Paris, France

Lester Mackey
University of California
Berkeley, California, USA

Daniel Manrique-Vallier
Indiana University
Bloomington, Indiana, USA

David Mimno
Cornell University
Ithaca, New York, USA

Shakir Mohamed
Google London
London, United Kingdom

Thomas Brendan Murphy
University College Dublin
Dublin, Ireland

David Newman
Google Los Angeles
Venice, California, USA

John Paisley
Columbia University
New York, New York, USA

Adler Perotte
Columbia University
New York, New York, USA

Hanhuai Shan
University of Minnesota
Minneapolis, Minnesota, USA

Suyash Shringarpure
Stanford University
Stanford, California, USA

Burton H. Singer
University of Florida
Gainesville, Florida, USA

Tracy M. Sweet
University of Maryland
College Park, Maryland, USA

Andrew C. Thomas
Carnegie Mellon University
Pittsburgh, Pennsylvania, USA

Greg Ver Steeg
University of Southern California
Marina del Rey, California, USA

David Weiss
University of Pennsylvania
Philadelphia, Pennsylvania, USA

Frank Wood
Oxford University
Oxford, United Kingdom

Eric P. Xing
Carnegie Mellon University
Pittsburgh, Pennsylvania, USA

Jun Zhu
Tsinghua University
Beijing, China

List of Figures

List of Tables

Part I

Mixed Membership: Setting the Stage

1

Introduction to Mixed Membership Models and Methods

Edoardo M. Airoldi

Department of Statistics, Harvard University, Cambridge, MA 02138, USA

David M. Blei

Departments of Statistics and Computer Science, Columbia University, New York, NY 10027, USA

Elena A. Erosheva

Department of Statistics, University of Washington, Seattle, WA 98195, USA

Stephen E. Fienberg

Department of Statistics, Heinz College, and Machine Learning Department, Carnegie Mellon University, Pittsburgh, PA 15213, USA

CONTENTS

Mixed membership models have emerged over the past 20 years as a flexible cluster-like modeling tool for unsupervised analyses of high-dimensional multivariate data where the assumption that an observational unit belongs to a single cluster, or principal component, is violated. Instead, one assumes that every unit partially belongs to all clusters, according to an individual membership vector. Mixed membership models were introduced essentially independently in a number of different statistical application settings: (1) survey data (Berkman et al., 1989; Erosheva, 2002; Erosheva et al., 2007), (2) population genetics (Pritchard et al., 2000b; Rosenberg et al., 2002), (3) text analysis (Blei et al., 2003; Erosheva et al., 2004; Airoldi et al., 2010), and then later on in (4) image processing and annotation (Barnard et al., 2003; Fei-Fei and Perona, 2005), and (5) molecular biology (Segal et al., 2005; Airoldi et al., 2006; 2007; 2013).

1.1 Historical Developments

This volume chronicles recent developments in the area of mixed membership modeling. Mixed membership models are used to characterize complex multivariate data such as those arising in studies of genetic build-up of biological organisms, patterns in disease and disability manifestations, combinations of topics covered by text documents, political ideology or electorate voting patterns, or heterogeneous relationships in networks. Early applications of mixed membership modeling included the admixture model in genetics (Pritchard et al., 2000a), the Grade of Membership model in medical classification studies (Manton et al., 1994b), and the latent Dirichlet allocation model in machine learning (Blei et al., 2003).

In contrast to the finite mixture or parametric clustering models (McLachlan and Peel, 2000), mixed membership models assume that individuals or observational units may only partly belong to population mixture categories, referred to in various fields as *topics*, *extreme profiles*, *pure* or *ideal types*, *states*, or *subpopulations*. The degree of membership then is a vector of continuous non-negative latent variables that add up to 1 (in mixture models, membership is a binary indicator). The original idea for a mixed membership type of modeling goes back to at least the 1970s when the Grade of Membership (GoM) model was developed by mathematician Max Woodbury to allow for "fuzzy" classifications in medical diagnosis problems (Woodbury et al., 1978). The model had not received a lot of attention from statisticians in the early years, and was later characterized by seemingly controversial statements regarding the nature of the compositional data implied by the GoM model (Haberman, 1995). It was not until the early 2000s, with the widespread use of Bayesian methods and a better explanation of the duality between the discrete and continuous nature of latent structure in the GoM model, that a new Bayesian approach to the GoM model had been developed (Erosheva, 2003). The almost simultaneous and independent development of the admixture model in genetics (Pritchard et al., 2000a) and the latent Dirichlet allocation (LDA) model in computer science (Blei et al., 2003) also relied on the use of Bayesian estimation or approximate Bayesian estimation techniques, as in the case of LDA. This class of mixed membership models (Erosheva, 2002) unifies the LDA, GoM, and admixture models in a common framework and provides ways to construct other individual-level mixture models by varying assumptions on the population, sampling unit and latent variable levels, and the sampling scheme.

The word *mixed* in the name *mixed membership* comes from the alternative latent class specification of the models where each attribute is generated according to its distribution in a certain basis category (Erosheva et al., 2007). For example, each word in an article corresponds to a particular topic, whereas the article's composition as a whole corresponds to the author's intention to cover a selection of topics. Thus, the multivariate collection of outcomes for each sampling unit is composed of a mix of attributes that originate from the basis categories, e.g., words within a document that are generated from topics covered by that document. In the case of discrete data, the latent topic indicators for each word do not necessarily have to be the latent variables in the model. An alternative data-generating process that results in the same likelihood can be based on the latent degrees of membership controlling the proportions of attributes originating from each basis category (Erosheva, 2005). For this reason, mixed membership models have been occasionally referred to as partial membership models (e.g., Erosheva, 2004); however, that name has not gained widespread use and the name mixed membership remains the most commonly used descriptor (Erosheva and Fienberg, 2005).

1.2   A General Formulation for Mixed Membership Models

The general mixed membership model relies on four levels of assumptions: population, subject, latent variable, and sampling scheme. Population level assumptions describe the general structure of the population that is common to all subjects. Subject level assumptions specify the distribution of observed responses given individual membership scores. Membership scores are usually unknown and hence can also be viewed as latent variables. The next assumption specifies whether the membership scores are treated as unknown fixed quantities or as random quantities in the model. Finally, the last level of assumptions specifies the number of distinct observed characteristics (attributes) and the number of replications for each characteristic. We describe each set of assumptions formally in turn.

Population Level

Assume there are K original or basis subpopulations in the populations of interest. For each subpopulation k, denote by $f(x_j|\theta_{kj})$ the probability distribution for response variable j, where θ_{kj} is a vector of parameters. Assume that within a subpopulation, responses to observed variables are independent.

Subject Level

For each subject, membership vector $\lambda = (\lambda_1, \ldots, \lambda_K)$ provides the degrees of a subject's membership in each of the subpopulations. The probability distribution of observed responses x_j for each subject is fully defined by the conditional probability $Pr(x_j|\lambda) = \sum_k \lambda_k f(x_j|\theta_{kj})$, and the assumption that response variables x_j are independent, conditional on membership scores. In addition, given the membership scores, observed responses from different subjects are independent.

Latent Variable Level

With respect to the latent variables, one could either assume that they are fixed unknown constants or that they are random realizations from some underlying distribution.

1. If the membership scores λ are fixed but unknown, the conditional probability of observing x_j, given the parameters θ and membership scores, is

$$Pr(x_j|\lambda; \boldsymbol{\theta}) \quad = \quad \sum_{k=1}^{K} \lambda_k f(x_j|\theta_{kj}). \tag{1.1}$$

2. If membership scores λ are realizations of latent variables from some distribution D_α, parameterized by vector α, then the probability of observing x_j given the parameters is:

$$Pr(x_j|\alpha, \boldsymbol{\theta}) \quad = \quad \int \left(\sum_{k=1}^{K} \lambda_k f(x_j|\theta_{kj}) \right) dD_\alpha(\lambda). \tag{1.2}$$

Sampling Scheme

Suppose R independent replications of J distinct characteristics are observed for one subject, $\{x_1^{(r)}, \ldots, x_J^{(r)}\}_{r=1}^{R}$. Then, if the membership scores are treated as realizations from distribution D_α, the conditional probability is

$$Pr\left(\{x_1^{(r)}, \ldots, x_J^{(r)}\}_{r=1}^{R} | \alpha, \boldsymbol{\theta} \right) = \int \left(\prod_{j=1}^{J} \prod_{r=1}^{R} \sum_{k=1}^{K} \lambda_k f(x_j^{(r)}|\theta_{kj}) \right) dD_\alpha(\lambda). \tag{1.3}$$

When the latent variables are treated as unknown constants, the conditional probability for observing R replications of J variables can be derived analogously. In general, the number of observed characteristics J need not be the same across subjects, and the number of replications R need not be the same across observed characteristics.

One can obtain a number of mixed membership models using this general set up by specifying different choices of J and R, and different latent variable assumptions. For instance, the Grade of Membership model of Manton et al. (1994b) assumes polytomous responses are observed to J survey questions without replications and uses the fixed-effects assumption for the membership scores. Potthoff et al. (2000) employ a variation of the Grade of Membership model by treating the membership scores as Dirichlet random variables; the authors refer to the resulting model as a *Dirichlet generalization of latent class models*. In genetics, Pritchard et al. (2000a) use a *clustering model with admixture*, which they labeled as structure. For diploid individuals the clustering model assumes that $R = 2$ replications (genotypes) are observed at J distinct locations (loci), treating the proportions of a subject's genome that originated from each of the basis subpopulations as random Dirichlet realizations. Variations of mixed membership models for text documents called *probabilistic latent semantic analysis* (Hofmann, 2001) and *latent Dirichlet allocation* (Blei et al., 2003) both assume that a single characteristic (word) is observed a number of times for each document, but the former model considers the membership scores as fixed unknown constants, whereas the latter treats them as random Dirichlet realizations.

The mixed membership model framework presented above unifies several specialized models that have been developed independently in the social sciences, genetics, and text mining applications.

1.3   Advantages of Mixed Membership Models in Applied Statistics

Mixed membership models have had a significant impact on applied statistics. Over the past decade, the data that statisticians analyze have become more diverse and structured, and with this complexity comes the opportunity to model individual data points as belonging to multiple groups. Indeed, for many modern datasets—such as large-scale text documents and complex networks—we believe that there is rarely a case for the simpler models. Statisticians need mixed membership models or alternatives to them, and this is the reason to study them.

The main areas to which mixed membership models have been applied are reflected in the contents of this volume.

Document Collections

Mixed membership models are widely applied to document collections (Blei et al., 2003; Blei, 2012). In document collections, the mixed membership assumptions naturally capture the heterogeneity of language, where documents each exhibit multiple themes and to different degree. When modeling documents as data, each document is a collection of words from a vocabulary. (These are grouped as categorical data.) Mixed membership models allow each document to exhibit multiple components, where each component is a distribution over words. Conditioned on a collection, inspecting the posterior of the components reveals the "topics" inherent in the documents, i.e., the significant patterns of words associated under a single theme. For this reason, mixed membership models of text are often called *topic models*.

Mixed membership models for text have been extended in a myriad of ways and developed for many text-based applications. As examples, they have been developed into time series (Blei and Lafferty, 2006), into further hierarchicalized models of word contagion (Doyle and Elkan, 2009), into Bayesian nonparametric variants (Teh et al., 2006), and into models of interconnected documents (Chang and Blei, 2009). In some ways, mixed membership models of text have become a benchmark for new innovations in mixed membership modeling.

Network Data

Another central application of mixed membership models is for the analysis of network data. A network consists of a population of units and their relationships, represented via a graph with a set of nodes and edges between them. Networks arise naturally in sociological settings, co-author analysis, and a variety of biological problems. A classical latent-variable model of networks is the *stochastic blockmodel* (Wang and Wong, 1987), which assumes that each node belongs to a community, and that its assigned community mediates its connection to other nodes. While these assumptions may have been appropriate for small scale network analysis, modern networks are heterogenous. Nodes belong to *multiple* communities, and each node's connections reflect its particular signature of community memberships. This is a natural setting for mixed membership models.

Airoldi et al. (2008) developed the mixed membership extension of the stochastic blockmodel. Each node possesses an associated membership vector containing community proportions; each edge (present or absent) is associated with a community assignment drawn from the corresponding nodes' proportions. Note that modeling networks is fundamentally different from modeling documents because the observations are by definition intertwined. (We typically assume that documents, in contrast, are conditionally independent.) Mixed membership network models remain an active area of research. Further innovations include modeling dynamic networks (Ho et al., 2011) and including node attributes in modeling (Kim and Leskovec, 2011; Azari and Airoldi, 2012; Azizi et al., 2014). More broadly, networks are a type of *dyadic data*—data with entries indexed by a row and column—for which we can conceive more general mixed membership models (Mackey et al., 2010).

Social and Health Sciences Applications

The earliest mixed membership model, the Grade of Membership model (GoM) was developed by the statistician Max Woodbury (Woodbury et al., 1978), in the context of a medical classification problem where subsets of symptoms were observed on each patient. The goal was to identify and characterize sub-patterns of illness in a particular disease such as depression (Davidson et al., 1989), schizophrenia (Manton et al., 1994a), and Alzheimer's (Corder and Woodbury, 1993). GoM model analysis has been applied extensively to disability survey data—to analyze patters in binary indicators of basic and instrumental activities of daily living—in a frequentist (Berkman et al., 1989; Manton et al., 1991) and Bayesian framework (Erosheva et al., 2007). Mixed membership methodologies have been extended to longitudinal settings to capture heterogeneous pathways of disability and cognitive trajectories at the later portion of life (this volume: Manrique-Vallier, 2014 and Lecci, 2014). In political science, researchers have used mixed membership models to analyze politically-oriented beliefs, values, and attitudes from survey data (this volume: Gross and Manrique-Vallier, 2014) and have developed mixed membership models for rank data to analyze votes in Irish elections (Gormley and Murphy, 2009). Other applications of mixed membership models include assessing the risk of privacy violations in databases (Manrique-Vallier and Reiter, 2012), and even reconstructing the contents of a city based on sparse archeological evidence (Mimno, 2011).

Population Genetics

In computational biology, mixed membership models have had a tremendous impact, most notably following the structure model of Pritchard et al. (2000a). In this setting, we observe a collection of human genomes in which each is a collection of alleles (A,G,C,T) measured at different locations. The model assumes that there are ancestral populations, groups of original humans that share a unique genetic signature, which migrated around the world and mixed. The observed genomes—the data we are analyzing—reflect the results of that mixing. Each genome exhibits the populations with different proportions, and each population is characterized by its allele probabilities across genome locations. Posterior inference of the proportions and populations reveals the latent genetic structure of modern humans.

This kind of analysis has been used in two ways. First, as for networks and text, it is useful for exploring genetic patterns and forming hypotheses about our genetic history. Second, it is important for correcting analyses that seek to find associations between genes and traits. Patterns in ancestral populations, though not observed, are a confounder to making such associations; inferences from mixed membership models are useful in accounting for them. In this volume Shringarpure and Xing (2014) discuss some interesting variants on the original Pritchard et al. (2000a).

1.4　Theoretical Issues with Mixed Membership Models

The early examples of original mixed membership models described above were developed for discrete data, involving multivariate binary data, multinomial data, and ranks, and researchers using them considered responses to survey questions, counts of words in a document, sequences of genotypes, presence or absence of interactions between units, etc. Even though the general formulation of mixed membership models allows for combining outcomes of different types in a single omnibus model (Erosheva, 2002), the theoretical properties of mixed membership models applied to continuous data and data of mixed outcomes and applications of mixed membership for such problems is quite limited, e.g., see the discussion in Heller et al. (2008) and the the analysis of gene expression data by Rogers et al. (2005).

Extending mixed membership models to continuous data and data of mixed types is nontrivial. In this volume Galyardt (2014) demonstrates that the two interpretations—mixed attributes (the 'switching' interpretation) and partial memberships (the 'between' interpretation)—which are typically assumed as equivalent interpretations of mixed membership models, can not be taken for granted in the presence of continuous data. In fact, the 'between' interpretation no longer applies. Gruhl and Erosheva (2014) consider a broader class of individual-level mixture models and compare two members of this class—the mixed membership and the partial membership model (Heller et al., 2008)—for analyzing continuously-valued data. In essence, given individual-specific weights reflecting membership, mixed membership models assume that data are generated from individual-specific distributions that are weighted arithmetic averages of the subpopulation distributions, and partial membership models assume that individual-specific data are generated from a weighted geometric average of the subpopulation distributions. They explain that multivariate data my not provide researchers with a clear signal about the preferred type of individual-level mixture model. However, in this volume, analyzing a player statistics dataset from the National Basketball Association, Gruhl and Erosheva (2014) argue that the use of partial membership in that specific context is more appropriate. Partial membership models also happen to be more computationally convenient. Galyardt (2014) and Gruhl and Erosheva (2014) raise a number of issues for future work with individual-level mixture models for continuous data; some of these issues bear a clear connection to the large body of statistical literature on mixture models in general and on mixtures of normals in particular (McLachlan and Peel, 2000).

1.4.1　General Issues Inherent to Mixtures

While applications for mixed membership models especially in the form of extensions of topic models for text are widespread, these models suffer from a number of theoretical difficulties they inherit from mixture models. A lack of understanding of such issues may impact the validity of empirical analyses based on mixed membership models. Below we list a few key issues, borrowing material from a blog post on the topic by Wasserman (2012).

These issues are best illustrated in the context of a simple mixture model. Consider a finite mixture of Gaussians,

$$p(x; \psi) = \sum_{j=1}^{k} w_j \, \phi(x; \mu_j, \Sigma_j),$$

where $\phi(x; \mu_j, \Sigma_j)$ denotes a Gaussian density with mean vector μ_j and covariance matrix Σ_j. The weights $w_1, \ldots, w_k$ are non-negative and sum to 1. The entire set of parameters is $\psi = (\mu_1, \ldots, \mu_k, \Sigma_1, \ldots, \Sigma_k, w_1, \ldots, w_k)$. One can also consider k, the number of components, to be another parameter.

Now lets consider some of the weird things that can happen.

Infinite Likelihood. The likelihood function (for the Gaussian mixture) is infinite at some points in the parameter space. This is not necessarily bad, since the infinities are at the boundary and one can use the largest (finite) maximum in the interior as an estimator. But the infinities can cause numerical problems.

Multi-modality of the Likelihood. In fact, the likelihood has many modes (Richards and Buot, 2006). Finding the global (but not infinite) mode is a difficult. The EM algorithm only finds local modes. In this sense, the MLE is not really a well-defined estimator because it cannot be found. In the machine learning literature, there have been a number of papers trying to establish estimators for mixture models that can be found in polynomial time. For example, see Kalai et al. (2012).

Multi-modality of the Density. One may naïvely think that a mixture of k Gaussians would have k modes. But, in fact, it can have less than k or more than k. See Carreira-Perpinan and Williams (2003) and Edelsbrunner, Fasy, Rote (2012).

Non-identifiability. Recall that a model $\{p(x; \theta) : \theta \in \Theta\}$ is identifiable if

$$\theta_1 \neq \theta_2 \quad \text{implies} \quad p(x; \theta_1) \neq p(x; \theta_2).$$

Mixture models are non-identifiable in two different ways. First, there is non-identifiability due to permutation of labels. This is a nuisance, and there are strategies to deal with it (Stephens, 2000). A bigger issue is local non-identifiability. Suppose that

$$p(x; \eta, \mu_1, \mu_2) = (1 - \eta)\phi(x; \mu_1, 1) + \eta\phi(x; \mu_2, 1).$$

When $\mu_1 = \mu_2 = \mu$, we have that $p(x; \eta, \mu_1, \mu_2) = \phi(x; \mu)$. The parameter η has disappeared. Similarly, when $\eta = 1$, the parameter μ_2 disappears. This means that there are subspaces of the parameter space where the family is not identifiable. The result is that all the usual theory about the distribution of the MLE, the distribution of the likelihood ratio statistic, the properties of BIC, and so on, becomes very complicated.

Irregularity. Mixture models do not satisfy the usual regularity conditions that make parametric models easy to deal with. Consider the following example from Chen (1995). Let

$$p(x; \theta) = \frac{2}{3}\phi(x; -\theta, 1) + \frac{1}{3}\phi(x; 2\theta, 1).$$

Then $I(0) = 0$ where $I(\theta)$ is the Fisher information. Moreover, no estimator of θ can converge faster than $n^{-1/4}$. Compare this to a Normal family $\phi(x; \theta, 1)$ where the Fisher information is $I(\theta) = n$ and the maximum likelihood estimator converges at rate $n^{-1/2}$.

Non-intuitive Group Membership. Mixtures are often used for finding clusters. Suppose that

$$p(x) = (1 - \eta)\phi(x; \mu_1, \sigma_1^2) + \eta\phi(x; \mu_2, \sigma_2^2)$$

with $\mu_1 < \mu_2$. Let $Z = 1, 2$ denote the two components. We can compute $P(Z = 1|X = x)$ and $P(Z = 2|X = x)$ explicitly. We can then assign an x to the first component if $P(Z = 1|X = x) > P(Z = 2|X = x)$. It is easy to check that, with certain choices of σ_1, σ_2, all large values of x get assigned to component 1 (i.e., the leftmost component). Technically this is correct, yet it seems to be an unintended consequence of the model.

Improper Posteriors. Suppose we have a sample from the simple mixture

$$p(x; \mu) = \frac{1}{2}\phi(x; 0, 1) + \frac{1}{2}\phi(x; \mu, 1).$$

Then any improper prior on μ yields an improper posterior for μ regardless of how large the sample size is. Also, Wasserman (2012) shows that the only priors that yield posteriors in close agreement to frequentist methods are data-dependent priors.

These issues are often exacerbated in more complex mixed membership models. They should be taken seriously. In most applications, however, available additional information can be used to mitigate, and sometimes resolve, the problems listed above. The papers we have collected in this volume provide good examples, and explain why we do not share Wasserman's negative assessment: "that mixtures, like tequila, are inherently evil and should be avoided at all costs."

References

Airoldi, E. M., Blei, D. M., Fienberg, S. E., and Xing, E. P. (2008). Mixed membership stochastic blockmodels. *Journal of Machine Learning Research* 9: 1981–2014.

Airoldi, E. M., Blei, D. M., Fienberg, S. E., and Xing, E. P. (2006). Mixed membership stochastic block models for relational data with application to protein-protein interactions. In *Proceedings of the International Biometrics Society Annual Meeting*.

Airoldi, E. M., Erosheva, E. A., Fienberg, S. E., Joutard, C., Love, T., and Shringarpure, S. (2010). Reconceptualizing the classification of PNAS articles. *Proceedings of the National Academy of Sciences* 107: 20899–20904.

Airoldi, E. M., Fienberg, S. E., and Xing, E. P. (2007). Mixed membership analysis of genome-wide expression studies. Unpublished manuscript.

Airoldi, E. M., Wang, X., and Lin, X. (2013). Multi-way blockmodels for analyzing coordinated high-dimensional responses. *Annals of Applied Statistics* 7: 2431–2457.

Azari, H. and Airoldi, E. M. (2012). Graphlet decomposition of a weighted network. *Journal of Machine Learning Research* : 54–63.

Azizi, E., Airoldi, E. M., and Galagan, J. E. (2014). Learning modular structures from network data and node variables. *Journal of Machine Learning Research, W&CP.* In press.

Barnard, K., Duygulu, P., de Freitas, N., Forsyth, D., Blei, D. M., and Jordan, M. I. (2003). Matching words and pictures. *Journal of Machine Learning Research* 3: 1107–1135.

Berkman, L., Singer, B. H., and Manton, K. G. (1989). Black/white differences in health status and mortality among the elderly. *Demography* 26: 661–678.

Blei, D. M. (2012). Probabilistic topic models. *Communications of the ACM* 55: 77–84.

Blei, D. M. and Lafferty, J. D. (2006). Dynamic topic models. In *Proceedings of the 23rd International Conference on Machine Learning (ICML '06)*. New York, NY, USA: 113–120.

Blei, D. M., Ng, A. Y., and Jordan, M. I. (2003). Latent Dirichlet allocation. *Journal of Machine Learning Research* 3: 993–1022.

Carreira-Perpinan, M. A. and Williams, C. K. I. (2003). On the number of modes of a Gaussian mixture. In *Scale Space Methods in Computer Vision, Proceedings of the 4th International Conference on Scale Space*. Springer, 625–640.

Chang, J. and Blei, D. M. (2009). Relational topic models for document networks. In *Proceedings of the 12th International Conference on Artificial Intelligence and Statistics (AISTATS 2009)*. *Journal of Machine Learning Research – Proceedings Track* 5, 81–88.

Chen, J. (1995). Optimal rate of convergence for finite mixture models. *Annals of Statistics* 23: 221–233.

Corder, E. H. and Woodbury, M. A. (1993). Genetic heterogeneity in Alzheimer's disease: A Grade of Membership analysis. *Genetic Epidemiology* 10: 495–499.

Davidson, J. R., Woodbury, M. A., Zisook, S., and Giller, E. L., Jr. (1989). Classification of depression by Grade of Membership: A confirmation study. *Psychological Medicine* 19: 987–998.

Doyle, G. and Elkan, C. (2009). Accounting for burstiness in topic models. In *Proceedings of the 26th International Conference on Machine Learning (ICML '09)*. New York, NY, USA: ACM, 281–288.

Edelsbrunner, H., Fasy, B. T., and Rote, G. (2012). Add isotropic Gaussian kernels at own risk: More and more resiliant modes in higher dimensions. In *Proceedings of the ACM Symposium on Computational Geometry (SoCG 2012)*. New York, NY, USA: ACM, 91–100.

Erosheva, E. A. (2002). Grade of Membership and Latent Structure Models with Application to Disability Survey Data. Ph.D. thesis, Carnegie Mellon University, Pittsburgh, Pennsylvania, USA.

Erosheva, E. A. (2003). Bayesian estimation of the Grade of Membership model. In Bernardo, J., Bayarri, M., Berger, J., Dawid, A., Heckerman, D., Smith, A., and West, M. (eds), *Bayesian Statistics 7*. New York, NY: Oxford University Press, 501–510.

Erosheva, E. A. (2004). Partial membership models with application to disability survey data. In *Statistical Data Mining and Knowledge Discovery*. Chapman & Hall/CRC, 117–134.

Erosheva, E. A. (2005). Comparing latent structures of the Grade of Membership, Rasch, and latent class models. *Psychometrika* 70: 619–628.

Erosheva, E. A. and Fienberg, S. E. (2005). Bayesian mixed membership models for soft clustering and classification. In Weihs, C. and Gaul, W. (eds), *Classification – The Ubiquitous Challenge*. Springer, 11–26.

Erosheva, E. A., Fienberg, S. E., and Joutard, C. (2007). Describing disability through individual-level mixture models for multivariate binary Data. *Annals of Applied Statistics* 1: 502–537.

Erosheva, E. A., Fienberg, S. E., and Lafferty, J. D. (2004). Mixed-membership models of scientific publications. *Proceedings of the National Academy of Sciences* 97: 11885–11892.

Galyardt, A. (2014). Interpreting mixed membership: Implications of Erosheva's representation theorem. In Airoldi, E. M., Blei, D. M., Erosheva, E. A., and Fienberg, S. E. (eds), *Handbook of Mixed Membership Models and Its Applications*. Chapman & Hall/CRC.

Gormley, I. C. and Murphy, T. B. (2009). A Grade of Membership model for rank data. *Bayesian Analysis* 4: 265–296.

Gruhl, J. and Erosheva, E. A. (2014). A tale of two (types of) memberships. In Airoldi, E. M., Blei, D. M., Erosheva, E. A., and Fienberg, S. E. (eds), *Handbook of Mixed Membership Models and Its Applications*. Chapman & Hall/CRC.

Haberman, S. J. (1995). Book review of 'Statistical Applications Using Fuzzy Sets', by Kenneth G. Manton, Max A. Woodbury, and Larry S. Corder. *Journal of the American Statistical Association* 90(431): 1131–1133.

Heller, K. A., Williamson, S., and Ghahramani, Z. (2008). Statistical models for partial membership. In *Proceedings of the 25th International Conference on Machine Learning (ICML '08)*. New York, NY, USA: ACM, 392–399.

Ho, Q., Song, L., and Xing, E. P. (2011). Evolving cluster mixed-membership blockmodel for time-evolving networks. In *Proceedings of the 14th International Conference on Artifical Intelligence and Statistics (AISTATS 2011)*. Palo Alto, CA, USA: AAAI.

Hofmann, T. (2001). Unsupervised learning by probabilistic latent semantic analyis. *Machine Learning* 42: 177–196.

Kalai, A., Moitra, A., and Valiant, G. (2012). Disentangling Gaussians. *Communications of the ACM* 55: 113–120.

Kim, M. and Leskovec, J. (2011). Modeling social networks with node attributes using the multiplicative attribute graph model. *arXiv preprint*, arXiv:1106.5053.

Fei-Fei, L. and Perona, P. (2005). A Bayesian hierarchical model for learning natural scene categories. In *Proceedings of the 10th IEEE Computer Vision and Pattern Recognition (CVPR 2005)*. San Diego, CA, USA: IEEE Computer Society, 524–531.

Gross, J. H. and Manrique-Vallier, D. (2014). A mixed membership approach to political ideology. In Airoldi, E. M., Blei, D. M., Erosheva, E. A., and Fienberg, S. E. (eds), *Handbook of Mixed Membership Models and Its Applications*. Chapman & Hall/CRC.

Lecci, F. (2014). An analysis of development of dementia through the Extended Trajectory Grade of Membership model. In Airoldi, E. M., Blei, D. M., Erosheva, E. A., and Fienberg, S. E. (eds), *Handbook of Mixed Membership Models and Its Applications*. Chapman & Hall/CRC.

Mackey, L., Weiss, D. and Jordan, M. I. (2010). Mixed membership matrix factorization. In Fürnkranz, J. and Joachims, T. (eds), *Proceedings of the 27th International Conference on Machine Learning (ICML '10)*. Omnipress, 711–718.

Manrique-Vallier, D. (2014). Mixed membership trajectory models. In Airoldi, E. M., Blei, D. M., Erosheva, E. A., and Fienberg, S. E. (eds), *Handbook of Mixed Membership Models and Its Applications*. Chapman & Hall/CRC.

Manrique-Vallier, D. and Reiter, J. (2012). Estimating identification disclosure risk using mixed membership models. *Journal of the American Statistical Association* 107: 1385–1394.

Manton, K. G., Stallard, E., and Woodbury, M. A. (1991). A multivariate event history model based upon fuzzy states: Estimation from longitudinal surveys with informative nonresponse. *Journal of Official Statistics* 7: 261–293.

Manton, K. G., Woodbury, M. A., Anker, M., and Jablensky, A. (1994a). Symptom profiles of psychiatric disorders based on graded disease classes: An illustration using data from the WHO International Pilot Study of Schizophrenia. *Psychological Medicine* 24: 133–144.

Manton, K. G., Woodbury, M. A., and Tolley, H. D. (1994b). *Statistical Applications Using Fuzzy Sets*. Wiley-Interscience.

McLachlan, G. and Peel, D. (2000). *Finite Mixture Models*. Wiley Series in Probability and Statistics. Wiley-Interscience, 1st edition.

Mimno, D. (2011). Reconstructing Pompeian households. In *Proceedings of the 27th Conference on Uncertainty in Artificial Intelligence (UAI 2011)*. Corvallis, OR, USA: AUAI Press, 506–513.

Potthoff, R. G., Manton, K. G., Woodbury, M. A., and Tolley, H. D. (2000). Dirichlet generalizations of latent-class models. *Journal of Classification* 17: 315–353.

Pritchard, J. K., Stephens, M., and Donnelly, P. (2000a). Inference of population structure using multilocus genotype data. *Genetics* 155: 945–959.

Pritchard, J. K., Stephens, M., Rosenberg, N. A., and Donnelly, P. (2000b). Association mapping in structured populations. *American Journal of Human Genetics* 67: 170–181.

Richards, D. and Buot, M. -L. G. (2006). Counting and locating the solutions of polynomial systems of maximum likelihood equations. *I. Journal of Symbolic Computation* 41: 234–244.

Rogers, S., Girolami, M., Campbell, C., and Breitling, R. (2005). The latent process decomposition of cDNA microarray data sets. *IEEE/ACM Transactions on Computational Biology and Bioinformatics* 2: 143–156.

Rosenberg, N. A., Pritchard, J. K., Weber, J. L., Cann, H. M., Kidd, K. K., Zhivotovsky, L. A., and Feldman, M. W. (2002). Genetic structure of human populations. *Science* 298: 2381–2385.

Segal, E., Pe'er, D., Regev, A., Koller, D., and Friedman, N. (2005). Learning module networks. *Journal of Machine Learning Research* 6: 503–556.

Shringarpure, S. and Xing, E. P. (2014). Population stratification with mixed membership models. In Airoldi, E. M., Blei, D. M., Erosheva, E. A., and Fienberg, S. E. (eds), *Handbook of Mixed Membership Models and Its Applications*. Chapman & Hall/CRC.

Stephens, M. (2000). Bayesian analysis of mixture models with an unknown number of components—An alternative to reversible jump methods. *Annals of Statistics* 2: 40–74.

Teh, Y. W., Jordan, M. I., Beal, M. J., and Blei, D. M. (2006). Hierarchical Dirichlet processes. *Journal of the American Statistical Association* 101: 1566–1581.

Wang, Y. and Wong, G. (1987). Stochastic block models for directed graphs. *Journal of the American Statistical Association* 82: 8–19.

Wasserman, L. (2000). Asymptotic inference for mixture models by using data-dependent priors. *Journal of the Royal Statistical Society: Series B* 62(1): 159–180.

Wasserman, L. (2012). Mixture models: The twilight zone of statistics. Blog post, http://normaldeviate.wordpress.com/2012/08/04/mixture-models-the-twilight-zone-of-statistics/. Accessed on December 12, 2013.

Woodbury, M. A., Clive, J., and Garson A., Jr. (1978). Mathematical typology: A Grade of Membership technique for obtaining disease definition. *Computers and Biomedical Research* 11: 277–298.

2

A Tale of Two (Types of) Memberships: Comparing Mixed and Partial Membership with a Continuous Data Example

Jonathan Gruhl

Department of Statistics, University of Washington, Seattle, WA 98195, USA

Elena A. Erosheva

Department of Statistics, University of Washington, Seattle, WA 98195, USA

CONTENTS

Mixed membership models such as the Grade of Membership and latent Dirichlet allocation models have primarily focused on the analysis of binary and categorical data. In this chapter, we will focus on exploring the performance of two different types of membership models with continuous data: one that has a classic mixed membership structure and one that has a partial membership structure. The Bayesian partial membership model was recently proposed by Heller et al. (2008) as a promising alternative to mixed membership motivated by continuous data. The Bayesian partial membership model based on exponential family distributions allows for computationally efficient modeling of a variety of data types. Heller et al. (2008) demonstrated a partial membership analysis of a discrete dataset. In this work, we use a dataset that has a collection of continuous variables describing NBA (National Basketball Association) players and their playing styles as a motivating example. Although NBA players are typically assigned to one of five player positions, the language used to describe players and playing styles is often suggestive of individual-level mixtures. In this chapter, we compare the exponential family form of the Bayesian partial membership model with the general mixed membership model on simulated binary and continuous data. We then extend the partial membership framework to account for correlated membership scores. Based on the properties of the two types of models and the nature of the NBA data, we argue for choosing a partial membership model over a mixed membership model in this case. We show how the NBA players can be modeled as individual-level mixtures using the correlated partial membership model. To our knowledge, this is the first individual-level mixture analysis of continuous data.

2.1 Introduction

Mixture models provide a model-based approach to clustering. Population-level mixture models describe a population as a collection of subpopulations where each individual (or observational unit) belongs exclusively to one of the subpopulations (Lazarsfeld and Neil, 1968). Individual-level mixture models, on the other hand, allow each individual to belong to multiple subpopulations at once, with varying degrees of membership among individuals (Woodbury et al., 1978; Pritchard et al., 2000; Blei et al., 2003; Erosheva, 2002). Because the instance of individuals belonging exclusively to one subpopulation is a special case of individuals belonging simultaneously to multiple subpopulations, individual-level mixture models can be viewed as a relaxation of population-level mixtures such as finite mixture or latent class models.

The family of mixed membership models constitutes the predominant means of employing individual-level mixture models. At a high level, the mixed membership model assumes that data arise from individual-specific distributions that are arithmetic averages of the subpopulation distributions with individual-specific weights. Heller et al. (2008) formulated an alternative structure for individual-level mixtures, the Bayesian partial membership model, where the data can be viewed as arising from a (normalized) weighted geometric average of the subpopulation distributions with individual-specific weights.

When the subpopulation distributions are of exponential family form, the partial membership model allows for computationally efficient, individual-level mixture modeling of a variety of data types. In this chapter, we concentrate on the exponential family form of the partial membership model and compare this model to corresponding mixed membership models for the binary data case and the continuous data case. We highlight the differences in the data-generating behavior between the two types of models which have connection to the work of Galyardt (2014).

To demonstrate an individual-level mixture model analysis with continuous data, we use (NBA) National Basketball Association player statistics from the 2010–11 season (Hoopdata, 2012). The case of continuous data is of particular interest as existing individual-level mixture models have given less attention to continuous data. Even though the general class of mixed membership models (Erosheva, 2002) can accommodate any type of outcome (discrete or continuous), or even a mix of different types of outcomes in a model, the early independent developments have been motivated by discrete data problems whether in genetics, medicine, or computer science. Likewise, existing applications of mixed membership models primarily focus on binary, multinomial, and rank data: medical classification based on observed symptoms (Woodbury et al., 1978), counts of words in documents (Blei et al., 2003), responses to binary or multiple choice survey items such as disability manifestations (Erosheva et al., 2007), voter rankings of political candidates (Gormley and Murphy, 2009), counts of features present in an image (Wang et al., 2009), presence or absence of interactions between units (Airoldi et al., 2008), etc. A mixed membership analysis of continuous gene expression data with the latent process decomposition model by Rogers et al. (2005) is an exception. One reason for the continued focus of mixed membership models on discrete data is that little is known about this type of modeling for continuous data, and that basic examples of mixed membership for continuous data do not seem realistic.

The rest of the chapter is organized as follows. We provide more background on the NBA player data in Section 2.2. Section 2.3 provides a review of the mixed membership model and the partial membership model (no implied connection to the partial membership model in Erosheva, 2004). We compare the models and their applications to binary and continuous data in Section 2.4. Our comparison of these models for the continuous data case helps explicate our decision to use the partial membership for the NBA data analysis. In Section 2.5, we introduce an extension of the partial membership model that allows us to accommodate correlations among class membership scores. In Section 2.6, we analyze NBA playing styles using the correlated partial membership model.

2.2 Compositional Playing Styles of NBA Players

In the New York Times basketball blog *Off The Dribble*, Joshua Brustein highlighted NBA-related research presented at the 2012 MIT Sloan Sports Analytics Conference (Brustein, 2012). Team chemistry and construction were recurring themes in the research with the intent of understanding how team chemistry and construction might relate to winning. In understanding the team construction process, comparing it across teams, and ultimately relating it to game outcomes, it is helpful to be able to group players by playing style and/or ability.

Typically, basketball players are assigned to one of five positions: point guard (PG), shooting guard (SG), small forward (SF), power forward (PF), and center (C). Some players may play multiple positions. For instance, some players may play both the point guard and shooting guard positions or both the small forward and power forward positions. NBA observers may commonly use a more informal typology of players with three categories that consolidate the above positions by physical attributes and function on the court: point guard, wings (shooting guards and small forwards), and bigs (power forwards and centers). However, current positions and player assignments to those positions may not fully reflect the variety of playing styles (Lutz, 2012). To classify players based on their playing style as reflected in their statistics, Lutz (2012) carried out a model-based cluster analysis of players based on their season statistics. We would like to take a different approach and, rather than assign players to strictly one playing style or identify clusters that are themselves mixtures of more pure clusters, assume that players themselves demonstrate compositions of different pure playing styles. This assumption is intuitively plausible. For instance, the term "combo guard" is regularly used to describe a player who combines the skills and the playing style of a typical point guard and a typical shooting guard. As a result, we would like to use an individual-level mixture model for our analysis of the NBA data.

To characterize players, we consider 13 different statistics from the 2010–11 NBA season available on hoopdata.com (Hoopdata, 2012). Our dataset is composed of 332 players who had played 30 or more games and averaged 10 or more minutes per game. We selected 13 statistics that characterize different elements of players' styles; these largely overlap with the statistics used by Lutz (2012) in a model-based cluster analysis of similar data.

The variables in our dataset include: minutes played per game, percent of made field goals that are assisted, assist rate, turnover rate, offensive rebound rate, defensive rebound rate, steals per 40 minutes, blocks per 40 minutes, and number of shots attempted per 40 minutes at each of the following locations: at the rim, from 3–9 feet, from 10–15 feet, from 16–23 feet, and beyond the 3-point line. All of the variables are continuous, but some, such as minutes played per game (maximum of 48) or percent of field goals made (0–100), are restricted in their range.

In addition to these variables, Lutz (2012) also included the number of games played as another statistic in the cluster analysis. We elected not to use this variable as it is likely to be influenced by events such as injuries that may have little connection to a player's style. Table 2.1 lists the variables, their abbreviations, and formulas of calculated statistics in our dataset.

Figures 2.1 and 2.2 display two bivariate scatterplots for selected player statistics. The data pat-

TABLE 2.1

Variables, abbreviations, and formulas (if calculated).

Variable	Description and Formula
Min	Minutes played per game
% Ast	Percent of made field goals that are assisted $\frac{\text{field goals that are assisted}}{\text{total made field goals}}$
AR	Assist Ratio $\frac{\text{Assists} \times 100}{\text{FGA} + (\text{FTA} \times .44) + \text{Turnovers}}$
TOR	Turnover Ratio $\frac{\text{Turnovers} \times 100}{\text{FGA} + (\text{FTA} \times .44) + \text{Turnovers}}$
ORR	Offensive Rebound Rate $\frac{100 \times (\text{Player ORebs} \times (\text{Team Min}/5))}{(\text{Player Min} \times (\text{Team ORebs} + \text{Opp DRebs}))}$
DRR	Defensive Rebound Rate $\frac{100 \times (\text{Player DRebs} \times (\text{Team Min}/5))}{(\text{Player Min} \times (\text{Team DRebs} + \text{Opp ORebs}))}$
Rim	Attempted field goals at the rim per 40 minutes
Close	Attempted field goals from 3-9 feet per 40 minutes
Medium	Attempted field goals from 10-15 feet per 40 minutes
Long	Attempted field goals from 16-23 feet per 40 minutes
3s	3-point field goals attempted per 40 minutes
Stls	Steals per 40 minutes
Blks	Blocks per 40 minutes

terns presented in Figures 2.1 and 2.2 are typical of other bivariate scatterplots in this dataset (not shown). The shapes of the plotted points indicate the player's position. In the list of positions in the legend, the positions 'G', 'GF,' and 'F' are listed in addition to the five main positions listed earlier. Hoopdata.com uses these designations in their positional assignments to describe players who regularly play multiple positions. G (guard) is typically used to describe a player who plays both point guard and shooting guard, GF (guard-forward) to describe a player who plays both shooting guard and small forward, and F (forward) to describe a player who plays both small forward and power forward.

Figure 2.1 plots the assist and turnover ratios of the players. The data appear to fan out from the lower left corner, adopting an almost triangular shape. Within this shape, we can see some patterns. Players designated as point guards and guards dominate the points in the upper right. Players manning the forward, power forward, and center positions generally appear to have low assist ratios and span the range of turnover ratios, comprising the points lining the left side of the plot.

Figure 2.2 presents the corresponding plot for defensive rebound rate and 3-point field goals attempted per 40 minutes. We see a different pattern in this data with a clear cluster of players comprised of forwards, power forwards, and centers that rarely attempt 3-point field goals. Separate from this cluster is a cloudlike structure of points that tends to shoot some 3-pointers. Within the cloud, we see that point guards tend to have lower defensive rebound rates while forwards, power forwards, and centers tend to have higher rebound rates.

Our first step with individual-level mixture modeling for these data is to identify which compositional representation of continuous data is better suited for analyzing NBA player statistics. Next, we will present formulations of mixed membership and partial membership models and examine the data-generative capabilities of these models for both discrete and continuous data.

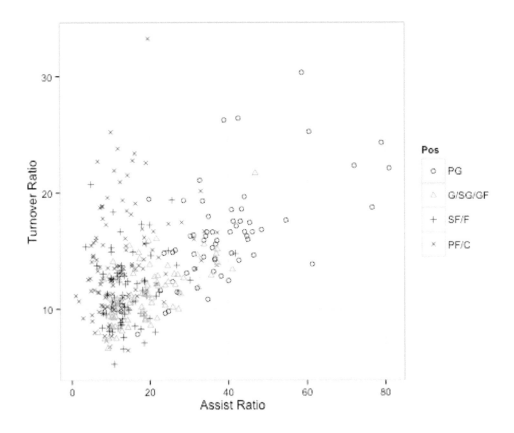

FIGURE 2.1
Bivariate scatterplot of players' assist ratio and turnover ratio. The symbols of the points represent the different positions of each player.

2.3 Two Types of Membership

In this section, we introduce the mixed membership and the partial membership models. For each of these two individual-level mixture models, we first consider a standard population-level mixture model formulation and then present each individual-level mixture model as a relaxation of the population-level mixture. Heller et al. (2008) used Bayesian methods to estimate the partial membership model. Similarly, Bayesian methods are frequently employed with mixed membership models. As we will see, the hierarchical Bayesian representations of the two models have many features in common.

2.3.1 Mixed Membership Model

Let $\mathbf{y}_i$ be a vector of p outcomes for the ith individual or observational unit. We use K to denote the number of pure types or mixture components. Let $p_k(\cdot)$ specify the density particular to pure type k, and let $\boldsymbol{\theta}_k$ represent the parameters characterizing $p_k(\cdot)$ for pure type k. The population-level mixture model with K components assumes the existence of K membership indicator variables, π_{ik}, for each individual i that designate the cluster or pure type to which the individual belongs.

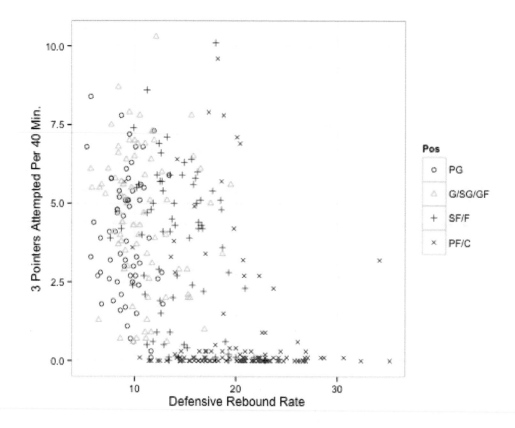

FIGURE 2.2
Bivariate scatterplot of players' defensive rebound rate and 3-point field goals attempted per 40 minutes. The symbols of the points represent the different positions of each player.

As such, $\pi_{ik} \in \{0,1\}$ with the restriction $\sum_k \pi_{ik} = 1$. The probability density for $\mathbf{y}_i$, given a collection of parameters $\Theta = (\boldsymbol{\theta}_1, \dots, \boldsymbol{\theta}_K)$ for all K pure types and given the latent pure type membership indicator π_{ik} for pure type k and individual i, is

$$p(\mathbf{y}_i|\Theta, \boldsymbol{\pi}_i) = \sum_k^K \pi_{ik} p_k(\mathbf{y}_i|\boldsymbol{\theta}_k). \tag{2.1}$$

For the mixed membership model, one replaces π_{ik} with a membership score g_{ik}. Instead of being restricted to either 0 or 1, the membership score g_{ik} is allowed to range continuously between 0 and 1, subject to the constraint $\sum_k g_{ik} = 1$. The mixed membership then takes the form

$$p(\mathbf{y}_i|\Theta, \mathbf{g}_i) = \prod_j^J \sum_k^K g_{ik} p_{jk}(y_{ij}|\boldsymbol{\theta}_{jk}). \tag{2.2}$$

Here, conditional on the membership vector $\mathbf{g}_i = (g_{i1}, \dots, g_{iK})$, the observations $\mathbf{y}_i$ are assumed to be independent.

In the Bayesian representation of the model introduced, $\mathbf{g}_i \sim D_g(\alpha, \boldsymbol{\rho})$, where D_g is a prior suitable for compostional parameters and $\alpha, \boldsymbol{\rho}$ are hyperparameters. As $\mathbf{g}_i$ lies in the $K-1$ probability simplex, the most common choices for D_g are the Dirichlet (Blei et al., 2003) and logistic normal (Blei and Lafferty, 2007) distributions. For the class-specific and outcome-specific parameters, θ_{jk}, a conjugate prior, is typically assumed:

$$\theta_{jk} \sim \text{Conj}(\boldsymbol{\lambda}, \boldsymbol{\nu}), \tag{2.3}$$

where $\boldsymbol{\lambda}, \boldsymbol{\nu}$ are hyperparameters. The mixed membership model has a latent class representation that suggests a data augmentation approach for estimation (Erosheva, 2003). This approach adds an additional level of hierarchy to the model by including latent classification variables.

2.3.2   Partial Membership Model

An alternative means of specifying Equation (2.1) is through the product of the densities:

$$p(\mathbf{y}_i | \boldsymbol{\Theta}, \boldsymbol{\pi}) = \prod_k^K p_k(\mathbf{y}_i | \boldsymbol{\theta}_k)^{\pi_{ik}}. \tag{2.4}$$

We specify the partial membership model by relaxing Equation (2.4) so that

$$p(\mathbf{y}_i | \boldsymbol{\Theta}, \mathbf{g}) = \frac{1}{c} \prod_k^K p_k(\mathbf{y}_i | \boldsymbol{\theta}_k)^{g_{ik}}, \tag{2.5}$$

where $g_{ik} \in [0, 1]$ and c is a normalizing constant. Heller et al. (2008) further highlights the case where p_k is an exponential family density (denoted $\text{Exp}(\cdot)$):

$$p_k(\mathbf{y}_i | \boldsymbol{\psi}_k) = \text{Exp}(\boldsymbol{\psi}_k). \tag{2.6}$$

Here, $\boldsymbol{\psi}_k$ denotes the natural parameters for pure type k. Let $\boldsymbol{\Psi}$ denote the collection of the natural parameters for all pure types.

Substituting exponential family densities for p_k in Equation (2.5), we obtain

$$p(\mathbf{y}_i | \boldsymbol{\Psi}, \mathbf{g}) = \text{Exp}\left(\sum_k g_k \boldsymbol{\psi}_k\right). \tag{2.7}$$

In addition, let the natural parameters for each pure type follow a conjugate prior distribution

$$\boldsymbol{\psi}_k \sim \text{Conj}(\boldsymbol{\lambda}, \boldsymbol{\nu}), \tag{2.8}$$

where $\boldsymbol{\lambda}, \boldsymbol{\nu}$ are hyperparameters. As with the mixed membership model, we assume $\mathbf{g}_i \sim D_g(\alpha, \boldsymbol{\rho})$ where D_g is a prior suitable for compostional parameters and $\alpha, \boldsymbol{\rho}$ are hyperparameters.

Conditional on the membership scores, $\mathbf{y}_i$ is distributed according to the same exponential family distribution as the pure types but with natural parameters that are a convex combination of the natural parameters of the pure type distributions. The use of the exponential family distributions allows one to model a variety of outcome types. Going forward, we focus on this particular case of the Bayesian partial membership model.

2.4   Comparison of Partial and Mixed Membership

In this section, we compare and contrast the partial membership model with the mixed membership model using simulated data. Figure 2.3 provides a graphical comparison of the models' data generative processes. Although the generative structures are otherwise very similar, we see that whereas the mixed membership model assumes local independence (i.e., the outcomes are conditionally independent given the pure type memberships), the partial membership model makes no such assumption. What we can not see in Figure 2.3 is how the pure type parameters and membership scores are combined together mathematically to define the individual-level distributions of the outcomes. To explore this, we examine scatterplots for data generated by the two types of models when the data are continuous and probabilities of success generated by the respective models when the data are binary. Understanding the differences in the continuous data case will help us select an appropriate model for the NBA player data introduced in Section 2.2.

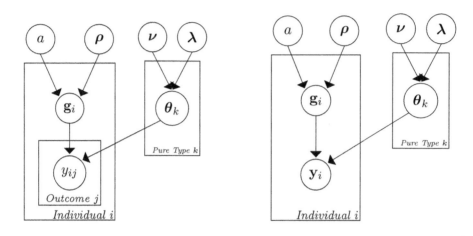

FIGURE 2.3
Graphical representations of the mixed membership (left) and partial membership models (right).

2.4.1   Continuous Data

For both the partial and mixed membership models, we begin by assuming that the pure type densities are normal. We allow the means of the normal distributions to vary by pure type. Similar to model-based clustering with the mixtures of normals (Fraley et al., 2012), different specifications are possible for the variances. For this work, we will focus on two cases. In the first case, we consider variance specifications to be the same across the pure types. In the second case, we allow the variances to differ across pure types. While the mixed membership model uses a local independence assumption, the partial membership model does not. Hence, for the partial membership model, we additionally consider two cases of variance specification: the case where the outcomes are correlated, conditional on the membership scores, and the case where the outcomes are uncorrelated, conditional on the membership scores. Next, we specify mixed membership and partial membership models for continuous data with normally distributed pure types before examining scatterplots of simulated data under the different scenarios of variance specification.

Mixed Membership

Under the mixed membership model and a local independence assumption, each outcome y_{ij}, conditional on the pure type memberships for individual i, is distributed

$$y_{ij}|\mathbf{g}_i, \Theta \sim \sum_k g_{ik} \mathrm{N}\left(\mu_{jk}, \sigma_{jk}^2\right). \qquad (2.9)$$

If we restrict $\sigma_{jk}^2 = \sigma_j^2$ so that the variances do not differ by pure type, then the model formulation remains the same. As we will see, in the case of the partial membership model, the model formulation can be simplified under the same restriction.

Partial Membership

In the case of the partial membership model, we may assume multivariate normal densities as we are not restricted to the conditional independence assumption. The observed data for individual i, $\mathbf{y}_i$ will also be multivariate normally distributed conditional on the pure type membership for individual i and the pure type parameters (recall Equation 2.7). The natural parameters of a multivariate normal distribution are $\Sigma^{-1}\mu$ and $-\frac{1}{2}\Sigma^{-1}$, where μ and Σ are the mean and covariance matrix of a multivariate normal distribution. Let $\Theta = \{\mu_k, \Sigma_k, k = 1, \ldots, K\}$ denote the collection of pure type means and covariance matrices. As a result, the natural parameters of $p(\mathbf{y}_i|\mathbf{g}_i, \Theta)$ are $\sum_k g_{ik}\Sigma_k^{-1}\mu_k$ and $-\frac{1}{2}\sum_k g_{ik}\Sigma_k^{-1}$.

Using the standard parameterization for the multivariate normal distribution, the vector of observed data, $\mathbf{y}_i$, is conditionally distributed

$$\mathbf{y}_i|\mathbf{g}_i, \Theta \sim \mathrm{N}\left(\left(\sum_k g_{ik}\Sigma_k^{-1}\right)^{-1}\left(\sum_k g_{ik}\Sigma_k^{-1}\mu_k\right), \left(\sum_k g_{ik}\Sigma_k^{-1}\right)^{-1}\right). \tag{2.10}$$

If we restrict $\Sigma_1 = \cdots = \Sigma_K = \Sigma$, then

$$\mathbf{y}_i|\mathbf{g}_i, \Theta \sim \mathrm{N}\left(\sum_k g_{ik}\mu_k, \Sigma\right). \tag{2.11}$$

Finally, if we assume the outcomes $\mathbf{y}_i$ are conditionally independent given the pure type memberships (local independence), each outcome y_{ij} conditional on the pure type memberships for individual i is distributed

$$y_{ij}|\mathbf{g}_i, \Theta \sim \mathrm{N}\left(\left(\sum_k g_{ik}\sigma_{jk}^{-2}\right)^{-1}\left(\sum_k g_{ik}\sigma_{jk}^{-2}\mu_{jk}\right), \left(\sum_k g_{ik}\sigma_{jk}^{-2}\right)^{-1}\right), \tag{2.12}$$

where σ_{jk}^2 is the j-th diagonal element of Σ_k, now a diagonal matrix.

Simulated Data Scenarios

We now compare data generated by each of the two models. Consider three pure types with two normally distributed outcomes. We present the means for each pure type in Table 2.2.

For the variance specifications, we explore two scenarios, one where the variances for each outcome are the same across pure types and a second where the variances differ across pure types. For each scenario, we consider three models: a mixed membership with a local independence assumption, a partial membership with a local independence assumption, and a partial membership model with no restrictions on dependence.

Table 2.3 summarizes Scenario 1 for which we assume the variance for the first outcome is 4 for all pure types and 9 for the second outcome for all pure types. Because of the local independence assumption used in the mixed membership model, there is no correlation between the two outcomes. As a means of comparison, we consider a corresponding partial membership model that employs the local independence assumption and hence also has the correlation between the two outcomes restricted to 0. Finally, the partial membership model without a local independence assumption assumes a correlation of 0.4.

Table 2.4 presents the corresponding information for Scenario 2 where the covariances may vary by pure type. For the partial membership model with full covariance matrix, the correlations by pure type were set to 0.4, -0.4, and 0.7.

TABLE 2.2
Pure Type Means.

	Pure Type		
Outcome	1	2	3
1	10	25	40
2	25	40	10

TABLE 2.3
Covariance matrices under Scenario 1.

Model	Pure Types 1–3
Mixed Membership	$\begin{pmatrix} 4 & 0 \\ 0 & 9 \end{pmatrix}$
Partial Membership (Uncorrelated)	$\begin{pmatrix} 4 & 0 \\ 0 & 9 \end{pmatrix}$
Partial Membership (Correlated)	$\begin{pmatrix} 4 & 2.4 \\ 2.4 & 9 \end{pmatrix}$

TABLE 2.4
Covariance matrices under Scenario 2.

Model	Pure Type		
	1	2	3
Mixed Membership	$\begin{pmatrix} 4 & 0 \\ 0 & 16 \end{pmatrix}$	$\begin{pmatrix} 9 & 0 \\ 0 & 1 \end{pmatrix}$	$\begin{pmatrix} 9 & 0 \\ 0 & 4 \end{pmatrix}$
Partial Membership (Uncorrelated)	$\begin{pmatrix} 4 & 0 \\ 0 & 16 \end{pmatrix}$	$\begin{pmatrix} 9 & 0 \\ 0 & 1 \end{pmatrix}$	$\begin{pmatrix} 9 & 0 \\ 0 & 4 \end{pmatrix}$
Partial Membership (Correlated)	$\begin{pmatrix} 4 & 3.2 \\ 3.2 & 16 \end{pmatrix}$	$\begin{pmatrix} 9 & -1.2 \\ -1.2 & 1 \end{pmatrix}$	$\begin{pmatrix} 9 & 4.2 \\ 4.2 & 4 \end{pmatrix}$

Scenario 1: Same Variances Across Pure Types

Under Scenario 1, we assume the pure type covariances are common across all three pure types. We keep the population parameters constant and vary the distribution of membership scores to produce scatterplots of observed data.

We generated 1000 random membership vectors from a Dirichlet $(a\rho)$ distribution with $a = 1$ and $\rho = (1/3, 1/3, 1/3)$. Using these membership scores, we simulated 1000 bivariate outcomes. The results are depicted in Figure 2.4(b). The left plot shows the mixed membership model and the center plot displays the corresponding partial membership model with a diagonal covariance matrix (i.e., local independence was assumed as in the case of the mixed membership model). The right plot shows partial membership model results with a full covariance matrix where the variances of the outcomes are the same as the previous two cases but the correlation between the outcomes is set to 0.4.

In Figure 2.4(b), the mixed membership model generates points in three columns. Looking more closely, each column can be divided horizontally into three parts corresponding to the means for each pure type for y_{i2}. Dividing the columns in this manner produces $K^2 = 9$ clusters of points, consistent with the latent class representation described by Erosheva (2006) and the more extreme depiction presented in Figure 4 in Heller et al. (2008). The partial membership model, in both the diagonal and full covariance matrix cases, generates points in a more cloud-like structure. One can see that the partial membership model with the full covariance matrix generates a set of points that is "rotated," albeit slightly, as compared to the set generated by the partial membership model with a diagonal covariance matrix.

By varying the values of a, we can further compare the models. If we set $a = 10$, the membership scores will fluctuate more closely around 1/3 than $a = 1$. Figure 2.4(c) presents 1000 generated data points with membership scores generated from a Dirichlet $(a\rho)$ distribution with $a = 10$ and $\rho = (1/3, 1/3, 1/3)$. In the case of the mixed membership model, the K^2 clusters become slightly more apparent while the data generated by the partial membership models reduce to single clusters with less variation. If we set $a = 1/10$, the membership scores tend to be closer to the extremes 0 or 1. Figure 2.4(a) presents the simulated data from each model with this set of membership scores. The three plots now appear largely similar. The primary differences are that the set of points generated by the partial membership model with full covariance matrix is "rotated" as compared to the other two and that the mixed membership model appears to show greater variation in points on the periphery.

Scenario 2: Different Variances Across Pure Types

We subsequently generate data points from each individual-level mixture model according to Scenario 2. Again, the pure type covariances for Scenario 2 are listed in Table 2.4. Figure 2.5(b) presents the data generated by the mixed membership model, the partial membership model with diagonal covariance, and the partial membership model with unrestricted covariance for $a = 1$. The sets of points generated by the mixed membership and partial membership model with diagonal covariance appear rectangular in shape. The set of points from the partial membership model with diagonal covariance is more densely populated in the center while one can faintly make the clusters in the set of points generated by the mixed membership model. The partial membership model with full covariance matrices on the other hand is more triangular in structure.

Figures 2.5(c) and 2.5(a) provide the corresponding plots for membership vectors generated by $a = 10$ and $a = 1/10$, respectively. With $a = 10$, we again see the greater concentration of points into a single cluster for the partial membership models while the different clusters become a little more apparent for the mixed membership model. In the case of $a = 1/10$, the mixed membership and partial membership with diagonal covariance models again appear very similar. The full covariance partial membership model, however, displays a triangular boundary with an empty center.

Overall, while the mixed and partial membership models can produce scatterplots that look

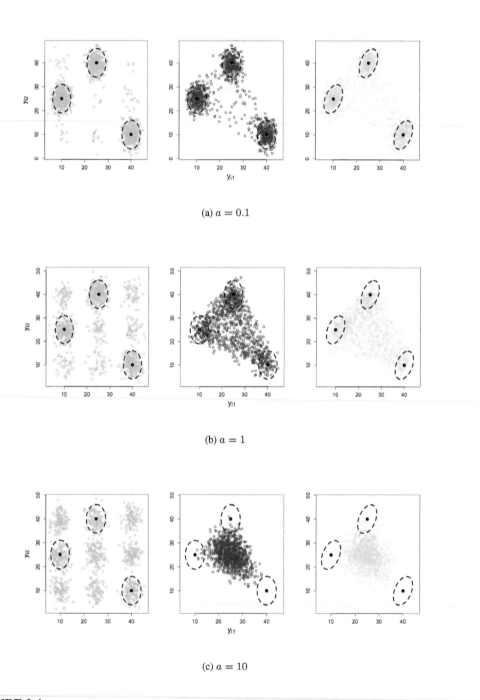

(a) $a = 0.1$

(b) $a = 1$

(c) $a = 10$

FIGURE 2.4
Simulated data according to different individual-level mixture models assuming variances are the same across pure types. Each panel contains: mixed membership (left), partial membership with local independence assumption (center), partial membership with full covariance matrix (right). The solid points represent the pure type centers and the dashed ellipses represent 2SD contours.

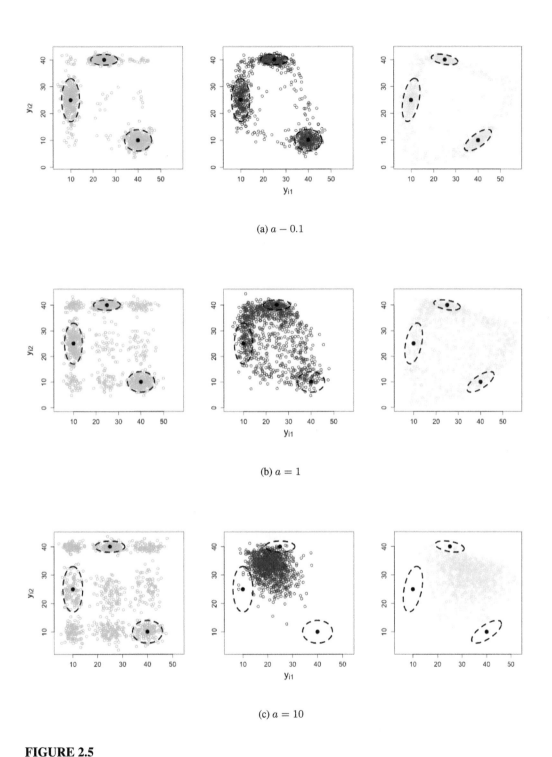

(a) $a - 0.1$

(b) $a = 1$

(c) $a = 10$

FIGURE 2.5

Simulated data according to different individual-level mixture models for the case where the variances are different across pure types. Each panel contains: mixed membership (left), partial membership with local independence assumption (center), partial membership with full covariance matrix (right). The solid points represent the pure type centers and the dashed ellipses represent 2SD contours.

very similar for some special cases of the distribution of the membership scores, we observe that the partial membership models generate scatterplots that are more contiguous. At the same time, we emphasize that placements of pure type means and variances, as well as the selection of the distribution of the membership scores, can create different patterns that would not be as easy to recognize as either mixed or partial membership. To investigate this further, one could consider a template for variance specification as provided by model-based clustering with Gaussian clusters (Fraley et al., 2012).

2.4.2 Binary Data

We now examine the mixed and partial membership models for binary data. We follow a geometric approach (Erosheva, 2005) where we keep the population parameters constant and examine population heterogeneity manifolds obtained by letting subject-level parameters vary over their natural range.

In the case of binary data, we compare the models by examining the probability of a positive response, $p(y_{ij} = 1|\mathbf{g}_i, \Theta)$, for outcome j and individual i, conditional on the pure type membership of individual i. Let θ_{jk} denote the probability of a positive response for pure type k and outcome j. Then,

$$\theta_{ij} = p(y_{ij} = 1|\mathbf{g}, \Theta) = \sum_k g_{ik}\theta_{jk}, \qquad (2.13)$$

so that $y_{ij}|\mathbf{g}, \Theta$ has a Bernoulli distribution where the probability of a positive response is a weighted arithmetic mean of the pure type response probabilities.

In the case of the partial membership model, $y_{ij}|\mathbf{g}, \Theta$ also has a Bernoulli distribution but where the natural parameter is a convex combination of the pure type natural parameters, $\sum_k g_{ik}\ln[\theta_{jk}/(1 - \theta_{jk})]$. As a result,

$$\theta_{ij} = p(y_{ij} = 1|\mathbf{g}, \Theta) = \frac{\prod_k \theta_{jk}^{g_{ik}}}{\prod_k \theta_{jk}^{g_{ik}} + \prod_k (1 - \theta_{jk})^{g_{ik}}}. \qquad (2.14)$$

In the case of the partial membership model, the probability of a positive response (Equation 2.14) is a normalized weighted geometric mean of the pure type response probabilities.

We now examine how these differences in the mixed membership and partial membership models for binary data manifest themselves for different pure type membership and parameter values. We consider $K = 2$ pure types and $p = 2$ outcomes. Let g_i denote the degree of membership for an arbitrary individual in the first pure type; the degree of membership in the second pure type is then $1 - g_i$. We examine θ_{ij}, the marginal probability of a positive response for outcome j, and individual i given by Equations (2.13) and (2.14) for the two types of models, respectively.

TABLE 2.5
Pure type response probabilities.

Scenario	θ_{11}	θ_{12}	θ_{21}	θ_{22}
1	0.1	0.8	0.3	0.6
2	0.05	0.8	0.3	0.95
3	0.01	0.8	0.3	0.99
4	0.001	0.8	0.3	0.999
5	0	0.8	0.3	1

Table 2.5 presents five sets of the pure type response probabilities, θ_{jk}; the corresponding

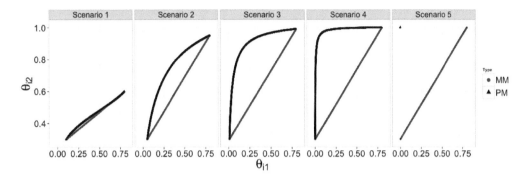

FIGURE 2.6
Marginal probability plots for Scenarios 1–5 in Table 2.5 obtained with the partial membership model (darker) and the mixed membership model (lighter).

marginal probability plots appear in Figure 2.6. Treating the pure type response probabilities as constant, we examine population heterogeneity manifolds obtained by letting membership scores g_i vary over their natural range from 0 to 1. The darker points indicate the population heterogeneity manifolds obtained with the partial membership model for given θ_{i1} and θ_{i2}, whereas the lighter points indicate the corresponding manifolds for the mixed membership model.

For Scenario 1, we see that the heterogeneity manifold for the partial membership model is a nonlinear path that closely resembles the heterogeneity manifold for the mixed membership model. As the pure type response probability θ_{11} decreases and θ_{22} increases over the five scenarios, the paths of points increasingly diverge. Finally, for Scenario 5, the partial membership model produces the heterogeneity manifold that takes only three pairs of values, sitting at the corners of the marginal probability space. At $g_i = 0$ and $g_i = 1$, the partial membership model produces θ_{ij} values equivalent to the mixed membership model. For values of g_i, $0 < g_i < 1$, in Scenario 5, $\theta_{i1} = 0$, and $\theta_{i2} = 1$ under the partial membership model. Consistent with the geometric mean representation in Equation (2.14), in scenarios where one of the pure type conditional response probabilities equals 1, any partial membership in the pure type implies that that individual's probability for that outcome must be 1. Similarly, when one of the pure type conditional response probabilities equals 0, any partial membership in that pure type implies that the probability for that outcome must be 0. We do not observe this property in the mixed membership model that employs the arithmetic mean to derive individual-specific marginal probabilities (Equation 2.13). Moreover, as one of the pure type probabilities decreases to 0 or increases to 1, the population heterogeneity manifolds obtained under the partial membership and mixed membership models increasingly diverge, as shown in Figure 2.6.

Overall, we have demonstrated that the partial and mixed membership models exhibit different data-generating behavior. In the case of continuous data, the partial membership model generates data in more contiguous patterns that may be more natural for some applications. However, except for some special cases, it may not be possible to tell the nature of individual-level mixing from scatterplots. Hence, data mechanisms need to be considered.

Our decision for the analysis of the NBA data is to use a partial membership model. We believe that a partial membership model could better describe the types of data patterns displayed in Figures 2.1 and 2.2 than a mixed membership model. However, an equally important factor in our decision is the nature of individual-level mixing in the data. The NBA player data contain variables that themselves are summary statistics as opposed to individual player's actions. While mixed membership modeling should be more appropriate for the latter type of data that could exhibit changes in (latent) pure type assignments for each variable, we find the partial membership representation to be more consistent with the averages reported over an NBA season. These considerations are akin to the switching and blending interpretations discussed in Galyardt (2014).

2.5   A Correlated Partial Membership Model for Continuous Data

Before analyzing the NBA player style data with a partial membership model for continuous data, we develop an extension of the partial membership model that allows for correlated membership scores. We subsequently discuss estimation of the correlated partial membership model.

2.5.1   Correlated Memberships

One limitation of the partial membership model as originally formulated is its inability to flexibly accommodate correlations among an individual's membership in the pure types. The Dirichlet prior induces a small negative correlation among the pure type memberships in individuals. Blei and Lafferty (2007) addressed this shortcoming in mixed membership topic models by replacing the Dirichlet prior for individual membership scores with a logistic normal prior. Under this model, draws from the multivariate normal are transformed to map the probability simplex so that the values are positive and constrained to add to 1,

$$\boldsymbol{\eta}_{\mathbf{g}_i} \sim \mathrm{N}\left(\boldsymbol{\rho}, \boldsymbol{\Sigma}\right), \tag{2.15}$$

$$g_{ik} = \frac{\exp(\eta_{g_{ik}})}{\sum_l \exp(\eta_{g_{il}})}. \tag{2.16}$$

Because of the constraints that $\sum_k g_{ik} = 1$, we fix the Kth element of $\boldsymbol{\eta}_{\mathbf{g}_i}$ to 0 so that the vector contains only $K - 1$ free elements and $\boldsymbol{\rho}$ and $\boldsymbol{\Sigma}$ have dimensions $K - 1$ and $(K - 1) \times (K - 1)$, respectively. Atchison and Shen (1980) discuss properties and uses of the logistic normal, including a comparison with the Dirichlet distribution. They suggest that the logistic normal can suitably approximate the Dirichlet distribution so that little, if anything, would be lost if we applied the logistic normal in cases where a Dirichlet prior would be appropriate.

2.5.2   A Correlated Partial Membership Model

To model the continuous data in the NBA example, we assume the observed data points for individual i, $\mathbf{y}_i$ are conditionally independent given the pure type memberships for the individual, $\mathbf{g}_i$. Equation (2.9) gives the distribution of y_{ij} under this assumptions. Now let $\tau_{jk} = \sigma_{jk}^{-2}$ and let $\alpha_{jk} = \sigma_{jk}^{-2} \mu_{jk}$ in Equation (2.9) so that τ_{jk} and ϕ_{jk} correspond closely to the natural parameters of a normal distribution. Moreover, let $\Theta = \{g_{ik}, \phi_{jk}, \tau_{jk}, \rho_k, j = 1, \ldots, J, k = 1, \ldots, K\}$.

For α_{jk} and τ_{jk}, we specify normal and gamma prior distributions, respectively. The elements of the mean vector of the untransformed pure type memberships, ρ_k, are also specified to have normal prior distributions. For the covariance matrix for the untransformed pure type memberships, $\boldsymbol{\Sigma}$, we use an inverse Wishart prior distribution. Fully stated, the correlated partial membership model for continuous data is

$$y_{ij}|\mathbf{g}_i, \boldsymbol{\Theta} \sim \mathrm{N}\left(\left(\sum_k g_{ik}\tau_{jk}\right)^{-1}\left(\sum_k g_{ik}\alpha_{jk}\right), \left(\sum_k g_{ik}\tau_{jk}\right)^{-1}\right), \tag{2.17}$$

$$\alpha_{jk} \sim \mathrm{N}\left(m_{\alpha_{jk}}, s_{\alpha_{jk}}^2\right), \tag{2.18}$$

$$\tau_{jk} \sim \text{Gamma}\left(\nu_{\tau_{jk}}, \phi_{\tau_{jk}}\right), \tag{2.19}$$

$$g_{ik} = \frac{\exp(\eta_{g_{ik}})}{\sum_l \exp(\eta_{g_{il}})}, \tag{2.20}$$

$$\boldsymbol{\eta_{g_i}} \sim \text{N}\left(\boldsymbol{\rho}, \boldsymbol{\Sigma}\right), \tag{2.21}$$

$$\rho_k \sim \text{N}\left(m_{\rho_k}, s_{\rho_k}^2\right), \tag{2.22}$$

$$\boldsymbol{\Sigma} \sim \text{Inv. Wishart}\left(\nu_\Sigma, S_\Sigma\right). \tag{2.23}$$

In order to obtain posterior samples of μ_{jk} and σ_{jk}^2 rather than α_{jk} and τ_{jk}, we may transform the posterior samples of α_{jk} and τ_{jk} to μ_{jk} and σ_{jk}^2.

2.5.3 Estimation

Let Ω denote the set of hyperparameters for the prior distributions of the parameters in Θ as specified in Equation (2.17). The joint probability of $\mathbf{Y}$ and Θ conditional upon Ω, Σ is

$$
\begin{aligned}
p\left(\mathbf{Y}, \Theta | \boldsymbol{\Sigma}, \Omega\right) = &\prod_i^I \prod_j^J (2\pi)^{-1/2} \left(\sum_k^K g_{ik}\tau_{jk}\right)^{1/2} \\
&\cdot \exp\left(-\frac{\sum_k^K g_{ik}\tau_{jk}}{2}\left[y_{ij} - \left(\sum_k^K g_{ik}\tau_{jk}\right)^{-1}\sum_k^K g_{ik}\alpha_{jk}\right]^2\right) \\
&\prod_j^J \prod_k^K \left(2\pi s_{\alpha_{jk}}^2\right)^{-1/2} \exp\left(-\frac{1}{2s_{\alpha_{jk}}^2}\left(\alpha_{jk} - m_{\alpha_{jk}}\right)^2\right) \\
&\prod_j^J \prod_k^K \frac{\phi_{\tau_{jk}}^{\nu_{\tau_{jk}}}}{\Gamma\left(\nu_{\tau_{jk}}\right)} \tau_{jk}^{\nu_{\tau_{jk}}-1} \exp\left(-\phi_{\tau_{jk}}\tau_{jk}\right) \\
&\prod_i^I (2\pi)^{-(K-1)/2} |\boldsymbol{\Sigma}|^{-1/2} \exp\left(-\frac{1}{2}\left(\boldsymbol{\eta}_i - \rho\right)^T \boldsymbol{\Sigma}^{-1}\left(\boldsymbol{\eta}_i - \rho\right)\right) \\
&\prod_k^{K-1} \left(2\pi s_{\rho_k}^2\right)^{-1/2} \exp\left(-\frac{1}{2s_{\rho_k}^2}\left(\rho_k - m_{\rho_k}\right)^2\right).
\end{aligned}
\tag{2.24}
$$

As Heller et al. (2008) noted, all of the parameters in Θ are continuous and, moreover, we may take the derivatives of the log of the above probability expression. As a result, the problem of Bayesian estimation for this model lends itself to Hybrid (Hamiltonian) Monte Carlo. Hybrid Monte Carlo uses the derivative of the log joint probability to inform its proposals. As a result, in high dimensions, this algorithm may outperform more traditional algorithms such as Metropolis-Hastings or Gibbs sampling. For a thorough introduction to Hybrid Monte Carlo, see Neal (2010). In order to avoid the imposition of non-negativity restrictions on τ_{jk} in the Hybrid Monte Carlo algorithm, we employ the transformation $\eta_{\tau_{jk}} = \log(\tau_{jk})$ so that the parameter may take values unrestricted over the real line.

We do not rely on Hybrid Monte Carlo to draw Σ but rather draw Σ in a separate Gibbs step for the correlated partial membership model. Thus, to sample (Θ, Σ), we apply a Gibbs sampling algorithm where the first step involves sampling Θ via Hybrid Monte Carlo and then Σ from its full conditional distribution,

$$\boldsymbol{\Sigma} \sim \text{Inv. Wishart}\left(\nu_\Sigma + n, S_\Sigma + \left(\mathbf{H}_G - \mathbf{1}_n\rho^T\right)^T\left(\mathbf{H}_G - \mathbf{1}_n\rho^T\right)\right), \tag{2.25}$$

where $\mathbf{H}_G$ is a $n \times K - 1$ matrix of the untransformed membership scores.

2.6   Application to the NBA Player Data

We now apply the correlated membership model to NBA player data from the 2010–11 season. We considered models with 4, 5, and 6 pure types. We employed posterior predictive model checks to examine the fit of the model-based marginal distributions and rank correlations to the observed data. We ultimately settled on a model with 5 pure types as this model had the smallest number of classes that still provided sufficient fit to the data. The 5 pure type correlated partial membership model resulted in easily interpretable classes from a substantive viewpoint. Also, each pure type had at least one membership score above 0.20, meaning that at least one player had 1/5 or more of their membership in that type.

We ran the Gibbs sampling algorithm with a Hybrid Monte Carlo step for 80,000 iterations, keeping every 20th draw. We discarded the first 1000 of the retained draws as burn-in, leaving us with 3000 samples from the posterior distribution. To asses convergence, we examined trace plots and used the Geweke (Geweke, 1992) and Raftery-Lewis (Raftery and Lewis, 1995) diagnostic tests.

In examining the posterior estimates for the pure type specific means, μ_{jk}, presented in Table 2.6, we notice that some of the posterior means take negative values when all of the statistics recorded are strictly positive. For example, in the case of the % Ast statistic (the percentage of made field goals that are assisted), the range of the data is [0, 100], yet only one of the estimated pure type means lies inside this range. This observation is not worrisome by itself as it could be that no individual has high membership values in the pure types with negative means. We are more concerned with the associated predictive distributions for the observed data that are directly related to model fit. Nonetheless, when a pure type is characterized by values outside the range of observed data, the interpretation of this pure type is more complicated than of those pure types that can in principle be achievable in the population.

Figure 2.7 presents a posterior predictive model check that compares the marginal distribution of the percent of made fields goals assisted (% Ast) statistic against the replicated values for the statistic. The histogram depicts the observed data while the black points represent the posterior predictive mean count of replicated values falling in the corresponding bin. The black segment represents the 95% credible interval. We observe in Figure 2.7 that the model fits the marginal distribution of the data well; we obtained similar findings for other variables (not shown).

Although the model provides a good fit to the observed data, the shortcoming of this model is that it still places (small) non-zero predictive density in the improbable region of the data. This shortcoming will naturally arise when we use a normal distribution to model range-restriced data.

Examining the ordering of the posterior means can provide us with a way to characterize the pure types in relation to one another. Table 2.6 illustrates that pure type 1 comprises players who play a high number of minutes (Min), have a high percentage of their shots assisted (% Ast), shoot mid- and long-range jumpers (Medium, Long, 3s), and have a low steals rate (Stls). We refer to this pure type as the "high minute shooters." A high percentage of shots assisted (% Ast) and high volume of 3-point shots (3s) also describes pure type 2, but members of this pure type have fewer shots at all other distances (Rim, Close, Medium, Long) and a lower number of minutes played. We refer to this pure type as the "3-point specialists." The posterior means for the 3rd pure type are high relative to those for the other pure types across almost all variables except for the the mid- to long-range jumpers (Medium, Long, 3s). We use the term "active player" for this pure type. Low minutes played (Min) and high offensive rebound rates (ORR) are the most distinguishing features of pure type 4 which we refer to as the "limited big men" pure type. High assist (AR) and turnover (TOR) ratios, high steals per 40 minutes, a low percentage of shots assisted and low blocked shots per 40 minutes mark the final pure type. We refer to this pure type as the "ball handlers" pure type.

Figure 2.8 presents the mean posterior memberships of the players in these different pure types. The points' symbols denote their assigned position recorded in the original dataset. Here, we can see

TABLE 2.6

Posterior means for pure type mean parameters, μ_{jk}.

Var.	Pure Type				
	1	2	3	4	5
Min	42.40	13.36	416.04	18.09	29.43
% Ast	132.09	108.08	-179.33	70.81	-12.47
AR	4.36	131.68	916.56	9.20	607.91
TOR	6.55	-132.22	210.90	16.96	194.49
ORR	-29.30	-5.90	1552.00	9.34	1.70
DRR	228.25	7.51	1077.20	17.26	8.01
Rim	0.26	-0.64	116.07	3.73	6.03
Close	14.73	-0.13	238.88	1.38	2.07
Medium	340.24	0.08	10.87	1.17	3.35
Long	77.21	0.72	-33.99	3.05	3.81
3s	10.80	15.21	22.61	0.02	2.14
Stls	-3.51	0.96	54.93	0.81	1.82
Blks	12.81	0.25	61.76	1.84	0.21

that high membership in some pure types corresponds to certain positional assignments. Thus, the highest memberships in the limited big men pure type (pure type 4) are obtained by centers (C) and power forwards (PF) while the ball handlers pure type (pure type 5) is dominated by point guards (PG). Membership in pure types 1–3 does not have a close correspondence with specific assigned positions. For pure types 1–3, and to a lesser extent for pure type 5, no players come close to being fully represented by the pure type. This explains why the model performs well for predicting the marginal probability for the % Ast outcome despite having posterior means for pure types 1–3 and 5 to be out of bounds on that variable.

In contrast to the original partial membership model with Dirichlet membership scores (Heller et al., 2008), the correlated partial membership model allows for a more flexible correlation structure among components of the membership vector. Table 2.7 presents the posterior mean correlations of the pure type memberships that range from -0.664 to 0.410. The limited big man pure type (pure type 4) shows low to moderate negative correlations with all other pure types. The active player pure type, on the other hand, shows low to moderate positive correlations with the high minute shooter and 3-point specialist pure types and small negative correlations with the limited big man and ball handler pure types. We note that it is impossible to observe positive correlations under the Dirichlet type models. This suggests that our decision to allow for more flexible modeling of the pure type membership correlations was appropriate for the data.

TABLE 2.7

Posterior mean correlations of membership scores.

	1	2	3	4	5
1	1.000	0.081	0.410	-0.528	0.160
2	0.081	1.000	0.223	-0.553	-0.080
3	0.410	0.223	1.000	-0.235	-0.328
4	-0.528	-0.553	-0.235	1.000	-0.664
5	0.160	-0.080	-0.328	-0.664	1.000

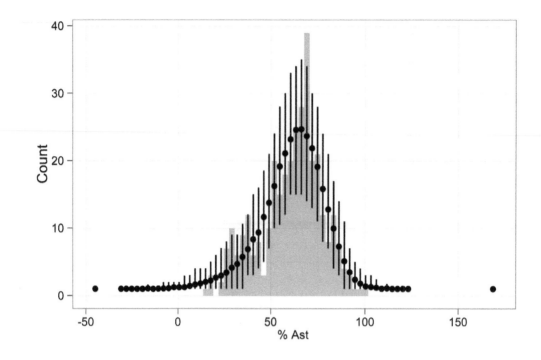

FIGURE 2.7
Histogram of the observed values for the % Ast statistic. The black points indicate the mean count across replicated datasets for each score. The black vertical segment indicates the interval from the 2.5% to 97.5% quantiles across replicated datasets.

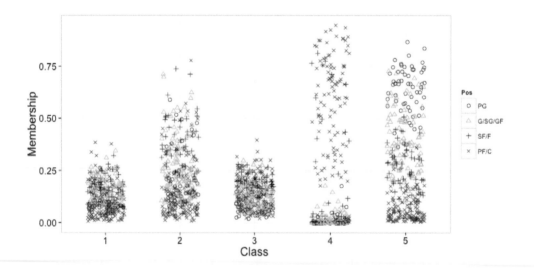

FIGURE 2.8
The mean posterior memberships of the players by pure type. The shapes of the points represent the different positions of each player.

To explore the compositional styles of NBA players further, consider the posterior mean membership scores and the corresponding credible intervals for three NBA "combo guards": Mario Chalmers, Steve Blake, and Rudy Fernandez, as identified by Lutz (2012). As a point of contrast, we examine the corresponding quantities for Chris Paul, who is generally considered to be an example of a pure point guard (Figure 2.9). We observe that 80% of Chris Paul's membership is in the ball handlers pure type. For the other three players, their membership is largely split between the ball handlers pure type and the 3-point specialists. Thus, we see that the correlated partial membership model describes the combo guard players using a mixture of pure types. This result stands in contrast to the results of the cluster analysis performed by Lutz (2012), where the combo guards comprised their own cluster, entirely separate from the other 12 clusters found in that analysis. Our correlated partial membership model uses only 5 pure types but characterizes the heterogeneity in individual playing styles as combinations of these pure types.

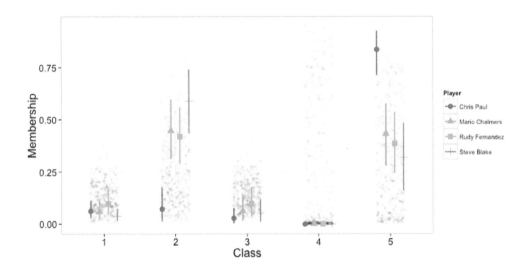

FIGURE 2.9
The mean posterior memberships and 95% posterior credible intervals of Chris Paul, Mario Chalmers, Rudy Fernandez, and Steve Blake. The grey points represent the posterior mean memberships of the other players in the data.

2.7 Summary and Discussion

In this chapter, we explored two individual-level mixture models for latent compositional data, namely, the mixed and partial memberships models. We found that the partial membership model has better potential for producing realistic representations of contiguous data patterns. However, we note that high-dimensional multivariate distributions of real data typically present even more complexity than the simulated examples considered here, which could easily mask the soft clustering nature of the underlying process. In such cases, one should consider a plausible interpretation for the latent compositional data at hand. For example, we point out that the partial membership formulation is consistent with the blending interpretation of mixed membership models as proposed by Galyardt (2014), because the NBA player dataset is primarily composed of continuous summary statistics. By contrast, in the binary data case, depending on the placement of pure type response probabilities, we observe that the partial membership model may result in a very particular behavior where fewer outcome combinations are possible compared to the Grade of Membership model. The implication of this finding for individual-level mixture models with binary data is that partial membership may not be appropriate for all binary data cases.

We modified the partial membership model to incorporate a logistic normal distribution for pure type memberships, similar to the correlated topic model extension (Blei and Lafferty, 2007) of the latent Dirichlet allocation models (Blei et al., 2003). This approach gave us more flexibility in specifying the dependence structure among the pure type memberships. We have illustrated the use of a partial membership model on continuous data using NBA player statistics. The NBA dataset provided an illustrative example where pure type membership scores exhibited both negative and positive correlations. We note that it is not possible to obtain positive correlations when one employs a Dirichlet distribution for the membership scores.

Although our partial membership analysis of the NBA player data resulted in a good fit as measured by the posterior predictive model checks, the limitation of using Gaussian pure type distributions is that the predicted values may lie outside of the allowable data intervals for variables that are constrained in their range. While it may be possible to specify other distributions for the pure types that can produce suitably constrained predicted values, a more general semiparametric approach that can accommodate not only range-restricted variables but also mixed data with both discrete and continuous outcomes could be more beneficial going forward (Gruhl et al., 2013). Examples of mixed outcome data are increasingly common in medicine and the social sciences, and the development of individual-level mixture models could be helpful for characterizing patterns in multivariate mixed outcomes.

References

Airoldi, E. M., Blei, D. M., Fienberg, S. E., and Xing, E. P. (2008). Mixed membership stochastic blockmodels. *The Journal of Machine Learning Research* 9: 1981–2014.

Atchison, J. and Shen, S. (1980). Logistic-normal distributions: Some properties and uses. *Biometrika* 67: 261–272.

Blei, D. M. and Lafferty, J. D. (2007). A correlated topic model of Science. *Annals of Applied Statistics* 1: 17–35.

Blei, D. M., Ng, A. Y., and Jordan, M. I. (2003). Latent Dirichlet allocation. *The Journal of Machine Learning Research* 3: 993–1022.

Brustein, J. (2012). Data Crunchers Look to Quantify Chemistry in N.B.A. Off The Dribble: The New York Times N.B.A. Blog.

Erosheva, E. A. (2002). Grade of Membership and Latent Structure Models with Application to Disability Survey Data. Ph.D. thesis, Carnegie Mellon University, Pittsburgh, Pennsylvania, USA.

Erosheva, E. A. (2003). Bayesian estimation of the Grade of Membership model. In Bernardo, J. M., Bayarri, M. J., Berger, J. O., Dawid, A. P., Heckerman, D., Smith, A. F. M., and West, M. (eds), *Bayesian Statistics 7*. New York, NY: Oxford University Press, 501–510.

Erosheva, E. A. (2004). Partial membership models with application to disability survey data. In *Statistical Data Mining and Knowledge Discovery*. Chapman & Hall/CRC, 117–134.

Erosheva, E. A. (2005). Comparing latent structures of the Grade of Membership, Rasch, and latent class models. *Psychometrika* 70: 619–628.

Erosheva, E. A. (2006). Latent Class Representation of the Grade of Membership Model. Tech. report 492, University of Washington.

Erosheva, E. A., Fienberg, S. E., and Joutard, C. (2007). Describing disability through individual-level mixture models for multivariate binary data. *Annals of Applied Statistics* 1: 502–537.

Fraley, C., Raftery, A. E., Murphy, T. B., and Scrucca, L. (2012). MCLUST Version 4 for R: Normal Mixture Modeling for Model-Based Clustering, Classification, and Density Estimation. Tech. report 597, University of Washington.

Galyardt, A. (2014). Interpreting mixed membership: Implications of Erosheva's representation theorem. In Airoldi, E. M., Blei, D. M., Erosheva, E. A., and Fienberg, S. E. (eds), *Handbook of Mixed Membership Models and Its Applications*. Chapman & Hall/CRC.

Geweke, J. (1992). Evaluating the accuracy of sampling-based approaches to calculating posterior moments. In Bernardo, J., Berger, J., Dawid, A., and Smith, J. (eds), *Bayesian Statistics 4*. Oxford, UK: Oxford University Press, 169–193.

Gormley, I. C. and Murphy, T. B. (2009). A Grade of Membership model for rank data. *Bayesian Analysis* 4: 265–296.

Gruhl, J., Erosheva, E. A., and Crane, P. (2013). A semiparametric approach to mixed outcome latent variable models: Estimating the association between cognition and regional brain volumes. *Annals of Applied Statistics* To appear.

Heller, K. A., Williamson, S., and Ghahramani, Z. (2008). Statistical models for partial membership. In *Proceedings of the 25th International Conference on Machine Learning (ICML '08)*. New York, NY, USA: ACM, 392–399.

Hoopdata (2012). NBA Player Statistics. www.hoopdata.com.

Lazarsfeld, P. F. and Neil, H. W. (1968). *Latent Structure Analysis*. Boston, MA: Houghton Mifflin.

Lutz, D. (2012). A cluster analysis of NBA players. In *Proceedings of the MIT Sloan Sports Analytics Conference*. Boston, MA, USA.

Neal, R. M. (2010). MCMC using Hamiltonian dynamics. In Brooks, S., Gelman, A., Jones, G., and Meng, X. -L. (eds), *Handbook of Markov Chain Monte Carlo*. Chapman & Hall/CRC, 113–162.

Pritchard, J., Stephens, M., and Donnelly, P. (2000). Inference of population structure using multi-locus genotype data. *Genetics* 155: 945–959.

Raftery, A. E. and Lewis, S. (1995). The number of iterations, convergence diagnostics and generic metropolis algorithms. In Gilks, W. R., Richardson, S., and Spiegelhalter, D. (eds), *Markov Chain Monte Carlo in Practice*. London, U.K.: Chapman and Hall.

Rogers, S., Girolami, M., Campbell, C., and Breitling, R. (2005). The latent process decomposition of cDNA microarray data sets. *IEEE/ACM Transactions on Computational Biology and Bioinformatics* 2: 143–156.

Wang, C., Blei, D. M., and Fei-Fei, L. (2009). Simultaneous image classification and annotation. In *Proceedings of the IEEE Conference on Computer Vision and Pattern Recognition (CVPR 2009)*. Los Alamitos, CA, USA: IEEE Computer Society, 1903–1910.

Woodbury, M. A., Clive, J., and Garson, A., Jr. (1978). Mathematical typology: A Grade of Membership technique for obtaining disease definition. *Computers and Biomedical Research* 11: 277–298.

3

Interpreting Mixed Membership Models: Implications of Erosheva's Representation Theorem

April Galyardt

Department of Educational Psychology, University of Georgia, Athens, GA 30602, USA

CONTENTS

The original three mixed membership models all analyze categorical data. In this special case there are two equivalent interpretations of what it means for an observation to have mixed membership. Individuals with mixed membership in multiple profiles may be considered to be 'between' the profiles, or they can be interpreted as 'switching' between the profiles. In other variations of mixed membership, the between interpretation is inappropriate. This chapter clarifies the distinction between the two interpretations and characterizes the conditions for each interpretation. I present a series of examples that illustrate each interpretation and demonstrate the implications for model fit. The most counterintuitive result may be that no change in the distribution of the membership parameter will allow for a between interpretation.

3.1 Introduction

The idea of mixed membership is a simple, intuitive idea. Individuals in a population may belong to multiple subpopulations, not just a single class. A news article may address multiple topics rather than fitting neatly in a single category (Blei et al., 2003). Patients sometimes get multiple diseases at the same time (Woodbury et al., 1978). An individual may have genetic heritage from multiple subgroups (Pritchard et al., 2000; Shringarpure, 2012). Children may use multiple strategies in mathematics problems rather than sticking to a single strategy (Galyardt, 2012).

The problem of how to turn this intuitive idea into an explicit probability model was originally solved by Woodbury et al. (1978) and later independently by Pritchard et al. (2000) and Blei et al. (2003). Erosheva (2002) and Erosheva et al. (2004) then built a general mixed membership framework to incorporate all three of these models.

Erosheva (2002) and Erosheva et al. (2007) also showed that every mixed membership model has an equivalent finite mixture model representation. The proof in Erosheva (2002) shows that the relationship holds for categorical data; Erosheva et al. (2007) indicates that the same result holds in general.

The behavior of mixed membership models is best understood in the context of this representation theorem. The shape of data distributions, the difference between categorical and continuous data, possible interpretations, and identifiability all flow from the finite mixture representation (Galyardt, 2012). This chapter describes the general mixed membership model and then explores the implications of Erosheva's representation theorem.

3.2 The Mixed Membership Model

Due to the history of mixed membership models, and the fact that they were independently developed multiple times, there are now two common and equivalent ways to define mixed membership models. The generative model popularized by Blei et al. (2003) is more intuitive so we will discuss it first, followed by the the general model (Erosheva, 2002; Erosheva et al., 2004).

3.2.1 The Generative Process

The generative version of mixed membership is the more common representation in the machine learning community. This is due largely to the popularity of latent Dirichlet allocation (LDA) (Blei et al., 2003), which currently has almost 5000 citations according to Google Scholar. LDA has inspired a wide variety of mixed membership models, e.g., see Fei-Fei and Perona (2005), Girolami and Kaban (2005), and Shan and Banerjee (2011), though these models still fit within the general mixed membership model of Erosheva (2002) and Erosheva et al. (2004).

The foundation of the mixed membership model is the assumption that the population consists of K profiles, indexed $k = 1, \ldots, K$, and that each individual $i = 1, \ldots, N$ belongs to the profiles in different degrees. If the population is a corpus of documents, then the profiles may represent the topics in the documents. If we are considering the genetic makeup of a population of birds, then the profiles may represent the original populations that have melded into the current population. In image analysis, the profiles may represent the different categories of objects or components in the images, such as *mountain, water, car*, etc. When modeling the different strategies that students use to solve problems, each profile can represent a different strategy.

Each individual has a membership vector, $\theta_i = (\theta_{i1}, \ldots, \theta_{iK})$, that indicates the degree to which they belong to each profile. The term *individual* here simply refers to a member of the population and could refer to an image, document, gene, person, etc. The components of θ are non-negative and sum to 1, so that θ can be treated as a probability vector. For example, if student i used strategies 1 and 2, each about half the time, then this student would have a membership vector of $\theta_i = (0.5, 0.5, 0, \ldots, 0)$. Similarly, if an image was 40% water and 60% mountain then this would be indicated by θ_i.

Each observed variable X_j, $j = 1, \ldots, J$ has a different probability distribution within each profile. For example, in an image processing application, the water profile has a different distribution of features than the mountain profile. In another application, such as an assessment of student learning, different strategies may result in different response times on different problems. Note that X_j may be univariate or be multidimensional itself, and that we may observe $r = 1, \ldots, R_{ij}$ replications of X_j for each individual i, denoted X_{ijr}. The distribution of X_j within profile k is given by the cumulative distribution function (cdf) F_{kj}.

We introduce the indicator vector Z_{ijr} to signify which profile individual i followed for replication r of the jth variable. For example, in textual analysis, Z_{ijr} would indicate which topic the rth word in document i came from. In genetics, Z_{ijr} indicates which founding population individual i inherited the rth copy of their jth allele from.

The membership vector θ_i indicates how much each individual belongs to each profile so that $Z_{ijr} \sim Multinomial(\theta_i)$. We will write Z_{ijr} in the form that, if individual i followed profile k for replication r of variable j, then $Z_{ijr} = k$. The distribution of X_{ijr} given Z_{ijr} is then

$$X_{ijr}|Z_{ijr} = k \quad \sim \quad F_{kj}. \tag{3.1}$$

The full data generating process for individual i is then given by:

1. Draw $\theta_i \sim D(\theta)$.

2. For each variable $j = 1, \ldots, J$:

    (a) For each replication $r = 1, \ldots, R_{ij}$:
        i. Draw a profile $Z_{ijr} \sim Multinomial(\theta_i)$.
        ii. Draw an observation $X_{ijr} \sim F_{Z_{ijr},j}(x_j)$ from the distribution of X_j associated with the profile Z_{ijr}.

3.2.2 General Mixed Membership Model

The general mixed membership model (MMM) makes explicit the assumptions that are tacit within the general model. These assumptions are collected into four layers of assumptions: population level, subject level, sampling scheme, and latent variable level.

The ***population level*** assumptions are that there are K different profiles within the population, and each has a different probability distribution for the observed variables F_{kj}.

The ***subject level*** assumptions begin with the individual membership parameter θ_i that indicates which profiles individual i belongs to. We then assume that the conditional distribution of X_{ij} given θ_i is:

$$F(x_j|\theta_i) = \sum_{k=1}^{K} Pr(Z_{ijrk} = 1|\theta_i)F(x_j|Z_{ijrk} = 1), \tag{3.2}$$

$$= \sum_{k=1}^{K} \theta_{ik} F_{kj}(x_j). \tag{3.3}$$

Equation (3.3) is the result of combining Steps 2(a)i and 2(a)ii in the generative process. Z_{ijr} is simply a data augmentation vector, and we can easily write the distribution of the observed data without it. Notice that Step 2 of the generative process assumes that the X_{ijr} are independent given θ_i. In psychometrics this is known as a local independence assumption. This exchangeability assumption allows us to write the joint distribution of the response vector $X_i = (X_{i1}, \ldots, X_{iJ})$, conditional on θ_i as

$$F(x|\theta_i) = \prod_{j=1}^{J} \left[\sum_{k=1}^{K} \theta_{ik} F_{kj}(x_j) \right]. \tag{3.4}$$

This conditional independence assumption also contains the assumption that the profile distributions are themselves factorable. If an individual belongs exclusively to profile k (for example, an image contains only water), then $\theta_{ik} = 1$, and all other elements in the vector θ_i are zero. Thus,

$$F(x|\theta_{ik} = 1) = \prod_j F_{kj}(x_j) = F_k(x). \tag{3.5}$$

The *sampling scheme level* includes the assumptions about the observed replications. Step 2(a) of the generative process assumes that replications are independent given the membership vector θ_i. Thus the individual response distribution becomes:

$$F(x|\theta_i) = \prod_{j=1}^{J} \prod_{r=1}^{R_{ij}} \left[\sum_{k=1}^{K} \theta_{ik} F_{kj}(x_{j_r}) \right]. \tag{3.6}$$

Note that Equations (3.3), (3.4), and (3.6) vary for each individual with the value of θ_i. It is in this sense that MMM is an individual-level mixture model. The distribution of variables for each profile, the F_{kj}, is fixed at the population level, so that the components of the mixture are the same, but the proportions of the mixture change individually with the membership parameter θ_i.

The *latent variable level* corresponds to Step 1 of the generative process. We can treat the membership vector θ as either fixed or random. If we wish to treat θ as random, then we can integrate Equation (3.6) over the distribution of θ, yielding:

$$F(x) = \int \prod_{j=1}^{J} \prod_{r=1}^{R_{ij}} \left[\sum_{k=1}^{K} \theta_{ik} F_{kj}(x_j) \right] d\mathrm{D}(\theta). \tag{3.7}$$

The final layer of assumptions about the latent variable θ is crucial for purposes of estimation, but it is unimportant for the discussion of mixed membership model properties in this chapter. All of the results presented here flow from the exchangeability assumption in Equation (3.4), and hold whether we use Equation (3.6) or (3.7) for estimation.

3.3　The Development of Mixed Membership

Independently, Woodbury et al. (1978), Pritchard et al. (2000), and Blei et al. (2003) developed remarkably similar mixed membership models to solve problems in three very different content areas.

Grade of Membership Model

The Grade of Membership model (GoM) is by far the earliest example of mixed membership (Woodbury et al., 1978). The motivation for creating this model came from the problem of designing a system to help doctors diagnose patients. The problems with creating such a system are numerous: Patients may not have all of the classic symptoms of a disease, they may have multiple diseases, relevant information may be missing from a patient's profile, and many diseases have similar symptoms.

In this setting, the mixed membership profiles represent distinct diseases. The observed data X_{ij} are categorical levels of indicator j for patient i. The profile distributions $F_{kj}(x_j)$ indicate which level of indicator j is likely to be present in disease k. Since X_{ij} is categorical, and there is only one measurement of an indicator for each patient, the profile distributions are multinomial with $n = 1$. In this application, the individual's disease profile is the object of inference, so that the likelihood in Equation (3.4) is used for estimation.

Population Admixture Model

Pritchard et al. (2000) models the genotypes of individuals in a heterogeneous population. The profiles represent distinct populations of origin from which individuals in the current population have inherited their genetic makeup.

The variables X_j are the genotypes observed at J locations, and for diploid individuals two replications are observed at each location ($R_j = 2$). Across a population, a finite number of distinct alleles are observed at each location j, so that X_j is categorical and F_{kj} is multinomial for each sub-population k.

In this application, the distribution of the membership parameters θ_i is of as much interest as the parameters themselves. The parameters θ_i are treated as random realizations from a symmetric Dirichlet distribution. It is important to note that a symmetric Dirichlet distribution will result in an identifiability problem that is not present when θ has an asymmetric distribution (Galyardt, 2012).

One interesting feature of the admixture model is that it includes the possibility of both unsupervised and supervised learning. Most mixed membership models are estimated as unsupervised models. That is, the models are estimated with no information about what the profiles may be and no information about which individuals may have some membership in the same profiles. Pritchard et al. (2000) considers the unsupervised case, but also considers the case where there is additional information. In this application, the location where an individual bird was captured means that it is likely a descendent of a certain population with a lower probability that it descended from an immigrant. This information is included with a carefully constructed prior on θ, which also incorporates rates of migration.

Latent Dirichlet Allocation

Latent Dirichlet allocation (Blei et al., 2003) is in some ways the simplest example of mixed membership, as well as the most popular. LDA is a textual analysis model, where the goal is to identify the topics present in a corpus of documents. Mixed membership is necessary because many documents are about more than one topic.

LDA uses a "bag-of-words" model, where only the presence or absence of words in a document is modeled and word order is ignored. The individuals i are the documents. The profiles k represent the topics present in the corpus. LDA models only one variable, the words present in the documents ($J = 1$). The number of replications R_{ij} is simply the number of words in document i. The profile distributions are multinomial distributions over the set of words: $F_{kj} = Multinomial(\lambda_k, n = 1)$, where λ_{kw} is the probability of word w appearing in topic k. LDA uses the integrated likelihood in Equation (3.7). The focus here is on estimating the topic profiles, and the distribution of membership parameters, rather than the θ_i themselves. LDA also uses a Dirichlet distribution for θ, however it does not use a *symmetric* Dirichlet, and so it avoids the identifiability issues that are present in the admixture model (Galyardt, 2012).

3.3.1 Variations of Mixed Membership Models

Variations of mixed membership models fall into two broad groups: The first group alters the distribution of the membership parameter θ, the second group alters the profile distributions F_{kj}.

Membership Parameters

The membership vector θ is non-negative and sums to 1 so that it lies within a $K - 1$ dimensional simplex. The two most popular distributions on the simplex are the Dirichlet and the logistic-normal.

Both LDA and the population admixture model use a Dirichlet distribution as the prior for the membership parameter. This is the obvious choice when the data is categorical, since the Dirichlet distribution is a conjugate prior for the multinomial. However, the Dirichlet distribution introduces a strong independence condition on the components of θ subject to the constraint $\sum_k \theta_{ik} = 1$ (Aitchison, 1982).

In many applications, this strong independence assumption is a problem. For example, an article with partial membership in an evolution topic is more likely to also be about genetics than astronomy. In order to model an interdependence between profiles, Blei and Lafferty (2007) uses a logistic-normal distribution for θ. Blei and Lafferty (2006) takes this idea a step further and creates a dynamic model where the mean of the logistic-normal distribution evolves over time.

Fei-Fei and Perona (2005) analyzes images, where the images contain different proportions of the profiles *water, sky, foliage*, etc. However, images taken in different locations will have a different underlying distribution for the mixtures of each of these profiles. For example, rural scenes will have more foliage and fewer buildings than city scenes. Fei-Fei and Perona (2005) addresses this by giving the membership parameters a distribution that is a mixture of Dirichlets.

Profiles

In all three of the original models, the data are categorical and the profile distributions F_{kj} are multinomial. More recently, we have seen a variety of mixed membership models for data that is not categorical, with different parametric families for the F_k distributions.

Latent process decomposition (Rogers et al., 2005) describes the different processes that might be responsible for different levels of gene expression observed in microarray datasets. In this application, X_{ij} measures the expression level of the jth gene in sample i, a continuous quantity. This leads to profile distributions $F_{kj} = N(\mu_{kj}, \sigma_{kj})$.

The simplical mixture of Markov chains (Girolami and Kaban, 2005) is a mixed membership model where each profile is characterized by a Markov chain transition matrix. The idea is that over time an individual may engage in different activities, and each activity is characterized by a probable sequence of actions.

The mixed membership naive Bayes model (Shan and Banerjee, 2011) is another extension of LDA which seeks to define a 'generalization' of LDA. This model simply requires the profile distributions F_{kj} to be exponential family distributions. This is a subset of models that falls within Erosheva's general mixed membership model (Erosheva et al., 2004). Moreover, other exponential family profile distributions will not have the same properties as the multinomial profiles used in LDA (Galyardt, 2012). The main contribution of Shan and Banerjee (2011) is a comparison of different variational estimation methods for particular choices of F_{kj}.

3.4   The Finite Mixture Model Representation

Before we discuss the relationship between mixed membership models (MMM) and finite mixture models (FMM), we will briefly review FMM.

3.4.1   Finite Mixture Models

Finite mixture models (FMM) go by many different names, such as "latent class models" or simply "mixture models," and they are used in many different applications from psychometrics to clustering and classification.

The basic assumption is that within the population there are different subgroups, $s = 1, \ldots, S$, which may be called clusters or classes depending on the application. Each subgroup has its own distribution of data, $F_s(x)$, and each subgroup makes up a certain proportion of the population, π_s.

The distribution of data across the population is then given by:

$$F(x) = \sum_{s=1}^{S} \pi_s F_s(x). \tag{3.8}$$

For reference, the distribution of data over the population in a MMM, given by Equation (3.7), is:

$$F(x) = \int \prod_{j=1}^{J} \prod_{r=1}^{R_{ij}} \left[\sum_{k=1}^{K} \theta_{ik} F_{kj}(x_j) \right] d\mathrm{D}(\theta). \tag{3.9}$$

Finite mixture models can be considered a special case of mixed membership models. In a mixed membership model, the membership vector θ_i indicates how much individual i belongs to each of the profiles k, thus θ lies in a $K-1$ dimensional simplex. If the distribution of the membership parameter θ is restricted to the corners of the simplex, then θ_i will be an indicator vector and Equation (3.9) will reduce to the form of Equation (3.8). So a finite mixture model is a special case of mixed membership with a particular distribution of θ.

3.4.2 Erosheva's Representation Theorem

Even though FMM is a special case of MMM, every MMM can be expressed in the form of an FMM with a potentially much larger number of classes. Haberman (1995) suggests this relationship in his review of Manton et al. (1994). Erosheva et al. (2007) shows that it holds for categorical data and indicates that the same result holds in the general case as well. Here the theorem is presented in a general form.

Before we consider the formal version of the theorem, we can build some intuition based on the generative version of MMM. In the generative process, to generate the data point X_{ijr} for individual i's replication r of variable j, we first draw an indicator variable $Z_{ijr} \sim Multinomial(\theta_i)$ that indicates which profile X_{ijr} will be drawn from. Let us write Z_{ijr} in the form: $Z_{ijr} = k$, if X_{ijr} was drawn from profile k. Effectively, Z indicates that individual i 'belongs' to profile k for observation j_r.

The set of all possible combinations of Z defines a set of FMM classes, which we shall write as $\mathcal{Z} = \{1, \ldots, K\}^R$, where R is the total number of replications of all variables. For individual i, let $\zeta_i = (Z_{i11}, \ldots, Z_{iJR_J}) \in \mathcal{Z}$. So ζ_i indicates which profile an individual belongs to for each and every observed variable.

Representation Theorem. *Assume a mixed membership model with J features and K profiles. To account for any replications in features, assume that each feature j has R_j replications, and let $R = \sum_{j=1}^{J} R_j$. Write the profile distributions as*

$$F_k(x) = \prod_{r=1}^{R} F_{kr}(x_r).$$

Then the mixed membership model can be represented as a finite mixture model with components indexed by $\zeta \in \{1, \ldots, K\}^R = \mathcal{Z}$, where the classes are

$$F_\zeta^{FMM}(x) = \prod_{r=1}^{R} F_{\zeta_r, r}(x_r) \tag{3.10}$$

and the probability associated with each class ζ is

$$\pi_\zeta = \mathbb{E} \left[\prod_{r=1}^{R} \theta_{\zeta_r} \right]. \tag{3.11}$$

Proof. Begin with the individual mixed membership distribution, conditional on θ_i.

$$F(x|\theta_i) \quad = \quad \prod_r \sum_k \theta_{ik} F_{kr}(x_r), \tag{3.12}$$

$$= \quad \sum_{\zeta \in \mathcal{Z}} \prod_r \theta_{i\zeta_r} F_{\zeta_r r}(x_r). \tag{3.13}$$

Equation (3.13) reindexes the terms of the finite sum when Equation (3.12) is expanded. Distributing the product over r yields Equation (3.14):

$$F(x|\theta_i) \quad = \quad \sum_{\zeta \in \mathcal{Z}} \left(\left[\prod_r \theta_{i\zeta_r} \right] \left[\prod_r F_{\zeta_r r}(x_r) \right] \right), \tag{3.14}$$

$$= \quad \sum_{\zeta \in \mathcal{Z}} \pi_{i\zeta} F_\zeta(x). \tag{3.15}$$

Integrating Equation (3.15) yields the form of a finite mixture model:

$$F(x) = \mathbb{E}_\theta \left[\sum_{\zeta \in \mathcal{Z}} \pi_{i\zeta} F_\zeta(x) \right] = \sum_{\zeta \in \mathcal{Z}} \pi_\zeta F_\zeta(x). \tag{3.16}$$

$\square$

Erosheva's representation theorem states that if a mixed membership model needs K profiles to express the diversity in the population, an equivalent finite mixture model will require K^R components. In addition, if we compare Equation (3.15) to Equation (3.16), then we see that each individual's distribution is also a finite mixture model, with the same components as the population FMM but with individual mixture proportions.

The mixed membership model is a much more efficient representation for high-dimensional data—we need only K profiles instead of K^R. However, there is a tradeoff in the constraints on the shape of the data distribution (Galyardt, 2012). The rest of this chapter will explore some of these constraints.

3.5   A Simple Example

A finite mixture model is described by the components of the mixture F_ζ and the proportion associated with each component, π_ζ. The representation theorem tells us that when a MMM is expressed in FMM form, the components are completely determined by MMM profiles (Equation 3.10), and that the proportions are completely determined by the distribution of the membership vector θ (Equation 3.11).

We can think of the MMM profiles F_{kj} as forming a *basis* for the FMM components F_ζ. Consider a very simple example with two dimensions ($J = 2$) and two profiles ($K = 2$). Suppose that the first profile has a uniform distribution on the unit square and the second profile has a concentrated normal distribution centered at (0.3, 0.7):

$$F_1(x) \quad = \quad F_{11}(x_1) \times F_{12}(x_2) \quad = \quad Unif(0,1) \times Unif(0,1), \tag{3.17}$$

$$F_2(x) \quad = \quad F_{21}(x_1) \times F_{22}(x_2) \quad = \quad N(0.3, 0.1) \times N(0.7, 0.1). \tag{3.18}$$

From a generative perspective, an individual with membership vector $\theta_i = (\theta_{i1}, \theta_{i2})$ will have $Z_{i1} = 1$ with probability θ_{i1} and $Z_{i1} = 2$ with probability θ_{i2}, so that $X_{ij} \sim Unif(0, 1)$ with probability θ_{i1}, and $X_{ij} \sim N(0.3, 0.1)$ with probability θ_{i2}. Similarly, for variable $j = 2$, with probability θ_{i1}, $Z_{i2} = 1$, and with probability θ_{i2}, $Z_{i2} = 2$. In total, there are $K^J = 4$ possible combinations of $\zeta_i = (Z_{i1}, Z_{i2})$:

$$X_i|\zeta_i = (1, 1) \quad \sim \quad Unif(0, 1) \times Unif(0, 1), \tag{3.19}$$

$$X_i|\zeta_i = (1, 2) \quad \sim \quad Unif(0, 1) \times N(0.7, 0.1), \tag{3.20}$$

$$X_i|\zeta_i = (2, 1) \quad \sim \quad N(0.3, 0.1) \times Unif(0, 1), \tag{3.21}$$

$$X_i|\zeta_i = (2, 2) \quad \sim \quad N(0.3, 0.1) \times N(0.7, 0.1). \tag{3.22}$$

Equations (3.19)–(3.22) are the four FMM components for this MMM model, F_ζ (Figure 3.1), and they are formed from all the possible combinations of the MMM profiles F_{kj}. It is in this sense that the MMM profiles form a basis for the data distribution.

The membership parameter θ_i governs how much individual i 'belongs' to each of the MMM profiles. If $\theta_{i1} > \theta_{i2}$, then $\zeta_i = (1, 1)$ is more likely than $\zeta_i = (2, 2)$. Notice, however, that since multiplication is commutative, $\theta_{i1}\theta_{i2} = \theta_{i2}\theta_{i1}$, so that $\zeta_i = (1, 2)$ always has the same probability as $\zeta_i = (2, 1)$.

Figure 3.2 shows the data distribution of this MMM for two different distributions of θ. The change in the distribution of θ affects only the probability associated with each component. Thus the MMM profiles define the modes of the data, and the distribution of θ controls the height of the modes.

Alternate Profiles

Consider an alternate set of MMM profiles, G:

$$G_1(x) \quad = \quad Unif(0, 1) \times N(0.7, 0.1), \tag{3.23}$$

$$G_2(x) \quad = \quad N(0.3, 0.1) \times Unif(0, 1). \tag{3.24}$$

The G profiles are essentially a rearrangement of the F profiles, and will generate exactly the same FMM components as the F profiles (Figure 3.3). For any MMM model, there are $K!^{(J-1)}$ sets of basis profiles which will generate the same set of components in the FMM representation (Galyardt, 2012). The observation that multiple sets of MMM basis profiles can generate the same FMM components has implications for the identifiability of MMM, which is explored fully in Galyardt (2012).

Multivariate X_j

The same results hold when X_j is multivariate. Consider an example where each profile F_{kj} is a multivariate Gaussian, as used in the GM-LDA model in Blei and Jordan (2003). Then we can write the profiles as:

$$F_1(x) \quad = \quad F_{11}(x_1) \times F_{12}(x_2) \quad = \quad MvN(\mu_{11}, \Sigma_{11}) \times MvN(\mu_{12}, \Sigma_{12}),$$

$$F_2(x) \quad = \quad F_{21}(x_1) \times F_{22}(x_2) \quad = \quad MvN(\mu_{21}, \Sigma_{21}) \times MvN(\mu_{22}, \Sigma_{22}).$$

The corresponding FMM components are then:

$$X_i|\zeta_i = (1, 1) \quad \sim \quad MvN(\mu_{11}, \Sigma_{11}) \times MvN(\mu_{12}, \Sigma_{12}), \tag{3.25}$$

$$X_i|\zeta_i = (1, 2) \quad \sim \quad MvN(\mu_{11}, \Sigma_{11}) \times MvN(\mu_{22}, \Sigma_{22}), \tag{3.26}$$

$$X_i|\zeta_i = (2, 1) \quad \sim \quad MvN(\mu_{21}, \Sigma_{21}) \times MvN(\mu_{12}, \Sigma_{12}), \tag{3.27}$$

$$X_i|\zeta_i = (2, 2) \quad \sim \quad MvN(\mu_{21}, \Sigma_{21}) \times MvN(\mu_{22}, \Sigma_{22}). \tag{3.28}$$

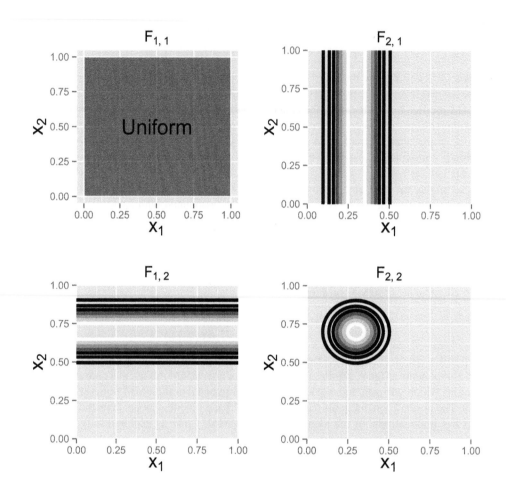

FIGURE 3.1

Each of the four boxes shows the contour plot of an FMM component in Equations (3.19)–(3.22). They correspond to the MMM defined by the F profiles in Equations (3.17)–(3.18). X_1 and X_2 are the two observed variables. Lighter contour lines indicate higher density.

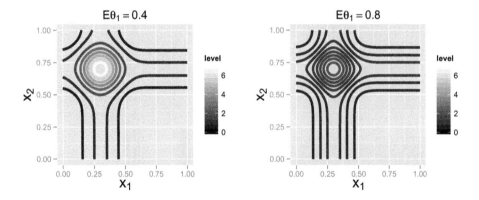

FIGURE 3.2
Contour plot of the MMM defined by the profiles in Equations (3.17)–(3.18) with two different distributions of θ. X_1 and X_2 are the two observed variables. Lighter contour lines indicate higher density; the scale is the same for both figures.

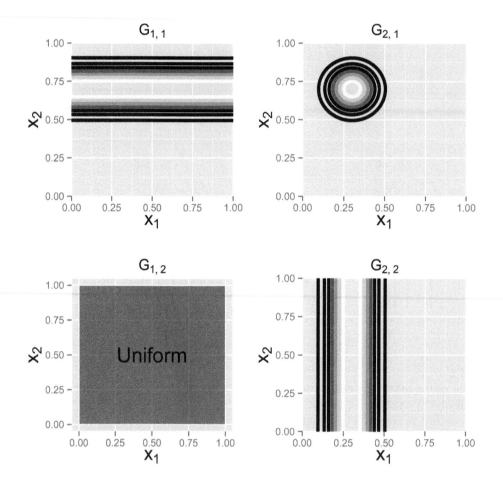

FIGURE 3.3
Each of the four boxes shows the contour plot of an FMM component corresponding to the MMM defined by the G profiles in Equations (3.23)–(3.24). Note that these are the same components as those defined by the F profiles in Figure 3.1 and Equations (3.19)–(3.22), simply re-indexed. X_1 and X_2 are the two observed variables. Lighter contour lines indicate higher density.

There are still K^R FMM components; the only difference is that these clusters are not in an R-dimensional space but a higher-dimensional space, depending on the dimensionality of the X_j.

3.6 Categorical vs. Continuous Data

All three of the original mixed membership models, and a majority of the subsequent variations, were built for categorical data. This focus on categorical data can lead to intuitions about mixed membership models which do not hold in the general case. Since every mixed membership model can be expressed as a finite mixture model, the best way to understand the difference between continuous and categorical data in MMM is to focus on how different data types behave in FMM.

Let us begin by considering the individual distributions conditional on profile membership (Equation 3.3):

$$F(x_j|\theta_i) = \sum_{k=1}^{K} \theta_{ik} F_{kj}(x_j).$$

In general, this equation does not simplify, but in the case of categorical data, it does. This is the key difference between categorical data and any other type of data.

If variable X_j is categorical, then we can represent the possible values for this variable as $\ell_1, \ldots, \ell_{L_j}$. We represent the distribution for each profile as $F_{kj}(x_j) = Multinomial(\lambda_{kj}, n = 1)$, where λ_{kj} is the probability vector for profile k on feature j, and n is the number of multinomial trials. The probability of observing a particular value l within basis profile k is written as:

$$Pr(X_j = l|\theta_k = 1) = \lambda_{kjl}. \tag{3.29}$$

The probability of individual i with membership vector θ_i having value l for feature j is then

$$Pr(X_{ij} = l|\theta_i) = \sum_{k=1}^{K} \theta_{ik} Pr(X_j = l|\theta_k = 1) = \sum_{k=1}^{K} \theta_{ik} \lambda_{kjl}. \tag{3.30}$$

Consider LDA as an example. Assume that document i belongs to the *sports* and *medicine* topics. The two topics each have a different probability distribution over the lexicon of words, say $Multinomial(\lambda_s)$ and $Multinomial(\lambda_m)$. The word *elbow* has a different probability of appearing in each topic, $\lambda_{s,e}$ and $\lambda_{m,e}$, respectively. Then the probability of the word *elbow* appearing in document i is given by $\lambda_i = \theta_{is}\lambda_{s,e} + \theta_{im}\lambda_{m,e}$. Since the vector θ_i sums to 1, the individual probability λ_i must be between $\lambda_{s,e}$ and $\lambda_{m,e}$. The individual probability is *between* the probabilities in the two profiles.

We can simplify the mathematics further if we collect the λ_{kj} into a matrix by rows and call this matrix λ_j. Then $\theta_i^T \lambda_j$ is a vector of length L_j where the lth entry is individual i's probability of value l on feature j, as in Equation (3.30).

We can now write individual i's probability vector for feature j as

$$\lambda_{ij} = \theta_i^T \lambda_j. \tag{3.31}$$

The matrix λ_j defines a linear transformation from θ_i to λ_{ij}, as illustrated in Figure 3.4. Since θ_i is a probability vector and sums to 1, λ_{ij} is a convex combination of the the profile probability vectors λ_{kj}. Thus the individual λ_{ij} lies within a simplex where the extreme points are the λ_{kj}. In other words, the individual response probabilities lie between the profile probabilities. This leads Erosheva et al. (2004) and others to refer to the profiles as "extreme profiles." For categorical data, the parameters of the profiles form the extremes of the individual parameter space.

Moreover, since the mapping from the individual membership parameters θ_i to the individual feature probabilities λ_{ij} is linear, the distribution of individual response probabilities is effectively the same as the population distribution of membership parameters (Figure 3.4).

Thus, when feature X_j is categorical, an individual with membership vector θ_i has a probability distribution of

$$F(x_j|\theta_i) = Multinomial(\theta_i^T \lambda_j, n = 1). \tag{3.32}$$

This is the property that makes categorical data special. When the profile distributions are multinomial with $n = 1$, the individual-level mixture distributions are also multinomial with $n = 1$. Moreover, we also have that the parameters of the individual distributions, the $\theta_i^T \lambda_j$, are convex combinations of the profile parameters, the λ_{kj}. In this sense, when the data are categorical, an individual with mixed membership in multiple profiles is effectively *between* those profiles.

In general, this between relationship does not hold. The general interpretation is a *switching* interpretation, and is clearly captured by the indicator variable Z_{ijr} in the generative model. Z_{ijr} indicates which profile distribution k generated the observation X_{ijr}. Thus, Z indicates that an individual switched from profile k for the jth variable to profile k' for the $j + 1^{st}$ variable.

The between interpretation for categorical data only holds in the multinomial parameter space: λ_i is between the profile parameters λ_k. The behavior in data space is the same switching behavior as defined in the general case. Individuals may only give responses that are within the support of at least one of the profiles.

Consider LDA as an example. The observation X_{ir} is the rth word appearing in document i; each profile is a multinomial probability distribution over the set of words. "Camel" may be a high probability word in the *zoo* topic, while "cargo" has high probability in the *transportation* topic. For a document with partial membership in the zoo and transportation topics, the word camel will have a probability of appearing that is between the probability of camel in the zoo topic and its probability in the transportation topic. Similarly for the word cargo. However, it doesn't make sense

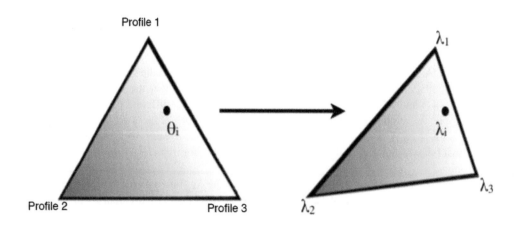

FIGURE 3.4

The membership parameter θ_i lies in a $K - 1$ simplex. When the mixed membership profiles are $F_{kj} = Multinomial(\lambda_{kj}, n = 1)$, the membership parameters are mapped linearly onto response probabilities (Equation 3.31), indicated by the arrow. The density, indicated by the shading, is preserved by the linear mapping. This mapping allows us to interpret individual i's position in the θ-simplex as equivalent to their response probability vector.

to talk about the word "cantaloupe" being between camel and cargo. With categorical data, there is no 'between' in the data-space. The between interpretation only holds in the parameter space.

Consider another example: suppose that we are looking at response times for a student taking an assessment, where X_{ij} is the response time of student i on item j and each profile represents a particular strategy. Suppose that one strategy results in a response time with a distribution $N(10, 1)$ and another less effective strategy has a response time distribution of $N(20, 2)$. In the mixed membership model, an individual with membership vector $\theta_i = (\theta_{i1}, \theta_{i2})$ then has a response time distribution of $\theta_{i1} N(10, 1) + \theta_{i2} N(20, 2)$. This individual may use strategy 1 or strategy 2, but a response time of 15 has a low probability under both strategies and in the mixture. The individual may switch between using strategy 1 and strategy 2 on subsequent items, but a response time between the two distributions is never likely, no matter the value of θ. Moreover, the individual distribution is no longer normal but a mixture of normals (Titterington et al., 1985). Thus, for this continuous data, we can use a switching interpretation, but a between interpretation is unavailable.

3.6.1 Conditions for a 'Between' Interpretation

The between interpretation arises out of a special property of the multinomial distribution: the individual probability distributions are in the same parametric family as the profile distributions, multinomial with $n = 1$, and the individual parameters are between the profile parameters (Equation 3.31 and Figure 3.4).

For the between interpretation to be available, this is the property we need to preserve. The individual distributions $F(x|\theta_i)$ must be in the same parametric family as each profile distribution F_k. Additionally, if F is parameterized by ϕ, then the individual parameters ϕ_i must lie between the profile parameters ϕ_k.

Thus, the property we are looking for is that an individual with membership parameter θ_i would have an individual data distribution of $F(X; \theta_i^T \phi)$, so that for each variable j we would have:

$$X_{ij}|\theta_i \sim \sum_k \theta_{ik} F_{kj}(X_j; \phi_{kj}) = F_j(X_j; \theta_i^T \phi_{.j}). \tag{3.33}$$

In other words, the between interpretation is only available if the profile cumulative distribution functions (cdfs) are linear transformations of their parameters. The only exponential family distribution with this property is the multinomial distribution with $n = 1$. Thus, it is the only common profile distribution which allows a between interpretation (Galyardt, 2012).

The partial membership models in Gruhl and Erosheva (2013) and Mohamed et al. (2013) use a likelihood that is equivalent to Equation (3.33) in the general case. This fundamentally alters the mixed membership exchangeability assumption for the distribution of $X_{ij}|\theta_i$ and preserves the between interpretation in the general case.

Example

We will focus on a single variable j, omitting the subscript j within this example for simplicity. Let the profile distributions be Gaussian mixture models with proportions $\beta_k = (\beta_{k1}, \ldots, \beta_{kS})$ and fixed means c_s. If we denote the cdf of the standard normal distribution as Φ, then we can write the profiles as

$$F_k(x) = F(x; \beta_k) = \sum_s \beta_{ks} \Phi(x - c_s). \tag{3.34}$$

Define $\beta_{is} = \theta_i^T (\beta_{1s}, \ldots, \beta_{Ks})$. Then the individual distributions, conditional on the membership vector θ_i, are

$$X|\theta_i \sim \sum_k \theta_{ik} \left[\sum_s \beta_{ks} \Phi(x - c_s) \right], \tag{3.35}$$

$$= \sum_s \beta_{is} \Phi(x - c_s), \tag{3.36}$$

$$= F(x; \beta_i). \tag{3.37}$$

Thus, the individual parameter β_i is in between the profile parameters β_k.

Now let us change the profile distributions slightly. Suppose the means are no longer fixed constants but are also variable parameters:

$$F_k^\star(x) = F^\star(x; \beta_k, \mu) = \sum_s \beta_{ks} \Phi(x - \mu_s). \tag{3.38}$$

In this case the individual conditional distributions are given by

$$X|\theta_i \sim \sum_k \theta_{ik} \left[\sum_s \beta_{ks} \Phi(x - \mu_s) \right], \tag{3.39}$$

$$= F^\star(x; \beta_i, \mu). \tag{3.40}$$

Figure 3.5 shows three example profiles of this form and the distribution of $X|\theta_i$ for two individuals. Here, the between interpretation does not hold in the entire parameter space. Individual data distributions are the same form as the profile distributions—both are in the $F^\star$ parametric family. However, $F^\star$ has two parameters, β and μ. The individual mixing parameter β_i will lie in a simplex defined by the profile parameters β_k, since $\beta_{is} = \theta_i^T(\beta_{1s}, \dots, \beta_{Ks})$.

The fact that the individual mixing parameter β_i is literally 'between' the profile mixing parameters β_k allows us to interpret individuals as a 'blend' of the profiles. The same is not true for the μ parameter. We only have the between interpretation when considering the β parameters.

Now, let's make another small change to the profile distributions. Suppose that the standard deviation of the mixture components is not the same for each profile:

$$F_k^\dagger(x) = F^\dagger(x; \beta_k, \mu, \sigma_k) = \sum_s \beta_{ks} \Phi\left(\frac{x - \mu_s}{\sigma_k}\right). \tag{3.41}$$

Now the conditional individual distributions are

$$X|\theta_i \sim \sum_k \theta_{ik} F_k^\dagger(x), \tag{3.42}$$

$$= \sum_k \theta_{ik} \left[\sum_s \beta_{ks} \Phi\left(\frac{x - \mu_s}{\sigma_k}\right) \right], \tag{3.43}$$

$$= \sum_k \sum_s \theta_{ik} \beta_{ks} \Phi\left(\frac{x - \mu_s}{\sigma_k}\right). \tag{3.44}$$

Equation (3.44) does not simplify in any way. The conditional individual distribution is no longer of the $F^\dagger$ form and as such does not have parameters that are between the profile parameters. Figure 3.6 is an analog of Figure 3.5 and shows three $F^\dagger$ profiles and the distribution of $X|\theta_i$ for two individuals.

This example is analogous to the model of genetic variation, *mStruct* (Shringarpure, 2012). In this model, the population is comprised of K ancestral populations, and each member of the current population has mixed membership in these ancestral populations. *mStruct* also accounts for the fact that the current set of alleles may contain mutations from the ancestral set of alleles.

Each ancestral population has different proportions $\beta_k = (\beta_{k1}, \dots, \beta_{kS})$ of the set of founder

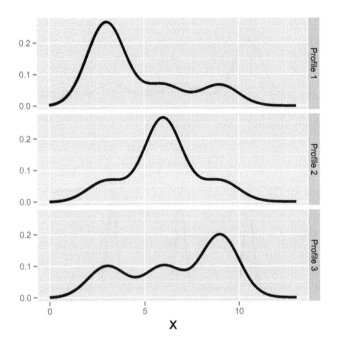

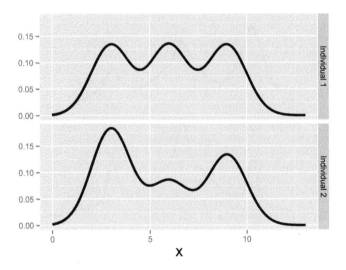

FIGURE 3.5

Non-multinomial profile distributions that preserve the 'between' interpretation. The top graph shows three profiles of the form $F_k^* = \sum_s \beta_{ks} \Phi(x - \mu_s)$ (Equation 3.38). The mixture means μ_s and the standard deviations are the same for each profile. The lower graph shows two individual distributions where $X|\theta_i \sim F^*(x; \beta_i, \mu)$ (Equation 3.40).

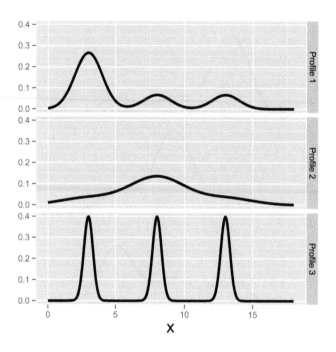

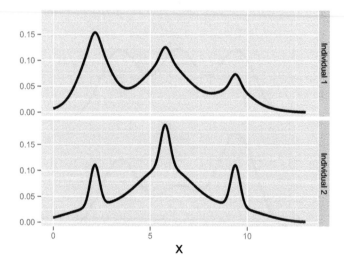

FIGURE 3.6
Profile distributions that do not preserve the 'between' interpretation. The top graph shows three
profiles of the form $F_k^\dagger = \sum_s \beta_{ks} \Phi\left(\frac{x - \mu_s}{\sigma_k}\right)$ (Equation 3.41). The mixture means μ_s are the same
for each profile, but the standard deviations σ_k are different. The lower graph shows two individual
distributions with $X|\theta_i \sim \sum_k \sum_s \theta_{ik} \beta_{ks} \Phi\left(\frac{x - \mu_s}{\sigma_k}\right)$ (Equation 3.44).

alleles at locus j: $\mu_j = (\mu_{j1}, \ldots, \mu_{jS})$. The observed allele for individual i at locus j, X_{ij}, will have mutated from the founder alleles according to some probability distribution $P(\cdot|\mu_{js}, \delta_{kj})$, with the mutation rate δ_{kj} differing depending on the ancestral population. Thus, the profile distributions are

$$F_{kj}(x_j) \quad = \quad F(x_j; \beta_{kj}, \mu_j, \delta_{kj}) \quad = \quad \sum_{s=1}^{S} \beta_{kjs} P(x|\mu_{js}, \delta_{kj}). \tag{3.45}$$

The individual probability distribution of alleles at locus j, conditional on their membership in the ancestral profiles is then given by

$$X_{ij}|\theta_i \sim \sum_k \theta_{ik} \left[\sum_s \beta_{kjs} P(x|\mu_{js}, \delta_{kj}) \right]. \tag{3.46}$$

In the same way that the conditional individual distributions in the $F^\dagger$ model (Equation 3.44) do not simplify, the individual distributions in the *mStruct* do not simplify.

3.7 Contrasting Mixed Membership Regression Models

In this section, we compare and contrast two mixed membership models which are identical in the exchangeability assumptions and the structure of the models. The only difference is that in one case the data is categorical, and in the other case it is continuous. In the categorical case, the between interpretation holds and mixed membership is a viable way to model the structure of the data. In the continuous case, the between interpretation does not hold and mixed membership cannot describe the variation that is present in the data.

Let us suppose that in addition to the variables X_{ij} we also observe a set of covariates T_{ij}. For example, T may be the date a particular document was published or the age of a participant at the time of the observation. In this case, we may want the MMM profiles to depend on these covariates: $F_k(x|t)$. There are many ways to incorporate covariates into F, but perhaps the most obvious is a regression model.

Every regression model, whether linear, logistic, or nonparametric is based on the same fundamental assumption: $\mathbb{E}[X|T = t] = m(t)$. When X is binary, $X|T = t \sim \text{Bernoulli}(m(t))$. When X is continuous, we most often use $X|T = t \sim N(m(t), \sigma^2)$. In general, we tend not to treat these two cases as fundamentally different, they are both just regression. The contrast between these two mixed membership models is inspired by an analysis of the National Long Term Care Survey (Manrique-Vallier, 2010) and an analysis of children's numerical magnitude estimation (Galyardt, 2010; 2012). In Manrique-Vallier (2010), X is binary and T is continuous, so that the MMM profiles are

$$F_k(x|t) = \text{Bernoulli}(m_k(t)). \tag{3.47}$$

In Galyardt (2010), both X and T are continuous, so that the MMM profiles are

$$F_k(x|t) = N(m_k(t), \sigma_k^2). \tag{3.48}$$

Note, however, that for the reasons explained here and detailed in Section 3.7.2, a mixed membership analysis of the numerical magnitude estimation data was wildly unsuccessful (Galyardt, 2010). An analysis utilizing functional data techniques was much more successful (Galyardt, 2012).

The interesting question is why an MMM was successful in one case and unsuccessful in the other. At the most fundamental level, the answer is that a mixture of Bernoullis is still Bernoulli, and a mixture of normals is not normal. This is a straightforward application of Erosheva's representation theorem.

To simplify the comparison, let us suppose that we observe a single variable ($J = 1$), with replications at points T_r, $r = 1, \ldots, R$. For example, X_{ir} may be individual i's response to a single survey item observed at different times T_{ir}. To further simplify, we will use only $K = 2$ MMM profiles with distributions $F(x; m_k(t))$. Thus for an individual with membership parameter θ_i, the conditional data distribution is:

$$X_i|T_i, \theta_i \ \sim \ \prod_r \left[\sum_k \theta_{ik} F(X_{ir}; m_k(T_{ir})) \right]. \tag{3.49}$$

3.7.1   Mixed Membership Logistic Regression

When the MMM profiles are logistic regression functions (Equation 3.47), then the conditional data distribution for an individual with membership parameter θ_i becomes

$$X_i|T_i, \theta_i \ \sim \ \prod_r \left[\sum_k \theta_{ik} Bernoulli(m_k(T_{ir})) \right], \tag{3.50}$$

with

$$m_k(t) = \text{logit}^{-1}(\beta_{0k} + \beta_{1k}t). \tag{3.51}$$

Equation (3.50) is easily rewritten as

$$X_i|T_i, \theta_i \ \sim \ \prod_r \left[Bernoulli \left(\sum_k \theta_{ik} m_k(T_{ir}) \right) \right]. \tag{3.52}$$

In this case, we can write an individual regression function,

$$m_i(t) = \sum_k \theta_{ik} m_k(t). \tag{3.53}$$

This individual regression function m_i does not have the same loglinear form as m_k, so we cannot talk about individual β parameters being between the profile parameters. However, it is a single smooth regression function that summarizes the individual's data, and m_i will literally be between the m_k. Figure 3.7 shows an example with two such logistic regression profile functions and a variety of individual regression functions specified by this mixed membership model.

3.7.2   Mixed Membership Regression with Normal Errors

When the MMM profiles are regression functions with normal errors (Equation 3.48), the conditional distribution for individual i's data is given by

$$X_i|T_i, \theta_i \ \sim \ \prod_r \left[\sum_k \theta_{ik} N \left(m_k(T_{ir}), \sigma_k^2 \right) \right]. \tag{3.54}$$

Since a mixture of normal distributions is not normal, Equation (3.54) does not simplify. In this case it is impossible to write a smooth regression function m_i. Figure 3.8 demonstrates this by showing two profile regression functions and contour plots of the density for two individuals, $X_i|T_i, \theta_i$.

It can be tempting to suggest that a change in the distribution of the membership parameter θ may

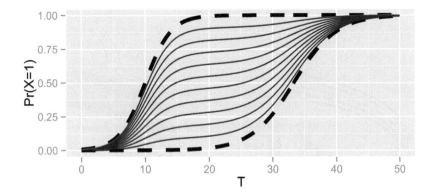

FIGURE 3.7
Profile and individual regression functions in a mixed membership logistic regression model. The thick dashed lines indicate the profile regression functions $m_k(t)$. The thin lines show individual regression functions $m_i(t)$ for a range of values of θ_i.

resolve this issue. However, according to Erosheva's representation theorem, the profile distributions F_k control *where* the data is and θ only controls how much data is in each location (Equations 3.10, 3.11, and Section 3.5). Figure 3.9 illustrates the result of making θ_i a function of t, $\theta_i(t)$.

If the profile distributions F are linear transformations of their parameters (Equation 3.33), then a mixed membership regression model with profiles $F(m(x))$ will have individual regression functions $m_i(x)$. Otherwise a mixed membership model will not produce continuous individual regression functions.

Functional data are a class of data of the form $X_{ij} = f_i(t_{ij}) + \epsilon_{ij}$, where f_i is an individual smooth function, but we only observe a set of noisy measurements X_{ij} and t_{ij} for each individual (Ramsay and Silverman, 2005; Serban and Wasserman, 2005). For example, suppose we observe the height of children at different ages, or temperature at discrete intervals over a period of time. In this type of data analysis, the functions f_i and the similarities and variation between them are the primary objects of inference.

The examples in this section demonstrate that without fundamentally altering the exchangeability assumption of the general mixed membership model (Equation 3.4), a MMM cannot fit functional data. Equation (3.54) will never produce smooth individual regression functions. Galyardt (2012), Gruhl and Erosheva (2013), and Mohamed et al. (2013) suggest a way in which the exchangeability assumption might be altered to model individual regression functions as lying between the profile functions.

3.7.3 Children's Numerical Magnitude Estimation

The mixed membership regression model with normal errors is based on an analysis of the strategies and representations that children use to estimate numerical magnitude. This has been an active area of research in recent years (Ebersbach et al., 2008; Moeller et al., 2008; Siegler and Booth, 2004; Siegler and Opfer, 2003; Siegler et al., 2009). The primary task in experiments studying numerical magnitude estimation is a number line task. The experimenter presents each child with a series of number lines which have only the endpoints marked. The scale of the number lines is most often 0 to 100, or 0 to 1000. The child estimates a number by marking the position where they think the number 'belongs.' Each child will estimate a series of numbers, with a single number line on each page.

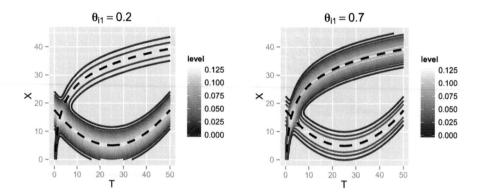

FIGURE 3.8
Mixed membership regression model with normal errors. The two plots show contours of the data
distribution for two different values of θ_i. The thick dashed lines indicate the profile regression
functions. Lighter contour lines indicate higher density. Note that there is no individual regression
function m_i, which can summarize data from this distribution.

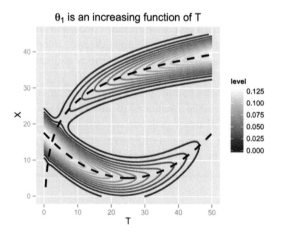

FIGURE 3.9
Mixed membership regression model with normal errors. Contour plot of an individual data distribution where $\theta_{i1}(t)$ is an increasing function of T. The thick dashed lines indicate the profile
regression functions. Lighter contour lines indicate higher density. We cannot summarize data from
this distribution with any smooth regression function $m_i(t)$.

There are competing theories as to how children represent numerical magnitude and the strategies that they use to estimate numbers (Ebersbach et al., 2008; Galyardt, 2012; Moeller et al., 2008; Opfer and Siegler, 2007; Siegler et al., 2009). This argument is not our primary concern. We will focus on the aspect of performance that all of the studies agree upon: there is an immature pattern and a mature pattern. Older children are able to accurately and linearly estimate numerical magnitude. That is, if T_{ir} is the rth number you ask child i to estimate, then their estimates X_{ir} can be modeled as $X_{ir} = T_{ir} + \epsilon_{ir}$.

Young children consistently overestimate small numbers. For example, a kindergardener estimating on the 0–100 scale may place the number 23 three-quarters of the distance from 0 to 100, near a position of 75. These children also appear to not differentiate well between larger quantities, so that they might place both 56 and 84 near a position of 90. The estimate from a child displaying the immature pattern will follow $X_{ir} = m(X_{ir}) + \epsilon_{ir}$. The exact functional form of $m(x)$ is disputed; Opfer and Siegler (2007) and Siegler et al. (2009) suggest that it is logarithmic; Ebersbach et al. (2008) and Moeller et al. (2008) suggest that it is piece-wise linear.

At this point, it seems natural to model children who are learning the mature representation as having mixed membership in both representations (Galyardt, 2010). We can represent each strategy with a MMM profile and use the membership parameter to indicate the degree to which a child has learned the mature strategy. Thus the profiles are mixed membership regression functions with normal errors, as in Equation (3.54). The distribution of individual data predicted by this model would be similar to the distributions shown in Figure 3.8. This mixed membership model would embody a 'switching' interpretation; sometimes the child uses the mature strategy and sometimes the child uses the immature strategy.

This is where the difference between the switching and blending interpretations becomes critical. Children using the immature strategy will estimate the number 30 near the position 80, while those using the mature strategy will estimate the position accurately at 30. If a child is blending the two strategies, then a model should predict an estimate at a position between 30 and 80. On the other hand, if a child is switching between the mature and the immature strategy, then a model should predict estimates near these two points and have lower probability in the middle.

Figure 3.10 shows data from a number line estimation task for six representative individuals. We can see immediately that this is functional data. Each child's strategy can be represented by a single smooth curve, f_i.

Some children clearly display the immature pattern, some children display the mature pattern. The interesting patterns belong to the children between the two extremes. Yet the mixed membership regression model cannot capture this variation, even with the addition of more profiles. The profiles are normal, and since mixtures of normals are not normal, the individual distributions will not be normal. Therefore the exchangeability assumptions in Equation (3.54) will not produce a smooth regression function for each individual.

In this kind of application, we want to model where each individual lies between the two extremes. A mixed membership model cannot capture the patterns of variation that are present in this data. As one measure of model misfit, an attempt to use the mixed membership model with normal errors (Equation 3.54) on this data resulted in estimates of $\sigma > 30$, with data on a scale of 0–100 (Galyardt, 2010). One way to solve this problem is to apply functional data analysis tools, the approach successfully used in Galyardt (2012). Another approach is to alter the exchangeability assumption to allow for a 'between' interpretation (Gruhl and Erosheva, 2013; Mohamed et al., 2013).

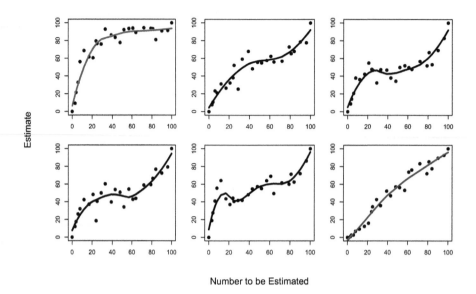

Number to be Estimated

FIGURE 3.10
Each box displays data from a single child participant in Siegler and Booth (2004). Individuals were selected to display the range of strategies observed in the data. The immature and mature patterns are present, but other intermediate patterns are present as well.

3.8   Discussion

Everything presented in this chapter is a straightforward observation based on Erosheva's representation theorem (Erosheva et al., 2007). Every mixed membership model can be expressed as a finite mixture model with a much larger number of classes. Therefore, the best way to understand how mixed membership models behave and how we should interpret them is by focusing on the relationship with finite mixture models.

Categorical data and the multinomial distribution have a unique behavior within the family of finite mixture models. Therefore categorical data have a unique behavior within the family of mixed membership models.

In general, individuals with mixed membership in multiple profiles should be interpreted as switching between the profiles. For example, a student who uses one strategy on one problem and switches to another strategy for the next problem; or one segment of an image from the water profile that then switches its next segment to the tree profile. This switching interpretation is inherent in the exchangeability assumption that observed variables are independent conditional on the individual's membership parameter.

Only in a small set of special cases, including the multinomial distribution, can we interpret mixed membership as individuals being between the profiles. In these cases, the general switching interpretation is also accurate. Think of an individual who has mixed heritage. In the between interpretation, we can consider this individual as blending the two heritages together. Whereas in the switching interpretation, one gene may come from one heritage while the next gene comes from another heritage. In this special case, both interpretations work.

Changing the distribution of the membership parameters has no effect on which interpretations are available. Whether or not the profile distributions are linear transformations of their parameters is the only thing that determines whether the between interpretation is available. The same property is at work in the more complicated regression examples as in the simple examples.

Mixed membership models individuals switching between profiles. Partial membership (Galyardt, 2012; Gruhl and Erosheva, 2013; Mohamed et al., 2013) models individuals blending profiles. Only in very special cases do the two interpretations overlap.

References

Aitchison, J. (1982). The statistical analysis of compositional data. *Journal of the Royal Statistical Society. Series B (Methodological)* 44: 139–177.

Blei, D. M. and Jordan, M. I. (2003). Modeling annotated data. In *Proceedings of the 26th Annual International ACM SIGIR Conference on Research and Development in Information Retrieval (SIGIR '03)*. New York, NY, USA: ACM, 127–134.

Blei, D. M. and Lafferty, J. D. (2006). Dynamic topic models. In *Proceedings of the 23rd International Conference on Machine Learning (ICML '06)*. New York, NY, USA: ACM, 113–120.

Blei, D. M. and Lafferty, J. D. (2007). A correlated topic model of Science. *Annals of Applied Statistics* 1: 17–35.

Blei, D. M., Ng, A. Y., and Jordan, M. I. (2003). Latent Dirichlet allocation. *Journal of Machine Learning Research* 3: 993–1022.

Ebersbach, M., Luwel, K., Frick, A., Onghena, P., and Verschaffel, L. (2008). The relationship between the shape of the mental number line and familiarity with numbers in 5- to 9-year old children: Evidence for a segmented linear model. *Journal of Experimental Child Psychology* 99: 1–17.

Erosheva, E. A. (2002). Grade of Membership and Latent Structure Models with Application to Disability Survey Data. Ph.D. thesis, Department of Statistics, Carnegie Mellon University, Pittsburgh, Pennsylvania, USA.

Erosheva, E. A., Fienberg, S. E., and Lafferty, J. D. (2004). Mixed-membership models of scientific publications. *Proceedings of the National Academy of Sciences* 101: 5220–5227.

Erosheva, E. A., Fienberg, S. E., and Joutard, C. (2007). Describing disability through individual-level mixture models for multivariate binary data. *Annals of Applied Statistics* 1: 502–537.

Fei-Fei, L. and Perona, P. (2005). A Bayesian hierarchical model for learning natural scene categories. In *Proceedings of the 10th IEEE Computer Vision and Pattern Recognition (CVPR 2005)*. San Diego, CA, USA: IEEE Computer Society, 524–531.

Galyardt, A. (2010). Mixed membership models for continuous data. In *Proceedings of the 75th International Meeting of the Psychometric Society (IMPS 2010)*. Athens, GA, USA.

Galyardt, A. (2012). Mixed Membership Distributions with Applications to Modeling Multiple Strategy Usage. Ph.D. thesis, Department of Statistics, Carnegie Mellon University, Pittsburgh, Pennsylvania, USA.

Girolami, M. and Kaban, A. (2005). Sequential activity profiling: Latent Dirichlet allocation of Markov chains. *Data Mining and Knowledge Discovery* 10: 175–196.

Gruhl, J. and Erosheva, E. A. (2013). A tale of two (types of) memberships: Comparing mixed and partial membership with a continuous data example. In Airoldi, E. M., Blei, D. M., Erosheva, E. A., and Fienberg, S. E. (eds), *Handbook of Mixed Membership Models and Its Applications*. Chapman & Hall/CRC.

Haberman, S. J. (1995). Book review of statistical applications using fuzzy sets. *Journal of the American Statistical Association* 90: 1131–1133.

Manrique-Vallier, D. (2010). Longitudinal Mixed Membership Models with Applications to Disability Survey Data. Ph.D. thesis, Department of Statistics, Carnegie Mellon University, Pittsburgh, Pennsylvania, USA.

Manton, K. G., Woodbury, M. A., and Tolley, H. D. (1994). *Statistical Applications Using Fuzzy Sets*. New York, NY: John Wiley & Sons.

Moeller, K., Pixner, S., Kaufmann, L., and Nuerk, H. -C. (2008). Children's early mental number line: Logarithmic or decomposed linear? *Journal of Experimental Child Psychology* 103: 503–515.

Mohamed, S., Heller, K. A., and Ghahramani, Z. (2013). A simple and general exponential family framework for partial membership and factor analysis. In Airoldi, E. M., Blei, D. M., Erosheva, E. A., and Fienberg, S. E. (eds), *Handbook of Mixed Membership Models and Its Applications*. Chapman and Hall/ CRC Press.

Opfer, J. E. and Siegler, R. S. (2007). Representational change and children's numerical estimation. *Cognitive Psychology* 55: 169–195.

Pritchard, J. K., Stephens, M., and Donnelly, P. (2000). Inference of population structure using multilocus genotype data. *Genetics* 155: 945–959.

Ramsay, J. and Silverman, B. W. (2005). *Functional Data Analysis*. Springer Series in Statistics. New York, NY: Springer, 2nd edition.

Rogers, S., Girolami, M., Campbell, C., and Breitling, R. (2005). The latent process decomposition of cDNA microarray data sets. *IEEE/ACM Transactions on Computational Biology and Bioinformatics* 2: 143–156.

Serban, N. and Wasserman, L. (2005). CATS: Clustering after transformation and smoothing. *Journal of the American Statistical Association* 100: 990–999.

Shan, H. and Banerjee, A. (2011). Mixed-membership naive Bayes models. *Data Mining and Knowledge Discovery* 23: 1–62.

Shringarpure, S. (2012). Statistical Methods for Studying Genetic Variation in Populations. Ph.D. thesis, Machine Learning Department, Carnegie Mellon University, Pittsburgh, Pennsylvania, USA.

Siegler, R. S. and Booth, J. L. (2004). Development of numerical estimation in young children. *Child Development* 75: 428–444.

Siegler, R. S. and Opfer, J. E. (2003). The development of numerical estimation: Evidence for multiple representations of numerical quantitiy. *Psychological Science* 14: 237–243.

Siegler, R. S., Thompson, C. A., and Opfer, J. E. (2009). The logarithmic-to-linear shift: One learning sequence, many tasks, many time scales. *Mind, Brain and Education* 3: 142–150.

Titterington, D. M., Smith, A. F. M., and Makov, U. E. (1985). *Statistical Analysis of Finite Mixture Distributions*. Chichester, UK: John Wiley & Sons.

Woodbury, M. A., Clive, J., and Garson, A., Jr. (1978). Mathematical typology: A Grade of Membership technique for obtaining disease definition. *Computers and Biomedical Research* 11: 277–298.

4

A Simple and General Exponential Family Framework for Partial Membership and Factor Analysis

Zoubin Ghahramani

Department of Engineering, University of Cambridge, Cambridge CB2 1TN, United Kingdom

Shakir Mohamed

Google London, London SW1W 9TQ, United Kingdom

Katherine Heller

Department of Statistical Science, Duke University, Durham, NC 27708, USA

CONTENTS

In this chapter we show how mixture models, partial membership models, factor analysis, and their extensions to more general mixed membership models, can be unified under a simple framework using the exponential family of distributions and variations in the prior assumptions on the latent variables that are used. We describe two models within this common latent variable framework: a Bayesian partial membership model and a Bayesian exponential family factor analysis model. Accurate inferences can be achieved within this framework that allow for prediction, missing value imputation, and data visualization, and importantly, allow us to make a broad range of insightful probabilistic queries of our data. We emphasize the adaptability and flexibility of these models for a wide range of tasks, characteristics that will continue to see such models used at the core of modern data analysis paradigms.

4.1 Introduction

Latent variable models are ubiquitous in machine learning and statistics and are core components of many of the most widely-used probabilistic models, including mixture models (Newcomb, 1886; Bishop, 2006), factor analysis (Bartholomew and Knott, 1999), probabilistic principal components analysis (Tipping and Bishop, 1997; Bishop, 2006), mixed membership models (Erosheva et al., 2004), and matrix factorization (Lee and Seung, 1999; Salakhutdinov and Mnih, 2008), amongst others. The use and success of latent variables lies in that they provide us a mechanism with which to achieve many of the desiderata of modern data modeling: robustness to noise, allowing for accurate predictions of future events, the ability to handle and impute missing data, and providing insights into the phenomena underlying our data. For example, in mixture models the latent variables represent the membership of data points to one of a set of underlying classes; in topic models the latent variables allow us to represent the distribution of topics captured within a set of documents.

The broad applicability of mixed membership models is expanded upon throughout this volume, and here we shall focus on simpler instances of the general mixed membership modeling framework to emphasize this wide applicability. In this chapter, we show how mixture models, factor analysis, and partial membership models and their generalization to mixed membership models can be unified under a common modeling framework. Moreover, we show how exponential family likelihoods can be used to provide a very general tool for modeling diverse data types, such as binary, count, or non-negative data, etc. Specifically, we will develop two models: a Bayesian partial membership model (BPM) (Heller et al., 2008) and a Bayesian exponential family factor analysis (EXFA) (Mohamed et al., 2008), and demonstrate the power of these models for accurate prediction and interpretation of data.

As a case study, we will use an analysis of recorded votes: data that lists the names of those voting for or against a motion. In particular, we will focus on the roll call of the U.S. senate and demonstrate the different perspectives of the data that can be obtained, including the types of probabilistic queries that can be made with an accurate model of the data. Recorded votes are stored as a binary matrix and we describe a general approach for handling this type of data, and generally, any data that can be described by members of the exponential family of distributions. We develop two probabilistic models: the first is a model for *partial memberships* that allows us to describe senators on a scale of fully-allegiant Democrats to fully-allegiant Republicans. This is a natural way of thinking about such data, since senators are often grouped into blocs depending on their degree of membership to these two groups, such as moderate Democrats, Republican majority, etc. Secondly, we develop *factor models* that provide a means of representing the underlying factors or traits that senators use in their decision making. These two models will be shown to arise naturally from the same probabilistic framework, allowing us to explore different assumptions on the underlying structure of the data.

We begin our exposition by providing the required background on conjugate-exponential family models (Section 4.2.1). We then show that by considering a relaxation of standard mixture models we arrive naturally at two useful model classes: latent Dirichlet models and latent Gaussian models (which we expand upon in Section 4.4). In Section 4.2.3, we show that the assumption of Dirichlet distributed latent variables allows us to develop a model that quantifies the partial membership of objects to clusters, and that the assumption of continuous, unconstrained latent variables in Section 4.2.4 leads to an exponential family factor analysis. We focus on Markov chain Monte Carlo methods for learning in both models in Section 4.3. Whereas many types of mixed membership models focus on representing the data at two levels (e.g., a subject and a population level), here we operate at one level (subject level) only, and we describe the relationship between our approach and other mixed membership models such as latent Dirichlet allocation and mixed membership matrix factorization (Blei et al., 2003; Erosheva et al., 2004; Mackey et al., 2010) in Section 4.4. We provide some experimental results and explore the roll call data in Section 4.5.

Notation. Throughout this chapter we represent observed data as an $N \times D$ matrix $\mathbf{X} = [\mathbf{x}_1, \dots, \mathbf{x}_N]^\top$, with an individual data point $\mathbf{x}_n = [x_{n1}, \dots, x_{nD}]$. N is the number of data points and D is the number of input features. $\boldsymbol{\Theta}$ is a $K \times D$ matrix of model parameters with rows $\boldsymbol{\theta}_k$. $\mathbf{V}$ is a $N \times K$ matrix $\mathbf{V} = [\mathbf{v}_1, \dots, \mathbf{v}_N]^\top$ of latent variables with rows $\mathbf{v}_n = [v_{n1}, \dots, v_{nK}]$, which are K-dimensional vectors of continuous values in $\mathbb{R}$. K is the number of latent factors representing the dimensionality of the latent variable.

4.2 Membership Models for the Exponential Family

4.2.1 The Exponential Family of Distributions

The choice of likelihood function $p(\mathbf{x}|\boldsymbol{\eta})$ for parameters $\boldsymbol{\eta}$ and observed data $\mathbf{x}$ is central to the models we describe here. In particular, we would like to model data of different types, i.e., data that may be binary, categorical, real-valued, etc. To achieve this objective, we make use of the exponential family of distributions, which is an important family of distributions that emphasizes the shared properties of many standard distributions, including the binomial, Poisson, gamma, beta, multinomial, and Gaussian distributions (Bickel and Doksum, 2001). The exponential family of distributions allows us to provide a singular discussion of the inferential properties associated with members of the family and thus, to develop a modeling framework generalized to all members of the family.

In the exponential family of distributions, the conditional probability of $\mathbf{x}$ given parameter value $\boldsymbol{\eta}$ takes the following form:

$$p(\mathbf{x}|\boldsymbol{\eta}) = \exp\{s(\mathbf{x}_n)^\top \boldsymbol{\eta} + h(\mathbf{x}_n) - g(\boldsymbol{\eta})\}, \tag{4.1}$$

where $s(\mathbf{x}_n)$ are the sufficient statistics, $\boldsymbol{\eta}$ is a vector of natural parameters, $h(\mathbf{x}_n)$ is a function of the data, and $g(\boldsymbol{\eta})$ is the cumulant or log-partition function. For this chapter, the natural representation of the exponential family likelihood is used such that $s(\mathbf{x}) = \mathbf{x}$. For convenience, we shall represent a variable $\mathbf{x}$ that is drawn from an exponential family distribution using the notation $\mathbf{x} \sim \text{Expon}(\boldsymbol{\eta})$, with natural parameters $\boldsymbol{\eta}$.

Probability distributions that belong to the exponential family also have corresponding conjugate prior distributions $p(\boldsymbol{\eta})$, for which both $p(\boldsymbol{\eta})$ and $p(\mathbf{x}|\boldsymbol{\eta})$ have the same functional form. The conjugate prior distribution for the exponential family distribution of Equation (4.1) is:

$$p(\boldsymbol{\eta}) \propto \exp\{\boldsymbol{\lambda}^\top \boldsymbol{\eta} - \nu g(\boldsymbol{\eta}) + f(\boldsymbol{\lambda})\}, \tag{4.2}$$

where $\boldsymbol{\lambda}$ and ν are hyperparameters of the prior distribution. We use the shorthand $\boldsymbol{\eta} \sim \text{Conj}(\boldsymbol{\lambda}, \nu)$ to denote draws from a conjugate distribution.

As an example, consider binary data, for which an appropriate data distribution is the Bernoulli distribution and the corresponding conjugate prior is the Beta distribution. The Bernoulli distribution has the form $p(x|\mu) = \mu^x (1 - \mu)^{1-x}$, with μ in [0,1]. The exponential family form, using the terms in Equation (4.1), is described using $h(x) = 0$, $\eta = \ln(\frac{\mu}{1-\mu})$ and $g(\eta) = \ln(1 + e^\eta)$. The natural parameters can be mapped to the parameter values of the distribution using the link function, which is the logistic sigmoid in the case of the Bernoulli distribution. The terms of the conjugate distribution can also be derived easily.

4.2.2   Beyond Mixture Models

Mixture models are a common approach for assigning membership of observations to a set of distinct clusters. For a finite mixture model with K mixture components, the probability of a data observation $\mathbf{x}_n$ given parameters $\boldsymbol{\Theta}$ is

$$p(\mathbf{x}_n|\boldsymbol{\Theta}) = \sum_{k=1}^{K} \rho_k p_k(\mathbf{x}_n|\boldsymbol{\theta}_k), \tag{4.3}$$

where $p_k(\cdot)$ is the probability distribution of mixture component k, and ρ_k is the mixing proportion. We can express this using indicator variables $\mathbf{v}_n = [v_{n1}, v_{n2}, \dots, v_{nK}]$ as

$$p(\mathbf{x}_n|\boldsymbol{\Theta}) = \sum_{\mathbf{v}_n} p(\mathbf{v}_n) \prod_{k=1}^{K} \left(p_k(\mathbf{x}_n|\boldsymbol{\Theta}_k)\right)^{v_{nk}}, \tag{4.4}$$

where $v_{nk} \in \{0, 1\}$, $\sum_k v_{nk} = 1$, and $p(v_{nk} = 1) = \rho_k$. If $v_{nk} = 1$, then observation n belongs to cluster k, and therefore v_{nk} indicates the membership of observations to clusters.

We now consider a relaxation of this model: relaxing the constraint that $v_{nk} \in \{0, 1\}$ to instead be continuous-valued and removing the sum-to-one constraint. The probability in Equation (4.4) must now be modified, and becomes

$$p(\mathbf{x}_n|\boldsymbol{\Theta}) = \int_{\mathbf{v}_n} p(\mathbf{v}_n) \frac{1}{Z(v_n, \boldsymbol{\Theta})} \prod_{k=1}^{K} \left(p_k(\mathbf{x}_n|\boldsymbol{\Theta}_k)\right)^{v_{nk}} d\mathbf{v}_n, \tag{4.5}$$

where we have integrated over the continuous latent variables rather than summing, and have introduced the normalizing constant Z, which is a function of $\mathbf{v}_n$ and $\boldsymbol{\Theta}$, to ensure normalization.

By substituting the exponential family distribution (4.1) into Equation (4.5), the likelihood can be expressed as

$$\mathbf{x}_n|\mathbf{v}_n, \boldsymbol{\Theta} \sim \text{Expon}\left(\sum_{k} v_{nk}\boldsymbol{\theta}_k\right), \tag{4.6}$$

which is obtained by combining terms in log-space and requiring the resulting distribution to be normalized. The computation of the normalizing constant Z in Equation (4.5) is thus always tractable. Thus, we see that the observed data can be described by an exponential family distribution with natural parameters that are given by the linear combination of the coefficients $\boldsymbol{\theta}_k$ weighted by the latent variables v_{nk}.

We consider two types of constraints on the latent variables, which give rise to two important model classes. These are:

Partial membership models. The latent variables can take any value in the range $v_{nk} \in [0, 1]$. It is with this relaxation that we are able to represent data points that can belong partially to a cluster. Such ideas are found in fuzzy set theory, mixed membership, and topic modeling.

Factor models. The latent variables are allowed to take any continuous value $v_{nk} \in \mathbb{R}$. Popular models that stem from this assumption include factor analysis (FA) (Bartholomew and Knott, 1999), probabilistic principal components analysis (PCA) (Tipping and Bishop, 1997), and probabilistic matrix factorization (PMF) (Salakhutdinov and Mnih, 2008), amongst others. The latent variables form a continuous, low-dimensional representation of the input data. For easier interpretation, one can restrict the latent variables to be nonnegative, allowing for a parts-based explanation of the data (Lee and Seung, 1999).

Thus, we obtain a unifying framework for many popular latent variable models, whose key difference lies in the nature of the latent variables used. Table 4.1 summarizes this insight and lists some of the models that can arise from this framework.

TABLE 4.1
Models which can be derived from the unifying framework for latent variable models.

Model	Domain
mixture models	$v_{nk} \in \{0, 1\}$
partial membership models (Heller et al., 2008)	$v_{nk} \in [0, 1]$
exponential family PCA (Collins et al., 2002; Mohamed et al., 2008)	$v_{nk} \in \mathbb{R}$
nonnegative matrix factorization (Lee and Seung, 1999)	$v_{nk} \in \mathbb{R}^+$

4.2.3 Bayesian Partial Membership Models

We consider a model for partial membership that we refer to as the Bayesian partial membership model (BPM) (Heller et al., 2008). The BPM is a model in which we consider observations have partial membership each of K classes. Consider political affiliations as an example: an individual's political leaning is not wholly socialist or wholly conservative, but may have partial membership in both these political schools.

At the outset it is important to note the distinction between partial membership and uncertain membership. Responsibilities in mixture models are representations of the uncertainty in assigning full membership to a cluster, and this uncertainty can often be reduced with more data. Partial membership represents a fractional membership in multiple clusters, such as a senator with moderate views in between that of being fully Republican or fully Democrat.

Figure 4.1(a) is a graphical representation of the generative process for the Bayesian partial membership model. The plate notation represents replication of variables and the shaded node represents observed variables. We denote the K-dimensional vector of positive hyperparameters by α. The generative model is: Draw mixture weights ρ_k from a Dirichlet distribution with

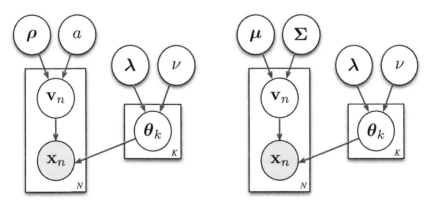

(a) Bayesian partial membership   (b) Exponential family factor analysis

FIGURE 4.1
Graphical models representing the relationship between latent variables, parameters, and observed data for exponential family latent variable models.

hyperparameters $\boldsymbol{\alpha}$, and a positive scaling factor a from an exponential distribution with hyper-parameter $\beta > 0$; then draw a vector of partial memberships $\mathbf{v}_n$ from a Dirichlet distribution, representing the extent to which the observation belongs to each of the K clusters.

$$\boldsymbol{\rho} \sim \text{Dir}(\boldsymbol{\alpha}); \qquad a \sim \text{Exp}(\beta), \tag{4.7}$$

$$\mathbf{v}_n \sim \text{Dir}(a\boldsymbol{\rho}). \tag{4.8}$$

Each cluster k is characterized by an exponential family distribution with natural parameters $\boldsymbol{\theta}_k$ that are drawn from a conjugate exponential family distribution, with hyperparameters $\boldsymbol{\lambda}$ and ν. Given the latent variables and parameters, each data point is drawn from a data-appropriate exponential family distribution:

$$\boldsymbol{\theta}_k \sim \text{Conj}\left(\boldsymbol{\lambda}, \nu\right), \tag{4.9}$$

$$\mathbf{x}_n \sim \text{Expon}\left(\sum_k v_{nk}\boldsymbol{\theta}_k\right). \tag{4.10}$$

We denote $\boldsymbol{\Omega} = \{\mathbf{V}, \boldsymbol{\Theta}, \boldsymbol{\rho}, a\}$ as the set of unknown parameters with hyperparameters $\boldsymbol{\Psi} = \{\boldsymbol{\alpha}, \beta, \boldsymbol{\lambda}, \nu\}$. Given this generative specification, the joint-probability is:

$$p(\mathbf{X}, \boldsymbol{\Omega}|\boldsymbol{\Psi}) = p(\mathbf{X}|\mathbf{V}, \boldsymbol{\Theta})p(\mathbf{V}|a, \boldsymbol{\rho})p(\boldsymbol{\Theta}|\boldsymbol{\lambda}, \nu)p(\boldsymbol{\rho}|\boldsymbol{\alpha})p(a|\beta)$$

$$= \prod_{n=1}^{N} p(\mathbf{x}_n|\mathbf{v}_n, \boldsymbol{\Theta})p(\mathbf{v}_n|a, \boldsymbol{\rho}) \prod_{k=1}^{K} p(\boldsymbol{\theta}_k|\boldsymbol{\lambda}, \nu)p(\boldsymbol{\rho}|\boldsymbol{\alpha})p(a|\beta). \tag{4.11}$$

Substituting the forms for each distribution, the log joint probability is:

$$\ln p(\mathbf{X}, \boldsymbol{\Omega}|\boldsymbol{\Psi}) = \sum_{n=1}^{N} \left\{ \left(\sum_k v_{nk}\boldsymbol{\theta}_k\right)^{\top} \mathbf{x}_n + h(\mathbf{x}_n) + g\left(\sum_k v_{nk}\boldsymbol{\theta}_k\right) \right\} \tag{4.12}$$

$$+ \sum_{k=1}^{K} \left[\boldsymbol{\lambda}^T \boldsymbol{\theta}_k + \nu g(\boldsymbol{\theta}_k) + f(\boldsymbol{\lambda})\right]$$

$$+ N \ln \Gamma\left(\sum_k a\rho_k\right) - N \sum_k \ln \Gamma\left(a\rho_k\right) + \sum_n \sum_k (a\rho_k - 1) \ln v_{nk}$$

$$+ \ln \Gamma\left(\sum_k \alpha_k\right) - \sum_k \ln \Gamma(\alpha_k) + \sum_k (\alpha_k - 1) \ln \rho_k + \ln b - ba.$$

We arrive at the BPM model using a continuous latent variable relaxation of the mixture model. As a result, the BPM reduces to mixture modeling when $a \to 0$ with mixing proportions $\boldsymbol{\rho}$, and follows from the limit of Equation (4.8). The BPM bears interesting relationships to several well-known models, including latent Dirichlet allocation (LDA) (Blei et al., 2003), mixed membership models (Erosheva et al., 2004), discrete components analysis (DCA) (Buntine and Jakulin, 2006), and exponential family PCA (Collins et al., 2002; Moustaki and Knott, 2000), which we discuss in Section 4.2.4. Unlike LDA and mixed membership models that capture partial memberships in the form of attribute-specific mixtures, the BPM does not assume a factorization over attributes and provides a general way of combining exponential family distributions with partial membership.

4.2.4   Exponential Family Factor Analysis

We now consider a Bayesian model for exponential family factor analysis (EXFA) (Mohamed et al., 2008). We can think of an exponential family factor analysis as a method of decomposing an observed data matrix $\mathbf{X}$, which can be of any type supported by the exponential family of distributions, into two matrices $\mathbf{V}$ and $\boldsymbol{\Theta}$; we define the product matrix $\mathbf{P} = \mathbf{V}\boldsymbol{\Theta}$. Since the likelihood depends only on $\mathbf{V}$ and $\boldsymbol{\Theta}$ through their product $\mathbf{P}$, this can also be seen as a model for matrix factorization. In traditional factor analysis and probabilistic PCA, the elements of the matrix $\mathbf{P}$, which are the means of Gaussian distributions, lie in the same space as that of the data $\mathbf{X}$. In the case of EXFA and similar methods for non-Gaussian PCA such as EPCA (Collins et al., 2002; Moustaki and Knott, 2000), this matrix represents the natural parameters of the exponential family distribution of the data.

The generative process for the EXFA model is described by the graphical model of Figure 4.1(b). Let $\mathbf{m}$ and $\mathbf{S}$ be hyperparameters representing a K-dimensional vector of initial mean values and an initial covariance matrix, respectively. Let α and β be the hyperparameters corresponding to the shape and scale parameters of an inverse-gamma distribution. We begin by drawing $\boldsymbol{\mu}$ from a Gaussian distribution and the elements σ_k^2 of the diagonal matrix $\boldsymbol{\Sigma}$ from an inverse-gamma distribution. For each data point n of the factor score matrix $\mathbf{V}$, we draw a K-dimensional Gaussian latent variable $\mathbf{v}_n$:

$$\boldsymbol{\mu} \sim \mathcal{N}(\boldsymbol{\mu}|\mathbf{m}, \mathbf{S}); \qquad \sigma_k^2 \sim i\mathcal{G}(\alpha, \beta) \tag{4.13}$$

$$\mathbf{v}_n \sim \mathcal{N}(\mathbf{v}_n|\boldsymbol{\mu}, \boldsymbol{\Sigma}). \tag{4.14}$$

The data is described by an exponential family distribution with natural parameters given by the product of the latent variables $\mathbf{v}_n$ and parameters $\boldsymbol{\theta}_k$. The exponential family distribution modeling the data and the corresponding prior over the model parameters is:

$$\boldsymbol{\theta}_k \sim \mathrm{Conj}\,(\boldsymbol{\lambda}, \nu) \tag{4.15}$$

$$\mathbf{x}_n|\mathbf{v}_n, \boldsymbol{\Theta} \sim \mathrm{Expon}\left(\textstyle\sum_k v_{nk}\boldsymbol{\theta}_k\right). \tag{4.16}$$

We denote $\boldsymbol{\Omega} = \{\mathbf{V}, \boldsymbol{\Theta}, \boldsymbol{\mu}, \boldsymbol{\Sigma}\}$ as the set of unknown parameters with hyperparameters $\boldsymbol{\Psi} = \{\mathbf{m}, \mathbf{S}, \alpha, \beta, \boldsymbol{\lambda}, \nu\}$. Given this specification, in Equations (4.13)–(4.16), the log joint probability distribution is:

$$p(\mathbf{X}, \boldsymbol{\Omega}|\boldsymbol{\Psi}) = p(\mathbf{X}|\mathbf{V}, \boldsymbol{\Theta})p(\boldsymbol{\Theta}|\boldsymbol{\lambda}, \nu)p(\mathbf{V}|\boldsymbol{\mu}, \boldsymbol{\Sigma})p(\boldsymbol{\mu}|\mathbf{m}, \mathbf{S})p(\boldsymbol{\Sigma}|\alpha, \beta)$$

$$\ln p(\mathbf{X}, \boldsymbol{\Omega}|\boldsymbol{\Psi}) = \sum_{n=1}^{N}\left[\left(\sum_k v_{nk}\boldsymbol{\theta}_k\right)^\top \mathbf{x}_n + h(\mathbf{x}_n) + g\left(\sum_k v_{nk}\boldsymbol{\theta}_k\right)\right] \tag{4.17}$$

$$+ \sum_{k=1}^{K}\left[\boldsymbol{\lambda}^\top\boldsymbol{\theta}_k + \nu g(\boldsymbol{\theta}_k) + f(\boldsymbol{\lambda})\right]$$

$$+ \sum_{n=1}^{N}\left[-\frac{K}{2}\ln(2\pi) - \frac{1}{2}\ln|\boldsymbol{\Sigma}| - \frac{1}{2}(\mathbf{v}_n - \boldsymbol{\mu})^T\boldsymbol{\Sigma}^{-1}(\mathbf{v}_n - \boldsymbol{\mu})\right]$$

$$- \frac{K}{2}\ln(2\pi) - \frac{1}{2}\ln|\boldsymbol{S}| - \frac{1}{2}(\boldsymbol{\mu} - \boldsymbol{m})^T\boldsymbol{S}^{-1}(\boldsymbol{\mu} - \boldsymbol{m})$$

$$+ \sum_{i=1}^{K}\left[\alpha\ln\beta - \ln\Gamma(\alpha) + (\alpha - 1)\ln\sigma_i^2 - \beta\sigma_i^2\right],$$

where the functions $h(\cdot)$, $g(\cdot)$, and $f(\cdot)$ correspond to the functions of the chosen conjugate-exponential family distribution for the data.

Whereas mixture models represent membership to a single cluster, and the BPM represents partial membership to the set of clusters, EXFA explains the data using linear combinations of all latent classes (an all-membership). EXFA thus provides a natural way of combining different exponential family distributions and producing a shared latent embedding of the data using Gaussian latent variables.

4.3 Prior to Posterior Analysis

For both the Bayesian partial membership (BPM) model and exponential family factor analysis (EXFA), typical tasks include prediction, missing data imputation, dimensionality reduction, and data visualization. To achieve this, we must infer the posterior distribution $p(\Omega|\mathbf{X}, \mathbf{\Psi})$, by which we can visualize the structure of the data and compute predictive distributions. Due to the lack of conjugacy, analytic computation of the posterior is not possible. Although many approximation methods exist for computing posterior distributions, we focus on Markov chain Monte Carlo (MCMC) because it provides a simple, powerful, and often surprisingly scalable family of methods. Using MCMC involves representing the posterior distribution by a set of samples, following which we use these samples for analysis, prediction, and decision making.

4.3.1 Markov Chain Monte Carlo

Markov chain Monte Carlo (MCMC) methods are a general class of sampling methods based on constructing a Markov chain with the desired posterior distribution as the equilibrium distribution of the Markov chain. MCMC methods are popular in machine learning and Bayesian statistics and include widely-known methods such as Gibbs sampling, Metropolis-Hastings, and slice sampling (Robert and Casella, 2004; Gilks et al., 1995). For sampling in the models of Sections 4.2.3 and 4.2.4, we make use of a general purpose MCMC algorithm known as Hybrid (or Hamiltonian) Monte Carlo (HMC) sampling.

Hybrid Monte Carlo (HMC), which was first described by Duane et al. (1987), is based on the simulation of Hamiltonian dynamics as a way of exploring the sample space of the posterior distribution. Consider the task of generating samples from the distribution $p(\Omega|\mathbf{\Psi}, \mathbf{X})$, with $\mathbf{\Psi}$ being any relevant hyperparameters; we denote $\mathbf{u}$ as an auxiliary variable. Intuitively, HMC combines auxiliary variables with gradient information from the joint-probability to improve mixing of the Markov chain, with the gradient acting as a force that results in more effective exploration of the sample space. HMC can be used to sample from continuous distributions for which the density function can be evaluated (up to a known constant). This makes HMC particularly amenable to sampling in non-conjugate settings where the full conditional distributions required for Gibbs sampling cannot be derived, but for which the joint probability density and its derivatives can be computed. These properties make HMC well-suited to sampling from the BPM and EXFA models, since these models do not have a conjugate structure and all unknown variables Ω are continuous and differentiable, making it possible to exploit available gradient information.

For HMC, a potential energy function and a kinetic energy function are defined, whose sum forms the Hamiltonian energy:

$$\mathcal{H}(\Omega, \mathbf{u}) = \mathcal{E}(\Omega|\mathbf{\Psi}) + \mathcal{K}(\mathbf{u}), \qquad \text{(Hamiltonian Energy)} \qquad (4.18)$$

$$\mathcal{E}(\Omega|\mathbf{\Psi}) = -\ln p(\Omega, \mathbf{X}|\mathbf{\Psi}), \qquad \text{(Potential Energy)} \qquad (4.19)$$

$$\mathcal{K}(\mathbf{u}) = -\tfrac{1}{2}\mathbf{u}^\top \mathbf{M}\mathbf{u}. \qquad \text{(Kinetic Energy)} \qquad (4.20)$$

The Hamiltonian can be seen as the log of an augmented distribution to be sampled from: $p(\mathbf{X}, \Omega, \mathbf{u}|\mathbf{\Psi}) = p(\mathbf{X}, \Omega|\mathbf{\Psi})\mathcal{N}(\mathbf{u}|0, \mathbf{M})$, where $\mathbf{M}$ is a preconditioning matrix often referred to as a mass matrix, which in the simplest case is set to the identity matrix. The gradient of the potential energy is defined as $\Delta(\Omega) = \frac{\partial \mathcal{E}(\Omega)}{\partial \Omega}$. We defer further details of the physical underpinnings describing Hamiltonian dynamics and its appropriateness for MCMC to the work of Neal (2010) and Neal (1993).

We present the full algorithm for HMC in Algorithm 1. Each iteration of HMC has two steps. In the first step, we assume that an initial sample (state) for Ω is given and we generate a Gaussian

---

Evaluate Gradient $\mathbf{g} = \Delta(\boldsymbol{\theta})$ with initial $\boldsymbol{\theta}$ //*
[f]g = gradE(theta)
Evaluate Energy $E = \mathcal{E}(\boldsymbol{\theta}|\boldsymbol{\psi})$ //*
[f]E = findE(theta)
for L *iterations* **do**
    Initialize new momentum $\mathbf{u}$ drawn from a Gaussian
    Calculate: $\mathcal{K}(\mathbf{u}) = \frac{1}{2}\mathbf{u}^\top\mathbf{u}$ and $\mathcal{H} = \mathcal{E}(\boldsymbol{\theta}|\boldsymbol{\psi}) + \mathcal{K}(\mathbf{u})$
    $\boldsymbol{\theta}^{new} \leftarrow \boldsymbol{\theta}; \qquad \mathbf{g}^{new} \leftarrow \mathbf{g};$
    for L *leapfrog steps* **do**
        $\mathbf{u} \leftarrow \mathbf{u} - \frac{\epsilon}{2}\mathbf{g}$ //*
        [f]Make half-step in u
        $\boldsymbol{\theta}^{new} \leftarrow \boldsymbol{\theta}^{new} + \epsilon\mathbf{u}$ //*
        [f]Make a step in theta
        $\mathbf{g}^{new} \leftarrow \Delta(\boldsymbol{\theta}^{new})$ //*
        [f]gradE(thetaNew)
        $\mathbf{u} \leftarrow \mathbf{u} - \frac{\epsilon}{2}\mathbf{g}^{new}$ //*
        [f]make half step in u

    end
    $E^{new} = \mathcal{E}(\boldsymbol{\theta}^{new}|\boldsymbol{\psi})$//*
    [f]Enew = findE(thetaNew)
    Calculate $\mathcal{K}(\mathbf{u}) = \frac{1}{2}\mathbf{u}^\top\mathbf{u}$
    Hamiltonian $\mathcal{H}^{new} \leftarrow E^{new} + \mathcal{K}(\mathbf{u})$
    if $rand() < \exp(-(\mathcal{H}^{new} - \mathcal{H}))$ **then**
        Accept $\leftarrow$ True
        $\mathbf{g} \leftarrow \mathbf{g}^{new}; \quad \boldsymbol{\theta} \leftarrow \boldsymbol{\theta}^{new}; \quad E \leftarrow E^{new}$

    else
        Accept $\leftarrow$ False
    end
end

Algorithm 1: Hybrid Monte Carlo (HMC) Sampling (MacKay, 2003).

variable $\mathbf{u}$ for the momentum (line 4, Algorithm 1). In the second step, we simulate Hamiltonian dynamics, which follow the equations of motion to move the current sample and momentum to a new state. The Hamiltonian dynamics must be discretized for implementation and the most popular discretization is known as the leapfrog method (lines 7–11). The leapfrog approximation is simulated for L steps using a step-size ϵ. The samples $\boldsymbol{\Omega}^*$ and $\mathbf{u}^*$ at the end of the leapfrog steps form the proposed state, which is accepted using the Metropolis criterion (line 15):

$$\min\left(1, \exp(-\mathcal{H}(\boldsymbol{\Omega}^*, \mathbf{u}^*) + \mathcal{H}(\boldsymbol{\Omega}, \mathbf{u}))\right). \tag{4.21}$$

Finally, marginal samples from $p(\boldsymbol{\Omega})$ are obtained by ignoring $\mathbf{u}$.

Aspects of Implementation

To implement HMC correctly we must adjust the energy function to account for variables that may be constrained, such as variables that are nonnegative or bound between [0,1]. We make use of the following transformations:

    BPM with $a > 0$, $\sum_k \pi_k = 1$ **and** $\sum_k v_{nk} = 1$:

$$a = \exp(\eta); \qquad \pi_k = \frac{\exp(r_k)}{\sum_{k'} \exp(r_{k'})}; \qquad v_{nk} = \frac{\exp(\omega_{nk})}{\sum_{k'} \exp(\omega_{nk'})}.$$

EXFA with $\sigma_k^2 > 0$: $\qquad \sigma_k^2 = \exp(\xi_k).$

The use of these transformations requires the inclusion of the determinant of the Jacobian of the change of variables, as well as consistent application of the chain rule for differentiation taking into account the change of variables.

HMC has two tunable parameters: the number of leapfrog steps L and the step-size ϵ. In general, the step-size should be chosen to ensure that the sampler's rejection rate is between 25% and 35%, and to use a large number of leapfrog steps. Here we generally make use of L between 80 and 100. The tuning of these parameters can be challenging in some cases, and we show ways in which these choices can be explored in the experimental section. We fix the mass matrix to the identity but this can also be tuned, and we discuss aspects of this in Section 4.6. Analysis of the optimal acceptance rates for HMC is discussed in Beskos et al. (2010); Neal (2010) provides a great deal of guidance in tuning HMC samplers.

Many datasets contain missing values, and we can account for this missing data in a principled manner by dividing the data into the set of observed and missing entries $\mathbf{X} = \{\mathbf{X}^{obs}, \mathbf{X}^{missing}\}$ and conditioning on the set $\mathbf{X}^{obs}$ during inference. In practice, the pattern of missing data is represented by a masking matrix, which is the indicator matrix of elements that are observed versus missing. Probabilities are then computed using elements of the masking matrix set to 1.

4.4 Related Work

Mixed Membership Models and LDA. In general, mixed membership models (Erosheva et al., 2004) organize the data in two levels using an admixture structure (mixture-of-mixtures model). Latent Dirichlet allocation (LDA) (Blei et al., 2003), as an instance of a mixed membership model, organizes the data at the level of words and then documents, expressing this data likelihood as a mixture of multinomials. LDA combines this mixture-likelihood with a K-dimensional Dirchlet-distributed latent variable $\mathbf{v}$ as a distribution over topics. The BPM is a similar *latent Dirichlet model*, but the latent variable represents partial memberships, and instead of a two-level structure, the BPM indexes the data directly using an exponential family likelihood. LDA assumes that each data attribute (i.e., words) of an observation (i.e., document) is drawn independently from a mixture distribution given the membership vector for the data point, $x_{nd} \sim \sum_k v_{nk} p(x|\theta_{kd})$. As a result, LDA makes the most sense when the observations (documents) being modeled constitute bags of exchangeable sub-objects (words). Furthermore, for both LDA and mixed membership models, there is a discrete latent variable for every sub-object, corresponding to which mixture component that sub-object was drawn from. This large number of discrete latent variables makes MCMC sampling potentially much more expensive than sampling in the exponential family models we describe here. A more detailed discussion and comparison of mixed and partial membership models is in the chapter by Gruhl and Erosheva (Gruhl and Erosheva, 2013, §2.4) in this volume and complements this discussion.

Latent Gaussian Models. EXFA employs a K-dimensional Gaussian latent variable $\mathbf{v}$ and is thus an example of a *latent Gaussian model*. This is one of the most established classes of models and includes generalized linear regression models, nonparametric regression using Gaussian processes, state-space and dynamical systems, unsupervised latent variable models such as PCA, factor analysis (Bartholomew and Knott, 1999), probabilistic matrix factorization (Salakhutdinov and Mnih, 2008), and Gaussian Markov random fields. In generalized linear regression (Bickel and Doksum, 2001), the latent variables v_n are the predictors formed by the product of covariates and regression coefficients; in Gaussian process regression (Rasmussen and Williams, 2006), the latent variables $\mathbf{v}$ are drawn jointly from a correlated Gaussian using a mean function and a covariance function formed using the covariates; and in probabilistic PCA and factor analysis (Tipping and Bishop, 1997; Bartholomew and Knott, 1999), latent variables $\mathbf{v}_n$ are Gaussian with isotropic or diagonal covariances, respectively.

EXFA also follows as a Bayesian interpretation of exponential family PCA (Collins et al., 2002) and generalized latent trait models (Moustaki and Knott, 2000). Instead of fully Bayesian inference, these related models specify an objective function that is optimized to obtain the MAP solution. Similarly to the BPM, in EXFA the data is indexed directly using an exponential family distribution rather than through an admixture structure. With this realization though, it is easy to see the connection and extension of EXFA to a generalized mixed membership matrix factorization (MMMF) model by instead considering a two-level representation of the data similar to that described by Mackey et al. (2010).

Both the BPM and EXFA model the natural parameters of an exponential family distribution. This makes them different from other latent variable models, such as nonnegative matrix factorization (NMF) (Lee and Seung, 1999; Buntine and Jakulin, 2006), since these alternative approaches model the mean parameters of distributions rather than their natural parameters. The use of natural parameters allows for easier learning of model parameters, since these are often unconstrained, unlike learning for NMF which requires special care in handling constraints, e.g., leading to the multiplicative updates required for learning in NMF.

Fuzzy Clustering. Partial membership is a cornerstone of fuzzy theory, and the notion that probabilistic models are unable to handle partial membership is used to argue that probability is a sub-theory, or different in character from fuzzy logic (Zadeh, 1965; Kosko, 1992). With the BPM, we are able to demonstrate that probabilistic models *can* be used to describe partial membership. Rather than using a mixture model for clustering, an alternative is given by fuzzy set theory and fuzzy k-means clustering (Bezdek, 1981). Fuzzy k-means clustering (Gasch and Eisen, 2002) iteratively minimizes the objective function: $J = \sum_n \sum_k v_{nk}^{\gamma_f} D^2(\mathbf{x}_n, \mathbf{c}_k)$, where $\gamma_f > 1$ is the fuzzy exponent parameter, v_{nk} represents the degree of membership of data point n to cluster k, where $\sum_k v_{nk} = 1$ and $D^2(\mathbf{x}_n, \mathbf{c}_k)$ is a squared distance between the observation $\mathbf{x}_n$ and the cluster centre $\mathbf{c}_k$. By varying γ_f, it is possible to attain different degrees of partial membership, with $\gamma_f = 1$ being k-means with no partial membership.

We compare fuzzy clustering and the BPM in Section 4.5 and find that the two approaches achieve very similar results, with the advantage of probabilistic models being that we obtain estimates of uncertainty, are able to deal with missing data, and can combine these models naturally with the wider set of probabilistic models. Thus, we hope that this work demonstrates that, contrary to the common misconception, fuzzy set theory is not needed to represent partial membership in probabilistic models, and that this can be achieved with established approaches for probabilistic modeling.

4.5 Experimental Results

We demonstrate the effectiveness of the models presented in this chapter using synthetic datasets as well as a real-world case study: roll call data from the U.S. Senate. We evaluate the performance of the methods by computing the negative log predictive probability (NLP) on test data. The test sets are created by setting 10% of the elements of the data matrix as missing data in the training set and then learning in the presence of this missing data. We provide Matlab code to reproduce all the results in this section online.[1]

4.5.1 Synthetic Binary Data

Noisy Bit Patterns

We evaluate the behaviour of EXFA using a synthetic binary dataset. The synthetic data was generated by creating three 16-bit prototype vectors, with each bit being set with probability 0.5. Each of the three prototypes is replicated 100 times, resulting in a dataset of 300 observations. Noise is then added to the data by flipping bits in the dataset with probability of 0.1 (Tipping, 1999; Mohamed et al., 2008). We use HMC to generate 5000 samples from the EXFA model with $K = 3$ factors, and demonstrate the evolution of the sampler in Figure 4.2.

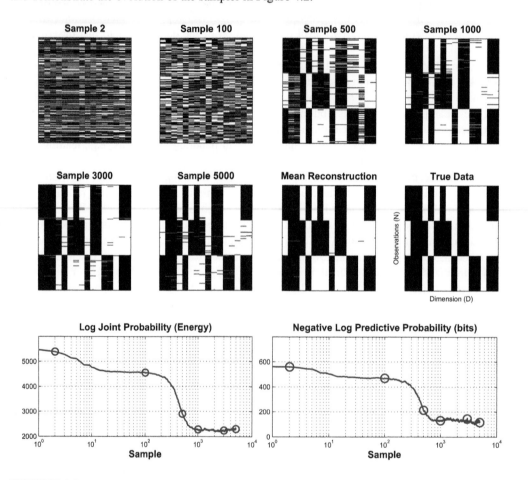

FIGURE 4.2
Reconstruction of data samples at various stages of the sampling in EXFA. Top two rows: Greyscale reconstructions at various samples and the true, noise-free data. Bottom row: Change in the energy function (using training data) and the corresponding predictive probability (using test data). We show circular markers at samples for which the reconstructions are shown above.

---

[1] See www.shakirm.com/code/EFLVM/.

Since the sampler is initialized randomly, we see that the initial samples have no discernible structure. As the sampling proceeds, the energy rapidly decreases (the energy is the negative log joint probability, meaning lower is better), and useful structure can be seen after the 500th sample. By the end of the sampling, the samples correctly capture the true data, as seen by comparing the mean reconstruction computed using the last 1000 samples, and the true data in Figure 4.2. The predictive probability of the test data computed for every sample also decreases as the sampler progresses, indicating that the correct latent structure has been inferred, allowing for accurate imputation of the missing data. The random predictor would have an $NLP = 10\% \times 300 \times 16 = 480$ bits, and we can see that the NLP we obtain is much lower than this. The maximum likelihood estimation of EXFA has $NLP = 1148$ bits, which is significantly worse than the Bayesian prediction. This is a well-known problem, since maximum likelihood estimation in this model suffers from severe overfitting, highlighting an important advantage of Bayesian methods over optimization methods (Mohamed et al., 2008).

Plots such as Figure 4.2 are also useful as tools for tuning an HMC sampler. For a fixed K, the region of high energy is fixed, so this can be used to choose a step-size and the number of leapfrog steps that allow us to rapidly reach this region. We fix $L = 80$ and tune ϵ by monitoring the progression of the sampler.

In practice, we can choose the number of latent factors K by cross-validation. To do this, we create 10 replications of our data and for each dataset we set 10% of the elements of the matrix as missing, using these elements as a held-out dataset. We then generate samples from the model over a range of K, and use the reconstruction error on the held-out data to choose the K that gives the best performance. We compare the negative log predictive probability (NLP) for K in the range of 2 to 20. We show the performance on the training and testing data in terms of root mean squared error (RMSE) as well as predictive probability (NLP) in Figure 4.3. We also compare the performance of the fully Bayesian approach using HMC that we presented, and the performance of maximum likelihood estimation in this model. The maximum likelihood estimators experience severe overfitting as shown by the RMSE on the training data. Since we would prefer a simpler model to a more complex one, we choose $K = 3$ vased on the graphs of RMSE and NLP on the test data. We discuss this issue of selecting K, and in particular, automatic methods for its selection in Section 4.6.

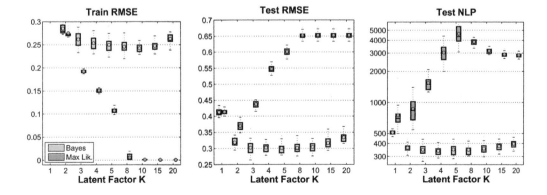

FIGURE 4.3
Choosing the number of latent factors K by cross-validation. We find that $K = 3$ is an appropriate number of latent factors.

Simulated Data from the BPM

We generated a synthetic binary dataset from the BPM consisting of $N = 50$ points, each being a $D = 32$ dimensional vector using $K = 3$ clusters. We ran HMC for 4000 iterations, using the first half as burn-in. To compare the true partial memberships $\mathbf{V}_T$ to the inferred memberships $\mathbf{V}_L$, we computed $\mathbf{U}_T = \mathbf{V}_T \mathbf{V}_T^\top$ and $\mathbf{U}_L = \mathbf{V}_L \mathbf{V}_L^\top$, which is a measure of the degree of shared membership between pairs of observations for the true and inferred partial memberships, respectively (Heller et al., 2008). This measure is invariant to permutations of the cluster labels, and the range of entries is between [0,1]. We show image-maps of these matrices in Figure 4.4. The difference between entries of the true and inferred shared memberships $|\mathbf{U}_T - \mathbf{U}_L|$ is shown in the histogram. The two matrices are highly similar, with 90% of entries being different from the true value by less than 0.2, showing that the sampler was able to learn the true partial memberships.

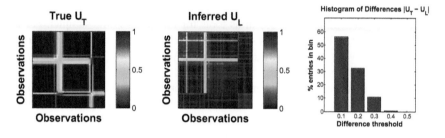

FIGURE 4.4
Image maps showing true shared partial memberships $\mathbf{U}_T$ and inferred shared membership $\mathbf{U}_L$ for synthetic data generated from the BPM model. The histogram shows the percentage of entries in $|\mathbf{U}_T - \mathbf{U}_L|$ that fall within a given difference threshold.

4.5.2   Senate Roll Call Data

Having evaluated the behavior of the BPM and EXFA on synthetic data, we demonstrate their use in exploring membership behavior from the U.S. Senate roll call as a case study. Specifically, we analyze the roll call from the 107th U.S. Congress (2001–2002) (Jakulin, 2002). The data consists of 99 senators (one senator died in 2002, and neither he nor his replacement are included), by 633 votes. It also includes the outcome of each vote, which we treat as an additional data point (like a senator who always voted the actual outcome). The matrix contains binary features for yea and nay votes, and abstentions are recorded as missing values. For the perspective of a political scientist analyzing such data, see the chapter by Gross and Manrique-Vallier (Gross and Manrique-Vallier, 2013).

We analyze the data using the BPM with $K = 2$ clusters, and show results of this analysis in Figure 4.5. Since there are two clusters and the amount of membership always sums to 1 across clusters, the figure looks the same regardless of whether we look at the 'Democrat' or 'Republican' cluster. The cyan line in Figure 4.5 indicates the partial membership assigned to each of the senators with their names overlaid. We can see that most Republicans and Democrats are clustered together in the flat regions of the line (with partial memberships very close to 0 or 1), but that there is a fraction of senators (around 20%) that lie somewhere in-between. Interesting properties of this figure include the location of Senator Jeffords (in magenta) who left the Republican party in 2001 to become an Independent who caucused with the Democrats. Also, Senator Chafee who is known as a moderate Republican and who often voted with the Democrats (for example, he was the only Republican to vote against authorizing the use of force in Iraq), and Senator Miller, a conservative Democrat who supported George Bush over John Kerry in the 2004 U.S. Presidential election. Lastly, it is interesting to note the location of the outcome data point, which is very much in the middle. This

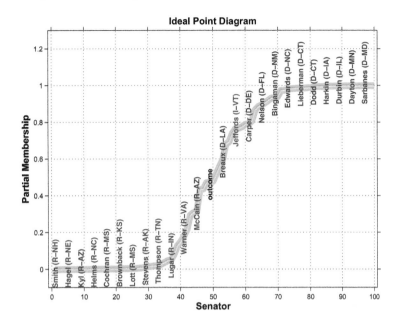

FIGURE 4.5
Analysis of the partial memberships for the 107th U.S. Senate roll call using BPM. The line shows
the amount of membership in the 'Democrat' cluster with the names of Democrat senators overlaid
in blue and Republican senators in red.

makes sense since the 107th Congress was split 50-50 (with Republican Vice President Dick Cheney
breaking ties), until Senator Jeffords became an Independent, at which point the Democrats had a
one seat majority.

We also analyzed the data using fuzzy k-means clustering, which found very similar rankings
of senators to the 'Democrat' cluster. Fuzzy k-means was very sensitive to the exact ranking and
degree of partial membership, since it is highly sensitive to the fuzzy exponent parameter γ_f, which
is typically set by hand. Figure 4.6 shows the change in partial membership for the outcome of
the most-allegiant Democrat and Republican senator (using the result of Figure 4.5), for a range of
values for the fuzzy exponent. The graph shows that the assigned partial membership can vary quite
dramatically depending on the choice of γ_f. This type of sensitivity to parameters does not exist in
the Bayesian models we present here, since they can be inferred automatically.

The BPM provides a very natural representation of the membership of individuals in this data
to political leanings. An alternative viewpoint can be obtained using EXFA. With EXFA, the latent variables do not have an interpretation as a degree of membership, but rather provide a low-
dimensional embedding of the data, which for the case of two latent factors, can be used to provide
a spatial visualization of senators. We show the results of analyzing the roll call data with EXFA in
Figure 4.8 using $K = 2$ latent factors, producing 4000 samples from the HMC sampler and using
the first half as burn-in. The latent embedding in Figure 4.8 is color-coded blue for Democrats and
red for Republicans, and shows that there is a natural separation of the data into these two groups.
Similarly to the BPM, we observe that most senators are clustered into a Democrat or Republican
cluster, with a percentage who straddle the boundary between these two groups. Again, we see the
effect of the independent candidate and the outcome. It is also important to note the connection
between both BPM and EFA to ideal point models in political science (Bafumi et al., 2005), which
aim to spatially represent political preferences on a left-to-right scale. Using the BPM and EFA,

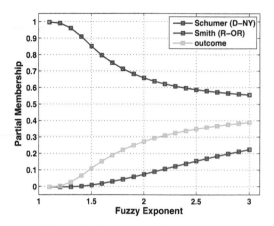

NLP	BPM	EFA	DPM
Mean	192	188	196
Min	100	92	112
Median	173	171	178
Max	428	412	412
Outcome	230	183	245

FIGURE 4.7
Comparison of negative log predictive probabilities (in bits) across senators for BPM, EFA, and DPM.

FIGURE 4.6
Sensitivity of partial memberships in fuzzy k-means with respect to the fuzzy exponent.

we have with Figures 4.5 and 4.8 shown Bayesian approaches of producing 1D and 2D ideal point representations, respectively.

As a further comparison of the BPM and EXFA, we also analyze the roll call data using a Dirichlet Process mixture model (DPM). We ran the DPM for 1000 Gibbs sampling iterations, sampling both assignments and concentration parameters. The DPM confidently finds four clusters: one cluster consists soley of Democrats, another solely of Republicans, a third cluster contains 9 moderate Republicans and Democrats as well as the outcome, and the last cluster consists of a single senator (Hollings (D-SC)).

We calculate the negative log-predictive probability (NLP, in bits) across senators for the BPM, EXFA, and DPM (Figure 4.7). We present the mean, minimum, median, and maximum NLP over all senators, which represents the number of bits needed to encode a senator's voting behavior. We also show the outcome separately. Except for the maximum, the BPM is able to produce a more compressed representation for each senator than the DPM, showing the sensibility of inferring partial memberships for this data, rather than assignments to clusters. EXFA produces the most compressed representation, since it used unconstrained latent variables and thus has greater modeling flexibility. These two approaches emphasize the tradeoff between modeling efficiency and interpretability that must be considered when analyzing such data.

The BPM gives an intuitive numerical quantity to the degree of membership, whereas EXFA gives an intuitive spatial understanding of this membership. Factor models are also often used to model the covariance structure of data and provide further insight into the data. For Gaussian data, this covariance is given by $\Theta\Theta^\top$. For non-Gaussian data, we can compute the marginal covariance $p(x_i = x_j), i \neq j$, by Monte Carlo integration using the posterior samples obtained. We show this in Figure 4.8(b) for the first 30 votes. The figure shows that there are many roll calls that are highly correlated, e.g., the first 14 entries represent the opening of the congress and are votes for chairs of various committees. Often not being votes of contention, there is highly correlated voting for these motions. Analysis of this matrix gives insight into the evolution of votes in the congress and provides an example of some of the probabilistic queries that can be made once the posterior samples are obtained. Other interesting probabilistic queries of this nature include examining the similarity of senators using the KL-distance between their latent posterior distributions, or examining the influence of senators to the voting outcomes using the marginal likelihood each senator contributes to the total probability.

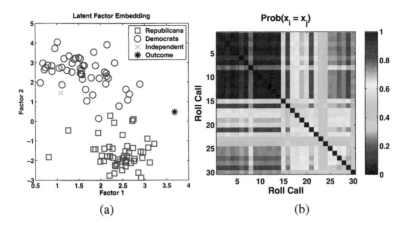

FIGURE 4.8

Analysis of the partial memberships for the 107th U.S. Senate roll call using EXFA. (a) The left plot shows the latent embedding produced using two latent factors. (b) The right plot shows the marginal covariance between votes.

4.6 Discussion

Having gained an understanding of exponential family latent variable models and their behavior, we now consider some of the questions that affect our ability to use such models in practice. Questions that arise include: how to decide between competing models, methods for choosing the latent dimensionality K, difficulty in tuning the MCMC samplers, and obstacles in applying these models to large datasets. We expand on these questions and discuss the ways in which our models can be extended to address them.

Choice of Model. In this chapter we have considered mixture models, the Bayesian partial membership model, exponential family factor analysis, and mixed membership models. The choice of one model type over another depends on the whether the modeling assumptions made match our beliefs regarding the process that generated the data, as well as the aim of our modeling effort, whether for visualization, predictive, or explanatory purposes. The BPM and EXFA are models with a single layer of latent variables that we showed are relaxations of K-component mixture models. These models thus make use of a single layer of latent variables, and we demonstrated in the experiments that the models allowed for de-noising of data, effective imputation of missing data, and are useful tools for visualization of high-dimensional data. The structure of the models proved to be intuitive and flexible, and appropriate for the tasks we presented.

More flexible versions of these models can be obtained by considering the mixed membership analogues of the BPM and EXFA, such as Grade of Membership models (Erosheva et al., 2007; Gross and Manrique-Vallier, 2013) and mixed membership matrix factorization (Mackey et al., 2010), respectively. In addition, other prior assumptions may be needed; sparsity is one such prior assumption that has gained importance and the inclusion of sparsity in the models discussed here is described by Mohamed et al. (2012). Galyardt (2013) in this volume shows that mixed membership models have an equivalent representation as a mixture model, with a number of mixture components polynomial in K, thus providing a highly efficient representation of high-dimensional data. Inference in these more complex models is harder due to the increased number of latent and assignment variables, making the factors affecting our choice of model based on the tradeoff between simplicity, flexibility, and the computational complexity of the available models. A formal model comparison would rely on Bayesian model selection, in which the 'best' model is chosen based on the evaluation of the marginal likelihood or model evidence (Carlin and Chib, 1995).

Choosing the Latent Dimensionality. In Section 4.5 we used cross-validation to determine the appropriate dimensionality of the latent variables. Ideally, we would wish to learn K automatically using the training data only. An alternative approach to cross-validation is by Bayesian model selection where we evaluate and compare the marginal likelihood or evidence for various models, e.g., as described by Minka (2001) for probabilistic PCA. The models we have described can also be adapted to include the determination of K as part of the learning algorithm. Bishop (1999) exploited sparsity by employing automatic relevance determination (ARD), which uses a large number of latent factors and sets to zero any factors that are not supported by the data; K is then the number of non-zero columns at convergence of the algorithm. It is also possible to specify the dimensionality of the latent variables as part of our model construction. This approach requires an efficient means of sampling in spaces with changing dimensionality, most often achieved by trans-dimensional MCMC, such as the approach described by Lopes and West (2004). More recent approaches have focused on the construction of nonparametric latent factor models using the Indian buffet process or other nonparametric priors to automatically adapt the dimensionality of latent variables (Knowles and Ghahramani, 2010; Bhattacharya and Dunson, 2011).

Tuning MCMC Samplers. We made use of the standard approach for Hybrid Monte Carlo (HMC) sampling here, but this can be improved to increase the number of uncorrelated samples obtained. We used an identity mass matrix, but adaptively estimating the mass matrix using the empirical covariance or Hessian of the log joint probability from the samples during the burn-in phase can be used, reducing sensitivity to the choice of step-size ϵ (Atchade et al., 2011).

Using an appropriate mass matrix allows proposals to be made at an appropriate scale, thus allowing for larger step-sizes during sampling. But estimation of the mass matrix (and computing its inverse) can add significantly to the computation involved in HMC. Adaptive tuning of the mass matrix was also shown using the Riemann geometry of the joint-probability by Girolami and Calderhead (2011). Another way of improving HMC was proposed in Shahbaba et al. (2011), and involves splitting the Hamiltonian in a way that allows much of the movement around the state-space to be done at low computational cost. Tuning the HMC parameters can be challenging, especially for the non-expert, and methods now exist for the automatic tuning of HMC's parameters (Hoffman and Gelman, 2011; Wang et al., 2013). Any of these approaches removes the need for tuning HMC and have the promise of making the application of HMC much more general purpose.

Deterministic Approximations for Large-scale Learning. With the increasing size of datasets, the availability of scalable inference is an important factor in the practial use of many models. MCMC methods can be shown to scale well to large datasets (Salakhutdinov and Mnih, 2008). Deterministic approximations are increasingly used in the development of scalable algorithms and can allow better exploitation of the distributed nature of modern computing environments. Variational inference for LDA was described by Teh et al. (2007), and such an approach can be applied to the BPM. For latent Gaussian models, approximate inference methods such as integrated nested Laplace approximations (INLA) (Rue et al., 2009) have been proposed. INLA is effective for models whose latent variables are controlled by a small number of hyperparameters, limiting the application of this approach for learning in EFA. Variational methods for EXFA have also been successfully explored (Khan et al., 2010).

4.7 Conclusion

In this chapter, we have described a principled Bayesian framework for latent variable modeling that is generalized to the exponential family of distributions. We began with the widely-used mixture model and showed that a relaxation of the assumption that each data point belongs to one and only one cluster allows us to explore different aspects of the structure underlying the data. We obtained the Bayesian partial membership (BPM) model by allowing the latent variables to represent fractional membership in multiple clusters, and obtained exponential family factor analysis (EXFA) by considering continuous latent variables (which explain contributions to the data using a linear combination from all clusters). By framing these models in the same latent variable framework, we exploited the continuous nature of the unknown parameters and demonstrated how Hybrid Monte Carlo can be implemented and tuned for such models. We also described the connection to other latent variable and mixed membership models. Using both synthetic and real-world data, we demonstrated the use of these models for visualization and predictive tasks and the wide range of insightful probabilistic queries that can be made using these models.

References

Atchade, Y., Fort, G., Moulines, E., and Priouret, P. (2011). Adaptive Markov chain Monte Carlo: Theory and methods. In Barber, D., Cemgil, A. T., and Chiappa, S. (eds), *Bayesian Time Series Models*. Cambridge, UK: Cambridge University Press.

Bafumi, J., Gelman, A., Park, D. K., and Kaplan, N. (2005). Practical issues in implementing and understanding Bayesian ideal point estimation. *Political Analysis* 13: 171–187.

Bartholomew, D. J. and Knott, M. (1999). *Latent Variable Models and Factor Analysis: Kendall's Library of Statistics 7*. Wiley, 2nd edition.

Beskos, A., Pillai, N., Roberts, G. O., Sanz-Serna, J. -M., and Stuart, A. M. (2010). Optimal tuning of hybrid Monte Carlo. http://arxiv.org/abs/1001.4460.

Bezdek, J. C. (1981). *Pattern Recognition with Fuzzy Objective Functions Algorithms*. Norwell, MA: Kluwer Academic Publishers.

Bhattacharya, A. and Dunson, D. B. (2011). Sparse Bayesian infinite factor models. *Biometrika* 98: 291–306.

Bickel, P. J. and Doksum, K. A. (2001). *Mathematical Statistics: Basic Ideas and Selected Topics I*. Prentice Hall, 2nd edition.

Bishop, C. M. (1999). Bayesian PCA. In Kearns, M. S., Solla, S. A., and Cohn, D. A. (eds), *Advances in Neural Information Processings Systems 11*. Cambridge, MA: The MIT Press, 382–388.

Bishop, C. M. (2006). *Pattern Recognition and Machine Learning*. Information Science and Statistics. Springer.

Blei, D. M., Ng, A. Y., and Jordan, M. I. (2003). Latent Dirichlet allocation. *Journal of Machine Learning Research* 3: 993 –1022.

Buntine, W. and Jakulin, A. (2006). Discrete components analysis. In *Subspace, Latent Structure and Feature Selection*, vol. 3940 of *Lecture Notes in Computer Science*. Springer, 1–33.

Carlin, B. P. and Chib, S. (1995). Bayesian model choice via Markov chain Monte Carlo methods. *Journal of Royal Statistical Society Series B* : 473–484.

Collins, M., Dasgupta, S., and Schapire, R. E. (2002). A generalization of principal component analysis to the exponential family. In Dietterich, T. G., Becker, S., and Ghahramani, Z. (eds), *Advances in Neural Information Proceeding Systems 14*. Cambridge, MA: The MIT Press, 617–624.

Duane, S., Kennedy, A. D., Pendleton, B. J., and Roweth, D. (1987). Hybrid Monte Carlo. *Physics Letters B* 195: 216–222.

Erosheva, E. A., Fienberg, S. E., and Lafferty, J. D. (2004). Mixed membership models of scientific publications. *Proceedings of the National Academy of Sciences* 101: 5220–5227.

Erosheva, E. A., Fienberg, S. E., and Joutard, C. (2007). Describing disability through individual-level mixture models for multivariate binary data. *Annals of Applied Statistics* 1: 502–537.

Galyardt, A. (2014). Interpreting mixed membership models: Implications of Erosheva's representation theorem. In Airoldi, E. M., Blei, D. M., Erosheva, E. A., and Fienberg, S. E. (eds), *Handbook of Mixed Membership Models and Its Applications*. Chapman & Hall/CRC.

Gasch, A. and Eisen, M. (2002). Exploring the conditional coregulation of yeast gene expression through fuzzy *k*-means clustering. *Genome Biology* 3 : 1–22.

Gilks, W. R., Richardson, S., and Spiegelhalter, D. J. (eds) (1995). *Markov Chain Monte Carlo in Practice: Interdisciplinary Statistics (Chapman & Hall/CRC Interdisciplinary Statistics)*. Chapman & Hall/CRC.

Girolami, M. and Calderhead, B. (2011). Riemann manifold Langevin and Hamiltonian Monte Carlo methods. *Journal of Royal Statistical Society Series B* 73: 123–214.

Gross, J. H. and Manrique-Vallier, D. (2013). A mixed-membership approach to the assessment of political ideology from survey responses. In Airoldi, E. M., Blei, D. M., Erosheva, E. A., and Fienberg, S. E. (eds), *Handbook of Mixed Membership Models and Its Applications*. Chapman & Hall/CRC.

Gruhl, J. and Erosheva, E. A. (2013). A tale of two (types of) mixed memberships: Comparing mixed and partial membership with a continuous data example. In Airoldi, E. M., Blei, D. M., Erosheva, E. A., and Fienberg, S. E. (eds), *Handbook of Mixed Membership Models and Its Applications*. Chapman & Hall/ CRC.

Heller, K. A., Williamson, S., and Ghahramani, Z. (2008). Statistical models for partial membership. In *Proceedings of the 25th International Conference on Machine Learning (ICML '08)*. New York, NY, USA: ACM, 392–399.

Hoffman, M. D. and Gelman, A. (2011). The No-U-Turn Sampler: Adaptively Setting Path Lengths in Hamiltonian Monte Carlo. Tech. report, http://arxiv.org/abs/1111.4246.

Jakulin, A., Buntine, W., La Pira, T. M., and Brasher, H. (2002). Analyzing the U.S. Senate in 2003: Similarities, clusters, and blocs. *Political Analysis* 17 : 291–310.

Khan, M. E., Marlin, B. M., Bouchard, G., and Murphy, K. P. (2010). Variational bounds for mixed-data factor analysis. In Lafferty, J. D., Williams, C. K. I., Shawe-Taylor, J., Zemel, R. S., and Culotta, A. (eds), *Advances in Neural Information Proceeding Systems 23*. Red Hook, NY: Curran Associates, Inc., 1108–1116.

Knowles, D. and Ghahramani, Z. (2011). Nonparametric Bayesian sparse factor models with application to gene expression modeling. *Annals of Applied Statistics* 5 : 1534–1552.

Kosko, B. (1992). *Neural Networks and Fuzzy Systems*. Prentice Hall.

Lee, D. D. and Seung, H. S. (1999). Learning the parts of objects by non-negative matrix factorization. *Nature* 401: 788 –791.

Lopes, H. F. and West, M. (2004). Bayesian model assessment in factor analysis. *Statistica Sinica* 14: 41–68.

MacKay, D. J. C. (2003). *Information Theory, Inference & Learning Algorithms*. Cambridge, UK: Cambridge University Press.

Mackey, L., Weiss, D., and Jordan, M. I. (2010). Mixed membership matrix factorization. In Fürnkranz, J. and Joachims, T. (eds), *Proceedings of the 27th International Conference on Machine Learning (ICML '10)*. Omnipress, 711–718.

Minka, T. P. (2001). Automatic choice of dimensionality for PCA. In *Advances in Neural Information Proceeding Systems 11*. Cambridge, MA: The MIT Press, 598–604.

Mohamed, S., Heller, K. A., and Ghahramani, Z. (2008). Bayesian exponential family PCA. In Koller, D., Schuurmans, D., Bengio, Y., and Bottou, L. (eds), *Advances in Neural Information Proceeding Systems 21*. Red Hook, NY: Curran Associates, Inc., 1089–1096.

Mohamed, S., Heller, K. A., and Ghahramani, Z. (2012). Bayesian and L1 approaches for sparse unsupervised learning. In *Proceedings of the 29th International Conference on Machine Learning (ICML '12)*. Omnipress.

Moustaki, I. and Knott, M. (2000). Generalized latent trait models. *Psychometrika* 65: 391–411.

Neal, R. M. (1993). Probabilistic Inference Using Markov Chain Monte Carlo Methods. Tech. report CRG-TR-93-1, University of Toronto.

Neal, R. M. (2010). MCMC using Hamiltonian dynamics. In Brooks, S., Gelman, A., Jones, G., and Meng, X. -L. (eds), *Handbook of Markov Chain Monte Carlo*. Chapman & Hall/CRC, 113–162.

Newcomb, S. (1886). A generalized theory of the combination of observations so as to obtain the best result. *American Journal of Mathematics* 8: 343–366.

Rasmussen, C. E. and Williams, C. K. I. (2006). *Gaussian Processes for Machine Learning*. Cambridge, MA: The MIT Press.

Robert, C. P. and Casella, G. (2004). *Monte Carlo Statistical Methods*. Springer, 2nd edition.

Rue, H., Martino, S., and Chopin, N. (2009). Approximate Bayesian inference for latent Gaussian models by using integrated nested Laplace approximations. *Journal of Royal Statistical Society Series B* 71: 319–392.

Salakhutdinov, R. and Mnih, A. (2008). Bayesian probabilistic matrix factorization using Markov chain Monte Carlo. In *Proceedings of the 25th International Conference on Machine Learning (ICML '08)*. New York, NY, USA: ACM, 880–887.

Shahbaba, B., Lan, S., Johnson, W. O., and Neal, R. M. (2011). Split Hamiltonian Monte Carlo. Tech. report, University of California, Irvine. http://arxiv.org/abs/1106.5941.

Teh, Y. W., Newman, D. and Welling, M. (2007). A collapsed variational Bayesian inference algorithm for latent Dirichlet allocation. In Schölkopf, B., Platt, J., and Hofmann, T. (eds), *Advances in Neural Information Processing Systems 19*. Red Hook, NY: Curran Associates, Inc., 1343–1350.

Tipping, M. E. (1999). Probabilistic visualization of high dimensional binary data. In Kearns, M. J., Solla, S. A., and Cohn, D. A. (eds), *Advances in Neural Information Proceeding Systems 11*. Cambridge, MA: The MIT Press, 592–598.

Tipping, M. E. and Bishop, C. M. (1997). Probabilistic Principal Components Analysis. Tech. report NCRG/97/010, Neural Computing Research Group, Aston University.

Wang, Z., Mohamed, S., and de Freitas, N. (2013). Adaptive Hamiltonian and Riemann manifold Monte Carlo. In *Proceedings of the 30th International Conference on Machine Learning (ICML '13). Journal of Machine Learning Research W&CP* 28 : 1462–1470.

Zadeh, L. (1965). Fuzzy sets. *Information and Control* 8 : 338–353.

5

Nonparametric Mixed Membership Models

Daniel Heinz

Department of Mathematics and Statistics, Loyola University of Maryland, Baltimore, MD 21210, USA

CONTENTS

One issue with parametric latent class models, regardless of whether or not they feature mixed memberships, is the need to specify a bounded number of classes a priori. By contrast, nonparametric models use an unbounded number of classes, of which some random number are observed in the data. In this way, nonparametric models provide a method to infer the correct number of classes based on the number of observations and their similarity.

The following chapter seeks to provide mathematical and intuitive understanding of nonparametric mixed membership models, focusing on the hierarchical Dirichlet process mixture model (HDPM). This model can be understood as a nonparametric extension of the Grade of Membership model (GoM) described by Erosheva et al. (2007). To elucidate this relationship, the Dirichlet mixture model (DM) and Dirichlet process mixture model (DPM) are first reviewed; many of the interesting properties of these latent class models carry over to the GoM and HDPM models.

After describing these four models, the HDPM model is further explored through simulation studies, including an analysis of how the model parameters affect the model's clustering behavior.

An overview of inference procedures is also provided with a focus on Gibbs sampling and variational inference. Finally, some example applications and model extensions are briefly reviewed.

5.1 Introduction

Choosing the appropriate model complexity is a problem that must be solved in almost any statistical analysis, including latent class models. Simple models efficiently describe a small set of behaviors, but are not flexible. Complex models describe a wide variety of behaviors, but are subject to overfitting the training data. For latent class models, complexity refers to the number of groups used to describe the distribution of observed and/or predicted data. One strategy is to fit multiple models of varying complexity then decide among them with a post-hoc analysis (e.g., penalized likelihood). Nonparametric mixture models provide an alternate strategy which bypasses the need to choose the correct number of classes. The hallmark of nonparametric models is that their complexity increases stochastically as more data are observed. The rate of accumulation is determined by various tuning parameters and the similarity of observed data.

One of the best-known examples of nonparametric Bayesian inference is the Dirichlet process mixture model (DPM), a nonparametric version of the Dirichlet mixture model (DM). The DM model assumes that the population consists of a fixed and finite number of classes and it therefore bounds the number of classes used to represent any sample. By contrast, a DPM posits that the population consists of an infinite number of classes. Of these, some finite but *unbounded* number of classes are observed in the data. Because the number of classes is unbounded, the model always has a positive probability of assigning a new observation to a new class.

Both Dirichlet mixtures and Dirichlet process mixtures assume that observations are fully exchangeable. Extensions for both models exist for situations in which full exchangeability is inappropriate. This may be the case when multiple measurements are made for individuals in the sample. For example, one may consider a survey analysis in which each individual responds to several items. In this case, one expects two responses to be more similar if they come from the same individual. The Grade of Membership model (GoM) adapts the DM model for partial exchangeability (Erosheva et al., 2007). In the GoM model, two responses are exchangeable if and only if they are measured from the same individual. Like the DM model, it bounds the number of classes. The GoM model is known as a *mixed membership model* or *individual-level mixture model* because each individual in the sample is associated with unique mixing weights for the various classes.

The hierarchical Dirichlet process mixture model (HDPM) extends the GoM model in the same way that the Dirichlet process mixture model extends the DM model. As with the GoM model, the HDPM model assumes that responses are exchangeable if and only if they come from the same individual. Whereas the GoM assumes that the population consists of a fixed and finite number of classes, the HDPM posits an infinite number of classes. Thus, the HDPM model does not bound the number of classes used to represent the sample.

All four models (DM, DPM, GoM, and HDPM) cluster observations into various classes where class memberships are unobserved. They are distinguished by the type of exchangeability (full or partial) and whether or not the number of classes in the population is bounded a priori.

Nonparametric mixture models, such as the DPM and HDPM models, have several intuitive advantages. Because the number of classes is not fixed, they provide a posterior distribution over the model complexity. Posterior inference includes a natural weighting of high-probability models of varying complexity. Hence, uncertainty about the "true" number of classes is measurable. Furthermore, because the number of classes is unbounded, nonparametric models always include a positive probability that the next observation belongs to a previously unobserved class. This property is especially nice when considering predictive distributions. If the number of classes is unknown, it is possible that the next observation will be unlike any of the previous observations.

This chapter aims to provide an intuitive understanding of the hierarchical Dirichlet mixture model. Sections 5.2 and 5.3 begin with the fully exchangeable models, showing how the DM model is built into the nonparametric HDPM model by removing the bound on the number of classes. Properties of these mixtures are illustrated and compared using the Chinese restaurant process. This relationship forms the foundation for exploring properties of the GoM and HDPM models in Sections 5.4 and 5.5. In Section 5.6, the role of tuning parameters for the HDPM model is explored intuitively and illustrated through simulations. Section 5.7 provides an overview of inference strategies for DPM and HDPM models. Section 5.8 reviews some example applications with Section 5.9 devoted to brief descriptions of some model extensions.

5.2 The Dirichlet Mixture Model

Suppose a sample contains n observations, $(x_1, \ldots, x_n)$, where x_i is possibly vector-valued. A latent class model assumes that each observation belongs to one of K possible classes, where K is a finite constant. Observations are conditionally independent given their class memberships, but dependence arises because class memberships are not observed. As a generative model, each x_i is drawn by randomly choosing a class, say z_i, then sampling from the class-specific distribution.

Denote the population proportions of these K classes by $\pi = (\pi_1, \ldots, \pi_K)$ and the distribution of class k by F_k. For simplicity, assume that these distributions belong to some parametric family, $\{F(\cdot|\theta) : \theta \in \Theta\}$. Therefore, $F_k = F(\cdot|\theta_k)$, where $\theta_k \in \Theta$ denotes the class-specific parameter for class k. Given the class proportions and parameters, the latent class model is described by a simple hierarchy:

$$z_i|\pi \overset{\text{i.i.d.}}{\sim} \text{Mult}(\pi) \qquad i = 1 \ldots n.$$
$$x_i|z_i, \theta \sim F(\cdot|\theta_{z_i}) \qquad i = 1 \ldots n.$$

Here, $\text{Mult}(\pi)$ is the multinomial distribution satisfying $\mathbb{P}(z_i = k) = \pi_k$ for $k = 1 \ldots K$.

Inferential questions include learning the mixing proportions (π_k), class parameters (θ_k), and possibly the latent class assignments (z_i). Uncertainty about the class proportions and parameters may be expressed through prior laws. In the Dirichlet mixture model (DM), the class proportions have a symmetric Dirichlet prior, $\pi \sim \text{Dir}(\alpha/K)$. This distribution is specified by the precision $\alpha > 0$ and has the density function

$$f(\pi|\alpha) = \frac{\Gamma(\alpha)}{[\Gamma(\alpha/K)]^K} \prod_{k=1}^{K} \pi_k^{\alpha/K-1}, \tag{5.1}$$

wherever π is a K-dimensional vector whose elements are non-negative and sum to 1. The range of possible values for π is known as the $(K-1)$-*dimensional simplex* or more simply the $(K-1)$-simplex. The expected value of the symmetric Dirichlet distribution is the uniform probability vector $E[\pi] = \left(\frac{1}{K}, \ldots, \frac{1}{K}\right)$. The precision specifies the concentration of the distribution about this mean, with larger values of α translating to less variability.

More generally, an asymmetric Dirichlet distribution is defined by a precision $\alpha > 0$ and a mean vector $\pi_0 = E[\pi] = (\pi_{01}, \ldots, \pi_{0K})$ in the $(K-1)$-simplex. If $\pi \sim \text{Dir}(\alpha, \pi_0)$, then its distribution function is

$$f(\pi|\alpha, \pi_0) = \frac{\Gamma\left(\sum_{k=1}^{K} \alpha\pi_{0k}\right)}{\prod_{k=1}^{K} \Gamma(\alpha\pi_{0k})} \prod_{k=1}^{K} \pi_{0k}^{\alpha\pi_{0k}-1}. \tag{5.2}$$

Note that the symmetric Dirichlet $\text{Dir}(\alpha/K)$ is equivalent to the distribution $\text{Dir}\left(\alpha, \frac{1}{K}\mathbf{1}\right)$, where $\mathbf{1}$ denotes the vector of ones.

The symmetric Dirichlet prior influences the way in which the DM model groups observations into the K possible classes. To finish specifying the model, each class is associated with its class parameter, θ_k. The class parameters are assumed to be i.i.d. from some prior distribution $H(\lambda)$, where λ is a model-specific hyperparameter. This results in the following model:

Dirichlet Mixture Model

$$\pi|\alpha \sim \text{Dir}\left(\frac{\alpha}{K}\right).$$

$$\theta_k|\lambda \overset{\text{i.i.d.}}{\sim} H(\lambda) \qquad\qquad k = 1\ldots K.$$

$$z_i|\pi \overset{\text{i.i.d.}}{\sim} \text{Mult}(\pi) \qquad\qquad i = 1\ldots n.$$

$$x_i|z_i, \theta \sim F(\cdot|\theta_{z_i}) \qquad\qquad i = 1\ldots n.$$

Because class memberships are dependent only on α, the Dirichlet precision fully specifies the prior clustering behavior of the model. The hyperparameter λ only influences class probabilities during posterior inference. A priori, observations are expected to be uniformly dispersed across the K classes and α measures how strongly the prior insists on uniformity.

5.2.1 Finite Chinese Restaurant Process [1]

Imagine a restaurant with an infinite number of tables, each of which has infinite capacity. Observations are represented by customers and class membership is defined by the customer's choice of dish. All customers at a particular table eat the same dish. When a customer sits at an unoccupied table, he selects one of the K possible dishes for his table with uniform probabilities of $1/K$. Because multiple tables may serve the same dish, the class membership of an observation must be defined by the customer's *dish* rather than his table.

The first customer sits at the first table and randomly chooses one of the K dishes. The second customer joins the first table with probability $1/(1+\alpha)$ or starts a new table with probability $\alpha/(1+\alpha)$. As subsequent customers enter, the probability that they join an occupied table is proportional to the number of people already seated there. Alternatively, they may choose a new table with probability proportional to α.

Mathematically, let T denote the number of occupied tables when the nth customer arrives and let t_n denote the table that he chooses. Given the seating arrangement of the previous customers, the probability function for t_n is

$$f(t_n|\alpha, \mathbf{t}_{\widetilde{n}}) \propto \begin{cases} \sum_{i<n} \mathbf{1}\,(t_i = t_n), & t_n \leq T \\ \alpha, & t_n = T+1 \end{cases}, \tag{5.3}$$

where $\mathbf{t}_{\widetilde{n}}$ denotes the table assignments of all but the nth customer and $\mathbf{1}\,(t_i = t_n)$ is the indicator function, which is equal to 1 if $t_i = t_n$ and 0 otherwise.

Since all customers at a table eat the same dish, if a customer joins an occupied table, he eats

---

[1]The typical Chinese restaurant process, as described in Section 5.3.2, illustrates the clustering behavior of the Dirichlet process mixture model after integrating out the unknown vector of class proportions (π). Here, the modified finite version describes a Dirichlet mixture by fixing the number of possible dishes at $K < \infty$.

whatever dish was previously chosen for that table. When a customer starts a new table, he must select a dish for that table by randomly choosing one of the K menu items with uniform probabilities. Therefore, the distribution for the nth customer's dish is

$$f(z_n|\alpha, \mathbf{z}_{\tilde{n}}) = \frac{\sum_{i<n} \mathbf{1}(z_i = z_n) + \alpha/K}{n - 1 + \alpha} \qquad k \leq K, \tag{5.4}$$

where z_i is the dish (class membership) for the ith customer and $\mathbf{z}_{\tilde{n}}$ is the vector of dishes for all but the nth customer.

The Chinese restaurant analogy depicts the clustering behavior of the Dirichlet mixture model. Notably, tables with many customers are more likely to be chosen by subsequent customers. This creates a "rich-get-richer" effect. The marginal distribution of z_n is uniform due to the symmetric Dirichlet prior, but the conditional distribution given $(z_1, \ldots, z_{n-1})$ is skewed toward the class memberships of previous customers. Let $f_{\text{EDF}}(k) = \frac{\sum_{i<n} \mathbf{1}(z_i=k)}{n-1}$ denote the empirical distribution of the first $n - 1$ class memberships. Equation 5.4 can be written as a weighted combination of the empirical distribution and the uniform prior:

$$f(z_n|\alpha, \mathbf{z}_{\tilde{n}}) \propto (n - 1)f_{\text{EDF}}(z_n) + \alpha\frac{1}{K}. \tag{5.5}$$

Equation (5.5) shows the smoothing behavior of the Dirichlet mixture when the class proportions (π) are integrated out. Specifically, the class weights for the nth observation are smoothed toward the average value of $\frac{1}{K}$. The Dirichlet precision α controls the degree of smoothing. It has the effect of adding α prior observations spread evenly across all K classes. Because the class memberships are fully exchangeable, this equation expresses the conditional distribution for any z_i based on the other class memberships by treating x_i as the last observation.

Recall that the customers' dishes represent their class memberships. To completely specify the mixture distribution, each dish is associated with a parameter value, $\theta_k \overset{\text{i.i.d.}}{\sim} H$. In other words, each customer that eats dish k represents an observation from the kth class with class parameter θ_k. The class parameters are mutually independent and independent of the latent class memberships.

An important property of the DM model is that z_i is bounded by K. At most, K classes will be used to represent the n observations in the sample. The next section explores the behavior of this model when the bound is removed.

5.3 The Dirichlet Process Mixture Model

Consider the issue of deciding how many classes are needed to represent a given sample. One method is to fit latent class models for several values and use diagnostics to compare the fits. Such methods include, among others, cross-validation techniques (Hastie et al., 2009) and penalized likelihood scores such as Akaike information criterion (AIC) (Akaike, 1973) and Bayesian information criterion (BIC) (Schwarz, 1978). Though AIC and BIC are popular choices, their validity for latent class models has been criticized (McLachlan and Peel, 2000). Instead of choosing the single best model complexity, one can use a prior distribution over the number of classes to calculate posterior probabilities (Roeder and Wasserman, 1997). Given a suitable prior, reversible jump Markov chain Monte Carlo techniques can sample from a posterior distribution which includes models of varying dimensionality (Green, 1995; Giudici and Green, 1999). An alternate strategy is to assume that the number of latent classes in the population is unbounded. The Dirichlet process mixture model (DPM) arises as the limiting distribution of Dirichlet mixture models when K approaches infinity. This limit uses a Dirichlet process as the prior for class proportions and parameters. This

section reviews properties of the the Dirichlet process and its relationship to the finite Dirichlet mixture model. These properties elucidate the hierarchical Dirichlet process (Section 5.5), which uses
multiple Dirichlet process priors to construct a nonparametric mixed membership model.

5.3.1 The Dirichlet Process

The Dirichlet process is a much-publicized nonparametric process formally introduced by Ferguson (1973). It is a prior law over probability distributions whose finite-dimensional marginals have
Dirichlet distributions. Dirichlet processes have been used for modeling Gaussian mixtures when
the number of components is unknown (Escobar and West, 1995; MacEachern and Müller, 1998;
Rasmussen, 1999), survival analysis (Kim, 2003), hidden Markov models with infinite state-spaces
(Beal et al., 2001), and evolutionary clustering in which both data and clusters come and go as time
progresses (Xu et al., 2008).

    The classical definition of a Dirichlet process constructs a random measure P in terms of finite-
dimensional Dirichlet distributions (Ferguson, 1973). Let α be a positive scalar and let H be a
probability measure with support Θ. If $P \sim \mathrm{DP}(\alpha, H)$ is a Dirichlet process with precision α and
base measure H, then for any natural number K,

$$\big(P(A_1), \ldots, P(A_K)\big) \sim \mathrm{Dir}\big(\alpha H(A_1), \ldots, \alpha H(A_K)\big), \tag{5.6}$$

whenever $(A_k)_{k=1}^K$ is a measurable finite partition of Θ.

    Sethuraman (1994) provides a constructive definition of P based on an infinite series of inde-
pendent beta random variables. Let $\phi_k \stackrel{\mathrm{i.i.d.}}{\sim} \mathrm{Beta}(1, \alpha)$ and $\theta_k \stackrel{\mathrm{i.i.d.}}{\sim} H$ be independent sequences.
Define $\pi_1 = \phi_1$, and set $\pi_k = \phi_k \prod_{j=1}^{k-1}(1 - \phi_j)$ for $k > 1$. The random measure $P = \sum_{k=1}^{\infty} \pi_k \delta_{\theta_k}$
has distribution $\mathrm{DP}(\alpha, H)$, where δ_x is the degenerate distribution with $f(x) = 1$. This definition
of the Dirichlet process is called a stick-breaking process. Imagine a stick of unit length which is
divided into an infinite number of pieces. The first step breaks off a piece of length $\pi_1 = \phi_1$. After
$k - 1$ steps, the remaining length of the stick is $\prod_{i=1}^{k-1}(1 - \phi_i)$. The kth step breaks off a fraction
ϕ_k of this length, which results in a new piece of length π_k.

    The stick-breaking representation shows that $P \sim \mathrm{DP}(\alpha, H)$ is discrete with probability 1.
The measure P is revealed to be a mixture of an infinite number of point masses. Hence, there is
a positive probability that a finite sample from P will contain repeated values. This leads to the
clustering behavior of the following Dirichlet process mixture model (Antoniak, 1974):

$$P \sim \mathrm{DP}(\alpha, H).$$

$$\theta_i^* | P \stackrel{\mathrm{i.i.d.}}{\sim} P \qquad\qquad\qquad\qquad i = 1 \ldots n.$$
$$x_i | \theta_i^* \sim F(\cdot | \theta_i^*) \qquad\qquad\qquad\qquad i = 1 \ldots n.$$

    Let $(\theta_1, \ldots, \theta_K)$ denote the unique values of the sequence $(\theta_1^*, \ldots, \theta_n^*)$, where K is the random
number of unique values. Set z_i such that $\theta_i^* = \theta_{z_i}$. Given z_i and θ, the distribution of the ith
observation is

$$F(x_i | z_i, \theta) = F(\cdot | \theta_{z_i}). \tag{5.7}$$

By comparing Equation (5.7) to the Dirichlet mixture model, one can interpret z_i as a class membership, θ as the class parameters, and K as the number of classes represented in the sample. Note that
K is random in this model, whereas it is a constant in the DM model. Therefore, the DPM model
provides an implicit prior over the number of classes in the sample. Antoniak (1974) specifies this
prior explicitly.

    To make direct comparisons between Dirichlet process mixtures and Dirichlet mixtures, it is
useful to disentangle the distributions of π and θ_k. Let $\pi \sim \mathrm{SBP}(\alpha)$ denote the vector of weights

based on the stick-breaking process. Extend the notation Mult(π) to include infinite multinomial distributions such that $\mathbb{P}(z_i = k) = \pi_k$ for all positive integers k. The DPM model is equivalent to the following hierarchy:

Dirichlet Process Mixture Model

$$\pi|\alpha \sim \text{SBP}(\alpha).$$

$$\theta_k|\lambda \overset{\text{i.i.d.}}{\sim} H(\lambda) \qquad\qquad k = 1, 2, \ldots.$$

$$z_i|\pi \overset{\text{i.i.d.}}{\sim} \text{Mult}(\pi) \qquad\qquad i = 1 \ldots n.$$

$$x_i|z_i, \theta \sim F(\cdot|\theta_{z_i}) \qquad\qquad i = 1 \ldots n.$$

The above hierarchy directly shows the relationship between the Dirichlet mixture model and the DP mixture model. Where the Dirichlet mixture uses a symmetric Dirichlet prior for π, the DP mixture uses the stick-breaking process to generate an infinite sequence of class weights. In fact, Ishwaran and Zarepour (2002) shows that the marginal distribution induced on $x_1, \ldots, x_n$ by the DM model approaches that of the DPM model as the number of classes increases to infinity. Thus, the Dirichlet process mixture model may be interpreted as the infinite limit of finite Dirichlet mixture models.

5.3.2 Chinese Restaurant Process

A Chinese restaurant process illustrates the clustering behavior of a Dirichlet process mixture model when the unknown class proportions (π) are integrated out (Aldous, 1985). Customers arrive and choose tables as in the finite version for Dirichlet mixtures, but the menu in the full Chinese restaurant process contains an infinite number of dishes.

Recall that in the finite Chinese restaurant process, a customer who sits at an empty table chooses one of the K available dishes using uniform probabilities. For a DP mixture, the menu has an unlimited number of dishes. The discrete uniform distribution is not defined over infinite sets, but this technicality can be sidestepped. Class parameters are assigned independently of each other and the enumeration of the dishes is immaterial. Therefore, dishes do not need to be labeled until after they are sampled. Whatever dish happens to be selected first can be labeled 1, the second dish to be chosen can be labeled 2, and so on. In other words, when sampling a dish, there is no need to distinguish between any of the unsampled dishes. Because there are finitely many sampled dishes and infinitely many unsampled dishes, a "uniform" distribution implies that, with probability 1, the customer selects a new dish from the distribution $H(\lambda)$. Note that if the distribution H has any points with strictly positive probability, there is a chance that the "new" dish chosen by the customer will be the same as an already observed dish. To avoid this technicality and simplify wording, one may assume that H is continuous. The mathematics are the same in either case.

In the Chinese restaurant process, the first customer sits at the first table and randomly chooses a random dish, which is labeled 1. The second customer joins the first table with probability $1/(1+\alpha)$ or starts a new table with probability $\alpha/(1+\alpha)$. As subsequent customers enter, the probability that they join an occupied table is proportional to the number of people already seated there. Alternatively, they may choose a new table with probability proportional to α.

Suppose that there are T occupied tables when the nth customer enters. The probability distribution for the nth customer's table, t_n, is the same as in the finite Chinese restaurant process (Equation 5.3). In contrast, the distribution for his dish, z_n, is slightly different because each table has a unique dish. Let K be the current number of unique dishes:

$$\mathbb{P}(z_n = k|\alpha, \mathbf{z}_{\tilde{n}}) = \begin{cases} \frac{\sum_{i < n} \mathbb{1}(z_i = k)}{n - 1 + \alpha}, & k \leq K \\ \frac{\alpha}{n - 1 + \alpha}, & k = K + 1 \end{cases}. \tag{5.8}$$

Again, $\mathbf{z}_{\tilde{n}}$ denotes the vector of dishes (class assignments) for all but the nth customer.

To finish specifying the mixture, each dish is associated with a class parameter drawn independently from the base measure H. As in the finite Chinese restaurant process, the class parameter for the nth observation can also be written as a weighted combination of the current empirical distribution and the prior distribution:

$$F(\theta_n) \propto (n-1)f_{\text{EDF}} + \alpha H, \tag{5.9}$$

where $f_{\text{EDF}} = \sum_{i<n} \delta_{\theta_{z_i}} / (n-1)$ denotes the empirical distribution of the first $n-1$ class memberships. Note that the only difference from the finite Chinese restaurant is that Equation (5.9) uses the prior distribution H in place of the prior uniform probability $\frac{1}{K}$ of Equation (5.5).

The Chinese restaurant process illustrates how the population (restaurant) has an infinite number of classes (dishes), but only a finite number are represented (ordered by a customer). Note that each customer selects a random table with probabilities that depend on the precision α, but not the base measure H. Hence, the choice of α amounts to an implicit prior on the number of classes. Antoniak (1974) specifies this prior explicitly. Notably, the number of classes increases stochastically with both n and α. In the limit, as α approaches infinity, each customer chooses an unoccupied table. As a result, there is no clustering and each observation belongs to its own unique class. The distribution of $(\theta_{z_1}, \ldots, \theta_{z_n})$ approaches an i.i.d. sample from H. In the other extreme, as α approaches zero, each customer chooses the first table, resulting in a single class. In effect, the population distribution is no longer a mixture distribution.

5.3.3   Comparison of Dirichlet Mixtures and Dirichlet Process Mixtures

Both the DM model and the DPM model assume that observed data are representatives of a finite number of latent classes. The chief difference being that the DM model places a bound on the number of classes while the DPM model does not. Ishwaran and Zarepour (2002) makes this relationship explicit: as the number of classes increases to infinity, the DM model converges in distribution to the DPM model. Because the number of classes is unbounded, there is always a positive probability that the next response represents a previously unobserved class. The DPM model is an example of a nonparametric Bayesian model, which allows model complexity to increase as more data are observed.

A comparison of Equations (5.8) and (5.4) reveals how the nonparametric DPM model differs from the bounded-complexity DM model. In both models, the distribution of an observation's class is simply the empirical distribution of the previous class memberships, plus additional α prior observations. However, the prior weight is distributed differently. In the DM model, the α prior observations are placed uniformly over the K classes. Once all K classes have been observed, there is no chance of observing a novel class. In the DPM model, the α prior observations are placed on the next unoccupied table, which will serve a new dish with probability 1 (if the base measure H is continuous.) Hence, there is always a non-zero probability that the next observation belongs to a previously unobserved class, though this probability decreases as the sample size increases. While the DPM model allows greater flexibility in clustering, both models yield the same distribution for the observations and class parameters when conditioned on the vector of class memberships.

DP mixtures, and other nonparametric Bayesian models, are one strategy for determining the appropriate model complexity given a set of data. The theory behind these mixtures states that there is the possibility that some classes have not been encountered yet. For prediction, as opposed to estimation, this flexibility may be especially attractive since the new observation may not fit well into any of the current classes.

Recall that the precision α amounts to a prior over the number of classes. The posterior distribution of the latent class memberships provides a way to learn about the complexity from the observations that does not require choosing a specific value. Furthermore, it is possible to expand the DPM model to include a hyperprior for α (Escobar and West, 1995).

5.4 Mixed Membership Models

In the mixture models of Sections 5.2 and 5.3, each individual is assumed to belong to one of K underlying classes in the population. In a mixed membership model, each individual may belong to multiple classes with varying degrees of membership. In other words, each observation is associated with a *mixture* of the K classes. For this reason, mixed membership models are also called *individual-level mixture models*. Mixed membership models have been used for survey analysis (Erosheva, 2003; Erosheva et al., 2007), language models (Blei et al., 2003; Erosheva et al., 2004), and analysis of social and protein networks (Airoldi et al., 2008). This section focuses on models where K is a finite constant, which bounds the number of classes that may be observed.

Consider a population with K classes. Let $H(\lambda)$ denote the prior over class parameters where λ is a hyperparameter. In the DM model, individual i has membership in a single class, denoted z_i. Alternatively, the ith individual's class may be represented as the K-dimensional vector $\pi_i = (\pi_{i1}, \ldots, \pi_{iK})$, where π_{ik} is 1 if $z_i = k$ and 0 otherwise. By contrast, a mixed membership model allows π_i to be any non-negative vector whose elements sum to 1. The range of π_i is called the $(K-1)$-simplex. Geometrically, the simplex is a hyper-tetrahedron in $\mathbb{R}^K$. A mixed membership model allows π_i to take any value in the simplex while the DM model constrains π_i to be one of the K vertices.

The Grade of Membership model (GoM) extends the Dirichlet mixture model to allow for mixed membership (Erosheva et al., 2007). Both models can be understood as mixtures of the K possible classes. The DM model has a single population-level mixture for all individuals. In the GoM model, the population-level mixture provides typical values for the class weights, but the actual weights vary between individuals. As with the DM model, the population-level mixture in the GoM model has a symmetric Dirichlet prior. This mixture serves as the expected value for the individual-level mixtures, which also have a symmetric Dirichlet distribution. Denote the Dirichlet precision at the population level by α_0 and the precision at the individual level by α. The GoM model can be expressed by the following hierarchy:

$$\pi_0 | \alpha_0 \sim \text{Dir}(\alpha_0/K).$$

$$\theta_k | \lambda \overset{\text{i.i.d.}}{\sim} H(\lambda) \qquad\qquad k = 1 \ldots K.$$

$$\pi_i | \alpha, \pi_0 \overset{\text{i.i.d.}}{\sim} \text{Dir}(\alpha \pi_{01}, \ldots, \alpha \pi_{0K}) \qquad\qquad i = 1 \ldots n.$$

$$x_i | \pi_i, \theta \sim \sum_{k=1}^{K} \pi_{ij} F(\cdot | \theta_k) \qquad\qquad i = 1 \ldots n.$$

This model is the same as the DM model, except for the individual-level mixture proportions. The Dirichlet mixture model constrains π_{ik} to zero for all but one class, so the distribution of x_i is a "mixture" of one randomly selected component. By contrast, in a mixed membership model, π_i can take any value in the $(K-1)$-simplex resulting in a true mixture of the K components.

Clearly, the GoM model generalizes the K-dimensional DM model by allowing more flexibility in individual-level mixtures. Conversely, the GoM model can also be described as a special case of a larger DM model. Suppose each individual is measured across J different items. (For simplicity of notation, assume J is constant for all individuals; removing this restriction is trivial.) Erosheva et al. (2007) provides a representation theorem to express a K-class GoM model as a constrained DM model with K^J classes. Therefore, this theorem will assist in building the GoM model into a nonparametric model in much the same way that Section 5.3 built the DM model into the nonparametric DPM model.

Erosheva et al. describe their representation theorem in the context of survey analysis. In this case, a sample of n people each respond to a series of J survey items and an observation is the collection of one person's responses to all of the items. Let $\pi_i = (\pi_{i1}, \ldots, \pi_{iK})$ be the membership vector and let $\mathbf{x}_i = (x_{i1}, \ldots, x_{iJ})$ be the response vector for the ith individual. (Henceforth, the ith observation shall be explicitly denoted as a vector because scalar observations do not naturally fit with the representation theorem.) According to the GoM model, the distribution of $\mathbf{x}_i$ is a mixture of the K class distributions with mixing proportions given by π_i. Alternatively, the GoM model can be interpreted as a Dirichlet mixture model in which individuals can move among the classes for each response. That is, individual i may belong to class z_{ij} in response to item j, but belong to a different class z_{ij^*} in response to item j^*. The probability that individual i behaves as class k for a particular item is π_{ik}. Note that this probability depends on the individual, but is constant across all items. Let z_{ij} denote the class membership for the ith individual in response to the jth item. The distribution of the ith individual's response is determined by $\mathbf{z}_i = (z_{i1}, \ldots, z_{iJ})$ and the class parameters (θ). Therefore, individual i may be considered a member of the latent class $\mathbf{z}_i$ with class parameter $(\theta_{z_{i1}}, \ldots, \theta_{z_{iJ}})$. Each of the J components takes on one of K possible classes, making the GoM model a constrained DM model with K^J possible classes. The constraints arise because π_i is constant across all items in the GoM model, whereas the DM model allows all probabilities to vary freely. Thus, the probability of class $\mathbf{z}_i$ is constant under permutation of its elements in the GoM model but not the DM model.

The representation theorem suggests augmenting the GoM model with the collection of latent individual-per-item class memberships:

Grade of Membership Model

$$\pi_0 | \alpha_0 \sim \text{Dir}(\alpha_0/K).$$

$$\theta_k | \lambda \overset{\text{i.i.d.}}{\sim} H(\lambda) \qquad\qquad k = 1 \ldots K.$$

$$\pi_i | \alpha, \pi_0 \overset{\text{i.i.d.}}{\sim} \text{Dir}(\alpha\pi_{01}, \ldots, \alpha\pi_{0K}) \qquad i = 1 \ldots n.$$

$$z_{ij} | \pi_i \sim \text{Mult}(\pi_i) \qquad\qquad i = 1 \ldots n, \; j = 1 \ldots J.$$

$$x_{ij} | z_{ij}, \theta \sim F(\cdot | \theta_{z_{ij}}) \qquad\qquad i = 1 \ldots n, \; j = 1 \ldots J.$$

As with the DM model, each measurement (x_{ij}) is generated by randomly choosing a latent class membership then sampling from the class-specific distribution. In both models, responses are assumed to be independent given the class memberships. The responses in the DM model are fully exchangeable because each one uses the same vector of class weights. The GoM model includes individual-level mixtures that allow for the individual's class weights to vary from the population average. Therefore, responses are exchangeable only if they belong to the same individual. Note that z_{ij} is a positive integer less than or equal to K. The next section builds this latent class representation into a nonparametric model by removing this bound on the value of z_{ij}.

5.5 The Hierarchical Dirichlet Process Mixture Model

Table 5.1 illustrates two analogies that may help elucidate the hierarchical Dirichlet process mixture model (HDPM). Comparing the columns reveals that the relationship between the HDPM model and the DPM model is similar to the relationship between the GoM model and the DM model. Recall that the GoM model introduces mixed memberships to the DM model by introducing priors for individual-level mixtures that allow them to vary from the overall population mixture. In the same way, the HDPM adds mixed memberships to the DPM model through individual-level priors. In both the GoM and HDPM models, the population-level mixture provides the expected value for the individual-level mixtures. Comparing the *rows* of Table 5.1 shows that the relationship between the HDPM and GoM models is similar to the relationship between the DPM and DM models. In both cases, the former model is a nonparametric version of the latter model that arises as a limiting distribution when the number of classes is unbounded. The HDPM and DPM models are specified mathematically by replacing the symmetric Dirichlet priors of the GoM and DM models with Dirichlet *process* priors.

	Number of Classes	
Exchangeability	Bounded	Unbounded
Full	DM	DPM
Partial	GoM	HDPM

TABLE 5.1
The relationship among the four main models of this chapter.

The hierarchical Dirichlet process mixture model (HDPM) incorporates a Dirichlet process for each individual, $P_i \sim \text{DP}(\alpha, P_0)$, where the base measure P_0 is itself drawn from a Dirichlet process, $P_0 \sim \text{DP}(\alpha_0, H)$. Thus, the model is parametrized by a top-level base measure, H, and two precision parameters, α_0 and α.

$$P_0|\alpha_0, H \sim \text{DP}(\alpha_0, H).$$
$$P_i|\alpha, P_0 \overset{\text{i.i.d.}}{\sim} \text{DP}(\alpha, P_0) \qquad i = 1 \ldots n.$$
$$\theta_{ij}^*|P_i \sim P_i \qquad i = 1 \ldots n, \ j = 1 \ldots J.$$
$$x_{ij}|\theta_{ij}^* \sim F(\cdot|\theta_{ij}^*) \qquad i = 1 \ldots n, \ j = 1 \ldots J.$$

Note that $E[P_i] = P_0$. Thus the population-level mixture, P_0, provides the expected value for the individual-level mixtures and the precision α influences how closely the P_is fall to this mean.

A stick-breaking representation of the HDPM model allows it to be expressed as a latent class model. Since P_0 has a Dirichlet process prior, it can be written as a random stick-breaking measure $P_0 = \sum_{k=1}^{\infty} \pi_{0k}\delta_{\theta_{0k}}$, where $\pi_0 \sim \text{SBP}(\alpha_0)$ and θ_0 is an infinite i.i.d. sample with distribution H. Likewise, each individual mixture P_i can be expressed as $P_i = \sum_{k=1}^{\infty} \pi_{ik}^*\delta_{\theta_{ik}}$, where $\pi_i^* \sim \text{SBP}(\alpha)$ and θ_i is an infinite i.i.d. sample from P_0. Because $\theta_{ik} \sim P_0$, it follows that each $\theta_{ik} \in \theta_0$. Therefore, $P_i = \sum_{k=1}^{\infty} \pi_{ik}\delta_{\theta_{0k}}$, where $\pi_{ik} = \sum_{j=1}^{\infty} \pi_{ij}^*\mathbf{1}(\theta_{ij} = \theta_{0k})$. P_0 specifies the set of possible class parameters and the expected class proportions; P_i allows individual variability in class proportions. Since the class parameters are the same for all individuals, the notation θ_{0k} may be replaced by the simpler θ_k. While π_0 may be generated using the same stick-breaking procedure used in Section 5.3, the individual-level π_is require a different procedure given by Teh et al. (2006). Given π_0 and θ, let $\phi_{ik} \sim \text{Beta}\left(\alpha\pi_{0k}, \alpha\left(1 - \sum_{j=1}^{k} \pi_{0j}\right)\right)$. Define $\pi_{i1} = \phi_{i1}$, and set $\pi_{ik} = \phi_{ik}\prod_{j=1}^{k-1}(1 - \phi_{0j})$ for $k > 1$. The random measure $P_i = \sum_{k=1}^{\infty} \pi_{ik}\delta_{\theta_k}$ has distribution $\text{DP}(\alpha, P_0)$. Denote the conditional distribution of $\pi_i|\pi_0$ by $\text{SBP}_2(\alpha, \pi_0)$. The latent class representation of the HDPM model is as follows:

Hierarchical DP Mixture Model

$$\pi_0|\alpha_0, H \sim \text{SBP}(\alpha_0).$$
$$\theta_k|\lambda \overset{\text{i.i.d.}}{\sim} H(\lambda) \qquad k = 1 \ldots K.$$
$$\pi_i|\alpha, \pi_0 \overset{\text{i.i.d.}}{\sim} \text{SBP}_2(\alpha, \pi_0) \qquad i = 1 \ldots n.$$
$$z_{ij}|\pi_i \sim \text{Mult}(\pi_i) \qquad i = 1 \ldots n, \ j = 1 \ldots J.$$

$$x_{ij}|z_{ij}, \theta \sim F(\cdot|\theta_{z_{ij}}) \qquad\qquad i = 1 \ldots n, \ j = 1 \ldots J.$$

The three lowest levels in the HDPM model (pertaining to π_i, z_{ij}, and x_{ij}) represent the item-level distribution. Individual i, in response to item j, chooses a class according to its unique mixture: $z_{ij} \sim \text{Mult}(\pi_i)$. Its response is then given according to the distribution for that class: $x_{ij} \sim F(\cdot|\theta_{z_{ij}})$. This behavior is the same as the Grade of Membership model. Each individual is associated with a unique mixture over a common set of classes. The individual's mixture defines the probability that individual i behaves as class k in response to item j. Since this probability does not depend on j, two responses are exchangeable if and only if they come from the same individual. Unlike the GoM model, the HDPM model does not bound the number of classes a priori.

5.5.1 Chinese Restaurant Franchise

Teh et al. (2006) uses the analogy of a Chinese restaurant *franchise* to describe the clustering behavior of the hierarchical Dirichlet process. As each individual has a unique class mixture, each is represented by a distinct restaurant. Each of the customers represents one of the individual's features. For example, in the context of survey analysis, a customer is the individual's response to one of the survey items. The restaurants share a common menu with an infinite number of dishes to represent the various classes in the population. Each restaurant operates as an independent Chinese restaurant process with respect to seating arrangement, but dishes for each table are chosen by a different method which depends on the entire collection of restaurants.

Let x_{ij} denote the jth customer at the ith restaurant. When the customer enters, he chooses a previous table based on how many people are sitting there, or else starts a new table. The distribution for the table choice is the same as the Chinese restaurant process. It is given by Equation (5.3), taking t_n to denote the new customer's table and $t_1, \ldots, t_{n-1}$ to denote the previous customers' tables, where the numbering is confined to tables at restaurant i.

If the customer sits at a new table, he must select a dish. As with table choice, the customer will choose a previously selected dish with a probability that depends on how popular it is. Specifically, the probability is proportional to the number of other tables currently serving the dish across the entire franchise. Alternatively, with probability proportional to α_0, the customer will choose a new dish.

Suppose there are T tables currently occupied in the entire franchise when a customer decides to sit at a new table, becoming the first person at table $T + 1$. Denote the dish served at table t by d_t and let K denote the current count of unique dishes. If a new table is started, the distribution for the next dish, d_{T+1} is

$$\mathbb{P}(d_{T+1} = k|d_1, \ldots, d_T) = \begin{cases} \frac{\sum_{t=1}^{T} \mathbf{1}(d_t=k)}{T+\alpha_0}, & k \leq K \\ \frac{\alpha_0}{T+\alpha_0}, & k = K + 1 \end{cases}. \qquad (5.10)$$

Note that the customer has three choices: he may join an already occupied table (i.e., choose locally from the dishes already being served at restaurant i); start a new table and choose a previous dish (i.e., choose a dish from the global menu); or start a new table and select a new dish. Let z_{iJ} denote the dish chosen by the Jth customer at restaurant i. The Chinese restaurant franchise shows that the distribution of z_{iJ} is comprised of three components:

$$\mathbb{P}(z_{iJ} = k|\alpha_0, \alpha, \mathbf{z}_{\widetilde{iJ}}) = \begin{cases} 5\frac{\sum_{j=1}^{J-1} \mathbf{1}(z_{ij}=k)}{J-1+\alpha} + \frac{\alpha}{J-1+\alpha}\frac{\sum_{t=1}^{T} \mathbf{1}(d_t=k)}{T+\alpha_0}, & k \leq K \\ \frac{\alpha}{J-1+\alpha} \cdot \frac{\alpha_0}{T+\alpha_0}, & k = K + 1 \end{cases}, \qquad (5.11)$$

where T is the current number of occupied tables, K is the current number of distinct dishes across the entire franchise, and $\mathbf{z}_{\widetilde{iJ}}$ is the set of all dish assignments except for z_{iJ}. Note that the weight for a dish $k \leq K$ is the sum of the number of *customers at restaurant* i eating that dish plus α

times the number of *tables in the franchise* serving that dish. In other words, dishes are chosen according to popularity, but popularity within restaurant i is weighted more heavily. In practical terms, measurements from the same individual share information more strongly than measurements from multiple individuals. The precision α specifies the relative importance of the local-level and global-level mixtures.

To finish specifying the HDPM model, each dish is associated with a class parameter drawn independently from the base measure H. The distribution of $\theta_{z_{iJ}}$ is a three-part mixture. Let $F_{\text{Ind}} = \sum_{j<J} \delta_{\theta_{z_{iJ}}}/(J-1)$ be the empirical distribution of class parameters based on the customers at restaurant i. Let $F_{\text{Pop}} = \sum_t \delta_{\theta_{d_t}}/T$ be the empirical distribution based on the proportion of tables serving each dish across the entire restaurant. The distribution of $\theta_{z_{iJ}}$ is

$$F\left(\theta_{z_{iJ}} | \alpha, \alpha_0, \theta_{z_{\widetilde{iJ}}}\right) \propto (J-1)F_{Ind} + \alpha T F_{Pop} + \alpha\alpha_0 H, \tag{5.12}$$

where $\theta_{z_{\widetilde{iJ}}}$ denotes the class parameters for all customers except for the Jth customer of the ith restaurant. As with the other three mixture models, DM, DPM, and GoM, the responses are assumed to be independent given the class memberships. Hence, Equations (5.11) and (5.12) can be applied to any customer by treating him as the last one.

Equation (5.12) illustrates the role of the two precision parameters. Larger values of α place more emphasis on the population-level mixture, so that individual responses will tend to be closer to the overall mean. Meanwhile, larger values of α_0 place more emphasis on the base measure H. This specifies the prior distribution of $\theta_{z_{iJ}}$. After observing the response x_{iJ}, the class weights are updated by the likelihood of observing x_{iJ} within each class. Unfortunately, this model requires fairly complicated bookkeeping as to track the number of customers at each table and the number of tables serving each menu item. Blunsom et al. (2009) proposes a strategy to reduce this overhead. Inference is discussed more fully in Section 5.7.

5.5.2 Comparison of GoM and HDPM models

The relationship between the HDPM and GoM models is very similar to the relationship between the DP and DPM models described in Section 5.3.3. Both the GoM and HDPM models assume that observed data are representatives of a finite number of classes. Whereas the DM and DPM models assume that the observations are fully exchangeable, the GoM and HDPM models are mixed membership models that treat some observations as more similar than others. For example, in survey analysis, two responses from one individual are assumed to be more alike than responses from two different individuals. In text analysis, the topics within one document are assumed to be more similar than topics contained in different documents.

The HDPM model is similar to the GoM model as both combine individual-level and population-level mixtures. The chief difference between the GoM and HDPM models is that the GoM model bounds the number of classes a priori while the HDPM model does not. Teh et al. (2006) makes this relationship explicit. It shows that the HDPM model is the limiting distribution of GoM models when the number of classes approaches infinity. Because the number of classes is unbounded, there is always a positive probability that the next response represents a previously unobserved class.

The clustering behaviors of the HDPM and GoM models are very similar. In the GoM model, the individual-level mixtures (π_i) are shrunk toward the overall population-level mixture (π_0), which is itself shrunk toward the uniform probability vector, $\left(\frac{1}{K}, \ldots, \frac{1}{K}\right)$. The Dirichlet process priors in the HDPM model exhibit a similar property, where the individual-level P_is are shrunk toward the overall population-level mixture, P_0. Whereas the GoM model shrinks π_0 toward the uniform prior over the K classes, the HDPM model shrinks P_0 toward a prior base measure H. The result is that the HDPM model always maintains a strictly positive probability that a new observation is assigned to a new class. This is illustrated by the Chinese restaurant franchise, with the exact probability of a new class given by Equation (5.11). In other words, the DM and GoM models place a finite bound on the observed number of classes, but the DPM and HDPM models are nonparametric models that allow the number of classes to grow as new data are observed. While the HDPM allows greater flexibility in clustering than the GoM model, both models yield the same distribution for the observations and class parameters, when conditioned on the vector of class memberships.

Recall that the precision α in the DPM model amounts to a prior over the number of observed classes. In the HDPM model, this prior is specified by two precision parameters: α_0 at the population level and α at the individual level. The posterior distribution of the latent class memberships provides a way to learn about the complexity of the data without choosing a specific value. Teh et al. (2006) extends the HDPM model with hyperpriors for both α_0 and α to augment the model's ability to infer complexity from the data. Section 5.6 explores the role of α_0 and α intuitively with simulation studies for illustration.

5.6   Simulated HDPM Models

The Chinese restaurant process can be used to construct simple simulations of HDPM models. Such simulations can reveal how the the model is affected by changes in the tuning parameters or sample size. Specifically, the simulations in this section illustrate the behavior of mixtures at both the population level and individual level, as well as the similarity between different individuals.

5.6.1   Size of the Population-Level Mixture

Figure 5.1 shows how the number of population classes in the HDPM model is affected by sample size and the two precision parameters. The values result from simulation of a Chinese restaurant franchise in which each restaurant receives 16 customers (e.g., each individual responds to 16 survey items.) With α and n held fixed, the average size (number of components) of the population mixture increases as α_0 increases from 1 to 100. The average mixture also increases with α when n and α_0 are fixed, but the difference is not significant except when $\alpha_0 = 100$. Thus, both precisions affect the expected number of classes, but α_0 may limit or dampen the effect of α. Intuitively, a large value of α causes more customers to choose a new table, but a low value for α_0 means that they frequently choose a previously ordered dish. Hence, new classes will be encountered infrequently. Indeed, as α_0 approaches 0, the limit in the number of classes is 1, regardless of the value of α. Standard errors in mixture size were also estimated by repeating each simulation 100 times. The effect of α_0 and α on the standard error is similar to the effect on the mean mixture size (see Figure 5.2).

5.6.2   Size of Individual-Level Mixtures

Figures 5.3–5.4 show that the effect of the HDPM parameters on the size of the individual mixtures is similar to their effect on the population mixture size. As seen in Figure 5.3, the average number of classes in each individual mixture increases with both α and α_0. The first third of the chart comes from simulations in which each individual responds to 16 survey items. For the second third, there are 64 items per individual, and there are 100 items per individual in the last third. There is a clear interaction between the precision parameters and the number of survey items. The size of individual mixtures increases with the number of responses per individual. In terms of the Chinese restaurant franchise, more tables are needed as more customers enter each restaurant. Interestingly, this effect is influenced by the two precision parameters. The mixture size increases more dramatically for large values of α and α_0. Figure 5.4 shows how the standard deviation among the size of the individual mixtures is affected by the number of survey items and the precision parameters. The effect on variability is similar to, but weaker than, the effect on average mixture size. In most cases, the variability in the size of individual mixtures is quite small compared to the average. Note that this analysis compares only the *number* of classes represented by the individuals. This does not take into account the number of shared dishes between restaurants nor the variability in their class proportions (the π_is).

5.6.3 Similarity among Individual-Level Mixtures

In the HDPM model, each individual is associated with a unique mixture of the population classes. The similarity among the individuals is influenced by both α_0 and α, though in opposite manners. Large values of α lead to individual mixtures being closer to the population, and hence, each other. Therefore, similarity among the individuals tends to increase with α. On the other hand, similarity tends to decrease as α_0 increases. Intuitively, the number of classes in the entire sample tends to be small when α_0 is small. Hence, individual mixtures select from a small pool of potential classes. This leads to high similarity between individuals. When α is large, individual mixtures select from a large pool of potential classes and tend to be less similar. Indeed, as α_0 approaches infinity, the class parameters tend to behave as i.i.d. draws from the base measure H. If H is non-atomic, then the individual mixtures will not have any components in common. In the other extreme, as α_0 approaches zero, the number of classes in the sample tends to 1. This results in every individual "mixture" being exactly the same, with 100% of the weight on the sole class.

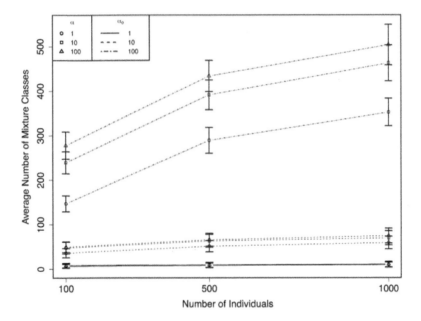

FIGURE 5.1

The effect of prior precisions and the number of individuals on the expected population mixture size in HDPM models. The population mixture size is the number of classes represented across all individuals in response to $J = 16$ measurements. Error bars represent two standard errors. Estimates are based on 100 simulations of each model.

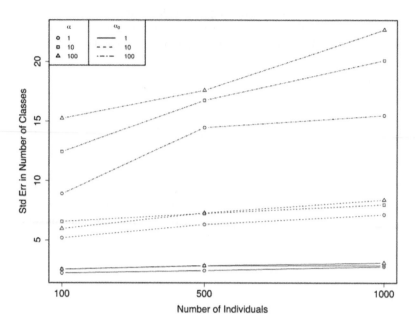

FIGURE 5.2
Standard errors for the population mixture size for various sample sizes and precisions in the HDPM model, based on 100 simulations per model.

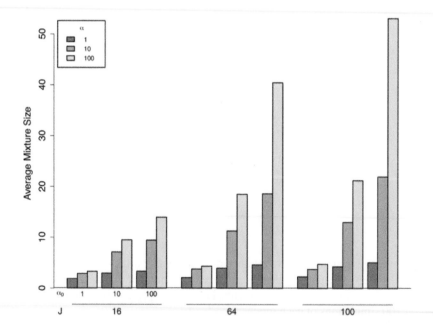

FIGURE 5.3
The effect of prior precisions and number of measurements per individual on the average individual-level mixture size. The mixture size represents the number of classes an individual represented during J measurements. Based on 100 simulations per model.

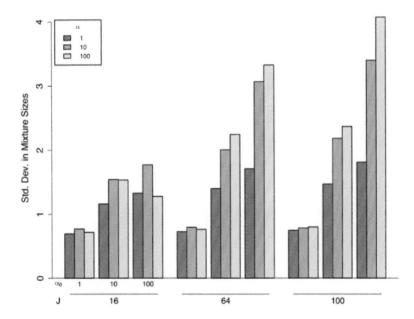

FIGURE 5.4
Standard deviation in individual-level mixture size for various HDPM models. Mixture size measures the number of classes an individual represented during J measurements. Based on 100 simulations per model.

The effect of the two precision parameters on similarity is shown in Figure 5.5 based on 100 simulations of $n = 50$ individuals responding to $J = 16$ items. There are several reasonable choices for measuring similarity. Here, the similarity between two individuals is defined by

$$\text{Sim}(i, i') = \frac{\sum_{k=1}^{K} \min(n_{ik}, n_{i'k})}{J}, \tag{5.13}$$

where J is the number of items per individual (16 in this case), K is the total number of classes in the sample, and n_{ik} is the number of times individual i responded to a survey item as a member of class k. In effect, $\text{Sim}(i, i')$ counts the number of times both individuals represented the same class, after arranging the second individual's responses to maximize the overlap with the first individual. Rearrangement is valid because responses from each individual are exchangeable.

Note that none of the heatmaps in Figure 5.5 exhibit any strong structure. This is expected under the HDPM model since the individuals are conditionally independent given the population-level mixture. On the other hand, one can easily see that the similarity among individuals in a particular model increases as α increases and as α_0 decreases. This exactly matches the intuition explained above.

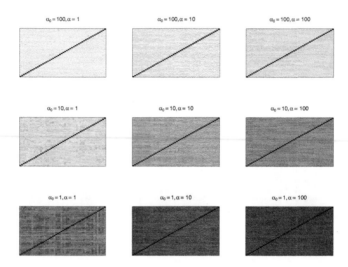

FIGURE 5.5
The effect of HDPM precision parameters on the similarity of individual-level mixtures. Darker areas correspond to higher similarity. Similarity is averaged across 100 simulations.

5.7   Inference Strategies for HDP Mixtures

Two broad categories of inference strategies for nonparametric mixture models are Markov chain Monte Carlo (MCMC) sampling and variational inference. Sampling techniques have the advantage of converging to the correct answer, at least under certain circumstances. Unfortunately, these techniques often require a great deal of computation time and it can be very difficult to assess convergence. Convergence for variational inference can be achieved quickly and assessed easily, but at the cost of some bias. Some simulation experiments have shown that the bias is not too drastic, at least when H is in the exponential family (Blei and Jordan, 2006).

5.7.1 Markov Chain Monte Carlo Techniques

Escobar and West (1995) demonstrate a Gibbs sampling scheme to estimate the posterior distribution for the DPM model, including inference for the precision term α. They directly sample from $f(\theta_{z_n}|\theta_{z_{\tilde{n}}}, \alpha)$, where $\theta_{z_{\tilde{n}}}$ denotes the class parameters for all observations except the nth one. Unfortunately, Markov chains built on this representation are slow to converge. In order for a class parameter to change, each member of that class must move to a new or different class one at a time. Thus, in order to remove a class or create a new one, there are low-probability intermediate states in which observations are in their own class. A more efficient strategy is to represent θ_{z_n} as the class parameters (θ_k) and class memberships (z_i) (MacEachern, 1994). This strategy is sometimes called the "collapsed" Gibbs sampler. Class assignments can be updated by combining the prior probabilities from the Chinese restaurant process with the likelihood of x_i given the class parameters. Let K denote the current number of classes in the mixture. For $k \leq K + 1$, let $f_{CRP}(k)$ be the probability that $z_n = k$ conditioned on the rest of the class memberships under the Chinese restaurant process (Equation 5.8). The probability that observation n should be assigned to class $k \leq K$ is

$$\mathbb{P}(z_n = k|\alpha, \mathbf{z}_{\tilde{n}}) \propto f_{CRP}(k)f(x_n|\theta_k), \tag{5.14}$$

where $\mathbf{z}_{\tilde{n}}$ denotes all class assignments except for z_n. The probability that x_n should be assigned to a new class is

$$\mathbb{P}(z_n = K + 1|\alpha, \mathbf{z}_{\tilde{n}}) \propto f_{CRP}(K + 1) \int_{\Theta} f(x_i|\theta)dH(\theta). \tag{5.15}$$

Since the observations are exchangeable, these equations can be used for any z_i by treating x_i as the last observation. Once the class membership vectors are updated, the class parameter θ_k can be updated from the posterior distribution given the prior H and the set of observations currently assigned to class k, denoted by $A_k = \{i : z_i = k\}$:

$$f(\theta_k) \propto \prod_{x_i \in A_k} f(x_i|\theta_k)dH(\theta_k). \tag{5.16}$$

In cases where two or more classes share similar structure, Jain and Neal (2004) proposes a "split-merge algorithm" that allows larger jumps in MCMC updates. This algorithm uses a Metropolis-Hastings step to potentially split one class into two or merge two classes into one. For the DPM model, MCMC sampling is fairly straightforward if the base measure $H(\theta)$ is conjugate to $F(\cdot|\theta)$. This conjugacy is important for two reasons. First, the probability of moving x_i to a new class depends on the integral $\int_{\Theta} f(x_i|\theta)dH(\theta)$. Second, in the collapsed Gibbs sampler, conjugacy leads to simple updates of θ_k given the observations in class k. Strategies for non-conjugate H include the "no gaps" algorithm, which augments the latent class representation with empty classes (MacEachern and Müller, 1998), and a split-merge algorithm for non-conjugate base measures (Jain and Neal, 2007).

For the HDPM model, Gibbs sampling is more complex due to the larger amount of bookkeeping required. In order to update z_{ij}, it is necessary to keep track of how many tables have dish k, how many customers are at each of those tables, and which restaurant the tables are in. This can lead to heavy memory requirements in large datasets. Blunsom et al. (2009) proposes a more efficient representation based on the idea of histograms. For each dish k and each positive integer m, they simply maintain a count of how many tables with m customers are serving dish k. This representation takes advantage of the exchangeability properties of the HDPM model. Due to the fact that responses are independent given the latent classes, it does not matter which table a customer actually sits at. When a customer joins a table, the appropriate bin count is decremented and the bin above is incremented. For example, if a customer is assigned to table 9, which has two previous customers, then there are now three customers at table 9. Thus, there is one fewer table with two customers and one more table with three customers. When a customer leaves a table, the opposite happens. The appropriate bin is decremented and the bin below is incremented.

Once the mechanism for implementing the Chinese restaurant franchise is decided, MCMC sampling can proceed as in the DPM model. That is, the latent class assignments (z_{ij}) and class parameters (θ_k) can be alternately updated. Let K be the current number of classes. For $k \leq K + 1$, let $f_{\text{CRF}}(k)$ be the probability that $z_n = k$ given the rest of the class assignments under the Chinese restaurant franchise (Equation 5.11). The probability that observation n should be assigned class $k \leq K$ is

$$\mathbb{P}(z_n = k | \alpha, \alpha_0, z_{\widetilde{ij}}) \propto f_{\text{CRF}}(k) f(x_n | \theta_k), \tag{5.17}$$

where $z_{\widetilde{ij}}$ denotes all class assignments except for z_{iJ}. The probability that x_n should be assigned to a new class is

$$\mathbb{P}(z_n = K + 1 | \alpha, \alpha_0, z_{\widetilde{ij}}) \propto f_{\text{CRF}}(k) \int_{\Theta} f(x_n | \theta) dH(\theta). \tag{5.18}$$

Note that the updates are the same as in the DPM model, except that $f_{\text{CRP}}(k)$ is replaced by $f_{\text{CRF}}(k)$.

Since the observations are independent given the latent class assignments, these equations can be used for any z_{ij} by treating x_{ij} as the last observation. Once the class parameters are updated, the class parameter θ_k can be updated in the same way as in the DPM model. Namely, the new value of θ_k is randomly generated from its posterior distribution given the prior H and the set of observations assigned to class k as in Equation (5.16).

5.7.2 Variational Inference

Variational inference can be viewed as an extension of the expectation maximization algorithm (EM) (Beal, 2003). Whereas EM uses an iterative approach to find a point estimate for some vector of unobserved variables (e.g., latent variables and parameters), variational inference attempts to approximate their entire posterior distribution.

Let θ and $\mathbf{z}$ be the sets of model parameters and latent variables. In DPM and HDPM models, direct calculation of $f(\theta, \mathbf{z} | \mathbf{x})$ is impractical due to the intractable calculation of the data marginal. The intractability arises from the complex interactions among parameters and latent variables. The variational approach is to constrain the posterior to some simpler family of *variational functions* that treat these values as independent. The posterior is approximated by finding the variational function closest to the true posterior (e.g., in KL divergence). Because the variational functions break the dependence between some variables, it is possible to minimize the divergence by iteratively optimizing one piece of the function at a time, given the rest of the function. For example, one may constrain $f(\theta, \mathbf{z} | \mathbf{x})$ to be of the form $q_\theta(\theta | \mathbf{x}) \cdot q_z(\mathbf{z} | \mathbf{x})$. This can be optimized using coordinate ascent by iteratively updating q_θ and q_z based on the value of the other function.

Blei and Jordan (2006) provides an explicit algorithm for DP mixtures when the base measure H is exponential family . Teh et al. (2008) describes a variational approach for hierarchical models that can be used for mixed membership models. The latent variables in the DP mixture are the class proportions (π_k), class parameters (θ_k), and class assignments (z_k). Rather than work with the class proportions, Blei and Jordan work directly with (ϕ_k), the beta random variables from the stick-breaking process. In order to update the variational functions, they also limit the number of components in the variational function to a finite number, say T. However, they optimize the KL-divergence between this truncated stick-breaking measure and the full DP posterior with infinite components. This yields a set of variational functions parametrized by:

$$q(\phi, \theta, \mathbf{z}) = \prod_{t=1}^{T-1} q_{\gamma_t}(\phi_t) \prod_{t=1}^{T} q_{\tau_t}(\theta_t) \prod_{i=1}^{n} q_{\rho_i}(z_i), \tag{5.19}$$

where each q_{γ_t} is a beta distribution, each q_{τ_t} is in the same family as the prior H, and each $q_{\rho_i}(z_i)$ is multinomial. Notice that the variational function for each variable is the same family as its marginal under the true posterior, however the variational function treats all variables as independent.

The variational function updates proceed like posterior updates given the data and the current value of the other functions. Blei and Jordan (2006) provide explicit updates for each function in the case where H is exponential family. They compared this variational algorithm to the collapsed Gibbs sampler and found that the log-probability of held-out data was similar, but that the variational approach required less computation time. Furthermore, the computation time for variational inference did not increase dramatically in the range of 5- to 40-dimensional observations.

Variational inference is even more efficient if some dimensions of the parameter space can be integrated out. For example, if inferential goals do not include recovering the full mixture posterior, it is possible to integrate out the mixing proportions (Kurihara et al., 2007). This still allows posterior analysis of class membership and parameters as well as calculation of a lower bound for the data marginal. Teh et al. (2008) extends this *collapsed* algorithm to hierarchical Dirichlet process mixtures.

One of the advantages of nonparametric models is that they allow the complexity of the model to grow as new data are observed. This property may be especially advantageous for *streaming* applications, for which new data continually arrive. Online variational inference algorithms have been developed for mixed membership models (Canini et al., 2009; Hoffman et al., 2010; Rodriguez, 2011) including the HDPM model (Wang et al., 2011).

5.7.3 Hyperparameters

The parameters for the DPM and HDPM models include the precisions for Dirichlet processes at each level of mixing and possible hyperparameters for the base distribution H. For example, if H is a normal distribution, a hyperprior may be used to learn about its mean and variance. Typically, hyperpriors are at least used for the precision parameters α_0 and α, since inference can be sensitive to these choices. For example, in one of the first practical applications of the DPM model, Escobar and West (1995) show that the posterior distribution over K is quite sensitive although the predictive distribution is robust. To decrease sensitivity, they recommend using diffuse gamma hyperpriors for precision parameters. Gamma hyperpriors are convenient because the induced posterior for α given the data and latent variables depends only on the number of classes. Thus, the value of α can be updated efficiently based on the current value of the other latent and observed variables.

5.8 Example Applications of Hierarchical DPs

Erosheva et al. (2007) applies the GoM model to data from the National Long Term Care Survey. Alternatively, the HDPM model provides a nonparametric approach to the same data. For each individual, the survey contains binary outcomes on 6 "Activities of Daily Living" (ADL) and 10 "Instrumental Activities of Daily Living" (IADL). ADL items include basic activities required for personal care, such as eating, dressing, and bathing. IADL items include basic activities necessary to reside in the community such as doing laundry, cooking, and managing money. Positive responses (disabled) to each item signify that during the past week the activity was not completed or not expected to be completed without the assistance of another person or equipment. Each survey response is regarded as an independent Bernoulli random variable: $X_{ij} \sim \text{Bern}(\theta_{ij})$, where X_{ij} is the response of the ith individual on the jth item and θ_{ij} is the probability of a positive response. In this context, a mixture model asserts that the population consists of various sub-groups with varying probabilities of a positive (disabled) response. For example, the population may contain healthy, mildly disabled, and disabled cohorts with increasing probabilities of positive responses.

The GoM model combines mixture models for both the population and individual level. The individual-level mixture asserts, for example, that an individual may behave as a member of the healthy cohort in response to item 1, but behave as a member of the disabled cohort in response to item 2. Each individual is associated with unique mixture probabilities. The population mixture defines the overall proportions of the various cohorts across all individuals and items.

Replacing the GoM model with the HDPM model yields a similar structure, except that the number of classes does not need to be specified a priori.

Blei et al. (2003) presents a mixed membership model for modeling documents called latent Dirichlet allocation (LDA). The classes are various topics (e.g., computer science, operating systems, and machine learning). Each topic is considered a multinomial distribution over some finite vocabulary. The class parameters are the multinomial proportions, which are smoothed using a Dirichlet prior. Each document in the sample is associated with a unique mixture of topic proportions. A word is generated by selecting a topic from the document-level mixture, then choosing a word from the topic-specific multinomial. As with the Grade of Membership model, the number of classes (topics) must be specified a priori. Alternatively, one can use a hierarchical Dirichlet process mixture, in which the number of potential topics is countably infinite (Teh et al., 2006). Under this nonparametric mixed membership model, each new word has a positive probability of belonging to a new topic. Hoffman et al. (2008) uses a similar model to measure musical similarity, where the documents are musical pieces and the "topics" are features.

5.8.1 The Infinite Hidden Markov Model

In the hidden Markov model, a sequence of observations $(x_1, x_2, \ldots, x_n)$ are explained by a second sequence of latent variables $(y_1, y_2, \ldots, y_n)$. The latent sequence is modeled by a Markov chain and the observation (or *emission*) at time t is assumed to depend only on the state of the chain at time t. Hidden Markov models assume fixed finite numbers for both the number of latent states and the number of possible emissions. Each state s is associated with a vector of transition probabilities, $\pi_s^T = (\pi_{s1}^T, \ldots, \pi_{sK}^T)$, where $\pi_{sk}^E = \mathbb{P}(y_{t+1} = k | y_t = s)$; and a vector of emission probabilities, $\pi_s^E = (\pi_{s1}^T, \ldots, \pi_{sV}^E)$, where $\pi_{sv}^E = \mathbb{P}(x_t = v | y_t = s)$.

A hidden Markov model can be specified as a mixed membership model by taking the latent states as the possible classes. The vectors π_s^T and π_s^E define mixtures over the state-space and emission space; since y_{t+1} and x_t are conditionally independent given y_t, one may consider each mixture separately. Denoting the number of possible states by K, a mixed membership model for the transitions can be defined by Dirichlet priors:

$$\pi_0^T \sim \text{Dir}(\alpha_0^T / K).$$

$$\pi_s^T \overset{\text{i.i.d.}}{\sim} \text{Dir}(\alpha^T \cdot \pi_0^T) \qquad\qquad s = 1 \ldots K.$$

$$y_{t+1} | y_t \sim \text{Mult}(\pi_{y_t}^T) \qquad\qquad t = 1 \ldots n.$$

The state-dependent vectors, π_s^T, allow each state to have unique transition probabilities, which are shrunk toward the population-level weights, π_0^T. Separate Dirichlet priors can be used to define a mixed membership model for emissions, with V denoting the number of possible values:

$$\pi_0^E \sim \text{Dir}(\alpha_0^E/V).$$

$$\pi_s^E \overset{\text{i.i.d.}}{\sim} \text{Dir}(\alpha^E \cdot \pi_0^E) \qquad\qquad s = 1 \ldots K.$$

$$x_t | y_t \sim \text{Mult}(\pi_{y_t}^E) \qquad\qquad t = 1 \ldots n.$$

As with the transition vectors, each state has unique emission probabilities, which are shrunk toward the population averages. Beal et al. (2001) developed a nonparametric version of HMMs by replacing the Dirichlet priors with Dirichlet process priors. As there are a countable infinite number of potential states and emissions, they call this model the *infinite* hidden Markov model (iHMM). The authors apply this model to a language processing problem. The latent state y_t denotes a topic which specifies a multinomial distribution for the tth word. Because both the transition and emission models are nonparametric, there is a non-zero probability that the Markov chain transitions to a new topic, or that a topic produces a previously unobserved word.

5.9 Other Nonparametric Mixed Membership Models

5.9.1 Multiple-Level Hierarchies

The four main models in this chapter: DM in Section 5.2, DPM in Section 5.3, GoM in Section 5.4, and HDPM in Section 5.5 all produce exchangeability within any given mixture. The individuals ($\mathbf{x}_i$) are exchangeable in all models and the per-item responses (x_{ij}) are also exchangeable in the GoM and HDPM models. If this exchangeability structure is unrealistic or undesired, one way to introduce dependence is to include multiple levels of hierarchy. For example, in the National Long Term Care Survey, some responses concern "Activities of Daily Living" and others concern "Instrumental Activities of Daily Living." In theory, an individual's class membership probabilities could vary depending on the sub-category. This can be modeled by including an extra layer of Dirichlet process mixing with S denoting the number of sub-categories:

$$P_0 \sim \text{DP}(\alpha_0, H).$$

$$P_i | P_0 \overset{\text{i.i.d.}}{\sim} \text{DP}(\alpha_1, P_0) \qquad\qquad i = 1 \ldots n.$$

$$P_{is} | P_i \sim \text{DP}(\alpha_2, P_{ij}) \qquad\qquad s = 1 \ldots S.$$

$$\theta_{isj} | P_{is} \sim P_{is} \qquad\qquad i = 1 \ldots n, \; s = 1 \ldots S, \; j = 1 \ldots J.$$

$$X_{isj} | \theta_{isj} \sim F(X | \theta_{isj}) \qquad\qquad i = 1 \ldots n, \; s = 1 \ldots S, \; j = 1 \ldots J.$$

This model includes mixtures at the population level (P_0), at the individual level (P_i), and at the sub-category level for each individual (P_{is}). The degree to which any two responses share information is determined by how many hierarchy levels separate them (as well as the relevant precision parameters). As before, responses are fully exchangeable within each mixture.

A double hierarchy may also be appropriate if individuals come from multiple sub-populations. For example, one could divide individuals based on type of residence: apartment, house, or nursing home. In this case, the HDPM model would include mixtures at the population level, sub-population level (type of residence), and individual level. Furthermore, if items are divided into various categories, then it is possible to include a fourth level of the hierarchy to account for this. In theory, any number of levels is possible, although the number of latent variables needed to represent a mixture model grows with each new level. Teh et al. (2006) provides an example application of a three-level HDPM model that they use to analyze articles from the proceedings of the *Neural Information Processing Systems (NIPS)* conference from the years 1988 to 1999. The articles are divided into nine sections, such as "algorithms and architectures" and "applications." Documents from the same section are expected to have a similar distribution of topics. Their model incorporates topic mixtures at the document level, section level, and population level. Here, the population is the entire collection of documents. The section level mixture allows a document to share information about topics more closely with documents within the same section than with documents in other sections.

5.9.2   Dependent Dirichlet Processes

In some cases, the problem is not to create a desired exchangeability structure but to induce correlation between different mixture components. This may be the case when covariates are measured for each observation. Exchangeability implies that there is no a priori difference among the possible covariate values. This may be appropriate for nominal variables such as gender or ethnicity. In this case, the effects of the covariate can be accounted for using additional hierarchy levels as described above. On the other hand, exchangeability may not be appropriate for ordinal or continuous variables, such as years of experience or age. Dependent Dirichlet processes have been developed for these types of covariates. For example, spatial Dirichlet processes have been used when the "individuals" are points in space (Gelfand et al., 2005; Duan et al., 2007). Such models produce Dirichlet process mixtures at each point, such that the mixtures are more similar when points are closer together. Temporal versions of dependent Dirichlet processes have also been developed which allow a nonparametric mixture to evolve over time (Xu et al., 2008; Ahmed and Xing, 2008). Although applications of dependent Dirichlet processes have focused on extensions to the DPM model, they provide potential sources for new nonparametric mixture models when hierarchical versions are developed.

Exchangeability may also be undesirable if one believes that certain classes tend to co-occur more often than other classes. Teh et al. (2006) uses HDPM models to describe documents (the observations) as a mixture of various topics (the classes). In sufficiently broad collections of documents, one may find that certain topics often appear together. For example, a document that focuses on the topic "politics" may be more likely to include the topic "economics" and less likely to include the topic "baseball." In other words, the occurrence of politics and economics may be positively correlated whereas the occurrence of politics and baseball may be negatively correlated. Unfortunately, the exchangeability property of the HDPM model prevents it from explicitly describing this correlation. Paisley et al. (2012) replaces the hierarchical Dirichlet process with a prior that they call the discrete infinite logistic normal distribution. This prior produces a mixed membership model that is able to explicitly describe correlated topics. Paisley et al. uses this prior to model a collection of 10,000 documents from Wikipedia.

5.9.3   Pitman-Yor Processes

The Pitman-Yor process (Pitman and Yor, 1997), or two-parameter Poisson-Dirichlet process, provides more flexibility in the clustering behavior of Dirichlet process mixture models. In addition to the base measure (H) and precision (α), there is a discount parameter, $0 \leq d \leq 1$. The Pitman-Yor process allows negative values for α provided that $\alpha > -d$.

The Pitman-Yor process can be illustrated using a more general version of the Chinese restaurant process. Consider a hierarchical model with $\theta_1, \theta_2, \ldots$ being a sequence of i.i.d. random variables with random distribution P, where P has a Pitman-Yor process prior. Similar to the Chinese restaurant process, when a customer arrives he either joins an existing table or begins a new table. Let K be the current number of occupied tables, z_i the dish for the ith customer, and $\mathbf{z}_{\tilde{n}}$ the vector of dishes except for z_n:

$$\mathbb{P}(z_n = k | \alpha, d, \mathbf{z}_{\widetilde{n}}) = \left\{ \begin{array}{ll} \frac{\sum_{i<n} \mathbf{1}(z_i=k)-d}{n-1+\alpha}, & k \leq K \\ \frac{\alpha-dK}{n-1+\alpha}, & k = K+1 \end{array} \right. \quad (5.20)$$

Notice that the discount parameter, d, reduces the clustering effect. The number of previous customers is reduced by d, and this weight is instead placed on the probability of a new table. For the limiting case with $d = 1$, $\theta_1, \ldots, \theta_n$ is an i.i.d. sample from H. On the other extreme, if $d = 0$, then the result is a Dirichlet process. As Teh (2006) shows, the number of unique values increases stochastically with both d and α. Recall that the stick-breaking process for the Dirichlet process constructs class proportions by setting $\pi_k = \phi_k \prod_{r=1}^{k-1}(1 - \phi_r)$, where $\phi_k \overset{\text{i.i.d.}}{\sim} \text{Beta}(1, \alpha)$. The Pitman-Yor process has a similar constructive definition using different beta marginals: $\phi_k \overset{\text{i.i.d.}}{\sim} \text{Beta}(1-d, \alpha+id)$ instead. It produces heavier tails than the Dirichlet process: π_k decreases stochastically with k for both processes, but this effect is more extreme with the Dirichlet process. Pitman-Yor processes have been used in applications such as natural language processing (Teh, 2006; Goldwater et al., 2006; Wallach et al., 2008) and image processing (Sudderth and Jordan, 2008).

Due to the stick-breaking construction, strategies for using Dirichlet processes can be adapted to Pitman-Yor processes. For example, it is straightforward to specify a hierarchical Pitman-Yor process by analogy to the hierarchical Dirichlet process. Teh (2006) constructs a MCMC sampling scheme for the hierarchical Pitman-Yor process, while Sudderth and Jordan (2008) develops a variational inference algorithm. The variational function updates for the Pitman-Yor process are similar to the Dirichlet process updates, since the stick-breaking proportions still have beta distributions. An open problem is to develop a more efficient collapsed strategy that integrates over the class proportions.

5.10 Conclusion

Nonparametric mixtures have been an active area of research since Sethuraman (1994) provided the seminal stick-breaking representation of the Dirichlet process. The Dirichlet process mixture model and its extensions have been used in many domains for modeling a population with an unbounded number of classes. The hierarchical Dirichlet process applies the same strategy for mixed membership models. Individual-level Dirichlet processes provide nonparametric mixtures for each individual, while a population-level Dirichlet process enables individuals to share statistical information. Such models have been used for survey analysis, document modeling, music models, and image analysis.

References

Ahmed, A. and Xing, E. P. (2008). Dynamic non-parametric mixture models and the recurrent Chinese restaurant process: With applications to evolutionary clustering. In Wang, W. (ed), *Proceedings of the 2008 SIAM Conference on Data Mining (SDM '08)*. Philadelphia, PA, USA: SIAM, 219–230.

Airoldi, E. M., Blei, D. M., Fienberg, S. E., and Xing, E. P. (2008). Mixed-memberhsip stochastic blockmodels. *Journal of Machine Learning Research* 9: 1823–1856.

Akaike, H. (1973). Information theory and an extension of the maximum likelihood principle. In

Petrov, B. N. and Csaki, F. (eds), *Proceedings of the 2ⁿᵈ International Symposium on Information Theory*. Budapest, Hungary: Akadèmiai Kiadò, 267–281.

Aldous, D. (1985). Exchangeability and related topics. In *École d'Été de Probabilités de Saint-Flour XIII*. Berlin: Springer, 1–198.

Antoniak, C. E. (1974). Mixtures of Dirichlet processes with applications to nonparametric problems. *Annals of Statistics* 2: 1152–1174.

Beal, M. J. (2003). Variational Algorithms for Approximate Bayesian Inference. Ph.D. thesis, Gatsby Computational Neuroscience Unit, University College London.

Beal, M. J., Ghahramani, Z., and Rasmussen, C. E. (2001). The infinite hidden Markov model. In Dietterich, T. G., Becker, S., and Ghahramani, Z. (eds), *Advances in Neural Information Processings Systems 14*. Cambridge, MA: The MIT Press, 577–584.

Blei, D. M. and Jordan, M. I. (2006). Variational inference for Dirichlet process mixtures. *Bayesian Analysis* 1: 121–144.

Blei, D. M., Ng, A. Y., and Jordan, M. I. (2003). Latent Dirichlet allocation. *Journal of Machine Learning Research* 3: 993–1022.

Blunsom, P., Cohn, T., Goldwater, S., and Johnson, M. (2009). A note on the implementation of hierarchical Dirichlet processes. In *Proceedings of the 47ᵗʰ Annual Meeting of the Association for Computational Linguistics and 4ᵗʰ International Joint Conference on Natural Language Processing of the Asian Federation of Natural Language Processing (ACL-IJCNLP 2009)*. Singapore: Association for Computational Linguistics, 337–340.

Canini, K. R., Shi, L., and Griffiths, T. L. (2009). Online inference of topics with latent Dirichlet allocation. In van Dyk, D. and Welling, M. (eds), *Proceedings of the 12ᵗʰ International Conference on Artificial Intelligence and Statistics (AISTATS 2009)*. *Journal of Machine Learning Research – Proceedings Track* 5 : 65–72.

Duan, J. A., Guindani, M., and Gelfand, A. E. (2007). Generalized spatial Dirichlet process models. *Biometrika* 94: 809–825.

Erosheva, E. A., Fienberg, S. E., and Lafferty, J. D. (2004). Mixed-membership models of scientific publications. *Proceedings of the National Academy of Sciences* 101: 5220–5227.

Erosheva, E. A. (2003). Partial membership models with application to disability survey data. In Bozdogan, H. (ed), *Statistical Data Mining and Knowledge Discovery*. Chapman & Hall/CRC, 117–134.

Erosheva, E. A., Fienberg, S. E., and Joutard, C. (2007). Describing disability through individual-level mixture models for multivariate binary data. *Annals of Applied Statistics* 1: 502–537.

Escobar, M. D. and West, M. (1995). Bayesian density estimation and inference using mixtures. *Journal of the American Statistical Association* 90: 577–588.

Ferguson, T. S. (1973). A Bayesian analysis of some nonparametric problems. *Annals of Statistics* 1: 209–230.

Gelfand, A. E., Kottas, A., and MacEachern, S. N. (2005). Bayesian nonparametric spatial modeling with Dirichlet processes mixing. *Journal of the American Statistical Association* 100: 1021–1035.

Giudici, P. and Green, P. J. (1999). Decomposable graphical Gaussian model determination. *Biometrika* 86: 785–801.

Goldwater, S., Griffiths, T. L., and Johnson, M. (2006). Interpolating between types and tokens by estimating power-law generators. In Weiss, Y., Schölkopf, B., and Platt, J. (eds), *Advances in Neural Information Processing Systems 18*. Cambridge, MA: The MIT Press, 459–466.

Green, P. J. (1995). Reversible jump Markov chain Monte Carlo computation and Bayesian model determination. *Biometrika* 82: 711–732.

Hastie, T., Tibshirani, R., and Friedman, J. (2009). *The Elements of Statistical Learning: Data Mining, Inference, and Prediction*. New York, NY: Springer, 2nd edition.

Hoffman, M. D., Blei, D. M., and Bach, F. (2010). Online learning for latent Dirichlet allocation. In Lafferty, J. D., Williams, C. K. I., Shawe-Taylor, J., Zemel, R., and Culotta, A. (eds), *Advances in Neural Information Processing Systems 23*. Red Hook, NY: Curran Associates, Inc., 856–864.

Hoffman, M. D., Blei, D. M., and Cook, P. (2008). Content-based musical similarity computation using the hierarchical Dirichlet process. In *Proceedings of the 9th International Conference on Music Information Retrieval (ISMIR 2008)*. ISMIR, 349–354.

Ishwaran, H. and Zarepour, M. (2002). Exact and approximate sum representations for the Dirichlet process. *The Canadian Journal of Statistics* 30: 269–283.

Jain, S. and Neal, R. M. (2004). A split-merge Markov chain Monte Carlo procedure for the Dirichlet process mixture model. *Journal of Computational and Graphical Statistics* 13: 158–182.

Jain, S. and Neal, R. M. (2007). Splitting and merging components of a nonconjugate Dirichlet process mixture model. *Bayesian Analysis* 2: 445–472.

Kim, Y. (2003). On the posterior consistency of mixtures of Dirichlet process priors with censored data. *Scandinavian Journal of Statistics* 30: 535–547.

Kurihara, K., Welling, M., and Teh, Y. W. (2007). Collapsed variational Dirichlet process mixture models. In *Proceedings of the 20th International Joint Conference on Artificial Intelligence (IJCAI '07)*. IJCAI, 2796–2801.

MacEachern, S. N. (1994). Estimating normal means with a conjugate style Dirichlet process prior. *Communications in Statistics - Simulation and Computation* 23: 727–741.

MacEachern, S. N. and Müller, P. (1998). Estimating mixture of Dirichlet process models. *Journal of Computational and Graphical Statistics* 7: 223–238.

McLachlan, G. J. and Peel, D. (2000). *Finite Mixture Models*. New York, NY: Wiley.

Paisley, J., Wang, C., and Blei, D. M. (2012). The discrete infinite logistic normal distribution. *Bayesian Analysis* 7 : 997–1034.

Pitman, J. and Yor, M. (1997). The two-parameter Poisson-Dirichlet distribution derived from a stable subordinator. *Annals of Statistics* 25: 855–900.

Rasmussen, C. E. (1999). The infinite Gaussian mixture model. In Solla, S. A., Leen, T. K., and Müller, K.-R. (eds), *Advances in Neural Information Processings Systems 12*. Cambridge, MA: The MIT Press, 554–560.

Rodriguez, A. (2011). On-line learning for the infinite hidden Markov model. *Communications in Statistics - Simulation and Computation* 40: 879–893.

Roeder, K. and Wasserman, L. (1997). Practical Bayesian density estimation using mixtures of normals. *Journal of the American Statistical Association* 92: 894–902.

Schwarz, G. E. (1978). Estimating the dimension of a model. *Annals of Statistics* 6: 461–464.

Sethuraman, J. (1994). A constructive definition of Dirichlet priors. *Statistica Sinica* 4: 639–650.

Sudderth, E. B. and Jordan, M. I. (2008). Shared segmentation of natural scenes using dependent Pitman-Yor processes. In Koller, D., Schuurmans, D., Bengio, Y., and Bottou, L. (eds), *Advances in Neural Information Processing Systems 21*. Red Hook, NY: Curran Associates, Inc., 1585–1592.

Teh, Y. W. (2006). A hierarchical Bayesian language model based on Pitman-Yor processes. In *ACL-44: Proceedings of the 21st International Conference on Computational Linguistics and the 44th Annual Meeting of the Association for Computational Linguistics*. Morristown, NJ, USA: Association for Computational Linguistics, 985–992.

Teh, Y. W., Jordan, M. I., Beal, M. J., and Blei, D. M. (2006). Hierarchical Dirichlet processes. *Journal of the American Statistical Association* 101: 1566–1581.

Teh, Y. W., Kurihara, K., and Welling, M. (2008). Collapsed variational inference for HDP. In Platt, J., Koller, D., Singer, Y., and Roweis, S. (eds), *Advances in Neural Information Processing Systems 20*. Red Hook, NY: Curran Associates, Inc., 1481–1488.

Wallach, H. M., Sutton, C., and McCallum, A. (2008). Bayesian modeling of dependency trees using hierarchical Pitman-Yor priors. In *Proceedings of the Workshop on Prior Knowledge for Text and Language (held in conjunction with ICML/UAI/COLT)*. Helsinki, Finland, 15–20.

Wang, C., Paisley, J., and Blei, D. M. (2011). Online variational inference for the hierarchical Dirichlet process. In *Proceedings of the 14th International Conference on Artificial Intelligence and Statistics (AISTATS 2011)*. Palo Alto, CA, USA: AAAI, 752–760.

Xu, T., Zhang, Z. M., Yu, P. S., and Long, B. (2008). Evolutionary clustering by hierarchical Dirichlet process with hidden Markov state. In *Proceedings of the 8th IEEE International Conference on Data Mining (ICDM '08)*. Los Alamitos, CA, USA: IEEE Computer Society, 658–667.

Part II

The Grade of Membership Model and Its Extensions

6

A Mixed Membership Approach to the Assessment of Political Ideology from Survey Responses

Justin H. Gross

Department of Political Science, University of North Carolina at Chapel Hill, Chapel Hill, NC 27599, USA

Daniel Manrique-Vallier

Department of Statistics, Indiana University, Bloomington, IN 47408, USA

CONTENTS

Political scientists have long observed that members of the public do not tend to exhibit highly constrained patterns of political beliefs and values in the way that partisan elites often do. In answering survey questions designed to measure latent ideology, they may act as if drawing responses randomly from different perspectives. In the American context, survey respondents frequently defy easy categorization as prototypical liberal or conservative, and yet their response patterns reflect structure that may be characterized by reference to such ideal types. We propose a mixed membership approach to survey-based measurement of ideology. Modeling survey respondents as partial members of a small number of ideological classes allows us to interpret the "mixed signals" they seem to send as a natural consequence of their competing inclinations. We illustrate our approach by reanalyzing data from a classic study of core beliefs and values (Feldman, 1988) and find that the most dramatic difference between prototypical members of the two main ideologies identified is not their vision of what society *should be* but rather their belief in what American society *actually is*.

6.1   Introduction

A rich and important tradition in political science involves the analysis of patterns of political ideology.[1] Initially, the identification and characterization of such patterns had been performed in a rather ad-hoc manner, albeit based on careful philosophical and qualitative considerations within a theoretical framework established by leading scholars on the subject. This approach reached its peak with what many consider its best and most influential example, *The American Voter*, by Campbell et al. (1960, see especially ch. 9, "Attitude Structure and the Problem of Ideology"). Beginning with Converse (1964), a wave of critical rethinking on the subject emerged. Along the way, various researchers have applied modern empirical tools of survey analysis and statistical inference to revisit previously held assumptions about the structure of American political attitudes and beliefs (e.g., Marcus et al. (1974); Achen (1975); Stimson (1975); Feldman (1988); Conover and Feldman (1984); Zaller (1992); Pew Research Center (2011); Ellis and Stimson (2012)). In essence, this constitutes a measurement problem: we treat data, e.g., responses to survey questions, as manifest indicators of respondents' latent dispositions on politics and policy. A number of different analytical tools have been utilized in approaching the problem, with factor analysis being the most frequently employed and item response models gaining in popularity. The basic goal of these endeavors is typically to understand the structure of—or constraints on—beliefs, values, and attitudes at two levels: the individual and the population at large. Thus, the objects of interest include configurations of views that may be expected to coexist within a single person and the relative frequency with which these particular sets of views are held (or called upon in responding to survey items).

A basic task in the study of ideology is the construction of typologies. Simple examples of typologies are the well-known distinctions between "left" and "right" or "liberal" and "conservative." Types such as "liberal" or "conservative" correspond to particular configurations of attitudes, values, and beliefs that hypothetical adherents to these ideological types are supposed to exhibit. Of course nothing precludes the construction of typologies with more than two classes, as long as they offer meaningful analytical distinctions. For example, in the most recent (as of 2012) of a series of reports from the Pew Center for People and the Press, survey response data, grouped using clustering techniques,[2] revealed two distinct groups of Republican-leaning respondents (labeled post facto as *Staunch Conservatives* and *Main Street Republicans*); three categories of people inclined toward the Democratic party (*New Coalition Democrats*, *Hard-Pressed Democrats*, and *Solid Liberals*; and three so-called "Middle Groups" (*Libertarians*, *Disaffecteds*, and *Post-Moderns*) (Pew Research Center, 2011).[3] In previous reports in the Pew series, issued in 1987, 1994, 1999, and

---

[1]Except where noted, we use the term *ideology* somewhat generically to include any patterns of political and policy-oriented beliefs, values, and attitudes. Within political science, psychology, and the scholarly study of public opinion, the term is more narrowly defined as a highly constrained special case of this.

[2]The full Pew report provides little detail on the clustering procedure employed, indicating only that scales were developed using factor analysis, with clustering carried out on responses measured along the resulting scales. The report does not specify the particular clustering algorithm employed: "The typology groups are created using a statistical procedure called 'cluster analysis' which accounts for respondents' scores on all nine scales as well as party identification to sort them into relatively homogeneous groups." Several competing cluster solutions were then compared, "evaluated for their effectiveness in producing cohesive groups that were sufficiently distinct from one another, large enough in size to be analytically practical, and substantively meaningful." Other than citing the reliance on both statistical and substantive criteria, no additional detail is provided on the selection of the particular clustering appearing in the report. Aside from modeling assumptions implicit in the first-stage factor analysis, the overall approach is not model-based; certainly the clustering stage is model-free.

[3]Worth noting is that those identified as middle-groups were not necessarily neatly placed on a scale from liberal to conservative, as is typically done for "moderates" not clearly identifiable as liberal or conservative in contemporary terms. For example, Libertarians were those who tended to express strong views in favor of reduced government in all aspects of life, social and economic, leading to positions more associated with Republicans on economic issues and with Democrats on a number of social issues including support for political secularism. Meanwhile, Disaffecteds would not be fruitfully described in terms of a continuum between liberal and conservative, as they were typified more by their cynicism regarding politics and voting in general, yet they did tend to be somewhat more likely to consider themselves Republicans. While

2005, somewhat different typologies emerged from cluster analysis. It is unclear whether identical clustering techniques have been used in all Pew typology studies.

A straightforward means of employing typologies in describing ideological inclinations is to use the types as direct descriptions of individuals. This allows us to partition the population of interest into a few disjoint homogeneous subsets, whose members share the same configuration of attitudes, beliefs, and values—for example, "liberals" and "conservatives"—or perhaps a larger set, as in the Pew report. This approach, however, requires an important trade-off between interpretability and ability of accounting for individual heterogeneity. On the one hand, we want to be able to rely on as few types as possible, each specifying meaningful distinctions pertaining to its members. On the other, we want to avoid oversimplifying the ideological phenomenon, thereby creating too coarse of a partition to adequately account for reality.

A different approach, which we consider more natural for the study of political ideology, is the the use of a mixed membership (MM) framework. Mixed membership allows us to specify partial membership in multiple reference types and to quantify the strength of these memberships. This enables us to describe ideology in terms of a reduced number of prototypical configurations, such as liberal and conservative, while allowing these configurations to coexist within a single individual. In this way we could, for instance, describe an individual's ideology as a combination of "14% conservative and 86% liberal."

In the remainder of this introductory section, we consider the challenge of measuring individuals' ideologies or political belief systems, typical methods for handling the task, and what a mixed membership modeling approach may offer political scientists wishing to answer key problems in the study of ideology. We also reflect briefly upon the notion of individuals as partial adherents to more than one ideological profile and look a bit more closely at why MM models provide such a suitable empirical counterpart to this analytical framework. The data with which we illustrate an application of this measurement model are described in Section 6.2. Next, in Section 6.3, we present a general mixed membership model for ideology, which treats survey respondents as if they were drawing upon partial membership in different latent ideological prototypes (or *extreme profiles*) in order to determine their responses. Some details regarding model fit are offered in Section 6.4, after which we discuss our results in Section 6.5. Finally, we conclude with a brief examination of how scholars of political psychology and public opinion stand to benefit more broadly from a mixed membership approach to their investigations, and offer a candid assessment of the limitations of the current model and possible remedies to be pursued in future work.

6.1.1 Understanding Ideology and the Structure of Politically-Oriented Beliefs, Values, and Attitudes

Among scholars who wish to understand how members of the public reach evaluations about parties, policies, and candidates, or simply about what they see on the evening news, there are a variety of approaches that may be taken. What is common to most of these is an assumption that people have certain dispositions, outlooks, or "basic orientations" (Feldman, 1988) upon which they rely in making such evaluations. Important debates have revolved around the question of whether ordinary people seem to apply "abstract ideological principles, sweeping ideas about how government and society should be organized" (Kinder, 1983, p. 390) in order to reach opinions on a variety of issues. For some, the notion of *ideology* itself is inherently unidimensional, a "general left-right scheme...organizing a wide range of fairly disparate concerns" (Zaller, 1992, p. 26). In this narrow sense of "ideology," as the term is typically employed by political scientists, the observation that

---

the Disaffected category had been identified in all three previous reports, Post-Moderns, on the other hand, were a newly emergent type. Young and heavily Democratic in party membership, they agreed with staunch liberals on such issues as the environment, immigration, and separation of church and state, yet were more wary of New Deal and Great Society policies. (Pew Research Center, 2011, pp. 20–21).

most people do not rely on such a unified structuring of political views and perceptions has long been of great interest (Campbell et al., 1960; Converse, 1964). And yet, even though members of the mass public do not think about most political issues using a left-right scheme to nearly the extent that political elites do, they neither approach each new object of evaluation independently, nor do they think or care enough about politics and policy to be able to do so (Page and Shapiro, 1992; Lupia and McCubbins, 1998). Thus, the basic orientations people use to make sense of the political world may be somewhat varied and not well captured by a single—or possibly even a small number—of dimensions.

We start from the widely shared assumption that such latent structure does exist in individuals and that it drives, probabilistically, their responses to survey questions, as proposed by Zaller (1992). It is this latent structure that we refer to here as *ideology*. As a prominent example of fundamental latent structures that do not meet the classical notion of a left-right overarching scale, some have suggested that particular nations or cultures have a few prominent core beliefs and values (e.g., communitarian or individualist orientations) that may have a high degree of popularity, but which may be of more or less importance in individuals' psyches. Converse (1964, p. 211) posits that "psychological constraints" may be at play, whereby "a few crowning postures—like premises about survival of the fittest in the spirit of Social Darwinism—serve as a sort of glue to bind together many more specific attitudes and beliefs, and these postures are of prime centrality in the belief system as a whole."

Feldman (1988) and others follow up on this by examining the *core beliefs* and *core values* that may provide just this sort of psychological constraint. More recently, Ellis and Stimson (2012) make similar ideas the centerpiece of their "alternative conception of 'ideology,' ... defined by citizens' specific beliefs and values regarding what governments should and should not be doing." This "operational ideology" is distinguished from a "symbolic ideology" based on a person's self-identification or one based on more vague sentiments about " 'government' or 'government programs' broadly framed."

Our own conceptualization of ideology here follows that of Ellis and Stimson in its reliance on fundamental values and beliefs, especially as related to the appropriate role and obligations of government. Note that there is nothing inherent in such a definition that requires ideology to be unidimensional, although left-right orientation can certainly be a useful heuristic and is the focus of these authors' own discussion. If ideology is instead conceptualized as the degree to which certain values and beliefs are salient for individuals as they evaluate political objects such as candidates and policy proposals, it may be less than ideal to measure ideology using a continuous, unbounded interval. For instance, we might expect that the vast majority of Americans will embrace values of reward for hard work, equality of opportunity, or freedom from governmental interference. Different kinds of people may find some of these considerations more compelling than others, but it would be surprising to find, for example, a group of Americans openly hostile to the notion of equal opportunity. Thus, we would like our measurement tools to be able to reflect this, by allowing us to distinguish groups of respondents not only by the values and beliefs which most starkly divide them, but also by how consistently they embrace those values and beliefs that are widely held.

6.1.2   Measuring Ideology with Survey Data

Regardless of the details of a latent structure approach to ideology (whether, for example, we treat latent and observed variables as continuously varying, ordinal, or measurable in terms of unordered levels), a key assumption is that all variation in survey responses can in fact be explained by the underlying latent structure. Survey responses will thus be conditionally independent given one's ideology (i.e., belief/value structure). The matter of how to conceptualize the latent space, whether as a multidimensional continuum or a typology with multiple possible latent classes, is largely a pragmatic question about what best reveals an otherwise invisible structure to the researcher in a manner appropriate to the questions being asked. It may be that certain renderings of this space (as, say, a unidimensional continuum) are rather limited in what they can tell us about how opinions are generated, but a choice among different representations will quite properly hinge upon what best allows lucid communication of findings. A discrete multivariate approach to the survey responses themselves makes sense, since such a treatment reflects the actual structure of Likert scale items typically found in public opinion surveys, allowing the researcher to avoid the false assumption of a continuous scale and comparable units separating response levels.

The main approaches one may take in measuring ideology as a latent construct, inferred from responses to carefully selected survey questions, can be divided into heuristic or purely descriptive techniques on one hand, and principled, model-based approaches on the other. Among the former are basic principal components analysis (PCA),[4] multidimensional scaling (MDS) (Marcus et al., 1974), Q-analysis (Conover and Feldman, 1984), and correspondence analysis (CORA), a categorical analogue to PCA. The latter includes factor analysis (FA) (Feldman, 1988) and other forms of latent structure analysis such as latent trait analysis/item-response theory (IRT) (Treier and Hillygus, 2009) and latent class analysis (LCA) (Taylor, 1983; Feldman and Johnston, 2009), as well as mixed membership/Grade of Membership models (MM/GoM), which may be thought of as either a sort of discrete factor analysis (Erosheva, 2002, pp. 16–20) or an extension of latent class analysis.

Although there are a number of different options for handling the measurement of ideology (and beliefs, attitudes, values, etc.), the most common approach is some form of factor analysis. As the oldest latent variable measurement technique, and the most deeply ingrained in the habits of social scientists, it has the advantage of being easily related to ordinary regression techniques, and dominates the early literature on mass belief structures. Converse sets the precedent of actually equating the factors discovered or confirmed via FA with dimensions of belief structure, generating political evaluations much as Spearman's general intelligence quotient g generates responses to IQ test items (Spearman, 1904). "Factor analysis is *the* statistical technique designed to reduce a number of correlated variables to a more limited set of *organizing dimensions*" (Converse, 1964, our emphasis). One reason that factor analysis became the dominant approach to measuring latent ideological structure was that it was the earliest to be implemented in standard statistical computing packages. The representation of individuals' ideals and beliefs located in a low-dimensional continuous space also conformed well with evocative metaphors, adapted from economics, by which voters were considered to occupy a location in ideological space (or representing preferred tradeoffs among various competing public goods) and should be expected to prefer candidates located nearby (or with similar ideal balance among policy priorities) (Downs, 1957). Although the application of factor analysis in such situations is a deeply entrenched tradition in the study of ideology and public opinion, and is not an unreasonable approach, it is more appropriate for continuous multivariate data than discrete multivariate data typically found in survey responses.

6.1.3 Citizens as Partial Adherents to Distinct Ideologies

Converse (1964) set forth a research agenda, carried out in various forms over the decades since, aimed at understanding the "constraints" on patterns of belief which people may simultaneously hold. He refers to the "combinations" and "permutations" of "idea-elements" actually observed for individuals. When the constraints are severe enough, the resulting packet of beliefs to which a set of people subscribes is considered an *ideology*, in its strict sense as a psychological term of art. As Kinder (1983, p. 390) puts it, the notion of ideology upon which scholars once focused, but which guides the political evaluations of few actual citizens, consisted of "abstract ideological principles, sweeping ideas about how government and society should be organized." In this previously

---

[4]Social scientists regularly use the term "principal components analysis" interchangeably with (exploratory) factor analysis—and PCA is treated as a special case of FA in statistical computer packages—but we are referring to its standard statistical meaning, a process by which orthogonal basis vectors of the reduced-dimensional space are chosen to maximize variance accounted for with each additional dimension included. The goal of PCA, as with the other descriptive approaches, is simply dimension reduction, not modeling or inference regarding the data generation process.

dominant understanding of ideology, answers to a variety of public opinion questions could be thought to follow logically from a highly rigid, overarching outlook. Of course, if ideology operated as a purely deductive process among adherents, survey responses would be generated deterministically, and we should see certain beliefs and opinions always occurring together and others never co-occurring.[5]

Given such a narrow view of ideology, it is easy to look at actual patterns of response as evidence that people are haphazard in their thinking about politics and policy. Zaller (1992) countered this by developing a highly influential theory of how individuals formulate their responses to public opinion polls by randomly sampling from a number of privately held "considerations" relevant to the question at hand. Such a formulation helps account for a number of puzzling observations, such as the tendency for particular individuals to give different answers to the same question on different occasions.

Latent class statistical modeling (Lazarsfeld and Henry, 1968; Goodman, 1974) corresponds fairly well to Zaller's theoretical model, as each member of a distinct class responds by drawing a particular response from a distribution associated with that type. However, such an approach has the intrinsic limitation of assuming that individuals belong exclusively to just one ideological group and that each such group is homogeneous. This characterization leaves out the possibility of individuals who do not fully conform to any of the categories of a typology, but rather respond as something of a hybrid.

Mixed membership models offer a conceptually attractive way of overcoming this limitation. Under mixed membership analysis, we still try to identify and characterize typical ideological classes. However, we regard individuals not as full members of those classes, but as *partial members*. This way, we take individuals' responses as arising from all the distributions associated with the classes, weighted according to individually specified membership in all of them.

Mixed membership models help formalize the idea of people as being partial adherents to different recognizable ideologies. Some—especially political leaders or "elites"—may adhere vigorously to a particular ideology and this would be reflected by full or nearly full membership in one group to the exclusion of the others. Others, perhaps the vast majority of the mass public, will draw on more widely dispersed vectors of partial membership in each. This offers a nice compromise between a continuous Euclidean latent space on one hand and a categorical latent space on the other; patterns in the population and in individuals themselves are described in terms of easily understood prototypical distributions over categorical responses, and yet individuals are treated as combinations of the various prototypes, with their partial memberships allowed to vary continuously. The generating process of responses may indeed be thought of hierarchically: in encountering a survey item, the respondent first randomly draws an ideological profile based upon his or her relative degree of membership in each extreme profile and then randomly draws a response from that profile's response distribution.

6.2   Application: The American National Election Survey

The data we analyze here come from a pilot study for the 1984 American National Election Study (NES), conducted by the Center for Political Studies of the Institute of Social Research at the University of Michigan during the summer of 1983. The study's purpose was to introduce and test new survey items, including a number of questions on core values that we will be using to illustrate

---

[5]Empirically, not only do people hold sets of beliefs and values that do not logically follow from one another, but we simultaneously hold beliefs and values that are logically inconsistent; the commonly held trio of preferences for more government spending, lower taxes, and a reduced deficit is but one prominent example.

our mixed membership modeling approach to measuring ideology. The complete data consist of reinterviews with 314 randomly selected respondents to the earlier 1982 National Election Study. We reanalyze the same 19 items investigated by Feldman (1988),[6] using only the 279 complete responses.

The initial national sample, obtained for the 1982 NES and from which the individuals in the 1983 pilot study considered here were subsampled, consisted of 1,418 respondents living within the primary areas of the survey's county-based sampling frame. These areas were all located within the 48 contiguous states (not including military bases). They include 12 major metropolitan areas, 32 other standard metropolitan statistical areas, and 30 counties or county-groups representing the rural subpopulation. Stratification was implemented independently within each of the four major geographical regions of the United States, as recognized at the time: northeast, north central (loosely, the midwest), south, and west, with each represented in proportion to population. The population under study included only United States citizens 18 years or older on Election Day, 1982.

According to Feldman's review of the existing literature at the time, three political attitudes dominated the American political psyche: "belief in equality of opportunity, support for economic individualism, and support for the free enterprise system," and the 19 items to be analyzed were all developed with the intent to measure these components of ideology. We have labeled all of these with our own variable names, which are intended to capture the spirit of the questions and distinguish similar items from one another based on the subtle wording differences. (See the Appendix for a complete list of the wording of questions.) Seven of the 19 items are intended to measure support for or belief in what Feldman calls "equal opportunity," including statements that attribute inequality to inherent individual differences (*natural inequality 1*, *natural inequality 2*, and *equality goal misguided*), one that claims a key role for society—perhaps understood as "government" by some respondents—in ensuring equal opportunity for success (*equal opportunity–society's responsibility*), assessments of whether inequality is a serious problem (*equal treatment* and *inequality big problem*), and one expressing support for the ideal of shared governance (*democracy*). The items dealing with "economic individualism" are closely related to one another and differ mostly in subtle ways, as indicated in our choice of variable names: *hard work optimism*, *hard work realism*, *hard work idealism*, *ambition pessimism*, *effort pessimism*, and *individual responsibility for failure*. Finally, the "free enterprise" items (*less intervention is better*, *intervention populism*, *laissez-faire capitalism*, *regulations not a threat to freedom*, *intervention causes problems*, and *free enterprise not intrinsic feature of gov't*) allow respondents to weigh possible tradeoffs between positive and negative consequences of governmental regulations. All items consist of statements to which the respondents may say that they "agree strongly," "agree but not strongly," "can't decide," "disagree but not strongly," or "disagree strongly." In order to avoid overparametrization for such a small sample yet capture the main qualitative differences in responses, we collapse these responses into three categories: agree, can't decide, or disagree.

6.3 Methods

We apply a technique known as the Grade of Membership model (GoM) to the study of political ideology. GoM models (Woodbury et al., 1978; Manton et al., 1994; Erosheva et al., 2007) are a sub-family of mixed membership models (Erosheva and Fienberg, 2005). They are well-suited to obtaining low-dimensional representations of high-dimensional multivariate unordered categorical data, such as those that are generated by opinion surveys. Similar to other MM techniques, GoM

---

[6]Feldman served on the ANES planning committee and was apparently directly involved with formulation of these questions, intended to measure three particular core values and beliefs of Americans.

models represent individuals as individually weighted combinations of a small number of "ideal individuals" or "extreme profiles" and use the data to estimate both the extreme profiles themselves and each subject's membership structure. The Bayesian version of the GoM model, which we introduce here and use throughout this chapter, was introduced by Erosheva et al. (2007) and applied to the study of disability in elders.

6.3.1 Grade of Membership Models

We consider a sample of N individuals. Each individual $i = 1, 2, \ldots, N$ has a corresponding J dimensional vector of manifest variables, $X_i = (X_{i1}, \ldots, X_{iJ})$, that collects the outcomes of interest. We assume that the components of the outcome's vector are unordered categorical variables with n_j levels each $(j = 1, \ldots, J)$. In our application, these outcomes are the answers to each of the J questions of the survey. For convenience, we label component's levels using consecutive numbers, $X_{ij} \in \{1, 2, \ldots, n_j\}$. We assume that there is only one response vector per individual.

GoM models assume the existence of a specific number, K, of "extreme profiles" or "pure types." These are idealized versions of individuals that we use as reference types for specifying the response distributions of actual individuals. We assume that real individuals are combinations of these extremes types. To formalize this, we endow each individual with his or her own *membership vector*, $g_i = (g_{i1}, \ldots, g_{ik}, \ldots, g_{iK})$. Each component of g_i, g_{ik} for $k = 1, \ldots, K$ specifies the degree of membership of individual i in the corresponding extreme profile among all K. We restrict membership vectors so that $g_i \in \Delta_{K-1} = \{(g_1, \ldots, g_K) : g_k \geq 0, \sum_{k=1}^{K} g_k = 1\}$, where Δ_{K-1} is the $K - 1$-dimensional simplex. Ideal individuals of the kth extreme profile have a membership vector whose kth component is $g_{ik} = 1$ and the rest, zeros.

We characterize the extreme profiles as follows: For any individual that is a *full member* of the kth extreme class (i.e., such that its membership vector has $g_{ik} = 1$ and $g_{ik'} = 0$ for $k' \neq k$), we assume that the response distribution of the jth entry of the manifest variables vector is a simple discrete distribution:

$$\Pr\left(X_{ij} = l | g_{ik} = 1\right) = \lambda_{jk}(l), \tag{6.1}$$

where $l \in \{1, 2, \ldots, n_j\}$ and $\lambda_{jk} = (\lambda_{jk}(1), \ldots, \lambda_{jk}(n_j)) \in \Delta_{n_j-1}$.

For generic individuals with membership vector g_i, we characterize their component-wise response distribution as the convex combination

$$\Pr\left(X_{ij} = l | g_i\right) = \sum_{k=1}^{K} g_{ik} \lambda_{jk}(l).$$

Geometrically, this specification means that the individual response distributions are located within the convex hull defined by the extreme profiles.

We further assume that the item responses j are conditionally independent given membership vectors. This local independence assumption (Holland and Rosenbaum, 1986) expresses the idea that the membership vector g completely explains the dependence structure among the J binary manifest variables. By making this assumption, we can construct the conditional joint distribution of responses:

$$\Pr\left(X_i = x_i | g_i\right) = \prod_{j=1}^{J} \sum_{k=1}^{K} g_{ik} \lambda_{jk}(x_{ij}).$$

Assuming further that the individuals are randomly sampled from the population, we finally obtain

$$\Pr\left(X = x|g\right) = \prod_{i=1}^{N}\prod_{j=1}^{J}\sum_{k=1}^{K} g_{ik}\lambda_{jk}(x_{ij}). \tag{6.2}$$

Membership vectors are unobserved latent quantities. In order to derive an unconditional expression for the joint distribution of observed responses, we assume that membership vectors are sampled from a common distribution, G_{α}, with support in Δ_{K-1}; whereby

$$\Pr\left(X = x\right) = \prod_{i=1}^{N}\int_{\Delta_{K-1}}\prod_{j=1}^{J}\sum_{k=1}^{K} g_{k}\lambda_{jk}(x_{ij})G(dg). \tag{6.3}$$

An interesting perspective on the GoM model results from considering the following equivalent data generation process, which generates N variates $x_i = (x_{i1}, \ldots, x_{iJ})$ for $i = 1, \ldots, N$, according to a GoM model with K extreme profiles (Haberman, 1995; Erosheva et al., 2007):

GoM Data Generation Process

For each $i = 1, 2, \ldots, N$

    Sample $g_i = (g_{i1}, \ldots, g_{iK}) \sim G$;

    For each $j = 1, 2, \ldots, J$

        Sample $z_{ij} \sim \mathbf{Discrete}_{1:K}(g_{i1}, g_{i2}, \ldots, g_{iK})$;

        Sample $y_{ij} \sim \mathbf{Discrete}_{1:n_j}\left(\lambda_{jz_{ij}}(1), \ldots, \lambda_{jz_{ij}}(n_j)\right)$.

According to this process, we can understand the generation of individual GoM variates as arising from a two-step procedure: (1) Given a membership vector g_i, we obtain the components of the response vector one by one. (2) For each of the J components, we determine an effective extreme profile—which is allowed to vary from component to component—by sampling it with probabilities given by g_i. Next, we sample the actual response as if the individual were a full member of that extreme profile for that question. The multiple membership is reflected by the fact that the individual answers to each question are generated according to different extreme profiles.

6.3.2 Full Bayesian Specification

For this application we closely follow Erosheva et al. (2007). We complete the specification of the GoM in a full Bayesian fashion by choosing the distribution of membership vectors, G_{α}, and a prior distribution for all parameters of our model.

For the membership vectors distribution G_{α}, we specify their common distribution as

$$g_i \stackrel{iid}{\sim} \mathbf{Dirichlet}(\alpha),$$

with $\alpha = (\alpha_0 \cdot \xi_1, \ldots, \alpha_0 \cdot \xi_K)$, $\alpha_0 > 0$ and $\xi = (\xi_1, \ldots, \xi_K) \in \Delta_{K-1}$. Parameter ξ is the expected value of distribution G_{α}. Using the generative process interpretation from the previous section, each component of ξ represents the expected proportion of item responses generated by each of the extreme profiles; thus we can understand it, informally, as the relative importance of each extreme profile in the population. Parameter α_0 is a concentration parameter that expresses how concentrated the probability distribution is about its expected value (as α_0 increases) or near the vertices of the simplex Δ_{K-1} (as α_0 decreases). When $\alpha_0 = K$ and $\xi = (1/K, \ldots, 1/K)$ so that $\alpha = \mathbf{1}_K$, distribution G_α becomes uniform over Δ_{K-1}.

We specify the hyperpriors of G_α as $\alpha_0 \sim \mathbf{Gamma}(1, 2)$ (in shape/rate parametrization) and $\xi \sim \mathbf{Dirichlet}(\mathbf{1_K})$. These choices specify a priori ignorance about the relative importance of each extreme profile in the population and a slight preference (although not so strong) for small values of α_0, with individuals likely to be relatively pure adherents to one or another extreme profile unless the data provided suggest otherwise.

Each conditional response distribution for item j and extreme profile k, $\lambda_{jk}(\cdot)$ consists of n_j scalar parameters restricted to the simplex Δ_{n_j}. For these parameters, we choose the prior distribution

$$\lambda_{jk} = (\lambda_{jk}(1), \ldots, \lambda_{jk}(n_j)) \overset{iid}{\sim} \mathbf{Dirichlet}(\mathbf{1}_{n_j}),$$

or a uniform distribution over $\Delta_{n_j - 1}$.

6.4 Fitting the Models

We have employed an MCMC algorithm to obtain samples from the posterior distribution of parameters given the data. The algorithm is an extension for multilevel variables of the sampler presented in Erosheva et al. (2007), originally developed for binary variables. This sampler is based on a data augmentation strategy using the equivalent generative process outlined in Section 6.3. We have fitted models with $K = 2, 3, 4$, and 5 extreme profiles using the prior distributions described in Section 6.3.2.

Similar to other latent structure models, GoM models are invariant to permutations of the extreme profile labels. For this reason we have re-labeled extreme profiles according to the decreasing sequence of the posterior estimates (posterior means) of the components of ξ. This ordering makes comparisons easier.

Table 6.1 shows posterior estimates (posterior means and standard deviations) for the population-level distribution of membership vectors α_0 and ξ for models with $K = 2, 3, 4, 5$ extreme profiles. In all cases, posterior estimates of α_0 are relatively small. This causes most membership vectors in the population to be dominated by a single extreme profile. However, α_0 is large enough so that the mixed membership becomes an important structural feature. For instance, for model $K = 3$ the probability that a single individual's responses to different questions are drawn from more than one extreme profile is approximately 0.65. Not surprisingly, given the scarcity of data ($n = 279$), posterior dispersions are rather large.

Investigating the posterior estimates of ξ we see that for $K \geq 3$, all models feature two dominant

K	α_0	ξ_1	ξ_2	ξ_3	ξ_4	ξ_5
2	0.510 (0.236)	0.971 (0.009)	0.029 (0.009)			
3	0.765 (0.233)	0.604 (0.129)	0.373 (0.129)	0.023 (0.007)		
4	0.772 (0.238)	0.591 (0.131)	0.384 (0.131)	0.016 (0.010)	0.010 (0.010)	
5	0.827 (0.229)	0.613 (0.122)	0.359 (0.121)	0.013 (0.010)	0.011 (0.011)	0.003 (0.003)

TABLE 6.1
Posterior estimates of parameters α_0 and ξ for models with $K = 2, 3, 4$, and 5 extreme profiles. Numbers between parenthesis are posterior standard deviations.

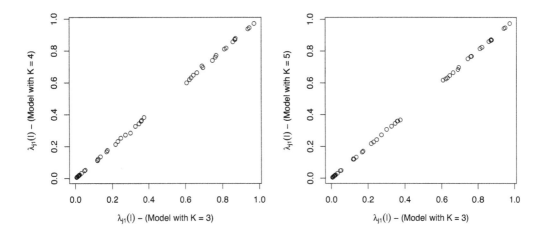

FIGURE 6.1
Comparison of posterior estimates of the first extreme profile, $\lambda_{j1}(l)$, for model with $K = 3$ versus models with $K = 4$ and $K = 5$ extreme profiles.

extreme profiles, with $\xi_1 \approx 0.60$, $\xi_2 \approx 0.36$, and $K - 2$ profiles with very small values of ξ_k. Closer inspection reveals that those two dominant extreme profiles ($k = 1$ and $k = 2$) are very similar for all models with $K \geq 3$. Plots in Figure 6.1 show the posterior estimates of $\lambda_{j1}(l)$ (parameters of the first extreme profile) for the model with $K = 3$ extreme profiles versus their counterparts for models with $K = 4$ and $K = 5$ extreme profiles. We see that all points lie almost perfectly on the main diagonal. The situation is similar for the second extreme profile ($k = 2$, not shown).

Based on these observations and the qualitative inspection of the estimates, we have selected the model with $K = 3$ for our inferences. This was the smallest model for which the two dominant extreme profiles appear, and any more complex model basically gives us the same information, supplemented only by additional extreme profiles with very small values of ξ_k. Interestingly, the estimate of the conditional response distribution, λ, for the dominant extreme profile in model $K = 2$ is almost numerically equal to the weighted sum (by ξ_1 and ξ_2) of the estimates of the two dominant extreme profiles of our selected model ($K = 3$, but also for $K = 4$ and $K = 5$). This suggests that the two dominant extreme profiles in our chosen model are basically a split of the first dominant profile of model $K = 2$. Our attempts to perform more formal evaluations failed to produce anything illuminating. Posterior predictive counts were difficult to produce and analyze due to the small sample size and large number of variables. We also evaluated the Advances in Computational Mathematics (AICM) index (Raftery et al., 2007; Erosheva et al., 2007), which selected a model with $K = 2$ extreme profiles.

6.5 Results and Discussion

6.5.1 Results

The multinomial conditional probabilities of responses given full membership, λ_{jk}, define the extreme profiles. Two of the three extreme profiles, $k = 1, 2$, account for 98% of the item responses, while profile $k = 3$ generates around 2%, and seems to be associated with a high probability of responding "can't decide" ($l = 2$) to the survey items (estimated as anywhere from around 10% to 83% for different items when $k = 3$).

Analyzing the posterior membership estimates of the 279 respondents, the vast majority have partial membership of less than .01 in $k = 3$. Just five members of the population would answer survey questions as primarily a member of this class (from .71 to .91 membership), and only 23 of the 279 have greater than 2% membership in this "neutral" prototype.

In order to better characterize the two dominant ideal types identified, let us examine each simply in terms of the probability of agreeing or disagreeing with each item when answers are based on considerations[7] rooted in one or the other dominant profile. For convenience—and in order to connect our findings with standard writings on American ideology—we use the term *conservative* as shorthand for $k = 1$ and *liberal* for $k = 2$. For reasons that will become apparent shortly, we might more accurately refer to these as, respectively, something like *Individualist/Believers in Realized American Ideals* and *Social Responsibility-Oriented/Still Waiting for American Ideals to be Realized*. Given the clumsiness of such labels, we will stick with the more common ideological identifiers, but consider them to be best understood in terms of response distributions associated with survey items, which we are about to examine.

6.5.2 Analyzing the Extreme Profiles: Americans' Core Values vs. Core Beliefs

In considering the estimated response distributions λ_k for the "conservative" ($k = 1$) and "liberal" ($k = 2$) ideal types (Table 6.2), one important thing to notice is the presence of certain high-valence items, enjoying the consensus one might expect of core values shared by most members of a society. For such items, the distinction between liberals and conservatives is not especially stark, but to the extent that one type is more predictably supportive of a statement than the other, the differences are in the direction that would be expected. For instance, both prototypical respondents would be unlikely to agree that our inherent differences should lead us to give up on the goal of equality ($j = 2$, *equality goal misguided*), but the prototypical conservative may have a greater probability of breaking with the norm: $\hat{\lambda}_{21}(1) = .272$ (*sd.* $= .045$) as opposed to $\hat{\lambda}_{22}(1) = .136$ (*sd.* $= .063$) for the prototypical liberal. Whether responding as a liberal or conservative, an individual would very likely support the democratic ideal of governance by all sorts of people—not only the most successful—($\sim .94$ or $.86$, respectively) as well as the notion that society has a responsibility to ensure equal opportunity of success for all ($\sim .89$ or $.82$, respectively). Yet, while a commitment to the ideal of equal opportunity in personal and public life is widely embraced, so too is the recognition that people are not equally well-suited to leadership positions (*natural inequality 1* ($\sim .87$ and $.76$) and *natural inequality 2* ($\sim .95$ and $.85$) among prototypical conservatives and liberals, respectively.)

---

[7]Here we intentionally use the term *considerations*, from Zaller (1992) in order to emphasize the connection between our measurement strategy and Zaller's theoretical framework. Just as Zaller depicts respondents drawing at random from an unobserved distribution of considerations in order to answer each question, we model such individuals as drawing an ideal type at random in proportion to their own latent membership vector, and then generating a response according to the distribution associated with the selected ideal type on the particular item.

		$\lambda_{jk}(l)$			
		Level: $l = 1$ (Agree)		$l = 3$ (Disagree)	
j	Question	$k = 1$	$k = 2$	$k = 1$	$k = 2$
1	*Equal treatment*	0.61 (0.10)	0.92 (0.05)	0.37 (0.10)	0.07 (0.05)
2	*Equality goal misguided*	0.27 (0.05)	0.14 (0.06)	0.7 (0.05)	0.83 (0.06)
3	*Equal opportunity society's responsibility*	0.82 (0.05)	0.89 (0.05)	0.17 (0.05)	0.10 (0.05)
4	*Natural inequality 1*	0.87 (0.04)	0.76 (0.07)	0.12 (0.04)	0.22 (0.07)
5	*Natural inequality 2*	0.95 (0.02)	0.85 (0.06)	0.05 (0.02)	0.14 (0.05)
6	*Democracy*	0.86 (0.04)	0.94 (0.04)	0.14 (0.04)	0.05 (0.04)
7	*Inequality big problem*	0.30 (0.14)	0.88 (0.07)	0.69 (0.14)	0.10 (0.07)
8	*Hard work optimism*	0.97 (0.02)	0.45 (0.17)	0.02 (0.02)	0.54 (0.17)
9	*Hard work realism*	0.12 (0.06)	0.47 (0.09)	0.87 (0.06)	0.52 (0.09)
10	*Individual responsibility for failure*	0.77 (0.06)	0.19 (0.12)	0.22 (0.06)	0.77 (0.12)
11	*Ambition pessimism*	0.76 (0.05)	0.88 (0.05)	0.23 (0.05)	0.11 (0.05)
12	*Hard work idealism*	0.64 (0.06)	0.21 (0.12)	0.35 (0.06)	0.78 (0.11)
13	*Effort pessimism*	0.75 (0.07)	0.95 (0.03)	0.25 (0.07)	0.04 (0.03)
14	*Less intervention is better*	0.81 (0.05)	0.42 (0.13)	0.17 (0.05)	0.55 (0.13)
15	*Intervention populism*	0.62 (0.06)	0.83 (0.06)	0.36 (0.05)	0.11 (0.06)
16	*Laissez-faire capitalism*	0.36 (0.05)	0.07 (0.07)	0.63 (0.05)	0.91 (0.07)
17	*Regulations not a threat to freedom*	0.33 (0.05)	0.49 (0.08)	0.66 (0.05)	0.49 (0.08)
18	*Intervention causes problems*	0.94 (0.04)	0.58 (0.13)	0.05 (0.03)	0.40 (0.13)
19	*Free enterprise not intrinsic feature of gov't*	0.12 (0.07)	0.41 (0.08)	0.87 (0.07)	0.58 (0.08)

TABLE 6.2
The two dominant extreme profiles for $K = 3$: Profile $k = 1$ (60.4% of responses) versus Profile $k = 2$ (37.3% of responses). Numbers in parentheses are posterior standard deviations. The grouping of items is based on Feldman (1988) and the original intent of the survey questionnaire design: the first concern *Equal Opportunity*; the second, *Economic Individualism*; and the third, *Free Enterprise*. The variable names, generic in the original, are our own.

In order to clarify which items are most important in defining each dominant extreme profile, we introduce the quantity

$$Cohes_{jk} = \frac{\max\limits_{l=1,...,n_j} \{\lambda_{jk}(l)\}}{\min\limits_{l=1,...,n_j} \{\lambda_{jk}(l)\}}, \tag{6.4}$$

or the *cohesion* of extreme profile k with respect to item j. The cohesion scores reflect the reliability with which each extreme type responds to an item. In Zaller's (1992) theory of survey response to opinion polling, this might correspond to individuals tending to answer a question predictably, perhaps because nearly all relevant considerations lead to the same response. This may alternatively be thought to measure the cohesiveness of hypothetical adherents to each extreme profile.

Additionally, we consider the hypotheses

$$DR_j : \arg\max_{l=1,...,n_j} \{\lambda_{j1}(l)\} \neq \arg\max_{l=1,...,n_j} \{\lambda_{j2}(l)\}, \tag{6.5}$$

for $j = 1, ..., J$. Hypothesis DR_j states that full adherents to the two dominant profiles have different modal responses to a given item j. Obtaining posterior estimates of the probability of DR_j enables us to draw inferences about how well different items distinguish the extreme profiles from one another. For example, a posterior probability value 0.01 for item $j = 7$, *inequality big problem*, $Pr[DR_7|Data]$, means that, given our data on 279 respondents, we find only a 1% chance that the most likely response to the item by the two types of pure respondents are the same; it is, rather, highly probable that the top response for *liberals* is recognition of inequality as a big problem while *conservatives* are more likely than not to deny inequality as a persistent issue.

Table 6.3 shows our posterior estimates (posterior means) of $Cohes_{jk}$ and our estimated posterior probabilities of hypotheses DR_j, for extreme profiles $k = 1$ (conservative) and $k = 2$ (liberal), and for every item in the survey ($j = 1, ..., 19$). Analyzing the cohesion scores we see that for certain items, both dominant extreme profiles are predictable and give identical responses (e.g., *intervention causes problems*); for others they reliably give opposite responses, i.e., one type is expected to agree and the other to disagree with an item (e.g., *inequality big problem*); for still other items, prototypical adherents to one profile are highly likely to give their modal response while prototypical adherents of the other are far less predictable (e.g., *laissez-faire capitalsm*).

For the first set of five items listed in Table 6.3, there is a greater than .50 posterior probability that pure liberals and pure conservatives will disagree in their favored responses. We can be highly confident ($> .99$) that prototypical liberals and conservatives—at least to the extent that these labels may be appropriately applied to $k = 1$ and $k = 2$—will tend to disagree when it comes to their reactions to *hard work idealism* ("If people work hard, they almost always get what they want"), *individual responsibility for failure* ("Most people who don't get ahead should not blame the system; they really have only themselves to blame"), and *inequality big problem* ("One of the big problems in this country is that we don't give everyone an equal chance"). Conservatives are also likely to disagree with liberals when it comes to the item most closely associated with contemporary definitions of American liberalism and conservatism: *less intervention is better* ("The less government gets involved with business and the economy, the better off this country will be"). Similarly, we expect them to disagree on an item that perhaps best captures a quintessential American belief that hard work pays off: *hard work optimism* ("Any person who is willing to work hard has a good chance of succeeding"). For the remaining 14 items, the bulk of our posterior probability is placed on identical modal responses for liberals and conservatives, though they may differ substantially in how predictable they are in choosing this modal response. For example, while both are more likely to agree with the statement that "There are many goods and services that would never be available to ordinary people without governmental intervention" (*intervention populism*), the mean posterior cohesion score for a prototypical liberal is 58, in contrast to around only 2 for a prototypical conservative. For eight items, we can be virtually certain that both dominant extreme profiles share a modal response.

If we look closely at the three items that most clearly distinguish our prototypical liberals from conservatives, two have been identified by Feldman (1988) as measures of the core belief in *Economic Individualism* and one as a measure of the belief in *Equal Opportunity*. All three, however, tap into beliefs about what *is* rather than what *should be*:

- *Hard work idealism* ($j = 12$): If people work hard, they almost always get what they want.

- *Individual responsibility for failure* ($j = 10$): Most people who don't get ahead should not blame the system; they really have only themselves to blame.

- *Inequality big problem* ($j = 7$): One of the big problems in this country is that we don't give everyone an equal chance.

Indeed, much of what seems to separate the response distributions for the two dominant ideal types has to do with how well respondents view the United States as actually living up to the ideals shared by many in both camps. In order to appreciate this, a distinction should be drawn between *beliefs* and *values*.

According to Glynn et al. (1999), "Values are ideals. Beliefs represent our understanding of the way things are, but values represent our understanding of the way things should be" (p. 105). The difference between beliefs and values is not always well delineated and, in fact, some survey questions may capture aspects of both. In certain cases, what is presented as a value may imply some belief about the way things actually are, and this may affect the responses of some individuals surveyed. For example, while a majority of individuals answering from either principal extreme profile claim a belief that equal treatment leads to fewer problems (*equal treatement*), pure liberals are nearly in uniform agreement with the statement, but conservatives have around a 38% chance of disagreeing with the sentiment. Why might there be resistance among conservatives, who otherwise generally embrace the goal of equality and the notion that society has a responsibility to ensure equal opportunity, according to their responses to other survey questions? Hidden within the question is an implied belief about the way things actually are: "If people were treated more equally in this country, we would have many fewer problems [*than we have now*]," with the italicized words as implied subtext. So if one believes that inequality leads to problems, but also that people already are treated equally and, perhaps, that commonly advocated programs aimed at the issue (e.g., affirmative action) are misguided, one might disagree with the survey item. Thus, one's national pride and a tendency to view the nation as having already realized the ideals of equal opportunity are considerations of prototypical conservatives that have a non-trivial probability of being primed by the choice of wording here.

Of the five items on which pure liberals and conservatives are expected to differ on their most likely responses, four address the locus of responsibility for individual success and failure. On all four measures, conservatives embrace the notion of individual responsibility for success and failure, while liberals are less convinced. Conservatives are unified in their belief that hard work has a "good chance" of yielding success, while liberals tend to disagree (albeit only at a 3:2 ratio). On a similar question, phrased differently, liberals largely reject an idealistic view of hard work, with a .79 probability of disagreeing that it will "almost always" lead to satisfying results, while conservatives have a .63 of *embracing* such idealism. On one of the most divisive questions, prototypical conservatives agree three to one that individuals should blame themselves if they "don't get ahead," while liberals find "the system" more at fault, by more than four to one! When it comes to the explicit assertion that inequality—specifically a lack of equal opportunity—remains a "big problem" in the United States, there is again a clear distinction between the two dominant types of respondents; liberals identify inequality as a big problem at over seven to one, while conservatives are nearly two to one in the opposite direction.

In short, the results of our Grade of Membership analysis reveal hidden structure in the beliefs and values of survey respondents missing from the original factor analytic results in Feldman (1988). While several core values are widely embraced across extreme profiles, prototypical liberals tend to be more unified in their support of those ideals typically associated with them (equality and democratic principles), while prototypical conservatives tend to be more consistent in embracing values tied to their own central narratives (rewards of hard work, individual responsibility and self-reliance, and antipathy towards government intervention). Only a few survey items serve to starkly contrast the two dominant extreme profiles, and those that most clearly distinguish them involve beliefs about the United States in which they live rather than simply ideals about their nation as it could be.

		$Cohes_{jk}$		
j	Question	$k = 1$ (cons)	$k = 2$ (lib)	DR_j
12	*Hard work idealism*	1.95	22.29	1.00
10	*Individual responsibility for failure*	4.14	16.08	1.00
7	*Inequality big problem*	11.29	14.56	0.99
14	*Less intervention is better*	5.79	19.11	0.66
8	*Hard work optimism*	330.92	8.62	0.61
17	*Regulations not a threat to freedom*	2.13	14.48	0.46
9	*Hard work realism*	29.55	7.43	0.36
18	*Intervention causes problems*	67.89	11.28	0.27
1	*Equal treatment*	1.89	7.35	0.16
19	*Free enterprise not intrinsic feature of gov't*	70.77	58.83	0.14
15	*Intervention populism*	1.79	58.46	0.02
16	*Laissez-faire capitalism*	1.82	14.02	0.01
13	*Effort pessimism*	3.31	21.71	0.00
11	*Ambition pessimism*	3.53	5.60	0.00
6	*Democracy*	6.90	5.89	0.00
5	*Natural inequality 2*	49.45	7.53	0.00
4	*Natural inequality 1*	8.18	32.55	0.00
3	*Equal opportunity society's responsibility*	5.15	55.54	0.00
2	*Equality goal misguided*	2.66	17.90	0.00

TABLE 6.3
Extreme profiles for $K = 3$: Profile $k = 1$ (60.4% of responses) versus Profile $k = 2$ (37.3% of responses). Cohesion scores represent posterior means for the odds that a prototypical adherent will give the top response to the question.

6.6 Conclusion

Typologies are ubiquitous in political science, providing useful frameworks for understanding variation in ideology, beliefs, and values, as well as conceptual development of many other areas (e.g., comparative political systems and conflicts, policy analysis, and identity politics). Typically, the manner in which such typologies are developed is ad hoc rather than model-based, reflecting a priori qualitative judgments about how the political world is naturally partitioned. As we demonstrate, it is also possible to use a mixed membership approach to construct typologies from data in a principled, model-based way, with qualitative interpretation taking place only after model-based estimates have been drawn, and without imposing a possibly artificially crisp partition on the population of interest. In some cases, the resulting typology will correspond well to what we expect and in others it might offer surprises. In our illustration, we see a bit of both: on one hand, two dominant extreme profiles emerge, which correspond roughly to what might be identified as conservatives and liberals, but on the other hand, the nature of these profiles presents us with a more nuanced view of what these ideal types look like. Analyzing survey response data with reference to such prototypes, we maintain a measure of simplicity that promotes understanding, while accommodating the heterogeneity actually present in real-world populations.

Where some have previously analyzed public opinion in terms of types, they have used ad hoc clustering techniques (Pew Research Center, 2011) without justification for the particular algorithm or verification that the results are robust to other choices of clustering routine. Latent class analysis is a surprisingly underutilized technique in political science that improves upon this by assuming that data are generated from distributions associated with distinct latent types of respondents (Feldman and Johnston, 2009) . Our mixed membership approach may be thought of as a generalization of LCA, combining advantages of categorical data inference available in LCA with advantages of continuity assumptions from factor analysis. In fact, one way of thinking about what we are doing here is that we have taken what would be allocated to error in the case of LCA and have integrated it with the structural component of our model. In a latent class analysis, good model fit must reduce the "error" associated with individuals who mostly answer as if they belonged to one type, but respond anomalously on certain questions. Allowing for Grades of Membership in these classes (reconceptualized as ideal types) takes what would otherwise be considered noise and attributes it to a person's internal complexity.

We see great potential for mixed membership modeling in the study of political psychology, behavior, and public opinion. It offers a sort of compromise between the concreteness of classification by types and the flexibility of multidimensional continuous latent variables as in factor analysis. Informal and qualitative accounts of political behavior, for example, rely heavily on such classifications as the "likely voter," the "independent voter," and the "alienated working class voter"; for the most part, evocative labels such as these are replaced in quantitative analysis by measures further removed from the familiar and useful prototypes, but which utilize interval-level scores to reflect diversity across the population. Assuming that individuals manifest partial membership in multiple recognizable types lets researchers use prominent response patterns as familiar reference points without reducing people to stereotypes. Furthermore, it allows us to discover new ideal types, patterns that we might not have otherwise noticed.

Mixed membership is a general idea that can be implemented and exploited in many ways. The particular technique that we employed in this application, the Grade of Membership model, has a fairly simple structure and is an appropriate tool for the basic soft clustering that we presented here. However, in order to investigate a wider array of political science research questions and to better use the available data, we need to develop more tools. First, we would like to investigate the relationship between individual ideology and other relevant variables, like cohort or income. This can be achieved by incorporating covariates into the model. One possible approach, introduced by

Manrique-Vallier (2010), is to specify the population-level membership distribution conditional on the covariates, keeping the extreme profiles common to the whole population. Such an extension would allow to estimate the effect of the covariates into the membership composition of the individuals, enabling us to answer questions such as: "Are older generations more conservative than younger ones?"

Another useful direction would be to use individual estimates of membership as predictors for some dependent variable of interest, for example, one's position on a particular policy issue, or reaction to an experimental intervention. This is a common approach with factor analysis, where we use scores on estimated dimension-reduced scales as inputs into a regression-style model of substantive interest. We can also understand mixed membership analysis as a dimension reduction technique: in our example we reduced the 19-dimensional response vector into a three-component membership vector that lies in a two-dimensional unit simplex. Thus, performing a similar analysis with membership vectors instead of factor loadings as inputs would achieve a similar aim, but make for a more intuitive analysis of results. For example, we could replace statements such as "For every standard deviation increase on the economic individualism factor, we expect..."—where the meaning of this factor is obscured—with more appealing statements of the form "an extra 25% conservatism leads to..." The actual implementation of this idea carries some difficulties though. While it might be tempting to perform regular regression analysis conditional on posterior point estimates of the individual membership scores, we have to make sure to correctly reflect the inherent posterior uncertainty of these estimates in the regression. One possible approach is to set up comprehensive hierarchical models that include the mixed membership and the regression parts. Another possible approach is to obtain samples from the posterior distribution of individual membership vectors and use multiple imputation techniques (Rubin, 1987) to perform the combined analysis.

One limitation of GoM models, stemming from their simple, local independence structure, is that given membership, all answers to questions are taken to be essentially equivalent. However, researchers usually design and organize surveys so that questions belong to specific domains, such as "economic issues" or "social issues," and therefore illuminate different (often known) aspects of ideology. One can envision a hierarchical extension in which, in addition to the mixed membership structure, questions are organized into domains and interact with the membership in different ways. The structure could be such that we take the original classification of questions as prior information with some degree of uncertainty, and learn the rest from the data.

If we or others are to effectively extend mixed membership analysis in any of these potential directions, we would be well-advised to keep in mind the simple observation of Achen (1975), who reminds us: "The greater the distance from data to conclusions, the more opportunity for errors." While latent variable modeling techniques grant us a principled way to measure underlying hidden concepts only indirectly revealed through survey responses, this typically comes at the expense of transparency; the connection between abstractions such as factor loadings and the observed data is often obscured in the minds of researchers and their audience. Among the various advantages of the MM/GoM approach to survey data in seeking to better understand the structure of mass attitudes, one of the most compelling is that it allows us the luxury of abstraction while preserving the close connection to data.

Appendix: Survey Items

Equal Opportunity

- *Equal treatment*: If people were treated more equally in this country, we would have many fewer problems. (V2169/V3120)

- *Equality goal misguided*: We should give up on the goal of equality, since people are so different to begin with. (V2172/V3122)

- *Equal opportunity-society's responsibility*: Our society should do whatever is necessary to make sure that everyone has an equal opportunity to succeed. (V2175/V3123)

- *Natural inequality 1*: Some people are just better cut out than others for important positions in society. (V2178/V3121)

- *Natural inequality 2*: Some people are better at running things and should be allowed to do so. (V2250)

- *Democracy*: All kinds of people should have an equal say in running this country, not just those who are successful. (V2253, Not in wave 2)

- *Inequality big problem*: One of the big problems in this country is that we don't give everyone an equal chance. (V2256, V3125)

Economic Individualism

- *Hard work optimism*: Any person who is willing to work hard has a good chance of succeeding. (V2170)

- *Hard work realism*: Hard work offers little guarantee of success. (V2173)

- *Individual responsibility for failure*: Most people who don't get ahead should not blame the system; they really have only themselves to blame. (V2176)

- *Ambition pessimism*: Even if people are ambitious, they often cannot succeed. (V2251)

- *Hard work idealism*: If people work hard, they almost always get what they want. (V2254)

- *Effort pessimism*: Even if people try hard, they often cannot reach their goals. (V2257)

Free Enterprise

- *Less intervention is better*: The less government gets involved with business and the economy, the better off this country will be. (V2171)

- *Intervention populism*: There are many goods and services that would never be available to ordinary people without governmental intervention. (2174)

- *Laissez-faire capitalism*: There should be no government interference with business and trade. (V2177)

- *Regulations not a threat to freedom*: Putting government regulations on business does *not* endanger personal freedom. (V2252)

- *Intervention causes problems*: Government intervention leads to too much red tape and too many problems. (V2255)

- *Free enterprise not intrinsic feature of gov't*: Contrary to what some people think, a free enterprise system is not necessary for our form of government to survive. (V2258)

References

Achen, C. (1975). Mass political attitudes and the survey response. *The American Political Science Review* 69: 1218–1231.

Campbell, A., Converse, P., Miller, W., and Donald, E. (1960). *The American Voter*. New York, NY: John Wiley & Sons.

Conover, P. and Feldman, S. (1984). How people organize the political world: A schematic model. *American Journal of Political Science* : 95–126.

Converse, P. (1964). The nature of belief systems in mass publics. In Apter, D. (ed), *Ideology and its Discontents*. New York, NY: Free Press, chap. 6, 207–261.

Downs, A. (1957). *An Economic Theory of Democracy*. New York, NY: Harper and Row.

Ellis, C. and Stimson, J. A. (2012). *Ideology in America*. Cambridge, UK: Cambridge University Press.

Erosheva, E. A. (2002). Grade of Membership and Latent Structure Models with Application to Disability Survey Data. Ph.D. thesis, Department of Statistics, Carnegie Mellon University, Pittsburgh, Pennsylvania, USA.

Erosheva, E. A. and Fienberg, S. E. (2005). Bayesian mixed membership models for soft clustering and classification. In *Classification–the Ubiquitous Challenge: Studies in Classification, Data Analysis, and Knowledge Organization*. Berlin Heidelberg: Springer, 11–26.

Erosheva, E. A., Fienberg, S. E., and Joutard, C. (2007). Describing disability through individual-level mixture models for multivariate binary data. *Annals of Applied Statistics* 1: 502–537.

Feldman, S. (1988). Structure and consistency in public opinion: The role of core beliefs and values. *American Journal of Political Science* 32: 416–440.

Feldman, S. and Johnston, C. (2009). Understanding the determinants of political ideology: The necessity of a multi-dimensional conceptualization. In *Proceedings of the 2010 Annual Meeting of the Midwest Political Science Association (MPSA 2010)*. Bloomington, IN, USA: MPSA.

Glynn, C., Herbst, S., O'Keefe, G., Shapiro, R., and Lindeman, M. (1999). *Public Opinion*. Boulder, CO: Westview Press.

Goodman, L. A. (1974). Exploratory latent structure analysis using both identifiable and unidentifiable models. *Biometrika* 61: 215–231.

Haberman, S. J. (1995). Review: Statistical applications using fuzzy sets, by K. Manton, M. Woodbury and H. Tolley. *Journal of the American Statistical Association* 90: 1131–1133.

Holland, P. W. and Rosenbaum, P. R. (1986). Conditional association and unidimensionality in monotone latent variable models. *Annals of Statistics* 14: 1523–1543.

Kinder, D. (1983). Diversity and complexity in American public opinion. In Finifter, A. (ed), *Political Science: The State of the Discipline*. Washington, D.C.: American Political Science Association, 389–425.

Lazarsfeld, P. F. and Henry, N. W. (1968). *Latent Structure Analysis*. Boston, MA: Houghton Mifflin.

Lupia, A. and McCubbins, M. (1998). *The Democratic Dilemma: Can Citizens Learn What They Need to Know?* Chicago, IL: Cambridge University Press.

Manrique-Vallier, D. (2010). Longitudinal Mixed Membership Models with Applications to Disability Survey Data. Ph.D. thesis, Department of Statistics, Carnegie Mellon University, Pittsburgh, Pennsylvania, USA.

Manton, K. G., Woodbury, M. A., and Tolley, H. D. (1994). *Statistical Applications Using Fuzzy Sets.* New York, NY: John Wiley & Sons.

Marcus, G., Tabb, D., and Sullivan, J. (1974). The application of individual differences scaling to the measurement of political ideologies. *American Journal of Political Science* 18: 405–420.

Page, B. and Shapiro, R. (1992). *The Rational Public: Fifty Years of Trends in Americans' Policy Preferences.* New York, NY: University of Chicago Press.

Pew Research Center (2011). Beyond Red vs. Blue: Political Typology. Tech. report, Pew Research Center for the People and the Press, Washington, D.C. Available at: http://www.people-press.org/2011/05/04/beyond-red-vs-blue-the-political-typology/.

Raftery, A. E., Newton, M., Satagopan, J., and Krivitsky, P. (2007). Estimating the integrated likelihood via posterior simulation using the harmonic mean identity. In Bernardo, J. M., Bayarri, M. J., Berger, J. O., Dawid, A. P., Heckerman, D., Smith, A. F. M., and West, M. (eds), *Bayesian Statistics 8.* Oxford, UK: Oxford University Press.

Rubin, D. B. (1987). *Multiple Imputation for Nonresponse in Surveys.* New York, NY: John Wiley & Sons.

Spearman, C. (1904). "General intelligence," objectively determined and measured. *The American Journal of Psychology* 15: 201–292.

Stimson, J. (1975). Belief systems: Constraint, complexity, and the 1972 election. *American Journal of Political Science* 19: 393–417.

Taylor, M. (1983). The black-and-white model of attitude stability: A latent class examination of opinion and nonopinion in the American public. *American Journal of Sociology* 89: 373–401.

Treier, S. and Hillygus, D. (2009). The nature of political ideology in the contemporary electorate. *Public Opinion Quarterly* 73: 679–703.

Woodbury, M. A., Clive, J., and Garson A., Jr. (1978). Mathematical typology: A Grade of Membership technique for obtaining disease definition. *Computers in Biomedical Research* 11: 277–98.

Zaller, J. (1992). *The Nature and Origins of Mass Opinion.* Cambridge, UK: Cambridge University Press.

7

Estimating Diagnostic Error without a Gold Standard: A Mixed Membership Approach

Elena A. Erosheva

Department of Statistics, University of Washington, Seattle, WA 98195-4320, USA

Cyrille Joutard

Institut de Mathématiques et de Modélisation de Montpellier & Université Montpellier 3, Montpellier Cedex 5, France

CONTENTS

Evaluation of sensitivity and specificity of diagnostic tests in the absence of a gold standard typically relies on latent structure models. For example, two extensions of latent class models in the biostatistics literature, Gaussian random effects (Qu et al., 1996) and finite mixture (Albert and Dodd, 2004), form the basis of several recent approaches to estimating sensitivity and specificity of diagnostic tests when no (or partial) gold standard evaluation is available. These models attempt to account for additional item dependencies that cannot be explained with traditional latent class models, where the classes typically correspond to healthy and diseased individuals.

We propose an alternative latent structure model, namely, the extended mixture Grade of Membership (GoM) model, for evaluation of diagnostic tests without a gold standard. The extended mixture GoM model allows for test results to be dependent on latent degree of disease severity, while also allowing for the presence of some individuals with deterministic response patterns such as all-positive and all-negative test results. We formulate and estimate the model in a hierarchical Bayesian framework. We use a simulation study to compare recovery of true sensitivity and specificity parameters with the extended mixture GoM model, and the latent class, Gaussian random effects, and finite mixture models.

Our findings indicate that when the true generating model contains deterministic mixture components and the sample size is large, all four models tend to underestimate sensitivity and overestimate specificity parameters. These results emphasize the need for sensitivity analyses in real life applications when the data generating model is unknown. Employing a number of latent structure models and examining how the assumptions on latent structure affect conclusions about accuracy of diagnostic tests is a crucial step in analyzing test performance without a gold standard. We illustrate the sensitivity analysis approach using data on screening for *Chlamydia trachomatis*. This example demonstrates that the extended mixture GoM model not only provides us with new latent structure and the corresponding interpretation to mechanisms that give rise to test results, but also provides new insights for estimating test accuracy without a gold standard.

7.1   Introduction

We consider the problem of estimating sensitivity and specificity of diagnostic or screening tests when results are available from multiple fallible tests but not from gold standard. This could happen when gold standard assessment doesn't exist or when economic or ethical issues in administering the gold standard prevent one from doing so.

Latent class analysis (Lazarsfeld and Henry, 1968; Goodman, 1974) has been at the core of model-based methods for analyzing diagnostic errors in the absence of a gold standard (Hui and Zhou, 1998; Albert and Dodd, 2004; Pepe and Janes, 2007). Recently, two extensions of latent class models, known as Gaussian random effects (Qu et al., 1996) and finite mixture (Albert and Dodd, 2004), have produced several new approaches to estimating sensitivity and specificity of diagnostic tests when no (or partial) gold standard evaluation is available (Hadgu and Qu, 1998; Albert and Dodd, 2004; 2008; Albert, 2007b). See Hui and Zhou (1998) for a comprehensive review of the earlier literature on evaluating diagnostic tests without gold standards. Pepe and Janes (2007) criticize latent class models as a tool for analyzing diagnostic test performance because of the lack of links between biological mechanisms giving rise to test results and dependencies induced by a structure of the model. For example, they assert that most diseases are not dichotomous but occur in varying degrees of severity. Hence, latent class models that employ discrete disease status as a latent variable cannot account for additional correlations induced by disease severity such as occurrences of false negatives for persons with mild disease. One example that Pepe and Janes (2007) provide talks about detection of a particular substance in a biological sample where the amount of substance affects all test results.

The best way to evaluate the performance of tests with unknown characteristics is, undoubtedly, to have at least a partial gold standard assessment (Albert and Dodd, 2008; Albert, 2007a). In the absence of a gold standard, however, having an arsenal of model-based methods can be informative for evaluating sensitivity of scientific conclusions regarding accuracy of diagnostic and screening tests. Because the true data generating mechanism is typically not known, Albert and Dodd (2004) (p. 433) recommends performing sensitivity analysis by using different models: "Although biological plausibility may aid the practitioner in favoring one model over another, a range of estimates from various models of diagnostic error (as well as standard errors) should be reported."

We present an alternative latent structure model for the analysis of test performance when no gold standard is available. Our model is an extension of the Grade of Membership model. The GoM model employs a degree of disease severity as a latent variable, therefore inducing a mixed membership latent structure where individuals can be members of diseased and healthy classes at the same time. This type of latent structure addresses the concerns of Pepe and Janes (2007). We extend the GoM model to obtain the extended mixture GoM model, analogous to the extended finite mixture model by Muthen and Shedden (1999) and the finite mixture model by Albert and Dodd (2004). The extended mixture GoM model allows for a mixture of deterministic and mixed membership responses. For example, some truly positive individuals may have deterministic positive response on every test while others may be subject to diagnostic error according to the GoM model. A version of this model has previously been applied in disability studies (Erosheva et al., 2007), but the extended mixture GoM model is new to the literature on diagnostic testing.

The remainder of the chapter is organized as follows. In Section 7.2 we review latent class (Lazarsfeld and Henry, 1968; Goodman, 1974), latent class random effects (also known as

Gaussian random effects) (Qu et al., 1996), and finite mixture models (Albert and Dodd, 2004) that are commonly used for analysis of diagnostic and screening tests. In Section 7.3 we introduce the GoM model, develop the extended mixture GoM model that allows for deterministic responses, discuss a hierarchical Bayesian framework for estimation of model parameters, and derive sensitivity and specificity estimates. In Section 7.4 we conduct a simulation study examining recovery of specificity and sensitivity parameters under the latent class, the latent class random effects, the finite mixture, and the extended mixture GoM models. We investigate performance of each of these models when the true data-generating model is known, varying the true model among the four alternatives. Our findings further emphasize the need for sensitivity analyses when no gold standard is available. In Section 7.5 we illustrate such a sensitivity analysis using a publicly available dataset on screening for *Chlamydia trachomatis* (CT) (Hadgu and Qu, 1998) . Finally, in Section 7.6 we relate results from our analyses of simulated and real data to prior findings in the literature.

7.2 Overview of Existing Model-Based Approaches

Sensitivity and specificity are key accuracy parameters of diagnostic and screening tests. The general framework for estimating diagnostic errors without a gold standard starts by assuming a latent structure model and then deriving sensitivity and specificity parameters that correspond to the model formulation. This section introduces a common notation and presents a concise overview of latent class (Lazarsfeld and Henry, 1968; Goodman, 1974), latent class Gaussian random effects (Qu et al., 1996), and finite mixture models (Albert and Dodd, 2004) that are commonly used for analysis of diagnostic and screening tests. For simplicity of the exposition, we omit the subject index.

Let $x = (x_1, x_2, \ldots, x_J)$ be a vector of dichotomous variables, where x_j takes on values $l_j \in \mathcal{L}_j = \{0, 1\}, j = 1, 2, \ldots, J$. Let $\mathcal{X} = \prod_{j=1}^{J} \mathcal{L}_j$ be the set of all possible outcomes l for vector x. Denote a positive test result by $x_j = 1$ and a negative result by $x_j = 0$. Denote the disease indicator by δ, with $\delta = 1$ standing for the presence of the disease. Let $\tau = P(\delta = 1)$ denote the disease prevalence parameter for the population of interest.

The latent class approach assumes two classes, the healthy and the sick. The probability to observe response pattern l is the weighted sum of probabilities to observe l from each latent class:

$$P(x = l) = P(x = l|\delta = 1)P(\delta = 1) + P(x = l|\delta = 0)P(\delta = 0), \quad l \in \mathcal{X}.$$

The tests are assumed to be conditionally independent given the true disease status. Test result x_j is a Bernoulli random variable with class conditional probabilities $\lambda_{1j} = P(x_j = 1|\delta = 1)$ and $\lambda_{2j} = P(x_j = 1|\delta = 0)$ for a given true disease status. The conditional probabilities $\lambda_{1j}, \lambda_{2j}, j = 1, \ldots, J$ and the weight $P(\delta = 1) = 1 - P(\delta = 0) = \tau$ are the model parameters.

For the jth diagnostic test, its sensitivity is the probability of the positive test result given that the true diagnosis is positive, $P(x_j = 1|\delta = 1)$, and its specificity is the probability of a negative response given that the true diagnosis is negative, $P(x_j = 0|\delta = 0) = 1 - P(x_j = 1|\delta = 0)$. The sensitivity and specificity of test j implied by the latent class model are then simply

$$P(x_j = 1|\delta = 1) = \lambda_{1j}$$

and

$$P(x_j = 0|\delta = 0) = 1 - \lambda_{2j}.$$

The Gaussian random effects model of Qu et al. (1996) is an attempt to relax the assumption of independence conditional on the true disease status. This model assumes that test outcomes are independent Bernoulli realizations with probabilities given by the standard normal cdf $\Phi\left(\beta_{j\delta} + \sigma_{\delta}b\right)$, where $\beta_{j\delta}$, $\delta = 0, 1$; $j = 1, \ldots, J$ are latent class parameters and b is an individual-specific standard Normal random effect. Under this latent class Gaussian random effects model,

$$P(x = l | \delta) = \left\{ \int \prod_j \Phi \left(\beta_{j\delta} + \sigma_\delta b \right)^{l_j} \left(1 - \Phi \left(\beta_{j\delta} + \sigma_\delta b \right) \right)^{1-l_j} \right\} \phi(b) db,$$

where $\phi(b)$ is the standard normal density. The sensitivity and specificity for test j under the latent class Gaussian random effects model are then

$$P(x_j = 1 | \delta = 1) = \Phi \left(\frac{\beta_{j1}}{(1 + \sigma_1^2)^{1/2}} \right)$$

and

$$P(x_j = 0 | \delta = 0) = 1 - \Phi \left(\frac{\beta_{j0}}{(1 + \sigma_0^2)^{1/2}} \right),$$

respectively.

The finite mixture model (Albert and Dodd, 2004) also uses the two-class structure as its basis and adds two point masses for the combinations of all-zero and all-one responses. These point masses correspond to the healthiest and the most severely diseased patients that are always classified correctly. Let t be an indicator that denotes correct classification. Specifically, let $t = 0$ if a healthy subject is always classified correctly (i.e., has the all-zero response pattern with J tests), $t = 1$ if a diseased subject is always classified correctly, and let $t = 2$ otherwise. Thus, subjects are either always classified correctly, when either $t = 0$ or $t = 1$, or a diagnostic error is possible when $t = 2$. Denote the probabilities for correctly classifying diseased and healthy subjects by $\eta_1 = P(t = 1)$ and $\eta_0 = P(t = 0)$, respectively. Let also $w_j(\delta_i)$ denote the probability of the jth test making a correct diagnosis when $t = 2$.

The finite mixture model of Albert and Dodd (2004) assumes that the test results x_j are independent Bernouilli random variables, conditional on the true disease status and the classification indicator. Thus,

$$P(x_j = 1 | \delta, t) = \begin{cases} w_j(1), & \text{if } \delta = 1, \text{ and } t = 2 \\ 1, & \text{if } \delta = 1, \text{ and } t = 1 \\ 1 - w_j(0), & \text{if } \delta = 0, \text{ and } t = 2 \\ 0, & \text{if } \delta = 0, \text{ and } t = 0. \end{cases}$$

Note that $P(x_j = 1 | \delta = 1, t = 0) = P(x_j = 1 | \delta = 0, t = 1) = 0$. The specificity and sensitivity of the jth test under the finite mixture model are then

$$P(x_j = 1 | \delta = 1) = \eta_1 + (1 - \eta_1) w_j(1)$$

and

$$P(x_j = 0 | \delta = 0) = \eta_0 + (1 - \eta_0) w_j(0),$$

respectively.

7.3   A Mixed Membership Approach to Estimating Diagnostic Error

The GoM model can be thought of as a different extension of latent class models where random effects are individual-specific grades of membership (Erosheva, 2005). The extended GoM mixture model combines individuals of mixed membership with those of full membership who have pre-determined response patterns. Although the extended mixture GoM model allows for an arbitrary choice of the number and the nature of deterministic response patterns, it is reasonable to assume two deterministic responses in the medical testing context. Analogous to the approach of Albert and Dodd (2004), we use two deterministic components in the extended mixture GoM model to allow for inclusion of some healthy and diseased individuals who have deterministic responses with the all-zero and all-one patterns, respectively. However, to model tests' diagnostic errors for other subjects in the population, our approach is to use the GoM model while the finite mixture model of Albert and Dodd (2004) relies on using the two-class latent class model to model diagnostic errors.

Next, we describe the GoM model before introducing the extended mixture GoM model and deriving a Bayesian estimation algorithm for the extension.

7.3.1 The Grade of Membership Model

As before, let $x = (x_1, x_2, \ldots, x_J)$ be a vector of dichotomous variables, where x_j takes on values $l_j \in \mathcal{L}_j = \{0, 1\}$, $j = 1, 2, \ldots, J$. Let K be the number of mixture components (extreme profiles) in the GoM model. To preserve generality, we will provide notation and estimation algorithms for an arbitrary value of K. However, in the medical testing context, we will assume $K = 2$ to be consistent with the existing literature.

Let $g = (g_1, g_2, \ldots, g_K)$ be a latent partial membership vector of K nonnegative random variables that sum to 1. In what follows, we use notation $p(\ )$ to refer to both probability density and probability mass functions. Each extreme profile is characterized by a vector of conditional response probabilities for manifest variables, given that the kth component of the partial membership vector is 1 and the others are zero, $\lambda_{kj} = p(x_j = 1|g_k = 1)$, $k = 1, 2, \ldots, K$; $j = 1, 2, \ldots, J$. Given partial membership vector $g \in [0, 1]^K$, the conditional distribution of manifest variable x_j is given by a convex combination of the extreme profiles' response probabilities, i.e., $p(x_j = 1|g) = \sum_{k=1}^{K} g_k \lambda_{kj}$, $j = 1, 2, \ldots, J$. Let us denote the distribution of g by $D(g)$. The local independence assumption states that manifest variables are conditionally independent, given the latent variables. Using this assumption and integrating out latent variable g, we obtain the marginal distribution for response pattern l in the form of a continuous mixture

$$p(x = l) = \int \prod_{j=1}^{J} \left(\sum_{k=1}^{K} g_k \lambda_{kj}^{l_j} (1 - \lambda_{kj})^{1-l_j} \right) dD(g), \ l \in \mathcal{X},$$

where $\mathcal{X} = \prod_{j=1}^{J} \mathcal{L}_j$ is the set of all possible outcomes for vector x.

The latent class representation of the GoM model leads naturally to a data augmentation approach (Tanner, 1996). Denote by $\mathbf{x}$ the matrix of observed responses x_{ij} for all subjects. Let $\boldsymbol{\lambda}$ denote the matrix of conditional response probabilities. Augment the observed data for each subject with realizations of the latent classification variables $z_i = (z_{i1}, \ldots, z_{iJ})$. Denote by $\mathbf{z}$ the matrix of latent classifications z_{ij}. Let $z_{ijk} = 1$, if $z_{ij} = k$ and $z_{ijk} = 0$ otherwise.

We assume the distribution of membership scores is Dirichlet with parameters α. The joint probability model for the parameters and augmented data is

$$p(\mathbf{x}, \mathbf{z}, \mathbf{g}, \boldsymbol{\lambda}, \alpha) = p(\boldsymbol{\lambda}, \alpha) \prod_{i=1}^{N} \left[p(z_i|g_i) p(x_i|\boldsymbol{\lambda}, z_i) \cdot Dir(g_i|\alpha) \right],$$

where

$$p(z_i|g_i) = \prod_{j=1}^{J} \prod_{k=1}^{K} g_{ik}^{z_{ijk}}, \quad p(x_i|\boldsymbol{\lambda}, z_i) = \prod_{j=1}^{J} \prod_{k=1}^{K} \left(\lambda_{kj}^{x_{ij}} (1 - \lambda_{kj})^{1-x_{ij}} \right)^{z_{ijk}}$$

and
$$Dir(g_i|\alpha) = \frac{\Gamma(\sum_k \alpha_k)}{\Gamma(\alpha_1)...\Gamma(\alpha_K)} g_{i1}^{\alpha_1 - 1} \cdots g_{iK}^{\alpha_K - 1}.$$

We assume the prior on extreme profile response probabilities λ is independent of the prior on the hyperparameters α. We further assume that the prior distribution of extreme profile response probabilities treats items and extreme profiles as independent, hence $p(\lambda, \alpha) = p(\alpha) \prod_{k=1}^{K} \prod_{j=1}^{J} p(\lambda_{kj})$. We assume the prior $p(\lambda_{kj})$ is $Beta(1,1)$. Estimation of the GoM model can be done via a Metropolis-Hastings within Gibbs algorithm as described by Erosheva (2003).

7.3.2   The Extended Mixture GoM Model

To define the extended mixture GoM model, we assume two patterns of deterministic responses that correspond to the all-zero and all-one test results. Similar to Albert and Dodd (2004), we introduce the classification indicator variable t. Let $t = 0$ for the healthiest individuals ($\delta = 0$) who are always classified correctly; let $t = 1$ for the sick individuals ($\delta = 1$) who are always classified correctly; and let $t = 2$ for the other individuals whose distribution of test results is given by the GoM model with parameters α, λ. Denote the respective weights for the multinomial distribution of t by $\theta = (\theta_0, \theta_1, \theta_2)$. The interpretation of θ_0 and θ_1 is similar to that of η_0 and η_1 in the finite mixture model of Albert and Dodd (2004); we are using a different notation symbol to emphasize that the values of those parameters will be different due to differences between the models.

Note that parameter estimation for the extended mixture GoM model would be identical to the estimation for the standard GoM model if we could modify the observed counts for the all-zero and all-one responses by subtracting the numbers of individuals who are always classified correctly. However, these numbers are typically unknown which means that we have to estimate weights of the deterministic components.

To derive the Markov chain Monte Carlo (MCMC) sampling algorithm for the extended mixture GoM model, we further augment data with individual classification indicators. Let N be the total number of individuals in the sample, and let $n_0^{(m)}$ and $n_1^{(m)}$ be the expected values of the all-zero cell count and the all-one cell count, respectively, for the mixed membership individuals (with $t = 2$) at the m-th iteration. Denote the number of individuals with at least one positive and at least one zero response in their response pattern by n_{mix}. The total number of individuals with $t = 2$ at the mth iteration is then $n_{GoM}^{(m)} = n_0^{(m)} + n_1^{(m)} + n_{mix}$. Let the prior distribution for weights θ be uniform on the simplex and update θ at the end of the posterior step with:

$$\theta_0^{(m+1)} = \theta_0^{(m)} + \frac{n_0^{(m)} - n_0^{(m+1)}}{N}, \quad \theta_1^{(m+1)} = \theta_1^{(m)} + \frac{n_1^{(m)} - n_1^{(m+1)}}{N},$$

and
$$\theta_2^{(m+1)} = \frac{n_0^{(m+1)} + n_1^{(m+1)} + n_{mix}}{N} = 1 - \theta_0^{(m+1)} - \theta_1^{(m+1)}.$$

Given the number of individuals subject to classification error, $n_{GoM}^{(m)}$, the estimation of model parameters for the stochastic GoM compartment is identical to that used in the case of the standard GoM model. We use a reparameterization of $\alpha = (\alpha_1, \ldots, \alpha_K)$ with $\xi = (\xi_1, \ldots, \xi_K)$ and α_0, which reflect proportions of the item responses that belong to each mixture category and the spread of the membership distribution. The closer α_0 is to zero, the more probability is concentrated near the mixture categories; similarly, the larger α_0 is, the more probability is concentrated near the population average membership score. We assume that α_0 and ξ are independent since they govern two unrelated qualities of the distribution of the GoM scores. In the absence of a strong prior opinion about hyperparameters α_0 and ξ, we take the prior distribution $p(\xi)$ to be uniform on the simplex and $p(\alpha_0)$ to be a proper diffuse gamma distribution. We also assume that the prior distribution on the GoM scores is independent of the prior distribution on the structural parameters. The joint distribution of the parameters and augmented data for the mixed membership component is

$$p(\lambda)p(\alpha_0)p(\xi)\left(\prod_{i=1}^{n_{GoM}}D(g_i|\alpha)\right)\prod_{i=1}^{n_{GoM}}\prod_{j=1}^{J}\prod_{k=1}^{K}\left(g_{ik}\lambda_{kj}^{x_{ij}}(1-\lambda_{kj})^{1-x_{ij}}\right)^{z_{ijk}},$$

where z_{ijk} is the latent class indicator as before. We sample from the posterior distribution of $\xi = (\xi_1,\ldots,\xi_K)$ and α_0 by using the Gibbs sampler with two Metropolis-Hastings steps (see Erosheva, 2003). The modified sampling algorithm for the extended mixture GoM model can be easily generalized to a number of deterministic response patterns greater than two.

7.3.3 Sensitivity and Specificity with the Extended Mixture GoM Model

Here we derive sensitivity and specificity estimates under the extended mixture GoM model. As before, let us denote the true diagnosis of a subject by $\delta = 1$ or $\delta = 0$ for the presence or absence of the disease, respectively. If patient i has the disease, then, under the extended mixture GoM model, this patient either belongs to the deterministic compartment with the clear positive diagnosis, $t = 1$, or they belongs to the stochastic compartment with the classification indicator $t = 2$. In terms of probability, this translates into $P(\delta = 1) = \theta_1 + \theta_2\xi_1$. As a consequence, the sensitivity of item j can be expressed as follows :

$$\begin{aligned}P(x_j = 1|\delta = 1) &= P(x_j = 1, t = 1|\delta = 1) + Pr(x_j = 1, t = 2|\delta = 1)\\ &= \left[P(t = 1) + P(x_j = 1, t = 2, \delta = 1)\right]/P(\delta = 1).\end{aligned}$$

Noticing that

$$P(x_j = 1, t = 2, \delta = 1) = P(x_j = 1|t = 2, \delta = 1)P(\delta = 1|t = 2)P(t = 2),$$

we obtain a parametric form for sensitivity of item j under the extended mixture GoM model:

$$P(x_j = 1|\delta = 1) = \frac{\theta_1}{\theta_1 + \theta_2\xi_1} + \lambda_{1j}\frac{\theta_2\xi_1}{\theta_1 + \theta_2\xi_1}.$$

Similarly, the absence of the disease for patient i means that either i belongs to the deterministic compartment, $t = 0$ of a clear negative diagnosis, or he/she belongs to the stochastic compartment with classification indicator $t = 2$. Therefore, $P(\delta = 0) = \theta_0 + \theta_2\xi_2$ and we obtain

$$\begin{aligned}P(x_j = 1|\delta = 0) &= P(x_j = 1, t = 0|\delta = 0) + P(x_j = 1, t = 2|\delta = 0)\\ &= P(x_j = 1, t = 0|\delta = 0) + P(x_j = 1, t = 2, \delta = 0)/P(\delta = 0).\end{aligned}$$

Because $P(x_j = 1, t = 0|\delta = 0) = 0$, we have

$$P(x_j = 1|\delta = 0) = \lambda_{2j}\frac{\theta_2\xi_2}{\theta_0 + \theta_2\xi_2},$$

and the specificity estimate of item j under the extended mixture GoM model can be obtained as $1 - P(x_j = 1|\delta = 0)$.

7.4 Simulation Study

In this section, we present a simulation study with the primary aim to examine recovery of sensitivity and specificity parameters under the four different latent structure models: the latent class (Lazarsfeld and Henry, 1968; Goodman, 1974), the latent class random effects (Qu et al., 1996), the finite mixture (Albert and Dodd, 2004), and the extended mixture GoM models introduced earlier. We investigate performance of each of these models when the true model is known, varying the true model among the four alternatives under two sample sizes, $N = 1000$ and $N = 4000$. We also report the comparative fit of the models in each case, however, model fit was not a primary goal of our study. Earlier work demonstrated difficulties in distinguishing between models with different dependence structures (Albert and Dodd, 2004), and pointed out that even equally well-fitting models may result in different accuracy estimates (Begg and Metz, 1990).

For the simulations we considered $J = 6$ and set the true specificities and sensitivities to be the same for all 6 items. Specifically, we used the value of 0.9 for the sensitivity parameter and 0.95 for the specificity. This setting allowed us to examine the recovery of accuracy parameters for a given model by simply computing the respective average sensitivity and specificity estimates across all items. and the hyperparameter was set at $\alpha_0 = 0.25$

We selected data generating designs under each model to reflect important features of biomedical screening and diagnostic data. Most noticeably, contingency tables formed on the basis of this type of data often contain many zeros and small observed cell counts but also have several large cell counts. The large observed cell counts typically include the all-zero and the all-one response patterns. The following parameter choices produced simulated data with many zeros and large all-zero and all-one counts and items with 0.9 sensitivity and 0.95 specificity:

1. For data generated under the latent class model, we chose: $\tau = 0.1$ and $\lambda_{1j} = 0.9$, $\lambda_{2j} = 0.05$, for all j.

2. For data generated under the latent class random effects model, we chose: $\sigma_0 = \sigma_1 = 1.5$, $\tau = 0.1$, and $\beta_{j0} = -2.965$, $\beta_{j1} = 2.31$, for all j.

3. For data generated under the finite mixture model, we chose: $\tau = 0.1$, $\eta_0 = 0.2$, $\eta_1 = 0.5$, and $w_j(0) = 0.9375$, $w_j(1) = 0.8$, for all j.

4. For data generated under the extended mixture Grade of Membership model, we chose the following parameter values: $\theta = (0.85, 0.05, 0.10)$, $\alpha = (0.02, 0.06)$, and $\lambda_{1j} = 0.7$, $\lambda_{2j} = 0.6$, for all j.

Among the four models, the latent class is the least complex with 13 independent parameters; the latent class random effects and the finite mixture models both have 15 independent parameters, and the extended mixture GoM model is the most complex with 16 independent parameters. We used BUGS (Bayesian inference using Gibbs sampling) to estimate the latent class, latent class random effects, and finite mixture models. We used a C code for estimation of the extended mixture GoM model.

Tables 7.1–7.4 report posterior means and standard errors of the sensitivity and specificity parameters, averaged over the six items for each model value of the log-likelihood, as well as goodness-of-fit criteria. We report the log-likelihood, the G^2 likelihood ratio criteria (Bishop et al., 1975), and the truncated sum of squared Pearson residuals (SSPR) χ^2 (Erosheva et al., 2007) computed for observed counts larger than 1 (i.e., the sum did not include residuals for the cells with zero observed counts). The log-likelihood and the goodness-of-fit criteria were evaluated as the posterior means of the parameters for each model.

TABLE 7.1

Results for the LCM generating model.

Criterion	LCM	LCRE	FM	ExtM-GoM
N=1000				
SSPR χ^2_{tr}	23.44	22.68	18.47	22.14
G^2	30.24	31.11	34.59	54.54
Log-likelihood	-1528.19	-1528.62	-1530.36	-1545.39
Sensitivity	0.894 (0.032)	0.890 (0.033)	0.893 (0.032)	0.910 (0.029)
Specificity	0.951 (0.007)	0.951 (0.007)	0.950 (0.007)	0.953 (0.006)
N=4000				
Criterion	LCM	LCRE	FM	ExtM-GoM
SSPR χ^2_{tr}	49.42	44.61	40.90	178.23
G^2	55.73	61.70	56.19	280.96
Log-likelihood	-6418.15	-6421.13	-6418.38	-6525.79
Sensitivity	0.902 (0.015)	0.895 (0.016)	0.901 (0.016)	0.913 (0.009)
Specificity	0.948 (0.004)	0.949 (0.004)	0.948 (0.004)	0.953 (0.005)

TABLE 7.2

Results for the LCRE generating model.

Criterion	LCM	LCRE	FM	ExtM-GoM
N=1000				
SSPR χ^2_{tr}	308.75	54.98	36.45	34.55
G^2	232.28	62.60	62.02	59.57
Log-likelihood	-1301.07	-1216.24	-1215.95	-1221.95
Sensitivity	0.838 (0.031)	0.886 (0.040)	0.878 (0.039)	0.905 (0.010)
Specificity	0.969 (0.005)	0.958 (0.009)	0.971 (0.007)	0.976 (0.004)
N=4000				
Criterion	LCM	LCRE	FM	ExtM-GoM
SSPR χ^2_{tr}	710.79	60.92	65.74	61.13
G^2	514.76	73.25	76.32	62.11
Log-likelihood	-5283.37	-5062.62	-5064.15	-5069.51
Sensitivity	0.868 (0.016)	0.938 (0.013)	0.876 (0.019)	0.936 (0.008)
Specificity	0.973 (0.003)	0.955 (0.005)	0.970 (0.003)	0.976 (0.005)

TABLE 7.3
Results for the FM generating model.

	N=1000			
Criterion	LCM	LCRE	FM	ExtM-GoM
SSPR χ^2_{tr}	159.19	55.07	43.91	43.40
G^2	100.70	71.37	75.01	80.61
Log-likelihood	-1517.72	-1503.05	-1504.87	-1513.80
Sensitivity	0.928 (0.027)	0.898 (0.053)	0.907(0.034)	0.915 (0.021)
Specificity	0.949 (0.007)	0.953 (0.008)	0.950 (0.008)	0.954 (0.006)
	N=4000			
Criterion	LCM	LCRE	FM	ExtM-GoM
SSPR χ^2_{tr}	109.61	44.53	36.38	93.08
G^2	91.77	53.30	43.89	108.38
Log-likelihood	-5287.40	-5268.16	-5263.46	-5292.61
Sensitivity	0.838 (0.019)	0.841 (0.023)	0.842 (0.020)	0.864 (0.011)
Specificity	0.969 (0.003)	0.968 (0.003)	0.968 (0.003)	0.969 (0.002)

TABLE 7.4
Results for the extended mixture GoM generating model.

	N=1000			
Criterion	LCM	LCRE	FM	ExtM-GoM
SSPR χ^2_{tr}	174.34	48.77	30.94	35.70
G^2	138.55	80.85	66.67	62.61
Log-likelihood	-887.45	-858.59	-851.50	-854.46
Sensitivity	0.791 (0.033)	0.843 (0.050)	0.857 (0.067)	0.835 (0.023)
Specificity	0.998 (0.002)	0.998 (0.004)	0.974 (0.014)	0.979(0.002)
	N=4000			
Criterion	LCM	LCRE	FM	ExtM-GoM
SSPR χ^2_{tr}	394.06	193.19	42.32	41.57
G^2	328.31	270.28	53.14	51.48
Log-likelihood	-3463.84	-3443.51	-3326.26	-3330.02
Sensitivity	0.772 (0.018)	0.874 (0.019)	0.789 (0.024)	0.858 (0.006)
Specificity	0.999 (0.001)	0.969 (0.003)	0.991 (0.005)	0.962 (0.002)

A first remark concerning the simulation results is that the fit criteria do not always favor a generating model. For example, perhaps not surprisingly, the truncated sum of squares of Pearson's residuals tends to be better for the finite mixture and the extended mixture GoM model. These models provide perfect fit for the two largest observed counts, even when they are not the true data-generating models. In general, however, we observe that different latent structure models can produce similar fit; this finding confirms that it can be difficult to distinguish between models with different dependence structures (Albert and Dodd, 2004).

Examining bias for the sensitivity and specificity estimates, we observe that when the latent class model is used to generate the data, all of the models successfully recover the accuracy parameters for both cases, $N = 1000$ and $N = 4000$.

The success in recovery of the accuracy parameters is not that great when more complex models are used to generate the data. Thus, when the latent class random effects model generates the data, even the true model recovers only the specificity parameter, but not the sensitivity parameter. For the latent class random effects as the generating model, the extended mixture GoM does best in recovering the sensitivity for the smaller sample size, but the finite mixture model does best in recovering the sensitivity for the larger sample size. When the finite mixture model is the generating model, all models perform well in recovering the sensitivity and specificity for the $N = 1000$ case, however, all models perform poorly for the $N = 4000$ case. In the latter scenario, we see that all the models considered, including the true model, underestimate the sensitivity and overestimate the specificity parameters. When the extended mixture GoM model is the generating model, we observe that the latent class random effects model does better with recovering the true value of the sensitivity parameter, even though the fit of this model is not as good compared to others. We also observe that all models tend to overestimate the specificity parameter for data generated with the extended mixture GoM model.

Finally, we observe that in the most difficult cases, the sensitivity parameter was underestimated to various degree by all of the models and the specificity—overestimated by all of the models. In such cases, models that provide larger sensitivity estimates and smaller specificity estimates could be considered especially informative in a sensitivity analysis with respect to latent structure assumptions. We also note that the sensitivity and especially specificity estimates under the extended mixture GoM model had smaller standard errors as compared to those of the other models, independently of the true model.

7.5 Analysis of *Chlamydia trachomatis* Data

In this section, we provide sensitivity analysis for a dataset on testing for *Chlamydia trachomatis*, originally considered by Hadgu and Qu (1998). *Chlamydia trachomatis* (CT) is the most common sexually transmitted bacterial infection in the U.S. The data contain binary outcomes on $J = 6$ tests for 4,583 women where positive responses correspond to detection of the disease. The six tests are Syva-DFA, Syva-EIA, Abbott-EIA, GenProbe, Sanofi-EIA, and a culture test (for more information, see Hadgu and Qu, 1998). Table 7.5 provides response patterns with positive observed counts. In our analyses, we follow Hadgu and Qu (1998), who only retained complete response patterns and assumed that patterns with zero observed counts were sampling and not structural zeros.

We assumed two basis latent classes and analyzed the CT data with the latent class, the latent class random effects, the finite mixture, and the extended mixture GoM models. For comparative purposes, we also provide results obtained with the GoM model, although we did not expect it to perform well based on our earlier experience with disability data (Erosheva et al., 2007).

To obtain draws from the joint posterior distribution, we used the same prior and proposal distributions for the standard GoM as well as for the extended mixture GoM models. We chose Gamma as the prior distribution on α_0, with the shape and the inverse scale parameter equal to 1, and chose the uniform prior for ξ-parameters. We chose the shape parameter for the proposal distribution on α_0 to be equal to $C_1 = 100$, and the sum of the parameters of the proposal distribution for ξ to be equal to $C_2 = 1$. We set starting values for λ to the estimated conditional response probabilities from the latent class model with two classes, and selected the starting value for hyperparameters α to be $(0.001, 0.099)$.

We monitored the convergence of MCMC chains using Geweke convergence diagnostics (Geweke, 1992), and Heidelberger and Welch stationarity and interval halfwidth tests (Heidelberger and Welch, 1983). Furthermore, we visually examined plots of successive iterations. With our choices of starting values and parameters for prior and proposal distributions, all these methods indicated favorable convergence of MCMC chains for both the standard and the extended mixture GoM models.

Under the extended mixture GoM model, the estimated proportion θ_0 of women belonging to the deterministic compartment of a clear negative diagnosis was $\hat{\theta}_0 = 0.917$ (sd = 0.009). The observed probability of all-zero response was 0.944. The estimated proportion θ_1 of women belonging to the deterministic compartment of a clear positive diagnosis was $\hat{\theta}_1 = 0.017$ (sd=0.009), while the observed probability of all-one response was 0.019. Thus, consistently with the idea of screening tests in medicine, about 97% of individuals with negative results on all six tests were healthy with probability 1 while about 89% of individual with positive results on all six tests were diseased with probability 1. Table 7.6 shows the posterior means and standard deviation estimates of parameters λ_{kj}, ξ, α_0 for the GoM portion of the extended mixture GoM model. We observe that the extreme profile $k = 1$ represents women with likely positive CT diagnosis, while extreme profile $k = 2$ represents women with likely negative CT diagnosis.

Table 7.5 shows the observed and expected cell counts as well as the number of parameters, the degrees of freedom, and the values of the likelihood ratio statistic G^2 (see Bishop et al., 1975) for all five models considered. As expected, we observe that the fits of the standard GoM model and the latent class model are rather poor. The extended mixture GoM model provides a comparable performance in fit compared to the latent class random effects model of Hadgu and Qu (1998) and the finite mixture model of Albert and Dodd (2004), however, it has one less degree of freedom.

Table 7.7 provides the sensitivity and specificity estimates for the six tests under the five models considered. We see that the sensitivity and specificity estimates are rather high. Given our findings from the simulation study that pointed towards widespread overestimation of specificity and underestimation of sensitivity, we should pay particular attention to the smallest specificity and largest sensitivity estimates. For specificity, the extended mixture GoM model and the finite mixture model provide the smallest values that range from about 0.993 for the culture test to about 0.997 for the Syva-DFA test. For sensitivity, the extended mixture GoM model provides the largest values ranging from about 0.757 for the Abbott-EIA test to about 0.985 for the culture test. Thus, it appears that the extended mixture GoM model contributes new information beyond that available from other latent structure models, for estimating accuracy parameters of diagnostic tests.

TABLE 7.5

Observed and expected cell counts under LCM, LCRE, GoM, ExtM-GoM, and FM models. χ^2 is the truncated sum of squared Pearson residuals for observed cell counts > 1; n is the number of independent parameters; df is degrees of freedom.

	Response pattern	Observed counts	Expected LCM	Expected LCRE	Expected GoM	Expected ExtM-GoM	Expected FM
1	111111	87	53.4	87.30	437.31	87	87.18
2	111110	2	0.85	0.12	11.53	0.50	0.40
3	111101	9	17.26	8.19	146.20	7.64	7.07
4	111011	2	8.47	2.93	73.43	2.60	2.64
5	110111	9	21.2	11.79	167.21	10.27	9.52
6	110101	6	6.85	5.03	56.78	6.98	6.68
7	110011	1	3.36	1.98	29.64	2.42	2.5
8	110001	6	1.09	2.09	10.27	1.91	1.85
9	101111	7	11.57	4.91	91.33	4.44	4.05
10	101001	1	0.59	0.93	6.93	0.97	0.87
11	100111	4	4.59	3.15	36.12	4.08	3.84
12	100101	1	1.48	3.26	13.57	3.08	2.80
13	100011	2	0.73	1.32	8.89	1.20	1.11
14	100001	5	0.27	3.15	4.17	2.09	1.87
15	100000	7	7.50	6.92	6.27	6.91	7.23
16	011111	6	10.58	4.14	93.3	4.02	3.72
17	011101	1	3.42	1.93	33.37	2.80	2.62
18	011011	1	1.68	0.76	15.54	1.00	1.00
19	011001	1	0.54	0.81	5.88	0.97	0.87
20	011000	5	0.07	4.96	1.13	1.26	1.22
21	010111	1	4.20	2.70	36.34	3.72	3.53
22	010101	5	1.36	2.81	13.30	2.89	2.64
23	010001	2	0.28	2.71	3.93	2.59	2.41
24	010000	9	13.56	8.92	7.29	10.34	10.91
25	001111	2	2.29	1.21	20.76	1.65	1.51
26	001101	1	0.74	1.25	9.46	1.47	1.28
27	001001	1	0.20	1.18	2.78	2.54	2.49
28	001000	14	18.46	13.88	10.24	12.72	13.64
29	000111	2	0.91	1.77	9.53	1.82	1.61
30	000101	4	0.37	4.25	5.12	3.39	3.12
31	000100	16	16.54	15.82	11.46	12.45	13.15
32	000011	3	0.22	1.70	3.33	2.43	2.31
33	000010	15	15.73	14.85	9.84	11.40	11.98
34	000001	17	20.02	17.04	11.45	19.50	20.50
35	000000	4328	4321.11	4328.30	3123.48	4328	4322.71
χ^2			604.95	48.61	1529.89	43.29	48.70
G^2			179.68	42.25	1141.70	61.44	76.24
n			13	15	14	16	15
df			50	48	49	47	48

TABLE 7.6
Posterior mean (standard deviation) estimates for the diagnostic error component of the extended mixture GoM model with two extreme profiles (stochastic compartment).

	$k = 1$	$k = 2$
$\lambda_{k,1}$	0.750 (0.058)	0.050 (0.022)
$\lambda_{k,2}$	0.732 (0.058)	0.076 (0.027)
$\lambda_{k,3}$	0.534 (0.065)	0.093 (0.027)
$\lambda_{k,4}$	0.822 (0.057)	0.089 (0.029)
$\lambda_{k,5}$	0.608 (0.068)	0.084 (0.026)
$\lambda_{k,6}$	0.970 (0.021)	0.134 (0.042)
α_0	0.075 (0.057)	
ξ_k	0.278 (0.044)	0.722 (0.044)

TABLE 7.7
Sensitivity and specificity (standard deviation) of the diagnostic tests for LCM, LCRE, GoM, ExtM-GoM, and FM models.

Test	LCM	LCRE	GoM	ExtM-GoM	FM
Specificity					
Syva-DFA	0.998 (0.001)	0.999 (0.001)	0.999 (0.001)	0.998 (0.001)	0.997 (0.001)
Syva-EIA	0.997 (0.001)	0.997 (0.001)	0.999 (0.001)	0.996 (0.001)	0.996 (0.001)
Abbott-EIA	0.996 (0.001)	0.997 (0.001)	0.998 (0.001)	0.995 (0.001)	0.995 (0.001)
GenProbe	0.996 (0.001)	0.997 (0.001)	0.998 (0.001)	0.996 (0.001)	0.995 (0.001)
Sanofi-EIA	0.996 (0.001)	0.997 (0.001)	0.997 (0.001)	0.996 (0.001)	0.996 (0.001)
Culture	0.995 (0.001)	0.998 (0.001)	0.998 (0.001)	0.994 (0.001)	0.993 (0.001)
Sensitivity					
Syva-DFA	0.835 (0.030)	0.747 (0.048)	0.825 (0.034)	0.870 (0.030)	0.860 (0.030)
Syva-EIA	0.822 (0.031)	0.731 (0.048)	0.824 (0.034)	0.861 (0.030)	0.851 (0.032)
Abbott-EIA	0.716 (0.036)	0.636 (0.047)	0.721 (0.037)	0.757 (0.035)	0.748 (0.037)
GenProbe	0.863 (0.028)	0.777 (0.047)	0.864 (0.031)	0.907 (0.030)	0.892 (0.029)
Sanofi-EIA	0.756 (0.034)	0.678 (0.048)	0.751 (0.037)	0.796 (0.036)	0.786 (0.036)
Culture	0.984 (0.011)	0.935 (0.034)	0.978 (0.013)	0.985 (0.010)	0.980 (0.011)

7.6 Conclusion

We presented the extended mixture GoM model as an alternative latent structure model for analyzing accuracy of diagnostic and screening tests. For the medical testing case, the extended mixture GoM model accommodates two types of individuals, those whose diagnosis is certain, independently of the test, and those who are subject to diagnostic error. The latter individuals can be thought of as stochastic "movers" because they may change in their disease status depending on the diagnostic test considered; this idea is analogous to longitudinal mover-stayer models (Blumen et al., 1955). In the extended mixture GoM model, the "stayers" have predetermined response patterns that correspond to particular cells in a contingency table. The extended GoM mixture model can also be seen as a combination of latent class and GoM mixture modeling, analogous to the extended finite mixture model (Muthen and Shedden, 1999). Similar to the finite mixture model of Albert and Dodd (2004), the extended GoM mixture model allows for some individuals to always be diagnosed correctly, however, it relies on the assumption of disease severity for modeling diagnostic error for the other individuals while the finite mixture model of Erosheva (2005) relies on the assumption of two latent classes for the same purpose.

To estimate the extended mixture GoM model, we used a hierarchical Bayesian approach to estimation, drawing on earlier work (Erosheva, 2003). Although we did not use informative priors in our examples, Pfeiffer and Castle (2005) and Albert and Dodd (2004) specifically mention the promise of Bayesian approach in estimating diagnostic error without a gold standard when good prior information is available.

Our findings with the simulation study and with the *Chlamydia trachomatis* (Hadgu and Qu, 1998) data further emphasize the need of carrying out sensitivity analyses when no gold standard is available (Albert and Dodd, 2004) . When the underlying latent structure is unknown, it is important to examine sensitivity of scientific conclusions regarding the estimated accuracy of diagnostic and screening tests to the latent structure assumptions.

This could be done by comparing test accuracy estimates across a number of different latent structure models. Our results demonstrate that the extended mixture GoM model can provide us with new information on estimating test sensitivity and specificity beyond that provided by the latent class (Lazarsfeld and Henry, 1968; Goodman, 1974), latent class Gaussian random effects (Qu et al., 1996), and finite mixture models (Albert and Dodd, 2004). In addition, the extended mixture GoM model offers a plausible interpretation for diagnostic and screening test results by combining the idea of diagnostic error that depends on disease severity for some individuals with the idea of certain diagnosis for the other, typically the most healthy and the most sick individuals.

Finally, the flexible framework of mixed membership models can be used to modify the extended mixture GoM model and to address, for example, diagnostic cases when disease status is ordinal and test results do not come in binary form (Wang and Zhou, 2012). Drawing on the recent development in the class of mixed membership models that includes the GoM model as a special case, one can also modify the model to accommodate, for example, outcomes of mixed types, multiple basis categories in the latent structure, and correlations among membership scores.

References

Albert, P. S. (2007a). Imputation approaches for estimating diagnostic accuracy for multiple tests from partially verified designs. *Biometrics* 63: 947–957.

Albert, P. S. (2007b). Random effects modeling approaches for estimating ROC curves from repeated ordinal tests without a gold standard. *Biometrics* 63: 593–602.

Albert, P. S. and Dodd, L. E. (2004). A cautionary note on the robustness of latent class models for estimating diagnostic error without a gold standard. *Biometrics* 60: 427–435.

Albert, P. S. and Dodd, L. E. (2008). On estimating diagnostic accuracy from studies with multiple raters and partial gold standard evaluation. *Journal of the American Statistical Association* 103: 61–73.

Begg, C. B. and Metz, C. E. (1990). Consensus diagnoses and "gold standards." *Medical Decision Making* 10: 29–30.

Bishop, Y. M. M., Fienberg, S. E., and Holland, P. W. (1975). *Discrete Multivariate Analysis: Theory and Practice*. Cambridge, MA: The MIT press.

Blumen, J., Kogan, M., and Holland, P. W. (1955). *The Industrial Mobility of Labor as a Probability Process*. Cornell Studies of Industrial and Labor Relations 6. Ithaca, NY: Cornell Univesity Press.

Erosheva, E. A. (2003). Bayesian estimation of the Grade of Membership model. In Bernardo, J. M., Bayarri, M. J., Berger, J. O., Dawid, A. P., Heckerman, D., Smith, A. F. M., and West, M., (eds), *Bayesian Statistics 7*. New York, NY: Oxford University Press, 501–510.

Erosheva, E. A. (2005). Comparing latent structures of the Grade of Membership, Rasch, and latent class models. *Psychometrika* 70: 619–628.

Erosheva, E. A., Fienberg, S. E., and Joutard, C. (2007). Describing disability through individual-level mixture models for multivariate binary data. *Annals of Applied Statistics* 1: 502–537.

Geweke, J. (1992). Evaluating the accuracy of sampling-based approaches to the calculation of posterior moments. In Bernardo, J. M., Bayarri, M. J., Berger, J. O., Dawid, A. P., Heckerman, D., Smith, A. F. M., and West, M. (eds), *Bayesian Statistics 4*. Oxford, UK: Oxford University Press, 162–193.

Goodman, L. A. (1974). Exploratory latent structure analysis using both identifiable and unidentifiable models. *Biometrika* 61: 215–231.

Hadgu, Y. and Qu, A. (1998). A biomedical application of latent class model with random effects. *Journal of the Royal Statistical Society (Series C): Applied Statistics* 47: 603–613.

Heidelberger, P. and Welch, P. (1983). Simulation run length control in the presence of an initial transient. *Operations Research* 31: 1109–1144.

Hui, S. L. and Zhou, X. H. (1998). Evaluation of diagnostic tests without gold standards. *Statistical Methods in Medical Research* 7: 354–370.

Lazarsfeld, P. F. and Henry, N. W. (1968). *Latent Structure Analysis*. Boston, MA: Houghton Mifflin.

Muthén, B. and Shedden, K. (1999). Finite mixture modeling with mixture outcomes using the EM algorithm. *Biometrics* 55: 463–469.

Pepe, M. S. and Janes, H. (2007). Insights into latent class analysis of diagnostic test performance. *Biostatistics* 8: 474–484.

Pfeiffer, R. M. and Castle, P. E. (2005). With or without a gold standard. *Epidemiology* 16: 595–597.

Qu, Y., Tan, M., and Kutner, M. H. (1996). Random effects models in latent class analysis for evaluating accuracy of diagnostic tests. *Biometrics* 55: 258–263.

Tanner, M. A. (1996). *Tools for Statistical Inference. Methods for the Exploration of Posterior Distributions and Likelihood Functions*. New York, NY: Springer, 3rd edition.

Wang, Z. and Zhou, X. (2012). Random effects models for assessing diagnostic accuracy of traditional chinese doctors in absence of a gold standard. *Statistics in Medicine* 31: 661–671.

8

Interpretability Constraints and Trade-offs in Using Mixed Membership Models

Burton H. Singer

Emerging Pathogens Institute, University of Florida, Gainesville, FL 32610, USA

Marcia C. Castro

Harvard School of Public Health, Boston, MA 02115, USA

CONTENTS

Although shared membership of individuals in two or more categories of a classification scheme is a distinguishing feature of the family of mixed membership models, relatively few analyses using these models pay much attention to this special feature. Most published analyses to-date focus on identifying and interpreting the extreme, or ideal, types consistent with a given body of data, thereby in effect using mixed membership models as crisp clustering techniques. Getting into the domain of shared membership quickly places the investigator in a difficult position, as standard estimation strategies produce a large number of ideal profiles, almost always greater than six, that represent best fitting representations of the data, while at the same time making it impossible to interpret what membership in, say, four or more profiles actually means. This conflict between statistical goodness-of-fit and subject-matter-based interpretability of shared membership cannot usually be resolved using conventional mixed membership models. We show that by introducing separate mixed membership models, each containing a small number of ideal profiles, to describe a population according to responses focused on distinct subject matter domains, and at the same time producing a vector of correlated grade of membership scores for the individuals, interpretation of shared memberships across the distinct subject matter domains becomes feasible. Deciding on what constitutes a good model requires tradeoffs between statistical goodness-of-fit criteria and frequently non-quantifiable subject-matter-based interpretation. We illustrate these unavoidable tradeoffs in several epidemiological contexts.

8.1 Introduction

Mixed membership models are ideally suited for characterizing heterogeneous populations where many individuals have multiple characteristics of interest, no combination of which occur at high frequency in an overall population. Examples of this phenomena are: (i) representation of the health status of elderly populations at the community level (Berkman et al., 1989); (ii) prevalence of, and variation in, co-infection with tropical diseases at the district level (Keiser et al., 2002; Raso et al., 2006); (iii) characterization of environmental and behavioral risks for Chagas disease in Latin America (Chuit et al., 2001); (iv) representation of malaria risk in complex eco-systems (e.g., the Brazilian Amazon) (Castro et al., 2006); and (v) represention of variation in disability among the elderly in the United States (Manton et al., 2006). Although mixed membership models have been fit to data from the above and other settings, very little attention has been given to the severe lack of interpretability of 'shared membership' of characteristics of individuals among a family of extreme, or ideal, types. Indeed, most published analyses to-date using this class of models focus on inter-pretation of ideal type profiles, and neglect the more complex stories inherent in the phenomenon of shared membership. This is tantamount to using a mixed membership model as a crisp cluster-ing methodology and bypassing the analysis of shared membership, which is the key distinguishing feature of the technology.

Standard estimation strategies for mixed membership models, if used in an exploratory mode, produce a number K of ideal profiles that lead to the best fitting and parsimonious representations of the multidimensional data. The optimum value of K can be defined based on a measure of relative goodness-of-fit of models run for several K values (Manton et al., 1994; Airoldi et al., 2010; White et al., 2012; Suleman, 2013). However, interpreting these profiles can be rather complex. Assuming K ideal profiles, each individual is assigned a grade of membership (GoM) vector $\mathbf{g}$ that corresponds to coordinates in a simplex with K vertices, each of which represents an ideal, or extreme, type of individual. For example, in the case of two profiles ($K = 2$), non-zero entries in $\mathbf{g}$ define the individuals that belong to the k_1 profile, while those equal to zero determine who does not belong to the same k_1 profile, but to k_2 instead. These cases can be understood as vertices (endpoints) of a line. However, some elements do not lie on the vertices but on the line connecting them. They share characteristics of both profiles. If, for example, the vertices correspond to 'very high risk' and 'very low risk', respectively, then $\mathbf{g} = (g_1, 1 - g_1)$ with $0 < g_1 < 1$ identifies an individual with a degree of risk that is intermediate between the extremes corresponding to the vertices. In the case of three profiles ($K = 3$), however, the geographical representation moves from a line to a triangle, and individuals with shared membership will lie on either an edge of the triangle or in the interior. Individuals on edges share conditions with only two vertices, and degree of similarity to one or the other of them provides for straightforward interpretation. However, individuals in the interior of the triangle share conditions with all three vertices, and writing a coherent English sentence describing the shared conditions becomes a more substantial challenge.

More formally, the number of non-zero entries in $\mathbf{g}$ is the number of vertices with one or more associated conditions that represent the state of a given individual. For a given K, there are $\binom{K}{2}$ edges of the simplex that represent shared conditions among two vertices. There are $\binom{K}{3}$ faces of the simplex that represent shared conditions among three vertices, and a $\mathbf{g}$-vector with all non-zero entries corresponding to an individual who shares conditions with all K vertices. The central problem associated with values of $K \geq 4$ is writing an interpretable description of what it means to share conditions with four or more ideal types. This, of course, is not a statistical problem, but it presents a challenge to investigators that, to the best of our knowledge, has received almost no attention in the extant literature on mixed membership models.

This paper addresses this issue. Our purpose is to present examples of problems where mixed membership modeling with non-standard model specifications and/or vertex-edge-face aggregation schemes facilitates interpretability of shared conditions. Section 8.2 contains specifications of the mixed membership model used in Sections 8.3 and 8.4. In particular, a non-standard specification is introduced where the original response vector is partitioned into blocks of variables associated with distinct subject-matter domains and a mixed membership model is estimated for each domain, thereby generating a set of GoM vectors for each individual, one vector for each domain in the partitioning. This facilitates interpretability in characterizations of disease risk, as we indicate in the next section. In Section 8.3 we discuss an example of characterization of community risk of Chagas disease in rural Argentina, where the original response vector is partitioned into components focused on indices of blood availability and environmental characteristics. Thus, GoM score vectors are generated for each of these domains and interpreted in the context of overall community risk of transmission of Chagas disease. In Section 8.4, we illustrate a standard mixed membership model with 6 pure types, and introduce a vertex-edge aggregation strategy in a study of changes in disability over time in the U.S. population during the period 1982–1994. The aggregation scheme facilitates interpretation of shared membership in a setting where model complexity could, in principle, impair our ability to describe subtleties in the data. We conclude with a brief discussion of open methodological problems that are an outgrowth of our examples.

8.2 Mixed Membership Model Specifications

Let $X = (X_1, \ldots, X_J)$ be a vector whose components are variables which can each assume a finite number of possible values. We consider the data analytic task of mapping response vectors $\{X^{(i)}, 1 \leq i \leq N\}$ for N individuals into a unit simplex with K vertices (to be estimated) and a GoM vector, $\mathbf{g}^{(i)} = (g_1, \ldots, g_K)^{(i)}$ with $g_1 + \ldots + g_K = 1$ for each individual, specifying a location in the unit simplex. Each vertex of the simplex is associated with a set of levels on a subset of the variables in X. Each set of such levels is interpreted as an ideal, or pure type, set of characteristics. GoM vectors that have all components equal to zero except one of them, say the kth—which must be a 1—identify individuals with response vectors having all of the conditions in the kth pure type. GoM vectors with two or more non-zero entries identify individuals whose response vectors share conditions with the pure types corresponding to the non-zero entries. They exhibit mixed membership across the K pure types, and provide the rationale for the terminology, 'mixed membership models.'

In the conventional version of mixed membership models (Manton et al., 1994; Erosheva et al., 2007), we assume that the variables in the response vector X are independent, conditional on the GoM score vector $\mathbf{g}$. More formally, we let $X^{(g)}$ denote the response vector for an individual with GoM score vector $\mathbf{g}$. Then we introduce the probability model (Singer, 1989):

$$Pr(X^{(g)} = \ell) = \int_{S_K} Pr(X^{(g)} = \ell | \mathbf{g} = \gamma) d\mu(\gamma) = \int_{S_K} \prod_{j=1}^{J} Pr(X_j^{(g)} = \ell_j | \mathbf{g} = \gamma) d\mu(\gamma), \quad (8.1)$$

where $\mu(\gamma)$ is the distribution of GoM scores and $S_K = \{\gamma = (\gamma_1, \ldots, \gamma_K) : \gamma_k \geq 0, \sum \gamma_k = 1\}$ is the unit simplex with K vertices. The conditional probabilities in Equation (8.1) can be written as

$$Pr(X^{(g)} = \ell_j | \mathbf{g} = \gamma) = \sum_{k=1}^{K} \gamma_k \lambda_{k,j,\ell_j},$$

where λ_{k,j,ℓ_j} (called pure type probabilities) can be defined as the probability in profile k of observing level ℓ_j on variable X_j. They are subject to the following constraints (Woodbury and Manton, 1982):

$$\sum_{\ell_j \in L_j} \lambda_{k,j,\ell_j} = 1 \text{ for } 1 \leq k \leq K \text{ and } 1 \leq j \leq J,$$

and L_j is the set of possible levels of variable X_j. Values of these probabilities close to 0 or 1 imply that a distinguished level is either almost certain to appear, or almost never to appear, in the kth pure type. The estimation problem for specification (8.1) is to find K, the associated pure type probabilities, and the individual GoM score vector $\mathbf{g}$ such that the model provides a good numerical fit to the data and is also interpretable. The words 'good fit' are associated with numerical goodness-of-fit criteria that are context free. However, the word 'interpretable' is not connected to statistical methodology, but is directly linked to our ability to describe the position of an individual in the simplex, i.e., describe the GoM score vector **g** in coherent English sentences that make sense in the scientific setting of a given dataset. The fundamental problem under discussion in this chapter is numerical identification of values of K which can be anywhere from 4 to 60 or 70, depending on the dataset, and the inability of the investigator to write a coherent paragraph describing GoM vectors with four or more non-zero entries and the sharing of membership among many pure types.

Depending on the dataset, and with a large value of K associated with the best fitting model, we may find that nearly all individuals have GoM vectors that place them on a vertex or on an edge of the simplex, sharing conditions with at most 2 pure types. This is a relatively easy situation in which to provide interpretable descriptions of response vectors sharing conditions between the pure types. If 90% or more of the individuals share conditions with at most 3 pure types—i.e., they are, at worst, situated in a face of the simplex—interpretable summaries of the shared conditions can also frequently be produced. In a study of classification of scientific papers (Airoldi et al., 2010), mixed membership models with $K = 20$ were utilized on the basis of goodness-of-fit criteria, but nearly all papers in the classification exercise shared conditions with at most 5 pure types. Most of the papers with shared pure types involved only 2 or 3 such profiles. In the context of that study, even sharing of 3 or 4 pure types resulted in interpretable formulations of the shared conditions. However, GoM vectors with 5 non-zero entries seem to be close to an upper bound on shared condition interpretability. In Section 8.4, we will show an example with $K = 6$, but where a subject-matter-driven aggregation of edges in the unit simplex leads to interpretability of the set of all shared conditions among pairs of vertices.

In studies of disease risk, as illustrated in Section 8.3, the variables in X can frequently be partitioned into subsets, each of which is associated with a different domain of risk. This context also makes it desirable to use 2 pure type specifications corresponding to the extremes of high and low risk for each domain. What we frequently find with high-dimensional X is that 2 pure type models provide a poor fit to the data, but models with 3, 4, and 5 pure types do much better numerically, while paying a high price in losing gradation of risk interpretations of shared conditions. One route out of this dilemma is to change the conditioning structure of the mixed membership model and produce two or more sets of 2 pure type representations, one for each substantively defined risk domain. Then for the variables associated with each domain, we have high and low risk interpretations of 2 pure types and a separate GoM score vector for each individual, one vector for each domain. Essentially we are trading higher dimensionality in K, with a single GoM vector, for lower dimensionality in K (namely, $K = 2$), but with a set of correlated GoM vectors for each individual, one vector for each risk domain.

We describe a mixed membership model structure for a multiple domain specification in the simplest setting (two domains). Let $X = (X^{(1)}, X^{(2)})$ be a response vector with subsets of variables $X^{(i)}$ associated with the subject matter domains indexed by $i = 1, 2$. We introduce the pair of GoM vectors $\mathbf{g}^{(1)}$ and $\mathbf{g}^{(2)}$ and assume that the variables in each of $X^{(i)}$ for $i = 1, 2$ are independent, conditional on $\mathbf{g} = (\mathbf{g}^{(1)}, \mathbf{g}^{(2)})$. Then we set

$$
\begin{aligned}
Pr(X = 1|\mathbf{g} = \gamma) &= \prod_{i \in (1)} Pr(X_i^{(1)} = \ell_i^{(1)}|\mathbf{g} = \gamma) \prod_{i \in (2)} Pr(X_i^{(2)} = \ell_i^{(2)}|\mathbf{g} = \gamma), \\
&= \prod_{i \in (1)} \sum_{k \in K(1)} \gamma_k^{(1)} \lambda_{k,i,\ell_i}^{(1)} \prod_{i \in (2)} \sum_{k \in K(2)} \gamma_k^{(2)} \lambda_{k,i,\ell_i}^{(2)}.
\end{aligned}
\tag{8.2}
$$

Here, $\mathbf{g} = (\mathbf{g}^{(1)}, \mathbf{g}^{(2)})$, $K(j) =$ Number of pure types in group(j), $j = 1, 2$. In the context of the risk profiles mentioned above, we would impose $K(1) = K(2) = 2$. In Section 8.3, we present an analysis of Chagas disease risk where the representation (8.2) plays a central role.

8.3 Chagas Disease Risk in Rural Argentina

Rural communities that are endemic for Chagas disease usually consist of privately owned habitats containing a primary house and peridomestic structures that serve as animal pens/housing, crop storage areas, and tool sheds and storage areas for agricultural equipment. The physical characteristics of houses and peridomestic structures are highly variable, thereby creating great heterogeneity within a community in sources of blood meals for triatomine bugs. Dogs, cats, chickens, and young children are all part of the transmission system, and their physical proximity during the night is an important factor in attracting *Triatoma infestans* vectors, the primary transmitters of *Trypanosoma cruzi* parasites that are the causative agents of Chagas disease.

We consider a community with 445 habitats in rural Santiago del Estero, Argentina (Paulone et al., 1991), where at baseline, 99.6% (443/445) habitats were infested with *T. infestans* bugs. As part of the initial data collection, *T. infestans* were collected from 390 (88%) of the houses and from the peridomestic structures of 280 (63%) habitats. A total of 6,518 *T. infestans* were captured in the 390 infested bedroom areas. Of these, 2,249 bugs were examined for *T. cruzi* parasites, and 697 (31%) were found positive. On the human side of the transmission system, 2,153 (69%) of the 3,194 persons in the community were serologically tested. The prevalence rate of seropositivity against *T. cruzi* infection was 29.2% (630/2153). Age specific seropositivity ranged from 9.6% in children under the age of 5 years to 57.7% in persons aged 70 or more. For the age group of children aged 5–14 the seropositivity rate was 25.3%.

Despite the high overall seropositivity rate, there were sets of habitats with very low rates and other sets with high rates in children in the age range 5–14. This is a useful age range for assessing Chagas disease incidence in the relatively recent past, particularly in a community where there has been no active control activity against infestation prior to the study by Paulone et al. (1991). Further, it is not at all obvious which habitats are at highest or lowest risk for Chagas disease transmission on the basis of a walking tour of the community, or even a verbal estimate from the community health officer. Our analytical problem is to identify the highest and lowest risk habitats in the midst of a highly heterogeneous endemic community, with the longer term objective of adapting the features of the low risk habitats on a wider scale as a hopefully low cost means of preventing transmission of *T. cruzi*.

To this end, a mixed membership modeling exercise using specification (8.2) was carried out using variables from two distinct domains: indices of blood availability and characteristics of the physical environment (Chuit et al., 2001). These are delineated in Table 8.1. For each domain a 2-pure type model (interpreted as levels of high and low risk) was fit to the data from the 445 habitats. A GoM score vector associated with each domain was generated for each habitat. Then, with only two profiles for each domain, the GoM vectors $\mathbf{g}^{(1)} = (g_1, 1 - g_1)$ and $\mathbf{g}^{(2)} = g_2, 1g_2)$ associated with each habitat have g_i indicating the degree of similarity of the habitat characteristics to the high risk profile for $i = 1$ (blood availability) and $i = 2$ (environmental characteristics). Using tertile cut points, for each risk domain (blood availability and environmental characteristics) we define a habitat to be low risk if $0 \leq g_i \leq 0.20$; intermediate risk if $0.20 < g_i \leq 0.70$; and high risk if $g_i > 0.70$. Cross classifying the habitats by their scores g_i, for $i = 1, 2$, Table 8.2 shows the breakdown of seropositivity rates for children in the age range 5–14 years.

TABLE 8.1
Response variables used in the mixed membership model.

Indices of Blood Availability	Conditions
number of dogs	= 2; > 2
number of cats	= 2: > 2
number of persons	= 2; > 2
persons/room	= 2; > 2
people/bed	= 2; > 2
people/[structures in the habitat]	= 2; > 2
persons + dogs + cats	= 5; 6 - 8; > 8
[persons + dogs + cats]/[room in the house]	= 6; >6
[persons + dogs + cats]/bed	= 4; > 4
[persons + dogs + cats]/[structure in the habitat]	= 3; > 3
Environmental Variables	**Conditions**
seasonal migration	No; Yes
condition of interior roof	Good (cement, zinc, fibrocement); Bad (straw, jarilla, discard)
condition of interior walls	Good (cement, lasterwall, no cracks); Bad (unplastered mud or brick with cracks)
condition of gallery roof	Good; Bad
number of rooms	= 1; > 1
number of beds	= 3; > 3
corn storage area	No; Yes
kitchen	No: Yes
equipment store room	No; Yes
corral	No; Yes
pig pen	No; Yes
brick pile	No; Yes

Source: Chuit et al. (2001).

TABLE 8.2
Seropositivity rates (%) by habitat risk for children aged 5–14.

Risk Domain		Environmental Characteristics		
	Risk level	Low	Medium	High
Blood Availability — Low	Low	18.8	16.2	21.4
Blood Availability — Medium	Medium	20.5	22.0	9.0
Blood Availability — High	High	19.3	22.8	25.0

Estimation in the GoM model (8.2) was carried out by estimating the GoM score vectors and pure type probabilities for each domain (environmental and blood availability) separately via GoM model (8.1) with $K = 2$ in each case. The two sets of GoM scores, $\mathbf{g}^{(1)}$ and $\mathbf{g}^{(2)}$, with a pair of such scores for each habitat, are—not surprisingly–correlated. Determination of the association between GoM scores begins with a scatter plot of (g_1, g_2) for the full set of habitats. Division of GoM scores into tertiles for purposes of classifying habitats by gradations of risk was the result of judgment by the investigators that such coarse graining led to qualitatively different categories of risk in both the environmental and blood availability domains. Quartile, quintile, or finer divisions could have been used, but these do not lead to meaningfully distinct risk categories. We emphasize here that this is not a statistically driven categorization. It is based on subject matter interpretations of meaningfully different levels of risk.

Turning to the profiles per se, the high risk profile for blood availability is represented by the logical AND statement: [more than 2 dogs] AND [6 persons or dogs or cats per room] AND [3 persons or dogs or cats per structure at the habitat]. Low risk is characterized by [None of the adverse conditions in the blood availability section of Table 8.1]. The high risk profile for physical environmental characteristics is given by: [poor interior roof] AND [poor gallery roof] AND [presence of a pig pen] AND [presence of a brick pile]. Low risk for environmental characteristics is characterized by no adverse house or peridomicile conditions from the full list in Table 8.1 (Chuit et al., 2001).

If the pattern of seropositivity rates corresponds to the risk levels for the habitats, Table 8.2 should be a double-gradient table in the sense that the rates should increase in going from left to right across each row and from top to bottom down each column. Rows 1 and 3 and column 2 exhibit this pattern, but there is one exceptional cell (row 2, column 3) corresponding to medium risk on blood availability and high risk on environmental characteristics. This requires some explanation, which we provide below. Column 1 appears to have a violation in row 3; however, these rates (19.3% and 20.5%) are not statistically significantly different from each other at level 0.05.

The aberrant cell (row 2, column 3) is high risk on environmental characteristics. This particularly means that there is a poor interior and gallery roof. While these conditions characterized habitats with high risk environmental characteristics up to approximately 18 months—two years prior to serological data collection for the present study, the owners engaged in roof repair on their houses. The immediate effect was to eliminate localities that were previously hospitable to *T. infestans*. It is, therefore, not surprising that the incidence rate for new Chagas disease cases dropped precipitously at those sites. It is important to emphasize that all owners of houses in the community did not engage in roof repair. Indeed, examination of Table 8.2 provoked a deeper inquiry into why the $(2, 3)$ cell was so anomalous. Examining the full information set for the habitats in this cell revealed that the GoM analysis had isolated the locations in a highly heterogeneous community where roof repair was making a major difference in Chagas disease incidence.

With the extreme habitats identified—meaning those scoring high risk on both blood availability and environmental characteristics as well as those scoring low risk on both dimensions—a more in-depth analysis was carried out to characterize the most (and least) risky habitats. To this end, variables defining host availability and environmental conditions (shown in Table 8.1) were used to calculate the odds ratio (OR) comparing the proportion of habitats with the highest level of risk to the proportion having the lowest level of risk; 95% confidence interval (CI) on the odds ratio was also calculated. A condition was then defined to be extremely risky if the lower bound of the 95% CI on the odds ratio exceeded 3.5. Analogously, for the habitats classified as low risk on both domains in Table 8.2, the odds ratio comparing the lowest risk condition on each variable with the highest risk condition on that variable and its 95% CI was calculated. Now, a condition was defined to be extremely low risk if the lower bound of the 95% CI exceeded 3.5. Applying these stringent criteria, a new set of low and high risk conditions were specified. They are described in Table 8.3 together with the seropositivity rates for the subset of habitats satisfying them.

Identification of habitats with these very different seropositivity rates that were, nevertheless, embedded in an endemic community provided evidence that our mixed membership methodology,

TABLE 8.3
High and low risk conditions and associated seropositivity rates for children aged 5–14.

Level of risk	Conditions	% seropositive
Low	[# of peridomicilliary structures = 1, but no presence of a food storage area] AND [1 dog OR 1 cat]	7.7
High	[# of peridomicilliary structures > 2] AND [presence of food storage area] AND [> 1 dog OR > 1 cat OR both]	36.4

together with the second stage screening of habitats that were low and those that were high on risk variables from the two domains, deserves attention in other risk assessment settings. The low risk conditions in Table 8.3 are associated with a seropositivity rate among children aged 5–14 of 7.7%, significantly lower than the rate observed in the total population (25.3%). These conditions are, in fact, a basis for relatively simple and inexpensive restructuring of individual habitats to substantially reduce Chagas disease risk. Further, from a methodological perspective, the standard mixed membership model structure, which contains only a single GoM score vector, automatically masks over the risk domain distinctions achievable with the partitioned model of Equation system (8.2).

A final methodological point pertaining to this example concerns the bivariate distribution of the GoM scores (g_1, g_2). An empirical distribution is obtained via conditional likelihood calculation of GoM scores for each of the domains separately using the specification (8.1). There is currently no theoretical basis for a priori imposing a class of bivariate distributions to represent the GoM scores from two domains in the context of Chagas disease epidemiology. This situation could change with particular applications, but thus far there is not enough experience using specification 2, or models with three or more distinct domains, to warrant putting forth defensible classes of bivariate distributions for (g_1, g_2). Carrying the modeling into a Bayesian framework would require specification of defensible prior distributions on the GoM scores. Except for a nearly uniform prior on the unit square, we await the development of subject-matter-driven specification of more informative priors for use with model specification (8.2).

8.4   Disability Change in the U. S. Population: 1982–1994

Populations aged 65 and older at the level of communities contain many people who have multiple disabilities and chronic conditions, no combination of which occurs at high frequency. This makes classification of elderly populations into disability/chronic conditions groups particularly problematic. Simply describing the joint distribution of co-morbid conditions is an unwieldy and difficulty task. This setting, however, is precisely where mixed membership models can play a useful role in terms of representing the heterogeneity in elderly populations via interpretable sets of pure types and characterizations of shared membership between them among selected sub-populations. Berkman et al. (1989) put forth an initial analysis in this direction, focused on the elderly community of New Haven, CT.

Data used for the analysis was derived from the National Long Term Care Survey (NLTCS) list-based samples of approximately 20,000 persons age 65+ drawn from Medicare enrollment files in the years 1982, 1984, 1989, and 1994. To ensure a national sample of the age 65+ population at each survey date, a fresh supplementary list sample was drawn from Medicare enrollment files in 1984, 1989, and 1994. A detailed description of the NLTCS is given in Corder et al. (1993). The analysis examined sub-groups that contributed to an overall decline in disability between 1982 and 1994, and some that did not follow this general trend. Mixed membership models with the structure of Equation system (8.1) were fit for each of the survey years to response vectors whose coordinates described the ability, or not, of individuals to perform a diverse set of "Activities of Daily Living" (ADLs), tests of physical functioning, or both. A battery of 27 ADLs, "Instrumental Activities of Daily Living" (IADLs), and functional impairment measures were employed for this purpose. They are listed in Table 8.4.

TABLE 8.4

Activities of daily living and measures of physical functioning assessed in the National Long Term Care Survey.

ADL Items: need help with	IADL Items: need help with
Eating	Heavy work
Getting in/out of bed	Light work
Dressing	Laundry
Bathing	Cooking
Using a toilet	Traveling
Getting about outside	Grocery shopping
	Managing money
Are you	Taking medicine
Bedfast	Telephoning
Using a wheelchair	
Restricted to no inside activity	
Can you	
See well enough to read a newspaper	
How much difficulty do you have: none, some, very difficult, cannot at all	
Climbing 1 flight of stairs	
Bending for socks	
Holding a 10 lb. package	
Reaching over head	
Combing hair	
Washing hair	
Grasping small objects	

Source: Manton et al. (1998).

For the community population, the best fitting mixed membership models for each of the survey years, satisfying Equation system (8.1), had $K = 6$ pure types. There was very little variation in the pure types across the survey years. Independent of the model, a 7th pure type/profile was added for the elderly institutionalized population. This group was quite homogeneous, having an average of 4.8 ADLs chronically impaired. The full set of pure types is described in Table 8.5.

Although the pure types are clearly interpretable, sharing of conditions across sets of 2 and

TABLE 8.5
Disability pure types from mixed membership model with $K = 6$.

I	Active, no functional impairments.
II	Very modest impairment, some difficulty climbing stairs, lifting a 10 lb. package, and bending for socks (no ADL or IADL).
III	Moderate physical impairment, great difficulty climbing stairs, lifting 10 lb. package, reaching over head, etc. (no ADL or IADL).
IV	All IADLs, great difficulty climbing stairs, and lifting a 10 lb. package.
V	Some ADLs and IADLs, difficulty climbing stairs, cannot lift a 10 lb. package.
VI	All ADLs and all IADLs, and all tasks (high percentage in wheelchairs).
VII	Institutionalized – these are not included in the mixed membership model.

Source: Singer and Ryff (2001).

especially 3 pure types with only mild differences among some of the characteristics presents serious difficulties for differentiating among sub-groups. To resolve this difficulty, we introduce a context-specific strategy for aggregating vertices and edges of the simplex to create a coarser set of disability categories. For this, observe that pure types I–III represent persons who are generally functionally intact. In contrast, pure types IV–VI identify persons with significant physical or cognitive impairments. Heterogeneity within the functionally intact group is represented by persons who share conditions with pairs of pure types I–III. Such people have GoM score vectors at a given survey with non-zero entries for precisely 2 pure types. For example, $\mathbf{g} = (0.3, 0.7, 0, 0, 0, 0)$ is the GoM score vector for a person whose responses on ADL, IADL, and physical functioning are closer to pure type II (a weighting of 0.7) than to pure type I (a weighting of 0.3). We will denote the category of functionally intact persons by C(1 - 3). There are persons with response vectors at one of the pure types I, II, or III, supplemented by persons who share conditions with any pair of them. Geometrically, persons in C(1 - 3) are either at one of the vertices in the unit simplex labeled I, II, or III, or they are on one of the edges that link pairs of these vertices.

Heterogeneity in the severely disabled group, labeled C(4 - 6), is represented by persons at pure types IV, V, and VI, or by those who share conditions with any pair of them. A different form of heterogeneity, C(int), is designated for persons who are on edges connecting one of the vertices [I, II, III] to one of the vertices [IV, V, VI]. A more extreme form of heterogeneity, designated C(res) is represented by persons who share conditions with 3 or more pure types. Geometrically they are identified by points in the faces or further in the interior of the unit simplex with $K = 6$. The partitioning of population aged 65+ into the four disability categories defined above, augmented by the institutionalized population, identifies clearly distinct groups with qualitatively different interpretations of their mix of disabilities.

Returning to the issue of disability decline mentioned at the beginning of this section, we know that there was a decline of 1.5% per annum in the proportion of the age 65+ population that was chronically disabled over the time period 1982–1994. The classification scheme introduced above facilitates our getting a much better picture of the variation in prevalence of chronic disability according to our more refined classification of it. To this end, Table 8.6 shows the percent per annum changes in prevalence of chronic disability from 1982–1994 by disability category generated from the aggregation of vertices and edges of the unit simplex that gave rise to C(1 - 3), C(4 - 6), C(int), and C(res). The table also includes the category Inst, which refers to institutionalized individuals.

Prior to the production of Table 8.6, separate GoM models were run for the years 1982, 1984,

TABLE 8.6
Percent per annum changes in prevalence of chronic disabilities from 1982–1994 by age and gender.

Disability category	Men aged 65 - 84	Women aged 65 - 84	Men aged 85+	Women aged 85+
C(1 - 3)	+0.21	+0.30	+1.45	+0.25
C(4 - 6)	-2.78	-5.31	-2.65	-2.91
C(int)	-2.11	-0.74	+4.08	-0.16
C(res)	-1.05	-1.01	-3.13	+0.05
Inst.	-1.60	-1.71	-0.94	+0.16

Source: Singer and Ryff (2001).

1989, and 1994. The remarkable feature of these separate analyses was that the number of pure types and the conditions entering into them were invariant over this 12-year period. Thus, changes involving an individual were captured in changes in their GoM score vectors over time. Having a chronic disability means having at least one ADL or IADL, where such disability has lasted, or was expected to last, at least 90 days. Prevalence of this condition in each of the disability categories is the basis for calculation of changes between 1982 and 1994.

It is important to emphasize that the invariance in number of pure types and the conditions that enter into them that were an empirical fact of life in the present analysis is by no means generic to GoM modeling over time. Indeed the number of pure types could have varied between, for example, 3 and 7, depending upon the time of assessment. In addition, the conditions entering into pure types at each assessment time could be different. Under such a scenario, it would be impossible to discuss changes over time via a table as simple as Table 8.6.

In-depth interpretation of the category-specific changes in Table 8.6 requires the use of a much richer set of variables from the NLTCS then used for the present methodological discussion of mixed membership models. Extensive analysis of disability changes can be found in Manton et al. (1998), Manton et al. (2006), and Manton (2008).

8.5 Discussion and Open Problems

The major feature of mixed membership models that motivated their specification in the first place (Woodbury and Clive, 1974; Woodbury et al., 1978) was the empirical fact, arising in many studies, that crisp classification of individuals into well-defined categories was frequently difficult, if not impossible. Standard clustering methods do not provide a way out of this impasse, and the observation that shared membership among two or more categories for individuals in a wide variety of scientific contexts is conceptually meaningful paved the way for elaboration of formal models to capture this idea (Woodbury and Clive, 1974; Woodbury et al., 1978; Davidson et al., 1989). Although mixed membership models can be specified according to a priori theories and used in a hypothesis testing mode, by far the most extensive use of the methodology has been in exploratory studies where K, the number of pure types, and the structure of the pure types themselves, is estimated from the data. In terms of numerical goodness-of-fit criteria, best fitting mixed membership models have been obtained in many instances where K takes on values in the range 15–50 (Airoldi et al., 2010). Then interpretive reports are presented with a focus on the structure of the pure types themselves, with only minimal—if any—discussion and interpretation of shared membership across 2 or more pure types. For a notable exception to this practice, see Airoldi et al. (2010).

Our own attempts to consider shared membership (Berkman et al., 1989; Chuit et al., 2001; Castro et al., 2006) rather quickly highlighted the interpretability difficulties involved in simply writing coherent sentences to explain shared membership involving 4 or 5 pure types. This led to the alternative specification shown in Equation (8.2), which is the simplest example of partitioning response vectors by distinct subject-matter domains and the introduction of the assumption of conditional independence of variables in each domain separately given the set of GoM score vectors for all of them. The problem of interpreting shared conditions across multiple pure types is then transferred to one of providing interpretable explanations for the correlation structure in the set of GoM score vectors across domains. The latter situation turned out to be especially informative in our example of characterizing Chagas disease risk in rural Argentina. This special setting is also a generic one for disease risk assessment. In fact, we think it highly desirable to apply the strategy implied by Equation (8.2) to represent risk profiles in complex eco-epidemiological contexts quite generally. In particular, the process of health impact assessment (HIA) for large scale industrial projects in the tropics could benefit from this approach (Krieger et al., 2012; Winkler et al., 2011; 2012a;b).

The example of classification of disability in the U.S. population, discussed in Section 8.4, is a prototype for interpreting shared conditions where a multiplicity of pure types can meaningfully be aggregated into coarse categories of roughly similar conditions. It is unclear, in terms of scientific subject-matter, how to characterize the problems that lend themselves to this kind of aggregation methodology. However, we feel it would be useful to attempt such pure type, edge, and even face consolidations in exploratory data analyses using the mixed membership specification shown in Equation (8.1), when K is in the range 4–6, and certainly when $K > 10$.

In summary, we demonstrate alternative specifications of mixed membership models where an increase in dimensionality of grade of membership scores is traded for simplicity in the number and structure of ideal types, with clear payoff in terms of interpretability of model output. Alternatively, sub-sets of vertices and edges—or even faces—linking them can be aggregated to form interpretable categories of individuals, about which coherent descriptions can be formulated. The scientific subject matter must dictate which among these and other dimension-reducing strategies are to be employed for a particular problem. Although the statistical details of fitting mixed membership models to data lie outside the scope of the present paper, we direct the reader to the interesting Bayesian formulations in Airoldi et al. (2008; 2010) that focus on model specification (8.1). There is no analogous rigorous Bayesian methodology to-date for the class of specifications exemplified by Equation (8.2). Here is an important challenge worth taking up in the immediate future.

Acknowledgments

BHS thanks the Army Research Office Multidisciplinary Research Initiative for financial support under MURI Grant #58153-MA-MUR, Prime Award #W91 INF-11-1-0036. MCC thanks the National Institute of Allergy and Infectious Diseases, Award Number R03AI094401-01 (PI Castro). The content is solely the responsibility of the authors and does not necessarily represent the official views of the funders.

References

Airoldi, E. M., Blei, D. M., Fienberg, S. E., and Xing, E. P.(2008). Mixed membership stochastic blockmodels. *Journal of Machine Learning Research* 9: 1981–2014.

Airoldi, E. M., Erosheva, E. A., Fienberg, S. E., Joutard, C., Love, T., and Shringarpure, S. (2010). Reconceptualizing the classification of PNAS articles. *Proceedings of the National Academy of Sciences* 107: 20899–20904.

Berkman, L., Singer, B. H., and Manton, K. G. (1989). Black/white differences in health status and mortality among the elderly. *Demography* 26: 661–678.

Castro, M. C., Monte-Mór, R. L., Sawyer, D. O., and Singer, B. H. (2006). Malaria risk on the Amazon frontier. *Proceedings of the National Academy of Sciences* 103: 2452–2457.

Chuit, R., Gurtler, R. E., Mac Dougall, L., Segura, E. L., and Singer, B. H. (2001). Chagas disease – risk asssessment by an environmental approach in Northern Argentina. *Revista de Patologia Tropical* 30: 193–207.

Corder, L. S., Woodbury, M. A., and Manton, K. G. (1993). Health loss due to unobserved morbidity: A design based approach to minimize nonsampling error in active life expectation estimates. In *Proceedings of the 6th International Workshop on Calculation of Health Expectancies (INSERM)*. Paris, France: J. Libbey Eurotext, 217–232.

Davidson, J. R., Woodbury, M. A., Zisook, S., and Giller, Jr., E. L. (1989). Classification of depression by Grade of Membership: A confirmation study. *Psychological Medicine* 19: 987–998.

Erosheva, E. A., Fienberg, S. E., and Joutard, C. (2007). Describing disability through individual-level mixture models. *Annals of Applied Statistics* 1: 502–537.

Keiser, J., N'Goran, E., Traore, M., Lohourignon, K. L., and Singer, B. H. (2002). Polyparasitism in *Schistosoma mansoni*, geohelminths, and intestinal protozoa in rural Côte d'Ivoire. *Journal of Parasitology* 88: 461–466.

Krieger, G. R., Bouchard, M. A., de Sa, I. M., Paris, I., Balge, Z., Williams, D., Singer, B. H., Winkler, M. S., and Utzinger, J. (2012). Enhancing impact: Visualization of an integrated impact assessment strategy. *Geo-Spatial Health* 6: 303–306, + video.

Manton, K. G. (2008). Recent declines in chronic disability in the elderly U.S. population: Risk factors and future dynamics. *Annual Review of Public Health* 29: 91–113.

Manton, K. G., Gu, X., and Lamb, V. L. (2006). Change in chronic disability from 1982 to 2004/2005 as measured by long-term changes in function and health in the U.S. elderly population. *Proceedings of the National Academy of Sciences* 103: 18374–18379.

Manton, K. G., Stallard, E., and Corder, L. S. (1998). The dynamics of dimensions of age-related disability from 1982 to 1994 in the U.S. elderly population. *Journal of Gerontology: Biological Sciences* 53A: B59–B70.

Manton, K. G., Woodbury, M. A., and Tolley, H. D. (1994). *Statistical Applications Usings Fuzzy Sets*. New York, NY: John Wiley & Sons.

Paulone, I., Chuit, R., Pérez, A. C., Canale, D., and Segura, E. L. (1991). The status of transmission of *Trypanasoma cruzi* in an endemic area of Argentina prior to control attempts, 1985. *Annals of Tropical Medicine and Parasitology* 85: 489–497.

Raso, G., Vounatsou, P., Singer, B. H., N'Goran, E., Tanner, M., and Utzinger, J. (2006). An integrated approach for risk profiling and spatial prediction of *Schistosoma mansoni*—hookworm coinfection. *Proceedings of the National Academy of Sciences* 103: 6934–6939.

Singer, B. H. and Ryff, C. D. (2001). Person-centered methods for understanding aging: The integration of numbers and narratives. In Binstock, R. and George, L. K. (eds), *Handbook of Aging and the Social Sciences*. San Diego, CA: Academic Press, 44–65.

Singer, B. H. (1989). Grade of Membership representations: Concepts and problems. In Karlin, S., Anderson, T. W., Athreya, K. B., and Iglehart, D. L. (eds), *Probability, Statistics, and Mathematics: Papers in Honor of Samuel Karlin*. Boston, MA: Academic Press, 317–334.

Suleman, A. (2013). An empirical comparison between Grade of Membership and principal component analysis. *Iranian Journal of Fuzzy Systems* 10: 57–72.

White, A., Chan, J., Hayes, C., and Murphy, T. B. (2012). Mixed membership models for exploring user roles in online fora. In *Proceedings of the 6th International AAAI Conference on Weblogs and Social Media (ICWSM 2012)*. Palo Alto, CA, USA: AAAI.

Winkler, M. S., Divall, M. J., Krieger, G. R., Balge, M. Z., Singer, B. H., and Utzinger, J. (2011). Assessing health impacts in complex eco-epidemiological settings in the humid tropics: The centrality of scoping. *Environmental Impact Assessment Review* 31: 310–319.

Winkler, M. S., Divall, M. J., Krieger, G. R., Schmidlin, S., Magassouba, M. L., Knoblauch, A. M., Singer, B. H., and Utzinger, J. (2012a). Assessing health impacts in complex eco-epidemiological settings in the humid tropics: Modular baseline health surveys. *Environmental Impact Assessment Review* 33: 15–22.

Winkler, M. S., Krieger, G. R., Divall, M. J., Singer, B. H., and Utzinger, J. (2012b). Health impact assessment of industrial development projects: A spatio-temporal visualization. *Geo-Spatial Health* 6: 299–301, + video.

Woodbury, M. A. and Clive, J. (1974). Clinical pure types as a fuzzy partition. *Journal of Cybernetics* 4: 111–121.

Woodbury, M. A., Clive, J., and Garson, A., Jr. (1978). Mathematical typology: A Grade of Membership technique for obtaining disease definition. *Computers and Biomedical Research* 11: 277–298.

Woodbury, M. A. and Manton, K. G. (1982). A new procedure for analysis of medical classification. *Methods of Information in Medicine* 21: 210–220.

9

Mixed Membership Trajectory Models

Daniel Manrique-Vallier

Department of Statistics, Indiana University, Bloomington, IN 47408, USA

CONTENTS

We present a model in which individuals can have multiple membership into "pure types" defined by ways of evolving over time. This modeling strategy allows us to use longitudinal data on several subjects to isolate and characterize a few typical trajectories over time, and to soft cluster individuals with respect to them. We present these methods in the context of an application to the study of patterns of aging in American seniors.

9.1 Introduction

In this chapter we introduce a Bayesian technique for soft clustering units based on similarities on their temporal evolution using longitudinal data. Clustering based on evolution over time, or trajectories, is of interest in many areas. For example, criminology researchers are often interested in identifying types of "criminal careers," and in determining how a population of offenders distributes across them (Nagin and Land, 1993). Similarly, clinical psychologists might be interested in characterizing the developmental course of specific disorders like depression (Dekker et al., 2007) or post-traumatic stress disorder (Orcutt et al., 2004).

In general terms, the typical clustering problem consists of arranging a number of units, e.g., a sample of people, into a smaller number of classes based on the similarity of some observed attributes without assuming prior knowledge of the specific characteristics of the groups. From a model-based perspective, this is usually accomplished by fitting mixtures of the form

$$p(y|\pi) = \sum_{k=1}^{K} \pi_k f_k(y) \tag{9.1}$$

to multivariate data about one or more individual characteristics, say $y = (y_1, ..., y_J)$, that are expected to inform about the relevant differences between individuals. Models of this form are usually interpreted as partitioning the population into K disjoint sub-populations, each representing a fraction π_k of the population. The sub-populations themselves are characterized by (usually parametric) densities $f_k()$, which are also to be estimated from the data.

When dealing with phenomena that evolve over time it is occasionally of interest to cluster individuals based on similarities on their temporal evolution. This requires longitudinal data, i.e., repeated observations of the same individuals at different points in time. The most direct approach is to assume that there exists a number of sub-populations, each characterized by a particular trajectory, and that each individual belongs to one and only one of them. Such a setup corresponds to the general model-based clustering approach described in the previous paragraph. In particular, trajectories define specific forms of $f_k()$, the sub-population's joint distribution of the sequence of observations.

In some applications the requirement that each individual belong exclusively to just one sub-population can be too restrictive to be realistic. For example, in the study of political ideology it is common to use terms that describe pure extreme positions, like "liberal" or "conservative." However, actually assuming that every individual is either a liberal or a conservative is too broad a description to be useful, let alone accurate. In particular, it hides the fact that many individuals have opinions about different topics that correspond to more than one "pure" ideological position. A better alternative would be to describe individuals' ideologies as mixtures of the pure types.

Modeling these structures motivates the mixed membership approach. Mixed membership models relax the assumption of exclusive cluster membership by allowing units to belong to more than one group simultaneously. We call this type of arrangement a soft clustering.

In this chapter we extend the mixed membership approach to longitudinal structures based on trajectories. In particular, our technique allows us to construct soft-classifications based on the ways in which individuals evolve over time. This, in turn, allows us to isolate a few extreme or pure trajectories—which can be informative and easy to analyze—and to characterize units in the population as individually-mixed combinations of them. We introduce this approach in the context of studying the individual patterns of evolution of disability in the elder American population. In the next section we present the applied context, which will serve as our illustrative application. Then we introduce the general notion of *clustering based on trajectories*, and present our method, the *Trajectory Grade of Membership* model, as a mixed membership extension of this idea. We present a fully Bayesian specification and an estimation algorithm based on Markov chain Monte Carlo (MCMC) sampling. Finally, we demonstrate the method by analyzing patterns of evolution of disability using data from the National Long Term Care Survey. Additional details regarding the model and analysis can be found in Manrique-Vallier (2013; 2010)

9.1.1 Application: The National Long Term Care Survey

It is well known that elder Americans are living longer than in the past. Their absolute number and proportion are increasing rapidly (Connor, 2006). Older people often require some form of long-term care, especially in the presence of disabilities (Manton et al., 1997). Thus, efficient allocation of resources and overall cost prediction require information about typical patterns of disability, their progression over time, and their distribution over the population.

With these issues in mind, a group of researchers and policy makers created the National Long Term Care Survey, NLTCS (Clark, 1998). The NLTCS is a longitudinal survey designed to evaluate the state and progression of chronic disability among the senior population in the United States. This instrument tracks disability by recording each individual's capacity to perform a set of "Activities of Daily Living" (ADL) such as eating, bathing, or dressing, and "Instrumental Activities of Daily Living" (IADL) such as preparing meals or maintaining finances.

The NLTCS is comprised of six waves of interviews, administered in 1982, 1984, 1989, 1994, 1999, and 2004. Each wave includes interviews of about 20,000 people. Whenever possible, individuals are followed from wave to wave until death. However, due to the high mortality rate in the target population, each wave also includes a replacement sample of approximately 5,000 new subjects. The inclusion of these new individuals keeps each wave's sample size approximately constant and representative of the population at each given time (Clark, 1998). In aggregate, approximately 49,000 people have been interviewed between 1982 and 2004.

We represent individual-level NLTCS data as an array, $(y_{jt})_{J \times T}$, of J binary items (ADLs and IADLs) measured at T points in time (waves). Each entry of the array y_{jt} represents the presence $(y_{jt} = 1)$ or absence $(y_{jt} = 0)$ of impairments to perform ADL/IADL j at the time of wave t. Table 9.1 shows a hypothetical example of individual NLTCS data, considering only the ADLs $(J = 6$, EAT: Eating; DRS: Dressing; TLT: Toileting; BED: Getting in and out of bed; MOB: Inside mobility; BTH: Bathing). In this example we see that our hypothetical subject did not experience impediments in bathing until the 5th wave of the survey, in 1999. Similarly, by the time of the 6th wave, he/she had limitations in performing all the ADLs from the list. The NTLCS also records other complementary information, some of which is time-dependent (e.g., Age, in the example), and some of which is fixed (e.g. Date of Birth and Date of Death, in the example). In this application we will be concerned only with the binary responses to the ADL questions, and the age of each individual at each wave.

| | | | Wave (t) | | | | |
| | | 1 | 2 | 3 | 4 | 5 | 6 |
	Year	1982	1984	1989	1994	1999	2004
	EAT ($j = 1$)	0	0	0	0	1	1
	DRS ($j = 2$)	0	1	0	0	0	1
	TLT ($j = 3$)	0	0	0	1	1	1
ADL (j)	BED ($j = 4$)	1	1	0	1	1	1
	MOB ($j = 5$)	0	0	0	0	1	1
	BTH ($j = 6$)	0	0	0	0	1	1
	Age:	66	69	74	79	84	89
Other	DOB:	1916					
	DOD:	2005					

TABLE 9.1
Example of data structure for a single fictional individual. The individual itself is indexed by the letter $i \in \{1...N\}$.

We will introduce our methods by modeling the evolution of the probability of acquiring specific disabilities as a function of personal time (time in the system or age), using the NLTCS data.

9.2 Longitudinal Trajectory Models

9.2.1 Clustering Based on Trajectories

Multivariate data arising from longitudinal studies can sometimes be thought of as the expression of a time-continuous underlying process. For example, in the study of disability among elders, it is reasonable to assume that the sequence of discrete disability measurements of an individual is an observable expression of an underlying "aging process," that relates age to the probability of experiencing a disability. We can go further and assume that this process is such that the probability of experimenting a functional disability will tend to increase as the person ages.

In some applications it is possible to describe the evolutionary process that underlies the observed longitudinal data using parametric functions of time or of a time-dependent covariate. We call these functions *trajectories*. An example of a trajectory is a function of age that determines the probability of experimenting a disability.

When the population under study is, or may be expected to be, heterogeneous, we cannot assume that every individual in the population follows the same underlying trajectory. In our application, for example, that would force us to expect that all individuals age the same way. Instead, when modeling these situations, we need to allow distinct individuals to respond to different trajectories.

Adopting such a modeling scheme, where each individual is allowed to have his/her own trajectory, opens up the possibility of clustering the population based on similarities among trajectories. For instance we could try to cluster individuals from the NLTCS into classes defined by "types of aging." We call this clustering strategy *clustering based on trajectories*. Besides its intrinsic application domain interest, clustering based on trajectories has also the advantage of allowing us to incorporate additional knowledge about the trajectories (e.g., their expected shape) to complement the information already contained in the individual sequences of responses.

9.2.2 Hard Clustering: Group-Based Trajectory Models

A direct approach to clustering based on trajectories is given by group-based trajectory models (GBTM) (Nagin, 1999; Connor, 2006). These models assume the existence of a few homogeneous sub-populations whose members' responses follow the same trajectories over time. Thus, it enables a type of hard-clustering of the population of interest.

To see how a GBTM works, let us consider modeling the progression of a single binary response. Let $y = (y_1, ..., y_t, ..., y_T)$ be the sequence of binary measurements at times $t = 1, 2, ..., T$ for the same individual. We assume that the individual has been sampled from one of K sub-populations, with probability π_k $(k = 1, 2, ..., K)$. Then we specify the trajectory of the probability of a positive response for a member of group k, $\Phi_{\theta_k}(x)$ as some convenient function of a time-varying quantity x, indexed by parameters θ_k:

$$\Phi_{\theta_k}(x) = \Pr(y_t = 1 | x, \text{individual belongs to group } k).$$

Let $\mathbf{x} = (x_1, x_2, ..., x_T)$ be a vector containing the T measurements of the time-dependent quantity of interest, e.g., the age of the individual at each survey wave. Then, assuming that given group membership and x_t, responses are all independent, we have that

$$p(y|\mathbf{x}, \theta) = \sum_{k=1}^{K} \pi_k \prod_{t=1}^{T} f_{\theta_k}(y_t; x_t), \tag{9.2}$$

where $f_{\theta_k}(y_t | x) = \Phi_{\theta_k}(x)^{y_t}(1 - \Phi_{\theta_k}(x))^{1-y_t}$. The model in (9.2) is a discrete mixture that specifies the distribution of a response variable within each sub-population, conditional on a time-dependent covariate.

Connor (2006) proposed an extension for multivariate binary data consisting of J binary variables measured at T points in time, and applied it to the analysis of the NLTCS. His proposal extended (9.2) by assuming conditional independence between responses to different items at distinct points in time, given covariates and group membership. Let y_{jt} be the response to item j at measurement t (e.g., to the jth ADL of the NLTCS at wave t). Connor's model is

$$p(y|x, \theta) = \sum_{k=1}^{K} \pi_k \prod_{j=1}^{J} \prod_{t=1}^{T} f_{\theta_{jk}}(y_{jt}; x_t). \qquad (9.3)$$

This specification characterizes each sub-population based on J trajectories, each of them common to all of their members.

9.2.3 Soft Clustering: The Trajectory GoM Model

The requirement of group-based GBTM of within-cluster homogeneity can be too restrictive in some applications. For instance, in the NLTCS case it essentially requires us to assume that every individual within a sub-population follows the exact same aging process. This is not plausible. Furthermore, one might even wonder if such sub-populations exist at all (see e.g., Kreuter and Muthén, 2008).

One way of relaxing this strong assumption is by replacing the requirement of exclusive membership with a mixed membership structure—thus constructing a soft clustering based on trajectories. In such a case, we interpret latent groups not as sub-populations—as with group-based trajectory models—but as characterizations of extreme cases. We then model individual trajectories as mixtures of *extreme trajectories* in different individual degrees.

The rest of this section is devoted to presenting one such model, which we will call a *Trajectory Grade of Membership* (TGoM) model.

In longitudinal multivariate settings we are interested in studying the simultaneous progression of a number, J, of variables as a function of time. For now, assume that response variables are binary and that we have measurements of each variable at a number, T, of points in time. Call y_{jt} the value of the jth variable ($j = 1, ..., J$) at measurement time t ($t = 1, ..., T$) for a particular individual. In the NLTCS case, y_{jt} is the disability measurement j (jth ADL) at wave t.

Similar to a group-based trajectory model, we assume the existence of a small number, K, of ideal types of individuals or *extreme profiles*. However, instead of assuming that particular individuals belong completely to those classes, we endow them with membership vectors, $g = (g_1 ... g_K)$ ($g_k > 0$, $\sum_k g_k = 1$). Membership vectors are a characteristic of each individual. Their components, g_k, represent the degree of membership of an individual in each of the K extreme profiles. Ideal individuals of the kth type are individuals whose membership vector's kth component has a value $g_k = 1$, and the rest of the entries are zeros. For instance, an individual with membership vector $g = (0, 1, 0, 0)$ belongs exclusively to the extreme profile $k = 2$. An individual with membership vector $g = (0.1, 0.2, 0.7)$ has 10% membership in extreme profile $k = 1$, 20% in $k = 2$, and 70% in $k = 3$.

We specify the trajectory of a positive response for each response variable j and extreme profile k, $\Phi_{\theta_{jk}}(x)$ as a function of time with parameter θ_{jk}. These trajectories correspond to idealized progressions of the variables of interest over time, in the same way that trajectories in the developmental trajectory model represent the progression of variables for particular groups over time. Different from GBTM though, we do not regard individuals as being samples from the sub-population, but mixtures of them.

Using the membership vectors, we model the trajectory of variable j for an individual with

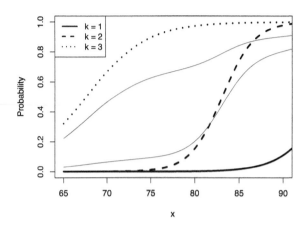

FIGURE 9.1
Example of extreme and individual trajectories. Extreme trajectories are drawn in thick lines. Individuals have membership vectors $(0.1, 0.2, 0.7)$ (top thin solid curve) and $(0.2, 0.7, 0.1)$ (bottom thin solid curve).

membership vector $g = (g_1, ..., g_K)$ as

$$p(y_{jt} | (g_1, ..., g_K), x, \theta) = \sum_{k=1}^{K} g_k f_{\theta_{jk}}(y_{jt}; x). \tag{9.4}$$

As an example with $K = 3$ extreme profiles, consider the situation in Figure 9.1. Curves in thick lines represent three extreme trajectories, $\Phi_{\theta_{j1}}(x)$, $\Phi_{\theta_{j2}}(x)$, $\Phi_{\theta_{j3}}(x)$, for an arbitrary ADL, j. According to (9.4), given an individual i whose membership vector is g_i, the probability of a positive response to item j is a weighted combination of the extremes, which defines an individualized trajectory, $\Phi^{(i)}(x) = \Pr(y_{ij} = 1 | g_i, x) = \sum_{k=1}^{K} g_{ik} \Phi_{jk}(x)$. Extreme trajectories—thick lines—correspond to (most likely fictional) individuals whose membership vectors are $(1, 0, 0)$, $(0, 1, 0)$, and $(0, 0, 1)$. The two individual trajectories—thin lines—in the picture correspond to individuals whose membership vectors are $(0.1, 0.2, 0.7)$ and $(0.2, 0.7, 0.1)$.

In order to characterize the joint distribution of individual responses, we introduce a local independence assumption: for a single individual, conditional on the value of the covariate of interest at time t, x_t, and its membership vector, g, the J responses at each of the T measurement times are mutually independent:

$$p(y | (g_1, ..., g_K), (x_1, ..., x_T), \theta) = \prod_{j=1}^{J} \prod_{t=1}^{T} \sum_{k=1}^{K} g_k f_{\theta_{jk}}(y_{jt}; x_t). \tag{9.5}$$

Moving to the sample, we assume that there are N individuals. We index them using the letter $i = 1, \ldots, N$ and add a corresponding sub-index to the individual-level quantities y_i, g_i, and x_i. Assuming that each individual has been randomly sampled from the population, we get the joint model for the whole sample $\mathbf{y}$, conditional on all the membership vectors $\mathbf{g}$, and all the time-varying covariates **x**:

$$p(\mathbf{y}\,|\mathbf{g},\mathbf{x},\theta) \;\; = \;\; \prod_{i=1}^{N}\prod_{j=1}^{J}\prod_{t=1}^{T}\sum_{k=1}^{K} g_{ik}f_{\theta_{jk}}(y_{ijt};x_{it}). \tag{9.6}$$

Finally, assuming that the membership vectors are i.i.d. samples from a common distribution, say F_α, we get the model

$$p(\mathbf{y}|\mathbf{x},\mathbf{g},\theta) \;\; = \;\; \prod_{i=1}^{N}\int_\Delta \prod_{j=1}^{J}\prod_{t=1}^{T}\sum_{k=1}^{K} g_k f_{\theta_{jk}}(y_{ijt};x_{it})F_\alpha(dg). \tag{9.7}$$

9.2.4 Latent Class Representation of the TGoM

Similar to the Grade of Memberhsip model (see Erosheva et al., 2007), the model in (9.5) admits an augmented data representation that makes it similar to the group-based multivariate developmental trajectory model in (9.3). A few algebraic manipulations on (9.5) (Erosheva, 2002) lead to the equivalence

$$\prod_{j=1}^{J}\prod_{t=1}^{T}\sum_{k=1}^{K} g_k f_{\theta_{jk}}(y_{jt};x_t) = \sum_{z\in Z}\prod_{j=1}^{J}\prod_{t=1}^{T} g_{z_j}f_{\theta_{jz_{jt}}}(y_{jt};x_t), \tag{9.8}$$

where $Z = \{1,2,...,K\}^{J\times T}$ is the set of all matrices (z_{jt}) whose entries take values in $\{1,2,...,K\}$. From here it follows that, after summing over all possible realizations of z, the model

$$p(y,z\,|x,\theta,g) \;\; = \;\; \prod_{j=1}^{J}\prod_{t=1}^{T}\prod_{k=1}^{K}\Big[g_k f_{\theta_{jk}}(y_{jl};x_t)\Big]^{I(z_{jt}=k)} \tag{9.9}$$

is equivalent to (9.5). For details applied to the case of the GoM model, see Erosheva et al. (2007). Considering that $g\sim G$ and integrating (9.9), we get the unconditional distribution

$$p(y|x) \;\; = \;\; \sum_{z\in Z}\pi_z \prod_{j=1}^{J}\prod_{t=1}^{T} f_{\theta_{jk}}(y_{jt};x_t), \tag{9.10}$$

where

$$\pi_z = E_G\left[\prod_{j=1}^{J}\prod_{t=1}^{T}\prod_{k=1}^{K} g_k^{I(z_{jt}=k)}\right]. \tag{9.11}$$

Equation (9.10) shows that the TGoM can be seen as a multivariate group-based DTM, just like (9.3), where the membership weights are restricted by the moments-based definition of π_z in (9.11). We can also see that the following generative process will produce N multivariate responses according to (9.10). Here we again add the individual index, $i\in\{1,...,N\}$, to y_{jt}, g, and z_{jt}.

Trajectory GoM Individual Response Generation Process

For each individual $i \in \{1, 2, ..., N\}$

    Sample $g_i = (g_{i1}, g_{i2}, ..., g_{iK}) \sim F_\alpha$.

    For each $j \in \{1...J\}$

        For each $t \in \{1...T\}$

            Sample $z_{ijt} \sim \textbf{Discrete}_{1:K}(g_i)$.

            Sample $y_{ijt} \sim \textbf{Bernoulli}(\Phi_{jz_{ijt}}(x_{it}))$.

9.2.5 Specifying the Trajectory Function

In our application it is reasonable to assume that the probability of presenting a disability should increase monotonically with age. We thus follow Connor (2006) in making $\theta_{jk} = (\beta_{0jk}, \beta_{1jk})$ and using the s-shaped function

$$\Phi_{\theta_{jk}}(x) = \frac{1}{1 + \exp(-\beta_{0jk} - \beta_{1jk}x)}, \tag{9.12}$$

where x is a scalar representing age.

In general, the choice of trajectory functions $\Phi_{jk}(\cdot)$ must be application-specific, as they encode assumptions about the nature of the underlying process. Thus, other applications would likely require different specifications.

9.2.6 Completing the Specification

To complete a full Bayesian specification of the TGoM model, we need to specify the membership distribution F_α and prior distributions for its parameter α and for the trajectory parameters θ_{jk}.

Following Erosheva et al. (2007), we assume $g_i | \alpha \sim \textbf{Dirichlet}(\alpha)$, where $\alpha = (\alpha_0 \cdot \xi_1, \alpha_0 \cdot \xi_2, ..., \alpha_0 \cdot \xi_K)$ with α_0 and $\xi_k > 0$ for all $k = 1, 2, ..., K$, and $\sum_{k=1}^{K} \xi_k = 1$. Under this parametrization, $\xi = (\xi_1, ..., \xi_K)$ is the expected value of the distribution. It also, more informally, represents the relative importance of profile k in the population. In turn, α_0 is a concentration parameter: the closer α_0 is to 0, the closer samples from F_α will be to the extreme profiles; conversely, the higher the value of α_0, the closer the samples from F_α will tend to be to their expected value, ξ. Thus, for ξ fixed, α_0 controls the amount of mixed membership. We also follow Erosheva et al. (2007) in specifying independent prior distributions $\alpha_0 \sim \textbf{Gamma}(\tau, \eta)$ and $\xi \sim \textbf{Dirichlet}(1_K)$.

Other specifications are possible, and in some problems they may be necessary. An important limitation of the Dirichlet distribution is its simple correlation structure. Regular Dirichlet distributions do not allow the capture of complex correlations between membership in different extreme profiles. This might be a limitation in applications where membership in some extreme profiles has non-trivial relationships with membership in other profiles. A natural extension that can be useful in such situations is the multinomial logistic normal prior (see e.g., Blei and Lafferty, 2007). Unfortunately this specification does not share the computational advantages of the Dirichlet distribution. In particular, it is not conjugate to the multinomial distribution. For the parameters of the extreme trajectories specified in (9.12), β, we specify independent prior distributions $\beta_{0jk} \overset{iid}{\sim} N(\mu_{\beta_0}, \sigma_{\beta 0}^2)$ and $\beta_{1jk} \overset{iid}{\sim} N(\mu_{\beta_1}, \sigma_{\beta 1}^2)$.

9.3 Estimation through Markov Chain Monte Carlo Sampling

Under the specification of extreme trajectories and priors outlined in this section, and following the augmented data representation in (9.9), the joint posterior distribution of parameters and augmented data is

$$p(\alpha, \beta, \mathbf{z}, \mathbf{g}| Data) \propto p(\beta)p(\alpha) \left(\prod_{i=1}^{N} p\left(g_i|\alpha\right) \right)$$

$$\times \prod_{i=1}^{N} \prod_{j=1}^{J} \prod_{t=1}^{T} \prod_{k=1}^{K} \left(g_{ik} \frac{\exp(y_{ijt}\beta_{0jk} + y_{ijt}\beta_{1jk}x_{it})}{1 + \exp(\beta_{0jk} + \beta_{1jk}x_{it})} \right)^{I(z_{ijt}=k)}.$$

Using the full Bayesian specification from Section 9.2.6 we have that

$$p(g_i|\alpha) = \textbf{Dirichlet}(g_i|\alpha_1, \alpha_2, ..., \alpha_k),$$
$$p(\alpha_0) = \textbf{Gamma}(\alpha_0|\tau, \eta),$$
$$p(\xi) = \textbf{Dirichlet}(\xi|\mathbf{1}_K) \quad \text{(Uniform on the } \Delta_{K-1}\text{)},$$

with $p(\alpha) = p(\alpha_0) \cdot p(\xi)$, where $\alpha_0 = \sum_k \alpha_k$ and $\xi = (\xi_1, \xi_2, ..., \xi_K)$ with $\alpha_k = \alpha_0 \cdot \xi_k$. Parameters τ and η are shape and inverse scale parameters, respectively.

Specifying an MCMC algorithm to obtain approximate realizations from this posterior distribution using the Gibbs sampling algorithm is just a matter of obtaining the full conditional distributions of each parameter and augmented data. An implementation of this algorithm follows.

1. **Sampling from** z: For every $i \in \{1 \ldots N\}, j \in \{1 \ldots J\}$, and $t \in \{1 \ldots T\}$, sample

$$z_{ijt}|... \sim \textbf{Discrete}(p_1, p_2, ..., p_K),$$

with $p_k \propto g_{ik} \exp(\beta_{0jk} + \beta_{1jk}x_{it})^{y_{ijt}} [1 + \exp(\beta_{0jk} + \beta_{1jk}x_{it})]^{-1}$, for all $k \in \{1, \ldots, K\}$.

2. **Sampling from** $(\beta_{0jk}, \beta_{1jk})$: Let $\rho_{it} = I(z_{ijt} = k)$. Then, the full joint conditional distribution of $(\beta_{0jk}, \beta_{1jk})$ is

$$p\left(\beta_{0jk}, \beta_{1jk}|...\right)$$

$$\propto \frac{\exp\left[-\frac{\beta_{0jk}^2}{2\sigma_0^2} + \beta_{0jk}\left(\frac{\mu_0}{\sigma_0} + \sum_{i,t} \rho_{it}y_{ijt} \right) - \frac{\beta_{1jk}^2}{2\sigma_1^2} + \beta_{1jk}\left(\frac{\mu_1}{\sigma_1} + \sum_{i,t} \rho_{it}x_{it}y_{ijt} \right) \right]}{\prod_{i,t} \left[1 + \exp\left(\beta_{0jk} + \beta_{0jk}x_{it}\right)\right]^{\rho_{it}}}.$$

This distribution does not have a recognizable form. Thus we use a random walk Metropolis step:

(a) Proposal step: Sample the proposal values

$$\beta_{0jk}^* \sim N(\beta_{0jk}, \sigma_{\beta 0}^2) \text{ and } \beta_{1jk}^* \sim N(\beta_{1jk}, \sigma_{\beta 1}^2),$$

where the values $\sigma_{\beta 0}^2$ and $\sigma_{\beta 1}^2$ are tuning parameters that we have to calibrate to achieve a good balance between acceptance of proposed values and exploration of the support of the target distribution.

(b) Acceptance step: Compute

$$r_M = \frac{p(\beta^*_{0jk}, \beta^*_{1jk} | \ldots)}{p(\beta_{0jk}, \beta_{1jk} | \ldots)}$$

$$= \prod_{i,t} \left[\frac{1 + \exp\left[\beta_{0jk} + \beta_{0jk} x_{it}\right]}{1 + \exp\left[\beta^*_{0jk} + \beta^*_{0jk} x_{it}\right]} \right]^{\rho_{it}} \tag{9.13}$$

$$\times \exp\left[-\frac{\beta^{*2}_{0jk} - \beta^2_{0jk}}{2\sigma_0^2} + (\beta^*_{0jk} - \beta_{0jk}) \left(\frac{\mu_0}{\sigma_0} + \sum_{i,t} \rho_{it} y_{ijt} \right) \right]$$

$$\times \exp\left[-\frac{\beta^{*2}_{1jk} - \beta^2_{1jk}}{2\sigma_1^2} + (\beta^*_{1jk} - \beta_{1jk}) \left(\frac{\mu_1}{\sigma_1} + \sum_{i,t} \rho_{it} y_{ijt} x_{it} \right) \right],$$

and make

$$(\beta_{0jk}, \beta_{1jk})^{(m+1)} = \begin{cases} (\beta^*_{0jk}, \beta^*_{1jk}) & \text{with probability } \min\{r_M, 1\} \\ (\beta_{0jk}, \beta_{1jk})^{(m)} & \text{with probability } 1 - \min\{r_M, 1\}. \end{cases}$$

3. **Sampling from** g_i**:** Since the Dirichlet distribution is conjugate to the multinomial, this expression is particularly simple:

$$g_i | \ldots \overset{indep.}{\sim} \textbf{Dirichlet}\left(\alpha_1 + \kappa_{i1}, \alpha_2 + \kappa_{i2}, \ldots, \alpha_K + \kappa_{iK}\right),$$

where $\kappa_{ik} = \sum_{j,t} I(z_{ijt} = k)$.

4. **Sampling from** α**:** For sampling from the full conditional distribution of α,

$$p(\alpha | \ldots) \propto \textbf{Gamma}(\alpha_0 | \tau, \eta) \times \textbf{Dirichlet}(\xi | 1_K) \times \prod_{i=1}^{N} \textbf{Dirichlet}(g_i | \alpha)$$

$$\propto \alpha_0^{\tau-1} \exp[-\alpha_0 \eta] \times \left[\frac{\Gamma(\alpha_0)}{\prod_{k=1}^{K} \Gamma(\alpha_k)} \right]^N \prod_{k=1}^{K} \left[\prod_{i=1}^{N} g_{ik} \right]^{\alpha_k}, \tag{9.14}$$

we use the Metropolis-Hastings within Gibbs step proposed by Manrique-Vallier and Fienberg (2008):

(a) (Proposal step) Sample $\alpha^* = (\alpha_1^*, \alpha_2^*, \ldots, \alpha_K^*)$, as independent lognormal variates from

$$\alpha_k^* \overset{indep.}{\sim} \text{lognormal}(\log \alpha_k, \sigma^2).$$

Again, σ is a tuning parameter that we have to calibrate.

(b) (Acceptance step) Let $\alpha_0^* = \sum_{k=1}^{K} \alpha_k^*$. Compute

$$r = \min\left\{ 1, \exp\left[-\tau(\alpha_0^* - \alpha_0)\right] \left(\prod_{k=1}^{K} \frac{\alpha_k^*}{\alpha_k} \right) \left(\frac{\alpha_0^*}{\alpha_0} \right)^{\tau-1} \right.$$

$$\left. \times \left[\frac{\Gamma(\alpha_0^*)}{\Gamma(\alpha_0)} \prod_{k=1}^{K} \frac{\Gamma(\alpha_k)}{\Gamma(\alpha_k^*)} \right]^N \prod_{k=1}^{K} \left(\prod_{i=1}^{N} g_{ik} \right)^{\alpha_k^* - \alpha_k} \right\},$$

and update the chain, from step m to step $m+1$ according to the rule

$$\alpha^{(m+1)} = \begin{cases} \alpha^* & \text{with probability } r \\ \alpha^{(m)} & \text{with probability } 1 - r. \end{cases}$$

To obtain samples from the posterior distribution of parameters, we simply cycle through Steps 1 to 4. Selection of tuning parameters, σ^2, $\sigma_{\beta 1}^2$, and $\sigma_{\beta 2}^2$, can be challenging. We present an automated procedure for choosing σ^2 in the next section.

9.3.1 Tuning the Population Proposal Distribution

The MCMC algorithm described in Section 9.3 requires that we select two sets of tuning parameters in order to balance good acceptance rates with an adequate exploration of the support of the posterior distribution. The effect of the tuning parameters for sampling β, $\sigma_{\beta 1}$, and $\sigma_{\beta 2}$ is fairly independent of sample size, and is also stable across models with different numbers of extreme profiles. Thus we tune it by trial and error. The situation is different for σ, the tuning parameter for sampling α. Acceptance rates for α are very sensitive to small changes in σ. Additionally, the same values of σ produce wildly different acceptance rates in models with different numbers of extreme profiles, and when dealing with different sample sizes.

In order to reduce the costly guesswork associated with choosing σ, we propose an automated procedure. With a pre-specified acceptance rate in mind, *acc*, we try different values of σ, and record whether the proposals for α are accepted or not. Then we gather these results and use logistic regression to pick a value of σ likely to achieve the target acceptance rate. Finally, we discard all the generated samples and run the chain, keeping σ fixed at the found value. We note that even though logistic regression assumptions are not satisfied in this case—in particular, observations are clearly not independent given predictors—we have empirically found this procedure to deliver excellent results.

In practice we use a two-phase search strategy. In the first phase we find an interval $[W_1, W_2]$ of values of σ that make the acceptance rate fall within a target interval $[acc_1, acc_2]$. The following algorithm implements this first step. It requires us to provide a reasonably wide starting interval for σ, $[W_1^s, W_2^s]$, and a number of steps, FS_1.

FIRST PASS: Reduce the interval $[W_1^s, W_2^s]$ so that $\Pr(acceptance) \in [acc_1, acc_2]$.

  Initialization: Let $\Delta = (\log(W_2^s) - \log(W_1^s))/FS_1$.

  For $n = 1, ..., FS_1$

    Update chain using $\log \sigma = \log(W_1^s) + \Delta n$.

    If α^* accepted, let $a_n = 1$. Otherwise, let $a_n = 0$.

  Fit a logistic regression model, $\mathrm{logit}(a_n) = \alpha + \beta \Delta n$. Get estimates $\hat{\alpha}$ and $\hat{\beta}$.

  Let $W_2 = \exp((\mathrm{logit}\, acc_1 - \hat{\alpha})/\hat{\beta})$ and $W_1 = \exp((\mathrm{logit}\, acc_2 - \hat{\alpha})/\hat{\beta})$.

We have found that starting values $W_1^s = 0.001$ and $W_2^s = 0.1$ work for most problems. Also, for a target of $acc = 30\%$, a good first-pass target interval is $acc_1 = 0.2$ and $acc_2 = 0.8$.

In the second phase, we search within the reduced interval $[W_1, W_2]$ for a single value of σ likely to attain the target acceptance rate, *acc*. The following algorithm also requires us to set up a number of iterations (FS_2) in advance. We have found that good choices for the number of iterations are $FS_1 = 700$ for the first phase and $FS_2 = 300$ for the second.

SECOND PASS: For $acc \in [W_1, W_2]$, find σ such that $\Pr(acceptance) = acc$.

   Initialization: Let $\Delta = (W_2 - W_1)/FS_2$.

   For $n = 1, ..., FS_2$

       Update chain using $\sigma = W_1 + \Delta n$.

       If α^* accepted, let $a_n = 1$. Otherwise, let $a_n = 0$.

   Fit a logistic regression model, $\text{logit}(a_n) = \alpha + \beta \log(\Delta n)$. Get estimates $\hat{\alpha}$ and $\hat{\beta}$.

   Let $\sigma = \exp((\log(acc/(1.0 - acc)) - \hat{\alpha})/\hat{\beta})$.

We reiterate the importance of discarding the values sampled during this calibration phase. The calibration operation uses all the acceptance outcomes generated during the adaptation phase to modify the kernel of the process. Thus, it renders the whole phase non-Markovian.

9.4   Using the TGoM

Now we return to our illustrative application. Our goal is to identify profiles of typical trajectories of progression into disability and to determine the structure of membership of the population into those profiles. For illustration purposes we have fit a TGoM model with $K = 3$ extreme profiles.

We used a sub-sample of the NLTCS that included $N = 39,323$ individuals measured on $T = 6$ waves. The response vector included the six ($J = 6$) binary coded ADLs shown in Table 9.1.

We chose the prior distribution for $\alpha = \alpha_0 \cdot \xi$ as independent $\alpha_0 \sim \textbf{Gamma}(1, 5)$ and $\xi \sim \textbf{Dirichlet}(1_K)$. This prior specification expresses the notion of complete ignorance about the relative importance of the extreme profiles in the population and preference for smaller values of the concentration parameter, α_0. The reasons behind the last choice are mostly interpretative: a Dirichlet distribution with small values of α_0 will produce individual realizations that are closer to one particular vertex of the simplex, with influence on the other vertices; and as α_0 goes all the way down to 0, a degenerate discrete distribution over the vertices. This arrangement allows us to talk about "dominant profiles" that are influenced by the others, easing the interpretation of the results while still allowing the mixed membership apparatus to handle a significant degree of heterogeneity. For the extreme trajectories parameters, β, we have chosen the relatively diffuse priors $\beta_{0jk}, \beta_{1jk} \sim N(0, 100)$.

We tuned the proposal distribution for α using the two-step algorithm described in Section 9.3.1. The resulting tuning parameter was $\sigma_\alpha = 0.011$. We set the remaining tuning parameters as $\sigma_{\beta 0} = 0.2$ and $\sigma_{\beta 1} = 0.02$.

The chain converges quickly, after around 15,000 iterations, but exhibits a rather high autocorrelation; for this reason, we had to perform long runs of 100,000 iterations. After that, we discarded the first 20,000 iterations and sub-sampled them, retaining one sample every 5 samples and discarding the rest. Figure 9.2 presents the trace plot of the parameter $\alpha_0 = \sum_{k=1}^{K} \alpha_k$.

TGoM models include two sets of directly interpretable parameters. The first group, α_0 and ξ, characterizes the common distribution of the individual mixed membership scores, F_α. Table 9.2 presents estimates (posterior means and standard deviations) of these parameters. From these summaries we see that the posterior distribution of α_0 is tightly concentrated around $\hat{\alpha}_0 = 0.261$. This value is small, but it still leaves room for a significant degree of mixed membership. In particular,

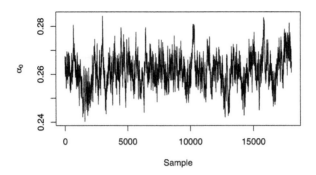

FIGURE 9.2
Trace plot of parameter α_0.

α_0	ξ_1	ξ_2	ξ_3
0.261	0.645	0.251	0.104
(0.006)	(0.004)	(0.004)	(0.002)

TABLE 9.2
Posterior estimates of population-level parameters for model with $K = 3$. Numbers between parenthesis are posterior standard deviations.

the posterior point estimates of α_0 and ξ imply that around 40% of the individuals have responses that are influenced by more than one extreme profile.

The extreme trajectory profiles, characterized by the β parameters, inform of typical progressions into disability as people get older. Figure 9.3 shows such trajectories for each ADL. The first extreme profile exhibits aging progressions where people remain basically healthy for most of their lives. As we consider the other extreme profiles ($k = 2$ and $k = 3$), we observe what we can describe as a decreasing gradation on the age of onset of disability: around 85 for profile $k = 2$ and around 70 for profile $k = 3$. This last profile describes a very early onset of disability, followed by a long decline. We note that extreme profiles are sorted according to their relative importance in the population (parameter ξ_k). This means that healthy aging trajectories are the most common in the population and that early onset of disability is not so prevalent.

To aid interpretation of the extreme profiles, we consider the quantity

$$Age_{0.5,jk} = -\frac{\beta_{0jk}}{\beta_{1jk}} + C, \tag{9.15}$$

which expresses the age at which an ideal individual of the extreme profile k reaches a 0.5 probability of being unable to perform ADL j. We take these numbers as indicative of the age of onset of disability in ADL-j for extreme profile k. We add the constant $C = 80$ because, before fitting the model, we have re-centered the original age data by subtracting 80 years, as a matter of computational convenience.

Table 9.3 shows posterior estimates of $Age_{0.5,jk}$ for our fitted model. We have sorted the ADLs according to the estimates of $Age_{0.5,jk}$ to give an idea of the sequence in which people start experimenting limitations. We note that the resulting sequence of ADLs remains the same on each extreme

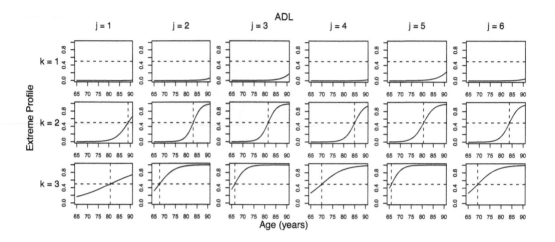

FIGURE 9.3
Extreme trajectories of disability over time for each ADL and extreme profile ($K = 3$). Vertical discontinuous lines mark parameter $Age_{0.5,jk}$

profile. Looking closer at the resulting sequence [inside mobility ($j = 5$) → toileting ($j = 3$) → dressing ($j = 2$) → bathing ($j = 6$) → getting in and out of bed ($j = 4$) → eating ($j = 1$)], we note that it corresponds to what we intuitively expect: the most severe disabilities are the latest to manifest. We also note that, due to the way we have specified individual trajectories in the model formulation, this sequence remains the same for the weighted individual trajectories.

ADL(j)	Extreme Profile-k (sd)					
	$k = 1$		$k = 2$		$k = 3$	
5 (BTH)	95.107	(0.155)	80.539	(0.093)	65.940	(0.164)
3 (MOB)	95.332	(0.139)	81.457	(0.084)	66.399	(0.155)
2 (BED)	97.824	(0.174)	83.156	(0.091)	67.674	(0.179)
6 (TLT)	99.538	(0.231)	83.731	(0.097)	69.315	(0.181)
4 (DRS)	100.210	(0.235)	84.873	(0.105)	69.959	(0.197)
1 (EAT)	104.768	(0.462)	88.933	(0.172)	80.725	(0.477)

TABLE 9.3
Posterior estimates of age of onset of disability (posterior means of parameter $Age_{0.5,jk}$) for model with $K = 3$ extreme profiles. Numbers between parenthesis are posterior standard deviations. ADLs are sorted increasingly according to estimates of $Age_{0.5,jk}$. Note that the sorted sequence of ADLs remains the same for every extreme profile.

9.5 Discussion and Extensions

Mixed membership models are powerful tools in situations in which we believe that a few prototypical or extreme cases can be isolated and analyzed, but we do not necessarily believe that units conform exactly to those cases.

In this chapter we have introduced a family of mixed membership models for longitudinal data, the Trajectory Grade of Membership models. These models characterize extreme profiles using functions that, with the help of time-dependent covariates, express the evolution of responses over time. Individuals have mixed membership on the extreme profiles, meaning that their evolution over time cannot be well described by a single extreme profile, but instead as combinations of the extremes, weighted by their individual membership. Through joint estimation of all the model's parameters from data, these methods allow us to infer the extreme profiles' characteristics (trajectories), the individual membership structure of units from the sample, and the distribution of the population with respect to the extreme profiles.

Our application to the study of disability and aging using data from the National Long Term Care Survey illustrates how TGoM models work. In this application, the extreme trajectories are simplified representations of prototypical ways of aging, expressed as the probability of becoming disabled as a function of age. The mixed membership structure represents the individual heterogeneity, by allowing individuals to follow individualized aging trajectories, described by weighted combinations of the extremes.

TGoM models conform to the general characterization of mixed membership models described in Erosheva et al. (2004) and Erosheva and Fienberg (2005). As such, they admit a number of natural extensions. First, we can expand the characterization of extreme profiles to include any other responses that might be reasonable to joint model. This may include discrete or continuous variables as well as other trajectories. For instance, analyzing the NLTCS, Manrique-Vallier (2010) modeled extreme profiles through the use of trajectories together with survival distributions. This way, extreme profiles did not only summarize typical ways of aging, but also typical survival patterns.

Another natural extension can be obtained by specifying the population-level distribution of individual membership vectors, F_α, conditional on individual-level covariates. Manrique-Vallier (2010; 2013) used this strategy to introduce cohort effects. Noting that as one considers younger cohorts, the distribution of individual membership vectors tends to be more concentrated towards extreme profiles characterized by healthy aging trajectories—to the detriment of other patterns. This allowed the detection of a steady improvement in the quality of aging for younger cohorts.

References

Blei, D. M. and Lafferty, J. D. (2007). A correlated topic model of Science. *Annals of Applied Statistics* 1: 17–35.

Clark, R. F. (1998). An Introduction to the National Long-Term Care Surveys. Office of Disability, Aging, and Long-Term Care Policy with the U.S. Department of Health and Human Services. http://aspe.hhs.gov/daltcp/reports/nltcssu2.htm.

Connor, J. T. (2006). Multivariate Mixture Models to Describe Longitudinal Patterns of Frailty in American Seniors. Ph.D. thesis, Department of Statistics, Carnegie Mellon University, Pittsburgh, Pennsylvania, USA.

Dekker, M. C., Ferdinand, R. F., van Lang, N. D., Bongers, I. L., Van Der Ende, J., and Verhulst,

F. C. (2007). Developmental trajectories of depressive symptoms from early childhood to late adolescence: Gender differences and adult outcome. *Journal of Child Psychology and Psychiatry* 48: 657–666.

Erosheva, E. A. (2002). Grade of Membership and Latent Structures with Application to Disability Survey Data, Ph.D. thesis, Department of Statistics, Carnegie Mellon University, Pittsburgh, Pennsylvania, USA.

Erosheva, E. A. and Fienberg, S. E. (2005). Bayesian mixed membership models for soft clustering and classification. In *Classification–the Ubiquitous Challenge: Studies in Classification, Data Analysis, and Knowledge Organization*. Berlin Heidelberg: Springer, 11–26.

Erosheva, E. A., Fienberg, S. E., and Joutard, C. (2007). Describing disability through individual-level mixture models for multivariate binary data. *Annals of Applied Statistics* 1: 502–537.

Erosheva, E. A., Fienberg, S. E., and Lafferty, J. D. (2004). Mixed-membership models of scientific publications. *Proceedings of the National Academy of Sciences* 101: 5220–5227.

Kreuter, F. and Muthén, B. (2008). Analyzing criminal trajectory profiles: Bridging multilevel and group-based approaches using growth mixture modeling. *Journal of Quantitative Criminology* 24: 1–31.

Manrique-Vallier, D. (2010). Longitudinal Mixed Membership Models with Applications to Disability Survey Data, Ph.D. thesis, Department of Statistics, Carnegie Mellon University, Pittsburgh, Pennsylvania, USA.

Manrique-Vallier, D. (2013). Longitudinal mixed membership trajectory models for disability survey data. *Pre-print*, http://arxiv.org/abs/1309.2324 [stat.AP] (under review).

Manrique-Vallier, D. and Fienberg, S. E. (2008). Population size estimation using individual level mixture models. *Biometrical Journal* 50: 1051–1063.

Manton, K. G., Corder, L., and Stallard, E. (1997). Chronic disability trends in elderly United States populations: 1982–1994. *Proceedings of the National Academy of Sciences* 94: 2593–2598.

Nagin, D. (1999). Analyzing developmental trajectories: A semiparametric, group-based approach. *Psychological Methods* 4: 139–157.

Nagin, D. and Land, K. (1993). Age, criminal careers, and population heterogeneity: Specification and estimation of a nonparametric, mixed Poisson model. *Criminology* 31: 327–362.

Orcutt, H. K., Erickson, D. J., and Wolfe, J. (2004). The course of PTSD symptoms among Gulf War veterans: A growth mixture modeling approach. *Journal of Traumatic Stress* 17: 195–202.

10

An Analysis of Development of Dementia through the Extended Trajectory Grade of Membership Model

Fabrizio Lecci

Department of Statistics, Carnegie Mellon University, Pittsburgh, PA 15213, USA

CONTENTS

Alzheimer's disease is the most frequent form of dementia in the elderly, and age is its most powerful risk factor. One idea is to model the probability of being diagnosed with dementia at different ages in order to construct trajectories for different categories of people. Mixed membership models constitute the most promising method for this problem. We develop a few ideas of Manrique-Vallier (2010) to extend the basic TGoM model. In particular, we propose a parametric dependence between the distribution of the membership vectors and a few time-invariant covariates that allows us to interpret their effect on the individual trajectories.

10.1 Introduction

The previous chapter by Manrique (Manrique-Vallier, 2013) introduced a family of mixed membership models, the Trajectory Grade of Membership models (TGoM), useful in analyzing longitudinal data, i.e., sequences of responses obtained from the same individuals at various points in time. Each of the N individuals in the analysis is represented by a trajectory, which describes the evolution of the probability of particular values of the response variables over time. The individual trajectories are modeled as weighted combinations of a small number K of typical trajectories, corresponding to K ideal types of individuals or *extreme profiles*. Considering only one response variable Y, an individual trajectory at time t can be written as

$$p(y_t|(g_1, \ldots, g_K), x_t, \theta) = \sum_{k=1}^{K} g_k f_{\theta_k}(y_t|x_t).$$

The membership vector $g = (g_1, \ldots, g_K)$ describes the degree of closeness of an individual to each extreme profile; x_t is the value of a time-dependent covariate (e.g., age), and $f_{\theta_k}(y_t|x_t)$ is a function of time with parameter θ_k describing the trajectory of an individual of extreme profile k.

In this chapter we develop a few ideas of Manrique-Vallier (2010) to extend the basic TGoM in two directions. First, we include the survival outcomes as a response: the presence of dementia is correlated with mortality in elderly years (Bowen et al., 1996; Brodaty et al., 2012; Mölsä et al., 1995) and information about survival times can help to explain disability patterns. Second, we add time-invariant covariates in the model. We propose a particular parametric dependence between the membership distribution G_α and time-invariant covariates that allows us to interpret their effect on the membership vector in a way that is similar to the effect of covariates in a simple logistic regression.

10.1.1   Application: The Cardiovascular Health Study—Cognition Study

Alzheimer's disease (AD) is the most common cause of dementia in the elderly, and age is the most important risk factor for the development of clinical dementia. The prevalence of AD increases exponentially between the ages of 65 and 85, approaching 50% in the oldest segment of the population (Evans et al., 1989; Fitzpatrick et al., 2004). After 90 years of age, the incidence of AD increases dramatically, from 12.7%/year in the 90–94 age group, to 21.2%/year in the 95–99 age group, and to 40.7%/year in those over 100 years old (Corrada et al., 2010). This risk of AD is further affected by the presence of the APOE*4 allele, male sex, lower education, and having a family history of dementia (Fitzpatrick et al., 2004; Launer et al., 1999; Tang et al., 1996). Medical risks include the presence of systemic hypertension, diabetes mellitus, cardiovascular disease, and cerebrovascular disease (Irie et al., 2005; Kuller et al., 2003; Luchsinger et al., 2001; Matsuzaki et al., 2010; Ohara et al., 2011; Skoog et al., 1996). Lifestyle factors affecting risk include physical and cognitive activity and diet (Erickson et al., 2010; Scarmeas et al., 2006; Verghese et al., 2003). It is the interactions among these risk factors and the pathobiological cascade of AD that determines the likelihood of a clinical expression of AD—either as dementia or Mild Cognitive Impairment (MCI) (Lopez et al., 2012).

The Cardiovascular Health Study—Cognition Study (CHS-CS) is a rich database of multiple metabolic, cardiovascular, cerebrovascular, and neuroimaging variables obtained over the past 20 years, as well as detailed cognitive assessments beginning in 1990–91(Saxton et al., 2004), 1998–99 (Lopez et al., 2003), 2002–03 (Lopez et al., 2007), and annually thereafter.

In 1992–94, 924 of the CHS participants in Pittsburgh underwent a structural MRI scan of the brain, and these individuals constitute the initial cohort of the Pittsburgh CHS-CS (Kuller et al., 2003). In our analysis we use data from the 652 individuals who were alive in 1998 and who agreed to genetic testing for APOE*4. We consider a single response variable Y that codes diagnosis for each individual at different ages:

$$Y = \begin{cases} 1 & \text{if dementia} \\ 2 & \text{if MCI} \\ 3 & \text{if normal.} \end{cases}$$

Age is the time dependent variable that defines the trajectories. In other words, we are interested in the probability of being diagnosed with MCI or dementia at different ages. We will also consider four time-invariant binary predictors: $X_1 =$ Race (White), $X_2 =$ Education (Beyond High School), $X_3 =$ Hypertension (Present), and $X_4 =$ APOE*4 (Present).

There are a variety of pathways or trajectories that individuals can take as part of the natural history of AD. In order to try to capture these different pathways, we adapt the work of Manrique-Vallier (2010) on modeling trajectories toward disability (Manrique-Vallier and Fienberg, 2009) that combines features of a version of the cross-sectional Grade of Membership model (Erosheva et al., 2007) with those of a longitudinal multivariate latent trajectory model (Connor, 2006). This technique allows our data to identify a small number of theoretically appealing 'canonical' trajectories to dementia or MCI and then express each individual's trajectory as a weighted combination of these canonical trajectories.

10.2 The Extended TGoM Model

In this section we present two extensions of the Trajectory Grade of Membership model. We start by recalling the basics of mixed membership models then gradually include survival outcomes and time-invariant predictors in our analysis.

Mixed membership models assume the existence of a small number of "typical classes" of individuals and model their evolution over time. They regard individuals as belonging to all of these classes in different degree by considering them as weighted combinations of the typical classes. It is possible to describe distinct general tendencies (the typical cases) while accounting for the individual variability.

Following the strategy described in the previous chapter, we start by assuming the existence of a specific number, K, of "typical classes" or "typical profiles" and we associate each individual, i, for $i \in \{1, \ldots, I\}$ (in our application $I = 652$), with its own membership vector $g_i = (g_{i1}, \ldots, g_{iK})$, representing the different degrees of closeness to each typical profile. Membership scores are restricted so that $g_{ik} > 0$ and $\sum_{k=1}^{K} g_{ik} = 1$ for any i. An individual with membership vector $g_i = (0, \ldots, 0, 1, 0, \ldots, 0)$, where 1 is in the kth position, is called an "ideal" (or extreme) individual of class k.

For any individual that is an ideal member of the kth typical class, we specify the distribution of the outcome variable Y_i to form a trajectory for the response variable. Therefore,

$$f_{\theta_k}(y_i|\text{Age}_i) = P(Y_i = y_i|\text{Age}_i, \ i\text{th individual in }k\text{th class})$$

indicates the probability of outcome y_i for an ideal individual of the kth class at a particular age.

We introduce the idea of mixed membership by setting the distribution of the outcome variable Y_i for each individual i as the convex combination

$$P(Y_i = y_i|(g_1, \ldots, g_K), \text{Age}_i) = \sum_{k=1}^{K} g_{ik} f_{\theta_k}(y_i|\text{Age}_i). \tag{10.1}$$

Then we assume that for a single individual, conditional on the age at time t, Age_{it}, and its membership vector, the responses at T measurement times are independent of each other:

$$P(\mathbf{Y_i} = \mathbf{y_i}|(g_1, \ldots, g_K), (\text{Age}_1, \ldots, \text{Age}_T)) = \prod_{t=1}^{T} \sum_{k=1}^{K} g_{ik} f_{\theta_k}(y_{it}|\text{Age}_{it}).$$

We further assume that the individuals are randomly sampled from the population and that the membership vectors are i.i.d. sampled from a common distribution G_α, with support Δ_{K-1} to obtain the unconditional expression

$$P(\mathbf{Y} = \mathbf{y}|\text{Age}) = \prod_{i=1}^{N} \int_{\Delta} \prod_{t=1}^{T} \sum_{k=1}^{K} g_k f_{\theta_k}(y_{it}|\text{Age}_{it}) G(dg).$$

10.2.1   Specifying the Trajectory Function

We must also specify a model for $f_{\theta_k}(y|\text{Age})$. Since in our application the outcome variable diagnosis has three ordered outcomes, we consider an ordered multinomial logit model (Gelman, 2007), described by the two following logistic regressions, for $k = 1, \ldots, K$:

$$P(Y > 1|\text{Age, individual in } k\text{th class}) = \text{logit}^{-1}(\beta_{0k} + \beta_{1k}\text{Age})$$
$$P(Y > 2|\text{Age, individual in } k\text{th class}) = \text{logit}^{-1}(\beta_{0k} + \beta_{1k}\text{Age} - c_k). \qquad (10.2)$$

We then compute the probabilities of individual outcomes using the formulas:

$$\begin{aligned}
P(Y = 1) &= 1 - P(Y > 1), \\
P(Y = 2) &= P(Y > 1) - P(Y > 2), \\
P(Y = 3) &= P(Y > 2).
\end{aligned} \qquad (10.3)$$

Therefore, the expression in (10.1) implicitly contains $\theta_k = (\beta_{0k}, \beta_{1k}, c_k)$, for $k = 1, \ldots, K$. The parameters c_k, which are called thresholds or cutpoints, are constrained to be positive, because the probabilities in (10.2) are strictly decreasing.

10.2.2   First Extension: Specifying the Dependency of Membership Vectors on Additional Covariates

Instead of attributing all variation over time to aging, we could place additional predictors in two different parts of the model. The first alternative is to place them at the level of the extreme profiles, as we have done for the variable Age. The second alternative is to model a dependency between the membership vectors and the new predictors. This is the strategy that we use, since it does not change the interpretation of the extreme profiles given in the previous chapter.

Suppose that, for each of the N individuals in our analysis, we have information about M binary time-invariant predictors $X_1, \ldots, X_M$. We evaluate the effect of these predictors on the proximity of individuals to the three trajectories by allowing the distribution of the membership vectors $g_i = (g_{i1}, g_{i2}, g_{i3})$ to depend on the predictors:

$$g_i|\alpha(\mathbf{x}_i) \sim \text{Dirichlet}(\alpha(\mathbf{x}_i)) \qquad \text{for } i = 1, \ldots, I, \qquad (10.4)$$

where

$$\begin{aligned}
\alpha(\mathbf{x}) = \big(&\exp(a_{01} + a_{11}x_1 + \cdots + a_{M1}x_M), \\
&\exp(a_{02} + a_{12}x_1 + \cdots + a_{M2}x_M), \\
&\cdots \\
&\exp(a_{0k} + a_{1k}x_1 + \cdots + a_{Mk}x_M)\big).
\end{aligned} \qquad (10.5)$$

Then by (10.5) and the properties of the Dirichlet distribution, we can see that

$$\log \frac{\mathbb{E}(g_{i1}|\mathbf{a}, \mathbf{x})}{\mathbb{E}(g_{i2}|\mathbf{a}, \mathbf{x})} = (a_{01} - a_{02}) + (a_{11} - a_{12})x_1 + \cdots + (a_{M1} - a_{M2})x_M,$$
$$\log \frac{\mathbb{E}(g_{i1}|\mathbf{a}, \mathbf{x})}{\mathbb{E}(g_{i3}|\mathbf{a}, \mathbf{x})} = (a_{01} - a_{03}) + (a_{11} - a_{13})x_1 + \cdots + (a_{M1} - a_{M3})x_M,$$
$$\log \frac{\mathbb{E}(g_{i2}|\mathbf{a}, \mathbf{x})}{\mathbb{E}(g_{i3}|\mathbf{a}, \mathbf{x})} = (a_{02} - a_{03}) + (a_{12} - a_{13})x_1 + \cdots + (a_{M2} - a_{M3})x_M,$$

so that we can interpret the difference $(a_{mk} - a_{mh})$ as the effect of variable X_m on the population log odds of the event "individual i has a trajectory near profile k" versus the event "individual i has a trajectory near profile h." Other specifications for the dependency of the distribution of the membership vectors on the time-invariant predictors are possible. See Manrique-Vallier (2010) for another parametric function of the covariates $\alpha(X)$ and Blei and Lafferty (2007); Galyardt (2012) for a logistic-normal prior that replaces the Dirichlet in (10.4).

10.2.3 Second Extension: Modeling Mortality

The presence of dementia is correlated with mortality. Patients with dementia are more likely to die than individuals of the same age without dementia (Bowen et al., 1996; Brodaty et al., 2012; Mölsä et al., 1995). Information about survival times can help to reconstruct certain regions of some trajectory patterns for which information about diagnoses is not sufficient. By design, all subjects in the CHS-CS are older than 65 years, therefore any reference to the distribution of survival time refers to the conditional version, given that the subjects have already lived more than 65 years. Within each canonical profile, we model the random survival time variable (s) in excess of 65 years using the Weibull distribution with inverse scale parameter λ_k and shape parameter δ_k, for $k = 1, \ldots, K$:

$$w(s; \lambda_k, \delta_k) = \delta_k \lambda_k^{\delta_k} s^{\delta_k - 1} e^{-(s\lambda_k)^{\delta_k}}.$$

Our objective is to understand the survival patterns and their effects on the trajectories to dementia. FollowingManrique-Vallier (2010), we make the following assumptions: 1) the canonical profiles specify both trajectories to dementia and mortality distributions; and 2) given the membership vector g_i, the survival time s and the Diagnosis Y are independent. Therefore, the joint model for dementia and mortality can be written as:

$$p(\mathbf{y_i}, s_i | g_i, \mathbf{Age}) = \left[\prod_{t=1}^{T} \sum_{k=1}^{K} g_{ik} f_{\theta_k}(y_{it} | \mathbf{Age}_{it}) \right] \left[\sum_{k=1}^{K} g_{ik} w(s_i | \lambda_k, \delta_k) \right],$$

where the first factor defines the trajectories for MCI and dementia, as described in the previous sections, and the second factor models the individual mortality patterns using the same number K of extreme profiles and membership vector g_i.

10.2.4 Full Bayesian Specification

We complete the Bayesian specification of the model by specifying uninformative priors for the trajectory parameters $\beta_{0k}, \beta_{1k}, c_k$ and the parameters a_{jk} of the membership distribution G_α:

$$\beta_{*k} \sim N(0, 100) \qquad \text{for } k = 1, 2, \ldots, K$$
$$c_k \sim N(0, 100) \qquad \text{for } k = 1, 2, \ldots, K$$
$$a_{jk} \sim N(0, 100) \qquad \text{for } j = 0, 1, \ldots, M \text{ and } k = 1, 2, \ldots, K.$$

We also specify the following priors for the parameters of the Weibull distribution used to model the survival outcomes

$$\delta_k \sim \text{Gamma}(1, 1) \qquad \text{for } k = 1, 2, \ldots, K$$
$$\lambda_k \sim \text{Gamma}(1, 0.1) \qquad \text{for } k = 1, 2, \ldots, K,$$

which are considered diffuse, but realistic to model human survival times in excess of 65 years.

10.3   Application to the CHS Data: Results

We fit the model described in the previous section to the CHS data using BUGS, a software package for Bayesian inference using Gibbs sampling (Lunn et al., 2009). The interested reader is referred to Manrique-Vallier (2010) for more details on an MCMC algorithm used in a similar setting. We report here primarily on a model with $K = 3$ canonical profiles (we discuss the selection of the number of profiles in Section 10.3.4 below). As described in Section 10.1.1 we recall that we consider the single outcome variable diagnoses (three levels: dementia, MCI, normal), the time-varying predictor Age, and four binary time-invariant predictors: X_1 = Race (White), X_2 = Education (Beyond High School), X_3 = Hypertension (Present), and X_4 = APOE*4 (Present).

10.3.1   The Trajectories Toward MCI and Dementia

Figure 10.1 shows the trajectories of the three canonical profiles, determined by the parameters, whose estimated posterior means and standard deviations are shown in Table 10.1. The probability of dementia as a function of age is shown in the left-hand panel, and the probability of MCI is shown in the right-hand panel. The bands around the three profiles are pointwise posterior 95% credible bands and describe the uncertainty related to the estimation of these trajectories. They are constructed using the MCMC draws of the parameters β_{0k} , β_{1k} , and c_k .

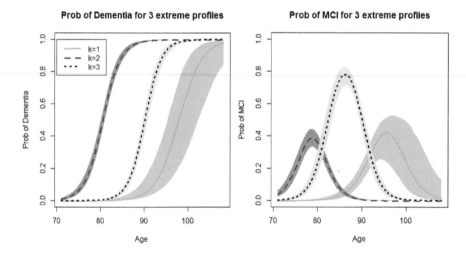

FIGURE 10.1

$K = 3$ typical trajectories for dementia and MCI with pointwise posterior 95% credible bands.

Profile 1 (continuous green curve), the 'healthy' profile, shows the typical or canonical trajectory of individuals whose peak probability of transitioning to MCI occurs between 95 and 100 years of age. This group has only a 50% probability of progressing to dementia by age 100. Profile 2 (dotted red curve), or 'unhealthy' profile, shows the typical or canonical trajectory of individuals who have a peak probability of progressing to MCI between the ages of 75 and 80, and a peak probability of progressing to dementia between the ages of 80 and 85. Finally, Profile 3 (dotted black curve), the 'intermediate' profile, shows the typical or canonical trajectory of individuals having a peak probability of progressing to MCI between 85 and 90 years of age, with a peak probability of progressing to dementia between 90 and 95 years. Figure 10.2 shows two individual trajectories as convex combinations of the canonical profiles as described by Equation (10.1). The trajectory closer to the unhealthy profile belongs to an individual with the following characteristics: non-white, less

Extreme: trajectory's parameter:	Estimate (sd)		
	k=1 (healthy)	k=2 (unhealthy)	k=3 (intermediate)
β_{0*}	38.011 (0.521)	38.904 (0.172)	47.913 (0.161)
β_{1*}	-0.388 (0.054)	-0.483 (0.027)	-0.531 (0.031)
c_*	1.799 (0.334)	1.647 (0.135)	4.197 (0.307)

TABLE 10.1
Posterior means and standard deviations for the parameters defining the three typical trajectories for dementia and MCI.

educated, hypertensive, ApoE4 present. The trajectory closer to the 'healthy' profile belongs to an individual with the opposite characteristics: white, education beyond high school, non-hypertensive, no-ApoE4.

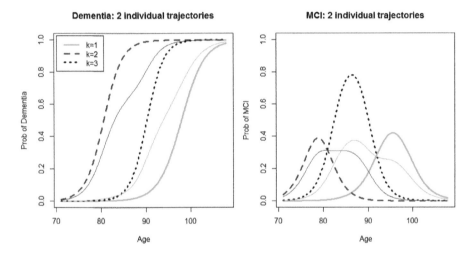

FIGURE 10.2
Two individual trajectories as weighted combinations of the three typical profiles. The trajectory closer to the 'unhealthy' profile belongs to an individual with the following characteristics: non-white, less educated, hypertensive, ApoE4 present. The trajectory closer to the 'healthy' profile belongs to an individual with the opposite characteristics: white, education beyond high school, non-hypertensive, no-ApoE4.

10.3.2 The Effect of Additional Covariates on the Membership Vectors

In order to understand the effects of the four time-invariant covariates on the closeness of an individual to each of the three canonical trajectories, it is necessary to examine the results in Table 10.2. The first three rows of the table show the effect of race on trajectory membership, and we see that for the comparison of Profiles 1 and 2, having race coded as white results in an increased probability of being near the healthy profile relative to being near the unhealthy profile (i.e., mean $a11 - a12 = 1.21$ with posterior 95% credible interval [0.82, 1.62]). In addition, race significantly increases the probability of being in the healthy profile relative to the intermediate profile. With regard to education, having more than a high school education resulted in increased closeness to the healthy profile relative to the unhealthy profile. However, education has no impact on the relative closeness of the intermediate profile to either the healthy or unhealthy profiles. Hypertension is associated with greater closeness to the unhealthy profile relative to the intermediate profile, while the presence of even a single copy of the APOE*4 allele increases the closeness of individuals to the unhealthy profile.

Effect of		Parameter:	Estimate [95% CI]
	Profile 1 Vs 2	$a_{11} - a_{12}$	1.21 [0.82, 1.62]
Race	Profile 1 Vs 3	$a_{11} - a_{13}$	0.39 [-0.15, 0.84]
	Profile 2 Vs 3	$a_{12} - a_{13}$	-0.83 [-1.26, -0.47]
	Profile 1 Vs 2	$a_{21} - a_{22}$	0.50 [0.10, 0.92]
Education	Profile 1 Vs 3	$a_{21} - a_{23}$	0.26 [-0.15, 0.81]
	Profile 2 Vs 3	$a_{22} - a_{23}$	-0.24 [-0.66, 0.17]
	Profile 1 Vs 2	$a_{31} - a_{32}$	-0.26 [-0.60, 0.07]
Hypertension	Profile 1 Vs 3	$a_{31} - a_{33}$	0.18 [-0.22, 0.62]
	Profile 2 Vs 3	$a_{32} - a_{33}$	0.43 [0.13, 0.79]
	Profile 1 Vs 2	$a_{41} - a_{42}$	-0.71 [-1.12, -0.26]
ApoE4	Profile 1 Vs 3	$a_{41} - a_{43}$	0.12 [-0.31, 0.60]
	Profile 2 Vs 3	$a_{42} - a_{43}$	0.83 [0.40, 1.23]

TABLE 10.2
Posterior means and 95% credible intervals for the parameters representing the effects of time-invariant predictors on the closeness of individual trajectories to the typical profiles.

10.3.3   The Survival Trajectories

We also estimated survival trajectories, shown in Figure 10.3, based on the results of Table 10.3. For Profiles 1 and 3 the survival curves are almost overlapping, indicating that for individuals close to these profiles, the probability of being alive is below 50% only after the age of 90. By contrast for individuals that are close to the unhealthy profile, the probability of being alive is below 50% before the age of 90. The difference in the age for a 50% probability of survival is approximately 5 years between the unhealthy profile, and the healthy and intermediate profiles. By contrast, the difference in the age at which the different profiles reach a 50% probability of dementia is approximately 10 years between each trajectory, and at least 20 years between the unhealthy and healthy profiles (See Figure 10.1).

Weibull's: parameter:	Estimate (sd)		
	k=1 (healthy)	k=2 (unhealthy)	k=3 (intermediate)
λ_*	3.887 (0.376)	4.050 (0.256)	5.397 (0.616)
δ_*	0.034 (0.001)	0.040 (0.001)	0.034 (0.001)

TABLE 10.3
Posterior means and standard deviations for the parameters defining the three typical survival trajectories.

10.3.4   Discussion: The Number of Typical Profiles

Finally, in order to evaluate our decision to include only three canonical trajectories in our model, we used the method of posterior predictive testing (Gelman, 2007) to compare the models with

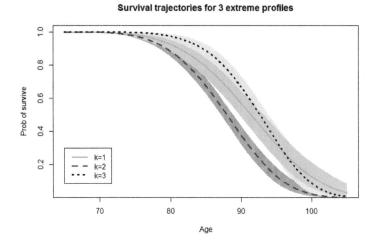

FIGURE 10.3

$K = 3$ typical survival trajectories with pointwise posterior 95% credible bands.

$K = 3$ and $K = 2$ canonical profiles; we found that there were systematic differences between the model with $K = 2$ canonical profiles and the data. Using the estimated posterior distribution of the parameters, we replicated the original diagnoses, obtaining 1000 different simulated datasets. The model with two canonical profiles systematically overestimates the number of individuals that are diagnosed with MCI at least once in their life. The histograms in Figure 10.4 show this test statistic for 1000 simulated datasets for the model with $K = 2$ canonical profiles and the model with $K = 3$ canonical profiles. The vertical bars indicate the true value of the test statistic: 338 individuals have been diagnosed with MCI at least once. Then we compared the original and simulated diagnoses using the proportions of individuals affected by MCI at every age. In Figure 10.5 the black lines represent the true proportion of individuals affected by MCI between the ages of 71 and 105, and the red lines represent the same proportions for 30 simulated datasets. There are some discrepancies between the true proportions and those that were replicated using the model with $K = 2$ canonical profiles, while the proportions simulated through the model with $K = 3$ canonical profiles show no apparent discrepancies. We also attempted a model with four canonical profiles, but the estimation process was very slow to converge, and produced a fourth additional canonical profile that essentially duplicated the healthier one. Based on these results we conclude that the model with $K = 3$ canonical profiles best fits the data.

10.4 Conclusion and Remarks

We reported here the results of an MMTM analysis of the natural history of the development of dementia among individuals over the age of 65. We investigated the relative merits of three separate trajectories, and then identified the effects of four time-invariant covariates on the nearness of individuals to each of these profiles. The results provide new insights into the natural history of AD and

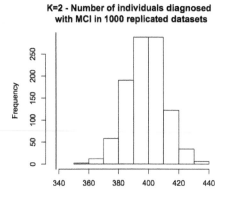
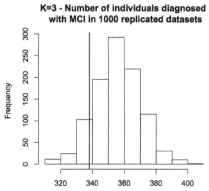

FIGURE 10.4
Posterior predictive check. Number of individuals diagnosed with MCI.

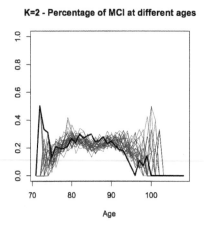
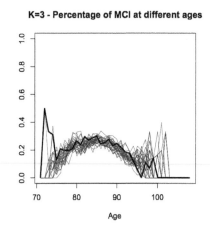

FIGURE 10.5
Posterior predictive check. Proportions of individuals affected by MCI for 30 simulated datasets (thinner red lines) and true proportions (thicker black lines).

related dementias, and may also provide evidence for a potential difference in the pathophysiology of the development of dementia as a function of age.

One of the important characteristics of MMTMs is that individual subjects are assumed to have weighted membership in each of the three canonical trajectories. Thus, while it is theoretically possible for an individual to be an ideal or perfect member of one trajectory, in fact, as shown in Figure 10.2, individuals actually share characteristics of all three profiles to varying degrees. The main extension of the TGoM presented in this chapter involves a particular dependency between the distribution of the membership vectors and the time-invariant predictors added in the model. Particular values of the new covariates help to explain the closeness of individuals to one or another of the ideal trajectories.

Our decision to include three canonical trajectories in our model was based on three separate factors: MCMC convergence time and cost, the model fit, and the interpretability of the trajectories. The three profiles' models not only provided us with a good cost-benefit ratio in terms of processing time and model fit (as assessed by posterior predictive model checking), but also provided interpretable trajectories—an unhealthy trajectory proceeding very rapidly through MCI to dementia, a slow trajectory that does not become apparent until after the age of 90, and an intermediate trajectory through MCI to dementia with a peak probability of a clinical syndrome in the late 80s. We can view the results of the analysis of survivorship as a kind of validation of the three profile model. Thus, the fact that the individuals with an 'unhealthy' trajectory are also the ones most likely to die sooner is consistent with the observation that demented individuals have a higher risk of death (Bowen et al., 1996; Mölsä et al., 1995).

If the use of MMTMs were extended to larger databases with appropriate follow-up and assessment schedules, we might be able to evaluate the relative contributions of other genetic factors, treatment history, and biomarkers on the natural history of dementia.

References

Blei, D. M. and Lafferty, J. D. (2007). A correlated topic model of Science. *Annals of Applied Statistics* : 17–35.

Bowen, J. D., Malter, A., Sheppard, L., Kukull, W., McCormick, W., Teri, L., and Larson, E. (1996). Predictors of mortality in patients diagnosed with probable Alzheimer's disease. *Neurology* 47: 433–439.

Brodaty, H., Heffernan, M., Kochan, N. A., Draper, B., Trollor, J. N., Reppermund, S., Slavin, M. J., and Sachdev, P. S. (2012). Mild cognitive impairment in a community sample: The Sydney memory and ageing study. *Alzheimer's & Dementia.* 9 : 310–317.

Connor, J. T. (2006). Multivariate Mixture Models to Describe Longitudinal Patterns of Frailty in American Seniors. Ph.D. thesis, Department of Statistics, Carnegie Mellon University, Pittsburgh, Pennsylvania, USA.

Corrada, M. M., Brookmeyer, R., Paganini-Hill, A., Berlau, D., and Kawas, C. H. (2010). Dementia incidence continues to increase with age in the oldest old: The 90+ study. *Annals of Neurology* 67: 114–121.

Erickson, K., Raji, C., Lopez, O. L., Becker, J. T., Rosano, C., Newman, A., Gach, H. M., Thompson, P., Ho, A., and Kuller, L. H. (2010). Physical activity predicts gray matter volume in late adulthood: The Cardiovascular Health Study. *Neurology* 75: 1415–1422.

Erosheva, E. A., Fienberg, S. E., and Joutard, C. (2007). Describing disability through individual-level mixture models for multivariate binary data. *Annals of Applied Statistics* 1: 346.

Evans, D. A., Funkenstein, H. H., Albert, M. S., Scherr, P. A., Cook, N. R., Chown, M. J., Hebert, L. E., Hennekens, C. H., and Taylor, J. O. (1989). Prevalence of Alzheimer's disease in a community population of older persons. *JAMA: The Journal of the American Medical Association* 262: 2551–2556.

Fitzpatrick, A. L., Kuller, L. H., Ives, D. G., Lopez, O. L., Jagust, W. J., Breitner, J., Jones, B., Lyketsos, C., and Dulberg, C. (2004). Incidence and prevalence of dementia in the Cardiovascular Health Study. *Journal of the American Geriatrics Society* 52: 195–204.

Galyardt, A. (2012). Mixed Membership Distributions with Applications to Modeling Multiple Strategy Usage. Ph.D. thesis, Department of Statistics, Carnegie Mellon University, Pittsburgh, Pennsylvania, USA.

Gelman, A. and Hill, J. (2007). *Data Analysis Using Regression and Multilevel/Hierarchical Models*. New York, NY: Cambridge University Press.

Irie, F., Fitzpatrick, A. L., Lopez, O. L., Peila, R., Kuller, L. H., Newman, A., and Launer, L. (2005). Type 2 diabetes (T2D), genetic susceptibility and the incidence of dementia in the Cardiovascular Health Study. *Neurology* 64: A316–A316.

Kuller, L. H., Lopez, O. L., Newman, A., Beauchamp, N. J., Burke, G., Dulberg, C., Fitzpatrick, A. L., Fried, L., and Haan, M. N. (2003). Risk factors for dementia in the Cardiovascular Health–Cognition Study. *Neuroepidemiology* 22: 13–22.

Launer, L., Andersen, K., Dewey, M., Letenneur, L., Ott, A., Amaducci, L., Brayne, C., Copeland, J., Dartigues, J. -F., Kragh-Sorensen, P. et al. (1999). Rates and risk factors for dementia and Alzheimer's disease results from EURODEM pooled analyses. *Neurology* 52: 78–78.

Lopez, O. L., Becker, J. T., Chang, Y. -F., Sweet, R. A., DeKosky, S. T., Gach, M. H., Carmichael, O. T., McDade, E., and Kuller, L. H. (2012). Incidence of mild cognitive impairment in the Pittsburgh Cardiovascular Health Study–Cognition Study. *Neurology* 79: 1599–1606.

Lopez, O. L., Jagust, W. J., DeKosky, S. T., Becker, J. T., Fitzpatrick, A. L., Dulberg, C., Breitner, J., Lyketsos, C., Jones, B., Kawas, C. et al. (2003). Prevalence and classification of mild cognitive impairment in the Cardiovascular Health Study–Cognition Study: Part 1. *Archives of Neurology* 60: 1385.

Lopez, O. L., Kuller, L. H., Becker, J. T., Dulberg, C., Sweet, R. A., Gach, H. M., and DeKosky, S. T. (2007). Incidence of dementia in mild cognitive impairment in the Cardiovascular Health Study–Cognition Study. *Archives of Neurology* 64: 416.

Luchsinger, J. A., Tang, M. -X., Stern, Y., Shea, S., and Mayeux, R. (2001). Diabetes mellitus and risk of Alzheimer's disease and dementia with stroke in a multiethnic cohort. *American Journal of Epidemiology* 154: 635–641.

Lunn, D., Spiegelhalter, D. J., Thomas, A., and Best, N. G. (2009). The BUGS project: Evolution, critique and future directions. *Statistics in Medicine* 28: 3049–3067.

Manrique-Vallier, D. (2010). Longitudinal Mixed Membership Models with Applications to Disability Survey Data. Ph.D. thesis, Department of Statistics, Carnegie Mellon University, Pittsburgh, Pennsylvania, USA.

Manrique-Vallier, D. (2013). Mixed membership trajectory models. In Airoldi, E. M., Blei, D. M., Erosheva, E. A., and Fienberg, S. E. (eds), *Handbook of Mixed Membership Models and Its Applications*. Chapman & Hall/CRC.

Manrique-Vallier, D. and Fienberg, S. E. (2009). Longitudinal mixed-membership models for survey data on disability. In *Longitudinal Surveys: From Design to Analysis: Proceedings of the XXV International Methodology Symposium*. Ottawa, QC, Canada: Statistics Canada.

Matsuzaki, T., Sasaki, K., Tanizaki, Y., Hata, J., Fujimi, K., Matsui, Y., Sekita, A., Suzuki, S., Kanba, S., Kiyohara, Y. et al. (2010). Insulin resistance is associated with the pathology of Alzheimer disease: The Hisayama study. *Neurology* 75: 764–770.

Mölsä, P., Marttila, R., and Rinne, U. (1995). Long-term survival and predictors of mortality in Alzheimer's disease and multi-infarct dementia. *Acta Neurologica Scandinavica* 91: 159–164.

Ohara, T., Doi, Y., Ninomiya, T., Hirakawa, Y., Hata, J., Iwaki, T., Kanba, S., and Kiyohara, Y. (2011). Glucose tolerance status and risk of dementia in the community: The Hisayama study. *Neurology* 77: 1126–1134.

Saxton, J., Lopez, O. L., Ratcliff, G., Dulberg, C., Fried, L., Carlson, M., Newman, A., and Kuller, L. H. (2004). Preclinical Alzheimer disease neuropsychological test performance 1.5 to 8 years prior to onset. *Neurology* 63: 2341–2347.

Scarmeas, N., Stern, Y., Tang, M. -X., Mayeux, R., and Luchsinger, J. A. (2006). Mediterranean diet and risk for Alzheimer's disease. *Annals of Neurology* 59: 912–921.

Skoog, I., Nilsson, L., Persson, G., Lernfelt, B., Landahl, S., Palmertz, B., Andreasson, L., Oden, A., and Svanborg, A. (1996). 15-year longitudinal study of blood pressure and dementia. *The Lancet* 347: 1141–1145.

Tang, M. -X., Jacobs, D., Stern, Y., Marder, K., Schofield, P., Gurland, B., Andrews, H., and Mayeux, R. (1996). Effect of oestrogen during menopause on risk and age at onset of Alzheimer's disease. *The Lancet* 348: 429–432.

Verghese, J., Lipton, R. B., Katz, M. J., Hall, C. B., Derby, C. A., Kuslansky, G., Ambrose, A. F., Sliwinski, M., and Buschke, H. (2003). Leisure activities and the risk of dementia in the elderly. *New England Journal of Medicine* 348: 2508–2516.

Part III

Topic Models: Mixed Membership Models for Text

11

Bayesian Nonnegative Matrix Factorization with Stochastic Variational Inference

John Paisley

Department of Electrical Engineering, Columbia University, New York, NY 10027, USA

David M. Blei

Departments of Statistics and Computer Science, Columbia University, New York, NY 10027, USA

Michael I. Jordan

Computer Science Division and Department of Statistics, University of California, Berkeley, CA 94720, USA

CONTENTS

We present stochastic variational inference algorithms for two Bayesian nonnegative matrix factorization (NMF) models. These algorithms allow for fast processing of massive datasets. In particular, we derive stochastic algorithms for a Bayesian extension of the NMF algorithm of Lee and Seung (2001), and a matrix factorization model called correlated NMF, which is motivated by the correlated topic model (Blei and Lafferty, 2007). We apply our algorithms to roughly 1.8 million documents from the *New York Times*, comparing with online LDA (Hoffman et al., 2010b).

11.1 Introduction

In the era of "big data," a significant challenge for machine learning research lies in developing efficient algorithms for processing massive datasets (Jordan, 2011). In several modern data-modeling environments, algorithms for mixed membership and other hierarchical Bayesian models no longer have the luxury of waiting for Markov chain Monte Carlo (MCMC) samplers to perform the tens of thousands of iterations necessary to approximately sample from the posterior, especially when per-iteration runtime is long in the presence of much data. Instead, stochastic optimization methods provide another non-Bayesian learning framework that is better suited to big data environments (Bottou, 1998).

This may seem unfortunate for Bayesian methods in machine learning, however, recent advances have combined stochastic optimization with hierarchical Bayesian modeling (Sato, 2001; Hoffman et al., 2010b; Wang et al., 2011), allowing for approximate posterior inference for "big data." Called *stochastic variational Bayes*, this method performs stochastic optimization on the objective function used in mean field variational Bayesian (VB) inference (Jordan et al., 1999; Sato, 2001; Hoffman et al., 2013). Like maximum likelihood (ML) and maximum a posteriori (MAP) inference methods, VB inference learns a point estimate that locally maximizes its objective function. But unlike ML and maximum MAP, which learn point estimates of a model's parameters, VB learns a point estimate on a set of probability distributions on these parameters.

Since maximizing the variational objective function minimizes the Kullback-Leibler divergence between the approximate posterior distribution and the true posterior (Jordan et al., 1999), variational Bayes is an approximate Bayesian inference method. Because it is an optimization algorithm, it can leverage stochastic optimization techniques (Sato, 2001). This has recently proven useful in mixed membership topic modeling (Hoffman et al., 2010b; Wang et al., 2011), where the number of documents constituting the data can be in the millions. However, the stochastic variational technique is a general method that can address big data issues for other model families as well.

In this paper, we develop stochastic variational inference algorithms for two nonnegative matrix factorization models, which we apply to text modeling. Integrating out the latent indicators of a probabilistic topic model results in a nonnegative matrix factorization problem, and thus the relationship to mixed membership models is clear. The first model we consider is a Bayesian extension of the well-known NMF algorithm of Lee and Seung (2001) with a KL penalty that has an equivalent maximum likelihood representation. This extension was proposed by Cemgil (2009), who derived a variational inference algorithm. We present a stochastic inference algorithm for this model, which significantly increases the amount of data that can be processed in a given period of time.

The second model we consider is motivated by the correlated topic model (CTM) of Blei and Lafferty (2007). We first present a new representation of the CTM that represents topics and documents as having latent locations in $\mathbb{R}^m$. In this formulation, the probability of any topic is a function of the dot-product between the document and topic locations, which introduces correlations among the topic probabilities. The latent locations of the documents have additional uses, which we show with a document retrieval example. We carry this idea into the nonnegative matrix factorization domain and present a stochastic variational inference algorithm for this model as well.

We apply our algorithms to 1.8 million documents from the *New York Times*. Processing this data in the traditional batch inference approach would be extremely expensive computationally since parameters for each document would need to be optimized before global parameters could be updated; MCMC methods are even less feasible. Using stochastic optimization, we show how stochastic VB can quickly learn the approximate posterior of these nonnegative matrix factorization models. Before deriving these inference algorithms, we give a general review of the stochastic VB approach.

We organize the chapter as follows: In Section 11.2 we review the latent indicator approach probabilistic topic modeling, which forms the jumping-off point for the matrix factorization models we consider. In Section 11.3 we review the Bayesian extension to NMF and present an alternate mixture representation of this model that highlights to relationship to existing models (Blei et al., 2003; Teh et al., 2007). In this section we also present correlated NMF, a matrix factorization model with similar objectives as the CTM. In Section 11.4 we review mean field variational inference in both its batch and stochastic forms. In Section 11.5 we present the stochastic inference algorithm for Bayesian NMF and correlated NMF. In Section 11.6 we apply the algorithm to 1.8 million documents from the *New York Times*.

11.2   Background: Probabilistic Topic Models

Probabilistic topic models assume a probabilistic generative structure for a corpus of text documents. They are an effective method for uncovering the salient themes within a corpus, which can help the user navigate large collections of text. Topic models have also been applied to a wide variety of data modeling problems, including those in image processing (Fei Fei and Perona, 2005) and political science (Grimmer, J., 2010(@), and are not restricted to document modeling applications, though modeling text will be the focus of this chapter.

A probabilistic topic model assumes the existence of an underlying collection of "topics," each topic being a distribution on words in a vocabulary, as well as a distribution on these topics for each document. For a K-topic model, we denote the set of topics as $\beta_k \in \Delta_V$, where β_{kv} is the probability of word index v given that a word comes from topic k. For document d, we denote the distribution on these K topics as $\theta_d \in \Delta_K$, where θ_{dk} is the probability that a word in document d comes from topic k.

For a corpus of D documents generated from a vocabulary of V words, let $w_{dn} \in \{1, \ldots, V\}$ denote the nth word in document d. In its most basic form, a latent-variable probabilistic topic model assumes the following hierarchical structure for generating this word,

$$w_{dn} \overset{ind}{\sim} \text{Discrete}(\beta_{z_{dn}}), \quad z_{dn} \overset{iid}{\sim} \text{Discrete}(\theta_d). \tag{11.1}$$

The discrete distribution indicates that $\text{Pr}(z_{dn} = i|\theta_d) = \theta_{di}$.

Therefore, to populate a document with words, one first selects the topic, or theme of each word, followed by the word-value itself using the distribution indexed by its topic. In this chapter, we work within the "bag-of-words" context, which assumes that the N_d words within document d are exchangeable; that is, the order of words in the document does not matter according to the model. We next review two bag-of-words probabilistic topic models.

Latent Dirichlet Allocation. A Bayesian topic model places prior distributions on β_k and θ_d. The canonical example of a Bayesian topic model is latent Dirichlet allocation (LDA) (Blei et al., 2003), which places Dirichlet distribution priors on these vectors,

$$\beta_k \overset{iid}{\sim} \text{Dirichlet}(c_0 \mathbf{1}_V/V), \quad \theta_d \overset{iid}{\sim} \text{Dirichlet}(a_0 \mathbf{1}_K). \tag{11.2}$$

The vector $\mathbf{1}_a$ is an a-dimensional vector of ones. LDA is an example of a conjugate exponential family model; all conditional posterior distributions are closed-form and in the same distribution family as the prior. This gives LDA a significant algorithmic advantage.

Correlated Topic Models. One potential drawback of LDA is that the Dirichlet prior on θ_d does not model correlations between topic probabilities. This runs counter to a priori intuition, which says that some topics are more likely to co-occur than others (e.g., topics on "politics" and "military" versus a topic on "cooking"). A correlated topic model (CTM) was proposed (Blei and Lafferty, 2007) to address this issue. This model replaces the Dirichlet distribution prior on θ_d with a logistic normal distribution prior (Aitchison, 1982),

$$\theta_{dk} = \exp\{y_{dk}\}/\sum_{j=1}^{K} \exp\{y_{dj}\}, \quad y_d \sim \text{Normal}(0, \Sigma). \tag{11.3}$$

The covariance matrix Σ contains the correlation information for the topic probabilities. To allow for this correlation structure to be determined by the data, the covariance matrix Σ has a conjugate inverse Wishart prior,

$$\Sigma \sim \text{invWishart}(A, m). \tag{11.4}$$

The correlated topic model can therefore "anticipate" co-occuring themes better than LDA, but since the logistic normal distribution is not conjugate to the multinomial, inference is not as straightforward.

Matrix Factorization Representations. As mentioned, these hierarchical Bayesian priors are presented within the context of latent indicator topic models. The distinguishing characteristic of this framework is the hidden data z_{dn}, which indicates the topic of word n in document d. Marginalizing out these random variables, one enters the domain of nonnegative matrix factorization (Lee and Seung, 2001; Gaussier and Goutte, 2005; Singh and Gordon, 2008). In this modeling framework, the data is restructured into a matrix of nonnegative integers, $X \in \mathbb{N}^{V \times D}$. The entry X_{vd} is a count of the number of times word v appears in document d. Therefore,

$$X_{vd} = \sum_{n=1}^{N_d} \mathbb{I}(w_{dn} = v). \tag{11.5}$$

Typically, most values of X can be expected to equal zero. Several matrix factorization approaches exist for modeling this representation of the data. In the next section, we discuss two NMF models for this data matrix.

11.3   Two Parametric Models for Bayesian Nonnegative Matrix Factorization

As introduced in the previous section, our goal is to factorize a $V \times D$ data matrix X of nonnegative integers. This matrix arises by integrating out the latent topic indicators associated with each word in a probabilistic topic model, thus turning a latent indicator model into a nonnegative matrix factorization model. The matrix to be factorized is not X, but an underlying matrix of nonnegative latent variables $\Lambda \in \mathbb{R}_+^{V \times D}$. Each entry of this latent matrix is associated with a corresponding entry in X, and we assume a Poisson data-generating distribution, with $X_{vd} \sim \text{Poisson}(\Lambda_{vd})$.

A frequently used model for X is simply called NMF, and was presented by Lee and Seung (1999). This model assumes Λ to be low-rank, the rank K being chosen by the modeler, and factorized into the matrix product $\Lambda = B\Theta$, with $B \in \mathbb{R}_+^{V \times K}$ and $\Theta \in \mathbb{R}_+^{K \times D}$. Lee and Seung (2001) presented optimization algorithms for two penalty functions; in this chapter we focus on the Kullback-Leibler (KL) penalty. This KL penalty has a probabilistic interpretation, since it results in an optimization program for NMF that is equivalent to a maximum likelihood approximation of the Poisson generating model,

$$\{B^*, \Theta^*\} = \max_{B,\Theta} P(X|B,\Theta) = \max_{B,\Theta} \prod_{v,d} \text{Poisson}(X_{vd}|(B\Theta)_{vd}).$$

A major attraction of the NMF algorithm is the fast multiplicative update rule for learning B and Θ (Lee and Seung, 2001). We next review the Bayesian extension of NMF (Cemgil, 2009). We then present a correlated NMF model that takes its motivation from the the latent-indicator correlated topic model.

11.3.1   Bayesian NMF

The NMF model with KL penalty was recently extended to the Bayesian setting under the name Bayesian NMF (Cemgil, 2009). This extension places gamma priors on all elements of B and Θ. The generative process of Bayesian NMF under our selected parameterization is

$$X_{vd} \sim \text{Poisson}(\sum_{k=1}^{K} \beta_{vk}\theta_{kd}), \tag{11.6}$$

$$\beta_{vk} \overset{iid}{\sim} \text{Gamma}(c_0/V, c_0), \quad \theta_{kd} \overset{iid}{\sim} \text{Gamma}(a_0, b_0). \tag{11.7}$$

Note that $\sum_v \beta_{vk} \neq 1$ with probability 1. We also observe that this is not a matrix factorization approach to LDA, though β and θ serve similar functions and have a similar interpretation, as discussed below. Therefore, it is still meaningful to refer to $\beta_{:k}$ as a "topic," and we adopt this convention below.

Just as the latent-variable probabilistic topic models discussed in Section 11.2 have nonnegative matrix factorization representations, the reverse direction holds for Bayesian NMF. The latent-variable representation of Bayesian NMF is insightful since it shows a close relationship with LDA. Using the data-generative structure given in Equation (11.1) (with an additional ˆ to distinguish from Equation (11.7)), the latent topics and distributions on these topics have the following generative process:

$$\hat{\beta}_k \overset{iid}{\sim} \text{Dirichlet}(c_0 \mathbf{1}_V/V), \quad \hat{\theta}_{dk} := \xi_{dk}\tilde{\theta}_{dk} \tag{11.8}$$

$$\xi_{dk} := \frac{e_k}{\sum_{j=1}^{K} \theta_{dj}e_j}, \quad \tilde{\theta}_d \overset{iid}{\sim} \text{Dirichlet}(a_0 \mathbf{1}_K), \quad e_k \overset{iid}{\sim} \text{Gamma}(c_0, c_0). \tag{11.9}$$

The vectors $\hat{\beta}_k$ and $\hat{\theta}_d$ correspond to the topics and document distributions on topics, respectively. Note that $\sum_k \hat{\theta}_{dk} = 1$. We see that when $e_k = 1$, LDA is recovered.[1] Thus, when the columns of B are restricted to the probability simplex, that is, when $e_k = 1$ with probability 1 for each k, one obtains the matrix factorization representation of LDA, also called GaP (Canny, 2004). Relaxing this constraint to gamma distributed random variables allows for a computationally simpler variational inference algorithm for the matrix factorization model, which we give in Section 11.5.1.

The representation in Equations (11.8) and (11.9) shows the motivation for parameterizing the gamma distributions on β as done in Equation (11.7). The desire is for ξ to be close to 1, which results in a model close to LDA. This parameterization gives a good approximation; since c_0 is commonly set equal to a fraction of V in LDA, for example $c_0 = 0.1V$, and because V is often on the order of thousands, the distribution of e_k is highly peaked around 1, with $\mathbb{E}[e_k] = 1$ and $\text{Var}(e_k) = 1/c_0$. Though this latent variable representation affords some insight into the relationship between Bayesian NMF and LDA, we derive a cleaner inference algorithm using the hierarchical structure in Equations (11.6) and (11.7).

11.3.2 Correlated NMF

We next propose a correlated NMF model, which we build on an alternate representation of the correlated topic model (CTM) (Blei and Lafferty, 2007). To derive the model, we first present the alternate representation of the CTM. Following a slight alteration to the prior on the covariance matrix Σ, we show how we can "unpack the information" in the CTM to allow for a greater degree of exploratory data analysis.

Recall that an inverse Wishart prior was placed on Σ, the covariance of the document-specific lognormal vectors, in Equation (11.4). Instead, we propose a Wishart prior,

$$\Sigma \sim \text{Wishart}(\sigma^2 I_K, m), \tag{11.10}$$

and assume a diagonal matrix parameter. Though this change appears minor, it allows for the prior to be expanded hierarchically in a way that allows the model parameters to contain more information that can aid in understanding the underlying dataset.

There are two steps to unpacking the CTM. For the first step, we observe that one can sample Σ from its Wishart prior by first generating a matrix $L \in \mathbb{R}^{m \times K}$, where each entry

---

[1]This additional random variable e_k arises out of the derivation by defining $e_k := \sum_{v=1}^{V} \beta_{vk}$, with β_{vk} drawn as in Equation (11.7). Nevertheless, e_k can be shown to be independent of all other random variables.

$L_{ij} \overset{iid}{\sim} \text{Normal}(0, \sigma^2)$, and then defining $\Sigma := L^T L$. It follows that Σ has the desired Wishart distribution.

Intuitively, with this expansion each topic now has a "location" ℓ_k, being the kth column in L. That is, column k of topic matrix B now has an associated latent location $\ell_k \in \mathbb{R}^m$, where $\ell_k \overset{iid}{\sim} \text{Normal}(0, \sigma^2 I_m)$. Note that when $m < K$, the covariance Σ is not full rank. This provides additional modeling flexibility to the CTM, which in previous manifestations required a full rank estimate of Σ (Blei and Lafferty, 2007).

The second step in unpacking the CTM is to define an alternate representation of $y_d \sim \text{Normal}(0, L^T L)$. We recall that this is the logistic normal vector that is passed to the softmax function in order to obtain a distribution on topics for document d, as described in Equation (11.3). We can again introduce Gaussian vectors, this time to construct y_d:

$$y_d := L^T u_d, \quad u_d \sim \text{Normal}(0, I_m). \tag{11.11}$$

The marginal distribution of y_d, or $p(y_d|L) = \int_{\mathbb{R}^m} p(y_d|L, u_d) p(u_d)\, du_d$, is a $\text{Normal}(0, L^T L)$ distribution, as desired. To derive this marginal, first let $y_d|L, u_d, \epsilon \sim \text{Normal}(L^T u_d, \epsilon)$, next calculate $p(y_d|L, \epsilon) = \text{Normal}(0, \epsilon I + L^T L)$, and finally let $\epsilon \to 0$. As with topic location ℓ_k, the vector u_d also has an interpretation as a location for document d. These locations are useful for search applications, as we show in Section 11.6.

For the latent variable CTM, this results in a new hierarchical prior for topic distribution θ_d. The previous hierarchical prior of Equation (11.3) becomes the following,

$$\theta_{dk} = \frac{\exp\{\ell_k^T u_d\}}{\sum_j \exp\{\ell_j^T u_d\}}, \quad \ell_k \overset{iid}{\sim} \text{Normal}(0, \sigma^2 I_m), \quad u_d \overset{iid}{\sim} \text{Normal}(0, I_m). \tag{11.12}$$

Transferring this into the domain of nonnegative matrix factorization, we observe that the normalization of the exponential is unnecessary. This is for a similar reason as with the random variables in Bayesian NMF, which made the transition from being Dirichlet distributed to gamma distributed. We also include a bias term α_d for each document. This performs the scaling necessary to account for document length.

The generative process for correlated NMF is similar to Bayesian NMF, with many distributions being the same. The generative process below for correlated NMF is

$$X_{vd} \sim \text{Poisson}(\textstyle\sum_{k=1}^K \beta_{vk} \exp\{\alpha_d + \ell_k^T u_d\}), \tag{11.13}$$

$$\beta_{vk} \overset{iid}{\sim} \text{Gamma}(c_0/V, c_0), \quad \ell_k \overset{iid}{\sim} \text{Normal}(0, \sigma^2 I_m), \quad u_d \overset{iid}{\sim} \text{Normal}(0, I_m).$$

The scaling performed by α_d allows the product $\ell_k^T u_d$ to only model random effects. We learn a point estimate of this parameter.

The latent locations introduced to the CTM and this model require the setting of the latent space dimension m. Since we are in effect modeling an m-rank covariance matrix Σ with these vectors, the variety of correlations decreases with m, and the model becomes more restrictive in the distributions on topics it can model. On the other hand, one should set $m \leq K$, since for $m > K$ there are $m - K$ redundant dimensions.

11.4 Stochastic Variational Inference

Text datasets can often be classified as a "big data" problem. For example, Wikipedia currently indexes several million entries, and the *New York Times* has published almost two million articles in the last 20 years. In other problem domains the amount of data is even larger. For example, a hyperspectral image can contain a hundred million voxels in a *single* data cube. With so much data, fast inference algorithms are essential. Stochastic variational inference (Sato, 2001; Hoffman et al., 2013) is a significant step in this direction for hierarchical Bayesian models.

Stochastic variational inference exploits the difference between *local* variables, or those associated with a single unit of data, and *global* variables, which are shared among an entire dataset. In brief, stochastic VB works by splitting a large dataset into smaller groups. These smaller groups can be quickly processed, with each iteration processing a new group of data. In the context of probabilistic topic models, the unit of data is a document, and the global variables are the topics (among other possible variables), while the local variables are document-specific and relate to the distribution on these topics.

Recent stochastic inference algorithms developed for LDA (Hoffman et al., 2010b), the HDP (Wang et al., 2011), and other models (e.g., in Paisley et al., 2012) have shown rapid speed-ups in inference for probabilistic topic models. Though mainly applied to latent-indicator topic models thus far, the underlying theory of stochastic VB is more general, and applies to other families of models. One goal of this chapter is to show how this inference method can be applied to nonnegative matrix factorization, placing the resulting algorithms in the family of online matrix factorization methods (Mairal et al., 2010). Specifically, we develop stochastic variational inference algorithms for the Bayesian NMF and correlated NMF models discussed in Section 11.3.

We next review the relevant aspects of variational inference that make deriving stochastic algorithms easy. Our approach is general, which will allow us to immediately derive the update rules for the stochastic VB algorithm for Bayesian NMF and correlated NMF. We focus on conjugate exponential models and present a simple derivation on a toy example—one for which online inference is not necessary, but which allows us to illustrate the idea.[2]

11.4.1   Mean Field Variational Bayes

Mean field variational inference is an approximate Bayesian inference method (Jordan et al., 1999). It approximates the full posterior of a set of model parameters $p(\Phi|X)$ with a factorized distribution $Q(\Phi) = \prod_i q(\phi_i)$ by minimizing their Kullback-Liebler divergence. This is done by maximizing the variational objective $\mathcal{L}$ with respect to the variational parameters Ψ of Q. The objective function is

$$\mathcal{L}(X, \Psi) = \mathbb{E}_Q[\ln p(X, \Phi)] + \mathbb{H}[Q]. \tag{11.14}$$

When the prior and likelihood of all nodes of the model falls within the conjugate exponential family, variational inference has a simple optimization procedure (Winn and Bishop, 2005). We illustrate this with the following example, which we extend to the stochastic setting in Section 11.4.2. This generic example gives the general form of the stochastic variational inference algorithm, which we later apply to Bayesian NMF and correlated NMF.

Consider D independent samples from an exponential family distribution $p(x|\eta)$, where η is the natural parameter vector. The data likelihood under this model has the standard form

$$p(X|\eta) = \left[\prod_{d=1}^{D} h(x_d)\right] \exp\left\{\eta^T \sum_{d=1}^{D} t(x_d) - DA(\eta)\right\}.$$

The sum of vectors $t(x_d)$ forms the sufficient statistic of the likelihood. The conjugate prior on η has a similar form

$$p(\eta|\chi, \nu) = f(\chi, \nu) \exp\left\{\eta^T \chi - \nu A(\eta)\right\}, \tag{11.15}$$

---

[2]Although Bayesian NMF is not in fact fully conjugate, we will show that a bound introduced for tractable inference modifies the joint likelihood such that the model effectively is conjugate. For correlated NMF, we will also make adjustments for non-conjugacy.

and conjugacy motivates selecting a q distribution in this same family,

$$q(\eta|\chi',\nu') = f(\chi',\nu')\exp\left\{\eta^T\chi' - \nu'A(\eta)\right\}. \tag{11.16}$$

After computing the variational lower bound given in Equation (11.14), which can be done explicitly for this example, inference proceeds by taking gradients with respect to variational parameters, in this case the vector $\psi := [\chi'^T, \nu']^T$, and then setting to zero to find their updated values. For conjugate exponential family models, this gradient has the general form

$$\nabla_\psi \mathcal{L}(X, \Psi) = - \begin{bmatrix} \frac{\partial^2 \ln f(\chi',\nu')}{\partial\chi'\partial\chi'^T} & \frac{\partial^2 \ln f(\chi',\nu')}{\partial\chi'\partial\nu'} \\ \frac{\partial^2 \ln f(\chi',\nu')}{\partial\nu'\partial\chi'^T} & \frac{\partial^2 \ln f(\chi',\nu')}{\partial\nu'^2} \end{bmatrix} \begin{bmatrix} \chi + \sum_{d=1}^{D} t(x_d) - \chi' \\ \nu + D - \nu' \end{bmatrix}, \tag{11.17}$$

and can be explicitly derived from the lower bound. Setting this to zero, one can immediately read off the variational parameter updates from the right vector, which in this case are $\chi' = \chi + \sum_{d=1}^{D} t(x_d)$ and $\nu' = \nu + D$. Though the matrix in Equation (11.17) is often very complicated, it is superfluous to batch variational inference for conjugate exponential family models. In the stochastic optimization of Equation (11.14), however, this matrix cannot be similarly ignored.

We show a visual representation of batch variational inference for Bayesian matrix factorization in Figure 11.1. The above procedure repeats for each variational Q distribution; first for all distributions of the right matrix, followed by those of the left. We note that, if conjugacy does not hold, gradient ascent can be used to optimize ψ.

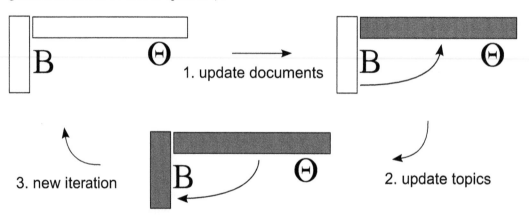

FIGURE 11.1
A graphic describing batch variational inference for Bayesian nonnegative matrix factorization. For each iteration, all variational parameters for document specific (local) variables are updated first. Using these updated values, the variational parameters for the global topics are updated. When there are many documents being modeled, i.e., when the number of columns is very large, step 1 in the image can have a long runtime.

11.4.2 Stochastic Optimization of the Variational Objective

Stochastic optimization of the variational lower bound involves forming a noisy gradient of $\mathcal{L}$ using a random subset of the data at each iteration. Let $C_t \subset \{1, \ldots, D\}$ index this subset at iteration t. Also, let ϕ_d be the model variables associated with observation x_d and Φ_X the variables shared among all observations. In Table 11.1, we distinguish the local from the global variables for Bayesian NMF and correlated NMF.

Model	Local variables	Global variables
Bayesian NMF	$\{\theta_{kd}\}_{k=1:K,d=1:D}$	$\{\beta_{vk}\}_{v=1:V,k=1:K}$
Correlated NMF	$\{u_d, \alpha_d\}_{d=1:D}$	$\{\beta_{vk}, \ell_k\}_{v=1:V,k=1:K}$

TABLE 11.1
Local and global variables for the two Bayesian nonnegative matrix factorization models considered in this chapter. Stochastic variational inference partitions the local variables into batches, with each iteration of inference processing one batch. Updates to the global variables follow each batch.

The *stochastic variational objective function* $\mathcal{L}_s$ is the noisy version of $\mathcal{L}$ formed by selecting a subset of the data,

$$\mathcal{L}_s(X_{C_t}, \Psi) = \frac{D}{|C_t|} \sum_{d \in C_t} \mathbb{E}_Q[\ln p(x_d, \phi_d | \Phi_X)] + \mathbb{E}_Q[\ln p(\Phi_X)] + \mathbb{H}[Q]. \tag{11.18}$$

This constitutes the objective function at step t. By optimizing $\mathcal{L}_s$, we are optimizing $\mathcal{L}$ in expectation. That is, since each subset C_t is equally probable, with $p(C_t) = \binom{D}{|C_t|}^{-1}$, and since $d \in C_t$ for $\binom{D-1}{|C_t|-1}$ of the $\binom{D}{|C_t|}$ possible subsets, it follows that

$$\mathbb{E}_{p(C_t)}[\mathcal{L}_s(X_{C_t}, \Psi)] = \mathcal{L}(X, \Psi). \tag{11.19}$$

Therefore, by optimizing $\mathcal{L}_s$ we are stochastically optimizing $\mathcal{L}$. Stochastic variational optimization proceeds by optimizing the objective in Equation (11.18) with respect to ψ_d, $d \in C_t$, followed by an update to Ψ_X that blends the new information with the old. For example, in the simple conjugate exponential model of Section 11.4.1, the update of the vector $\psi := [\chi'^T, \nu']^T$ at iteration t follows a gradient step,

$$\psi_t = \psi_{t-1} + \rho_t G \nabla_\psi \mathcal{L}_s(X_{C_t}, \Psi). \tag{11.20}$$

The matrix G is a positive definite preconditioning matrix and ρ_t is a step size satisfying $\sum_{t=1}^{\infty} \rho_t = \infty$ and $\sum_{t=1}^{\infty} \rho_t^2 < \infty$, which ensures convergence (Bottou, 1998).

The key to stochastic variational inference for conjugate exponential models is in selecting G. Since the gradient of $\mathcal{L}_s$ has the same form as Equation (11.17), the difference being a sum over $d \in C_t$ rather than the entire dataset, G can be set to the inverse of the matrix in (11.17) to allow for cancellation. An interesting observation is that this matrix is

$$G = -\left(\frac{\partial^2 \ln q(\eta|\psi)}{\partial \psi \partial \psi^T}\right)^{-1}, \tag{11.21}$$

which is the inverse Fisher information of the variational distribution $q(\eta|\psi)$. This setting of G gives the natural gradient of the lower bound, and therefore not only simplifies the algorithm, but gives an efficient step direction Amari (1998); Sato (2001). We note that this is the setting of G given in the stochastic variational algorithm of Sato (2001) and was used in Hoffman et al. (2010b) and Wang et al. (2011) for online LDA and HDP, respectively.

In the case where the prior-likelihood pair does not fall within the conjugate exponential family, stochastic variational inference still proceeds as described, instead using an appropriate G for the gradient step in Equation (11.20). The disadvantage of this regime is that the method truly is a gradient method, with the attendant step size issues. Using the Fisher information gives a clean and interpretable update.

This interpretability is seen by returning to the example in Section 11.4.1, where the stochastic variational parameter updates are

$$\chi_t' = (1 - \rho_t)\chi_{t-1}' + \rho_t \left\{ \chi + \frac{D}{|C_t|} \sum_{d \in C_t} t(x_d) \right\},$$

$$\nu' = \nu + D. \tag{11.22}$$

We see that, for conjugate exponential family distributions, each step of stochastic variational inference entails a weighted averaging of sufficient statistics from previous data with the sufficient statistics of new data *scaled up* to the size of the full dataset. We show a visual representation of stochastic variational inference for Bayesian matrix factorization in Figure 11.2.

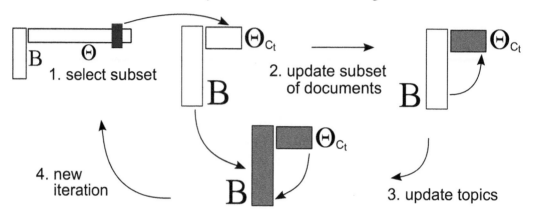

FIGURE 11.2
A graphic describing stochastic variational inference for Bayesian nonnegative matrix factorization. From the larger dataset, first select a subset of data (columns) uniformly at random, indexed by C_t at iteration t; for clarity we represent this subset as a contiguous block. Next, fully optimize the local variational parameters for each document. Because the subset is much smaller than the entire dataset, this step is fast. Finally, update the global topic variational parameters using a combination of information from the local updates and the previously seen documents, as summarized in the current values of these global variational parameters.

11.5   Variational Inference Algorithms

We present stochastic variational inference algorithms for Bayesian NMF and correlated NMF. Table 11.2 contains a list of the variational q distributions we use for each model. The variational objective functions for both models are given below. Since an expectation with respect to a delta function is simply an evaluation at the point mass, we write this evaluation for correlated NMF:

B-NMF: $\mathcal{L} = \mathbb{E}_q \ln p(X|B, \Theta, \boldsymbol{\alpha}) + \mathbb{E}_q \ln p(B) + \mathbb{E}_q \ln p(\Theta) + \mathbb{H}[Q]$

C-NMF: $\mathcal{L} = \mathbb{E}_q \ln p(X|B, L, U, \boldsymbol{\alpha}) + \mathbb{E}_q \ln p(B) + \ln p(L) + \ln p(U) + \mathbb{H}[Q].$

As is evident, mean field variational inference requires being able to take expectations of the log joint likelihood with respect to the predefined q distributions. As frequently occurs with VB inference, this is not possible here. We adopt the common solution of introducing a tractable lower bound for the problematic function, which we discuss next.

Model	Variational q distributions
Bayesian NMF	$q(\beta_{vk}) = \text{Gamma}(\beta_{vk}|g_{vk}, h_{vk})$
	$q(\theta_{kd}) = \text{Gamma}(\theta_{kd}|a_{kd}, b_{kd})$
Correlated NMF	$q(\beta_{vk}) = \text{Gamma}(\beta_{vk}|g_{vk}, h_{vk})$
	$q(\ell_k) = \delta_{\ell_k}, \; q(u_k) = \delta_{u_k}$

TABLE 11.2
The variational q distributions for Bayesian NMF and correlated NMF.

A Lower Bound of the Variational Objective Function

For both Bayesian NMF and correlated NMF, the variational lower bound contains an intractable expectation in the log of the Poisson likelihood. To speak in general terms about the problem, let ω_{kd} represent the document weights. This corresponds to θ_{kd} in Bayesian NMF and to $\exp\{\alpha_d + \ell_k^T u_d\}$ in correlated NMF.

The problematic expectation is $\mathbb{E}_q \ln \sum_{k=1}^{K} \beta_{vk}\omega_{kd}$. Given the concavity of the natural logarithm, we introduce a probability vector $p^{(vd)} \in \Delta_K$ for each (v, d) pair in order to lower-bound this function,

$$\ln\left(\sum_{k=1}^{K} \beta_{vk}\omega_{kd}\right) \geq \sum_{k=1}^{K} p_k^{(vd)} \ln(\beta_{vk}\omega_{kd}) - \sum_{k=1}^{K} p_k^{(vd)} \ln p_k^{(vd)}. \tag{11.23}$$

All expectations of this new function are tractable, and the vector $p^{(vd)}$ is an auxiliary parameter that we optimize with the rest of the model. After each iteration, we optimize this auxiliary probability vector to give the tightest lower bound. This optimal value is

$$p_k^{(vd)} \propto \exp\{\mathbb{E}_q[\ln \beta_{vk}] + \mathbb{E}_q[\ln \omega_{kd}]\}. \tag{11.24}$$

Section 11.5.1 contains the functional forms of these expectations.

11.5.1 Batch Algorithms

Given the relationship between batch and stochastic variational inference, we first present the batch algorithm for Bayesian NMF and correlated NMF, followed by the alterations needed to derive their stochastic algorithms. For each iteration of inference, batch variational inference cycles through the following updates to the parameters of each variational distribution.

Parameter Update for $q(\beta_{vk})$

The two gamma distribution parameters for this q distribution (Table 11.2) have the following updates,

$$g_{vk} = \frac{c_0}{V} + \sum_{d=1}^{D} X_{vd} p_k^{(vd)}, \tag{11.25}$$

$$h_{vk} = c_0 + \sum_{d=1}^{D} \mathbb{E}_q[\theta_{kd}], \qquad \text{(Bayesian NMF)} \tag{11.26}$$

$$h_{vk} = c_0 + \sum_{d=1}^{D} \exp\{\alpha_d + \ell_k^T u_d\}. \quad \text{(Correlated NMF)} \tag{11.27}$$

Expectations used in other parameter updates are $\mathbb{E}_q[\beta_{vk}] = g_{vk}/h_{vk}$ and $\mathbb{E}_q[\ln \beta_{vk}] = \psi(g_{vk}) - \ln h_{vk}$.

Parameter Update for $q(\theta_{kd})$ (Bayesian NMF)

The two gamma distribution parameters for this q distribution (Table 11.2) have the following updates,

$$a_{kd} = a_0 + \sum_{v=1}^{V} X_{vd}p_k^{(vd)}, \tag{11.28}$$

$$b_{kd} = b_0 + \sum_{v=1}^{V} \mathbb{E}_q[\beta_{vk}]. \tag{11.29}$$

Expectations used in other parameter updates are $\mathbb{E}_q[\theta_{kd}] = a_{kd}/b_{kd}$ and $\mathbb{E}_q[\ln \theta_{kd}] = \psi(a_{kd}) - \ln b_{kd}$.

Parameter Updates for $q(\ell_k)$ and $q(u_d)$ (Correlated NMF)

Since these parameters do not have closed-form updates, we use the steepest ascent gradient method for inference. The gradients of $\mathcal{L}$ with respect to ℓ_k and u_k are

$$\nabla_{\ell_k}\mathcal{L} = \sum_{v=1}^{V}\sum_{d=1}^{D}\left(X_{vd}p_k^{(vd)} - \mathbb{E}_q[\beta_{vk}]\exp\{\alpha_d + \ell_k^T u_d\}\right)u_d - \sigma^{-2}\ell_k, \tag{11.30}$$

$$\nabla_{u_d}\mathcal{L} = \sum_{v=1}^{V}\sum_{k=1}^{K}\left(X_{vd}p_k^{(vd)} - \mathbb{E}_q[\beta_{vk}]\exp\{\alpha_d + \ell_k^T u_d\}\right)\ell_k - u_d. \tag{11.31}$$

For each variable, we take several gradient steps to approximately optimize its value before moving to the next variable.

Paramter Update for α_d (Correlated NMF)

The point estimate for α_d has the following closed-form solution,

$$\alpha_d = \ln \sum_{v=1}^{V} X_{vd} - \ln \sum_{v,k} \mathbb{E}_q[\beta_{vk}]\exp\{\ell_k^T u_d\}. \tag{11.32}$$

We update this parameter after each step of u_d.

11.5.2   Stochastic Algorithms

By inserting the lower bound (11.23) into the log joint likelihood and then exponentiating, one can see that the likelihood β_{vk} is modified to form a conjugate exponential pair with its prior for both models. Hence, the discussion and theory of natural gradient ascent in Section 11.4.2 applies to both models with respect to the topic matrix B. For correlated NMF, this does not apply to the global variable ℓ. For this variable, we use the alternate gradient method discussed in Section 11.4.2.

After selecting a subset of the data using the index set $C_t \in \{1, \ldots, D\}$, stochastic inference starts by optimizing the local variables, which entails iterating between the parameter updates for θ_d and $p^{(vd)}$ for Bayesian NMF, and u_d and $p^{(vd)}$ for correlated NMF. Once these parameters have converged, we take a single step in the direction of the natural gradient to update the distributions on $\beta_{:,k}$ and use Newton's method in the step for ℓ_k. We use a step size of the form $\rho_t = (t_0 + t)^{-\kappa}$ for $t_0 > 0$ and $\kappa \in (.5, 1]$. This step size satisfies the necessary conditions for convergence discussed in Section 11.4.2 (Bottou, 1998). We also recall from Section 11.4.2 that D is the corpus size to which each batch C_t is scaled up.

Stochastic Update of $q(\beta_{vk})$

As with batch inference, this update is similar for Bayesian NMF and correlated NMF. In keeping with the generalization at the beginning of this section, we let ω_{kd} stand for θ_{kd} or $\exp\{\alpha_d + \ell_k^T u_d\}$, depending on the model under consideration. The update of the variational parameters of $q(\beta_{vk})$ is

$$
g_{vk}^{(t)} = (1 - \rho_t) g_{vk}^{(t-1)} + \rho_t \left\{ \frac{c_0}{V} + \frac{D}{|C_t|} \sum_{d \in C_t} X_{vd} p_k^{(vd)} \right\}, \tag{11.33}
$$

$$
h_{vk}^{(t)} = (1 - \rho_t) h_{vk}^{(t-1)} + \rho_t \left\{ c_0 + \frac{D}{|C_t|} \sum_{d \in C_t} \mathbb{E}_q[\omega_{kd}] \right\}. \tag{11.34}
$$

As expected from the theory, the variational parameters are a weighted average of their previous values and the sufficient statistics calculated using batch C_t.

Stochastic Update of $q(\ell_k)$ (Correlated NMF)

Stochastic inference for correlated NMF has an additional global variable in the location of each topic. The posterior of this variable—a point estimate—is not conjugate with the prior, and therefore we do not use the natural gradient stochastic VB approach discussed in Section 11.4.2. However, as pointed out in that section, we can still perform stochastic inference according to the general update given in Equation (11.20). With reference to this equation, we set the preconditioning matrix G to be the inverse negative Hessian and update ℓ_k at iteration t as follows,

$$
\ell_k^{(t)} = \ell_k^{(t-1)} + \rho_t G \nabla_{\ell_k} \mathcal{L}_s(X_{C_t}, U_{C_t}, \alpha_{C_t}, L, B), \tag{11.35}
$$

$$
G^{-1} = \sigma^{-2} I_m + \frac{D}{|C_t|} \sum_{d \in C_t} \sum_{v=1}^{V} \mathbb{E}_q[\beta_{vk}] \exp\{\alpha_d + \ell_k^T u_d\} u_d u_d^T.
$$

In batch inference, we perform gradient (steepest) ascent optimization as well. A key difference there is that we fully optimize each ℓ_k and u_d before moving to the next variable—indeed, for stochastic VB we still fully optimize the local variable u_d with steepest ascent during each iteration. For stochastic learning of ℓ_k, however, we only take *one* step in the direction of the gradient for the stochastic update of ℓ_k before moving on to a new batch of documents. In an attempt to take the best step possible, we use the Hessian matrix to construct a Newton step.

11.6 Experiments

We perform experiments using stochastic variational inference to learn the variational posteriors of Bayesian NMF and correlated NMF. We compare these algorithms with online LDA of Hoffman et al. (2010b). We summarize the dataset, parameter settings, experimental setup, and performance evaluation method below.

Dataset. We work with a dataset of 1,819,268 articles from the *New York Times* newspaper. The article dates range from January 1987 to May 2007. We use a dictionary of $V = 8000$ words learned from the data, and randomize the order of the articles for processing.

Parameter Settings. In all experiments, we set the parameter $c_0 = 0.05V$. When learning a K-topic model, for online LDA we set the parameter for the Dirichlet distribution on topic weights to $a_0 = 1/K$, and for Bayesian NMF we set the weight parameters to $a_0 = 1/K$ and $b_0 = 1/K$. For correlated NMF we use a latent space dimensionality of $m = 50$ for all experiments, and set $\sigma^2 = 1/m$.

Experimental setup. We compare stochastic inference for Bayesian NMF and correlated NMF with online LDA. For all models, we perform experiments for $K \in \{50, 100, 150\}$ topics. We also evaluate the inference method for several batch sizes using $|C_t| \in \{500, 1000, 1500, 2000\}$. We use a step size of $\rho_t = (1 + t)^{-0.5}$. Stochastic inference requires initialization of all global variational parameters. For topic-related parameters, we set the variational parameters to be the prior plus a Uniform(0,1) random variable that is scaled to the size of the corpus, similar to the scaling performed on the statistics of each batch. For correlated NMF, we sample $\ell_k \sim \text{Normal}(0, \sigma^2 I_m)$.

Performance Evaluation. To evaluate performance, we hold out every tenth batch for testing. On each testing batch, we perform threefold cross validation by partitioning each document into thirds. Using the current values of the global variational parameters, we then train the local variables on two-thirds of each document and predict the remaining third. For prediction, we use the mean of each variational q distribution. We average the per-word log-likelihoods of all words tested to quantify the performance of the model at the current step of inference. After testing the batch, stochastic inference proceeds as before, with the testing batch processed first—this doesn't compromise the algorithm since we make no updates to the global parameters during testing, and since every testing batch represents a new sample from the corpus.

Experimental Results. Figure 11.3 contains the log-likelihood results for the threefold cross validation testing. Each plot corresponds to a setting of K and $|C_t|$. From the plots, we can see how performance trends with these parameter settings. We first see that performance improves as the number of topics increases within our specified range. Also, we see that as the batch size increases, performance improves as well, but appears to reach a saturation point. At this point, increasing the batch size does not appear to significantly improve the direction of the stochastic gradient, meaning that the quality of the learned topics remains consistent over different batch sizes.

Performance is roughly the same for the three models considered. For online LDA and Bayesian NMF, this perhaps is not surprising given the similarity between the two models discussed in Section 11.3.1. Modeling topic correlations with correlated NMF does not appear to improve upon the performance of online LDA and Bayesian NMF. Nevertheless, correlated NMF does provide some additional tools for understanding the data.

In Figure 11.4 and Table 11.3, we show results for correlated NMF with $K = 150$ and $|C_t| = 1000$. In Table 11.3 we show the most probable words from the 40 most probable topics. In Figure 11.4 we show the correlations learned using the latent locations of the topics. The correlation between topic i and j is calculated as

$$\text{Corr}(\text{topic}_i, \text{topic}_j) = \ell_i^T \ell_j / \|\ell_i\|_2 \|\ell_j\|_j. \tag{11.36}$$

The learned correlations are meaningful. For example, two negatively correlated topics, topic 19 and topic 25, concern the legislative branch and football, respectively. On the other hand, topic 12, concerning baseball, correlates positively with topic 25 and negatively with topic 19. The ability to interpret topic meanings does not decrease as their probability decreases, as we show in Table 11.4.

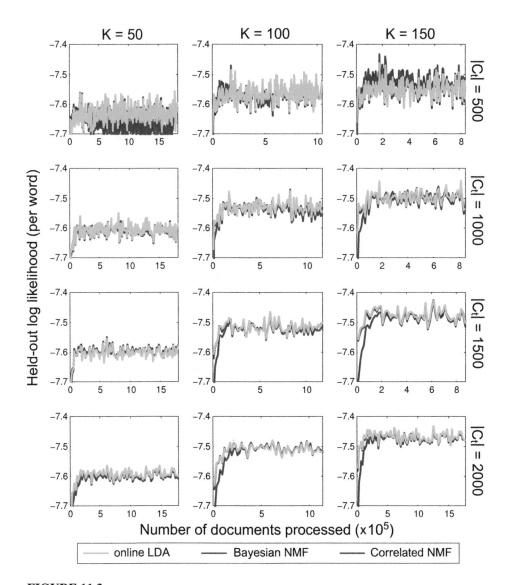

FIGURE 11.3
Performance results for Bayesian NMF, correlated NMF, and online LDA on the *New York Times* corpus. Results are shown for various topic settings and batch sizes. In general, performance is similar for all three models. Performance tends to improve as the number of topics increases. There appears to be a saturation level in batch size, that being the point where the increasing the number of documents does not significantly improve the stochastic gradient. Performance on this dataset does not appear to improve significantly as $|C_t|$ increases over 1000 documents.

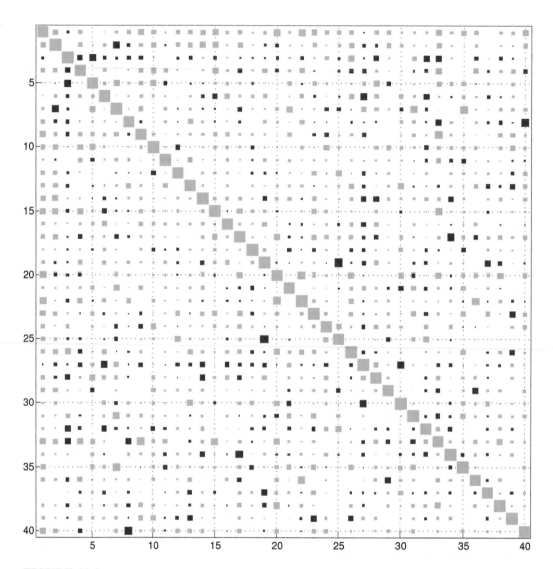

FIGURE 11.4
Correlations learned by correlated NMF for $K = 100$ and $|C_t| = 1000$. The figure contains correlations for the 40 most probable topics sorted by probability. Table 11.3 contains the most probable words associated with each topic in this figure. A green block indicates positive correlation, while a red block indicates negative correlations. The size of the block increases with an increasing correlation. The diagonal corresponds to perfect correlation, and is left in the figure for calibration.

Correlated NMF: Most probable words from the most probable topics

1. think, know, says, going, really, see, things, lot, got, didn
2. life, story, love, man, novel, self, young, stories, character, characters
3. wife, beloved, paid, notice, family, late, deaths, father, mother
4. policy, issue, debate, right, need, support, process, act, important
5. percent, prices, market, rate, economy, economic, dollar, growth, rose
6. government, minister, political, leaders, prime, officials, party, talks, foreign, economic
7. companies, billion, percent, corporation, stock, share, largest, shares, business, quarter
8. report, officials, department, agency, committee, commission, investigation, information, government
9. going, future, trying, likely, recent, ago, hopes, months, strategy, come
10. social, professor, society, culture, ideas, political, study, harvard, self
11. court, law, judge, legal, justice, case, supreme, lawyers, federal, filed
12. yankees, game, mets, baseball, season, run, games, hit, runs, series
13. trial, charges, case, prison, jury, prosecutors, federal, attorney, guilty
14. film, theater, movie, play, broadway, director, production, show, actor
15. best, need, better, course, easy, makes, means, takes, simple, free
16. party, election, campaign, democratic, voters, candidate, republican
17. town, place, small, local, visit, days, room, road, tour, trip
18. military, army, forces, troops, defense, air, soldiers, attacks, general
19. senate, bill, house, congress, committee, republicans, democrats
20. went, came, told, found, morning, away, saw, got, left, door
21. stock, investors, securities, funds, bonds, market, percent, exchange
22. asked, told, interview, wanted, added, felt, spoke, relationship, thought
23. restaurant, food, menu, cook, dinner, chicken, sauce, chef, dishes
24. school, students, education, college, teachers, public, campus
25. team, season, coach, players, football, giants, teams, league, game, bowl
26. family, father, mother, wife, son, husband, daughter, friends, life, friend
27. inc, net, share, reports, qtr, earns, sales, loss, corp, earnings
28. tax, budget, billion, spending, cuts, income, government, percent
29. art, museum, gallery, artists, show, exhibition, artist, works, paintings
30. police, officers, man, arrested, gun, shot, yesterday, charged, shooting
31. executive, chief, advertising, business, agency, marketing, chairman
32. game, points, knicks, basketball, team, nets, season, games, point, play
33. bad, far, little, hard, away, end, better, keep, break, worse
34. study, cancer, research, disease, tests, found, blood, test, cells
35. business, sold, market, buy, price, sell, selling, bought, sale, customers
36. public, questions, saying, response, criticism, attack, news, answer
37. street, park, avenue, west, east, side, village, neighborhood, central
38. feet, right, head, foot, left, side, body, eye, see, eyes
39. system, technology, research, program, industry, development, experts
40. music, band, songs, rock, jazz, song, pop, singer, album, concert

TABLE 11.3
Most probable words from the most probable topics for $K = 150$, $|C_t| = 1000$.

Correlated NMF: Most probable words from less probable topics
41. television, films, network, radio, cable, series, show, fox, nbc, cbs
44. minutes, salt, add, oil, pepper, cup, heat, taste, butter, fresh
48. computer, internet, technology, software, microsoft, computers, digital, electronic, companies, information
61. john, thomas, smith, scott, michael, james, lewis, howard, kennedy
83. iraq, iran, iraqi, hussein, war, saudi, gulf, saddam, baghdad, nations
84. mayor, governor, giuliani, council, pataki, bloomberg, cuomo, assembly
98. rangers, game, goal, devils, hockey, games, islanders, team, goals, season
111. israel, israeli, palestinian, peace, arab, arafat, casino, bank, west
114. china, chinese, india, korea, immigrants, immigration, asia, beijing
134. british, london, england, royal, prince, sir, queen, princess, palace
141. catholic, roman, irish, pope, ireland, bishop, priest, cardinal, john, paul

TABLE 11.4
Some additional topics not given in Table 11.3. Topics with less probability still capture coherent themes. Topics not shown were similarly coherent.

11.7   Conclusion

We have presented stochastic variational inference algorithms for two Bayesian nonnegative matrix factorization models: Bayesian NMF (Cemgil, 2009), a Bayesian extension of NMF (Lee and Seung, 1999); and correlated NMF, a new matrix factorization model that takes its motivation for the correlated topic model (Blei and Lafferty, 2007). Many other nonnegative matrix factorization models are candidates for stochastic inference, for example those based on Bayesian nonparametric priors such as the gamma process (Hoffman et al., 2010a) and the beta process Paisley et al. (2011).

References

Aitchison, J. (1982). The statistical analysis of compositional data. *Journal of the Royal Statistical Society, Series B* 44: 139–177.

Amari, S. (1998). Natural gradient works efficiently in learning. *Neural Computation* 10: 251–276.

Blei, D. M. and Lafferty, J. D. (2007). A correlated topic model of Science. *Annals of Applied Statistics* 1: 17–35.

Blei, D. M., Ng, A. Y., and Jordan, M. I. (2003). Latent Dirichlet allocation. *Journal of Machine Learning Research* 3: 993–1022.

Bottou, L. (1998). *Online Learning and Stochastic Approximations*. Cambridge, UK: Cambridge University Press.

Canny, J. (2004). GaP: A factor model for discrete data. In *Proceedings of the 27th Annual International ACM SIGIR Conference (SIGIR '04)*. New York, NY, USA: ACM, 122–129.

Cemgil, A. (2009). Bayesian inference in non-negative matrix factorisation models. *Computational Intelligence and Neuroscience*. Article ID 785152.

Fei-Fei, L. and Perona, P. (2005). A Bayesian hierarchical model for learning natural scene categories. In *Proceedings of the 10th IEEE Computer Vision and Pattern Recognition (CVPR 2005)*. San Diego, CA, USA: IEEE Computer Society, 524–531.

Gaussier, E. and Goutte, C. (2005). Relation between PLSA and NMF and implication. In *Proceedings of the 28th Annual International ACM SIGIR Conference on Research and Development in Information Retrieval (SIGIR '05)*. New York, NY, USA: ACM, 601–602.

Grimmer, J. (2010). A Bayesian hierarchical topic model for political texts: Measuring expressed agendas in senate press releases. *Political Analysis* 18: 1–35.

Hoffman, M. D., Blei, D. M., and Bach, F. (2010a). Online learning for latent Dirichlet allocation. In Lafferty, J. D., Williams, C. K. I., Shawe-Taylor, J., Zemel, R. S., and Culotta, A. (eds), *Advances in Neural Information Processing Systems 23*. Red Hook, NY: Curran Associates, Inc., 856–864.

Hoffman, M. D., Blei, D. M., and Cook, P. (2010b). Bayesian nonparametric matrix factorization for recorded music. In Fürnkranz, J. and Joachims, T. (eds), *Proceedings of the 27th International Conference on Machine Learning (ICML '10)*. Omnipress, 439–446.

Hoffman, M. D., Blei, D. M., Wang, C., and Paisley, J. (2013). Stochastic variational inference. *Journal of Machine Learning Research* 14: 1303–1347.

Jordan, M. I. (2011). Message from the president: The era of big data. *ISBA Bulletin* 18: 1–3.

Jordan, M. I., Ghahramani, Z., Jaakkola, T., and Saul, L. (1999). An introduction to variational methods for graphical models. *Machine Learning* 37: 183–233.

Lee, D. D. and Seung, H. S. (1999). Learning the parts of objects by non-negative matrix factorization. *Nature* 401: 788–791.

Lee, D. D. and Seung, H. S. (2001). Algorithms for non-negative matrix factorization. In Leen, T. K., Dietterich, T. G., and Tresp, V. (eds), *Advances in Neural Information Processing Systems 13*. Cambridge, MA: The MIT Press, 556–562.

Mairal, J., Bach, F., Ponce, J., and Sapiro, G. (2010). Online learning for matrix factorization and sparse coding. *Journal of Machine Learning Research* 11: 19–60.

Paisley, J., Carin, L., and Blei, D. M. (2011). Variational inference for stick-breaking beta process priors. In *Proceedings of the 28th International Conference on Machine Learning (ICML '11)*. Omnipress, 889–896.

Paisley, J., Wang, C., and Blei, D. M. (2012). The discrete logistic normal distribution. *Bayesian Analysis* 7: 235–272.

Sato, M. (2001). Online model selection based on the variational Bayes. *Neural Computation* 13: 1649–1681.

Singh, A. and Gordon, G. (2008). A unified view of matrix factorization models. In *Proceedings of the European Conference on Machine Learning and Knowledge Discovery in Databases – Part II (ECML/PKDD '08)*. Berlin, Heidelberg: Springer-Verlag, 358–373.

Teh, Y. W., Jordan, M. I., Beal, M. J., and Blei, D. M. (2007). Hierarchical Dirichlet processes. *Journal of the American Statistical Association* 101: 1566–1581.

Wang, C., Paisley, J., and Blei, D. M. (2011). Online learning for the hierarchical Dirichlet process. In *Proceedings of the 14ᵗʰ International Conference on Artificial Intelligence and Statistics (AISTATS 2011)*. Palo Alto, CA, USA: AAAI, 752–760.

Winn, J. and Bishop, C. M. (2005). Variational message passing. *Journal of Machine Learning Research* 6: 661–694.

12

Care and Feeding of Topic Models: Problems, Diagnostics, and Improvements

Jordan Boyd-Graber

Department of Computer Science, University of Colorado, Boulder, CO 80309, USA

David Mimno

Information Science, Cornell University, Ithaca, NY 14850, USA

David Newman

Google Los Angeles, Venice, CA 90291, USA

CONTENTS

Topic models are a versatile tool for understanding corpora, but they are not perfect. In this chapter, we describe the problems users often encounter when using topic models for the first time. We begin with the preprocessing choices users must make when creating a corpus for topic modeling for the first time, followed by options users have for running topic models. After a user has a topic model learned from data, we describe how users know whether they have a good topic model or not and give a summary of the common problems users have, and how those problems can be addressed and solved by recent advances in both models and tools.

12.1 Introduction

Topic models are statistical models for learning the latent structure in document collections, and have gained much attention in the machine learning community over the last decade. Topic models improve the ways users find and discover text content in digital libraries, search interfaces, and across the web, through their ability to automatically learn and apply subject tags to documents in a collection. However, this potential requires practitioners to overcome the problems often associated with topic models: when to use them, how to know when there are problems, how to fix those problems, and how to make topic models more useful.

Topic modeling is an increasingly popular framework for simultaneously soft clustering terms and documents into a fixed number of topics, which take the form of a multinomial distribution over terms in the document collection. Topic models are useful for a variety of research tasks and user-facing applications described below. We start by introducing notation for the original generative topic model, latent Dirichlet allocation (LDA) (Blei et al., 2003).

Latent Dirichlet allocation and its extensions form one popular class of topic models and will be the basis of discussion for this chapter. The LDA topic model is based on the assumption that documents have multiple topics.

In LDA topic modeling, each of D documents in the corpus is modeled as a discrete distribution over T latent topics, and each topic is a discrete distribution over the vocabulary of W words. In the LDA topic model, the number of topics T is fixed and specified by the modeler. For document d, the distribution over topics, $\theta_{t|d}$, is drawn from a Dirichlet distribution $\text{Dir}[\alpha]$, where α might either be a symmetric constant vector (say $\alpha_0 1$) or a hyperparameter with variable values (say $(\alpha_1, ..., \alpha_T)$) which can be estimated. Likewise, each distribution over words, $\phi_{w|t}$, is drawn from a Dirichlet distribution $\text{Dir}[\beta]$.

For the ith token in a document, a topic assignment z_{id} is drawn from $\theta_{t|d}$ and the word, x_{id}, is drawn from the corresponding topic, $\phi_{w|z_{id}}$. Hence, the generative process in LDA is given by

$$\theta_{t|d} \sim \text{Dir}[\alpha] \qquad\qquad \phi_{w|t} \sim \text{Dir}[\beta] \qquad\qquad (12.1)$$

$$z_{id} \sim \text{Mult}[\theta_{t|d}] \qquad x_{id} \sim \text{Mult}[\phi_{w|z_{id}}]. \qquad\qquad (12.2)$$

We can compute the posterior distribution of the topic assignments via Gibbs sampling or variational inference. Given samples from the posterior distribution we can compute point estimates of the document-topic proportions $\theta_{t|d}$ and the word-topic probabilities $\phi_{w|t}$. We will henceforth denote ϕ_t as the vector of word probabilities for a given topic t.

The original LDA topic model has been extended in dozens of ways. Most of the extensions are a result of addressing a potential limitation of LDA, or taking advantage of an opportunity made available by additional data. Some notable extensions include: the correlated topic model (Blei and Lafferty, 2005); the nonparametric topic model, or hierarchical Dirichlet process model (Teh et al., 2006); the hierarchical topic model (Blei et al., 2007); and the dynamic topic model (Blei and Lafferty, 2006). To a large extent, these particular extensions have not directly addressed some of the usability issues we focus on in this chapter.

Nevertheless, there has been a thriving cottage industry adding more and more information to topic models to correct some of the shortcomings we are interested in, either by modeling perspective (Paul and Girju, 2010; Lin et al., 2006), syntax (Wallach, 2006; Gruber et al., 2007), or authorship (Rosen-Zvi et al., 2004; Dietz et al., 2007). Similarly, there has been an effort to inject semantic knowledge into topic models (Boyd-Graber et al., 2007).

12.1.1 Using Topic Models

In the academic literature, topic modeling has been demonstrated to be highly effective in a wide range of research-oriented tasks, including multi-document summarization (Haghighi and Vanderwende, 2009), word sense discrimination (Brody and Lapata, 2009), sentiment analysis (Titov and McDonald, 2008), machine translation (Eidelman et al., 2012), information retrieval (Wei and Croft, 2006), discourse analysis (Purver et al., 2006; Nguyen et al., 2012), and image labeling (Fei-Fei and Perona, 2005). In these tasks the topics are used as features in some larger algorithm, and not as first-order outputs of interest.

Beyond these research-type tasks, topic modeling has been demonstrated in several user-facing applications. Here, the topics themselves are of direct interest. Applications range from search and discovery interfaces to other types of collection analysis interfaces. There are several noteworthy examples, including two from the U.S. funding agencies, NIH and NSF. The NIH Map ViewerTopic[1] is both a topic-based search interface and a map visualizing the research funded by NIH (Talley et al., 2011). The STAR METRICS Portfolio Explorer[2] features topics describing NSF-funded research. Another example is the topic model browser for the journal *Science*.[3]

The remainder of this chapter is organized as follows. In this section, we further introduce topic modeling: how one goes from raw data to a topic model. In Section 12.2, we talk about problems and issues with topic modeling. In Section 12.3, we discuss diagnostics that are useful for detecting and measuring these problems. Finally, in Section 12.4 we review new methods aimed at improving the performance and utility of topic models in addition to those aimed at addressing some of their problems.

12.1.2 Preprocessing Text Data

Topic models take documents that contain words as input. This seems simple enough, but often the process of going from a source document to a form that can be understood by topic models drastically changes the final output. Suppose, for example, that we wanted to build a topic model using Wikipedia as our data source. How would we turn that into a sequence of words that could be used as input to a topic model?

Readers experienced with data processing and natural language processing can safely skip to Section 12.1.3, where we assume that we have the necessary input data for topic modeling.

First, let's take a look at what an individual Wikipedia page looks like:[4]

```
<!DOCTYPE html PUBLIC "-//W3C//DTD XHTML 1.0 Transitional//EN"
"http://www.w3.org/TR/xhtml1/DTD/xhtml1-transitional.dtd">
<html lang="en" dir="ltr" class="client-nojs"
xmlns="http://www.w3.org/1999/xhtml"> <head> <title>Princess
Ida - Wikipedia, the free encyclopedia</title> <meta
http-equiv="Content-Type" content="text/html; charset=UTF-8"
/> <meta http-equiv="Content-Style-Type" content="text/css"
/> <meta name="generator" content="MediaWiki 1.18wmf1" />
```

[1] See https://app.nihmaps.org.

[2] See http://readidata.nitrd.gov/star/.

[3] See http://topics.cs.princeton.edu/Science/.

[4] For this example, we use the HTML representation of a Wikipedia article. This is because it's easy to inspect on the web, isn't restricted by copyrights, and has many of the problems that web corpora have. For real applications, you should **not** use HTML served by Wikipedia's web servers but instead download their XML dumps available at http://dumps.wikimedia.org. This will make your life easier (it lacks many of the problems that we address in this section) and will save both you and Wikipedia bandwidth.

Little in this raw format is what we would call a word, and being able to effectively use this as an input to topic models would require us to do substantial preprocessing. Once we remove extraneous material, we still have to determine what "words" we're going to use and how to extract them from the remaining text. We go through each of these steps to produce a document in a form that is usable for topic modeling.

Many times the files that comprise our corpus have extraneous information that do not add to the *content* of the data. With the *Princess Ida* example, HTML obscures what the underlying words are. We can remove them using a regular expression or a variety of text processing tools (e.g., using the *Natural Language Toolkit* (Bird et al., 2009)).

```
Princess Ida - Wikipedia, the free encyclopedia Princess Ida From
Wikipedia, the free encyclopedia Jump to: navigation , search
Princess Ida; or, Castle Adamant is a comic opera with music by
Arthur Sullivan and libretto by W. S. Gilbert. It was their eighth
operatic collaboration of fourteen.

   ⋮

Personal tools Log in / create account Namespaces Article Discussion
Variants Views Read Edit View history Actions Search Navigation Main
page Contents Featured content Current events Random article
Donate to Wikipedia Interaction Help About Wikipedia Community por-
tal Recent changes Contact Wikipedia Toolbox What links here Related
changes Upload file Special pages Permanent link Cite this page
Print/export Create a book Download as PDF Printable version
Languages Fran\xc3\xa7ais Italiano This page was last modified on 23
September 2011 at 23:59. Text is available under the Creative
Commons Attribution-ShareAlike License; additional terms may apply.
See Terms of use for details. Wikipedia&reg; is a registered trade-
mark of the Wikimedia Foundation, Inc., a non-profit organization.
Contact us Privacy policy About Wikipedia Disclaimers Mobile view
```

Now that we've removed some of the HTML that obscured the content, we can see content that is ofter referred to as boilerplate: text that is repeated verbatim across many documents. Many forms of boilerplate (Freedman, 2007) text appears on this Wikipedia page. Some of it fulfills a legal function ("Text is available under the Creative Commons"), a navigation function ("Search Navigation"), and some of it provides metadata ("last modified on").

While these data are useful and necessary for an HTML page, they do not tell us about the *content* of the document, which is the goal of topic modeling. Failing to remove this boilerplate material can result in the discovery of topics that include just this boilerplate text. Because such text is on many pages, this is often a suboptimal result.

Typically, boilerplate can be removed by heuristics (e.g., removal of the first or last N bytes), or failing that, methods that can discover boilerplate (Kohlschütter et al., 2010). Such text can take many forms: signatures from prolific posters in a newsgroup, legalese in advertisements, contact information in press releases, or quotes appearing at the start of book chapters.

Removing such boilerplate gives us:

Princess Ida; or, Castle Adamant is a comic opera with music by Arthur
Sullivan and libretto by W. S. Gilbert. It was their eighth operatic
collaboration of fourteen. Princess Ida opened at the Savoy Theatre on
January 5, 1884, for a run of 246 performances. The piece concerns a
princess who founds a women's university and teaches that women are
superior to men and should rule in their stead. The prince to whom she
had been married in infancy sneaks into the university, together with
two friends, with the aim of collecting his bride. They disguise themselves as women students but are discovered, and all soon face a
literal war between the sexes.

---

which is finally getting us the content we want. Now we can begin extracting words from the text. Recall that most topic models treat documents as a bag-of-words, so we can stop caring about the order of the tokens within the text and concentrate on how many times a particular word appears in the text.

With this in mind, below we show the sixty most frequent "words" sorted by frequency if we consider words to be anything delimited by whitespace.

the	and	of	to	in	a	The	[	]
Princess	Ida	Gilbert	that	Sullivan	,	%	was	his
(	by	is	with	.	for	as	Carte	at
D'Oyly	her	p.	on	£	).	King	not	she
Lady	had	Act	I	opera	edit	)	but	Opera
are	from	1884	Hilarion	London	Savoy	has	women's	you
were	Hilarion,	In	first	Company	Gama	W.	he	if

Many of these strings are not what we would consider to be words but are instead punctuation. In most applications of topic modeling, we do not care about the punctuation used, so we likely want to remove them. Many of these words are also not content words; words like "the," "and," "of," etc. are functional words that don't provide any information about what the article is about. Such terms are typically called stopwords.

In addition to including items that are not helping us understand what the document is about, we are also making distinctions between words that under most reasonable interpretations should be viewed as identical. For example, the words "Hilarion" and "Hilarion" are considered to be distinct. Similarly, "opera" and "Opera" are considered to be distinct. This suggests that we need to be more aggressive when separating words.

On the other hand, there are also clues that we need to be less aggressive in separating words. For example, there are multi-word expressions that we might want to treat as pseudowords—e.g., "gilbert and sullivan" might be a reasonable multiword expression to treat as a fixed unit, as would "princess ida" and "king gama."[5]

How do we address these issues? These problems are typically viewed as problems of stopword removal, normalization, tokenization, and collocation discovery. We discuss each of them in turn.

Stopword Removal

The most common way to remove words that do not contribute to the meaning of a document is to use a fixed list. Such lists are available in many languages and typically take care of most stopwords. However, such lists are not complete, and there are often corpus-specific stopwords that such lists

---

[5]There has been considerable interest in simultaneously discovering multiword expressions either *after* topic modeling (Blei and Lafferty, 2009) or as part of the process for discovering topics (Johnson, 2010; Hardisty et al., 2010) . However, we view it as a preprocessing step (which is much more efficient).

would never discover. For example, in the Wikipedia corpus, "edit" or "citation" might appear so often in the HTML pages of Wikipedia that they do not serve to differentiate documents.[6]

Rather than having a set list of stopwords, other approaches take an adaptive threshold for which words are stopwords. For example, one could compute the tf-idf (Salton, 1968) of each term in a document and only consider terms that are above some reasonably set threshold.

Normalization

Here, we use normalization in a very broad sense. For a particular concept, there may be many different character strings that can represent it in a language. For instance, "Dog," "Dogs," "dog," and "dogs" both refer to the same underlying concept, except that some are plural, and some are capitalized. For the purposes of topic modeling, we may wish to assume that these are actually the same word. Converting to lower case and applying a stemming algorithm (Porter, 1980) can convert all of these to a cannonical form, "dog."

For languages with a richer morphology (Taghva et al., 2005), this is particularly critical. Failing to do so can lead to an overly large vocabulary (which slows inference) and can lead to poorer topics, as identical words in slightly different syntactic contexts are treated as distinct. However, for English, this is more a matter of taste. When topics are designed for human inspection, many users prefer not to see stemmed words.

Tokenization

Tokenization (or segmentation) is the process for breaking a string of text into its constituent words. For English, whitespace is a good proxy for detecting word boundaries. However, it is not perfect (as we saw above), and there may be other conventions for breaking a string of text into constituent words. For example, Treebank tokenization (Marcus et al., 1993) separates "won't" into "wo" and "n't." Other languages with implicit word boundaries may require more involved preprocessing (Goldwater et al., 2006).

Collocation Discovery

Often, a word's meaning is constrained by its local context (Schemann and Knight, 1995). For example, "house" means one thing, but when it appears together with "white house," it means quite another. Discovering multi-word expressions is a common task in natural language processing (Manning and Schütze, 1999). Often, topic modeling is done while ignoring multiword expressions.

This can lead to suboptimal outcomes for a number of reasons. First, it can lead to topics that join together unrelated concepts. For example, by treating "soviet" and "union" as separate tokens, a topic model might group together documents on the soviet union and the civil war (Chang et al., 2009a) . Even when topic models don't make such errors, it can annoy savvy users who see obvious multi-word expressions separated or displayed in the wrong order (e.g., displaying a topic as "bush," "clinton," "house," and "white" as a topic).

Let us now return to our Wikipedia article on *Princess Ida*, where we identified bigrams scored by point-wise mutual information (PMI), removed stopwords, and tokenized based on all punctuation and whitespace. We did not perform any normalization beyond converting everything to lower case. This gives a much more reasonable list of the most frequent words (seen in the following table).

Note, however, that there are still some problems: "opera" and "operas" are still distinct, "d'oyly carte" was turned into "oyly carte", and "edit" (a wikipedia-specific stopword) are still present. If we believe that these were problematic (or if we saw such issues in the output), we could apply a

---

[6]In practice, one should use Wikipedia XML dump, which would avoid some of these issues; again, we're using the HTML version to give examples of some of the issues that might arise with web corpora.

princess_ida	sullivan	opera	princess	gilbert
chorus	ida	oyly_carte	gilbert_and_sullivan	edit
london	women	1884	first	hilarion
king_hildebrand	may	king_gama	university	act
college	hildebrand	company	lady_blanche	men
one	ainger	melissa	musical	new
piece	productions	florian	gently	lady_psyche
plot	production	recordings	richard	role
rollins_and_witts	savoy	savoy_theatre	tennyson	three
castle_adamant	cyril	early	gama	january_1884
john	man	music	operas	revival
1870	1954	also	although	arthur_sullivan

stemming algorithm that would strip terminal "s" on plurals (at the risk of diminishing interpretability), improve tokenization (at the risk of allowing spurious punctuation to enter words), or add to our stop list (at the risk of removing real content-bearing contributions to documents).

At this point, it's often helpful to look at the most frequent words summed over all documents. This often gives you an idea of where problems might lie. If the results look reasonable, then you can press ahead with inference.

12.1.3 Running Topic Models

There are many different implementations of topic modeling software available;[7] each has (or should!) have its own discussion of how to specifically run the models and prepare input. The goal of this section is not to describe how to run any particular implementation but to talk about what needs to happen to go from raw data to an inferred topic model.

Broadly, implementations fall into two general categories: those that use variational inference (Blei et al., 2003) or Gibbs sampling (Griffiths and Steyvers, 2004)While describing these techniques is outside the scope of this chapter, they both attempt to discover the latent variables that best explain a dataset.

Preparing Data

After completing the steps in Section 12.1.2, the data must be converted into a form that is efficiently readable by software. This takes two steps: selecting the vocabulary and representing the data.

Typically this is done by converting strings into integers (e.g., "opera" is 0, "princess_ida" is 1). Typically you do not want to create an integer for every unique string that appears as a type in your corpus. It increases the amount of memory and time needed to run inference and can also introduce errors from misspellings or tokenization errors. Because natural languages have a power-law distribution, many types only appear in a handful of documents (or one). Including such types is useless for topic models, which attempt to generalize across documents.

Next, the data are reduced to this integer form. There are two ways to do this: representing a document by a single array of integers, with each element in the array corresponding to one appearance of a word, or as two paired arrays a and b, where $a[i]$ represents the identity of a word and $b[i]$ represents the frequency of the word in a document. The former is more common for inference using sampling; the latter is more common for variational inference.

---

[7]For most uses, we suggest Mallet, http://mallet.cs.umass.edu. For particularly large datasets, we suggest Yahoo LDA (Narayanamurthy, 2011) or Mr. LDA (Zhai et al., 2012).

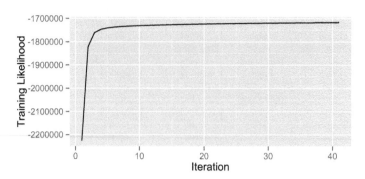

FIGURE 12.1
Training likelihood for variational inference. The shape of the curve shows that inference is increasing likelihood and is nearly converged. Other inference methods may have different convergence profiles, but it should have a similar shape.

Initialization

Both variational inference and Gibbs sampling can be viewed as a search over latent variables. Variational inference searches over variational parameters that induce a distribution over a model's latent variables, and states in the Markov chain for Gibbs sampling are direct assignments of latent variables. Thus, in either case, models must be initialized.

The most important aspect of initialization is to avoid local minima. Some initializations are 'good enough' so that inference will not want to leave the initial state. One common example of this is initializing the variational distributions as uniform distributions; this is a local optimum and will not allow inference to improve upon the initialized state (with boring, identical, uniform topics).

A better approach is to initialize randomly. In practice, this results in either perturbing the initial variational distributions from uniform slightly or, in a Gibbs sampler, setting topic assignments uniformly at random.

Another approach is to initialize the state in a way that might give your algorithm a boost to speed convergence. For example, one could initialize a topic model by initializing each topic with a single document. For other models, other initializations are also possible, but it is important to be aware of the possibility of falling into a local minimum. If inference is working correctly, your model should not be that sensitive to initialization.

Regardless of how you initialize your model and regardless of what inference technique you use, it's important to have many multiple starting points to inference. This guards against problems of local optima and allows you to make better estimates about the stability of your inferred latent variables.

Inference

Running inference itself is the most important step in the process; it produces a learned model from raw data. If you've implemented inference yourself, it is also likely that this aspect has taken the most time.

Typically, implementations work based on a series of iterations. Each iteration updates slightly the state of the algorithm, working slowly toward finding a local optimum. With each iteration, the model should estimate the data likelihood, i.e., given the current guess of the latent variables, what is the probability of recreating the data?

You should watch this quantity closely. If the quantity is consistently going down, it probably means you have a bug. If the quantity is improving steadily, it is a good sign that inference is making progress (although there could be other problems lurking underneath). It is difficult to say how many iterations are needed for inference; it depends on initialization, the data size, the complexity of the model, and what form of inference you're using. However, once the likelihood converges to a value, it is usually a sign that your inference has converged (although this is not always a sure-fire indicator, particularly for MCMC (Neal, 1993)).

Ready-made implementations should provide this information to you; even if you trust the code, you should still pay attention to verify that inference is progressing as it should.

Hyperparameter Updating

Hyperparameters in topic models are those that are not latent variables in the model but instead are the 'most basic' parameters of the topic model. Typically, these are the Dirichlet parameters that are assumed to have generated the per-document topic distributions and the per-topic distributions over words. More generally, these are any unknown parameters that govern latent variables (and are a part of any statistical model, not just topic models).

Particularly if you've derived inference for the model yourself, it's very tempting to set hyperparameters and forget them. After all, you're getting good results, the models are learning interesting things, and you've proved your point. At the risk of editorializing, we would encourage authors to explore sampling hyperparameters:

- It is not that hard, both from the programmer's perspective and from the amount of time it takes the computer;

- If you're using any kind of perplexity or likelihood-based evaluation, you will almost certainly lose to anything that does hyperparameter optimization (Wallach et al., 2009a) ; and

- It will improve the (qualitative) quality of the results.

12.1.4   Evaluation of Topic Models

One of the most important features of topic modeling is that it does not require 'supervision' in the form of annotations. In addition to text documents, many text mining and NLP tasks require additional information such as document-level labels for classification, word-level labels for part-of-speech tagging, phrase-structure trees for parsing, and relevance judgments for information retrieval. With the exception of classification and translation, document creators do not naturally produce such labels, and hiring experts to add annotations can be expensive and time-consuming. In contrast, topic models require only a segmentation of documents into word tokens. They can therefore be applied quickly to large volumes of data.

The benefit of supervised models, however, is that if we take the human-generated labels as a gold standard, measuring and comparing the performance of different methods is simple: we hold out a section of the labeled data as a testing set, train a model on the remaining data, and ask that model to predict labels for the testing set. If the predicted labels match the 'true' labels, the model is effectively learning the association between input data and output labels. In topic modeling, where the model is not trained to predict specific topics, there is no supervised gold standard.

Finding patterns in data is the central goal of topic modeling, but in order to make scientific statements, we must also be able to make predictions about future observations. As an alternative to predicting annotations given previously unseen documents, we can attempt to predict the unseen documents themselves. Simply generating documents and comparing them to a held-out set, however, is not feasible. In classification, there are a finite number of possible document labels. For a given testing document, even random guessing has a reasonably good probability of selecting the correct label. In contrast, the number of possible sequences of words from a vocabulary is exponential in the length of the document. Therefore, rather than measuring accuracy or some rank-based metric, we calculate the marginal probability of the held-out documents under the model. This metric measures the degree to which the model concentrates its probability mass on a relatively small set of 'sensible' documents rather than the vastly larger set of completely random documents.

If, given some held-out document set $\mathbf{w}$, some model A assigns greater marginal probability $p(\mathbf{w}|A)$ than some model B, we assume that model A has more effectively learned the language of the document set than model B. Model A is, in some sense, less 'surprised' by the real documents than model B. Borrowing a term from statistical language modeling, we refer to the negative log probability of the held-out set divided by the number of tokens $-\log p(\mathbf{w}|A)/|\mathbf{w}|$ as the *perplexity* of model A.

Unfortunately, even measuring the marginal probability of a document under a topic model is not computationally tractable due to the exponentially large number of possible topic assignments for words. Good approximations, however, can be evaluated tractably (Wallach et al., 2009b; Buntine, 2009).

Although measurements of held-out probability are important, they are not, by themselves, sufficient. There are several common problems:

- People use topic models to summarize the semantic components of a large document collection, but good predictive power does not necessarily mean that a model provides a meaningful representation of concepts.

- Users frequently distinguish between the quality of different topics: some are seen as coherent or pure, while others are seen as random or illogical. Marginal probability, however, depends on all topics, and therefore cannot be easily decomposed as a function of individual topics.

- Calculations of marginal probability can be sensitive to hyperparameter settings.

12.2 Problems

The topic model is based on the simple assumption that documents contain multiple topics. But is this assumption valid? An article on salary caps in the NFL may be about *sports* and *remuneration*, but do those two topics account for every word written in that article? And is the bag-of-words assumption (that word order is irrelevant) valid? In topic models, every word in a document is probabilistically assigned a topic label, and therefore topics need to explain or account for all words that appear. Is this a reasonable assumption?

Topic models are based on a generative model that clearly does not match the way humans write. However, topic models are often able to learn meaningful and sensible models. Of course, models are learned from the data—a collection of documents—so the quality of the model depends on the quality of the training data.

Most evaluation of topic models has focused on statistical measures of perplexity or likelihood of test data. But this type of evaluation has limitations. The perplexity measure does not reflect the semantic coherence of individual topics learned by a topic model, nor does perplexity necessarily indicate how well a topic model will perform in some end-user task. Recent research has shown potential issues with perplexity as a measure— Chang et al. (2009b) suggests that human judgments can be contrary to perplexity measures.

With this in mind, we pose the following overarching questions relating to evaluating topic models:

Q1 Are individual topics meaningful, interpretable, coherent, and useful?
Q2 Are assignments of topics to documents meaningful, appropriate, and useful?

Q3 Do topics facilitate better or more efficient document search, navigation, understanding, browsing?

While the final question is ultimately the most important for assessing the end-user utility of topic models, it is appropriate to address these questions in order. It doesn't make sense to talk about the quality of assignments of topics to documents if one can't agree on what a topic is about. Although topics themselves are not the end goal (the end goal is to use topics to improve some end-user task), the evaluation framework is built on the usability and usefulness of individual topics, and our focus in this chapter is primarily on the first of the three questions.

12.2.1 Categories of Poor Quality Topics

Before considering bad topics, it is helpful to consider what we are looking for in a topic. The following topic has several good, though not essential, properties:

> trout fish fly fishing water angler stream rod flies salmon...

It is specific. There is a clear focus on words related to the sport of trout fishing. It is coherent. All of the words are likely to appear near one another in a document. Some words (*water, fly*) are ambiguous and may occur in other contexts, but they are appropriate for this context. It is concrete. We can picture the angler with his rod catching trout in the stream. It is informative. Someone unfamiliar with the topic can work from general words (*fishing*) to learn about more unfamiliar words (*angler*). Relationships between entities can be inferred (trout and salmon both live in streams and can be caught in similar ways).

There are a variety of ways topics can be "bad," and we list some of them here. This value judgement is contextual: "good" or "bad" depends on a variety of factors that may involve the task, user, experience, etc. Here we take "bad" as some general idea of lack of usability, usefulness, utility, etc.

General and Specific Words

In any natural language, the most frequent words have less specific meaning, while rare words have very precise meanings. Stopwords such as *the, and, of* are the most extreme examples, but this gradient in specificity remains even after removing such words. For example, in a collection of publications from an artificial intelligence conference, words in the 99th percentile by token frequency might include *algorithm, model, estimation*. At the opposite end, there are large numbers of words that occur only once or twice, such as *dopaminergic* and *phytoplankton*.

> notion sense choice situation idea natural explicitly explicit definition refer...

> level significantly_higher significantly_lower lower higher_lever measured significantly_different investigate differ tended positive_correlation significantly_increased...

> might doesn't fact anyone does isn't mean anyway point quite...

> quite rather couple wasn't far seems less three however point...

Topic models often contain one or more topics consisting of frequent, non-specific words. Users perceive these topics as overly general and therefore not useful in understanding the divisions within a corpus. Such topics often consist of the most frequent words that were not removed as part of the stoplist.

Low-frequency words can also be problematic. Topics that contain many specific words are often perceived as unhelpful because they do not provide a general overview of the corpus. Such topics are also more vulnerable to random chance than topics containing more frequent words because they rely on words with small sample sizes.

Mixed and Chained Topics

Many topics are perceived as low quality by users because they are "mixed" or "chained."

> zinc migraine veterans zn headache magnesium military war zn2 csd affairs episodic deficiency...

A mixed topic can be defined as a set of words $\mathcal{T} = \{w_1, w_2, w_3, ..., w_N\}$ that do not make sense together, but that contains subsets $\mathcal{S}_1, \mathcal{S}_2, ...$, each of which individually form a sensible combination of words. For example, the words

> dog, cat, bird, honda, chevrolet, bmw

do not make sense together, but *dog, cat, bird* describe animals and *honda, chevrolet, bmw* describe makes of cars.

A chained topic is like a mixed topic: a set of words that is low quality overall but contains high quality subsets. The difference is that in a chained topic every high quality subset shares at least one word with another subset. For example, the set

> reagan, roosevelt, clinton, lincoln, honda, chevrolet, bmw

combines the names of U.S. presidents with makes of cars, but *lincoln* can be both categories. Chained topics can be caused by ambiguous words such as *lincoln*, but can also result from hierarchical relationships. A broader concept like *tax* may include several narrower concepts (*sales tax, property tax*). These more specific individual words (*sales, property*) may by themselves form non-sensical combinations.

Identical Topics

One common problem with the topic models learned on corpora is that the topics all look the same (or nearly so). Since topic models are meant to explain a corpus, having identical topics is clearly a suboptimal outcome. We discuss some of the possible causes of this outcome and how you can fix them.

> company customer market product business revenue companies software...

> market product company sale patent companies commercial cost...

One reason that topics might appear to be identical is that the *prior* topic distribution is being observed. Normally, the prior distribution is combined with data to produce a posterior conditioned on that data. However, the prior is still a model of text even without data, and most implementations will happily provide the prior distribution as the "result," even if it has not been supplied with data.

This result might be of particular concern if the inference took a suspiciously short amount of time or if inference chose not to use some of the topics available to it. Both problems are relatively easy to fix—perhaps preprocessing created empty documents or too many topics were chosen.

Incomplete Stopword List

In contrast, one of the symptoms of an incomplete stopword list is topics filled with highly frequent words (but the topics are not identical). Often, the topics discovered are perfectly reasonable, but buried underneath the convention of displaying the n most probable words in a topic.

> vii viii xiv xiii xii xvi xix xviii xvii xxix xxx xxi xxii xxiv xiii...

> david nick elizabeth brad kelsey ted drew theresa ricky russell...

This is often resolved by adding the most frequent words in the topics to the stopword list and then rerunning inference. Alternatively, one could adopt models that have asymmetric priors (Wallach et al., 2009a) or explicitly model syntax (Griffiths et al., 2005; Boyd-Graber and Blei, 2008).

Nonsensical Topics

Another possible problem is that the topics learned will be distinct, but otherwise inscrutable. This is often the result of preprocessing errors or providing the model with too much information.

> tree plum ink blossom **chp** branch bird paper...

Remember that topic models discover words that often appear together in documents. If your "documents" evince a structure that has similar correlation patterns between "words," it will gladly create a topic (we use scare quotes to highlight that the determination of what a document and word is is often subjective and is often impacted by preprocessing steps).

For example, if some documents are created by optical character recognition (OCR), frequent OCR errors will likely occur together; this can create a topic of such errors. Similarly, if metadata are included in the specification of a document, this also might create topics to model this boilerplate material (e.g., as we did in Section 12.1.2).

12.3 Diagnostics

Now we have topics, but how do we know how good the topics are? Traditionally in the literature, measurements have focused on measures based on held-out likelihood (Blei et al., 2003; Blei and Lafferty, 2005) or an external task that is independent of the topic space such as sentiment detection (Titov and McDonald, 2008) or information retrieval (Wei and Croft, 2006). This is true even for models engineered to have semantically coherent topics (Boyd-Graber et al., 2007).

For models that use held-out likelihood, Wallach et al. (2009b) provides a summary of evaluation techniques. These metrics borrow tools from the language modeling community to measure how well the information learned from a corpus applies to unseen documents. These metrics generalize easily and allow for likelihood-based comparisons of different models or selection of model parameters such as the number of topics. However, this adaptability comes at a cost: these methods only measure the probability of observations; the internal representation of the models is ignored.

However, not measuring the internal representation of topic models is at odds with their presentation and development. Most topic modeling papers display qualitative assessments of the inferred topics or simply assert that topics are semantically meaningful, and practitioners use topics for model-checking during the development process. Hall et al. (2008), for example, used latent topics deemed historically relevant to explore themes in the scientific literature. Even in production environments, topics are presented as themes: Rexa,[8] a scholarly publication search engine, displays the topics associated with documents.

In this section, we focus on metrics that *do* pay attention to the underlying topics either by asking individuals directly or by measuring the properties of the discovered topics.

12.3.1 Human Evaluation of Topics

Chang et al. (2009b) presented the following task to evaluate the latent space of topic models. In the word intrusion task, the subject is presented with six randomly ordered words. The task of the user is to find the word which is out of place or does not belong with the others, i.e., the *intruder*.

When the set of words minus the intruder makes sense together, then the subject should easily identify the intruder. For example, most people readily identify *apple* as the intruding word in the set: dog, cat, horse, apple, pig, cow because the remaining words: dog, cat, horse, pig, cow make sense together—they are all animals. For the set: car, teacher, platypus, agile, blue, Zaire, which lacks such coherence, identifying the intruder is difficult. People will typically choose an intruder at random, implying a topic with poor coherence.

In order to construct a set to present to the subject, they select a topic from the model. They then select the five most probable words from that topic. In addition to these words, an intruder word is selected at random from a pool of words with low probability in the current topic (to reduce the possibility that the intruder comes from the same semantic group) but high probability in some other topic (to ensure that the intruder is not rejected outright due solely to rarity). All six words are then shuffled and presented to the subject.

What Topics Make Sense?

The word intrusion task was applied to two corpora: The *New York Times* (Sandhaus, 2008) and Wikipedia,[9] two real-world corpora that are viewed by millions of people each day. Figure 12.2 shows the spectrum from incoherent to coherent topics.

An additional finding was that there was not a clear association between traditional measures of topic models, such as held-out log-likelihood and more intuitive measures such as the word intrusion task.

12.3.2 Topic Diagnostic Metrics

While the techniques described in the previous section are useful, they are time consuming and relatively expensive. Are there ways to measure topic quality without relying on human judgments? Fortunately, there are several useful topic diagnostic metrics that depend only on statistics of individual words in a topic without considering relationships between words or external knowledge sources. None of these metrics is conclusive by itself, but taken together they can provide a useful automated summary of topic quality. As a running example, we consider a model trained with 100 topics on a corpus of political blogs from the 2008 U.S. presidential election.

Topic Size

Most topic model inference methods work by assigning the word tokens in a corpus to one of K topics. We can add up the number of tokens or fractions of a token assigned to a given topic to get a measure of topic size, where the unit is the number of word tokens. There is a strong relation between this measure of topic size and perceived topic quality: very small topics are frequently

---

[8] See http://rexa.info.
[9] See www.wikipedia.org.

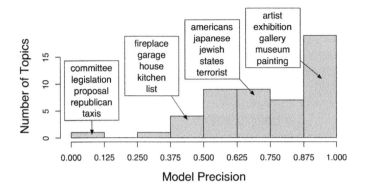

FIGURE 12.2

A histogram of the model precisions (the proportion of times users found the "intruder") on the *New York Times* corpus evaluated on a 50-topic LDA model. Example topics are shown for several bins; the topics in bins with higher model precision evince a more coherent theme.

bad (Talley et al., 2011). As an example, in the 2008 political blog model, the smallest topic by token count is *http player video window flag script false scriptalreadyrequested www*, with around 6,500 words (most topics lie between 15,000 and 20,000). This topic appears to represent URLs for embedded videos. Although it is arguably interpretable, it is not the sort of conceptual topic that many users may be looking for.

There are several possible explanations for this relationship. The most common topics in a corpus are usually well-represented in many documents. For the less-frequent topics, the model must estimate their word distribution from a smaller sample size. Smaller topics are also more vulnerable to become mixed with other topics because they do not 'own' their distribution as well.

Word Length

This metric measures the average length of the top N words in a topic. The usefulness of this metric varies by corpus, but in many cases it can be useful in picking up anomalous topics. The intuition is that words with more specific meaning tend to be longer, and vice versa. Examples include topics consisting of stopwords from a language other than the primary language of the corpus, and topics with many short acronyms, which are frequently ambiguous. In the political blog corpus, the topics with the smallest average word length are *legislator usmc aye nc ny fl pa oh ca tx va* (2.7 characters) and *re ll exit don doesn ve isn didn maverick guy* (4.15 characters). The *legislator* topic appears to represent abbreviations for U.S. states, perhaps related to legislative roll call voting. As with the previous metric, word length in this case does not necessarily indicate that a topic is uninterpretable, but it flags the fact that this is a different sort of cluster of co-occurring words. The *re* topic is more problematic, and indicates that there may be problems with tokenization of contractions such as *you're* or *don't*, possibly due to differences in character encodings for the apostrophe.

Distance from a Corpus Distribution

A topic is a probability distribution over the vocabulary of a corpus. We can define a "global" topic by counting the number of times each word is used in all documents and normalizing those counts. Topic distributions that are similar to this corpus-level distribution according to some measure of similarity between distributions, such as Jensen-Shannon distance or Hellinger distance, consist of the most common words in a corpus. These topics are often perceived as useless or overly general (AlSumait et al., 2009). The most common non-stopwords need to be assigned somewhere, so having a small number of these overly general topics may help to improve the quality of other topics, but it may not be necessary to display them to users.

Distance from the corpus distribution is most useful for documents that contain formulaic or administrative language, such as grant proposals. In corpora focused on a particular issue, this metric may be less useful. The most frequent words in the 2008 political blog corpus are *iraq war country states military security*, indicating that the corpus is dominated by discussion of the Iraq war. The closest topic to this overall distribution is *iraq troops war surge iraqi withdrawal security petraeus military forces*, which is a useful, coherent topic.

Difference between Token and Document Frequencies

We typically rank words within a topic by the number of word tokens (or fractional tokens) of a particular type that have been observed in the topic. We can also rank words within a topic by the number of documents that contain at least one token of a particular type in that topic. The difference between the token-based distribution over words and this document-based distribution is useful in identifying words that are prominent in a topic due to the burstiness of words. When a corpus contains many long documents, it is common for a word that is specific to a single document to occur often enough in that one document that it appears in the list of N top words for a topic. The highest ranking topic according to this metric is the *re* topic mentioned previously, where the tokens *re* and *ll* are the most bursty, possibly reflecting occasional use of second-person pronouns. The metric can also detect outlier words in otherwise more usable topics. The second most bursty topic is *financial crisis bailout fannie mortgage loans wall banks*, where the term *fannie* (referring to the U.S. financial entity known as Fannie Mae) is the most bursty.

Prominence within Documents

Topics often represent the major themes of a document, but they can also be clusters of "methodological" words, like words describing measurement (*larger, smaller, fast*) or days of the week. A good method for distinguishing between important topics and these more functional topics is to examine the proportion of documents assigned to a topic. The names of months may occur many times in a corpus, and more consistently with each other than any other words, but no documents are dominated by month names in the way that a document might be about molecular biology or a political debate. This property can be defined mathematically in several ways. One method is to count the number of documents such that the estimated probability of topic k $\tilde{\theta}_k$ is above some threshold, such as 0.2. Another is to count the number of documents for which topic k is the single largest topic. For example, the topic *meeting official officials conference visit senior reported event friday* is relatively large, with over 42,000 tokens, but it never appears as the single largest topic in any document. Meetings and conferences occur frequently, but are not by themselves worth discussing in great depth. In contrast, the topic *franken coleman ballots minnesota votes recount al board counted* has only a quarter of the total tokens of the *meeting* topic, but is the largest topic in 12.5% of the documents it appears in. This topic, about a contested senate election, refers to a specific event that is discussed in depth when it is discussed at all.

Burstiness

Many of the problems people observe in topic models are caused by the phenomenon of *burstiness* in natural language documents. This property states that within a context, for example a short document, there will be a small set of words that are globally rare but locally common.

Burstiness is related to, but distinct from, well-known power law properties of natural language. If we construct a list of all the distinct words in a corpus of documents and record, for each word, the number of documents that contain at least one instance of that word, the vast majority of those words will be rare, that is, they will occur in very few documents. The most common words, on the other hand, will make up roughly one half of the tokens in any given document. This relationship is known as Zipf's law.

Zipfian dynamics suggest that many of the words in a document will be rare, but burstiness describes an additional level of non-uniformity. It is not only likely that many of the tokens in a document will be rare, but it is also likely that many of them will be the *same rare word*. For example, assume you know the overall word-frequency statistics of a corpus. You can estimate the probability of every distinct word by dividing the number of occurrences of that word by the total number of tokens in the corpus $\mathcal{C}$. If you know nothing about a certain document, these corpus-level frequencies provide a reasonable estimator of the probability that a randomly selected word from that document will be, for example, *elephant*. For most words this probability will be a small number $p(w|\mathcal{C}) = \epsilon$. Once you have observed a particular word, however, the probability that the next word sampled at random from the same document will be of the same type is much larger than ϵ.

This phenomenon of burstiness violates the assumptions of a topic model, which assert that if we know the topic for a token position in a document, the probability that the word at that position is a particular type is independent of the document. When a topic is well represented in a corpus and most documents are short, the violations of this assumption may be averaged out. If there are long documents, however, the bursty words in those documents may have high prevalence in a topic despite not being representative of the central concept of a topic. Similarly, if a topic appears in only a few documents, each of which has its own bursty subset of the words that are associated with the topic, the topic may appear idiosyncratic or nonsensical.

When confronted with a bursty corpus, it may be useful to filter your documents so that documents are of similar length, perhaps by removing abnormally long documents or by breaking very long documents into smaller documents. It may also be worthwhile to consider particularly bursty words as stopwords to prevent them from dominating topics.

12.3.3 Topic Coherence Metrics

Our goal of answering whether individual topics are interpretable and coherent is partly addressed by the human evaluation of topics in Section 12.3.1. But how can we automatically measure topic coherence? And can we do this without disturbing the topic by adding intruder words? Earlier work presented an unsupervised approach to ranking topic significance and identifying what they call "junk" or "insignificant" topics (AlSumait et al., 2009). However, it was unclear to what extent their unsupervised approach and objective function agreed with human judgments, as they presented no user evaluations.

Subsequent work demonstrated that it is possible to automatically measure topic coherence with near-human accuracy (Newman et al., 2010a;b) using a topic coherence score based on pointwise mutual information of pairs of terms taken from topics. In both Newman et al. (2010a) and Newman et al. (2010b), 6000 human evaluations are used to show that their coherence score broadly agrees with human-judged topic coherence. Similar approaches further confirmed that humans agree with word-pair based topic coherence metrics (Mimno et al., 2011).

Topic coherence metrics are motivated by measuring word association between pairs of words in the list of the top-10 most likely topic words (here, top-10 is chosen arbitrarily as the typical number of terms displayed to a user; other settings such as top-20 could work equally as well). The intuition is that a topic will likely be judged as coherent if pairs of words from that topic are associated. Devising word association measures is a long-studied problem in computational linguistics. We opt for co-occurrence-based metrics that use corpus aggregates of the number of times two words are seen in a document. There are two flavors of counting term co-occurrences: either using a sliding window of fixed size (e.g., Do two terms appear in a window of 20 consecutive words?), or binarized at the document level (e.g., Does this document contain both these terms?). The former makes the metric more biased toward short-range dependencies.

For a final twist, we could either use the training corpus to count term co-occurrences, or we could opt for an *external* corpus to obtain these counts. The former is certainly easier, but one may be concerned that the training corpus is not representative—or may be polluted by unusual termwise statistics—as may happen in a text collection of blogs or tweets. In this case, the external corpus could come from a variety of sources, for example the entire collection of English Wikipedia articles.

Our topic coherence metrics take the form

$$\text{TC-f}(\mathbf{w}) = \sum_{i<j} f(w_i, w_j), i, j \in \{1 \ldots 10\}, \tag{12.3}$$

where $\mathbf{w} = \{w_1, w_2, \ldots, w_{10}\}$ are the top-10 most likely terms in a topic, and f is some function measuring the association between words w_i and w_j.

Let $N(w_i, w_j)$ be the number of times word w_i and w_j co-appear in a sliding window of fixed width (say 20 terms), applied to every document in the corpus used to obtain co-occurrence counts. Furthermore, $N(w_i)$ is the total count of times w_i appeared in that sliding window. Let $M(w_i, w_j)$ be the number of distinct documents where words w_i and w_j co-appear, and $M(w_i)$ is the total number of distinct documents that include term w_i. We create different metrics by using N or M to convert counts to probabilities, using the appropriate normalization. Two obvious quantities are pointwise mutual information (PMI) and log conditional probability (LCP). Note that PMI is symmetric, whereas LCP is one-sided.

$$\text{PMI}(w_i, w_j) = \log \frac{P(w_i, w_j)}{P(w_i)P(w_j)}, \tag{12.4}$$

$$\text{LCP}(w_i, w_j) = \log \frac{P(w_i, w_j)}{P(w_j)}. \tag{12.5}$$

Using these, we define the following three topic coherence metrics:

$$\text{TC-PMI}(\mathbf{w}) = \sum_{i<j} \text{PMI}(w_i, w_j), \tag{12.6}$$

$$\text{TC-LCP}(\mathbf{w}) = \sum_{i<j} \text{LCP}(w_i, w_j), \tag{12.7}$$

$$\text{TC-NZ}(\mathbf{w}) = \sum_{i<j} \mathbf{1}[N(w_i, w_j) = 0], \tag{12.8}$$

where all sums are over $i, j \in \{1 \ldots 10\}$. Note, we have added a third metric that simply counts the number of word pairs that are *never* observed in the reference corpus. These topic coherence metrics can be computed four different ways: using sliding window (N) or binarized (M) counts, obtained from training data or external data. For LCP, we can also do a symmetric metric instead of a one-sided metric by switching $i < j$ for $i \neq j$. When $N(w_i, w_j) = 0$, smoothing is required to compute a finite LCP, and for PMI we simply assume independence, PMI $= 0$.

Using TC-PMI computed with a 20-word sliding window on the entire 3M articles in English Wikipedia, Newman et al. (2010a;b) compared computed topic coherence to 6000 human-judged coherence scores, and obtained a Spearman rank correlation of $\rho = 0.8$, approximately the same as the inter-rater correlation computed on a leave-one-out basis. This topic coherence metric was used by Lau et al. (2010) for their best topic word task, and it performed well at detecting Chang et al.'s intruder word (Chang et al., 2009b).

We conclude this section by showing how these three different topic coherence metrics differ. Here, we focus on the metrics' ability to identify poor quality topics. We list sample topics learned from a collection of *New York Times* news articles, showing the lowest-scoring topics using the three metrics:

TC-PMI

```
why bad thing maybe doesn something does let isn really...
self sense often history yet power seems become itself perhaps...
came went told took later didn room began asked away...
need better problem must enough does likely less whether...
```

TC-LCP

```
space canadian station canada nasa mission air shuttle crew hughes...
fight lewis jones tyson vegas las boxing ring murphy elvis...
ball body wright arms watson puerto club rico hands swing...
blood thompson wilson cell test gladwin disease nixon gas sickle...
```

TC-NZ

```
eminem connor shea hanson mile daniels abbott seymour black trupia...
porter amin burke olsen omar horse horses martinez ruettgers botai...
hart hunter troy mack willis oxygen scooter terry chayes farrell...
greene weber sims fashion fairchild malley fletcher crosby sawyer
mccann...
```

The above examples show how PMI, LCP, and NZ-based topic coherence metrics identify different types of poor quality topics. TC-PMI tends to show poor quality topics that include terms that are more general and more frequent. TC-LCP shows topics that appear to relate to a nameable subject, but nevertheless are relatively incoherent. Finally, TC-NZ appears to do a good job at identifying the classic topic-of-names that is often learned by topic models.

12.4   Improving Topic Models

Now that we know what problems can appear in topic models and how to detect them, what can we do about them? At a high level, the problems can be interpreted as topics containing words that should not be together but are (e.g., "mixed" or "chained" topics) or distinct topics that should be together but aren't.

In this section, we discuss techniques to adapt the statistical formulating of topic models to incorporate these intuitive descriptions of problematic topics to create analysis of datasets that are more useful and more understandable. We also include a discussion on automatic topic labeling, another technique to improve the utility of topic models.

12.4.1   Interactive Topic Models

First, let's begin with a common-use case: a frustrated consumer of topic models staring at a collection of topics that do not make sense. In this section, we discuss interactive topic modeling (ITM), an in situ method for incorporating human knowledge into topic models.[10]

Recall that LDA views topics as distributions over words, and each document expresses an

---

[10]For full details, see Hu et al. (To Appear).

admixture of these topics. For "vanilla" LDA, these are symmetric Dirichlet distributions. A document is composed of observed words, which we call tokens, to distinguish specific observations from the word (type) associated with each token. Because LDA assumes a document's tokens are interchangeable, it treats the document as a bag-of-words, ignoring potential relations between words.

Constraints Change the Topics Discovered

This problem with vanilla LDA can be solved by encoding constraints, which will 'guide' different words into the same topic. Constraints change the underlying distribution by forcing words to either be positively or negatively correlated with each other. If a user sees two words that should appear in the same topic but do not, they can impose a positive correlation between the words. If the user sees two words that appear in a topic together but should not, they can impose a negative correlation between the words.

These correlations work by changing the underlying probabilistic model; while vanilla topic models assume that each topic is a distribution over words, we use tree-structured topics (Boyd-Graber et al., 2007; Andrzejewski et al., 2009). These models instead assume that topics first have a distribution over *concepts* and these concepts in turn have a distribution over words. By encoding word distributions as a tree, we can preserve conjugacy and relatively simple inference while encouraging correlations between words that are grouped together in concepts.

While these models can encourage words to be negatively or positively correlated, these constraints on the model must be added interactively as the user sees problems that must be corrected.

Interactively Adding Constraints

Interactively changing constraints can be accommodated in ITM, smoothly transitioning from unconstrained LDA to constrained LDA with one constraint, to constrained LDA with two constraints, etc.

A central tool that we use to transition between models is the strategic unassignment of states, which we call *ablation* (distinct from feature ablation in supervised learning). Gibbs sampling inference stores the topic assignment of each token. In the implementation of a Gibbs sampler, unassignment is done by setting a token's topic assignment to an invalid topic and decrementing any counts associated with that word.

The constraints created by users implicitly signal that words in constraints don't belong in a given topic. In other models, this input is sometimes used to 'fix,' i.e., deterministically hold constant topic assignments (Ramage et al., 2009). Instead, we change the underlying model, using the current topic assignments as a starting position for a new Markov chain with some states strategically unassigned; this is equivalent to performing online inference (Yao et al., 2009).

An alternative would be to not pursue this interactive strategy but instead restart inference from a new initialization. This, however, is counter to the goals of pursuing topic modeling interactively: restarting inference increases the latency users have to wait to see an updated model, restarting the model destroys any mental mapping of the model, and restarting the model could create additional problems into the model.

Merging Topics

To examine the viability of ITM, we begin with a qualitative demonstration that shows the potential usefulness of ITM. For this task, we used a corpus of about 2000 *New York Times* editorials from the years 1987 to 1996. We started by finding 20 initial topics with no constraints, as shown in Table 12.1 (left).

Notice that Topics 1 and 20 both deal with Russia. Topic 20 seems to be about the Soviet Union, with Topic 1 about the post-Soviet years. We wanted to combine the two into a single topic, so we created a constraint with all of the clearly Russian or Soviet words (*boris, communist, gorbachev, mikhail, russia, russian, soviet, union, yeltsin*). Running inference forward 100 iterations with the **Doc** ablation strategy yields the topics in Table 12.1 (right). The two Russia topics were combined into Topic 20. This combination also pulled in other relevant words that were not near the top of either topic before: "moscow" and "relations." Topic 1 is now more about elections in countries other than Russia. The other 18 topics changed little.

While we combined the Russian topics, other researchers analyzing large corpora might preserve the Soviet vs. post-Soviet distinction but combine topics about American government. ITM allows tuning for specific tasks.

Topic	Words	Topic	Words
1	election, yeltsin, russian, political, party, democratic, russia, president, democracy, boris, country, south, years, month, government, vote, since, leader, presidential, military	1	election, democratic, south, country, president, party, africa, lead, even, democracy, leader, presidential, week, politics, minister, percent, voter, last, month, years
2	new, york, city, state, mayor, budget, giuliani, council, cuomo, gov, plan, year, rudolph, dinkins, lead, need, governor, legislature, pataki, david	2	new, york, city, state, mayor, budget, council, giuliani, gov, cuomo, year, rudolph, dinkins, legislature, plan, david, governor, pataki, need, cut
3	nuclear, arms, weapon, defense, treaty, missile, world, unite, yet, soviet, lead, secretary, would, control, korea, intelligence, test, nation, country, testing	3	nuclear, arms, weapon, treaty, defense, war, missile, may, come, test, american, world, would, need, lead, get, join, yet, clinton, nation
4	president, bush, administration, clinton, american, force, reagan, war, unite, lead, economic, iraq, congress, america, iraqi, policy, aid, international, military, see	4	president, administration, bush, clinton, war, unite, force, reagan, american, america, make, nation, military, iraq, iraqi, troops, international, country, yesterday, plan
⋮		⋮	
20	soviet, lead, gorbachev, union, west, mikhail, reform, change, europe, leaders, poland, communist, know, old, right, human, washington, western, bring, party	20	soviet, union, economic, reform, yeltsin, russian, lead, russia, gorbachev, leaders, west, president, boris, moscow, europe, poland, mikhail, communist, power, relations

TABLE 12.1
Five topics from a 20-topic topic model on the editorials from the *New York Times* before adding a constraint (left) and after (right). After the constraint was added, which encouraged Russian and Soviet terms to be in the same topic, non-Russian terms gained increased prominence in Topic 1, and "Moscow" (which was not part of the constraint) appeared in Topic 20.

However, user constraints are not absolute. For example, in experiments some users attempted to merge topics about Apple computers and IBM-compatible personal computers discovered from the 20 Newsgroups corpus.[11] However, the model preferred to explain the data using two separate topics.

Separating Topics

Another possible imperfection in a topic model is that a single topic conflates two concepts that should be in distinct topics. This can be corrected by adding a constraint that two words cannot appear in the same topic. For example, in a collection of biomedical publications, a topic might be discovered that contains both words related to spinal cord and the urinary tract. Upon showing this to a domain expert—an NIH program manager—it was found that this was incorrect clustering. Introducing a constraint that these two words should not appear together results in the new topics in Table 12.2.

12.4.2 Generalized Pólya Urn Models

A topic model claims that, given topic assignments, the observed words are selected i.i.d. from a single set of topic distributions. If this assumption is true, then the expected number of documents that contain any pair of words w_i, w_j assigned to topic k should be a function of $p(w_i|k)$ and $p(w_j|k)$. Under this model, if those two probabilities are both large, it is unlikely that there will be no documents containing both words. Several of the topic quality metrics described in this chapter

---

[11] See http://people.csail.mit.edu/jrennie/20Newsgroups/.

Before	After	
bladder	spinal_cord	bladder
spinal_cord	spinal_cord_injury	women
sci	spinal	oc
spinal_cord_injury	injury	pelvic_floor
spinal	recovery	incontinence
urinary	motor	urinary_incontinence
urothelial	reflex	pelvic
cervical	urothelial	ui
injury	injured	prolapse
recovery	functional_recovery	ul
urinary_tract	plasticity	contraceptive
locomotor	locomotor	treatment
lumbar	cervical	stress
reflex	pathways	disorders

TABLE 12.2

Example of a topic being split using interactive topic modeling under the constraint that "bladder" and "spinal_cord_injury" should not be in the same topic. This results in "bladder" now being associated with incontinence.

measure mismatch between the theoretical co-occurrence implied by a model and actual word co-occurrence observed in documents.

The power that these simple metrics hold raises the question of why such topics should arise in the first place: if they are so easy to detect, why do they appear at all? The answer is that under standard specifications, topic models such as LDA cannot directly represent co-occurrence information. Mimno et al. (2011) presents an alternative model based on generalized Pólya urns (Mahmoud, 2008) that addresses this problem by encoding word co-occurrence information into the prior.

The generative process of a topic model is usually described in terms of discrete variables drawn from multinomial distributions that are themselves drawn from Dirichlet distributions. In this representation, the "meaning" of a topic is defined once and for all when the multinomial parameters for the topic-word distribution are sampled, and does not change no matter how many words are observed. An alternative generative model for LDA, which does not involve these intermediate multinomial parameters, is a standard Pólya urn process. Under this representation, the "meaning" of a topic evolves as words are sampled.

Consider an urn containing N balls, each with a single word written on it, such that N_w balls have word w written on them. If we draw and replace balls repeatedly, recording the word on each sampled ball, the frequency of each word in the resulting set of words is a distributed i.i.d. multinomial with $p(w) \propto N_w$.

If instead of replacing just the sampled ball we also add a new ball with the same word, the resulting set of words is distributed as a Dirichlet-compound multinomial. The DCM distribution is equivalent to a Dirichlet-multinomial hierarchical model with the parameters of the multinomial distribution integrated out. This model, the standard Pólya urn, is not i.i.d.: if we draw a ball with word w at time t, the probability that word w will appear on the next ball at $t + 1$ increases and the probability of all other words decreases. The model is, however, exchangeable, as the probability of a sequence of words is invariant to permutation of their order.

The Pólya urn process provides burstiness (a word, once seen, becomes more probable), but it cannot represent covariance since an increase in one word decreases the probability of all other words. The generalized Pólya urn extends the standard urn model by specifying a separate rule for adding new balls after sampling a ball of each type. For example, we might say that after sampling a ball with word w_2, we should replace it along with two new balls with word w_2, and one each of w_5, w_8, and w_{15}. In this way, w_2 would increase the probability of seeing w_2 again, but also increase the probability of the three other word types.

All three urn models can be represented by specifying a *schema* matrix $\mathbf{A}$, which defines the number of balls of each type to add after drawing a ball of each type. To define the simple sampling-with-replacement model we use a matrix of all zeros, indicating that no new balls will be added. For the standard Pólya urn, we use an identity matrix, which specifies that after seeing a ball of type w we add a single new ball of type w and nothing else. The generalized Pólya urn permits arbitrary values in the matrix (negative values are possible, corresponding to permanently removing balls, but can lead to instability). Mimno et al. (2011) defines a matrix with entries proportional to the co-document matrix used in the previously discussed evaluation metrics.

$$\begin{aligned} \boldsymbol{A}_{vv} &\propto \lambda_v D(v), \\ \boldsymbol{A}_{vw} &\propto \lambda_v D(w, v). \end{aligned} \tag{12.9}$$

As with the standard Pólya urn, the flexibility of the generalized Pólya urn comes at the cost of additional complexity. Specifically, the resulting distribution is no longer exchangeable, as the probability of a sequence depends on the order that words are observed. Nevertheless, the model can be effectively trained using a Gibbs sampler as if the distribution were exchangeable.

12.4.3 Regularized Topic Models

Topic models have the potential to improve search and discovery by extracting useful semantic themes from text documents. When learned topics are coherent and interpretable, they can be valuable for faceted browse, results set diversity, and document retrieval. However, when collections are made up of short documents or noisy text (e.g., web search result snippets or blog posts), learned topics can be less coherent, less interpretable, and less useful.

Predicated on recent evidence that a PMI-based topic coherence score is highly correlated with human-judged topic coherence (Newman et al., 2010a), Newman et al. (2011) proposed two Bayesian regularization formulations to improve topic coherence. Both methods use additional word co-occurrence data to improve the coherence and interpretability of learned topics, while still learning a faithful representation of the collection of interest, as measured by likelihood of test data. These *regularized topic models* are an alternative to the generalized Pólya urn models described in the previous section, and have similar objectives and goals.

To learn more coherent topic models for small or noisy collections, they introduced structured priors on ϕ_t based upon external data, which have a regularization effect on the standard LDA model. More specifically, the priors on ϕ_t depend on the structural relations of the words in the vocabulary as given by external data, which are characterized by the $W \times W$ "covariance" matrix $\mathbf{C}$. Intuitively, $\mathbf{C}$ is a matrix that captures the short-range dependencies between (i.e., co-occurrences of) words in the external data. One is only interested in relatively frequent terms from the vocabulary, so $\mathbf{C}$ is a sparse matrix and computations are still feasible.

Quadratic Regularizer. A standard quadratic form is used with a trade-off factor. Given a matrix of word dependencies $\mathbf{C}$, use the prior:

$$p(\phi_t | \mathbf{C}) \propto \left(\phi_t^T \mathbf{C} \phi_t\right)^\nu \tag{12.10}$$

for some power ν. The normalization factor is unknown but for MAP estimation we do not need it. Optimizing the log posterior with respect to $\phi_{w|t}$ subject to the usual constraints, one obtains the following fixed point update:

$$\phi_{w|t} \leftarrow \frac{1}{N_t + 2\nu} \left(N_{wt} + 2\nu \frac{\phi_{w|t} \sum_{i=1}^{W} C_{iw}\phi_{i|t}}{\phi_t^T C \phi_t} \right). \tag{12.11}$$

Unlike other topic models in which a covariance or correlation structure is used in the context of correlated priors for $\theta_{t|d}$, (as in the correlated topic model (Blei and Lafferty, 2005)), this method does not require the inversion of C, which would be impractical for even modest vocabulary sizes. (Interactive topic modeling, discussed in Section 12.4.1, also adds correlations without requiring this inversion because it preserves conjugacy.)

By using the update in Equation (12.11) we obtain the values for $\phi_{w|t}$. This means we no longer have conjugate priors for $\phi_{w|t}$ and thus the standard Gibbs-sample update

$$p(z_{id} = t | x_{id} = w, \mathbf{z}^{-i}) \propto \frac{N_{wt}^{-i} + \beta}{N_t^{-i} + W\beta}(N_{td}^{-i} + \alpha) \tag{12.12}$$

does not hold. Instead, at the end of each major Gibbs cycle, $\phi_{w|t}$ is re-estimated and the corresponding Gibbs update becomes:

$$p(z_{id} = t | x_{id} = w, \mathbf{z}^{-i}, \phi_{w|t}) \propto \phi_{w|t}(N_{td}^{-i} + \alpha). \tag{12.13}$$

Convolved Dirichlet Regularizer. Another approach to leveraging information on word dependencies from external data is to consider that each ϕ_t is a mixture of word probabilities ψ_t, where the coefficients are constrained by the word-pair dependency matrix C:

$$\phi_t \propto C\psi_t \qquad \text{where} \qquad \psi_t \sim \text{Dirichlet}(\gamma\mathbf{1}). \tag{12.14}$$

Each topic has a different ψ_t drawn from a Dirichlet, thus the model is a convolved Dirichlet. This means that we convolve the supplied topic to include a spread of related words. Optimizing the posterior and solving for $\psi_{w|t}$ one obtains:

$$\psi_{w|t} \propto \sum_{i=1}^{W} \frac{N_{it}C_{iw}}{\sum_{j=1}^{W} C_{ij}\psi_{j|t}} \psi_{w|t} + \gamma. \tag{12.15}$$

One follows the same semi-collapsed inference procedure used for the quadratic regularizer, with the updates in Equations (12.15) and (12.14) producing the values for $\phi_{w|t}$ to be used in the semi-collapsed sampler (12.13).

Using thirteen datasets from blog posts, news articles, and web searches, Newman et al. (2011) shows that both regularizers improve topic coherence and interpretability while learning a faithful representation of the collection of interest. Additionally, in an experiment involving 3,650 crowd-sourced topic comparisons, they show that humans judge the regularized topic models as being more coherent than LDA.

12.4.4 Automatic Topic Labeling

In user-facing applications that use topic models, topics are displayed to humans, typically using the top-10 or so terms in the topic. However, it can sometimes be difficult for end-users to interpret the rich statistical information encoded in the topics, or quickly getting the gist of a topic. One way of making topics more readily human interpretable is by annotating the topic with a short label. While this task is best done by a subject matter expert, recent work has shown that one can partially automate the generation of candidate labels for topics.

Short labels for topics are typically best expressed with multiword terms (for example STOCK

MARKET TRADING), or terms that might not be in the top-10 topic terms (for example, COLORS would be a good label for a topic of the form *red green blue cyan ...*). Lau et al. (2011) proposed a novel method for automatic topic labeling that first generates topic label candidates using English Wikipedia, and then ranks the candidates to select the best topic labels. Given the size and diversity of English Wikipedia, they posit that the vast majority of (coherent) topics or concepts are probably well represented by a Wikipedia article title.

Their method of predicting suitable candidate labels has two parts. They first have a system to generate a relatively long list of candidates. Then, they use lexical features of and association measures between candidate labels and topic terms in a support vector regression framework for ranking the labels.

Generating the list of candidates starts with querying Wikipedia using the top-10 topic terms. The top-ranked search results (article titles) returned constitute the initial set of *primary* candidates for each topic. Next we chunk parse the primary candidates using the OpenNLP chunker and extract out all noun chunks. For each noun chunk, we generate all component n-grams, out of which we remove all n-grams which are not in themselves article titles in English Wikipedia. For example, if the Wikipedia document title were the single noun chunk *United States Constitution*, we would generate the bigrams *United States* and *States Constitution*, and prune the latter; we would also generate the unigrams *United*, *States*, and *Constitution*, all of which exist as Wikipedia articles and are preserved.

Ranking candidate labels is premised on the idea that a good label should be strongly associated with the topic terms. To learn the association of a label candidate with the topic terms, we use several lexical association measures: pointwise mutual information (PMI), Student's t-test, Dice's coefficient, Pearson's χ^2 test, and the log-likelihood ratio. We also include conditional probability and reverse conditional probability measures based on the work of Lau et al. (2010). To calculate the association measures, we parse the full collection of English Wikipedia articles using a sliding window of width 20, and obtain term frequencies for the label candidates and topic terms. To measure the association between a label candidate and a list of topic terms, we average the scores of the top-10 topic terms.

These lexical features and association measures were used in a supervised model by training over topics where we have gold standard labeling of the label candidates using a support vector regression (SVR) model over all of the features. Table 12.3 shows examples of the top-ranked label candidate for four topics learned on four different corpora from diverse genres. We see that the top-ranked label candidate does a relatively good job of capturing the gist of each of the four topics.

china chinese olympics gold olympic team win beijing medal sport ...
Label: 2008 SUMMER OLYMPICS
church arch wall building window gothic nave side vault tower ...
Label: GOTHIC ARCHITECTURE
israel peace barak israeli minister palestinian agreement prime leader ...
Label: ISRAELI-PALESTINIAN CONFLICT
cell response immune lymphocyte antigen cytokine t-cell induce receptor ...
Label: IMMUNE SYSTEM

TABLE 12.3
A sample of topics and automatically generated topic labels.

12.5　Conclusion

While topic models are a popular technique for understanding large datasets, how to actually go from raw data to an effective topic analysis is often difficult for new users. This chapter discusses the iterative process for building topic models from preprocessing data to improving and understanding the results users can obtain from models. In time, this process can benefit from continued development by both tool builders and researchers.

However, tool builders will continue to make this process more straightforward by building unified interfaces that can seamlessly adjust tokenization, vocabulary, and topic models within a single interface. Improved visualization tools that can help users identify and correct topic modeling errors would also make the process of curating a topic model more straightwforward.

Researchers can improve the process by building models that are less sensitive to the seemingly arbitrary choices made by users. Models should be less sensitive to the vocabulary, should be able to segment overly long documents, and should detect when the data fail to meet the assumptions of topic models, such as when a corpus is in multiple languages or dialects. Finally, researchers can improve inference throughput and latency so that users can try more models more quickly.

Together, these advances will allow users to move from data to a final, quality model quickly and with minimal hassle.

References

AlSumait, L., Barbará, D., Gentle, J., and Domeniconi, C. (2009). Topic significance ranking of LDA generative models. In *Proceedings of the European Conference on Machine Learning and Knowledge Discovery in Databases: Part I (ECML PKDD '09)*. Berlin, Heidelberg: Springer-Verlag, 67–82.

Andrzejewski, D., Zhu, X., and Craven, M. (2009). Incorporating domain knowledge into topic modeling via Dirichlet Forest priors. In *Proceedings of the 26th International Conference of Machine Learning (ICML '09)*. New York, NY, USA: ACM, 25–32.

Bird, S., Klein, E., and Loper, E. (2009). *Natural Language Processing with Python*. O'Reilly Media.

Blei, D. M., Griffiths, T. L., and Jordan, M. I. (2007). The nested Chinese restaurant process and Bayesian nonparametric inference of topic hierarchies. *Journal of the ACM* 57 : 7.1–7.30.

Blei, D. M. and Lafferty, J. D. (2005). Correlated Topic Models. In Weiss, Y., Schölkopf, B., and Platt, J. (eds), *Advances in Neural Information Processing Systems 18*. Cambridge, MA: The MIT Press, 147–154.

Blei, D. M. and Lafferty, J. D. (2006). Dynamic topic models. In *Proceedings of the 23rd International Conference of Machine Learning (ICML '06)*. New York, NY, USA: ACM, 113–120.

Blei, D. M. and Lafferty, J. D. (2009). Visualizing topics with multi-word expressions. http://arxiv.org/abs/0907.1013 [stat.ML].

Blei, D. M., Ng, A. Y., and Jordan, M. I. (2003). Latent Dirichlet allocation. *Journal of Machine Learning Research* 3: 993–1022.

Boyd-Graber, J. and Blei, D. M. (2008). Syntactic Topic Models. In Koller, D., Schuurmans, D., Bengio, Y., and Bottou, L. (eds), *Advances in Neural Information Processing Systems 21*. Red Hook, NY: Curran Associates, Inc., 185–192.

Boyd-Graber, J., Blei, D. M., and Zhu, X. (2007). A topic model for word sense disambiguation. In *Proceedings of the 2007 Joint Conference on Empirical Methods in Natural Language Processing and Computational Natural Language Learning*. Stroudsburg, PA, USA: Association of Computational Linguistics, 1024–1033.

Brody, S. and Lapata, M. (2009). Bayesian word sense induction. In *Proceedings of the 12th Conference of the European Chapter of the Association for Computational Linguistics*. Stroudsburg, PA, USA: Association for Computational Linguistics, 103–111.

Buntine, W. (2009). Estimating likelihoods for topic models. In *Proceedings of the 1st Asian Conference on Machine Learning: Advances in Machine Learning*. Berlin, Heidelberg: Springer-Verlag, 51–64.

Chang, J., Boyd-Graber, J., and Blei, D. M. (2009a). Connections between the lines: Augmenting social networks with text. In *Proceedings of the 15th ACM SIGKDD International Conference on Knowledge Discovery and Data Mining (KDD '09)*. New York, NY, USA: ACM, 169–178.

Chang, J., Boyd-Graber, J., Wang, C., Gerrish, S., and Blei, D. M. (2009b). Reading tea leaves: How humans interpret topic models. In Bengio, Y., Schuurmans, D., Lafferty, J. D., Williams, C. K. I., and Culotta, A. (eds), *Advances in Neural Information Processing Systems 22*. Red Hook, NY: Curran Associates, Inc., 288–296.

Dietz, L., Bickel, S., and Scheffer, T. (2007). Unsupervised prediction of citation influences. In *Proceedings of the 24th Annual International Conference on Machine Learning (ICML '07)*. New York, NY, USA: ACM, 233-240.

Eidelman, V., Boyd-Graber, J., and Resnik, P. (2012). Topic models for dynamic translation model adaptation. In *Proceedings of the 50th Annual Meeting of the Association for Computational Linguistics: Short Papers - Vol. 2*. Stroudsburg, PA, USA: Association for Computational Linguistics, 115-119.

Fei-Fei, L. and Perona, P. (2005). A Bayesian hierarchical model for learning natural scene categories. In *Proceedings of the 10th IEEE Conference on Computer Vision and Pattern Recognition (CVPR 2005)*. Los Alamitos, CA,USA: IEEE Computer Society, 524–531.

Freedman, A. (2007). *The Party of the First Part: The Curious World of Legalese*. New York, NY: Henry Holt and Company.

Goldwater, S., Griffiths, T. L., and Johnson, M. (2006). Contextual dependencies in unsupervised word segmentation. In *Proceedings of the 21st International Conference on Computational Linguistics and the 44th Annual Meeting of the Association for Computational Linguistics*. Stroudsburg, PA, USA: Association for Comptuational Linguistics, 673–680.

Griffiths, T. L. and Steyvers, M. (2004). Finding scientific topics. *Proceedings of the National Academy of Sciences* 101: 5228–5235.

Griffiths, T. L., Steyvers, M., Blei, D. M., and Tenenbaum, J. B. (2005). Integrating topics and syntax. In Saul, L. K., Weiss, Y., and Bottou, L. (eds), *Advances in Neural Information Processing Systems 17*. Cambridge, MA: The MIT Press, 537–544.

Gruber, A., Rosen-Zvi, M., and Weiss, Y. (2007). Hidden topic Markov models. In *Proceedings of the 11ᵗʰ International Conference on Artificial Intelligence and Statistics (AISTATS 2007). Journal of Machine Learning Research – Proceedings Track* 2: 163–170.

Haghighi, A. and Vanderwende, L. (2009). Exploring content models for multi-document summarization. In *Proceedings of Human Language Technologies: The 2009 Annual Conference of the North American Chapter of the Association for Computational Linguistics*. Boulder, Colorado: Association for Computational Linguistics, 362–370.

Hall, D., Jurafsky, D., and Manning, C. D. (2008). Studying the history of ideas using topic models. In *Proceedings of the Conference on Empirical Methods in Natural Language Processing (EMNLP '08)*. Stroudsburg, PA, USA: Association for Computational Linguistics, 363–371.

Hardisty, E. A., Boyd-Graber, J., and Resnik, P. (2010). Modeling perspective using adaptor grammars. In *Proceedings of the 2010 Conference on Empirical Methods in Natural Language Processing (EMNLP '10)*. Stroudsburg, PA, USA: Association for Computational Linguistics, 284–292.

Hu, Y., Boyd-Graber, J., Satinoff, B., and Smith, A. (To Appear). Interactive topic modeling. *Machine Learning Journal*.

Johnson, M. (2010). PCFGs, topic models, adaptor grammars and learning topical collocations and the structure of proper names. In *Proceedings of the 48ᵗʰ Annual Meeting of the Association for Computational Linguistics*. Stroudsburg, PA, USA: Association for Computational Linguistics, 1148–1157.

Kohlschütter, C., Fankhauser, P., and Nejdl, W. (2010). Boilerplate detection using shallow text features. In *Proceedings of the 3ʳᵈ ACM International Conference on Web Search and Data Mining (WSDM '10)*. New York, NY, USA: ACM, 441–450.

Lau, J. H., Grieser, K., Newman, D., and Baldwin, T. (2011). Automatic Labelling of Topic Models. In *Proceedings of the Association for Computational Linguistics*, 1536–1545.

Lau, J. H., Newman, D., Karimi, S., and Baldwin, T. (2010). Best topic word selection for topic labelling. In *Proceedings of the 23ʳᵈ International Conference on Computational Linguistics (Coling 2010)*. Poster presented. Beijing, China, 605–613.

Lin, W. -H., Wilson, T., Wiebe, J., and Hauptmann, A. (2006). Which side are you on? Identifying perspectives at the document and sentence levels. In *Proceedings of the 10ᵗʰ Conference on Computational Natural Language Learning (CoNLL-X '06)*. Stroudsburg, PA, USA: Association for Computational Linguistics, 109–116.

Mahmoud, H. (2008). *Pólya Urn Models*. Chapman & Hall/CRC, 1st edition.

Manning, C. D. and Schütze, H. (1999). *Foundations of Statistical Natural Language Processing*. Cambridge, MA: The MIT Press.

Marcus, M. P., Santorini, B., and Marcinkiewicz, M. A. (1993). Building a large annotated corpus of English: The Penn treebank. *Computational Linguistics* 19: 313–330.

Mimno, D., Wallach, H. M., Talley, E. M., Leenders, M., and McCallum, A. (2011). Optimizing semantic coherence in topic models. In *Proceedings of the Conference on Empirical Methods in Natural Language Processing (EMNLP '11)*. Stroudsburg, PA, USA: Association for Computational Linguistics, 262–272.

Narayanamurthy, S. (2011). Yahoo! LDA project.

Neal, R. M. (1993). Probabilistic Inference Using Markov Chain Monte Carlo Methods. Tech. report CRG-TR-93-1, University of Toronto.

Newman, D., Baldwin, T., Cavedon, L., Huang, E., Karimi, S., Martinez, D., Scholer, F., and Zobel, J. (2010a). Visualizing search results and document collections using topic maps. *Web Semantics* 8: 169–175.

Newman, D., Bonilla, E., and Buntine, W. (2011). Improving topic coherence with regularized topic models. In Shawe-Taylor, J., Zemel, R. S., Bartlett, P., Pereira, F. C. N., and Weinberger, K. Q. (eds), *Advances in Neural Information Processing Systems 24*. Red Hook, NY: Curran Associates, Inc., 496–504.

Newman, D., Lau, J. H., Grieser, K., and Baldwin, T. (2010b). Automatic evaluation of topic coherence. In *Proceedings of Human Language Technologies: The 2010 Annual Conference of the North American Chapter of the Association for Computational Linguistics*. Stroudsburg, PA, USA: Association for Computational Linguistics, 100–108.

Nguyen, V. -A., Boyd-Graber, J., and Resnik, P. (2012). SITS: A hierarchical nonparametric model using speaker identity for topic segmentation in multiparty conversations. In *Proceedings of the 50th Annual Meeting of the Association for Computational Linguistics: Long Papers –Vol. 1*. Stroudsburg, PA, USA: Association for Computational Linguistics, 78–87.

Paul, M. and Girju, R. (2010). A two-dimensional topic-aspect model for discovering multi-faceted topics. In *Proceedings of the 24th AAAI Conference on Artificial Intelligence (AAAI 2010)*. Palo Alto, CA, USA: AAAI, 545–550.

Porter, M. F. (1980). An algorithm for suffix stripping. *Program: Electronic Library and Information Systems* 14: 130–137.

Purver, M., Körding, K., Griffiths, T. L., and Tenenbaum, J. B. (2006). Unsupervised topic modelling for multi-party spoken discourse. In *Proceedings of the 21st International Conference on Computational Linguistics and the 44th Annual Meeting of the Association for Computational Linguistics*. Stroudsburg, PA, USA: Association for Computational Linguistics, 17–24.

Ramage, D., Hall, D., Nallapati, R. M., and Manning, C. D. (2009). Labeled LDA: A supervised topic model for credit attribution in multi-labeled corpora. In *Proceedings of the 2009 Conference on Empirical Methods in Natural Language Processing: Vol. 1*. Stroudsburg, PA, USA: Association for Computational Linguistics, 248–256.

Rosen-Zvi, M., Griffiths, T. L., Steyvers, M., and Smyth, P. (2004). The author-topic model for authors and documents. In *Proceedings of the 20th Conference on Uncertainty in Artificial Intelligence*. Arlington, VA, USA: AUAI Press, 487–494.

Salton, G. (1968). *Automatic Information Organization and Retrieval*. McGraw Hill Text.

Sandhaus, E. (2008). The *New York Times* Annotated Corpus. Philadelphia, PA: Linguistic Data Consortium.

Schemann, H. and Knight, P. (1995). *German-English Dictionary of Idioms*. Oxford, UK: Routledge.

Taghva, K., Elkhoury, R., and Coombs, J. (2005). Arabic stemming without a root dictionary. In *Proceedings of the Internaional Conference on Information Technology: Coding and Computing – Vol. 1 (ITCC 2005)*. Los Alamitos, CA, USA: IEEE Computer Society, 152–157.

Talley, E. M., Newman, D., Mimno, D., Herr, B. W., Wallach, H. M., Burns, G. A. P. C., Leenders, M., and McCallum, A. (2011). Database of NIH grants using machine-learned categories and graphical clustering. *Nature Methods* 8: 443–444.

Teh, Y. W., Jordan, M. I., Beal, M. J., and Blei, D. M. (2006). Hierarchical Dirichlet processes. *Journal of the American Statistical Association* 101: 1566–1581.

Titov, I. and McDonald, R. (2008). A joint model of text and aspect ratings for sentiment summarization. In *Proceedings of the 46th Annual Meeting of the Association for Computational Linguistics: Human Language Technologies*. Stroudsburg, PA, USA: Association for Computational Linguistics, 308–316.

Wallach, H. M., Mimno, D., and McCallum, A. (2009a). Rethinking LDA: Why priors matter. In Bengio, Y., Schuurmans, D., Lafferty, J. D., Williams, C. K. I., and Culotta, A. (eds), *Advances in Neural Information Processing Systems 22*. Red Hook, NY: Curran Associates, Inc., 1973–1981.

Wallach, H. M. (2006). Topic modeling: Beyond bag-of-words. In *Proceedings of the 23rd International Conference of Machine Learning (ICML '06)*. New York, NY, USA: ACM, 977–984.

Wallach, H. M., Murray, I., Salakhutdinov, R., and Mimno, D. (2009b). Evaluation methods for topic models. In Bottou, L. and Littman, M. (eds), *Proceedings of the 26th Annual International Conference of Machine Learning (ICML '09)*. New York, NY, USA: ACM, 1105–1112.

Wei, X. and Croft, B. (2006). LDA-based document models for ad-hoc retrieval. In *Proceedings of the ACM SIGIR Conference on Research and Development in Information Retrieval (SIGIR '06)*. New York, NY, USA: ACM, 178–185.

Yao, L., Mimno, D., and McCallum, A. (2009). Efficient methods for topic model inference on streaming document collections. In *Proceedings of the 15th ACM SIGKDD International Conference on Knowledge Discovery and Data Mining*. New York, NY, USA: ACM, 937–946.

Zhai, K., Boyd-Graber, J., Asadi, N., and Alkhouja, M. (2012). Mr. LDA: A flexible large scale topic modeling package using variational inference in MapReduce. In *Proceedings of the 21st International Conference on World Wide Web*. New York, NY, USA: ACM, 879–888.

13

Block-LDA: Jointly Modeling Entity-Annotated Text and Entity-Entity Links

Ramnath Balasubramanyan

Language Technologies Institute, Carnegie Mellon University, Pittsburgh, PA 15213, USA

William W. Cohen

Machine Learning Department, Carnegie Mellon University, Pittsburgh, PA 15213, USA

CONTENTS

Identifying latent groups of entities from observed interactions between pairs of entities is a frequently encountered problem in areas like analysis of protein interactions and social networks. We present a model that combines aspects of mixed membership stochastic blockmodels and topic models to improve entity-entity link modeling by jointly modeling links and text about the entities that are linked. We apply the model to two datasets: a protein-protein interaction (PPI) dataset supplemented with a corpus of abstracts of scientific publications annotated with the proteins in the PPI dataset and an Enron email corpus. The induced topics' ability to help understand the nature of the data provides a qualitative evaluation of the model. Quantitative evaluation shows improvements in functional category prediction of proteins and in perplexity, using the joint model over baselines that use only link or text information. For the PPI dataset, the topic coherence of the emergent topics and the ability of the model to retrieve relevant scientific articles and proteins related to the topic are compared to that of a text-only approach that does not make use of the protein-protein interaction matrix. Evaluation of the results by biologists show that the joint modeling results in better topic coherence and improves retrieval performance in the task of identifying top related papers and proteins.

13.1 Introduction

The task of modeling latent groups of entities from observed interactions is a commonly encountered problem. In social networks, for instance, we might want to identify sub-communities. In the biological domain we might want to discover latent groups of proteins based on observed pairwise interactions. Mixed membership stochastic blockmodels (MMSB) (Airoldi et al., 2008; Parkkinen et al., 2009) approach this problem by assuming that nodes in a graph represent entities belonging to latent blocks with mixed membership, effectively capturing the notion that entities may arise from different sources and have different roles.

In another area of active research, models like latent Dirichlet allocation (LDA) (Blei et al., 2003) model text documents in a corpus as arising from mixtures of latent topics. In such models, words in a document are potentially generated from different topics using topic-specific word distributions. Extensions to LDA (Erosheva et al., 2004; Griffiths and Steyvers, 2004) additionally model other metadata in documents such as authors and entities by treating a latent topic as a set of distributions, one for each metadata type. For instance, when modeling scientific publications from the biological domain, a latent topic could have a word distribution, an author distribution, and a protein entity distribution. We refer to this model as *Link LDA* following the convention established by Nallapati et al. (2008). The different types of data that are contained in a document (e.g., words in the body, words in the title, authors, list of citations, etc.) are referred to as *entity types*.

In this chapter, we present a model, *Block-LDA*, that jointly generates text documents annotated with metadata about associated entities and external links between pairs of entities. This allows the model to use supplementary annotated text to influence and improve link modeling. The text documents are modeled as bags of entities of different types and the network is modeled as edges between entities of a source type to a destination type. Consider the example of a corpus of publications about the yeast organism and a network of protein-protein interactions in yeast. These publications are further annotated by experts with lists of proteins that are discussed in them. Therefore, each publication could be modeled as a collection of bags *vis a vis* bag of body-words, bag of authors, bag of proteins discussed in the paper, etc. Similarly, the network could be a collection of protein-protein interactions independently observed. The model merges the idea of latent topics in topic models with blocks in stochastic blockmodels. The joint modeling permits sharing of information about the latent topics between the network structure and text, resulting in more coherent topics. Co-occurrence patterns in entities and words related to them aid the modeling of links in the graph. Likewise, entity-entity links provide clues about topics in the text. We also propose a method to perform approximate inference in the model using a collapsed Gibbs sampler, since exact inference in the joint model is intractable.

We then use the model to organize a large collection of literature about yeast biology to enable topic-oriented browsing and retrieval from the literature. The analysis is performed using the mixed membership topic modeling to uncover latent structure in document corpora by identifying broad topics that are discussed in it. This approach complements traditional information retrieval tasks where the objective is to fulfill very specific information needs. By using joing modeling, we are able to use other sources of domain information related to the domain in addition to literature. In the case of yeast biology, an example of such a resource is a database of known protein-protein interactions (PPI) which have been identified using wetlab experiments. We perform data fusion by combining text information from articles and the database of yeast protein-protein interactions by using a latent variable model—Block-LDA (Balasubramanyan and Cohen, 2011), that jointly models the literature and PPI networks.

We evaluate the ability of the topic models to return meaningful topics by inspecting the top papers and proteins that pertain to them. We compare the performance of the joint model, i.e., Block-LDA, with a model that only considers the text corpora by asking a yeast biologist to evaluate the coherence of topics and the relevance of the retrieved articles and proteins. This evaluation serves to test the utility of Block-LDA on a real task as opposed to an internal evaluation (such as by using perplexity metrics). Our evaluaton shows that the joint model outperforms the text-only approach both in topic coherence and in top paper and protein retrieval as measured by precision@10 values.

The chapter is organized as follows: Section 15.2 introduces the model and presents a Gibbs sampling-based method for performing approximate inference with the model. Section 13.3 discusses related work, and Section 13.4 provides details of datasets used in the experiments. Sections 13.5.1 and 13.5.2 present the results of our experiments on two datasets from different domains. Finally, our conclusions are in Section 13.6.

13.2 Block-LDA

Variables in the model

K	-	the number of topics (therefore resulting in K^2 blocks in the network)
α_L	-	Dirichlet prior for the topic pair distribution for links
α_D	-	Dirichlet prior for document specific topic distributions
γ	-	Dirichlet prior for topic multinomials
π_L	-	multinomial distribution over topic pairs for links
θ_d	-	multinomial distribution over topics for document d
T	-	the number of types of entities in the corpus
$\beta_{t,z}$	-	multinomial over entities of type t for topic z
D	-	number of documents in the corpus
$z_{t,i}$	-	topic chosen for the i-th entity of type t in a document
$e_{t,i}$	-	the i-th entity of type t occurring in a document
N_L	-	number of links in the network
z_{i1} and z_{i2}	-	topics chosen for the two nodes participating in the i-th link
e_{i1} and e_{i2}	-	the two nodes participating in the i-th link

The Block-LDA model (plate diagram in Figure 13.1) enables sharing of information between the component on the left that models links between pairs of entities represented as edges in a graph with a block structure, and the component on the right that models text documents through shared latent topics. More specifically, the distribution over the entities of the type that are linked is shared between the blockmodel and the text model.

The component on the right, which is an extension of the LDA models, documents as sets of "bags of entities," with each bag corresponding to a particular type of entity. Every entity type has a topic-wise multinomial distribution over the set of entities that can occur as an instance of the entity type.

The component on the left is a generative model for graphs representing entity-entity links with an underlying block structure, derived from the sparse blockmodel introduced by Parkkinen et al. (2009). Linked entities are generated from topic-specific entity distributions conditioned on the topic pairs sampled for the edges. Topic pairs for edges (links) are drawn from a multinomial defined over the Cartesian product of the topic set with itself. Vertices in the graph representing entities therefore have mixed memberships in topics. In contrast to MMSB, only observed links are sampled, making this model suitable for sparse graphs.

Let K be the number of latent topics (blocks) we wish to recover. Assuming documents consist of T different types of entities (i.e., each document contains T bags of entities), and that links in the

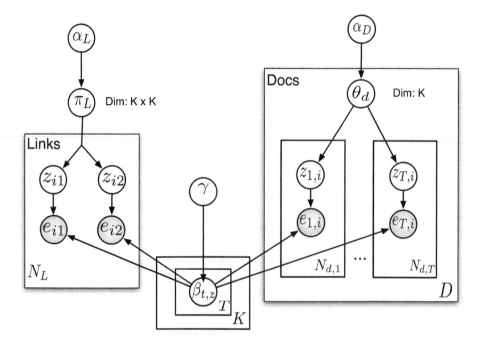

FIGURE 13.1
Block-LDA: plate diagram.

graph are between entities of type t_l and t_r, the generative process is as follows.

1. Generate topics:

   - For each type $t \in 1, \dots, T$, and topic $z \in 1, \dots, K$, sample $\beta_{t,z} \sim \text{Dirichlet}(\gamma)$, the topic specific entity distribution.

2. Generate documents. For every document $d \in \{1 \dots D\}$:

   - Sample $\theta_d \sim \text{Dirichlet}(\alpha_D)$ where θ_d is the topic mixing distribution for the document.
   - For each type t and its associated set of entity mentions $e_{t,i}$, $i \in \{1, \cdots, N_{d,t}\}$:
     - Sample a topic $z_{t,i} \sim \text{Multinomial}(\theta_d)$.
     - Sample an entity $e_{t,i} \sim \text{Multinomial}(\beta_{t,z_{t,i}})$.

   3. Generate the link matrix of entities of type t_l:

   - Sample $\pi_L \sim \text{Dirichlet}(\alpha_L)$ where π_L describes a distribution over the Cartesian product of topics for links in the dataset.
   - For every link $e_{i1} \rightarrow e_{i2}$, $i \in \{1 \cdots N_L\}$:
     - Sample a topic pair $\langle z_{i1}, z_{i2} \rangle \sim \text{Multinomial}(\pi_L)$.
     - Sample $e_{i1} \sim \text{Multinomial}(\beta_{t_l,z_{i1}})$.
     - Sample $e_{i2} \sim \text{Multinomial}(\beta_{t_r,z_{i2}})$.

Note that unlike the MMSB model introduced by Airoldi et al. (2008), this model generates only realized links between entities.

Given the hyperparameters α_D, α_L, and γ, the joint distribution over the documents, links, their topic distributions, and topic assignments is given by

$$p(\pi_L, \boldsymbol{\theta}, \boldsymbol{\beta}, \mathbf{z}, \mathbf{e}, \langle \mathbf{z_1}, \mathbf{z_2} \rangle, \langle \mathbf{e_1}, \mathbf{e_2} \rangle | \alpha_D, \alpha_L, \gamma) \propto \tag{13.1}$$

$$\prod_{z=1}^{K} \prod_{t=1}^{T} \mathrm{Dir}(\beta_{t,z} | \gamma_t) \times$$

$$\prod_{d=1}^{D} \mathrm{Dir}(\theta_d | \alpha_D) \prod_{t=1}^{T} \prod_{i=1}^{N_{d,t}} \theta_d^{z_{t,i}^{(d)}} \beta_{t,z_{t,i}^{(d)}}^{e_{t,i}} \times$$

$$\mathrm{Dir}(\pi_L | \alpha_L) \prod_{i=1}^{N_L} \pi_L^{\langle z_{i1}, z_{i2} \rangle} \beta_{t_l,z1}^{e_{i1}} \beta_{t_r,z2}^{e_{i2}}.$$

A commonly required operation when using models like Block-LDA is to perform inference on the model to query the topic distributions and the topic assignments of documents and links. Due to the intractability of exact inference in the Block-LDA model, a collapsed Gibbs sampler is used to perform approximate inference. It samples a latent topic for an entity mention of type t in the text corpus conditioned on the assignments to all other entity mentions using the following expression (after collapsing θ_D):

$$p(z_{t,i} = z | e_{t,i}, \mathbf{z}^{\neg \mathbf{i}}, \mathbf{e}^{\neg \mathbf{i}}, \alpha_D, \gamma) \tag{13.2}$$

$$\propto (n_{dz}^{\neg i} + \alpha_D) \frac{n_{zte_{t,i}}^{\neg i} + \gamma}{\sum_{e'} n_{zte'}^{\neg i} + |E_t|\gamma}.$$

Similarly, we sample a topic pair for every link conditional on topic pair assignments to all other links after collapsing π_L using the expression:

$$p(\mathbf{z_i} = \langle z_1, z_2 \rangle | \langle e_{i1}, e_{i2} \rangle, \mathbf{z}^{\neg \mathbf{i}}, \langle \mathbf{e_1}, \mathbf{e_2} \rangle^{\neg \mathbf{i}}, \alpha_L, \gamma) \tag{13.3}$$

$$\propto \left(n_{\langle z_1, z_2 \rangle}^{L \neg i} + \alpha_L \right) \times$$

$$\frac{\left(n_{z_1 t_l e_{i1}}^{\neg i} + \gamma \right) \left(n_{z_2 t_r e_{i2}}^{\neg i} + \gamma \right)}{\left(\sum_e n_{z_1 t_l e}^{\neg i} + |E_{t_l}|\gamma \right) \left(\sum_e n_{z_2 t_r e}^{\neg i} + |E_{t_r}|\gamma \right)}.$$

E_t refers to the set of all entities of type t. The ns refer to the number of topic assignments in the data.

- n_{zte}—the number of times an entity e of type t is observed under topic z.
- n_{zd}—the number of entities (of any type) with topic z in document d.
- $n_{\langle z_1, z_2 \rangle}^L$—count of links assigned to topic pair $\langle z_1, z_2 \rangle$.

The topic multinomial parameters and the topic distributions of links and documents are easily recovered using their MAP estimates after inference using the counts of observations:

$$\beta_{t,z}^{(e)} = \frac{n_{zte} + \gamma}{\sum_{e'} n_{zte'} + |E_t|\gamma}, \tag{13.4}$$

$$\theta_d^{(z)} = \frac{n_{dz} + \alpha_D}{\sum_{z'} n_{dz'} + K\alpha_D}, \tag{13.5}$$

$$\pi_L^{\langle z_1, z_2 \rangle} = \frac{n_{\langle z_1, z_2 \rangle} + \alpha_L}{\sum_{z_1', z_2'} n_{\langle z_1', z_2' \rangle} + K^2 \alpha_L}. \tag{13.6}$$

A de-noised form of the entity-entity link matrix can also be recovered from the estimated parameters of the model. Let B_t be a matrix of dimensions $K \times |E_t|$ where row $k = \beta_{t,k}$, $k \in \{1, \cdots, K\}$. Let Z be a matrix of dimensions $K \times K$ s.t $Z_{p,q} = \sum_{i=1}^{N_L} \mathbf{I}(z_{i1} = p, z_{i2} = q)$. The de-noised matrix M of the strength of association between the entities in E_{t_l} is given by $M = B_{t_l}^T Z B_{t_r}$.

13.3   Related Work

Link LDA, and many other extensions to LDA, model documents that are annotated with metadata. In a parallel area of research, various different approaches to modeling links between documents have been explored. For instance, pairwise-link-LDA (Nallapati et al., 2008) combines MMSB with LDA by modeling documents using LDA and generating links between them using MMSB. The relational topic model (Chang and Blei, 2009) generates links between documents based on their topic distributions. The copycat and citation influence models (Dietz et al., 2007) also model links between citing and cited documents by extending LDA and eliminating independence between documents. The latent topic hypertext model (LTHM) (Gruber et al., 2008) presents a generative process for documents that can be linked to each other from specific words in the citing document. These classes of models are different from the model proposed in this paper, Block-LDA, in that they model links between entities in the documents rather than links between documents.

The Nubbi model (Chang et al., 2009) tackles a related problem where entity relations are discovered from text data by relying on words that appear in the context of entities and entity pairs in the text. Block-LDA differs from Nubbi in that it models a document as bags of entities without considering the location of entity mentions in the text. The entities need not even be mentioned in the text of the document. The group-topic model (Wang et al., 2006) addresses the task of modeling events pertaining to pairs of entities with textual attributes that annotate the event. The text in this model is associated with events, which differs from the standalone documents mentioning entities considered by Block-LDA.

The author-topic model (AT) (Rosen-Zvi et al., 2004) addresses the task of modeling corpora annotated with the IDs of people who authored the documents. Every author in the corpus has a topic distribution over the latent topics, and words in the documents are drawn from topics drawn from the specific distribution of the author who is deemed to have generated the word. The author-recipient-topic model (ART) (McCallum et al., 2005) extends the idea further by building a topic distribution for every author-recipient pair. As we show in the experiments below, Block-LDA can also be used to model the relationships between authors, recipients, and words in documents by constructing an appropriate link matrix from known information about the authors and recipients of documents; however, unlike the AT and ART models which are primarily designed to model documents, Block-LDA provides a generative model for the links between authors and recipients in addition to documents. This allows Block-LDA to be used for additional inferences not possible with the AT or ART models, for instance, predicting probable author-recipient interactions. Wen and Lin (2010) describes an application of an approach that uses both content and network information to analyze enterprise data. While a joint modeling of the network and content is not used, LDA is used to study the topics in communications between people.

A summary of related models from prior work is shown in Table 13.1.

The Munich Institute for Protein Sequencing (MIPS) database (Mewes et al., 2004) includes a hand-crafted collection of protein interactions covering 8000 protein complex associations in yeast. We use a subset of this collection containing 844 proteins, for which all interactions were hand-curated (Figure 13.2(a)). The MIPS institute also provides a set of functional annotations for each protein which are organized in a tree, with 15 nodes at the first level (shown in Table 13.2). The 844

Model	Links	Documents
LDA	-	words
link LDA	-	words + entities
relational topic model	document-document	words + document IDs
pairwise-link-LDA, link-PLSA-LDA	document-document	words + cited document IDs
copycat, citation influence models	document-document	words + cited document IDs
latent topic hypertext model	document-document	words + cited document IDs
author-recipient-topic model	-	docs + authors + recipients
author-topic model	-	docs + authors
topic link LDA	document-document	words + authors
MMSB	entity-entity	-
sparse blockmodel (Parkkinen et al.)	entity-entity	-
Nubbi	entity-entity	words near entities or entity-pairs
group topic model	entity-entity	words about the entity-entity event
Block-LDA	entity-entity	words + entities

TABLE 13.1
Related work.

proteins participating in interactions are mapped to these 15 functional categories with an average of 2.5 annotations per protein.

We also use another dataset of protein-protein interactions in yeast that were observed as a result of wetlab experiments by collaborators. This dataset consists of 635 interactions that deal primarily with ribosomal proteins and assembly factors in yeast.

In addition to the MIPS PPI data, we use a text corpus that is derived from the repository of scientific publications at PubMed®. PubMed is a free, open-access, on-line archive of over 18 million biological abstracts, bibliographies, and citation lists for papers published since 1948 (U.S. National Library of Medicine, 2008). The subset we work with consists of approximately 40,000 publications

Metabolism
Cellular communication/signal transduction mechanism
Cell rescue, defense and virulence
Regulation of / interaction with cellular environment
Cell fate
Energy
Control of cellular organization
Cell cycle and DNA processing
Subcellular localisation
Transcription
Protein synthesis
Protein activity regulation
Transport facilitation
Protein fate (folding, modification, destination)
Cellular transport and transport mechanisms

TABLE 13.2
List of functional categories.

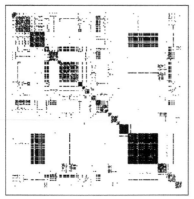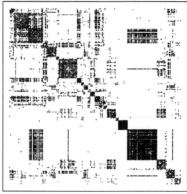

(a) MIPS interactions   (b) Co-occurences in text

FIGURE 13.2
Observed protein-protein interactions compared to thresholded co-occurrence in text.

about the yeast organism that have been curated in the Saccharomyces Genome Database (SGD) (Dwight et al., 2004) with annotations of proteins that are discussed in the publication. We further restrict the dataset to only those documents that are annotated with at least one protein from the MIPS database. This results in a MIPS-protein annotated document collection of 15,776 publications. The publications in this set were written by a total of 47,215 authors. We tokenize the titles and abstracts based on white space, lowercase all tokens, and eliminate stopwords. Low frequency (<5 occurrences) terms are also eliminated. The vocabulary contains 45,648 words.

13.4   Datasets

To investigate the co-occurrence patterns of proteins annotated in the abstracts, we construct a co-occurrence matrix. From every abstract, a link is constructed for every pair of annotated protein mentions. Additionally, protein mentions that occur fewer than 5 times in the corpus are discarded. Figure 13.2(b) shows that the resultant matrix looks very similar to the MIPS PPI matrix in Figure 13.2(a). This suggests that joint modeling of the protein-annotated text with the PPI information has the potential to be beneficial. The nodes representing proteins in Figures 13.2(a) and 13.2(b) are ordered by their cluster IDs, obtained by clustering them using k-means clustering, treating proteins as 15-bit vectors of functional category annotations.

The Enron email corpus (Shetty and Adibi, 2004) is a large publicly available collection of email messages subpoenaed as part of the investigation by the Federal Energy Regulatory Commission (FERC). The dataset contains 517,437 messages in total. Although the Enron Email Dataset contains the email folders of 150 people, two people appear twice with different usernames, and one user's emails consist solely of automated emails resulting in 147 unique people in the dataset. For the text component of the model, we use all the emails in the Sent[1] folders of the 147 users' mailboxes, resulting in a corpus of 96,103 emails. Messages are annotated with mentions of people from the set

---

[1]"Sent", "sent_items," and "_sent_mail" folders in users' mailboxes were treated as "Sent" folders.

of 147 Enron employees if they are senders or recipients of the email. Mentions of people outside of the 147 persons considered are dropped. While extracting text from the email messages, "quoted" messages are eliminated using a heuristic which looks for a "Forwarded message" or "Original message" delimiter. In addition, lines starting with a ">" are also eliminated. The emails are then tokenized after lowercasing the entire message, using whitespace and punctuation marks as word delimiters. Words occurring fewer than 5 times in the corpus are discarded. The vocabulary of the corpus consists of 32,880 words.

For the entity links component of the model, we build an email communication network by constructing a link between the sender and every recipient of an email message for every email in the corpus. Recipients of the emails include people directly addressed in the "To" field and people included in the "Cc" and "Bcc" fields. Similar to the text component, only links between the 147 Enron employees are considered. The link dataset generated in this manner has 200,404 links. Figure 13.3(a) shows the email network structure. The nodes in the matrix representing people are ordered by cluster IDs obtained by running k-means clustering on the 147 people. Each person s is represented by a vector of length 147, where the elements in the vector are normalized counts of the number of times an email is sent by s to the person indicated by the element.

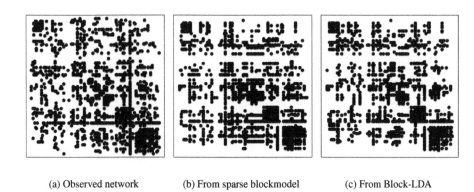

(a) Observed network   (b) From sparse blockmodel   (c) From Block-LDA

FIGURE 13.3
Enron network and its de-noised recovered versions.

13.5   Experimental Results

We present results from experiments using Block-LDA to model the yeast and Enron datasets described in Section 13.4.

13.5.1   Results from the Yeast Dataset

Perplexity and Convergence

First, we investigate the convergence properties of the Gibbs sampler used for inference in Block-LDA by observing link perplexity on held-out data at different epochs. Link perplexity of a set of links L is defined as

$$\exp\left(\frac{\sum_{e_1 \to e_2 \in L} \log\left(\sum_{\langle z_1,z_2\rangle} \pi^{\langle z_1,z_2\rangle} \beta^{(e_1)}_{t_l,z_1} \beta^{(e_1)}_{t_l,z_2}\right)}{|L|}\right). \tag{13.7}$$

Figure 13.4(a) shows the convergence of the link perplexity using Block LDA and a baseline model on the PPI+SGD dataset with 20% of the full dataset held-out for testing. The number of topics K is set at 15 since our aim is to recover topics that can be aligned with the 15 protein functional categories. α_D and α_L are sampled from Gamma$(0.1, 1)$. It can be observed that the Gibbs sampler burns-in after about 20 iterations.

Next, we perform two sets of experiments with the PPI+PubMed Central dataset. The text data has three types of entities in each document—words, authors, and protein annotations with the PPI data-linking proteins. In the first set of experiments, we evaluate the model using perplexity of held-out protein-protein interactions using increasing amounts of the PPI data for training.

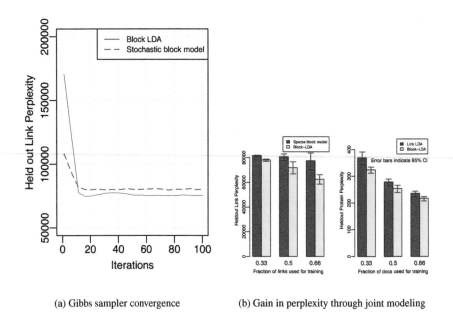

(a) Gibbs sampler convergence   (b) Gain in perplexity through joint modeling

FIGURE 13.4
Perplexity in the MIPS PPI+SGD dataset.

All 15,773 documents in the SGD dataset are used when textual information is used. When text is not used, the model is equivalent to using only the left half of Figure 13.1. Figures 13.5(a) and 13.5(b) show the posterior likelihood of protein-protein interactions recovered using the sparse blockmodel and Block-LDA, respectively. In the other set of experiments, we evaluate the model using protein perplexity in held-out text using progressively increasing amounts of text as training data. All the links in the PPI dataset are used in these experiments when link data are used. When link data are not used, the model reduces to Link LDA. In all experiments, the Gibbs sampler is run until the held-out perplexity stabilizes to a nearly constant value (≈ 80 iterations).

Figure 13.4(b) shows the gains in perplexity in the two sets of experiments with different amounts of training data. The perplexity values are averaged over 10 runs. In both sets of experiments, it can be seen that Block-LDA results in lower perplexities than using links/text alone.

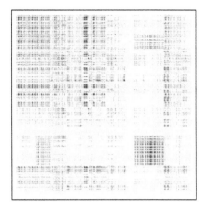
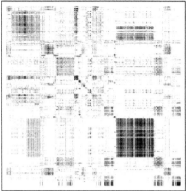

(a) Sparse blockmodel   (b) Block-LDA

FIGURE 13.5
Inferred protein-protein interactions.

These results indicate that co-occurrence patterns of proteins in text contain information about protein interactions, which Block-LDA is able to utilize through joint modeling. Our conjecture is that the protein co-occurrence information in text is a noisy approximation of the PPI data.

Table 13.3 shows the top words, proteins, and authors for sample topics induced by running Block-LDA over the full PPI+SGD dataset. These topics provide a qualitative feel for the topics that emerge using the model. The Gibbs sampling procedure was run until convergence (around 80 iterations) and the number of topics was set to 15. The topic tables were then analyzed, with a title and an analysis of the topic added after the inference procedure was completed. Details about proteins and yeast researchers were obtained from the SGD [2] website to understand the function of the top proteins in each topic and to get an idea of the research profile of the top authors mentioned.

Topic Coherence

A useful application of latent blockmodeling approaches is understanding the underlying nature of data.

We conduct three different evaluations of the emergent topics. First, we obtain topics from only the text corpus using a model that comprises the right half of Figure 13.1, which is equivalent to using the Link-LDA model. For the second evaluation, we use the Block-LDA model that is trained on the text corpus and the MIPS protein-protein interaction database. Finally, for the third evaluation, we replace the MIPS PPI database with the interaction obtained from the wetlab experiments. In all cases, we set K, the number of topics, to be 15. In each variant, we represent documents as three sets of entities, i.e., the words in the abstracts of the article, the set of proteins associated with the article as indicated in the SGD database, and the authors who wrote the article. Each topic therefore consists of three different multinomial distributions over the sets of the three kinds of entities described.

Topics that emerge from the different variants can possibly be assigned different indices even when they discuss the same semantic concept. To compare topics across variants, we need a method to determine which topic indices from the different variants correspond to the same semantic concept. To obtain the mapping between topics from each variant, we utilize the Hungarian algorithm

---

[2] See http://www.yeastgenome.org.

Words	mutant, mutants, gene, cerevisiae, growth, type, mutations, saccharomyces, wild, mutation, strains, strain, phenotype, genes, deletion
Proteins	rpl20b, rpl5, rpl16a, rps5, rpl39, rpl18a, rpl27b, rps3, rpl23a, rpl1b, rpl32, rpl17b, rpl35a, rpl26b, rpl31a
Authors	klis_fm, bussey_h, miyakawa_t, toh-e_a, heitman_j, perfect_jr, ohya_y_ws, sherman_f, latge_jp, schaffrath_r, duran_a, sa-correia_i, liu_h, subik_j, kikuchi_a, chen_j, goffeau_a, tanaka_k, kuchler_k, calderone_r, nombela_c, popolo_l, jablonowski_d, kim_j
Analysis	A common experimental procedure is to induce random mutations in the "wild-type" strain of a model organism (e.g., saccharomyces cerevisiae) and then screen the mutants for interesting observable characteristics (i.e. phenotype). Often the phenotype shows slower growth rates under certain conditions (e.g. lack of some nutrient). The RPL* proteins are all part of the larger (60S) subunit of the ribosome. The first two biologists, Klis and Bussey's research use this method.

(a) Analysis of mutations

Words	binding, domain, terminal, structure, site, residues, domains, interaction, region, subunit, alpha, amino, structural, conserved, atp
Proteins	rps19b, rps24b, rps3, rps20, rps4a, rps11a, rps2, rps8a, rps10b, rps6a, rps10a, rps19a, rps12, rps9b, rps28a
Authors	naider_f, becker_jm, leulliot_n, van_tilbeurgh_h, melki_r, velours_j, graille_m_s, janin_-j, zhou_cz, blondeau_k, ballesta_jp, yokoyama_s, bousset_l, vershon_ak, bowler_be, zhang_y, arshava_b, buchner_j, wickner_rb, steven_ac, wang_y, zhang_m, forgac_m, brethes_d
Analysis	Protein structure is an important area of study. Proteins are composed of amino-acid residues, functionally important protein regions are called domains, and functionally important sites are often "converved" (i.e., many related proteins have the same amino-acid at the site). The RPS* proteins all part of the smaller (40S) subunit of the ribosome. Naider, Becker, and Leulliot study protein structure.

(b) Protein structure

Words	transcription, ii, histone, chromatin, complex, polymerase, transcriptional, rna, promoter, binding, dna, silencing, h3, factor, genes
Proteins	rpl16b, rpl26b, rpl24a, rpl18b, rpl18a, rpl12b, rpl6b, rpp2b, rpl15b, rpl9b, rpl40b, rpp2a, rpl20b, rpl14a, rpp0
Authors	workman_jl, struhl_k, winston_f, buratowski_s, tempst_p, erdjument-bromage_h, kornberg_rd_a, svejstrup_jq, peterson_cl, berger_sl, grunstein_m, stillman_dj, cote_j, cairns_-br, shilatifard_a, hampsey_m, allis_cd, young_ra, thuriaux_p, zhang_z, sternglanz_r, krogan_nj, weil_pa, pillus_l
Analysis	In transcription, DNA is unwound from histone complexes (where it is stored compactly) and converted to RNA. This process is controlled by transcription factors, which are proteins that bind to regions of DNA called promoters. The RPL* proteins are part of the larger subunit of the ribosome, and the RPP proteins are part of the ribosome stalk. Many of these proteins bind to RNA. Workman, Struhl, and Winston study transcription regulation and the interaction of transcription with the restructuring of chromatin (a combination of DNA, histones, and other proteins that comprises chomosomes).

(c) Chromosome remodeling and transcription

TABLE 13.3
Top words, proteins, and authors: Topics obtained using Block-LDA on the PPI+SGD dataset.

(Kuhn, 1955) to solve the assignment problem where the cost of aligning topics together is determined using the Jensen-Shannon divergence measure.

Once the topics are obtained, we first obtain the proteins associated with the topic by retrieving the top proteins from the multinomial distribution corresponding to proteins. Then, the top articles corresponding to each topic are obtained using a ranked list of documents with the highest mass of their topic proportion distributions (θ) residing in the topic considered.

Manual Evaluation

To evaluate the topics, a yeast biologist who is an expert in the field was asked to mark each topic with a binary flag indicating if the top words of the distribution represented a coherent sub-topic in yeast biology. The top words of the distribution representing a topic were presented as a ranked list of words. This process was repeated for the three different variants of the model. The variant used to obtain results is concealed from the evaluator to remove the possibility of bias.

In the next step of the evaluation, the top articles and proteins assigned to each topic were presented in a ranked list and a similar judgment was requested to indicate if the article/protein was relevant to the topic in question. Similar to the topic coherence judgments, the process was repeated for each variant of the model. Screenshots of the tool used for obtaining the judgments can be seen in Figure 13.6. It should be noted that since the nature of the topics in the literature considered was highly technical and specialized, it was impractical to get judgments from multiple annotators.

To evaluate the retrieval of the top articles and proteins, we measure the quality of the results by computing the precision@10 score.

First, we evaluate the coherence of the topics obtained from the three variants described above. Table 13.4 shows that out of the 15 topics that were obtained, 12 topics were deemed coherent from the text-only model and 13 and 15 topics were deemed coherent from the Block-LDA models using the MIPS and wetlab PPI datasets, respectively.

Variant	Num. Coherent Topics
Only Text	12 / 15
Text + MIPS	13 / 15
Text + Wetlab	15 / 15

TABLE 13.4
Topic coherence evaluation.

Next, we study the precision@10 values for each topic and variant of the article retrieval and protein retrieval tasks (see Figures 13.7 or 13.8, respectively). The horizontal lines in the plots represent the mean of the precision@10 across all topics. It can be seen from the plots that for both the article and protein retrieval tasks, on average the joint models work better than the text-only model. For the article retrieval task, the model trained with the text + MIPS resulted in the higher mean precision@10 whereas for the protein retrieval task, the text + Wetlab PPI dataset returned a higher mean precision@10 value. For both the protein retrieval and paper retrieval tasks, the improvements shown by the joint models using either of the PPI datasets over the text-only model (i.e., the Link LDA model) were statistically significant at the 0.05 level using the paired Wilcoxon sign test. However, the difference in performance between the two joint models that used the two different PPI networks was insignificant, which indicates that there is no observable advantage in using one PPI dataset over the other in conjunction with the text corpus.

Functional Category Prediction

Proteins are identified as belonging to multiple functional categories in the MIPS PPI dataset, as described in Section 13.4. We use Block-LDA and baseline methods to predict proteins' functional

FIGURE 13.6
Screenshot: Article relevance annotation tool.

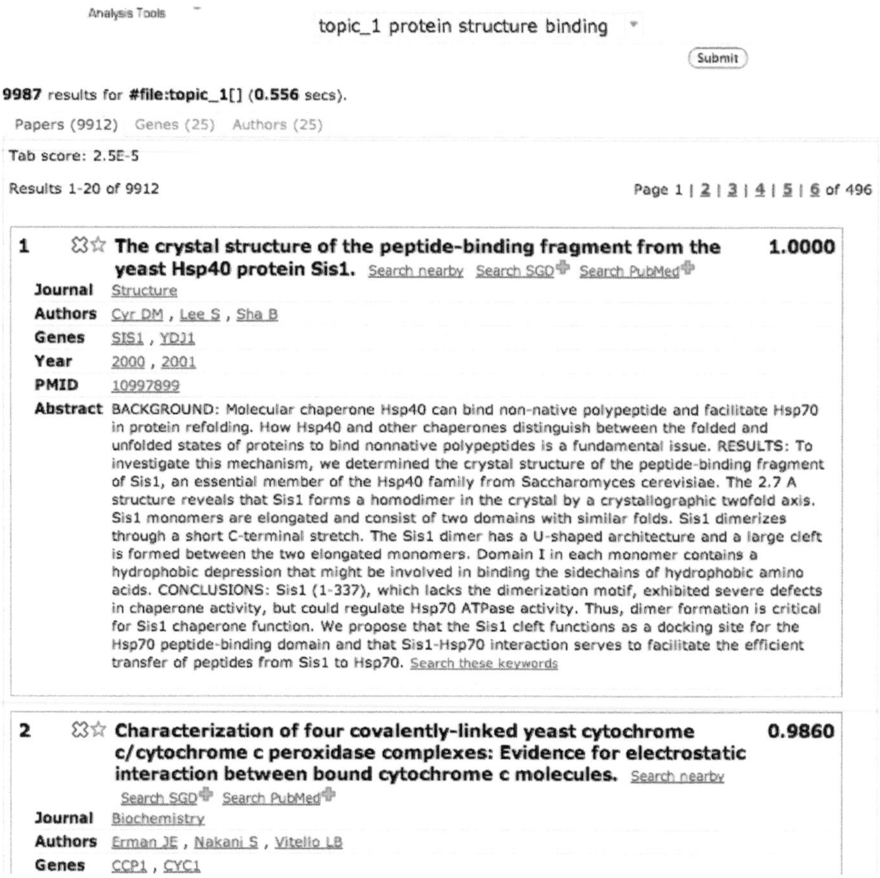

categories and evaluate them by comparing them to the ground truth in the MIPS dataset using the method presented in prior work (Airoldi et al., 2008). A model is first trained with K set to 15 topics to recover the 15 top-level functional categories of proteins. Every topic that is returned consists of a set of multinomials including β_{t_1}, the topic-wise distribution over all proteins. The values of β_{t_1} are thresholded such that the top $\approx 16\%$ (the density of the protein-function matrix) of entries are considered as such a positive prediction that the protein falls in the functional category corresponding to the latent topic. To determine the mapping of latent topic to functional category, 10% of the proteins are used in a procedure that greedily finds the alignment resulting in the best accuracy, as described in Airoldi et al. (2008). It is important to note that the true functional categories of proteins are completely hidden from the model. The functional categories are used only during evaluation of the resultant topics from the model.

The precision, recall, and F_1 scores of the different models in predicting the right functional categories for proteins are shown in Table 13.5. Since there are 15 functional categories and a protein has approximately 2.5 functional category associations, we expect only $\sim 1/6$ of protein-functional category associations to be positive. Precision and recall therefore depict a better picture of the predictions than accuracy. For the random baseline, every protein-functional category pair is randomly deemed to be 0 or 1 with the Bernoulli probability of an association being proportional to the ratio of 1s observed in the protein-functional category matrix in the MIPS dataset. In the MMSB approach, induced latent blocks are aligned to functional categories as described in Airoldi et al. (2008).

We see that the F_1 scores for the baseline sparse blockmodel and MMSB are nearly the same, and that combining text and links provides a significant boost to the F_1 score. This suggests that protein co-occurrence patterns in the abstracts contain information about functional categories that is also evidenced by the better than random F_1 score obtained using Link LDA, which uses only documents. All the methods considered outperform the random baseline.

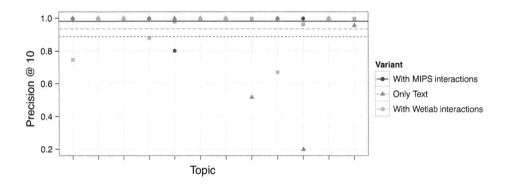

FIGURE 13.7
Retrieval performance - Article retrieval.

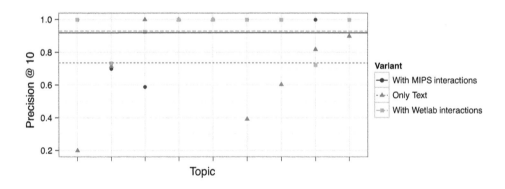

FIGURE 13.8
Retrieval performance - Protein retrieval.

13.5.2   Results from the Enron Email Corpus Dataset

As described in Section 13.4, the Enron dataset consists of two components—text from the sent folders and the network of senders and recipients of emails within the Enron organization. Each email is treated as a document and is annotated with a set of people consisting of the senders and recipients of the email. We first study the network reconstruction capability of the Block-LDA model. Block-LDA is trained using all 96,103 emails in the sent folders and the 200,404 links obtained from the full email corpus. Figures 13.3(a), 13.3(b), and 13.3(c) show the true communication

Method	F_1	Precision	Recall
Block-LDA	**0.249**	0.247	0.250
Sparse Blockmodel	0.161	0.224	0.126
Link LDA	0.152	0.150	0.155
MMSB	0.165	0.166	0.164
Random	0.145	0.155	0.137

TABLE 13.5
Functional category prediction.

matrix, the matrix reconstructed using the sparse mixed membership stochastic blockmodel and the matrix reconstructed using the Block-LDA model, respectively. The figures show that both models are approximately able to recover the communication network in the Enron dataset.

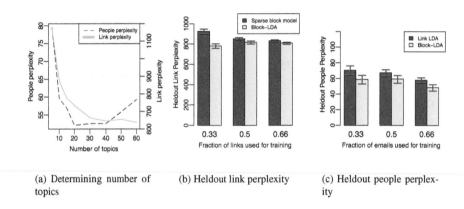

(a) Determining number of topics

(b) Heldout link perplexity

(c) Heldout people perplexity

FIGURE 13.9
Enron corpus: Perplexity.

Figure 13.9(a) shows the link perplexity and person perplexity in text of held-out data, as the number of topics is varied. Person perplexity is indicative of the surprise inherent in observing a sender or a recipient and can be used as a prior in tasks like predicting recipients for emails that are being composed. Link perplexity is a score for the quality of link prediction and captures the notion of social connectivity in the graph. It indicates how well the model is able to capture links between people in the communication network. The person perplexity in the plot decreases initially and stabilizes when the number of topics reaches 20. It eventually starts to rise again when the number of topics is raised above 40. The link perplexity on the other hand stabilizes at 20 and then exhibits a slight downward trend. For the remaining experiments with the Enron data, we set $K = 40$.

In the next set of experiments, we evaluate Block-LDA and other models by evaluating the person perplexity in held-out emails by varying the training and test set size. Similar to the experiments with the PPI data, the Gibbs sampler is run until the held-out perplexity stabilizes to a nearly constant value (≈ 80 iterations). The perplexity values are averaged over 10 runs. Figure 13.9(c) shows the person perplexity in text of held-out data as increasing amounts of the text data are used for training. The remainder of the dataset is used for testing. It is important to note that only Block-LDA uses the communication link matrix. A consistent improvement in person perplexity can be observed when email text data are supplemented with communication link data irrespective of the training set size. This indicates that the latent block structure in the links is beneficial while shaping latent topics from text.

Block-LDA is finally evaluated using link prediction. The sparse blockmodel, which serves as a baseline, does not use any text information. Figure 13.9(b) shows the perplexity in held-out data with varying amounts of the 200,404 edges in the network used for training. When textual information is used, all 96,103 emails are used. The histogram shows that Block-LDA obtains lower perplexities than the sparse blockmodel, which uses only links. As in the PPI experiments, using the text in the emails improves the modeling of the network of senders and recipients, although the effect is less marked when the number of links used for training is increased. The topical coherence in the latent topics induces better latent blocks in the matrix indicating a transfer of signal from the text to the network model.

13.6   Conclusion

We proposed a model that jointly models entity-entity links and entity-annotated text that permits co-occurence information in text to influence link modeling and vice-versa. Our experiments show that joint modeling outperforms approaches that use only a single source of information. Improvements are observed when the joint model is evaluated internally using perplexity in two different datasets and externally using protein functional category prediction in the yeast dataset. We also evaluated topics obtained from the joint modeling of yeast biology literature and protein-protein interactions in yeast and compared them to topics that were obtained from using only the literature. The topics were evaluated for coherence and by measuring the mean precision@10 score of the top articles and proteins that were retrieved for each topic. Evaluation by a domain expert showed that the joint modeling produced more coherent topics and showed better precision@10 scores in the article and protein retrieval tasks indicating that the model enabled information sharing between the literature and the PPI networks.

Acknowledgments

This work was funded by grant 1R101GM081293 from NIH, IIS-0811562 from NSF, and by a gift from Google. The opinions expressed in this paper are solely those of the authors.

References

Airoldi, E. M., Blei, D. M., Fienberg, S. E., and Xing, E. P. (2008). Mixed membership stochastic blockmodels. *Journal of Machine Learning Research* 9: 1981–2014.

Balasubramanyan, R. and Cohen, W. W. (2011). Block-LDA: Jointly modeling entity-annotated text and entity-entity links. In *Proceedings of the 2011 SIAM Conference on Data Mining (SDM '11)*. SIAM/Omnipress, 450–461.

Blei, D. M., Ng, A. Y., and Jordan, M. I. (2003). Latent Dirichlet allocation. *The Journal of Machine Learning Research* 3: 993–1022.

Chang, J. and Blei, D. M., (2009). Relational topic models for document networks. In *Proceedings* *of the of 12th International Conference on Artificial Intelligence and Statistics (AISTATS 2009). Journal of Machine Learning Research – Proceedings Track* 5, 81–88.

Chang, J., Boyd-Graber, J., and Blei, D. M. (2009). Connections between the lines: Augmenting social networks with text. In *Proceedings of the 15th ACM SIGKDD International Conference on Knowledge Discovery and Data Mining (KDD '09)*. New York, NY, USA: ACM, 169-178.

Dietz, L., Bickel, S., and Scheffer, T. (2007). Unsupervised prediction of citation influences. In *Proceedings of the 24th Annual International Conference on Machine Learning (ICML '07)*. New York, NY, USA: ACM, 233-240.

Dwight, S. S., Balakrishnan, R., Christie, K. R., Costanzo, M. C., Dolinski, K., Engel, S. R., Feierbach, B., Fisk, D. G., Hirschman, J., Hong, E. L., Issel-Tarver, L., Nash, R. S., Sethuraman, A., Starr, B., Theesfeld, C. L., Andrada, R., Binkley, G., Dong, Q., Lane, C., Schroeder, M., Weng, S., Botstein, D. and Cherry J., M. (2004). Saccharomyces genome database: Underlying principles and organisation. *Briefings in Bioinformatics* 5: 9.

Erosheva, E. A., Fienberg, S. E., and Lafferty, J. D. (2004). Mixed-membership models of scientific publications. *Proceedings of the National Academy of Sciences* 101: 5220.

Griffiths, T. L. and Steyvers, M. (2004). Finding scientific topics. *Proceedings of the National Academy of Sciences* 101 Suppl 1: 5228–5235.

Gruber, A., Rosen-Zvi, M., and Weiss, Y. (2008). Latent topic models for hypertext. In *Proceedings of the 24th Conference on Uncertainty in Artificial Intelligence (UAI 2008)*. Corvallis, OR, USA: AUAI Press, 230–239.

Kuhn, H. W. (1955). The Hungarian method for the assignment problem. *Naval Research Logistics Quarterly* 2: 83–97.

McCallum, A., Corrada-Emmanuel, A., and Wang, X. (2005). Topic and role discovery in social networks. *Proceedings of the 19th International Joint Conference on Artificial Intelligence (IJCAI '05)*. IJCAI, 786—791.

Mewes, H. -W., Amid, C., Arnold, R., Frishman, D., Gldener, U., Mannhaupt, G., Mnsterktter, M., Pagel, P., Strack, N., Stmpflen, V., Warfsmann, J., and Ruepp, A. (2004). MIPS: Analysis and annotation of proteins from whole genomes. *Nucleic Acids Research* 32: 41–44.

Nallapati, R. M., Ahmed, A., Xing, E. P., and Cohen, W. W. (2008). Joint latent topic models for text and citations. In *Proceeding of the 14th ACM SIGKDD International Conference on Knowledge Discovery and Data mining (KDD '08)*. New York, NY, USA: ACM, 542–550.

Parkkinen, J., Sinkkonen, J., Gyenge, A., and Kaski, S. (2009). A block model suitable for sparse graphs. In *Proceedings of the 7th International Workshop on Mining and Learning with Graphs (MLG 2009)*. Leuven, Belgium: poster presented.

Rosen-Zvi, M., Griffiths, T. L., Steyvers, M., and Smyth, P. (2004). The author-topic model for authors and documents. In *Proceedings of the 20th Conference on Uncertainty in Artificial Intelligence (UAI 2004)*. Arlington, VA, USA: AUAI Press, 487–494.

Shetty, J. and Adibi, J. (2004). The Enron Email Dataset Database Schema and Brief Statistical Report. Tech. report, Information Sciences Institute.

Wang, X., Mohanty, N., and McCallum, A. (2006). Group and topic discovery from relations and their attributes. In Weiss, Y., Schölkopf, B., and Platt, J. (eds), *Advances in Neural Information Processing Systems 18*. Cambridge, MA: The MIT Press, 1449—1456.

Wen, Z. and Lin, C. -Y. (2010). Towards finding valuable topics. In *Proceedings of the 2010 SIAM Conference on Data Mining (SDM '10)*. Philadelphia, PA, USA: SIAM, 720–731.

14

Robust Estimation of Topic Summaries Leveraging Word Frequency and Exclusivity

Jonathan M. Bischof

Department of Statistics, Harvard University, Cambridge, MA 02138, USA

Edoardo M. Airoldi

Department of Statistics, Harvard University, Cambridge, MA 02138, USA

CONTENTS

An ongoing challenge in the analysis of document collections is how to summarize content in terms of a set of inferred *themes* that can be interpreted substantively in terms of topics. However, the current practice in mixed membership models of text (Blei et al., 2003) of parametrizing the themes in terms of most frequent words limits interpretability by ignoring the differential use of words across topics. Words that are both common and exclusive to a theme are more effective at characterizing the topical content of such a theme. We consider a setting where professional editors have annotated documents to a collection of topic categories, organized into a tree, in which leaf-nodes correspond to the most specific topics. Each document is annotated to multiple categories, at different levels of the tree. We introduce hierarchical Poisson convolution (HPC) as a model to analyze annotated documents in this setting. The model leverages the structure among categories defined by professional editors to infer a clear semantic description for each topic in terms of words that are both frequent and exclusive. We develop a parallelized Hamiltonian Monte Carlo sampler that allows the inference to scale to millions of documents.

14.1 Introduction

A recurrent challenge in multivariate statistics is how to construct interpretable low-dimensional summaries of high-dimensional data. Historically, simple models based on correlation matrices, such as principal component analysis (Jolliffe, 1986) and canonical correlation analysis (Hotelling, 1936) have proven to be effective tools for data reduction. More recently, multilevel models have become a flexible and powerful tool for finding latent structure in high-dimensional data (McLachlan and Peel, 2000; Sohn and Xing, 2009; Blei et al., 2003; Airoldi et al., 2008). However, while interpretable statistical summaries are highly valued in applications, dimensionality reduction models are rarely optimized to aid qualitative discovery; there is no guarantee that the optimal low-dimensional projections will be understandable in terms of quantities of scientific interest that can help practitioners make decisions. Here, we design a model that targets scientific estimands of interest in text analysis and achieves a good balance between interpretability and dimensionality reduction.

We consider a setting in which we observe two sets of categorical data for each unit of observation: $w_{1:V}$, which live in a high-dimensional space, and $l_{1:K}$, which live in a structured low-dimensional space and provide a direct link to information of scientific interest about the sampling units. The goal of the analysis is twofold. First, we desire to develop a joint model for the observations $Y \equiv \{W_{D \times V}, L_{D \times K}\}$ that can be used to project the data onto a low-dimensional parameter space Θ in which interpretability is maintained by mapping categories in $\mathcal{L}$ to directions in Θ. Second, we would like the mapping from the original space to the low-dimensional projection to be scientifically interesting so that statistical insights about Θ can be understood in terms of the original inputs, $w_{1:V}$, in a way that guides future research.

In the application to text analysis that motivates this work, $w_{1:N}$ are the raw word counts observed in each document and $l_{1:K}$ are a set of labels created by professional editors that are indicative of topical content. Specifically, the words are represented as an unordered vector of counts, with the length of the vector corresponding to the size of a known dictionary. The labels are organized in a tree-structured ontology, from the most generic topic at the root of the tree to the most specific topic at the leaves. Each news article may be annotated with more than one label, at the editors' discretion. The number of labels is given by the size of the ontology and typically ranges from tens to hundreds of categories. In this context, the inferential challenge is to discover a low-dimensional representation of topical content, Θ, that aligns with the coarse labels provided by editors while at the same time providing a mapping between the textual content and directions in Θ in a way that formalizes and enhances our understanding of how low-dimensional structure is expressed in the space of observed words.

Recent approaches to this problem in the machine learning literature have taken a Bayesian hierarchical approach to this task by viewing a document's content as arising from a mixture of component distributions, commonly referred to as "topics," as they often capture thematic structure (Blei, 2012). As the component distributions are almost exclusively parameterized as multinomial distributions over words in the vocabulary, the loading of words onto topics is characterized in terms of the relative frequency of within-component usage. While relative frequency has proven to be a useful mapping of topical content onto words, recent work has documented a growing list of interpretability issues with frequency-based summaries: they are often dominated by contentless "stop" words (Wallach et al., 2009), sometimes appear incoherent or redundant (Mimno et al., 2011; Chang et al., 2009), and typically require post hoc modification to meet human expectations (Hu et al., 2011). Instead, we propose a new mapping for topical content that incorporates how words are used differentially across topics. If a word is common in a topic, it is also important to know whether it is common in many topics or relatively exclusive to the topic in question. Both of these summary statistics are informative: nonexclusive words are less likely to carry topic-specific content, while infrequent words occur too rarely to form the semantic core of a topic. We therefore look for the most frequent words in the corpus that are also likely to have been generated from the topic of interest to summarize its content. In this approach we borrow ideas from the statistical literature, in which models of differential word usage have been leveraged for analyzing writing styles in a supervised setting (Mosteller and Wallace, 1984; Airoldi et al., 2005; 2006; 2007a), and combine them with ideas from the machine learning literature, in which latent variable and mixture models based on frequent word usage have been used to infer structure that often captures topical content (McCallum et al., 1998; Blei et al., 2003; Canny, 2004; Airoldi et al., 2007b; 2010a). From a statistical perspective, models based on topic-specific distributions over the vocabulary cannot produce stable estimates of differential usage since they only model the relative frequency of words within topics. They cannot regularize usage across topics and naively infer the greatest differential usage for the rarest features (Eisenstein et al., 2011). To tackle this issue, we introduce the generative framework of hierarchical Poisson convolution (HPC) that parameterizes topic-specific word counts as unnormalized count variates whose rates can be regularized across topics as well as within them, making stable inference of both word frequency and exclusivity possible. HPC can be seen as a fully generative extension of sparse topic coding (Zhu and Xing, 2011) that emphasizes regularization and interpretability rather than exact sparsity. Additionally, HPC leverages hierarchical systems of topic categories created by professional editors in collections such as Reuters, the *New York Times*, *Wikipedia*, and *Encyclopedia Britannica* to make focused comparisons of differential use between neighboring topics on the tree and build a sophisticated joint model for topic memberships and labels in the documents. By conditioning on a known hierarchy, we avoid the complicated tasks of inferring hierarchical structure (Blei et al., 2004; Mimno et al., 2007; Adams et al., 2010) as well as the number of topics (Joutard et al., 2008; Airoldi et al., 2010a;b). We introduce a parallelized Hamiltonian Monte Carlo (HMC) estimation strategy that makes full Bayesian inference efficient and scalable.

Since the proposed model is designed to infer an interpretable description of human-generated labels, we restrict the topic components to have a one-to-one correspondence with the human-generated labels, as in Labeled LDA (Ramage et al., 2009). This *descriptive* link between the labels and topics differs from the *predictive* link used in Supervised LDA (Blei and McAuliffe, 2007; Perotte et al., 2012), where topics are learned as an optimal covariate space to predict an observed document label or response variable. The more restrictive descriptive link can be expected to limit predictive power but is crucial for learning summaries of individual labels. We then infer a description of these labels in terms of words that are both frequent and exclusive. We anticipate that learning a concise semantic description for any collection of topics implicitly defined by professional editors is the first step toward the semi-automated creation of domain-specific topic ontologies. Domain-specific topic ontologies may be useful for evaluating the semantic content of *inferred* topics, or for predicting the semantic content of new social media, including Twitter messages and Facebook wall posts.

14.2 A Mixed Membership Model for Poisson Data

The hierarchical Poisson convolution model is a data generating process for document collections whose topics are organized in a hierarchy, and whose topic labels are observed. We refer to the structure among topics interchangeably as a *hierarchy* or *tree* since we assume that each topic has exactly one parent and that no cyclical parental relations are allowed. Each document $d \in \{1, \ldots, D\}$ is a record of counts w_{fd} for every feature in the vocabulary, $f \in \{1, \ldots, V\}$. The length of the document is given by L_d, which we normalize by the average document length L to get $l_d \equiv \frac{1}{L} L_d$. Documents have unrestricted membership to any combination of topics $k \in \{1, \ldots, K\}$ represented by a vector of labels $\boldsymbol{I}_d$ where $I_{dk} \equiv I\{\text{doc } d \text{ belongs to topic } k\}$.

14.2.1   Modeling Word Usage Rates on the Hierarchy

The HPC model leverages the known topic hierarchy by assuming that words are used similarly in neighboring topics. Specifically, the log rate for a word across topics follows a Gaussian diffusion down the tree. Consider the topic hierarchy presented in the right panel of Figure 14.1. At the top level, $\mu_{f,0}$ represents the log rate for feature f overall in the corpus. The log rates $\mu_{f,1}, \ldots, \mu_{f,J}$ for first-level topics are then drawn from a Gaussian centered around the corpus rate with dispersion controlled by the variance parameter $\tau_{f,0}^2$. From first-level topics, we then draw the log rates for the second-level topics from another Gaussian centered around their mean $\mu_{f,j}$ and with variance $\tau_{f,j}^2$. This process is continued down the tree, with each parent node having a separate variance parameter to control the dispersion of its children.

The variance parameters τ_{fp}^2 directly control the local differential expression in a branch of the tree. Words with high variance parameters can have rates in the child topics that differ greatly from the parent topic p, allowing the child rates to diverge. Words with low variance parameters will have rates close to the parent and so will be expressed similarly among the children. If we learn a population distribution for the τ_{fp}^2 that has low mean and variance, it is equivalent to saying that most features are expressed similarly across topics a priori and that we would need a preponderance of evidence to believe otherwise.

14.2.2   Modeling the Topic Membership of Documents

Documents in the HPC model can contain content from any of the K topics in the hierarchy at varying proportions, with the exact allocation given by the vector $\boldsymbol{\theta}_d$ on the $K - 1$ simplex. The model assumes that the count for word f contributed by each topic follows a Poisson distribution

FIGURE 14.1
Graphical representation of hierarchical Poisson convolution (left panel) and detail on tree plate (right panel). For an introduction to this type of illustration, see Airoldi (2007).

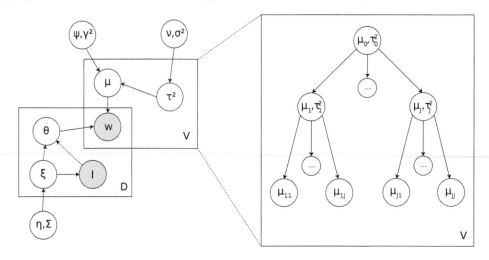

whose rate is moderated by the document's length and membership to the topic; that is, $w_{fdk} \sim$ $\text{Pois}(l_d \theta_{dk} \beta_{fk})$. The only data we observe is the total word count $w_{fd} \equiv \sum_{k=1}^{K} w_{fdk}$, but the infinite divisibility property of the Poisson distribution gives us that $w_{fd} \sim \text{Pois}(l_d \theta_d^T \beta_f)$. These draws are done for every word in the vocabulary (using the same θ_d) to get the content of the document.[1]

In labeled document collections, human coders give us an extra piece of information for each document I_d that indicates the set of topics that contributed its content. As a result, we know $\theta_{dk} = 0$ for all topics k where $I_{dk} = 0$, and only have to determine how content is allocated between the set of active topics.

The HPC model assumes that these two sources of information for a document are not generated independently. A document should not have a high probability of being labeled to a topic from which it receives little content and vice versa. Instead, the model posits a latent K-dimensional topic affinity vector $\xi_d \sim \mathcal{N}(\eta, \Sigma)$ that expresses how strongly the document is associated with each topic. The topic memberships and labels of the document are different manifestations of this affinity. Specifically, each ξ_{dk} is the log odds that topic label k is active in the document, with $I_{dk} \sim \text{Bernoulli}(\text{logit}^{-1}(\xi_{dk}))$. Conditional on the labels, the topic memberships are the relative sizes of the document's affinity for the active topics and zero for inactive topics: $\theta_{dk} \equiv e^{\xi_{dk}} I_{dk} / \sum_{j=1}^{K} e^{\xi_{dj}} I_{dj}$. Restricting each document's membership vectors to the labeled topics is a natural and efficient way to generate sparsity in the mixing parameters, stabilizing inference and reducing the computational burden of posterior simulation.

We outline the generative process in full detail in Table 14.1, which can be summarized in three steps. First, a set of rate and variance parameters are drawn for each feature in the vocabulary. Second, a topic affinity vector is drawn for each document in the corpus, which generate topic labels. Finally, both sets of parameters are then used to generate the words in each document. For simplicity of presentation we assume that each non-terminal node has J children and that the tree has only two levels below the corpus level, but the model can accommodate any tree structure.

14.2.3 Estimands for Text Analysis

In order to measure topical semantic content, we consider the topic-specific frequency and exclusivity of each word in the vocabulary. These quantities form a two-dimensional summary of each word's relation to a topic of interest, with higher scores in both frequency and exclusivity being positively related to topic specific content. Additionally, we develop a univariate summary of semantic content that can be used to rank words in terms of their semantic content. These estimands are simple functions of the rate parameters of HPC; the distribution of the documents' topic memberships is a nuisance parameter needed to disambiguate the content of a document between its labeled topics.

A word's topic-specific frequency, $\beta_{fk} \equiv \exp \mu_{fk}$, is directly parameterized in the model and is regularized across words (via hyperparameters ψ and γ^2) and across topics. A word's exclusivity to a topic, $\phi_{f,k}$, is its usage rate relative to a set of comparison topics $\mathcal{S}$: $\phi_{f,k} = \beta_{f,k} / \sum_{j \in \mathcal{S}} \beta_{f,j}$. A topic's siblings are a natural choice for a comparison set to see which words are overexpressed in the topic compared to a set of similar topics. While not directly modeled in HPC, the exclusivity parameters are also regularized by the τ_{fp}^2, since if the child rates are forced to be similar then the $\phi_{f,k}$ will be pushed toward a baseline value of $1/|\mathcal{S}|$. We explore the regularization structure of the model empirically in Section 14.4.

Since both frequency and exclusivity are important factors in determining a word's semantic content, a univariate measure of topical importance is a useful estimand for diverse tasks such as dimensionality reduction, feature selection, and content discovery. In constructing a composite measure, we do not want a high rank in one dimension to be able to compensate for a low rank in the

---

[1] This is where the model's name arises: the observed feature count in each document is the convolution of (unobserved) topic-specific Poisson variates.

TABLE 14.1
Generative process for hierarchical Poisson convolution.

Step	Generative process
Tree parameters	For feature $f \in \{1, \ldots, V\}$: • Draw $\mu_{f,0} \sim \mathcal{N}(\psi, \gamma^2)$ • Draw $\tau_{f,0}^2 \sim$ Scaled Inv-$\chi^2(\nu, \sigma^2)$ • For $j \in \{1, \ldots, J\}$ (first level of hierarchy): − Draw $\mu_{f,j} \sim \mathcal{N}(\mu_{f,0}, \tau_{f,0}^2)$ − Draw $\tau_{f,j}^2 \sim$ Scaled Inv-$\chi^2(\nu, \sigma^2)$ • For $j \in \{1, \ldots, J\}$ (terminal level of hierarchy): − Draw $\mu_{f,j1}, \ldots, \mu_{f,jJ} \sim \mathcal{N}(\mu_{f,j}, \tau_{f,j}^2)$ • Define $\beta_{f,k} \equiv e^{\mu_{f,k}}$ for $k \in \{1, \ldots, K\}$
Topic membership parameters	For document $d \in \{1, \ldots, D\}$: • Draw $\boldsymbol{\xi}_d \sim \mathcal{N}(\boldsymbol{\eta}, \boldsymbol{\Sigma} = \lambda^2 \boldsymbol{I}_K)$ • For topic $k \in \{1, \ldots, K\}$: − Define $p_{dk} \equiv 1/(1 + e^{-\xi_{dk}})$ − Draw $I_{dk} \sim$ Bernoulli(p_{dk}) − Define $\theta_{dk}(\boldsymbol{I}_d, \boldsymbol{\xi}_d) \equiv e^{\xi_{dk}} I_{dk} / \sum_{j=1}^K e^{\xi_{dj}} I_{dj}$
Data generation	For document $d \in \{1, \ldots, D\}$: • Draw normalized document length $l_d \sim \frac{1}{L}$Pois(v) • For every topic k and feature f: − Draw count $w_{fdk} \sim$ Pois$(l_d \boldsymbol{\theta}_d^T \boldsymbol{\beta}_f)$ • Define $w_{fd} \equiv \sum_{k=1}^K w_{fdk}$ (observed data)

other, since frequency or exclusivity alone are not necessarily useful. We therefore adopt the harmonic mean to pull the "average" rank toward the lower score. For word f in topic k, we define the $FREX_{fk}$ score as the harmonic mean of the word's rank in the distribution of $\phi_{.,k}$ and $\mu_{.,k}$:

$$FREX_{fk} = \left(\frac{w}{\text{ECDF}_{\phi_{.,k}}(\phi_{f,k})} + \frac{1-w}{\text{ECDF}_{\mu_{.,k}}(\mu_{f,k})} \right)^{-1},$$

where w is the weight for exclusivity (which we set to 0.5 as a default) and ECDF$_{x_{.,k}}$ is the empirical cdf function applied to the values x over the first index.

14.3 Scalable Inference via Parallelized HMC Sampler

We use a Gibbs sampler to obtain the posterior expectations of the unknown rate and membership parameters (and associated hyperparameters) given the observed data. Specifically, inference is conditioned on $\boldsymbol{W}$, a $D \times V$ matrix of word counts; $\boldsymbol{I}$, a $D \times K$ matrix of topic labels; $\boldsymbol{l}$, a D-vector of document lengths; and $\mathcal{T}$, a tree structure for the topics.

Creating a scalable inference method is critical since the space of latent variables grows linearly in the number of words and documents, with $K(D + V)$ total unknowns. Our model offers an advantage in that the parameters can be conceptually organized into two subsets, and the posterior distribution of each subset of parameters factors nicely, conditionally on the other subset of parameters. On one side, the conditional posterior of the rate and variance parameters $\{\boldsymbol{\mu}_f, \boldsymbol{\tau}_f^2\}_{f=1}^V$ factors by word given the membership parameters and the hyperparameters ψ, γ^2, ν, and σ^2. On the other, the conditional posterior of the topic affinity parameters $\{\boldsymbol{\xi}_d\}_{d=1}^D$ factors by document given the hyperparameters η and $\boldsymbol{\Sigma}$ and the rate parameters $\{\boldsymbol{\mu}_f\}_{f=1}^V$.

Therefore, conditional on the hyperparameters, we are left with two blocks of draws that can be broken into V or D independent threads. Using parallel computing software such as message passing interface (MPI), the computation time for drawing the parameters in each block is only constrained by resources required for a single draw. The total runtime need not significantly increase with the addition of more documents or words as long as the number of available cores also increases. Both of these conditional distributions are only known up to a constant and can be high-dimensional if there are many topics, making direct sampling impossible and random walk Metropolis inefficient. We are able to obtain uncorrelated draws through the use of Hamiltonian Monte Carlo (HMC) (Neal, 2011), which leverages the posterior gradient, and Hessian to find a distant point in the parameter space with high probability of acceptance. HMC works well for log densities that are unimodal and have relatively constant curvature. We give step-by-step instructions for our implementation of the algorithm in the Appendix.

After appropriate initialization, we follow a fixed Gibbs scan where the two blocks of latent variables are drawn in parallel from their conditional posteriors using HMC. We then draw the hyperparameters conditional on all the inputed latent variables.

14.3.1 A Blocked Gibbs Sampling Strategy

To set up the block Gibbs sampling algorithm, we derive the relevant conditional posterior distributions and explain how we sample from each.

Updating Tree Parameters

In the first block, the conditional posterior of the tree parameters factors by word:

$$p(\{\boldsymbol{\mu}_f, \boldsymbol{\tau}_f^2\}_{f=1}^V | \boldsymbol{W}, \boldsymbol{I}, \boldsymbol{l}, \psi, \gamma^2, \nu, \sigma^2, \{\boldsymbol{\xi}_d\}_{d=1}^D, \mathcal{T}) \propto$$
$$\prod_{f=1}^V \left\{ \prod_{d=1}^D p(w_{fd} | \boldsymbol{I}_d, l_d, \boldsymbol{\mu}_f, \boldsymbol{\xi}_d) \right\} \cdot p(\boldsymbol{\mu}_f, \boldsymbol{\tau}_f^2 | \psi, \gamma^2, \mathcal{T}, \nu, \sigma^2).$$

Given the conditional conjugacy of the variance parameters and their strong influence on the curvature of the rate parameter posterior, we sample the two groups conditional on each other to optimize HMC performance. Conditioning on the variance parameters, we can write the likelihood of the rate parameters as a Poisson regression where the documents are observations, the $\theta_d(\boldsymbol{I}_d, \boldsymbol{\xi}_d)$ are the covariates, and the l_d serve as exposure weights.

The prior distribution of the rate parameters is a Gaussian graphical model, so a priori the log rates for each word are jointly Gaussian with mean $\psi \mathbf{1}$ and precision matrix $\boldsymbol{\Lambda}(\gamma^2, \boldsymbol{\tau}_f^2, \mathcal{T})$, which has non-zero entries only for topic pairs that have a direct parent-child relationship.[2] The log conditional posterior is:

---

[2] In practice this precision matrix can be found easily as the negative Hessian of the log-prior distribution.

$$\log p(\boldsymbol{\mu}_f | \boldsymbol{W}, \boldsymbol{I}, \boldsymbol{l}, \{\tau_f^2\}_{f=1}^V, \psi, \gamma^2, \nu, \sigma^2, \{\boldsymbol{\xi}_d\}_{d=1}^D, \mathcal{T}) =$$

$$-\sum_{d=1}^D l_d \boldsymbol{\theta}_d^T \boldsymbol{\beta}_f + \sum_{d=1}^D w_{fd} \log(\boldsymbol{\theta}_d^T \boldsymbol{\beta}_f) - \frac{1}{2}(\boldsymbol{\mu}_f - \psi \mathbf{1})^T \boldsymbol{\Lambda}(\boldsymbol{\mu}_f - \psi \mathbf{1}).$$

We use HMC to sample from this unnormalized density. Note that the covariate matrix $\boldsymbol{\Theta}_{D \times K}$ is very sparse in most cases, so we speed computation with a sparse matrix representation.

We know the conditional distribution of the variance parameters due to the conjugacy of the inverse-χ^2 prior with the normal distribution of the log rates. Specifically, if $\mathcal{C}(\mathcal{T})$ is the set of child topics of topic k with cardinality J, then

$$\tau_{fk}^2 | \boldsymbol{\mu}_f, \nu, \sigma^2, \mathcal{T} \sim \text{Inv-}\chi^2\left(J + \nu, \frac{\nu\sigma^2 + \sum_{j \in \mathcal{C}}(\mu_{fj} - \mu_{fk})^2}{J + \nu}\right).$$

Updating Topic Affinity Parameters

In the second block, the conditional posterior of the topic affinity vectors factors by document:

$$p(\{\boldsymbol{\xi}_d\}_{d=1}^D | \boldsymbol{W}, \boldsymbol{I}, \boldsymbol{l}, \{\boldsymbol{\mu}_f\}_{f=1}^V, \boldsymbol{\eta}, \boldsymbol{\Sigma})$$

$$\propto \prod_{d=1}^D \left\{ \prod_{f=1}^V p(w_{fd} | \boldsymbol{I}_d, l_d, \boldsymbol{\mu}_f, \boldsymbol{\xi}_d) \right\} \cdot p(\boldsymbol{I}_d | \boldsymbol{\xi}_d) \cdot p(\boldsymbol{\xi}_d | \boldsymbol{\eta}, \boldsymbol{\Sigma}).$$

We can again write the likelihood as a Poisson regression, now with the rates as covariates. The log conditional posterior for one document is:

$$\log p(\boldsymbol{\xi}_d | \boldsymbol{W}, \boldsymbol{I}, \boldsymbol{l}, \{\boldsymbol{\mu}_f\}_{f=1}^V, \boldsymbol{\eta}, \boldsymbol{\Sigma}) =$$

$$-l_d \sum_{f=1}^V \boldsymbol{\beta}_f^T \boldsymbol{\theta}_d + \sum_{f=1}^V w_{fd} \log(\boldsymbol{\beta}_f^T \boldsymbol{\theta}_d) - \sum_{k=1}^K \log(1 + e^{-\xi_{dk}})$$

$$-\sum_{k=1}^K (1 - I_{dk}) \xi_{dk} - \frac{1}{2}(\boldsymbol{\xi}_d - \boldsymbol{\eta})^T \boldsymbol{\Sigma}^{-1}(\boldsymbol{\xi}_d - \boldsymbol{\eta}).$$

We use HMC to sample from this unnormalized density. Here the parameter vector $\boldsymbol{\theta}_d$ is sparse rather than the covariate matrix $\boldsymbol{B}_{V \times K}$. If we remove the entries of $\boldsymbol{\theta}_d$ and columns of $\boldsymbol{B}$ pertaining to topics k where $I_{dk} = 0$, then we are left with a low-dimensional regression where only the active topics are used as covariates, greatly simplifying computation.

Updating Corpus-Level Parameters

We draw the hyperparameters after each iteration of the block update. We put flat priors on these unknowns so that we can learn their most likely values from the data. As a result, their conditional posteriors only depend on the latent variables they generate.

The log corpus-level rates $\mu_{f,0}$ for each word follow a Gaussian distribution with mean ψ and variance γ^2. The conditional distribution of these hyperparameters is available in closed form:

$$\psi | \gamma^2, \{\mu_{f,0}\}_{f=1}^V \sim \mathcal{N}\left(\frac{1}{V}\sum_{f=1}^V \mu_{f,0}, \frac{\gamma^2}{V}\right),$$

$$\text{and} \quad \gamma^2 | \psi, \{\mu_{f,0}\}_{f=1}^V \sim \text{Inv-}\chi^2\left(V, \frac{1}{V}\sum_{f=1}^V (\mu_{f,0} - \psi)^2\right).$$

The discrimination parameters τ_{fk}^2 independently follow an identical scaled inverse-χ^2 with convolution parameter ν and scale parameter σ^2, while their inverse follows a Gamma($\kappa_\tau = \frac{\nu}{2}, \lambda_\tau = \frac{2}{\nu\sigma^2}$) distribution. We use HMC to sample from this unnormalized density. Specifically,

$$\log p(\kappa_\tau, \lambda_\tau | \{\boldsymbol{\tau}_f^2\}_{f=1}^V, \mathcal{T}) = (\kappa_\tau - 1) \sum_{f=1}^V \sum_{k \in \mathcal{P}} \log (\tau_{fk}^2)^{-1}$$

$$- |\mathcal{P}|V\kappa_\tau \log \lambda_\tau - |\mathcal{P}|V \log \Gamma(\kappa_\tau) - \frac{1}{\lambda_\tau} \sum_{f=1}^V \sum_{k \in \mathcal{P}} (\tau_{fk}^2)^{-1},$$

where $\mathcal{P}(\mathcal{T})$ is the set of parent topics on the tree. Each draw of $(\kappa_\tau, \lambda_\tau)$ is then transformed back to the (ν, σ^2) scale.

The document-specific topic affinity parameters $\boldsymbol{\xi}_d$ follow a multivariate normal distribution with mean parameter $\boldsymbol{\eta}$ and a covariance matrix parameterized in terms of a scalar, $\boldsymbol{\Sigma} = \lambda^2 \boldsymbol{I}_K$. The conditional distribution of these hyperparameters is available in closed form. For efficiency, we choose to put a flat prior on $\log \lambda^2$ rather than the original scale, which allows us to marginalize out $\boldsymbol{\eta}$ from the conditional posterior of λ^2:

$$\lambda^2 | \{\boldsymbol{\xi}_d\}_{d=1}^D \sim \text{Inv-}\chi^2 \left(DK - 1, \frac{\sum_d \sum_k (\xi_{dk} - \bar{\xi}_k)^2}{DK - 1} \right),$$

$$\text{and} \quad \boldsymbol{\eta} | \lambda^2, \{\boldsymbol{\xi}_d\}_{d=1}^D \sim \mathcal{N}\left(\bar{\boldsymbol{\xi}}, \frac{\lambda^2}{D} \boldsymbol{I}_K \right).$$

14.3.2 Estimation

As discussed in Section 14.2.3, our estimands are the topic-specific frequency and exclusivity of the words in the vocabulary, as well as the Frequency-Exclusivity (FREX) score that averages each word's performance in these dimensions. We use posterior means to estimate frequency and exclusivity, computing these quantities at every iteration of the Gibbs sampler and averaging the draws after the burn-in period. For the FREX score, we apply the ECDF function to the frequency and exclusivity posterior expectations of all words in the vocabulary to estimate the true ECDF.

14.3.3 Inference on Missing Document Categories

In order to classify unlabeled documents, we need to find the posterior predictive distribution of the membership vector $\boldsymbol{I}_{\tilde{d}}$ for a new document $\tilde{d}$. Inference is based on the new document's word counts $\boldsymbol{w}_{\tilde{d}}$ and the unknown parameters, which we hold constant at their posterior expectation. Unfortunately, the posterior predictive distribution of the topic affinities $\boldsymbol{\xi}_{\tilde{d}}$ is intractable without conditioning on the label vector, since the labels control which topics contribute content. We therefore use a simpler model where the topic proportions depend only on the relative size of the affinity parameters:

$$\theta_{dk}^*(\boldsymbol{\xi}_d) \equiv \frac{e^{\xi_{dk}}}{\sum_{j=1}^K e^{\xi_{dj}}} \quad \text{and} \quad I_{dk} \sim \text{Bern}\left(\frac{1}{1 + \exp(-\xi_{dk})} \right).$$

The posterior predictive distribution of this simpler model factors into tractable components:

$$p^*(\boldsymbol{I}_{\tilde{d}}, \boldsymbol{\xi}_{\tilde{d}} | \boldsymbol{w}_{\tilde{d}}, \boldsymbol{W}, \boldsymbol{I}) \approx p(\boldsymbol{I}_{\tilde{d}} | \boldsymbol{\xi}_{\tilde{d}}) \, p^*(\boldsymbol{\xi}_{\tilde{d}} | \{\hat{\boldsymbol{\mu}}_f\}_{f=1}^V, \hat{\boldsymbol{\eta}}, \hat{\boldsymbol{\Sigma}}, \boldsymbol{w}_{\tilde{d}})$$

$$\propto p(\boldsymbol{I}_{\tilde{d}} | \boldsymbol{\xi}_{\tilde{d}}) \, p^*(\boldsymbol{w}_{\tilde{d}} | \boldsymbol{\xi}_{\tilde{d}}, \{\hat{\boldsymbol{\mu}}_f\}_{f=1}^V) \, p(\boldsymbol{\xi}_{\tilde{d}} | \hat{\boldsymbol{\eta}}, \hat{\boldsymbol{\Sigma}}).$$

It is then possible to find the most likely $\boldsymbol{\xi}_{\tilde{d}}^*$ based on the evidence from $\boldsymbol{w}_{\tilde{d}}$ alone.

14.4 Empirical Analysis and Results

We analyze the fit of the HPC model to Reuters Corpus Volume I (RCV1), a large collection of newswire stories. First, we demonstrate how the variance parameters τ_{fp}^2 regularize the exclusivity with which words are expressed within topics. Second, we show that regularization of exclusivity has the greatest effect on infrequent words. Third, we explore the joint posterior of the topic-specific frequency and exclusivity of words as a summary of topical content, giving special attention to the upper right corner of the plot where words score highly in both dimensions. We compare words that score highly on the FREX metric to top words scored by frequency alone, the current practice in topic modeling. Finally, we compare the classification performance of HPC to baseline models.

14.4.1 An Overview of the Reuters Corpus

RCV1 is an archive of 806,791 newswire stories during a twelve-month period from 1996 to 1997.[3] As described in Lewis et al. (2004), Reuters staffers assigned stories into any subset of 102 hierarchical topic categories. In the original data, assignment to any topic required automatic assignment to all ancestor nodes, but we removed these redundant ancestor labels since they do not allow our model to distinguish intentional assignments to high-level categories from assignment to their offspring. In our modified annotations, the only documents we see in high-level topics are those labeled to them and none of their children, which maps onto general content. We preprocessed document tokens with the Porter stemming algorithm (leading to 300,166 unique stems) and chose the most frequent 3% of stems (10,421 unique stems, over 100 million total tokens) for the feature set.[4]

The Reuters topic hierarchy has three levels that divide the content into finer categories at each cut. At the first level, content is divided between four high-level categories: three that focus on business and market news (Markets, Corporate/Industrial, and Economics) and one grab bag category that collects all remaining topics from politics to entertainment (Government/Social). The second level provides fine-grained divisions of these broad categories and contains the terminal nodes for most branches of the tree. For example, the Markets topic is split between Equity, Bond, Money, and Commodity markets at the second level. The third level offers further subcategories where needed for a small set of second-level topics. For example, the Commodity markets topic is divided between Agricultural (soft), Metal, and Energy commodities. We present a graphical illustration of the Reuters topic hierarchy in Figure 14.2.

Many documents in the Reuters corpus are labeled to multiple topics, even after redundant ancestor memberships are removed. Overall, 32% of the documents are labeled to more than one node of the topic hierarchy. Fifteen percent of documents have very diverse content, being labeled to two or more of the main branches of the tree (Markets, Commerce, Economics, and Government/Social). Twenty-one percent of documents are labeled to multiple second-level categories on the same branch (for example, Bond markets and Equity markets in the Markets branch). Finally, 14% of documents are labeled to multiple children of the same second-level topic (for example, Metals trading and Energy markets in the Commodity markets branch of Markets). Therefore, a completely general mixed membership model such as HPC is necessary to capture the labeling patterns of the corpus. A full breakdown of membership statistics by topic is presented in Tables 14.2 and 14.3.

---

[3]Available upon request from the National Institute of Standards and Technology (NIST), http://trec.nist.gov/data/reuters/reuters.html.

[4]Including rarer features did not meaningfully change the results.

FIGURE 14.2
Topic hierarchy of Reuters corpus.

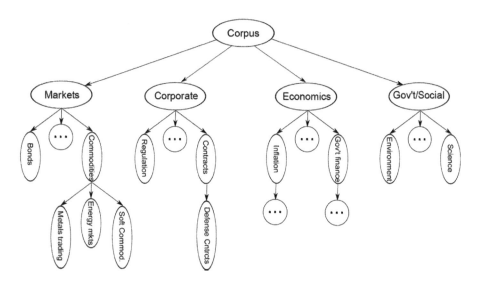

14.4.2   The Differential Usage Parameters Regulate Topic Exclusivity

A word can only be exclusive to a topic if its expression across the sibling topics is allowed to diverge from the parent rate. Therefore, we would only expect words with high differential usage parameters τ_{fp}^2 at the parent level to be candidates for highly exclusive expression ϕ_{fk} in any child topic k. Words with child topic rates that cannot vary greatly from the parent should have nearly equal expression in each child k, meaning $\phi_{fk} \approx \frac{1}{C}$ for a branch with C child topics. An important consequence is that, although the ϕ_{fk} are not directly modeled in HPC, their distribution is regularized by learning a prior distribution on the τ_{fp}^2.

This tight relation can be seen in the HPC fit. Figure 14.3 shows the joint posterior expectation of the differential usage parameters in a parent topic and exclusivity parameters across the child topics. Specifically, the left panel compares the rate variance of the children of Markets from their parent to exclusivity between the child topics; the right panel does the same with the two children of Performance, a second-level topic under the Corporate category. The plots have similar patterns. For low levels of differential expression, the exclusivity parameters are clustered around the baseline value, $\frac{1}{C}$. At high levels of child rate variance, words gain the ability to approach exclusive expression in a single topic.

14.4.3   Frequency Modulates the Regularization of Exclusivity

One of the most appealing aspects of regularization in generative models is that it acts most strongly on the parameters for which we have the least information. In the case of the exclusivity parameters in HPC we have the most data for frequent words, so for a given topic the words with low rates should be least able to escape regularization of their exclusivity parameters by our shrinkage prior on the parent's τ_{fp}^2.

TABLE 14.2

Topic membership statistics.

Topic code	Topic name	# docs	Any MM	CB L1 MM	CB L2 MM	CB L3 MM
CCAT	CORPORATE/INDUSTRIAL	2170	79.60%	79.60%	13.10%	0.80%
C11	STRATEGY/PLANS	24325	51.50	11.50	44.50	4.50
C12	LEGAL/JUDICIAL	11944	99.20	98.90	50.20	1.70
C13	REGULATION/POLICY	37410	85.90	55.60	61.40	4.50
C14	SHARE LISTINGS	7410	30.30	7.90	10.30	15.80
C15	PERFORMANCE	229	82.10	35.80	74.20	1.70
C151	ACCOUNTS/EARNINGS	81891	7.90	1.30	0.60	6.40
C152	COMMENT/FORECASTS	73092	18.90	4.80	1.60	13.50
C16	INSOLVENCY/LIQUIDITY	1920	66.70	31.50	54.60	3.60
C17	FUNDING/CAPITAL	4767	78.10	41.40	67.70	5.00
C171	SHARE CAPITAL	18313	44.60	3.20	1.70	41.50
C172	BONDS/DEBT ISSUES	11487	15.10	5.70	0.30	9.70
C173	LOANS/CREDITS	2636	24.70	8.50	3.60	15.60
C174	CREDIT RATINGS	5871	65.60	59.00	0.50	7.50
C18	OWNERSHIP CHANGES	30	76.70	23.30	76.70	3.30
C181	MERGERS/ACQUISITIONS	43374	34.40	6.50	4.80	26.90
C182	ASSET TRANSFERS	4671	28.30	4.70	5.70	21.00
C183	PRIVATISATIONS	7406	73.70	34.20	6.30	44.10
C21	PRODUCTION/SERVICES	25403	76.40	46.50	53.60	0.80
C22	NEW PRODUCTS/SERVICES	6119	55.00	15.30	49.10	0.40
C23	RESEARCH/DEVELOPMENT	2625	77.00	36.40	57.80	0.90
C24	CAPACITY/FACILITIES	32153	72.20	33.60	58.40	0.90
C31	MARKETS/MARKETING	29073	46.90	25.30	34.60	1.30
C311	DOMESTIC MARKETS	4299	80.60	73.70	9.50	18.70
C312	EXTERNAL MARKETS	6648	78.10	70.40	9.60	14.20
C313	MARKET SHARE	1115	39.70	10.30	5.10	27.80
C32	ADVERTISING/PROMOTION	2084	63.80	26.90	52.50	1.40
C33	CONTRACTS/ORDERS	14122	48.00	12.60	40.50	0.80
C331	DEFENCE CONTRACTS	1210	68.00	65.50	13.30	3.40
C34	MONOPOLIES/COMPETITION	4835	92.30	54.90	75.70	14.00
C41	MANAGEMENT	1083	75.60	52.10	59.90	2.00
C411	MANAGEMENT MOVES	10272	17.70	9.60	2.40	8.20
C42	LABOUR	11878	99.70	99.60	46.50	1.50
ECAT	ECONOMICS	621	90.50	90.50	9.70	1.40
E11	ECONOMIC PERFORMANCE	8568	43.00	24.20	29.10	5.10
E12	MONETARY/ECONOMIC	24918	81.70	75.40	17.90	13.70
E121	MONEY SUPPLY	2182	30.50	23.10	0.70	9.20
E13	INFLATION/PRICES	130	60.00	46.90	28.50	0.80
E131	CONSUMER PRICES	5659	24.70	15.60	6.00	12.00
E132	WHOLESALE PRICES	939	19.00	3.40	0.60	16.90
E14	CONSUMER FINANCE	428	73.80	43.20	61.00	1.60
E141	PERSONAL INCOME	376	75.00	63.80	9.60	22.30
E142	CONSUMER CREDIT	200	46.00	30.00	3.50	18.50
E143	RETAIL SALES	1206	27.50	19.70	2.40	10.20
E21	GOVERNMENT FINANCE	941	86.70	81.40	53.90	4.00
E211	EXPENDITURE/REVENUE	15768	78.20	72.40	16.10	13.80
E212	GOVERNMENT BORROWING	27405	32.70	29.60	2.70	4.50
E31	OUTPUT/CAPACITY	591	45.20	18.30	35.20	0.50
E311	INDUSTRIAL PRODUCTION	1701	17.70	9.80	3.10	9.30
E312	CAPACITY UTILIZATION	52	65.40	13.50	3.80	57.70
E313	INVENTORIES	111	26.10	10.80	0.00	16.20
E41	EMPLOYMENT/LABOUR	14899	100.00	100.00	49.40	2.20
E411	UNEMPLOYMENT	2136	92.00	90.60	10.40	12.00
E51	TRADE/RESERVES	4015	85.10	75.50	38.70	1.90
E511	BALANCE OF PAYMENTS	2933	63.80	43.70	8.20	25.70
E512	MERCHANDISE TRADE	12634	64.90	59.10	11.50	11.70
E513	RESERVES	2290	30.10	22.70	1.30	16.80
E61	HOUSING STARTS	391	51.70	47.80	13.80	0.80
E71	LEADING INDICATORS	5270	2.90	0.60	2.40	0.20

Key: MM = Mixed membership, CB Lx = Cross-branch MM at level x

TABLE 14.3
Topic membership statistics, continued.

Topic code	Topic name	# docs	Any MM	CB L1 MM	CB L2 MM	CB L3 MM
GCAT	GOVERNMENT/SOCIAL	24546	2.50	2.50	0.50	0.10
G15	EUROPEAN COMMUNITY	1545	16.10	6.90	14.60	0.00
G151	EC INTERNAL MARKET	3307	98.00	87.20	10.60	94.30
G152	EC CORPORATE POLICY	2107	96.70	90.70	40.30	50.30
G153	EC AGRICULTURE POLICY	2360	96.10	94.20	31.40	27.70
G154	EC MONETARY/ECONOMIC	8404	98.20	93.00	11.50	43.90
G155	EC INSTITUTIONS	2124	70.80	42.00	24.30	54.00
G156	EC ENVIRONMENT ISSUES	260	75.00	57.70	28.80	50.80
G157	EC COMPETITION/SUBSIDY	2036	100.00	99.80	60.20	32.50
G158	EC EXTERNAL RELATIONS	4300	80.70	62.80	27.00	24.80
G159	EC GENERAL	40	47.50	17.50	35.00	2.50
GCRIM	CRIME, LAW ENFORCEMENT	32219	79.50	41.60	59.40	0.90
GDEF	DEFENCE	8842	93.70	17.20	84.40	0.50
GDIP	INTERNATIONAL RELATIONS	37739	73.70	20.50	60.70	0.90
GDIS	DISASTERS AND ACCIDENTS	8657	75.70	40.10	52.20	0.20
GENT	ARTS, CULTURE, ENTERTAINMENT	3801	68.80	29.20	49.60	0.50
GENV	ENVIRONMENT AND NATURAL WORLD	6261	90.20	51.50	72.30	2.50
GFAS	FASHION	313	76.40	45.70	41.50	1.90
GHEA	HEALTH	6030	81.90	56.10	65.00	1.20
GJOB	LABOUR ISSUES	17241	99.60	99.40	44.60	3.30
GMIL	MILLENNIUM ISSUES	5	100.00	100.00	40.00	0.00
GOBIT	OBITUARIES	844	99.40	15.30	99.40	0.00
GODD	HUMAN INTEREST	2802	60.70	9.70	55.20	0.10
GPOL	DOMESTIC POLITICS	56878	79.60	29.70	63.00	1.80
GPRO	BIOGRAPHIES, PERSONALITIES, PEOPLE	5498	87.50	10.00	84.70	0.10
GREL	RELIGION	2849	86.10	6.60	84.30	0.10
GSCI	SCIENCE AND TECHNOLOGY	2410	55.20	22.20	45.10	0.30
GSPO	SPORTS	35317	1.30	0.60	0.90	0.00
GTOUR	TRAVEL AND TOURISM	680	89.60	69.70	34.70	3.40
GVIO	WAR, CIVIL WAR	32615	67.30	10.10	64.60	0.10
GVOTE	ELECTIONS	11532	100.00	13.30	100.00	1.30
GWEA	WEATHER	3878	73.90	46.80	46.40	0.10
GWELF	WELFARE, SOCIAL SERVICES	1869	95.40	75.50	74.10	3.40
MCAT	MARKETS	894	81.10	81.10	14.50	2.20
M11	EQUITY MARKETS	48700	16.30	12.30	3.90	2.90
M12	BOND MARKETS	26036	21.30	15.60	5.20	3.50
M13	MONEY MARKETS	447	65.80	51.90	23.30	1.60
M131	INTERBANK MARKETS	28185	15.10	9.40	0.70	6.40
M132	FOREX MARKETS	26752	36.90	24.70	3.10	16.10
M14	COMMODITY MARKETS	4732	18.00	16.70	2.30	0.10
M141	SOFT COMMODITIES	47708	24.10	22.80	5.50	2.00
M142	METALS TRADING	12136	34.70	19.30	4.10	16.10
M143	ENERGY MARKETS	21957	21.10	18.40	4.80	2.90

Key: MM = Mixed membership, CB Lx = Cross-branch MM at level x

Figure 14.4 displays words in terms of their frequency (on the X axis) and exclusivity (on the Y axis). The two panels correspond to two different topics, namely science and technology and research and development, and exclusivity scores are computed for each of these topics compared to their sibling topics in the topic hierarchy. We will refer to this plot as the FREX plot in the following. The left panel features the Science and Technology topic, a child in the grab bag Government/Social branch; the right panel features the Research/Development topic, a child in the Corporate branch. The overall shape of the joint posterior is very similar for both topics. On the left side of the plots, the exclusivity of rare words is unable to significantly exceed the $\frac{1}{C}$ baseline. This is because the model does not have much evidence to estimate usage in the topic, so the estimated rate is shrunk heavily toward the parent rate. However, we see that it is possible for rare words to be underexpressed in a topic, which happens if they are frequent and overexpressed in a sibling topic. Even though their rates are similar to the parent in this topic, sibling topics may have a much higher rate and account for most appearances of the word in the comparison group.

FIGURE 14.3
Exclusivity as a function of differential usage parameters.

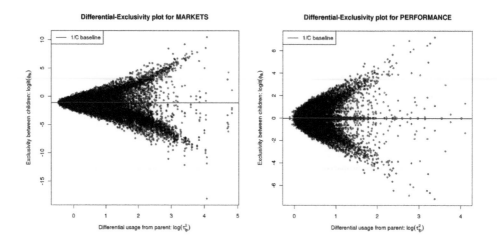

14.4.4   A Better Two-Dimensional Summary of Semantic Content

Words in the upper right of the FREX plot—those that are both frequent and highly exclusive—are of greatest interest. These are the most common words in the corpus that are also likely to have been generated from the topic of interest (rather than similar topics). We show words in the upper 5% quantiles in both dimensions for our example topics in Figure 14.5. These high-scoring words can help to clarify content even for labeled topics. In the Science and Technology topic, we see almost all terms are specific to the American and Russian space programs. Similarly, in the Research/Technology topic, almost all terms relate to clinical trials in medicine or to agricultural research.

We also compute the FREX score for each word-topic pair, a univariate summary of topical content that averages performance in both dimensions. In Table 14.4 we compare the top FREX words in three topics to a ranking based on frequency alone, which is the current practice in topic modeling. For context, we also show the immediate neighbors of each topic in the tree. The topic being examined is in bolded red, while the borders of the comparison set are solid. The Defense Contracts topic is a special case since it is an only child. In these cases, we use a comparison to the parent topic to calculate exclusivity.

By incorporating exclusivity information, FREX-ranked lists include fewer words that are used similarly everywhere (such as *said* and *would*) and fewer words that are used similarly in a set of related topics (such as *price* and *market* in the Markets branch). One can understand this result by comparing the rankings for known stopwords from the SMART list to other words. In Figure 14.6, we show the maximum ECDF ranking for each word across topics in the distribution of frequency (left panel) and exclusivity (right panel) estimates. One can see that while stopwords are more likely to be in the extreme quantiles of frequency, very few of them are among the most exclusive words. This prevents general and context-specific stopwords from ranking highly in a FREX-based index.

FIGURE 14.4
Frequency-Exclusivity (FREX) plots.

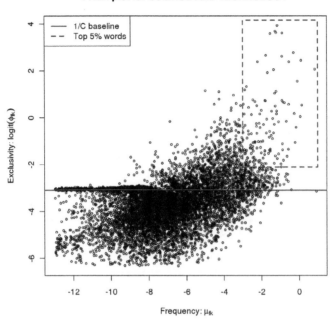

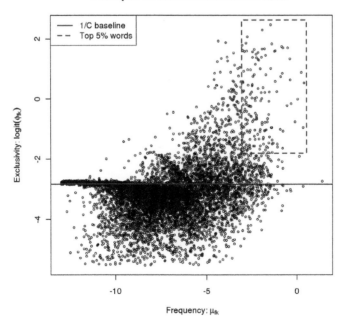

FIGURE 14.5
Upper right corner of FREX plot.

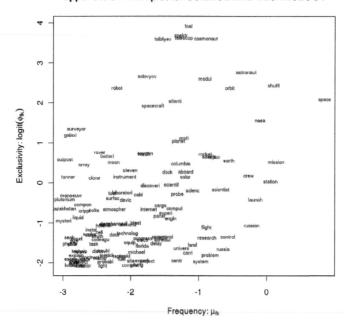

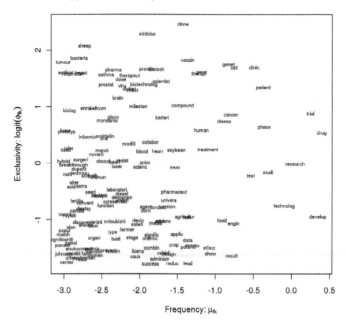

FIGURE 14.6
Comparison of FREX score components for SMART stopwords vs. regular words.

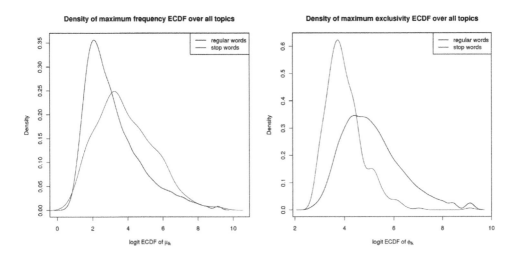

14.4.5 Classification Performance

We compare the classification performance of HPC with SVM and L2-regularized logistic regression. All methods were trained on a random sample of 15% of the documents using the 3% most frequent words in the corpus as features. These fits were used to predict memberships in the withheld documents, an experiment we repeated ten times with a new random sample as a training set. Table 14.5 shows the results of our experiment, using both micro averages (every document weighted equally) and macro averages (every topic weighted equally). While HPC does not dominate other methods, on average its performance does not deviate significantly from traditional classification algorithms.

HPC is not designed for optimizing predictive accuracy out-of-sample, rather it is designed to maximize interpretability of the label-specific summaries, in terms of words that are both frequent and exclusive. These results offer a quantitative illustration of the classical trade-off between predictive and explanatory power of statistical models (Breiman, 2001).

14.5 Concluding Remarks

Our thesis is that one needs to know how words are used differentially across topics as well as within them in order to understand topical content; we refer to these dimensions of content as word exclusivity and frequency. Topical summaries that focus on word frequency alone are often dominated by stopwords or other terms used similarly across many topics. Exclusivity and frequency can be visualized graphically as a latent space or combined into an index such as the FREX score to obtain a univariate measure of the topical content for words in each topic.

Naive estimates of exclusivity will be biased toward rare words due to sensitivity to small differences in estimated use across topics. Existing topic models such as LDA cannot regularize differential use due to topic normalization of usage rates; its symmetric Dirichlet prior on topic distributions

TABLE 14.4
Comparison of high FREX words (both frequent and exclusive) to most frequent words (featured topic name in underlined bold red; comparison set in solid ovals).

	High FREX	Most frequent
Metals Trading	copper	said
	aluminium	gold
	metal	price
	gold	copper
	zinc	market
	ounc	metal
	silver	trader
	palladium	tonn
	comex	trade
	platinum	close
	bullion	ounc
	preciou	aluminium
	nickel	london
	mine	dealer
Environment	greenpeac	said
	environment	would
	pollut	environment
	wast	year
	emiss	state
	reactor	nuclear
	forest	million
	speci	greenpeac
	environ	world
	eleph	water
	spill	group
	wildlif	govern
	energi	nation
	nuclear	environ
Defense Contracts	fighter	said
	defenc	contract
	missil	million
	forc	system
	defens	forc
	eurofight	defenc
	armi	would
	helicopt	aircraft
	lockhe	compani
	czech	deal
	martin	fighter
	militari	govern
	navi	unit
	mcdonnel	lockhe

TABLE 14.5

Classification performance for ten-fold cross-validation.

	SVM	L2-reg Logit	HPC
Micro-ave Precision	0.711 (0.002)	0.195 (0.031)	0.695 (0.007)
Micro-ave Recall	0.706 (0.001)	0.768 (0.013)	0.589 (0.008)
Macro-ave Precision	0.563 (0.002)	0.481 (0.025)	0.505 (0.094)
Macro-ave Recall	0.551 (0.006)	0.600 (0.007)	0.524 (0.093)

Standard deviation of performance over ten folds in parenthesis.

regularizes within, not between, topic usage. While topic-regularized models can capture many important facets of word usage, they are not optimal for the estimands used in our analysis of topical content.

HPC breaks from standard topic models by modeling topic-specific word counts as unnormalized count variates whose rates can be regularized both within and across topics to compute word frequency and exclusivity. It was specifically designed to produce stable exclusivity estimates in human-annotated corpora by smoothing differential word usage according to a semantically intelligent distance metric: proximity on a known hierarchy. This supervised setting is an ideal test case for our framework and will be applicable to many high value corpora such as the ACM library, IMS publications, the *New York Times* and Reuters, which all have professional editors and authors and provide multiple annotations to a hierarchy of labels for each document.

HPC offers a complex challenge for full Bayesian inference. To offer a flexible framework for regularization, it breaks from the simple Dirichlet-multinomial conjugacy of traditional models. Specifically, HPC uses Poisson likelihoods whose rates are smoothed across a known topic hierarchy with a Gaussian diffusion and a novel mixed membership model where document label and topic membership parameters share a Gaussian prior. The membership model is the first to create an explicit link between the distribution of topic labels in a document and of the words that appear in a document and allow for multiple labels. However, the resulting inference is challenging since, conditional on word usage rates, the posterior of the membership parameters involves Poisson and Bernoulli likelihoods of differing dimensions constrained by a Gaussian prior.

We offer two methodological innovations to make inference tractable. First, we design our model with parameters that divide cleanly into two blocks (the tree and document parameters) whose members are conditionally independent given the other block, allowing for parallelized, scalable inference. However, these factorized distributions cannot be normalized analytically and are the same dimension as the number of topics (102 in the case of Reuters). We therefore implement a Hamiltonian Monte Carlo conditional sampler that mixes efficiently through high-dimensional spaces by leveraging the posterior gradient and Hessian information. This allows HPC to scale to large and complex topic hierarchies that would be intractable for random walk Metropolis samplers. One unresolved bottleneck in our inference strategy is that the Markov chain Monte Carlo sampler mixes slowly through the hyperparameter space of the documents—the η and λ^2 parameters that control the mean and sparsity of topic memberships and labels. This is due to a large fraction of missing information in our augmentation strategy (Meng and Rubin, 1991). Conditional on all the documents' topic affinity parameters $\{\xi_d\}_{d=1}^{D}$, these hyperparameters index a normal distribution with D observations; marginally, however, we have much less information about the exact loading of each topic onto each document. While we have been exploring more efficient data augmentation strategies such as parameter expansion (Liu and Wu, 1999), we have not found a workable alternative to augmenting the posterior with the entire set of $\{\xi_d\}_{d=1}^{D}$ parameters.

14.5.1 Toward Automated Evaluation of Topic Models

While HPC was developed for the specific case of hierarchically labeled document collections, this framework can be readily extended to other types of document corpora. For labeled corpora where no hierarchical structure on the topics is available, one can use a flat hierarchy to model differential use. For document corpora where no labeled examples are available, a simple word rate model with a flat hierarchy and dense topic membership structure could be employed to get more informative summaries of inferred topics. In either case, the word rate framework could be combined with nonparameteric Bayesian models that infer hierarchical structure on the topics (Adams et al., 2010). We expect modeling approaches based on rates will play an important role in future work on text summarization.

The HPC model can also be leveraged to semi-automate the construction of topic ontologies targeted to specific domains, for instance, when fit to comprehensive human-annotated corpora such as Wikipedia, the *New York Times*, *Encyclopedia Britannica*, or databases such as JSTOR and the ACM repository. By learning a probabilistic representation of high quality topics, HPC output can be used as a gold standard to aid and evaluate other learning methods. Targeted ontologies have been a key factor in monitoring scientific progress in biology (Ashburner et al., 2000; Kanehisa and Goto, 2000). A hierarchical ontology of topics would lead to new metrics for measuring progress in text analysis. It would enable an evaluation of the semantic content of any collection of inferred topics, thus finally allowing for a *quantitative comparison* among the output of topic models. Current evaluations are qualitative, anecdotal, and unsatisfactory; for instance, authors argue that lists of most frequent words describing an arbitrary selection of topics inferred by a new model make sense intuitively, or that they are better then lists obtained with other models.

In addition to model evaluation, a news-specific ontology could be used as a prior to inform the analysis of unstructured text, including Twitter feeds, Facebook wall posts, and blogs. Unsupervised topic models infer a latent topic space that may be oriented around unhelpful axes, such as authorship or geography. Using a human-created ontology as a prior could ensure that a useful topic space is discovered without being so dogmatic as to assume that unlabeled documents have the same latent structure as labeled examples.

Appendix: Implementing the Parallelized HMC Sampler

Hamiltonian Monte Carlo Conditional Updates

Hamiltonian Monte Carlo (HMC) is the key tool that makes high-dimensional, non-conjugate updates tractable for our Gibbs sampler. It works well for log densities that are unimodal and have relatively constant curvature. We outline our customized implementation of the algorithm here; a general introduction can be found in Neal (2011).

HMC is a version of the Metropolis-Hastings algorithm that replaces the common Multivariate Normal proposal distribution with a distribution based on Hamiltonian dynamics. It can be used to make joint proposals on the entire parameter space or, as in this paper, to make proposals along the conditional posteriors as part of a Gibbs scan. While it requires closed form calculation of the posterior gradient and curvature to perform well, the algorithm can produce uncorrelated or negatively correlated draws from the target distribution that are almost always accepted.

A consequence of classical mechanics, Hamiltonian's equations can be used to model the movement of a particle along a frictionless surface. The total energy of the particle is the sum of its potential energy (the height of the surface relative to the minimum at the current position) and its kinetic energy (the amount of work needed to accelerate the particle from rest to its current velocity). Since energy is preserved in a closed system, the particle can only convert potential energy to kinetic energy (or vice versa) as it moves along the surface.

Imagine a ball placed high on the side of the parabola $f(q) = q^2$ at position $q = -2$. Starting out, it will have no kinetic energy but significant potential energy due to its position. As it rolls down the parabola toward zero, it speeds up (gaining kinetic energy), but loses potential energy to compensate as it moves to a lower position. At the bottom of the parabola the ball has only kinetic energy, which it then translates back into potential energy by rolling up the other side until its kinetic energy is exhausted. It will then roll back down the side it just climbed, completely reversing its trajectory until it returns to its original position.

HMC uses Hamiltonian dynamics as a method to find a distant point in the parameter space with high probability of acceptance. Suppose we want to produce samples from $f(q)$, a possibly unnormalized density. Since we want high probability regions to have the least potential energy, we parameterize the surface the particle moves along as $U(q) = -\log f(q)$, which is the height of the surface and the potential energy of the particle at any position q. The total energy of the particle, $H(p, q)$, is the sum of its kinetic energy, $K(p)$, and its potential energy, $U(q)$, where p is its momentum along each coordinate. After drawing an initial momentum for the particle (typically chosen as $p \sim \mathcal{N}(0, M)$, where M is called the *mass matrix*), we allow the system to evolve for a period of time—not so little that the there is negligible absolute movement, but not so much that the particle has time to roll back to where it started.

HMC will not generate good proposals if the particle is not given enough momentum in each direction to efficiently explore the parameter space in a fixed window of time. The higher the curvature of the surface, the more energy the particle needs to move to a distant point. Therefore the performance of the algorithm depends on having a good estimate of the posterior curvature $\hat{H}(q)$ and drawing $p \sim \mathcal{N}(0, -\hat{H}(q))$. If the estimated curvature is accurate and relatively constant across the parameter space, the particle will have high initial momentum along directions where the posterior is concentrated and less along those where the posterior is more diffuse.

Unless the (conditional) posterior is very well-behaved, the Hessian should be calculated at the log-posterior mode to ensure positive definiteness. Maximization is generally an expensive operation, however, so it is not feasible to update the Hessian every iteration of the sampler. In contrast, the log-prior curvature is very easy to calculate and well-behaved everywhere. This led us to develop the *scheduled conditional HMC sampler* (SCHMC), an algorithm for nonconjugate Gibbs draws that updates the log-prior curvature at every iteration but only updates the log-likelihood curvature in a strategically chosen subset of iterations. We use this algorithm for all non-conjugate conditional draws in our Gibbs sampler.

Specifically, suppose we want to draw from the conditional distribution $p(\boldsymbol{\theta}|\boldsymbol{\psi}_t, \boldsymbol{y}) \propto p(\boldsymbol{y}|\boldsymbol{\theta}, \boldsymbol{\psi}_t)p(\boldsymbol{\theta}|\boldsymbol{\psi}_t)$ in each Gibbs scan, where $\boldsymbol{\psi}$ is a vector of the remaining parameters and $\boldsymbol{y}$ is the observed data. Let $\mathcal{S}$ be the set of full Gibbs scans in which the log-likelihood Hessian information is updated (which always includes the first). For Gibbs scan $i \in \mathcal{S}$, we first calculate the conditional posterior mode and evaluate both the Hessian of the log-likelihood, $\log p(\boldsymbol{y}|\boldsymbol{\theta}, \boldsymbol{\psi}_t)$, and of the log-prior, $\log p(\boldsymbol{\theta}|\boldsymbol{\psi}_t)$, at that mode, adding them together to get the log-posterior Hessian. We then get a conditional posterior draw with HMC using the negative Hessian as our mass matrix. For Gibbs scan $i \notin \mathcal{S}$, we evaluate the log-prior Hessian at the current location and add it to our last evaluation of the log-likelihood Hessian to get the log-posterior Hessian. We then proceed as before. The SCHMC procedure is described in step-by-step detail in Algorithm 1.

SCHMC Implementation Details for the HPC Model

In the previous section we described our general procedure for obtaining samples from unnormalized conditional posteriors, the SCHMC algorithm. In this section, we provide the gradient and

---

input : $\boldsymbol{\theta}_{t-1}$, $\boldsymbol{\psi}_t$ (current value of other parameters), $\boldsymbol{y}$ (observed data), L (number of leapfrog steps), ϵ (stepsize), and $\mathcal{S}$ (set of full Gibbs scans in which the likelihood Hessian is updated)

output : $\boldsymbol{\theta}_t$

$\boldsymbol{\theta}_0^* \leftarrow \boldsymbol{\theta}_{t-1}$ /*Update conditional likelihood Hessian if iteration in schedule */

if $i \in \mathcal{S}$ **then**

$\hat{\boldsymbol{\theta}} \leftarrow \arg\max_{\boldsymbol{\theta}} \{\log p(\boldsymbol{y}|\boldsymbol{\theta}, \boldsymbol{\psi}_t) + \log p(\boldsymbol{\theta}|\boldsymbol{\psi}_t)\} \quad \hat{\boldsymbol{H}}_l(\boldsymbol{\theta}) \leftarrow \frac{\partial^2}{\partial\boldsymbol{\theta}\partial\boldsymbol{\theta}^T} \left[\log p(\boldsymbol{y}|\hat{\boldsymbol{\theta}}, \boldsymbol{\psi}_t)\right]|_{\boldsymbol{\theta}=\hat{\boldsymbol{\theta}}}$

end

/*Calculate prior Hessian and set up mass matrix */
$\hat{\boldsymbol{H}}_p(\boldsymbol{\theta}) \leftarrow \frac{\partial^2}{\partial\boldsymbol{\theta}\partial\boldsymbol{\theta}^T} [\log p(\boldsymbol{\theta}|\boldsymbol{\psi}_t)]|_{\boldsymbol{\theta}=\boldsymbol{\theta}_0^*} \quad \hat{\boldsymbol{H}}(\boldsymbol{\theta}) \leftarrow \hat{\boldsymbol{H}}_l(\boldsymbol{\theta}) + \hat{\boldsymbol{H}}_p(\boldsymbol{\theta}) \quad \boldsymbol{M} \leftarrow -\hat{\boldsymbol{H}}(\boldsymbol{\theta})$

/*Draw initial momentum */

Draw $\boldsymbol{p}_0^* \sim \mathcal{N}(\boldsymbol{0}, \boldsymbol{M})$ /*Leapfrog steps to get HMC proposal */

for $l \leftarrow 1$ *to* L **do**

$\boldsymbol{g}_1 \leftarrow -\frac{\partial}{\partial\boldsymbol{\theta}} [\log p(\boldsymbol{\theta}|\boldsymbol{\psi}_t, \boldsymbol{y})]|_{\boldsymbol{\theta}=\boldsymbol{\theta}_{l-1}^*} \quad \boldsymbol{p}_{l,1}^* \leftarrow \boldsymbol{p}_{l-1}^* - \frac{\epsilon}{2}\boldsymbol{g}_1 \quad \boldsymbol{\theta}_l^* \leftarrow \boldsymbol{\theta}_{l-1}^* + \epsilon(\boldsymbol{M}^{-1})^T\boldsymbol{p}_{l,1}^* \quad \boldsymbol{g}_2 \leftarrow$

$-\frac{\partial}{\partial\boldsymbol{\theta}} [\log p(\boldsymbol{\theta}|\boldsymbol{\psi}_t, \boldsymbol{y})]|_{\boldsymbol{\theta}=\boldsymbol{\theta}_l^*} \quad \boldsymbol{p}_l^* \leftarrow \boldsymbol{p}_{l,1}^* - \frac{\epsilon}{2}\boldsymbol{g}_2$

end

/*Calculate Hamiltonian (total energy) of initial position */
$K_{t-1} \leftarrow \frac{1}{2}(\boldsymbol{p}_0^*)^T\boldsymbol{M}^{-1}\boldsymbol{p}_0^* \quad U_{t-1} \leftarrow -\log p(\boldsymbol{\theta}_0^*|\boldsymbol{\psi}_t, \boldsymbol{y}) \quad H_{t-1} \leftarrow K_{t-1} + U_{t-1}$

/*Calculate Hamiltonian (total energy) of candidate position */

$K^* \leftarrow \frac{1}{2}(\boldsymbol{p}_L^*)^T\boldsymbol{M}^{-1}\boldsymbol{p}_L^* \quad U^* \leftarrow -\log p(\boldsymbol{\theta}_L^*|\boldsymbol{\psi}_t, \boldsymbol{y}) \quad H^* \leftarrow K^* + U^*$

/*Metropolis correction to determine if proposal accepted */

Draw $u \sim \text{Unif}[0, 1]$ $\log r \leftarrow H_{t-1} - H^*$ **if** $\log u < \log r$ **then**

$\boldsymbol{\theta}_t \leftarrow \boldsymbol{\theta}_L^*$

else

$\boldsymbol{\theta}_t \leftarrow \boldsymbol{\theta}_{t-1}$

end

Algorithm 1: Scheduled conditional HMC sampler for iteration i.

Hessian calculations necessary to implement this procedure for the unnormalized conditional densities in the HPC model, as well as strategies to obtain the maximum of each conditional posterior.

Conditional Posterior of the Rate Parameters

The log conditional posterior of the rate parameters for one word is:

$$\log p(\boldsymbol{\mu}_f|\boldsymbol{W}, \boldsymbol{I}, \boldsymbol{l}, \{\tau_f^2\}_{f=1}^V, \psi, \gamma^2, \nu, \sigma^2, \{\boldsymbol{\xi}_d\}_{d=1}^D, \mathcal{T})$$

$$= \sum_{d=1}^D \log \text{Pois}(w_{fd}|l_d\boldsymbol{\theta}_d^T\boldsymbol{\beta}_f) + \log \mathcal{N}(\boldsymbol{\mu}_f|\psi\boldsymbol{1}, \boldsymbol{\Lambda}(\gamma^2, \tau_f^2, \mathcal{T}))$$

$$= -\sum_{d=1}^D l_d\boldsymbol{\theta}_d^T\boldsymbol{\beta}_f + \sum_{d=1}^D w_{fd}\log(\boldsymbol{\theta}_d^T\boldsymbol{\beta}_f) - \frac{1}{2}(\boldsymbol{\mu}_f - \psi\boldsymbol{1})^T\boldsymbol{\Lambda}(\boldsymbol{\mu}_f - \psi\boldsymbol{1}).$$

Since the likelihood is a function of $\boldsymbol{\beta}_f$, we need to use the chain rule to get the gradient in $\boldsymbol{\mu}_f$ space:

$$\frac{\partial}{\partial \boldsymbol{\mu}_f}\left[\log p(\boldsymbol{\mu}_f|\boldsymbol{W}, \boldsymbol{I}, \boldsymbol{l}, \{\tau_f^2\}_{f=1}^V, \psi, \gamma^2, \{\boldsymbol{\xi}_d\}_{d=1}^D, \mathcal{T})\right]$$

$$= \frac{\partial l(\boldsymbol{\beta}_f)}{\partial \boldsymbol{\beta}_f}\frac{\partial \boldsymbol{\beta}_f}{\partial \boldsymbol{\mu}_f} + \frac{\partial}{\partial \boldsymbol{\mu}_f}\left[\log p(\boldsymbol{\mu}_f|\{\tau_f^2\}_{f=1}^V, \psi, \gamma^2, \mathcal{T})\right]$$

$$= -\sum_{d=1}^D l_d(\boldsymbol{\theta}_d^T \circ \boldsymbol{\beta}_f^T) + \sum_{d=1}^D \left(\frac{w_{fd}}{\boldsymbol{\theta}_d^T \boldsymbol{\beta}_f}\right)(\boldsymbol{\theta}_d^T \circ \boldsymbol{\beta}_f^T) - \boldsymbol{\Lambda}(\boldsymbol{\mu}_f - \psi\mathbf{1}),$$

where $\circ$ is the Hadamard (entrywise) product. The Hessian matrix follows a similar pattern:

$$\boldsymbol{H}(\log p(\boldsymbol{\mu}_f|\boldsymbol{W}, \boldsymbol{I}, \boldsymbol{l}, \{\tau_f^2\}_{f=1}^V, \psi, \gamma^2, \{\boldsymbol{\xi}_d\}_{d=1}^D, \mathcal{T})) = -\boldsymbol{\Theta}^T \boldsymbol{W} \boldsymbol{\Theta} \circ \boldsymbol{\beta}_f \boldsymbol{\beta}_f^T + \boldsymbol{G} - \boldsymbol{\Lambda},$$

where

$$\boldsymbol{W} = \text{diag}\left(\left\{\frac{w_{fd}}{(\boldsymbol{\theta}_d^T \boldsymbol{\beta}_f)^2}\right\}_{d=1}^D\right)$$

and

$$\boldsymbol{G} = \text{diag}\left(\frac{\partial l(\boldsymbol{\beta}_f)}{\partial \boldsymbol{\beta}_f} \circ \boldsymbol{\beta}_f^T\right) = \text{diag}\left(\frac{\partial l(\boldsymbol{\beta}_f)}{\partial \boldsymbol{\mu}_f}\right).$$

We use the BFGS algorithm with the analytical gradient derived above to maximize this density for iterations where the likelihood Hessian is updated; this quasi-Newton method works well since the conditional posterior is unimodal. The Hessian of the likelihood in β space is clearly negative definite everywhere since $\boldsymbol{\Theta}^T \boldsymbol{W} \boldsymbol{\Theta}$ is a positive definite matrix. The prior Hessian $\boldsymbol{\Lambda}$ is also positive definite by definition since it is the precision matrix of a Gaussian variate. However, the contribution of the chain rule term $\boldsymbol{G}$ can cause the Hessian to become indefinite away from the mode in μ space if any of the gradient entries are sufficiently large and positive. Note, however, that the conditional posterior is still unimodal since the logarithm is a monotone transformation.

Conditional Posterior of the Topic Affinity Parameters

The log conditional posterior for the topic affinity parameters for one document is:

$$\log p(\boldsymbol{\xi}_d|\boldsymbol{W}, \boldsymbol{I}, \boldsymbol{l}, \{\boldsymbol{\mu}_f, \tau_f^2\}_{f=1}^V, \boldsymbol{\eta}, \boldsymbol{\Sigma})$$

$$= l_d \sum_{f=1}^V \log \text{Pois}(w_{fd}|\boldsymbol{\beta}_f^T \boldsymbol{\theta}_d) + \log \text{Bernoulli}(\boldsymbol{I}_d|\boldsymbol{\xi}_d) + \log \mathcal{N}(\boldsymbol{\xi}_d|\boldsymbol{\eta}, \boldsymbol{\Sigma})$$

$$= -l_d \sum_{f=1}^V \boldsymbol{\beta}_f^T \boldsymbol{\theta}_d + \sum_{f=1}^V w_{fd} \log(\boldsymbol{\beta}_f^T \boldsymbol{\theta}_d) - \sum_{k=1}^K \log(1 + \exp(-\xi_{dk}))$$

$$- \sum_{k=1}^K (1 - I_{dk})\xi_{dk} - \frac{1}{2}(\boldsymbol{\xi}_d - \boldsymbol{\eta})^T \boldsymbol{\Sigma}^{-1}(\boldsymbol{\xi}_d - \boldsymbol{\eta}).$$

Since the likelihood of the word counts is a function of $\boldsymbol{\theta}_d$, we need to use the chain rule to get the gradient of the likelihood in $\boldsymbol{\xi}_d$ space. This mapping is more complicated than in the case of the $\boldsymbol{\mu}_f$ parameters since each ξ_{dk} is a function of all elements of $\boldsymbol{\theta}_d$:

$$\nabla l_d(\boldsymbol{\xi}_d) = \nabla l_d(\boldsymbol{\theta}_d)^T \boldsymbol{J}(\boldsymbol{\theta}_d \to \boldsymbol{\xi}_d),$$

where $J(\theta_d \to \xi_d)$ is the Jacobian of the transformation from θ space to ξ space, a $K \times K$ symmetric matrix. Let $S = \sum_{l=1}^{K} \exp \xi_{dl}$. Then

$$J(\theta_d \to \xi_d) = S^{-2} \begin{bmatrix} S \exp \xi_{d1} - \exp 2\xi_{d1} & \cdots & -\exp(\xi_{dK} + \xi_{d1}) \\ -\exp(\xi_{d1} + \xi_{d2}) & \cdots & -\exp(\xi_{dK} + \xi_{d2}) \\ \vdots & \ddots & \vdots \\ -\exp(\xi_{d1} + \xi_{dK}) & \cdots & S \exp \xi_{dK} - \exp 2\xi_{dK} \end{bmatrix}.$$

The gradient of the likelihood of the word counts in terms of θ_d is

$$\nabla l_d(\theta_d) = -l_d \sum_{f=1}^{V} \beta_f^T + \sum_{f=1}^{V} \frac{w_{fd}\beta_f^T}{\beta_f^T \theta_d}.$$

Finally, to get the gradient of the full conditional posterior, we add the gradient of the likelihood of the labels and of the normal prior on the ξ_d:

$$\frac{\partial}{\partial \xi_d} \left[\log p(\xi_d | W, I, l, \{\mu_f\}_{f=1}^{V}, \eta, \Sigma) \right]$$
$$= \nabla l_d(\theta_d)^T J(\theta_d \to \xi_d) + (1 + \exp \xi_d)^{-1} - (1 - I_d) - \Sigma^{-1}(\xi_d - \eta).$$

The Hessian matrix of the conditional posterior is a complicated tensor product that is not efficient to evaluate analytically. Instead, we compute a numerical Hessian using the analytic gradient presented above at minimal computational cost.

We use the BFGS algorithm with the analytical gradient derived above to maximize this density for iterations where the likelihood Hessian is updated. We have not been able to show analytically that this conditional posterior is unimodal, but we have verified this graphically for several documents and have achieved very high acceptance rates for our HMC proposals based on this Hessian calculation.

Conditional Posterior of the τ_{fk}^2 Hyperparameters

The variance parameters τ_{fk}^2 independently follow an identical scaled inverse-χ^2 with convolution parameter ν and scale parameter σ^2, while their inverse follows a Gamma$(\kappa_\tau = \frac{\nu}{2}, \lambda_\tau = \frac{2}{\nu\sigma^2})$ distribution. The log conditional posterior of these parameters is:

$$\log p(\kappa_\tau, \lambda_\tau | \{\tau_f^2\}_{f=1}^{V}, \mathcal{T}) = (\kappa_\tau - 1) \sum_{f=1}^{V} \sum_{k \in \mathcal{P}} \log (\tau_{fk}^2)^{-1}$$
$$- |\mathcal{P}|V\kappa_\tau \log \lambda_\tau - |\mathcal{P}|V \log \Gamma(\kappa_\tau) - \frac{1}{\lambda_\tau} \sum_{f=1}^{V} \sum_{k \in \mathcal{P}} (\tau_{fk}^2)^{-1},$$

where $\mathcal{P}(\mathcal{T})$ is the set of parent topics on the tree. If we allow $i \in \{1, \ldots, N = |\mathcal{P}|V\}$ to index all the f, k pairs and $l(\kappa_\tau, \lambda_\tau) = p(\{\tau_f^2\}_{f=1}^{V} | \kappa_\tau, \lambda_\tau, \mathcal{T})$, we can simplify this to

$$l(\kappa_\tau, \lambda_\tau) = (\kappa_\tau - 1) \sum_{i=1}^{N} \log \tau_i^{-2} - N\kappa_\tau \log \lambda_\tau - N \log \Gamma(\kappa_\tau) - \frac{1}{\lambda_\tau} \sum_{i=1}^{N} \tau_i^{-2}.$$

We then transform this density onto the $(\log \kappa_\tau, \log \lambda_\tau)$ scale so that the parameters are unconstrained, a requirement for standard HMC implementation. Each draw of $(\log \kappa_\tau, \log \lambda_\tau)$ is then transformed back to the (ν, σ^2) scale. To get the Hessian of the likelihood in log space, we calculate the derivatives of the likelihood in the original space and apply the chain rule:

$$H\left(l(\log \kappa_\tau, \log \lambda_\tau)\right) =$$

$$\begin{bmatrix} \kappa_\tau \frac{\partial l(\kappa_\tau, \lambda_\tau)}{\partial \kappa_\tau} + (\kappa_\tau)^2 \frac{\partial^2 l(\kappa_\tau, \lambda_\tau)}{\partial (\kappa_\tau)^2} & \kappa_\tau \lambda_\tau \frac{\partial^2 l(\kappa_\tau, \lambda_\tau)}{\partial \kappa_\tau \partial \lambda_\tau} \\ \kappa_\tau \lambda_\tau \frac{\partial^2 l(\kappa_\tau, \lambda_\tau)}{\partial \kappa_\tau \partial \lambda_\tau} & \lambda_\tau \frac{\partial l(\kappa_\tau, \lambda_\tau)}{\partial \lambda_\tau} + (\lambda_\tau)^2 \frac{\partial^2 l(\kappa_\tau, \lambda_\tau)}{\partial (\lambda_\tau)^2} \end{bmatrix},$$

where

$$\nabla l(\kappa_\tau, \lambda_\tau) = \begin{bmatrix} \sum_{i=1}^N \log \tau_i^{-2} - N \log \lambda_\tau - N \psi(\kappa_\tau) \\ -\frac{N \kappa_\tau}{\lambda_\tau} + \frac{1}{(\lambda_\tau)^2} \sum_{i=1}^N \tau_i^{-2} \end{bmatrix}$$

and

$$H\left(l(\kappa_\tau, \lambda_\tau)\right) = \begin{bmatrix} -N \psi'(\kappa_\tau) & -\frac{N}{\lambda_\tau} \\ -\frac{N}{\lambda_\tau} & \frac{N \kappa_\tau}{(\lambda_\tau)^2} - \frac{2}{(\lambda_\tau)^3} \sum_{i=1}^N \tau_i^{-2} \end{bmatrix}.$$

Following Algorithm 1, we evaluate the Hessian at the mode of this joint posterior. This is easiest to find on original scale following the properties of the Gamma distribution. The first order condition for λ_τ can be solved analytically:

$$\lambda_{\tau,MLE}(\kappa_\tau) = \arg \max_{\lambda_\tau} \left\{ l(\kappa_\tau, \lambda_\tau) \right\} = \frac{1}{\kappa_\tau N} \sum_{i=1}^N \tau_i^{-2}.$$

We can then numerically maximize the profile likelihood of κ_τ:

$$\kappa_{\tau,MLE} = \arg \max_{\kappa_\tau} \left\{ l(\kappa_\tau, \lambda_{\tau,MLE}(\kappa_\tau)) \right\}.$$

The joint mode in the original space is then $(\kappa_{\tau,MLE}, \lambda_{\tau,MLE}(\kappa_{\tau,MLE}))$. Due to the monotonicity of the logarithm function, the mode in the transformed space is simply $(\log \kappa_{\tau,MLE}, \log \lambda_{\tau,MLE})$. We can be confident that the conditional posterior is unimodal: the Fisher information for a Gamma distribution is negative definite, and the log transformation to the unconstrained space is monotonic.

References

Adams, R. P., Ghahramani, Z., and Jordan, M. I. (2010). Tree-structured stick breaking for hierarchical data. In Shawe-Taylor, J., Zemel, R., Lafferty, J. D., and Williams, C. (eds), *Advances in Neural Information Processing 23*. Red Hook, NY: Curran Associates, Inc., 19–27.

Airoldi, E. M. (2007). Getting started in probabilistic graphical models. *PLoS Computational Biology* 3: e252. doi:10.1371/journal.pcbi.0030252

Airoldi, E. M., Anderson, A. G., Fienberg, S. E., and Skinner, K. K. (2006). Who wrote Ronald Reagan's radio addresses? *Bayesian Analysis* 1: 289–320.

Airoldi, E. M., Blei, D. M., Fienberg, S. E., and Xing, E. P. (2008). Mixed membership stochastic blockmodels. *Journal of Machine Learning Research* 9: 1981–2014.

Airoldi, E. M., Cohen, W. W., and Fienberg, S. E. (2005). Bayesian models for frequent terms in text. *Proceedings of the Classification Society of North America and INTERFACE.*

Airoldi, E. M., Erosheva, E. A., Fienberg, S. E., Joutard, C., Love, T., and Shringarpure, S. (2010a). Reconceptualizing the classification of PNAS articles. *Proceedings of the National Academy of Sciences* 107: 20899–20904.

Airoldi, E. M., Fienberg, S. E., Joutard, C., and Love, T. (2010b). Hierarchical Bayesian mixed-membership models and latent pattern discovery. In Chen, M. -H., Müller, P., Sun, D., Ye, K., Dey, D. K. (eds), *Frontiers of Statistical Decision Making and Bayesian Analysis: In Honor of James O. Berger.* Springer, 360–376.

Airoldi, E. M., Fienberg, S. E., and Skinner, K. K. (2007a). Whose ideas? Whose words? Authorship of the Ronald Reagan radio addresses. *Political Science & Politics* 40: 501–506.

Airoldi, E. M., Fienberg, S. E., and Xing, E. P. (2007b). Mixed membership analysis of genome-wide expression data. Manuscript, http://arxiv.org/abs/0711.2520.

Ashburner, M., Ball, C. A., Blake, J. A., Botstein, D., Butler, H., Cherry, J. M., Davis, A. P., Dolinski, K., Dwight, S. S., Eppig, J. T., Harris, M. A., Hill, D. P., Issel-Tarver, L., Kasarskis, A., Lewis, S., Matese, J. C., Richardson, J. E., Ringwald, M., Rubin, G. M., and Sherlock, G. (2000). Gene ontology: Tool for the unification of biology. The gene ontology consortium. *Nature Genetics* 25: 25–29.

Blei., D. M. (2012). Introduction to probabilistic topic models. *Communications of the ACM* 55: 77–84.

Blei, D. M., Griffiths, T. L., Jordan, M. I., and Tenenbaum, J. B. (2004). Hierarchical topic models and the nested Chinese restaurant process. In Thrun, S., Saul, L., and Schölkopf, B. (eds), *Advances in Neural Information Processing Systems 16.* Cambridge, MA: The MIT Press, 17–32.

Blei, D. M. and McAuliffe, J. (2007). Supervised topic models. In Platt, J. C., Koller, D., Singer, Y., and Roweis, S. (eds), *Advances in Neural Information Processing Systems 20.* Cambridge, MA: The MIT Press, 121–128.

Blei, D. M., Ng, A. Y., and Jordan, M. I. (2003). Latent Dirichlet allocation. *Journal of Machine Learning Research* 3: 993–1022.

Breiman, L. (2001). Statistical modeling: The two cultures. *Statistical Science* 16: 199–231.

Canny, J. (2004). GaP: A factor model for discrete data. In *Proceedings of the 27th Annual International ACM SIGIR Conference on Research and Development in Information Retrieval (SIGIR '04).* New York, NY, USA: ACM, 122–129.

Chang, J., Boyd-Graber, J., Gerrish, S., Wang, C., and Blei, D. M. (2009). Reading tea leaves: How humans interpret topic models. In Bengio, Y., Schuurmans, D., Lafferty, J. D., Williams, C. K. I., and Culotta, A. (eds), *Advances in Neural Information Processing Systems 22.* Red Hook, NY: Curran Associates, Inc., 288–296.

Eisenstein, J., Ahmed, A., and Xing, E. P. (2011). Sparse additive generative models of text. In *Proceedings of the 28th International Conference on Machine Learning (ICML '11).* Omnipress, 1041–1048.

Hotelling, H. (1936). Relations between two sets of variants. *Biometrika* 28: 321–377.

Hu, Y., Boyd-Graber, J., and Satinoff, B. (2011). Interactive topic modeling. In *Proceedings of the 49th Annual Meeting of the Association for Computational Linguistics: Human Language Technologies - Volume 1*. Stroudsburg, PA, USA: Association for Computational Linguistics, 248–257.

Jolliffe, I. T. (1986). *Principal Component Analysis*. New York, NY: Springer-Verlag.

Joutard, C., Airoldi, E. M., Fienberg, S. E., and Love, T. (2008). Discovery of latent patterns with hierarchical Bayesian mixed-membership models and the issue of model choice. *Data Mining Patterns: New Methods and Applications*, 240–275.

Kanehisa, M. and Goto, S. (2000). KEGG: Kyoto encyclopedia of genes and genomes. *Nucleic Acids Research* 28: 27–30.

Lewis, D. D., Yang, Y., Rose, T. G., and Li, F. (2004). RCV1: A new benchmark collection for text categorization research. *Journal of Machine Learning Research* 5: 361–397.

Liu, J. S. and Wu, Y. N. (1999). Parameter expansion for data augmentation. *Journal of the American Statistical Association* 94: 1264–1274.

McCallum, A., Rosenfeld, R., Mitchell, T., and Ng, A. Y. (1998). Improving text classification by shrinkage in a hierarchy of classes. In *Proceedings of the 15th International Conference on Machine Learning (ICML '98)*. San Francisco, CA, USA: Morgan Kaufmann Publishers Inc., 359–367.

McLachlan, G. and Peel, D. (2000). *Finite Mixture Models*. New York, NY: John Wiley & Sons, Inc.

Meng, X. -L. and Rubin, D. B. (1991). Using EM to obtain asymptotic variance-covariance matrices: The SEM algorithm. *Journal of the American Statistical Association* 86: 899–909.

Mimno, D., Li, W., and McCallum, A. (2007). Mixtures of hierarchical topics with Pachinko allocation. In *Proceedings of the 24th International Conference on Machine Learning (ICML '07)*. New York, NY, USA: ACM. 633–640.

Mimno, D., Wallach, H. M., Talley, E. M., Leenders, M., and McCallum, A. (2011). Optimizing semantic coherence in topic models. In *Proceedings of the Conference on Empirical Methods in Natural Language Processings (EMNLP)*. Stroudsburg, PA, USA: Association for Computational Linguistics, 262–272.

Mosteller, F. and Wallace, D. (1984). *Applied Bayesian and Classical Inference: The Case of "The Federalist" Papers*. London, UK: Springer.

Neal, R. M. (2011). MCMC using Hamiltonian dynamics. In Brooks, S., Gelman, A., Jones, G. L., and Meng, X. -L. (eds), *Handbook of Markov Chain Monte Carlo*. Chapman & Hall/CRC, 113–162.

Perotte, A., Bartlett, N., Elhadad, N., and Wood, F. (2012). Hierarchically supervised latent Dirichlet allocation. In Shawe-Taylor, J., Zemel, R. S., Bartlett, P., Pereira, F., and Weinberger, K. Q. (eds), *Advances in Neural Information Processing Systems 24*. Red Hook, NY: Curran Associates, Inc., 2609–2617.

Ramage, D., Hall, D., Nallapati, R. M., and Manning., C. D. (2009). Labeled LDA: A supervised topic model for credit attribution in multi-labeled corpora. In *Proceedings of the Conference on Empirical Methods in Natural Language Processing - Volume 1 (EMNLP '09)*. Stroudsburg, PA: USA: Association for Computational Linguistics, 248–256.

Sohn, K.-A. and Xing, E. P. (2009). A hierarchical Dirichlet process mixture model for haplotype reconstruction from multi-population data. *Annals of Applied Statistics* 3: 791–821.

Wallach, H. M., Mimno, D., and McCallum, A. (2009). Rethinking LDA: Why priors matter. In Bengio, Y., Schuurmans, D., Lafferty, J. D., Williams, C. K. I., and Culotta, A. (eds), *Advances in Neural Information Processing Systems 22*. Red Hook, NY: Curran Associates, Inc., 1973–1981.

Zhu, J. and Xing, E. P. (2011). Sparse topical coding. In *Proceedings of the 27th Conference on Uncertainty in Artificial Intelligence (UAI 2011)*. Corvallis, OR, USA: AUAI Press, 831–838.

Part IV

Semi-Supervised Mixed Membership Models

15

Mixed Membership Classification for Documents with Hierarchically Structured Labels

Frank Wood

Department of Engineering, University of Oxford, Oxford, OX1 3PJ, UK

Adler Perotte

Department of Biomedical Informatics, Columbia University, New York, NY 10032, USA

CONTENTS

Placing documents within a hierarchical structure is a common task and can be viewed as a multi-label classification with hierarchical structure in the label space. Examples of such data include web pages and their placement in directories, product descriptions and associated categories from product hierarchies, and free-text clinical records and their assigned diagnosis codes. We present a model for hierarchically and multiply labeled bag-of-words data called hierarchically supervised latent Dirichlet allocation (HSLDA). Out-of-sample label prediction is the primary goal of this work, but improved lower-dimensional representations of the bag-of-words data are also of interest. We demonstrate HSLDA on large-scale data from clinical document labeling and retail product categorization tasks. We show that leveraging the structure from hierarchical labels improves out-of-sample label prediction substantially when compared to models that do not.

15.1 Introduction

Documents frequently come with additional information like labels or popularity ratings. Contemporary examples include the product ratings that accompany product descriptions, the number of "likes" that webpages have attracted, grades associated with assigned essays, and so forth.

This chapter covers one way to jointly model documents and the labels applied to them. In particular we focus on our own work on modeling documents with more complicated labels that themselves possess some kind of structural organization. Consider typical product catalogs. They usually contain text descriptions of products that have been organized into hierarchical product directories. The situation of a product into such a hierarchy (the path or paths in the product hierarchy that lead to it) can be thought of as a structured label. Jointly modeling the document and such a label is useful for automatically labeling new documents (corresponding in this example to automatically situating a new product in the product directory) and more.

Collections of hierarchically labeled documents abound, text and otherwise. We will consider hierarchically labeled patient clinical records in later sections. Applications like situating web-pages in hierarchical link directories are left for others to explore.

Because text documents are notoriously difficult to directly model we take an approach common to other chapters in this book. We use a mixed membership model of the text document to represent the document as a bag-of-words drawn from a document-specific mixture of topic distributions. The modeling choices we make in relating this representation to structured labels follows, as does its relationship to prior art.

15.1.1 Background

Mixed membership models, including the model upon which we build, latent Dirichlet allocation (LDA) (Blei et al., 2003), have been reviewed in other chapters. The key property we exploit for purposes of classification is that LDA provides a way to extract a latent, low-dimensional representation of text and other documents consisting of the frequency of word assignments to the topics that are assumed to have generated them. A topic is a distribution over words. Each document is a bag-of-words drawn from a document-specific mixture of topics.

Building a joint model of documents and labels using this representation is not new. It was first introduced by Blei and McAuliffe in a paper on "supervised" latent Dirichlet allocation (SLDA) (Blei and McAuliffe, 2008). SLDA built on LDA by incorporating "supervision" in the form of an observed exponential family response variable per document.

Latent Dirichlet Allocation

To explain both SLDA and to set the stage for our work, it helps to introduce our notation for LDA. Assume that there are K topics. Let $\phi_k \sim \text{Dir}_V(\gamma 1_V)$ be "topic" k, i.e., a distribution over a finite set of words. Here, V simply labels the variables to indicate that they have to do with the vocabulary. The vector 1_V consists of all ones and has length equal to the size of the vocabulary. The distribution Dir is the Dirichlet distribution. The constant γ controls the smoothness of the inferred topics. Larger values lead to smoother topic estimates. Let $\beta \mid \alpha' \sim \text{Dir}_K(\alpha' 1_K)$ be a "global" distribution over topics where K indicates that the distribution and variable sizes are equal to the number of topics K and α' controls the relative proportion of topics globally, large α' leading to all topics being roughly responsible for the same number of words. Intuitively, β is something like the average topic proportion independent of any particular document. Per document d, topic distributions $\theta_d \mid \beta, \alpha \sim \text{Dir}_K(\alpha\beta)$ are modeled as being deviations from the global distribution over topics where larger values of α result in all documents' topic distributions being more similar. We will use $z_{n,d} \mid \theta_d \sim \text{Multinomial}(\theta_d)$ to indicate which topic generated the nth word of document d. Drawing the nth word in document d from the indicated topic $w_{n,d} \mid z_{n,d}, \phi_{1:K} \sim$ Multinomial($\phi_{z_{n,d}}$) completes our notation of standard LDA.

Supervised Latent Dirichlet Allocation

Supervised LDA adds another per document observation, a label y_d. It is modeled as being generated by a generalized linear model (GLM) $y_d | \bar{\mathbf{z}}_d, \boldsymbol{\eta}, \delta \sim \text{GLM}(\bar{\mathbf{z}}_d, \boldsymbol{\eta}, \delta)$. Brushing aside exponential family and GLM link function generalities, what SLDA does is regress labels against the empirical distributions of assigned topic indicator variables $\bar{\mathbf{z}}_d = \{\bar{z}_{1,d}, \ldots, \bar{z}_{K,d}\}$, where $\bar{z}_{k,d} = \frac{\sum_n \mathbb{I}(z_{n,d}=k)}{\sum_n \sum_j \mathbb{I}(z_{n,d}=j)}$ is the fraction of words assigned to topic k and $\mathbb{I}(\cdot)$ is the indicator function that returns one if its argument is true. If the document labels are real-valued then one example choice for the regression relationship would be $y_d | \bar{\mathbf{z}}_d, \boldsymbol{\eta} \sim \mathcal{N}(\bar{\mathbf{z}}_d^T \boldsymbol{\eta}, \delta)$. Using a generalized linear model in the exponential family to parameterize this regression relationship allows for a wide variety of distributions over different kinds of label spaces to be represented in the same mathematical formalism. A variational expectation maximization algorithm was proposed in Blei and McAuliffe (2008) to learn model parameters. Experimental results in Blei and McAuliffe (2008) showed both excellent out-of-sample label prediction and improved topics. Topic improvement was measured by using the empirical topic proportions from SLDA as features for external, discriminative approaches to label prediction. Regression models built on SLDA topics outperformed the same built on LDA derived topics.

In one sense, SLDA is more general than the model we present in this chapter, namely, the labels need not be categorically valued. The main subject of this chapter, a generalization of SLDA called hierarchically supervised LDA (HSLDA) (Perotte et al., 2011) does not deal with real-valued labels, however, it is more general than SLDA in the case of categorical labels. The exponential family/GLM regression framework can theoretically account for multivariate labels and potentially even structured categorical labels. HSLDA is, however, a specific, practical way to model with structured categorical labels. Because we focus on hierarchically structured categorical labels, we refer to our model as a mixed membership hierarchical classification model.

15.2   Hierarchical Supervised Latent Dirichlet Allocation

This model (HSLDA) is designed to fit hierarchically, multiply-labeled, bag-of-word data. We call groups of bag-of-words data documents (unordered words in text documents, bag of visual feature representations of images, etc.). Let $w_{n,d} \in \Sigma$ be the nth observation in the dth document. Let $\mathbf{w}_d = \{w_{1,d}, \ldots, w_{1,N_d}\}$ be the set of N_d observations in document d. Let there be D such documents and let the size of the vocabulary be $V = |\Sigma|$.

Let the set of labels be $\mathcal{L} = \{l_1, l_2, \ldots, l_{|\mathcal{L}|}\}$. A label in HSLDA corresponds to a node in the graphical model in Figure 15.1. A label can either be observed or unobserved. Documents can be multiply labeled, meaning that subsets of label nodes in the graphical model in Figure 15.1 can have observed values. Each label $l \in \mathcal{L}$, except the root, has a parent $\text{pa}(l) \in \mathcal{L}$ also in the set of labels. We will, for exposition purposes, assume that this label set has hard "is-a" parent-child constraints (explained later), although this assumption can be relaxed at the cost of more computationally complex inference. Such a label hierarchy forms a multiply rooted tree. Readers may wish to consult Figure 15.2 or Figure 15.3, each of which is a label-tree graphical model.

In a label forest a node may be observed (the label was "applied" and, for instance, was observed to have value 1), unobserved and unknown, or unobserved and constrained to be either -1 or 1 by the structure of the label space. Without loss of generality we will consider a tree with a single root $r \in \mathcal{L}$. Each document has a variable $y_{l,d} \in \{-1, 1\}$ for every label which indicates whether the label applies to document d or not. In most cases $y_{l,d}$ will be unobserved; in some cases we will be able to fix its value because of constraints on the label hierarchy, and in the relatively minor remainder its value will be observed. In the applications we demonstrate, only positive labels are observed. This may not be true of all applications, however, positive-only label imbalance is a common problem. How we solve this problem will be discussed later.

The constraints imposed by an is-a label hierarchy are that if the lth label is applied to document d, i.e., $y_{l,d} = 1$, then all labels in the label hierarchy up to the root are also applied to document d, i.e., $y_{\text{pa}(l),d} = 1, y_{\text{pa}(\text{pa}(l)),d} = 1, \ldots, y_{r,d} = 1$. Conversely, if a label l' is marked as not applying to a document (i.e., $y_{l',d} = -1$) then no descendant label of that label can take value 1. We assume that at least one label is applied to every document. This is illustrated in Figure 15.1 where the root label is always applied but only some of the descendant labelings are observed as having been applied (diagonal hashing indicates that potentially some of the plated variables are observed).

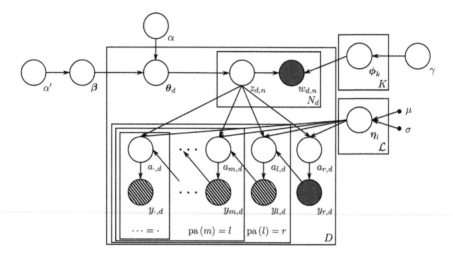

FIGURE 15.1
Hierarchically supervised latent Dirichlet allocation (HSLDA) graphical model.

In HSLDA, the bag-of-word document data is modeled using LDA with full, hierarchical topic estimation (i.e., global topic proportions are also estimated). Label responses are modeled using a conditional hierarchy of probit regressors and will be discussed next. The full HSLDA graphical model is given in Figure 15.1.

15.2.1   Generative Model

In the following box the HSLDA generative model is given for the "is-a hierarchy" set of label constraints. In the box and what follows in this chapter, K is the number of LDA "topics" (distributions over the elements of Σ), ϕ_k is a distribution over "words," θ_d is a document-specific distribution over topics, β is a global distribution over topics, $\text{Dir}_K(\cdot)$ is a K-dimensional Dirichlet distribution, $\mathcal{N}_K(\cdot)$ is the K-dimensional Normal distribution, $\mathbf{I}_K$ is the K dimensional identity matrix, $\mathbf{1}_d$ is the d-dimensional vector of all ones, and $\mathbb{I}(\cdot)$ is an indicator function that takes the value 1 if its argument is true and 0 otherwise.

HSLDA Generative Model

1. For each topic $k = 1, \ldots, K$

   - Draw a distribution over words $\phi_k \sim \text{Dir}_V(\gamma \mathbf{1}_V)$.

2. For each label $l \in \mathcal{L}$

   - Draw a label weight vector $\boldsymbol{\eta}_l \mid \mu, \sigma \sim \mathcal{N}_K(\mu \mathbf{1}_K, \sigma \mathbf{I}_K)$.

3. Draw the global topic proportions $\boldsymbol{\beta} \mid \alpha' \sim \text{Dir}_K(\alpha' \mathbf{1}_K)$.

4. For each document $d = 1, \ldots, D$

   - Draw topic proportions $\boldsymbol{\theta}_d \mid \boldsymbol{\beta}, \alpha \sim \text{Dir}_K(\alpha \boldsymbol{\beta})$.
   - For $n = 1, \ldots, N_d$
     - Draw topic assignment $z_{n,d} \mid \boldsymbol{\theta}_d \sim \text{Multinomial}(\boldsymbol{\theta}_d)$.
     - Draw word $w_{n,d} \mid z_{n,d}, \boldsymbol{\phi}_{1:K} \sim \text{Multinomial}(\boldsymbol{\phi}_{z_{n,d}})$.
   - Set $y_{r,d} = 1$.
   - For each label l in a breadth first traversal of $\mathcal{L}$ starting at the children of root r
     - Draw

$$
a_{l,d} \mid \bar{\mathbf{z}}_d, \boldsymbol{\eta}_l, y_{\text{pa}(l),d}
$$

$$
\sim \begin{cases} \mathcal{N}(\bar{\mathbf{z}}_d^T \boldsymbol{\eta}_l, 1), & y_{\text{pa}(l),d} = 1 \\ \mathcal{N}(\bar{\mathbf{z}}_d^T \boldsymbol{\eta}_l, 1)\mathbb{I}(a_{l,d} < 0), & y_{\text{pa}(l),d} = -1. \end{cases} \tag{15.1}
$$

     - Apply label l to document d according to $a_{l,d}$

$$
y_{l,d} \mid a_{l,d} = \begin{cases} 1 & \text{if } a_{l,d} > 0 \\ -1 & \text{otherwise.} \end{cases} \tag{15.2}
$$

Here $\bar{\mathbf{z}}_d^T = [\bar{z}_1, \ldots, \bar{z}_k, \ldots, \bar{z}_K]$ is the empirical topic distribution for document d, in which each entry is the percentage of the words in that document that come from topic k, $\bar{z}_k = N_d^{-1} \sum_{n=1}^{N_d} \mathbb{I}(z_{n,d} = k)$. As in Blei and McAuliffe (2008), the response variables are directly dependent on $\bar{\mathbf{z}}_d^T$ because this directly couples the topic assignments used to explain the words and the topic assignments used to explain the responses.

The second half of Step 4 is what is referred to as supervision in the supervised LDA literature. This is where the hierarchical classification of the bag-of-words data takes place and the is-a label constraints are enforced. For every label $l \in \mathcal{L}$, both the empirical topic distribution for document d and whether or not its parent label was applied (i.e., $\mathbb{I}(y_{\text{pa}(l),d} = 1)$) are used to determine whether or not label l is to be applied to document d as well. Equations (15.1) and (15.2) comprise a probit regression model in an auxiliary variable formulation (see Appendix). Note that in the case that the parent label is applied, i.e., $y_{\text{pa}(l),d} = 1$, the child label $y_{l,d}$ is applied with probability $P(\bar{\mathbf{z}}_d^T \boldsymbol{\eta}_l > 0)$. This is a conditional probit regression model for classification where $\boldsymbol{\eta}_l$ are the class-conditional regression parameters. The auxiliary variables $a_{l,d}$ make inference tractable but are not fundamental to the model—only the labels and regression parameters are actually of interest.

Note that $y_{l,d}$ can only be applied to document d (set to 1) if its parent label pa(l) is also applied (these expressions are specific to is-a constraints but could be modified to accommodate different constraints between labels). Note that multiple labels can be applied to the same document. The regression coefficients η_l which generate the labels are independent a priori, however, the hierarchical coupling in this model and conditional label dependency structure induces a posteriori dependence. The net effect of this conditional hierarchy of profit regressors is that child label predictors deeper in the label hierarchy are able to focus on finding features that distinguish label paths in the tree, conditioned on the fact that all the children of any particular node are by design members of some more general parent set. One can restrict this hierarchy to a depth of one hierarchy, recovering SLDA with probit link and univariate categorical labels. Also, one can nearly as easily make the conditional classification at each node multi-class rather than single-class if more than one label at each node is required. In many cases, however, a binary indicator along with a deeper or more complex tree is sufficient.

Note that the choice of variables $a_{l,d}$ and how they are distributed were driven at least in part by posterior inference efficiency considerations (see Appendix). In particular, choosing probit-style auxiliary variable distributions for the $a_{l,d}$'s yields conditional posterior distributions for both the auxiliary variables (15.5) and the regression coefficients (15.4), which are analytic. This simplifies posterior inference substantially. A review of probit regression can be found near the end of this chapter in the Appendix.

15.2.2 Dealing with Label Imbalance

In the common case where no negative labels are observed (like the example applications we consider in Section 15.4), the model must be explicitly biased towards generating negative labels in order to keep it from learning to only assign positive labels to all documents. This is a common problem in modeling with unbalanced labels. To see how this model can achieve this we draw the reader's attention to the μ parameter and, to a lesser extent, the σ parameter. Because $\bar{\mathbf{z}}_d$ is always positive, setting μ to a negative value results in a bias towards negative labelings, i.e., for large negative values of μ, all labels become a priori more likely to be negative ($y_{l,d} = -1$). We explore the effect of μ on out-of-sample label prediction performance in Section 15.4. In a very real way, μ is a knob that can be adjusted both before inference to induce a broad array of out-of-sample performance characteristics that vary along classical axes like specificity, recall, and accuracy. A similar but less principled solution can be effected by changing the decision boundary from 0 in (15.1) and (15.2). This technique can be used to vary out-of-sample label bias after learning.

15.2.3 Intuition

To help ground this abstract graphical model, recall the retail data example application. We asserted that retailers often have both a browseable product hierarchy and free-text descriptions for all products they sell. The situation of each product in a product hierarchy (often multiply situated) constitutes a multiple, hierarchical labeling $\mathbf{y}_d$ of the free-text product descriptions $\mathbf{w}_d$ for all products d. Note that a single product can be placed in the hierarchy in multiple places. This corresponds to multiple paths in the label hierarchy having labels that are all applied. HSLDA assumes that the free-text descriptions of all of the products in a particular node in the product hierarchy must be related. It also assumes that products deeper in the product hierarchy are described using language that is similar to that used to describe products in their parent classes. For instance, basketballs are probably described using language that is similar to that used to describe other basketballs, other balls, and more general sporting goods. In both the lay and technical senses, similar products should have product descriptions that share topics. If topic proportions are indicative of the text describing products that are grouped together, the key HSLDA assumption is that it should then be possible to use those proportions to decide (via probit classification) whether or not a particular product should be situated at a particular node in the product hierarchy. Conversely, that certain groups of products are known to be clustered together should inform the kinds of topics that are inferred from the product descriptions.

15.3 Inference

Our inference goal is to obtain a representation of the posterior distribution of the latent variables in the model. This posterior distribution can then be used for predictive inference of labels for held-out documents, among other things. Unfortunately, the posterior distribution we seek does not have a simple analytic form from which exact samples can be drawn. This is usually the case for posterior distributions of non-trivial probabilistic models and suggests approximating the posterior distribution by sampling.

In this section we derive the conditional distributions required to sample from the HSLDA posterior distribution using Markov chain Monte Carlo. The HSLDA sampler, like the collapsed Gibbs samplers for LDA (Griffiths and Steyvers, 2004), is itself a collapsed Gibbs sampler in which all of the latent variables that can be analytically marginalized are. Among others, the topic distributions $\phi_{1:K}$ and document-specific topic assignment distributions $\theta_{1:D}$ are analytically marginalized prior to deriving the following conditional distributions for sampling.

It will usually be the case that values $y_{l,d}$ will not be known for all labels $l \in \mathcal{L}$ in the space of possible labels. Values for $y_{l,d}$ that are enforced by label constraints and observed labels are set to their constrained values prior to inference and treated as observed. We will define $\mathcal{L}_d$ to be the subset of labels which have been observed (or observed via filling in from constraints) for document d. Marginalizing the probit regression auxiliary variables $a_{l',d}$ and $y_{l',d}$ for $l' \in \mathcal{L} \backslash \mathcal{L}_d$ is simple in the is-a hierarchy case because they can simply be ignored. The remaining latent variables (those that are not collapsed out) are the topic indicators $\mathbf{z} = \{z_{1:N_d,d}\}_{d=1,\dots,D}$, the probit regression parameters $\eta = \{\eta_l\}_{l \in \mathcal{L}}$, the auxiliary variables $\mathbf{a} = \{a_{l',d}\}_{l' \in \mathcal{L}_d, d=1,\dots,D}$, the global topic proportions β, and the concentration parameters α, α', and γ.

15.3.1 Gibbs Sampler

Let $\mathbf{a}$ be the set of all probit regression auxiliary variables, $\mathbf{w}$ the set of all words, η the set of all regression coefficients, and $\mathbf{z} \backslash z_{n,d}$ the set $\mathbf{z}$ with element $z_{n,d}$ removed.

First we consider the conditional distribution of $z_{n,d}$ (the assignment variable for each word $n = 1, \dots, N_d$ in documents $d = 1, \dots, D$). Following the factorization of the model (refer again to Figure 15.1), we can write

$$p\left(z_{n,d} \mid \mathbf{z}_d \backslash z_{n,d}, \mathbf{a}, \mathbf{w}, \eta, \alpha, \beta, \gamma\right)$$
$$\propto \prod_{l \in \mathcal{L}_d} p\left(a_{l,d} \mid \mathbf{z}, \eta_l\right) p\left(z_{n,d} \mid \mathbf{z}_d \backslash z_{n,d}, \mathbf{a}, \mathbf{w}, \alpha, \beta, \gamma\right).$$

The product is only over the subset of labels $\mathcal{L}_d$ which have been observed for document d. By isolating terms that depend on $z_{n,d}$ and absorbing all other terms into a normalizing constant as in Griffiths and Steyvers (2004) we find

$$p\left(z_{n,d} = k \mid \mathbf{z} \backslash z_{n,d}, \mathbf{a}, \mathbf{w}, \eta, \alpha, \beta, \gamma\right) \propto \qquad\qquad (15.3)$$
$$\left(c_{(\cdot),d}^{k,-(n,d)} + \alpha \beta_k\right) \frac{c_{w_{n,d},(\cdot)}^{k,-(n,d)} + \gamma}{\left(c_{(\cdot),(\cdot)}^{k,-(n,d)} + V\gamma\right)} \prod_{l \in \mathcal{L}_d} \exp\left\{-\frac{\left(\bar{\mathbf{z}}_d^T \eta_l - a_{l,d}\right)^2}{2}\right\},$$

where $c_{v,d}^{k,-(n,d)}$ is the number of words of type v in document d assigned to topic k omitting the

nth word of document d. The subscript $(\cdot)$ indicates to sum over the range of the replaced variable, i.e., $c_{w_{n,d},(\cdot)}^{k,-(n,d)} = \sum_d c_{w_{n,d},d}^{k,-(n,d)}$. Here, $\mathcal{L}_d$ is the set of labels which are observed for document d. We sample from (15.3) by first enumerating $z_{n,d}$ and then normalizing.

The conditional posterior distribution of the regression coefficients is given by

$$p(\boldsymbol{\eta}_l \mid \mathbf{z}, \mathbf{a}, \sigma) = \mathcal{N}(\hat{\boldsymbol{\mu}}_l, \hat{\boldsymbol{\Sigma}}), \tag{15.4}$$

$\boldsymbol{\eta}_l$ for $l \in \mathcal{L}$. Given that $\boldsymbol{\eta}_l$ and $a_{l,d}$ are distributed normally, the posterior distribution of $\boldsymbol{\eta}_l$ is normally distributed with mean $\hat{\boldsymbol{\mu}}_l$ and covariance $\hat{\boldsymbol{\Sigma}}$, where

$$\hat{\boldsymbol{\mu}}_l = \hat{\boldsymbol{\Sigma}} \left(\mathbf{1}\frac{\mu}{\sigma} + \bar{\mathbf{Z}}^T \mathbf{a}_l \right) \qquad \hat{\boldsymbol{\Sigma}}^{-1} = \mathbf{I}\sigma^{-1} + \bar{\mathbf{Z}}^T \bar{\mathbf{Z}}.$$

Here $\bar{\mathbf{Z}}$ is a $D \times K$ matrix such that row d of $\bar{\mathbf{Z}}$ is $\bar{\mathbf{z}}_d$, and $\mathbf{a}_l = [a_{l,1}, a_{l,2}, \ldots, a_{l,D}]^T$. The simplicity of this conditional distribution follows from the choice of probit regression (Albert and Chib, 1993); the specific form of the update is a standard result from Bayesian normal linear regression (Gelman et al., 2004). It also is a standard probit regression result that the conditional posterior distribution of $a_{l,d}$ is a truncated normal distribution (Albert and Chib, 1993) (see also the Appendix).

$$p\left(a_{l,d} \mid \mathbf{z}, \mathbf{Y}, \boldsymbol{\eta}\right)$$
$$\propto \begin{cases} \exp\left\{-\frac{1}{2}\left(a_{l,d} - \boldsymbol{\eta}_l^T \bar{\mathbf{z}}_d\right)\right\} \mathbb{I}\left(a_{l,d}y_{l,d} > 0\right) \mathbb{I}(a_{l,d} < 0), & y_{\mathrm{pa}(l),d} = -1 \\ \exp\left\{-\frac{1}{2}\left(a_{l,d} - \boldsymbol{\eta}_l^T \bar{\mathbf{z}}_d\right)\right\} \mathbb{I}\left(a_{l,d}y_{l,d} > 0\right), & y_{\mathrm{pa}(l),d} = 1. \end{cases}$$

HSLDA employs a hierarchical Dirichlet prior over topic assignments (i.e., β is estimated from data rather than fixed a priori). This has been shown to improve the quality and stability of inferred topics (Wallach et al., 2009). Sampling β, the vector of global topic proportions, can be done using the "direct assignment" method of Teh et al. (2006):

$$\boldsymbol{\beta} \mid \mathbf{z}, \alpha', \alpha \sim \mathrm{Dir}\left(m_{(\cdot),1} + \alpha', m_{(\cdot),2} + \alpha', \ldots, m_{(\cdot),K} + \alpha'\right). \tag{15.5}$$

Here, $m_{d,k}$ are additional auxiliary variables that are introduced by the direct assignment method to sample the posterior distribution of β. Their conditional posterior distribution is sampled according to

$$p\left(m_{d,k} = m \mid \mathbf{z}, \mathbf{m}_{-(d,k)}, \boldsymbol{\beta}\right) = \frac{\Gamma\left(\alpha\beta_k\right)}{\Gamma\left(\alpha\beta_k + c_{(\cdot),d}^k\right)} s\left(c_{(\cdot),d}^k, m\right)\left(\alpha\beta_k\right)^m, \tag{15.6}$$

where $s(n, m)$ denotes Stirling numbers of the first kind. The hyperparameters α, α', and γ are sampled using Metropolis-Hastings.

It remains now to show that HSLDA works. To do so we demonstrate results from modeling real-world datasets in the clinical and web retail domains. These results provide evidence that the two views (text and labels) mutually benefit multi-label classification. That is, modeling the joint is better than learning topic models and hierarchical classifiers independently.

15.4   Example Applications

15.4.1   Hospital Discharge Summaries and ICD-9 Codes

Despite the growing emphasis on meaningful use of technology in medicine, many aspects of medical record-keeping remain a manual process. In the U.S., diagnostic coding for billing and insurance purposes is often handled by professional medical coders who must explore a patient's extensive clinical record before assigning the proper codes.

A specific example of this involves labeling of hospital discharge summaries. These summaries are authored by clinicians to summarize patient hospitalization courses. They typically contain a record of patient complaints, findings, and diagnoses, along with treatment and hospital course. The kind of text one might expect to find in such a discharge summary is illustrated by this made-up snippet:

> History of Present Illness: Mrs. Carmen Sandiego is a 62-year-old female with a past medical history significant for diabetes, hypertension, hyperlipidemia, afib, status post MI in 5/2010 and cholecystectomy in 3/2009. The patient presented to the ED on 7/11/2011 with a right sided partial facial hemiparesis along with mild left arm weakness. The patient was admitted to the Neurology service and underwent a workup for stroke given her history of MI and many cardiovascular risk factors ...

For each hospitalization, trained medical coders review the information in the discharge summary and assign a series of diagnoses codes. Coding follows the ICD-9-CM controlled terminology, an international diagnostic classification for epidemiological, health management, and clinical purposes.[1] These ICD-9 codes are organized in a rooted-tree structure with each edge representing an is-a relationship between parent and child such that the parent diagnosis subsumes the child diagnosis. For example, the code for "Pneumonia due to adenovirus" is a child of the code for "Viral pneumonia," where the former is a type of the latter. A representative sub-tree of the ICD-9 code tree is shown in Figure 15.2. It is worth noting that the coding can be noisy. Human coders sometimes disagree (Cha, 2007), tend to be more specific than sensitive in their assignments (Birman-Deych et al., 2005), and sometimes make mistakes (Farzandipour et al., 2010).

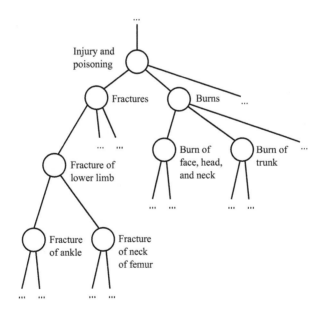

FIGURE 15.2
An illustration of a portion of the ICD9 hierarchy.

An automated process would ideally produce more complete and accurate diagnosis lists. The

---

[1] See http://www.cdc.gov/nchs/icd/icd9cm.htm.

task of automatic ICD-9 coding has been investigated in the clinical domain. Methods used to solve this problem (besides HSLDA) range from applying manually derived coding rules to applications of online rule learning approaches (Crammer et al., 2007; Goldstein et al., 2007; Farkas and Szarvas, 2008). Many classification schemes have been applied to this problem: K-nearest neighbor, naive Bayes, support vector machines, Bayesian Ridge Regression, as well as simple keyword mappings, all with promising results (Larkey and Croft, 1995; Ribeiro-Neto et al., 2001; Pakhomov et al., 2006; Lita et al., 2008).

The specific dataset we report results for in this chapter was gathered from the New York-Presbyterian Hospital clinical data warehouse. It consists of 6000 discharge summaries and their associated ICD-9 codes (7,298 distinct codes overall), representing a portion of the discharges from the hospital in 2009. All included discharge summaries had associated ICD-9 Codes. Summaries have 8.39 associated ICD-9 codes on average (std dev=5.01) and contain an average of 536.57 terms after preprocessing (std dev=300.29). We split our dataset into 5000 discharge summaries for training and 1000 for testing.

The text of the discharge summaries was tokenized with NLTK.[2] A fixed vocabulary was formed by taking the top 10,000 tokens with the highest document frequency (exclusive of names, places, and other identifying numbers). The study was approved by the Institutional Review Board and follows HIPAA (Health Insurance Portability and Accountability Act) privacy guidelines.

Here HSLDA is evaluated as a way to understand and model the relationship between a discharge summary and the ICD-9 codes that should be assigned to it. We show promising results for automatically assigning ICD-9 codes to hospital discharge records.

15.4.2 Product Descriptions and Catalogs

Many web-retailers store and organize their catalog of products in a mulitply-rooted hierarchy in addition to providing textual product descriptions for most products. Products can be discovered by users through free-text search and product category exploration. Top-level product categories are displayed on the front page of the website and lower-level categories can be discovered by choosing one of the top-level categories. Products can exist in multiple locations in the hierarchy.

Amazon.com is one such retailer. Its product categorization data is available as part of the Stanford Network Analysis Platform (SNAP) dataset (SNA, 2004). A representative sub-tree of the Amazon.com DVD product category tree is shown in Figure 15.3. Product descriptions were obtained separately from the Amazon.com website directly. Once such description is

> Winner of five Academy Awards, including Best Picture and Best Director, *The Deer Hunter* is simultaneously an audacious directorial conceit and one of the greatest films ever made about friendship and the personal impact of war. Like *Apocalypse Now*, it's hardly a conventional battle film—the soldier's experience was handled with greater authenticity in *Platoon*—but its depiction of war on an intimate scale packs a devastatingly dramatic punch ...

We study the collection of DVDs in the product catalog specifically. The resulting dataset contains 15,130 product descriptions for training and 1000 for testing. The product descriptions consist of 91.89 terms on average (std dev=53.08). Overall, there are 2,691 unique categories. Products are assigned on average 9.01 categories (std dev=4.91). The vocabulary consists of the most frequent 30,000 words omitting stopwords.

HSLDA is used here to understand and model the relationship between the product text description and the products' positioning in the product hierarchy. We show how to automatically situate a product in a hierarchical product catalog.

---

[2]See http://www.nltk.org.

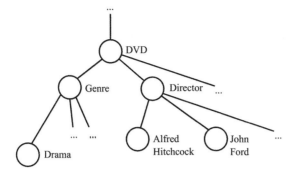

FIGURE 15.3
An illustration of a portion of the Amazon product hierarchy.

15.4.3   Comparison Models

We compare HSLDA to two closely related models. The comparison models are SLDA with independent regressors (hierarchical constraints on labels ignored, i.e., the regression is not conditional) and HSLDA fit by first performing LDA, then fitting probit regressors that respect the conditional label hierarchy (rather than jointly inferring the topics and the regression coefficients). These models were chosen because they are the strongest available competitors and because they highlight several pedagogical aspects of HSLDA, including performance in the absence of hierarchical constraints, the effect of the combined inference, and regression performance attributable solely to the hierarchical constraints.

SLDA with independent regressors is the most salient comparison model for our work. The distinguishing factor between HSLDA and SLDA is the additional structure imposed on the label space, a distinction that in developing HDSLA we hypothesized would result in a difference in predictive performance.

The second comparison model, HSLDA fit by performing LDA first followed by performing inference over the hierarchically constrained label space, does not allow the responses to influence the topics inferred by LDA. Combined inference has been shown to improve performance in SLDA (Blei and McAuliffe, 2008). This comparison model does not examine the value of utilizing the structured nature of the label space; instead, it highlights the benefit of combined inference over both the documents and the label space.

For all three models, particular attention was given to the settings of the prior parameters (μ, σ) for the regression coefficients (ν_l). These parameters implement an important form of regularization in HSLDA. In the setting where there are no negative labels, a Gaussian prior over the regression parameters with a negative mean implements a prior belief that missing labels are likely to be negative. Thus, we show model performance for all three models with a range of values for μ, the mean prior parameter for regression coefficients ($\mu \in \{-3, -2.8, -2.6, \ldots, 1\}$).

The number of topics for all models was set to $K = 50$, the prior distributions of $p(\alpha)$, $p(\alpha')$, and $p(\gamma)$ were all chosen to be gamma with a shape parameter of 1 and a scale parameter of 1000. Different values of K corresponding to different numbers of topics were explored, however, the results that we show in the following are not substantially changed in character. As is usual in mixed membership models, there is an ideal number of topics that should be used for out-of-sample prediction tasks, however, a full model-selection search varying topic cardinality was not performed for these datasets.

15.4.4   Evaluation and Results

We are particularly interested in the predictive performance on held-out data. Prediction performance was measured with standard metrics—sensitivity (true positive rate) and 1-specificity (false positive rate).

In each case the gold standard for testing was derived from the test data. To make the comparison as antagonistic to HSLDA as possible (relative to the other models), in evaluation only, ancestors of observed nodes in the label hierarchy were ignored, observed nodes were considered positive, and descendants of observed nodes were assumed to be negative. Note that this is different from our treatment of the observations during inference where we marginalize over possible settings of unobserved labels. For instance, as the SLDA model does not enforce hierarchical label constraints, when we consider only observed nodes we penalize HSLDA. This is because the is-a hierarchical constraints say that the ancestors of positively labeled nodes must also be positive, which the SLDA model cannot guarantee. Another antagonism of this gold standard is that it is likely to inflate the number of false positives because the labels applied to any particular document are usually not as complete as they could be. ICD-9 codes, for instance, are known to lack sensitivity and their use as a gold standard could lead to correctly positive predictions being labeled as false positives (Birman-Deych et al., 2005). However, given that the label space is often large (as in our examples), it is a reasonable assumption that erroneous false positives should not skew results significantly.

Predictive performance in HSLDA is evaluated by computing

$$p\left(y_{l,\hat{d}} \mid w_{1:N_{\hat{d}},\hat{d}}, w_{1:N_d,1:D}, y_{l\in\mathcal{L},1:D}\right)$$

for each test document $\hat{d}$ for each observed label $y_{l,\hat{d}}$ (given the test document words). For efficiency, the expectation of this probability distribution was approximated in the following way: Expectations of $\bar{z}_{\hat{d}}$ and η_l were estimated with samples from the posterior. Fixing these expectations, we performed Gibbs sampling over the hierarchy to acquire predictive samples for the documents in the test set. The true positive rate was calculated as the average expected labeling for gold standard positive labels. The false positive rate was calculated as the average expected labeling for gold standard negative labels.

As sensitivity and specificity can always be traded off, we examined sensitivity for a range of values for two different parameters—the prior means for the regression coefficients and the threshold for the auxiliary variables. The goal in this analysis was to evaluate the performance of these models subject to more or less stringent requirements for predicting positive labels. These two parameters have important related functions in the model. The prior mean in combination with the auxiliary variable threshold together encode the strength of the prior belief that unobserved labels are likely to be negative. Effectively, the prior mean applies negative pressure to the predictions and the auxiliary variable threshold determines the cutoff. For each model type, separate models were fit for each value of the prior mean of the regression coefficients. This is a proper Bayesian sensitivity analysis. In contrast, to evaluate predictive performance as a function of the auxiliary variable threshold, a single model was fit for each model type and prediction was evaluated based on predictive samples drawn subject to different auxiliary variable thresholds. These methods are significantly different since the prior mean is varied prior to inference, and the auxiliary variable threshold is varied following inference.

Figure 15.4(a) demonstrates the performance of the model on the clinical data as a ROC curve varying μ. For instance, a hyperparameter setting of $\mu = -1.6$ yields the following performance: the full HSLDA model had a true positive rate of 0.57 and a false positive rate of 0.13, the SLDA model had a true positive rate of 0.42 and a false positive rate of 0.07, and the HSLDA model where LDA and the regressions were fit separately had a true positive rate of 0.39 and a false positive rate of 0.08. These points are highlighted in Figure 15.4(a). Note that the figure is somewhat misleading because for any one value of μ, HSLDA outperforms the comparison models by a relatively large margin.

These results indicate that the full HSLDA model predicts more of the correct labels at a cost of an increase in the number of false positives relative to the comparison models. However, as shown in Figure 15.4(a), HSLDA outperforms no worse than the comparison models across the full range of specificities.

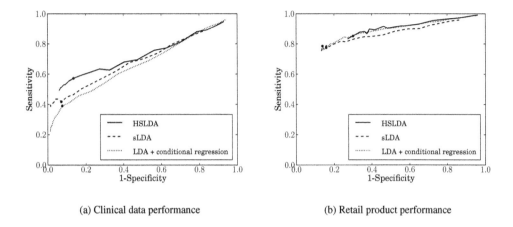

(a) Clinical data performance   (b) Retail product performance

FIGURE 15.4

ROC curves for HSLDA out-of-sample label prediction varying μ, the prior mean of the regression parameters. In both figures, solid is HSLDA, dashed are independent regressors + SLDA (hierarchical constraints on labels ignored), and dotted is HSLDA fit by running LDA first then running tree-conditional regressions.

Example topics (as word lists) learned for the discharge data are given below. These word lists are computed by sorting terms in decreasing order based on their probability under a given topic.

Topic 1	Topic 2
MASS	WOUND
CANCER	FOOT
RIGHT	CELLULITIS
BREAST	ULCER
CHEMOTHERAPY	LEFT
METASTATIC	ERYTHEMA
LEFT	PAIN
LYMPH	SWELLING
TUMOR	SKIN
BIOPSY	RIGHT
CARCINOMA	ABSCESS
LUNG	LEG
CHEMO	OSTEOMYELITIS
ADENOCARCINOMA	TOE
NODE	DRAINAGE

These topics closely correspond to common clinical concepts, namely cancers of the thorax and wounds common to diabetics suffering from poor peripheral circulation. Evaluations of the subject coherence of these topics relative to baselines are ongoing, but early results suggest positive findings similar to those reported for other supervised LDA models.

Figure 15.4(b) demonstrates the performance of the model on the retail product data as an ROC curve also varying μ. For instance, a hyperparameter setting of $\mu = -2.2$ yields the following performance: the full HSLDA model had a true positive rate of 0.85 and a false positive rate of 0.30, the SLDA model had a true positive rate of 0.78 and a false positive rate of 0.14, and the HSLDA model where LDA and the regressions were fit separately had a true positive rate of 0.77 and a false positive rate of 0.16. These results follow a similar pattern to the clinical data. These points are highlighted in Figure 15.4(b).

Example topics (as word lists) learned for the Amazon.com data are given below. These word lists were also computed by sorting terms in decreasing order based on their probability under a given topic.

Topic 1	Topic 2
SERIES	BASEBALL
EPISODES	TEAM
SHOW	GAME
SEASON	PLAYERS
EPISODE	BASKETBALL
FIRST	SPORT
TELEVISION	SPORTS
SET	NEW
TIME	PLAYER
TWO	SEASON
SECOND	LEAGUE
ONE	FOOTBALL
CHARACTERS	STARS
DISC	FANS
GUEST	FIELD

Figure 15.5 shows the predictive performance of HSLDA relative to the two comparison models on the clinical dataset as a function of the auxiliary variable threshold. For low values of the auxiliary variable threshold, the models predict labels in a more sensitive and less specific manner, creating the points in the upper right corner of the ROC curve. As the auxiliary variable threshold is increased, the models predict in a less sensitive and more specific manner, creating the points in the lower left hand corner of the ROC curve. HSLDA with full joint inference outperforms SLDA with independent regressors as well as HSLDA with separately trained regression.

15.5   Related Work

HSLDA does not, of course, stand alone. Models for structured labeling of bag-of-words data can be designed in a number of different ways.

As shown in Section 15.4, SLDA can be used to solve this kind of problem directly, however, doing so requires ignoring the hierarchical dependencies amongst the labels. Other models that incorporate LDA and supervision that could also be used to solve this problem include LabeledLDA (Ramage et al., 2009) and DiscLDA (Lacoste-Julien et al.). Various applications of these models to computer vision and document networks have been explored (Wang et al., 2009; Chang

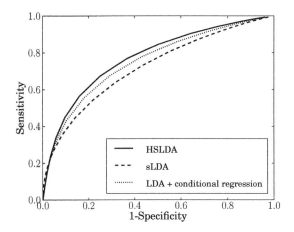

FIGURE 15.5
ROC curve for out-of-sample ICD-9 code prediction varying auxiliary variable threshold. $\mu = -1.0$ for all three models in this figure.

and Blei, 2010). None of these models, however, leverage dependency structure in the label space.

In other non-LDA-based related work, researchers have classified documents into hierarchies (a closely related task) using naive Bayes classifiers and support vector machines. Most of this work has been demonstrated on relatively small datasets and small label spaces, and has focused on single label classification without a model of documents such as LDA (McCallum et al., 1999; Dumais and Chen, 2000; Koller and Sahami, 1997; Chakrabarti et al., 1998).

15.6   Discussion

The SLDA model family, of which HSLDA is a member, can be understood in two different ways. One way is to see it as a family of topic models that improve on the topic modeling performance of LDA via the inclusion of observed supervision. An alternative, complementary way is to see it as a set of models that can predict labels for bag-of-word data. A large diversity of problems can be expressed as label prediction problems for bag-of-word data. A surprisingly large amount of data possess structured labels, either hierarchically constrained or otherwise. HSLDA directly addresses this kind of data and works well in practice. That it outperforms more straightforward approaches should be of interest to practitioners.

There are many kinds of problems that have the same characteristics as this: any data that consists of free text that has been partially or completely categorized by human editors; more specifically, any bag-of-words data that has been, at least in part, categorized. Examples include, but are not limited to, webpages and curated hierarchical directories of the same (DMO, 2002), product descriptions and catalogs, (e.g., AMA (2011) as available from SNA (2004)) and patient records and diagnosis codes assigned to them for bookkeeping and insurance purposes. The model we cover in this chapter shows one way to combine these two sources of information into a single model allowing one to categorize new text documents automatically, suggest labels that might be inaccurate, compute improved similarities between documents for information retrieval purposes, and more.

Extensions to this work include a nonparametric Bayesian extension with unbounded topic cardinality and relaxations to different kinds of label structure. Unbounded topic cardinality variants pose interesting inference challenges. Imposing different kinds of label structure constraints is possible within this framework but requires relaxing some of the assumptions we made in deriving the sampling distributions for HSLDA inference.

Appendix

Probit Regression

For reasons that are somewhat obscure, statisticians tend to use probit regression for binary classification whereas machine learners tend to use logistic regression. The "probit" function is the inverse of the normal cumulative distribution function (cdf). We denote the normal cdf function $\Phi(x; \mu, \sigma^2)$ with μ the mean, σ^2 the variance, and x the argument.

The range of the normal cdf is $(0, 1)$, which means that it can be interpreted as a probability. For instance, one can construct a generalized linear classification model (a "probit regression model") of the form

$$P(y_i = 1) = \Phi(x_i^T \beta; 0, \sigma^2). \tag{15.7}$$

Depending on convention (i.e., binary y_i represented as $\{1, 0\}$ or $\{1, -1\}$), the probability of y_i being labeled the opposite way is $P(y_i = -1)$ or $P(y_i = 0) = 1 - P(y_i = 1)$. Here x_i is a vector of covariates, β is a vector of weights, and y_i is a single, binary valued response. The close relationship between regression and classification is in full display here: probit regression is a "generalized linear *regression* model" as well as a "binary classifier."

In this model we would like to use labeled training data, $\{x_i, y_i\}_{i=1}^N$ to "learn" the value of β and then to use this value to predict the value of $y_{N+1}|x_{N+1}, \beta$. Being Bayesian about inference means that we will average over the posterior distribution of β when making predictions. This means that we want to draw samples from the posterior distribution of $\beta|\{x_i, y_i\}_{i=1}^N$. To do this efficiently one can introduce a set of auxiliary variables $\{u_i\}_{i=1}^N$.

By auxiliary variables we mean that such variables will be used as an intermediary for purposes of efficiency but will otherwise be uninteresting. They are variables introduced into a model in order to make inference easier but whose existence does not change the distribution of interest. Auxiliary variables for slice sampling are one particularly clever use of auxiliary variables. The auxiliary variable trick in probit regression is another.

For the purposes of exposition, forget about the i index and focus on a single instance y, x, and u. The argument we make will hold for all by simply reintroducing subscripts.

To start, let's propose a factorized joint distribution for these quantities

$$P(y, x, u) = P(y|u)P(u|x, \beta). \tag{15.8}$$

Straight away, one can see why this auxiliary variable scheme works. By the law of total probability we have

$$P(y, x) = \int P(y, x, u)du = \int P(y|u)P(u|x, \beta)du. \tag{15.9}$$

So, if by some means we generate S samples $\{u^{(s)}, y^{(s)}, x^{(s)}\}_{s=1}^S \sim P(y, x, u)$, we know that marginalizing u out (i.e., disregarding its value) we get samples $\{y^{(s)}, x^{(s)}\}_{s=1}^S \sim P(y, x)$.

We haven't specified the most important part of the auxiliary variable sampling scheme yet, namely, what $P(y|u)$ and $P(u|x, \beta)$ are. Let us try $y = \text{sign}(u)$ and $u \sim N(x^T \beta, \sigma^2)$. These choices are nice in a particular way. First let us verify that the marginalization of u out of this model results in the model specification in Equation (15.7):

$$
\begin{aligned}
P(y = 1 | x, \beta) &= \int P(y = 1 | u) P(u | x, \beta) du \\
&= \int \mathbb{I}(u > 0) N(u; x^T \beta, \sigma^2) du \\
&= \int_0^\infty N(u; x^T \beta, \sigma^2) du \\
&= 1 - \Phi(0; x^T \beta, \sigma^2) \\
&= \Phi(x^T \beta; 0, \sigma^2),
\end{aligned}
$$

where the last line comes from the fact that for symmetric distributions like the normal distribution, $\Phi(x^T \beta; 0, \sigma^2) = 1 - \Phi(-x^T \beta; 0, \sigma^2)$, and the mean of a normal cdf can be translated arbitrarily, i.e., $\Phi(-x^T \beta; 0, \sigma^2) = \Phi(0; x^T \beta, \sigma^2)$ (which comes from adding the offset $x^T \beta$ to the cdf argument and mean).

Having established the fact that for a particular sort of auxiliary variable choice, we get the same probit model as we wanted, why is this choice nice?

Well, it comes down to sampling β, u, and y. Generally, sampling β in the model without auxiliary variables will require hybrid Monte Carlo (HMC) or Metropolis-Hastings of some sort. Gibbs sampling often comes with substantial benefits. By making this choice of auxiliary variable, the conditional distribution of u_i given everything else is proportional to a truncated normal distribution, a distribution that is, by nature of its commonness, relatively straightforward to sample from. The big benefit, though, acrues from the posterior form for sampling β. With the u's "observed" (as they would be in a Gibbs sampler), the posterior distribution of β (for typical choices of prior) is precisely the same as that for linear regression, perhaps the most well-studied model in statistics. In that case, sampling β from its posterior distribution is quite simple usually, and certainly more so than sampling β without the u auxiliary variables.

The extension to the multivariate HSLDA setting is straightforward and follows this line of reasoning precisely. An extended discussion of the techniques suggested here and the multivariate generalization can be found in Gelman et al. (2004).

References

(2002). DMOZ open directory project. http://www.dmoz.org/.

(2004). Stanford network analysis platform. http://snap.stanford.edu/.

(2007). The computational medicine center's 2007 medical natural language processing challenge. http://www.computationalmedicine.org/challenge/previous.

(2011). Amazon, Inc. http://www.amazon.com/.

Albert, J. and Chib, S. (1993). Bayesian analysis of binary and polychotomous response data. *Journal of the American Statistical Association* 88: 669.

Birman-Deych, E., Waterman, A. D., Yan, Y., Nilasena, D. S., Radford, M. J., and Gage, B. F. (2005). Accuracy of ICD-9-CM codes for identifying cardiovascular and stroke risk factors. *Medical Care* 43: 480–5.

Blei, D. M. and McAuliffe, J. (2008). Supervised topic models. In Platt, J. C., Koller, D., Singer, Y., and Roweis, S. (eds), *Advances in Neural Information Processing Systems 20*. Cambridge, MA: The MIT Press, 121–128.

Blei, D. M., Ng, A. Y., and Jordan, M. I. (2003). Latent Dirichlet allocation. *Journal of Machine Learning Research* 3: 993–1022.

Chakrabarti, S., Dom, B., Agrawal, R., and Raghavan, P. (1998). Scalable feature selection, classification and signature generation for organizing large text databases into hierarchical topic taxonomies. *The VLDB Journal* 7: 163–178.

Chang, J. and Blei, D. M. (2010). Hierarchical relational models for document networks. *Annals of Applied Statistics* 4: 124–150.

Crammer, K., Dredze, M., Ganchev, K., Talukdar, P., and Carroll, S. (2007). Automatic code assignment to medical text. *Proceedings of the Workshop on BioNLP 2007: Biological, Translational, and Clinical Language Processing*. Stroudsburg, PA, USA: Association for Computational Linguistics, 129–136.

Dumais, S. and Chen, H. (2000). Hierarchical classification of web content. In *Proceedings of the 23rd Annual International ACM SIGIR Conference on Research and Development in Information Retrieval (SIGIR '00)*. New York, NY, USA: ACM, 256–263.

Farkas, R. and Szarvas, G. (2008). Automatic construction of rule-based ICD-9-CM coding systems. *BMC Bioinformatics* 9: S10.

Farzandipour, M., Sheikhtaheri, A., and Sadoughi, F. (2010). Effective factors on accuracy of principal diagnosis coding based on international classification of diseases, the 10th revision. *International Journal of Information Management* 30: 78–84.

Gelman, A., Carlin, J. B., Stern, H. S., and Rubin, D. B. (2004). *Bayesian Data Analysis*. Chapman and Hall/CRC, 2nd edition.

Goldstein, I., Arzumtsyan, A., and Uzuner, Ö. (2007). Three approaches to automatic assignment of ICD-9-CM codes to radiology reports. *AMIA Annual Symposium Proceedings* 2007: 279–283.

Griffiths, T. L. and Steyvers, M. (2004). Finding scientific topics. *Proceedings of the National Academy of Science (PNAS)* 101: 5228–5235.

Koller, D. and Sahami, M. (1997). Hierarchically Classifying Documents Using Very Few Words. Tech. report 1997-75, Stanford InfoLab, previous number = SIDL-WP-1997-0059.

Lacoste-Julien, S., Sha, F., and Jordan, M. I. (2008). DiscLDA: Discriminative learning for dimensionality reduction and classification. In Koller, D., Schuurmans, D., Bengio, Y., and Bottou, L. (eds), *Advances in Neural Information Processing Systems 21*. Red Hook, NY: Curran Associates, Inc., 897–904.

Larkey, L. and Croft, B. (1995). Automatic Assignment of ICD9 Codes to Discharge Summaries. Tech. report, University of Massachussets.

Lita, L. V., Yu, S., Niculescu, S., and Bi, J. (2008). Large scale diagnostic code classification for medical patient records. In *Proceedings of the 3rd International Joint Conference on Natural Language Processing (IJCNLP'08)*. Asian Federation for Natural Language Processing, 877–882.

McCallum, A., Nigam, K., Rennie, J. D. M., and Seymore, K. (1999). Building domain-specific search engines with machine learning techniques. In *Proceedings of the 16th International Joint Conference on Artifical Intelligence - Volume 2 (IJCAI '99)*. San Francisco, CA, USA: Morgan Kaufmann Publishers Inc., 662–667.

Pakhomov, S., Buntrock, J., and Chute, C. (2006). Automating the assignment of diagnosis codes to patient encounters using example-based and machine learning techniques. *Journal of the American Medical Informatics Association (JAMIA)* 13: 516–525.

Perotte, A., Barlett, N., Elhadad, N., and Wood, F. (2011). Hierarchically supervised latent Dirichlet allocation. In Shawe-Taylor, J., Zemel, R. S., Bartlett, P., Pereira, F. C. N., and Weinberger, K. Q. (eds), *Advances in Neural Information Processings Systems 24*. Red Hook, NY: Curran Associates, Inc., 2609–2617.

Ramage, D., Hall, D., Nallapati, R. M., and Manning, C. D. (2009). Labeled LDA: A supervised topic model for credit attribution in multi-labeled corpora. In *Proceedings of the 2009 Conference on Empirical Methods in Natural Language Processing: Vol. 1*. Stroudsburg, PA, USA: Association for Computational Linguistics, 248–256.

Ribeiro-Neto, B., Laender, A., and Lima, L. D. (2001). An experimental study in automatically categorizing medical documents. *Journal of the American society for Information science and Technology* 52: 391–401.

Teh, Y. W., Jordan, M. I., Beal, M. J., and Blei, D. M. (2006). Hierarchical Dirichlet processes. *Journal of the American Statistical Association* 101: 1566–1581.

Wallach, H. M., Mimno, D., and McCallum, A. (2009). Rethinking LDA: Why priors matter. In Bengio, Y., Schuurmans, D., Lafferty, J., Williams, C. K. I., and Culotta, A. (eds), *Advances in Neural Information Processing Systems 22*. Red Hook, NY: Curran Associates, Inc., 1973–1981.

Wang, C., Blei, D. M., and Fei-Fei, L. (2009). Simultaneous image classification and annotation. In *Proceedings of the 2009 IEEE Conference on Computer Vision and Pattern Recognition (CVPR 2009)*. Los Alamitos, CA, USA: IEEE Computer Society, 1903–1910.

16

Discriminative Mixed Membership Models

Hanhuai Shan

Department of Computer Science and Engineering, University of Minnesota: Twin Cities, Minneapolis, MN 55455, USA

Arindam Banerjee

Department of Computer Science and Engineering, University of Minnesota: Twin Cities, Minneapolis, MN 55455, USA

CONTENTS

Although mixed membership models have achieved great success in unsupervised learning, they have not been applied as widely to classification problems. In this chapter, we discuss a family of discriminative mixed membership (DMM) models. By combining unsupervised mixed membership models with multi-class logistic regression, DMM models can be used for classification. In particular, we discuss discriminative latent Dirichlet allocation (DLDA) for text classification and discriminative mixed membership naive Bayes (DMNB) for classification on general feature vectors. Two variation inference algorithms are considered for learning the models, including a fast inference algorithm which uses fewer variational parameters and is substantially more efficient than the standard mean field variational approximation. The efficacy of the models is demonstrated by extensive experiments on multiple datasets.

16.1 Introduction

In recent years, mixed membership (MM) models have found wide application in a variety of domains, such as topic modeling (Blei et al., 2003), bioinformatics (Airoldi et al., 2008), and social network analysis (Koutsourelakis and Eliassi-Rad, 2008). A key advantage of such models is that they provide a succinct and interpretable representation of otherwise large and high-dimensional datasets. However, one important restriction of most existing mixed membership models is that they are unsupervised models and cannot leverage class label information for classification. On the other hand, most of the popular classifiers, such as support vector machines (SVM) (Burges, 1998) and logistic regression (LR) (Pampel, 2000), are usually difficult to interpret. Therefore, an accurate discriminative classification model leveraging mixed membership models for interpretability is highly desirable.

This chapter discusses discriminative[1] mixed membership (DMM) models as a classification algorithm by combining multi-class logistic regression with unsupervised mixed membership models. In particular, two variants are considered in this chapter—discriminative latent Dirichlet allocation (DLDA) and discriminative mixed membership naive Bayes (DMNB). DLDA is applicable to text classification and uses LDA as the underlying mixed membership model (Blei et al., 2003). DMNB is applicable to non-text classification involving different types (e.g., numerical, categorical) of feature vectors and uses mixed membership naive Bayes (MNB) as the underlying mixed membership model (Shan and Banerjee, 2010).

Two variational inference algorithms, as well as corresponding variational EM algorithms are used to learn the model. The first inference algorithm is based on the ideas originally proposed in the context of LDA (Blei et al., 2003). The second algorithm uses a substantially smaller number of variational parameters, with no dependency on the dimensionality of the dataset. By design, the new algorithm has substantially smaller memory requirements, and is orders of magnitude faster, where the speedup times roughly increase with the dimensionality of data, i.e., the higher dimension of the data, the more computational achievements the algorithm gains.

The effectiveness of DMM models are established through extensive experiments on text data for DLDA and on UCI data for DMNB. The results show that DMM models achieve higher/competitive performance compared to state-of-the-art classification algorithms. More importantly, the new variational inference algorithm used in DMM is not only faster than the one used in Blei et al. (2003), but also leads to higher classification accuracy.

The rest of this chapter is organized as follows: Section 16.2 briefly reviews the related work. Sections 16.3 and 16.4 discuss DLDA and DMNB, respectively. Regular and fast variational inference algorithms are introduced in Section 16.5 and 16.6. We present the experimental results for DMM models in Section 16.7 and conclude in Section 16.8.

16.2 Related Work

This section gives a brief overview for unsupervised mixed membership models—latent Dirichlet allocation and mixed membership naive Bayes models, and then discusses the related work on incorporating supervised information into mixed membership models.

---

[1]"Discriminative" here does not mean a discriminative model, but a generative model used for classification instead of clustering.

16.2.1 Latent Dirichlet Allocation

Naive Bayes (NB) models, or mixture models (Redner and Walker, 1984; Banerjee et al., 2005b) in general, assume that for each data point $\mathbf{x}$, the latent component z is fixed across all features. While such an assumption is reasonable in certain domains, it puts a major restriction on the flexibility of naive Bayes models. Latent Dirichlet allocation (LDA)(Blei et al., 2003; Griffiths and Steyvers, 2004) is an elegant extension of standard mixture models that relaxes this assumption in the context of topic modeling, where each data point is a collection of tokens, e.g., a document with a collection of words. LDA assumes that each word in a document potentially comes from a separate topic z. If there are completely k topics, z can take value from 1 to k and is generated from a discrete distribution discrete(π) of this document, and all documents share a k-dimensional Dirichlet prior α. The generative process for each document $\mathbf{x}$ is as follows:

1. Choose a mixed membership vector $\pi \sim$ Dirichlet(α).

2. For each of m words $\{x_j, [j]_1^m\}$ ($[j]_1^m$ is defined as $\{j = 1, 2, \ldots, m\}$) in $\mathbf{x}$:

   (a) Choose a topic (component) $z_j = c \sim$ discrete(π).
   (b) Choose x_j from $p(x_j|\beta_c)$.

$\beta = \{\beta_c, [c]_1^k\}$ is a collection of parameters for k component distributions, where each β_c is a V-dimensional discrete distribution given V, the total number of words in the dictionary. The density function of a document $\mathbf{x}$ is

$$p(\mathbf{x}|\alpha, \beta) = \int_\pi p(\pi|\alpha) \left(\prod_{j=1}^m \sum_{c=1}^k p(z_j = c|\pi)p(x_j|\beta_c) \right) d\pi . \tag{16.1}$$

Computing the probability of a collection of documents is intractable, and several approximate inference techniques have been proposed to address the problem. The two most popular approaches include variational approximation (Blei et al., 2003) and Gibbs sampling (Geman and Geman, 1984; Griffiths and Steyvers, 2004).

16.2.2 Mixed Membership Naive Bayes

Although LDA achieves a good performance in topic modeling, it cannot deal with data points with numerical or real-valued features, or data points with heterogenous features. Mixed membership naive Bayes relaxes these limitations by introducing a separate exponential family distribution (Barndorff-Nielsen, 1978) for each feature. It is designed to deal with sparse and heterogenous feature vectors. Following MNB, the generative process for the data point $\mathbf{x}$ can be described as follows:

1. Choose a mixed membership vector $\pi \sim$ Dirichlet(α).

2. For each non-missing feature x_j of $\mathbf{x}$:

   (a) Choose a component $z_j = c \sim$ discrete(π) .
   (b) Choose a feature value $x_j \sim p_{\psi_j}(x_j|\theta_{jc})$.

Here, ψ_j and θ_{jc} jointly decide an exponential family distribution (Banerjee et al., 2005b; Barndorff-Nielsen, 1978) for feature j and component c. In particular,

$$p_{\psi_j}(x_j|\theta_{jc}) = \exp(x_j\theta_{jc} - \psi_j(\theta_{jc}))p_j(x_j) ,$$

where $\psi_j(\cdot)$ is the cumulant or the log-partition function, and $p_j(x_j)$ is a non-negative base measure. $\psi_j(\cdot)$ determines the exponential family model appropriate for feature j, e.g., Gaussian, Poisson, Bernoulli, etc., and θ_{jc} is the natural parameter corresponding to feature j and component c.

The density function for $\mathbf{x}$ is given by:

$$p(\mathbf{x}|\alpha, \Theta) = \int_\pi p(\pi|\alpha) \left(\prod_{\substack{j=1 \\ \exists x_j}}^{d} \sum_{c=1}^{k} p(z_j = c|\pi) p_{\psi_j}(x_j|\theta_{jc}) \right) d\pi , \qquad (16.2)$$

where $\exists x_j$ denotes any observed feature j for $\mathbf{x}$ and $\Theta = \{\theta_{jc}, [j]_1^d, [c]_1^k\}$.

16.2.3 Supervised LDA

Supervised latent Dirichlet allocation (SLDA) (Blei and McAuliffe, 2007) is an extension of LDA which accommodates the response variables as the supervised information other than the documents. The response variable is assumed to be generated from a normal linear model $N(w^T \bar{z}, \sigma^2)$, where w and σ^2 are the parameters and the covariates $\bar{z} = \sum_{j=1}^{M} z_j/M$ are the empirical average frequencies of each latent topic for the words in the document. If there are totally k components, each z_j is represented as a k-dimensional unit vector with only the cth entry being 1 if it denotes the cth component. The generative process for SLDA is as follows:

1. Choose a mixed membership vector $\pi \sim \text{Dirichlet}(\alpha)$.

2. For each of m words $\{x_j, [j]_1^m\}$ in $\mathbf{x}$:

   (a) Choose a topic (component) $z_j = c \sim \text{discrete}(\pi)$.

   (b) Choose x_j from $p(x_j|\beta_c)$.

3. Choose a response variable $y \sim N(w^T \bar{z}, \sigma^2)$.

The density function of SLDA is hence given by

$$p(\mathbf{x}|\alpha, \beta) = \int_\pi p(\pi|\alpha) \left(\prod_{j=1}^{m} \sum_{c=1}^{k} p(z_j = c|\pi) p(x_j|\beta_c) \right) p(y|\mathbf{z}, w, \sigma^2) d\pi .$$

Generally, SLDA is a combination of mixed membership models with generalized linear models to incorporate supervised information. In particular, the generalized linear model for SLDA to generate the response variable y is a univariate normal linear model. Therefore, SLDA is constrained to deal with one-dimensional, real-valued response variables.

Other than supervised LDA, recent years have seen quite a few extensions of incorporating supervised information into mixed membership models. Flaherty et al. (2005) proposed labeled latent Dirichlet allocation to incorporate functional annotation of known genes to guide gene clustering. Fei-Fei and Perona (2005) proposed a Bayesian model for natural scene categorization. Lacoste-Julien et al. (2008) proposed DiscLDA which determines document position on topic simplex with guidance of labels. Mimno and McCallum (2008) proposed a Dirichlet-multinomial regression which accommodates different types of metadata, including labels. Wang et al. (2008) proposed a correlated labeling model for multilabel classification. Wang et al. (2009) extended SLDA for image classification and annotation. Ramage et al. (2011) proposed partially labeled topic models which make use of unsupervised topic models but aligned some learned topics with labels. Wang and Blei (2011) proposed a collaborative topic regression model which uses ratings across different users as the supervised information for scientific articles and combines topic models and collaborative filtering together.

16.3 Discriminative LDA

SLDA (Blei and McAuliffe, 2007) incorporates the response variable into LDA (Blei et al., 2003), but it has two limitations preventing it from being used as a classification algorithm:

1. The response variables in SLDA are univariate real numbers assumed to be generated from a normal linear model, whereas the response variables, i.e., labels, are discrete categories in the classification setting. Although Blei and McAuliffe (2007) also gave a general framework for other types of response variables via generalized linear models, variational inference is not as straightforward as in SLDA. In particular, the Taylor expansion-based approach in Blei and McAuliffe (2007) forgoes the lower-bound guarantee of variational inference.

2. Like latent Dirichlet allocation, SLDA is designed for text data as a collection of homogeneous tokens. However, most non-text classification tasks, e.g., the UCI benchmark datasets, have features of heterogeneous types with measured values. SLDA is not designed for such data.

In this and the following sections, we discuss discriminative mixed membership models, which combine MM models with logistic regression for classification. The underlying MM models for DMM include LDA and MNB, yielding discriminative LDA and discriminative MNB for text and numerical data, respectively. We discuss DLDA first and introduce DMNB in the next section.

Assuming there are t classes and k components, the graphical model for DLDA is given in Figure 16.1(a). It is similar to LDA (Blei et al., 2003), except that it generates the label y other than the document $\mathbf{x}$ through logistic regression with parameter $\eta = \{\eta_1, \ldots, \eta_t\}$, where each η_h for $[h]_1^t$ is a k-dimensional vector and η_t is a zero vector by default. The generative process for each document $\mathbf{x}$ and label y is given as follows:

1. Choose a mixed membership vector $\pi \sim \text{Dirichlet}(\alpha)$.

2. For each of m words $\{x_j, [j]_1^m\}$ in the document $\mathbf{x}$,

   (a) Choose a component $z_j = c \sim \text{discrete}(\pi)$, $c \in \{1, 2, \ldots k\}$.

   (b) Choose a word $x_j \sim \text{discrete}(\beta_c)$.

3. Choose the label from a multi-class logistic regression $y \sim \text{LR}(\eta_1^T \bar{z}, \eta_2^T \bar{z}, \ldots, \eta_t^T \bar{z})$.

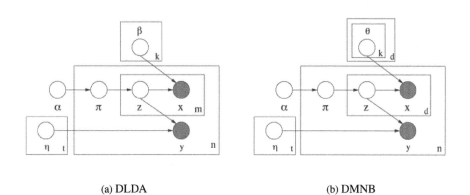

(a) DLDA   (b) DMNB

FIGURE 16.1
Graphical models for DLDA and DMNB.

As in SLDA, $\bar{z}$ is an average of $z_1 \ldots z_m$ over all observed words. LR $(\eta_1^T \bar{z}, \eta_2^T \bar{z}, \ldots, \eta_t^T \bar{z})$ denotes a logistic transformation on $[\eta_1^T \bar{z}, \eta_2^T \bar{z}, \ldots, \eta_t^T \bar{z}]$, which is equivalent to a discrete distribution $(p_1, \ldots p_{t-1}, 1 - \sum_{h=1}^{t-1} p_h)$ with $p_h = \frac{\exp(\eta_h^T \bar{z})}{1 + \sum_{h=1}^{t-1}(\eta_h^T \bar{z})}$ for $[h]_1^{t-1}$. In two-class classification, y is 0 or 1 generated from Bernoulli$(\frac{1}{1+\exp(-\eta_1^T \bar{z})})$, i.e., there is only one parameter η_1 to be estimated, η_2 is the zero vector by default.

DLDA could be considered as a variant of SLDA. In SLDA, the response variable is a real number generated from a normal linear model. In DLDA, the response variable is a classification label generated from a generalized linear model (McCullagh and Nelder, 1989), in particular, from the multivariate logistic regression.

From the generative model, the density function for $(\mathbf{x}, y)$ is given by:

$$p(\mathbf{x}, y | \alpha, \beta, \eta) = \int_\pi p(\pi | \alpha) \left(\prod_{j=1}^{m} \sum_{c=1}^{k} p(z_j = c | \pi) p(x_j | \beta_c) \right) p(y | \mathbf{z}, \eta) d\pi . \qquad (16.3)$$

The probability of the entire dataset of n documents and labels $(\mathcal{X} = \{\mathbf{x}_i, [i]_1^n\}, \mathcal{Y} = \{y_i, [i]_1^n\})$ is given by

$$p(\mathcal{X}, \mathcal{Y} | \alpha, \beta, \eta) = \prod_{i=1}^{n} \int_{\pi_i} p(\pi_i | \alpha) \left(\prod_{j=1}^{m_i} \sum_{c=1}^{k} p(z_{ij} = c | \pi_i) p(x_{ij} | \beta_c) \right) p(y_i | \mathbf{z}_i, \eta) d\pi_i . \qquad (16.4)$$

There are two important properties of DLDA and of discriminative mixed membership models in general: (1) The k-dimensional mixed membership $\bar{z}$ effectively serves as a low-dimensional representation of the original document. While $\bar{z}$ in LDA is inferred in an unsupervised way, it is obtained from a supervised dimensionality reduction in DLDA. (2) DLDA allows the number of classes t and the number of components k in the generative model to be different. If k was forced to be equal to t, for problems with a small number of classes, $\bar{z}$ would have been a rather coarse representation of the document. In particular, for two-class problems, $\bar{z}$ would lie on the 2-simplex, which may not be an informative representation for classification purposes. Decoupling the choice of k from t prevents such pathologies. In principle, one may find a proper k using a nonparametric Dirichlet process mixture model (Blei and Jordan, 2006).

In DLDA, following Blei and McAuliffe (2007), we have used $\bar{z}$ (the mean of z for all words) as an input to logistic regression. In principle, any other transformation of z could work, as long as it gives a reasonable representation of the original data point. The choice of $\bar{z}$ is due to the following: (1) Optimality: Given a set of data points, their best representative is always the mean according to a wide variety of divergence functions (Banerjee et al., 2005b; Banerjee, 2007). We also notice that $\eta_h^T \bar{z} = \eta_h^T E[z] = E[\eta_h^T z]$, which means that if we take the mean of $\eta_h^T z$ on each feature as the input to logistic transformation function, it is equivalent to using $\eta_h^T \bar{z}$ as the input to that function. (2) Simplicity: Since z is the latent variable, if we use a complicated transformation on z such as a non-linear function, it would greatly increase the difficulty in inference and learning.

16.4 Discriminative MNB

Discriminative MNB is similar to DLDA, but it is designed for non-text data with real-valued features and it keeps separate distributions for each feature as MNB. Given the graphical model in Figure 16.1(b), the generative process for the data point $\mathbf{x}$ and label y is as follows:

1. Choose a mixed membership vector $\pi \sim$ Dirichlet(α).

2. For each non-missing feature j in $\mathbf{x}$

    (a) Choose a component $z_j = c \sim \text{discrete}(\pi)$, $c \in \{1, 2, ...k\}$.

    (b) Choose a feature value $x_j \sim p_{\psi_j}(x_j | \theta_{jc})$.

3. Choose the label from a multi-class logistic regression $y \sim \text{LR}(\eta_1^T \bar{z}, \eta_2^T \bar{z}, \ldots, \eta_t^T \bar{z})$.

The density function for $(\mathbf{x}, y)$ is given by

$$p(\mathbf{x}, y | \alpha, \Theta, \eta) = \int_\pi p(\pi | \alpha) \left(\prod_{\substack{j=1 \\ \exists x_j}}^{d} \sum_{c=1}^{k} p(z_j = c | \pi) p_{\psi_j}(x_j | \theta_{jc}) \right) p(y | \mathbf{z}, \eta) d\pi . \tag{16.5}$$

The probability of the entire dataset of n documents and labels $(\mathcal{X} = \{\mathbf{x}_i, [i]_1^n\}, \mathcal{Y} = \{y_i, [i]_1^n\})$ is given by

$$p(\mathcal{X}, \mathcal{Y} | \alpha, \Theta, \eta)$$

$$= \prod_{i=1}^{n} \int_{\pi_i} p(\pi_i | \alpha) \left(\prod_{\substack{j=1 \\ \exists x_{ij}}}^{d} \sum_{c=1}^{k} p(z_{ij} = c | \pi_i) p_{\psi_j}(x_{ij} | \theta_{jc}) \right) p(y_i | \mathbf{z}_i, \eta) d\pi_i . \tag{16.6}$$

For a concrete exposition to MNB models, we will focus on two specific instantiations of such models based on univariate Gaussian and discrete distributions for each feature in each component, corresponding to real-valued features and discrete features, respectively. Note that although the two examples we give have the same family of distributions across all features, DMNB allows different features to have different distributions and parameters.

1. **DMNB-Gaussian:** Such models have Gaussian distributions for each feature, hence they are applicable to data with real-valued features. Given the model parameters α and $\{\boldsymbol{\mu}, \boldsymbol{\sigma}^2\} = \{(\mu_{jc}, \sigma_{jc}^2), [j]_1^d, [c]_1^k\}$, the density function is given by:

$$p(\mathbf{x}, y | \alpha, \boldsymbol{\mu}, \boldsymbol{\sigma}^2, \eta) \tag{16.7}$$

$$= \int_\pi p(\pi | \alpha) \left(\prod_{\substack{j=1 \\ \exists x_j}}^{d} \sum_{c=1}^{k} p(z_j = c | \pi) \frac{\exp\left(-\frac{(x_j - \mu_{jc})^2}{2\sigma_{jc}^2}\right)}{\sqrt{2\pi\sigma_{jc}^2}} \right) p(y | \mathbf{z}, \eta) d\pi.$$

2. **DMNB-Discrete:** Such models have discrete distributions for each feature, hence are applicable to data with categorical features. Assuming that feature j can take r_j possible values, each feature j and component c then has a discrete distribution $\{p_{jc}(r), [r]_1^{r_j}\}$, where $p_{jc}(r) \geq 0$ and $\sum_{r=1}^{r_j} p_{jc}(r) = 1$. Given the model parameters α and $\boldsymbol{p} = \{p_{jc}(r), [r]_1^{r_j}, [j]_1^d, [c]_1^k\}$, the density function is given by

$$p(\mathbf{x}, y | \alpha, \boldsymbol{p}, \eta) \tag{16.8}$$

$$= \int_\pi p(\pi | \alpha) \left(\prod_{\substack{j=1 \\ \exists x_j}}^{d} \sum_{c=1}^{k} p(z_j = c | \pi) p_{jc}(x_j) \right) p(y | \mathbf{z}, \eta) d\pi.$$

16.5   Inference and Estimation

For a given dataset $\{\mathcal{X}, \mathcal{Y}\}$, the learning task is to estimate the model parameters such that the likelihood of observing the whole dataset is maximized. A general approach for such a task is to use expectation maximization (EM) algorithms. However, the likelihood calculations in (16.3) and (16.5) are intractable, implying that a direct application of EM is not feasible. In this section, we introduce a variational inference method (Wainwright and Jordan, 2008), which alternates between obtaining a tractable lower bound to the true log-likelihood and choosing the model parameters to maximize the lower bound. To obtain a tractable lower bound, we consider an entire family of parameterized lower bounds with a set of free variational parameters, and pick the best lower bound by optimizing the lower bound with respect to the free variational parameters.

16.5.1   Variational Approximation

In most applications of the EM algorithm for mixture modeling, in the E-step, one can directly compute the latent variable distribution (Neal and Hinton, 1998; Banerjee et al., 2004), which is used to calculate the expectation of the likelihood; in the M-step, parameter estimation is done by maximizing the expectation of the complete likelihood. However, a direct computation of latent variable distribution $p(\pi, \mathbf{z}|\cdot)$ is not possible for DMM models. Hence, we introduce a tractable family of parameterized distributions $q_1(\pi, \mathbf{z}|\gamma, \phi)$ as an approximation to $p(\pi, \mathbf{z}|\cdot)$, where (γ, ϕ) are free variational parameters. In particular, following Blei et al. (2003), in DLDA

$$q_1(\pi, \mathbf{z}|\gamma, \phi) = q_1(\pi|\gamma) \prod_{j=1}^{m} q_1(z_j|\phi_j) , \qquad (16.9)$$

and in DMNB

$$q_1(\pi, \mathbf{z}|\gamma, \phi) = q_1(\pi|\gamma) \prod_{\substack{j=1 \\ \exists x_j}}^{d} q_1(z_j|\phi_j) . \qquad (16.10)$$

The plate diagram for q_1 is in Figure 16.2(a). For each data point, γ is a k-dimensional Dirichlet distribution parameter over π in both (16.9) and (16.10). $\phi = \{\phi_j, [j]_1^m\}$ in (16.9) are parameters for discrete distributions over the topics z of all m words, and $\phi = \{\phi_j, [j]_1^d, \exists x_j\}$ in (16.10) are parameters for discrete distributions over the latent components z for all m non-missing features out of d features in total.

Denoting the log-likelihood function with $\log p(\mathbf{x}, y|\alpha, \Lambda, \eta)$, where $\Lambda = \beta$ for DLDA and $\Lambda = \Theta$ for DMNB, applying Jensen's inequality (Blei et al., 2003) yields:

$$\log p(\mathbf{x}, y|\alpha, \Lambda, \eta) \geq E_{q_1}[\log p(\pi, \mathbf{z}, \mathbf{x}, y|\alpha, \Lambda, \eta)] + H(q_1(\pi, \mathbf{z}|\gamma, \phi)) . \qquad (16.11)$$

Therefore, (16.11) gives a lower bound to $\log p(\mathbf{x}, y|\alpha, \Lambda, \eta)$. For each $\{\mathbf{x}_i, y_i\}$, denoting the lower bound with $L(\gamma_i, \phi_i; \alpha, \Lambda, \eta)$, it can be expanded as

$$L(\gamma_i, \phi_i; \alpha, \Lambda, \eta) = E_{q_1}[\log p(\pi_i|\alpha)] + E_{q_1}[\log p(\mathbf{z}_i|\pi_i)] + E_{q_1}[\log p(\mathbf{x}_i|\mathbf{z}_i, \Lambda)]$$
$$- E_{q_1}[\log q_1(\pi_i|\gamma_i)] - E_{q_1}[\log q_1(\mathbf{z}_i|\phi_i)] + E_{q_1}[\log p(y_i|\mathbf{z}_i, \eta)] . \qquad (16.12)$$

It is easy to expand the first five terms in (16.12). For the last term, $E_{q_1}[\log p(y_i|\mathbf{z}_i, \eta)]$, there is no closed-form solution, but it could be lower-bounded after introducing a new parameter ξ. In particular, we have

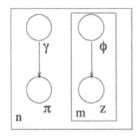

(a) q_1 for regular variational inference

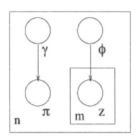

(b) q_2 for fast variational inference

FIGURE 16.2
Variational distributions for regular and fast variational inference.

$$E_{q_1}[\log p(y_i | \mathbf{z}_i, \eta)]$$
$$= E_{q_1}[\sum_{h=1}^{t-1} \eta_h^T \bar{z}_i y_{ih} - \log(1 + \sum_{h=1}^{t-1} \exp(\eta_h^T \bar{z}_i))]$$
$$= \sum_{h=1}^{t-1} \sum_{c=1}^{k} \eta_{hc} E_{q_1}[\bar{z}_{ic}] y_{ih} - E_{q_1}[\log(1 + \sum_{h=1}^{t-1} \exp(\eta_h^T \bar{z}_i))] . \qquad (16.13)$$

The second term of (16.13) could be expanded as follows:

$$- E_{q_1}[\log(1 + \sum_{h=1}^{t-1} \exp(\eta_h^T \bar{z}_i))]$$
$$\geq - \log(1 + \sum_{h=1}^{t-1} E_{q_1}[\exp(\sum_{c=1}^{k} \eta_{hc} \bar{z}_{ic})])$$
$$\geq - \log(1 + \sum_{h=1}^{t-1} E_{q_1}[\sum_{c=1}^{k} \bar{z}_{ic} \exp(\eta_{hc})])$$
$$= - \log(1 + \sum_{h=1}^{t-1} \sum_{c=1}^{k} E_{q_1}[\bar{z}_{ic}] \exp(\eta_{hc}))$$
$$\geq - \frac{1}{\xi_i} \sum_{h=1}^{t-1} \sum_{c=1}^{k} E_{q_1}[\bar{z}_{ic}] \exp(\eta_{hc}) + 1 - \frac{1}{\xi_i} - \log(\xi_i), \qquad (16.14)$$

where the first inequality is from Jensen's inequality, the second inequality is also from Jensen's inequality, noticing that $\bar{z}_i$ is actually a discrete distribution, and the third inequality is from $- \log(x) \geq 1 - \frac{x}{\xi} - \log(\xi)$ (Minka, 2003a), by introducing a new variational parameter $\xi > 0$. Given (16.14), the last term of (16.12) is lower-bounded by

$$E_{q_1}[\log p(y_i|\mathbf{z}_i, \eta)]$$

$$\geq \sum_{c=1}^{k} E_{q_1}[\bar{z}_{ic}] \sum_{h=1}^{t-1} \left(\eta_{hc} y_{ih} - \frac{1}{\xi_i} \exp(\eta_{hc}) \right) + 1 - \frac{1}{\xi_i} - \log(\xi_i) , \qquad (16.15)$$

where $E_{q_1}[\bar{z}_{ic}] = \frac{1}{m_i} \sum_{j=1}^{m_i} \phi_{ijc}$ in DLDA, and $E_{q_1}[\bar{z}_{ic}] = \frac{1}{m_i} \sum_{j=1, \exists x_j}^{d} \phi_{ijc}$ in DMNB.

Therefore, for DLDA, the six terms in (16.12) are given as follows:

$$E_{q_1}[\log p(\pi_i|\alpha)] = \log \Gamma \left(\sum_{c=1}^{k} \alpha_c \right) - \sum_{c=1}^{k} \log \Gamma(\alpha_c)$$

$$+ \sum_{c=1}^{k} (\alpha_c - 1) \left(\Psi(\gamma_{ic}) - \Psi \left(\sum_{l=1}^{k} \gamma_{il} \right) \right) , \qquad (16.16)$$

$$E_{q_1}[\log p(\mathbf{z}_i|\pi_i)] = \sum_{j=1}^{m_i} \sum_{c=1}^{k} \phi_{ijc} \left(\Psi(\gamma_{ic}) - \Psi \left(\sum_{l=1}^{k} \gamma_{il} \right) \right) , \qquad (16.17)$$

$$E_{q_1}[\log p(\mathbf{x}_i|\beta, \mathbf{z}_i)] = \sum_{j=1}^{m_i} \sum_{c=1}^{k} \sum_{v=1}^{V} \phi_{ijc} x_{ij}^v \log \beta_{cv} , \qquad (16.18)$$

$$E_{q_1}[\log q_1(\pi_i|\gamma_i)] = \log \Gamma \left(\sum_{c=1}^{k} \gamma_{ic} \right) - \sum_{c=1}^{k} \log \Gamma(\gamma_{ic})$$

$$+ \sum_{c=1}^{k} (\gamma_{ic} - 1) \left(\Psi(\gamma_{ic}) - \Psi \left(\sum_{l=1}^{k} \gamma_{il} \right) \right) , \qquad (16.19)$$

$$E_{q_1}[\log q_1(\mathbf{z}_i|\phi_i)] = \sum_{j=1}^{m_i} \sum_{c=1}^{k} \phi_{ijc} \log \phi_{ijc} , \qquad (16.20)$$

$$E_{q_1}[\log p(y_i|\mathbf{z}_i, \eta)] \geq \frac{1}{m_i} \sum_{c=1}^{k} \sum_{j=1}^{m_i} \sum_{h=1}^{t-1} \phi_{ijc} \left(\eta_{hc} y_{ih} - \frac{1}{\xi_i} \exp(\eta_{hc}) \right)$$

$$+ 1 - \frac{1}{\xi_i} - \log(\xi_i). \qquad (16.21)$$

For DMNB, the six terms in (16.12) are given as follows:

$$E_{q_1}[\log p(\pi_i|\alpha)] = \log \Gamma \left(\sum_{c=1}^{k} \alpha_c \right) - \sum_{c=1}^{k} \log \Gamma(\alpha_c)$$

$$+ \sum_{c=1}^{k} (\alpha_c - 1) \left(\Psi(\gamma_{ic}) - \Psi \left(\sum_{l=1}^{k} \gamma_{il} \right) \right) , \qquad (16.22)$$

$$E_{q_1}[\log p(\mathbf{z}_i|\pi_i)] = \sum_{\substack{j=1 \\ \exists x_{ij}}}^{d} \sum_{c=1}^{k} \phi_{ijc} \left(\Psi(\gamma_{ic}) - \Psi \left(\sum_{l=1}^{k} \gamma_{il} \right) \right) , \qquad (16.23)$$

$$E_{q_1}[\log p(\mathbf{x}_i|\mathbf{z}_i, \Theta)] = \sum_{\substack{j=1 \\ \exists x_{ij}}}^{d} \sum_{c=1}^{k} \phi_{ijc} \log p_{\psi_j}(x_{ij}|\theta_{jc}), \qquad (16.24)$$

$$E_{q_1}[\log q_1(\pi_i|\gamma_i)] = \log \Gamma \left(\sum_{c=1}^{k} \gamma_{ic} \right) - \sum_{c=1}^{k} \log \Gamma(\gamma_{ic})$$

$$+ \sum_{c=1}^{k} (\gamma_{ic} - 1) \left(\Psi(\gamma_{ic}) - \Psi \left(\sum_{l=1}^{k} \gamma_{il} \right) \right), \quad (16.25)$$

$$E_{q_1}[\log q_1(\mathbf{z}_i|\phi_i)] = \sum_{\substack{j=1 \\ \exists x_{ij}}}^{d} \sum_{c=1}^{k} \phi_{ijc} \log \phi_{ijc}, \quad (16.26)$$

$$E_{q_1}[\log p(y_i|\mathbf{z}_i,\eta)] \geq \frac{1}{m_i} \sum_{c=1}^{k} \sum_{\substack{j=1 \\ \exists x_{ij}}}^{d} \sum_{h=1}^{t-1} \phi_{ijc} \left(\eta_{hc} y_{ih} - \frac{1}{\xi_i} \exp(\eta_{hc}) \right)$$

$$+ 1 - \frac{1}{\xi_i} - \log(\xi_i). \quad (16.27)$$

After introducing ξ, the lower bound of the log-likelihood for each data point $(\mathbf{x}_i, y_i)$ can be represented as $L(\gamma_i, \phi_i, \xi_i; \alpha, \Lambda, \eta)$. The lower bound of the log-likelihood on the whole dataset $\{\mathcal{X}, \mathcal{Y}\}$ is simply the summation of $L(\gamma_i, \phi_i, \xi_i; \alpha, \Lambda, \eta)$ over all data points. The best lower bound can be computed by maximizing each $L(\gamma_i, \phi_i, \xi_i; \alpha, \Lambda, \eta)$ over the free parameters $(\gamma_i, \phi_i, \xi_i)$. A direct calculation gives the following update equations that iteratively maximize the lower bound. In particular, for DLDA, we have

$$\phi_{ijc} \propto \exp \left(\Psi(\gamma_{ic}) - \Psi \left(\sum_{l=1}^{k} \gamma_{il} \right) + \sum_{v=1}^{V} x_{ij}^v \log \beta_{cv} \right.$$

$$\left. + \frac{1}{m_i} \sum_{h=1}^{t-1} (\eta_{hc} y_{ih} - \exp(\eta_{hc})/\xi_i) \right), \quad (16.28)$$

$$\gamma_{ic} = \alpha_c + \sum_{j=1}^{m_i} \phi_{ijc}, \quad (16.29)$$

$$\xi_i = 1 + \frac{1}{m_i} \sum_{h=1}^{t-1} \sum_{c=1}^{k} \sum_{j=1}^{m_i} \phi_{ijc} \exp(\eta_{hc}) \cdot [i]_1^n, [j]_1^{m_i}, [c]_1^k. \quad (16.30)$$

For DMNB, we have

$$\phi_{ijc} \propto \exp \left(\Psi(\gamma_{ic}) - \Psi \left(\sum_{l=1}^{k} \gamma_{il} \right) + \log p_{\Psi_j}(x_{ij}|\theta_{jc}) \right.$$

$$\left. + \frac{1}{m_i} \sum_{h=1}^{t-1} (\eta_{hc} y_{ih} - \exp(\eta_{hc})/\xi_i) \right), \quad (16.31)$$

$$\gamma_{ic} = \alpha_c + \sum_{\substack{j=1 \\ \exists x_{ij}}}^{d} \phi_{ijc}, \quad (16.32)$$

$$\xi_i = 1 + \frac{1}{m_i} \sum_{h=1}^{t-1} \sum_{c=1}^{k} \sum_{\substack{j=1 \\ \exists x_{ij}}}^{d} \phi_{ijc} \exp(\eta_{hc}) \cdot [i]_1^n, [j]_1^d, [c]_1^k. \quad (16.33)$$

For a specific model, such as DMNB-Gaussian, the updating equation for ϕ_{ijc} could be obtained by replacing the corresponding distributions in place of $p_{\Psi_j}(x_{ij}|\theta_{jc})$ in (16.31). The form of the updates for γ_{ic} and ξ_i is independent of the exponential family being used.

16.5.2   Parameter Estimation

The goal of parameter estimation is to obtain (α, Λ, η) such that $\log p(\mathcal{X}, \mathcal{Y}|\alpha, \Lambda, \eta)$ is maximized. Since the log-likelihood is intractable, one can use the lower bound as a surrogate objective to be maximized. Note that for a fixed value of the variational parameters $(\gamma_i^*, \phi_i^*, \xi_i^*)$ for each $(\mathbf{x}_i, y_i)$, the lower bound of $\log p(\mathcal{X}, \mathcal{Y}|\alpha, \Lambda, \eta)$, i.e., $\sum_{i=1}^{n} L(\gamma_i^*, \phi_i^*, \xi_i^*; \alpha, \Lambda, \eta)$, is a function of the parameters (α, Λ, η). Maximizing $\sum_{i=1}^{n} L(\gamma_i^*, \phi_i^*, \xi_i^*; \alpha, \Lambda, \eta)$ with respect to (α, Λ, η) yields parameter estimate.

The update of α is independent of the specific model. Using the Newton-Raphson algorithm (Blei et al., 2003; Minka, 2003b) with line search, the updating equation is:

$$\alpha_c' = \alpha_c - \nu \frac{g_c - u}{h_c} \,, \ [c]_1^k \,, \tag{16.34}$$

where

$$
\begin{aligned}
g_c &= n\left(\Psi\left(\sum_{l=1}^{k} \alpha_l\right) - \Psi(\alpha_c) \right) + \sum_{i=1}^{n} \left(\Psi(\gamma_{ic}) - \Psi\left(\sum_{l=1}^{k} \gamma_{il}\right) \right) \\
h_c &= -n\Psi'(\alpha_c) \\
u &= \frac{\sum_{l=1}^{k} g_l/h_l}{w^{-1} + \sum_{l=1}^{k} h_l^{-1}} \\
w &= n\Psi'(\sum_{l=1}^{k} \alpha_l) \,.
\end{aligned}
$$

Since α has the constraint of $\alpha_c > 0$, by multiplying the second term of (16.34) by ν, we are performing a line search to prevent α_c to go out of the feasible range. At the beginning of each iteration, ν is set to 1. If the updated α_c falls into the feasible range, the algorithm goes on to the next iteration, otherwise, it reduces α by a factor of 0.5 until the updated α_c becomes valid.

For other model parameters, the update for η is given by

$$\eta_{hc} = \log \frac{\sum_{i=1}^{n} \sum_{j=1}^{m_i} y_{ih}\phi_{ijc}/m_i}{\sum_{i=1}^{n} \sum_{j=1}^{m_i} \phi_{ijc}/(m_i\xi_i)} \,, \ [c]_1^k \,, \ [h]_1^{t-1} .$$

The update equation for Λ is model dependent. For DLDA, the update equation for Λ, i.e., β, is given as follows:

$$\beta_{cv} \propto \sum_{i=1}^{n} \sum_{j=1}^{m_i} \phi_{ijc} x_{ij}^v \,, \ [c]_1^k \,, \ [v]_1^V \,. \tag{16.35}$$

For DMNB, following Redner and Walker (1984) and Banerjee et al. (2005b), the parameters Λ, i.e., Θ, can be estimated in a closed form for all exponential family distributions. From the Bregman divergence perspective, let τ_{jc} be the expectation parameter for the jth feature of the cth component, the estimation for τ_{jc} is given by

$$\tau_{jc} = \frac{\sum_{i=1, \exists x_{ij}}^{n} \phi_{ijc} s_{ij}}{\sum_{i=1, \exists x_{ij}}^{n} \phi_{ijc}} \,, \ [j]_1^d \,, \ [c]_1^k \,, \tag{16.36}$$

where s_{ij} is the sufficient statistic. The natural parameter θ_{jc} is given by conjugacy as

$$\theta_{jc} = \nabla f_j(\tau_{jc}) \,, \ [j]_1^d \,, \ [c]_1^k \,,$$

where $f_j(\cdot)$ is the conjugate of cumulant function ψ_j for each feature. We now give the parameter estimation for two special cases: DMNB-Gaussian and DMNB-Discrete.

DMNB-Gaussian: For Gaussians, by maximizing the lower bound, the exact update equations for μ_{jc} and σ_{jc}^2 can be obtained as

$$\mu_{jc} = \frac{\sum_{i=1,\exists x_{ij}}^n \phi_{ijc} x_{ij}}{\sum_{i=1,\exists x_{ij}}^n \phi_{ijc}}, \tag{16.37}$$

$$\sigma_{jc}^2 = \frac{\sum_{i=1,\exists x_{ij}}^n \phi_{ijc} (x_{ij} - \mu_{jc})^2}{\sum_{i=1,\exists x_{ij}}^n \phi_{ijc}}, \; [j]_1^d, \; [c]_1^k. \tag{16.38}$$

DMNB-Discrete: For a discrete distribution p_{jc} over $r = 1, \ldots, r_j$ values for feature j, the estimate of $p_{jc}(r)$ is given by

$$p_{jc}(r) \propto \sum_{i=1}^n \phi_{ijc} \mathbb{1}(x_{ij} = r), \; [j]_1^d, \; [r]_1^{r_j}, \tag{16.39}$$

where $\mathbb{1}(x_{ij} = r)$ is the indicator of observing value r for feature j in observation $\mathbf{x}_i$. While such a maximum likelihood (ML) estimate will give the maximizing parameters on an observed training set, there is the possibility of some probability estimates being zero. Such an eventuality does not pose a problem on the training set, but inference on unseen or test data may become problematic. If a feature in the test set takes a value that it has not taken in the entire training set, the model will assign a zero probability to the entire set of test observations. The standard approach to address the problem is to use smoothing, so that none of the estimated parameters is zero. In particular, we use Laplace smoothing, which results from a maximum a posteriori (MAP) estimate (DeGroot, 1970) assuming a Dirichlet prior over each discrete distribution, so that

$$p_{jc}(r) = \sum_{i=1}^n \phi_{ijc} \mathbb{1}(x_{ij} = r) + \epsilon, \; [c]_1^k, \; [j]_1^d, \; [r]_1^{r_j}, \tag{16.40}$$

for some $\epsilon > 0$.

16.5.3 Variational EM for DMNB

Based on the variational inference and parameter estimation updates, it is straightforward to construct a variational EM algorithm to estimate (α, Λ, η). Starting with an initial guess $(\alpha^{(0)}, \Lambda^{(0)}, \eta^{(0)})$, the variational EM algorithm alternates between two steps:

1. E-Step: Given $(\alpha^{(t-1)}, \Lambda^{(t-1)}, \eta^{(t-1)})$, for each data point $\mathbf{x}_i$, find the optimal variational parameters

$$(\gamma_i^{(t)}, \phi_i^{(t)}, \xi_i^{(t)}) = \underset{(\gamma_i, \phi_i, \xi_i)}{\arg\max} \, L(\gamma_i, \phi_i, \xi_i; \alpha^{(t-1)}, \Lambda^{(t-1)}, \eta^{(t-1)}).$$

$L(\gamma_i^{(t)}, \phi_i^{(t)}, \xi_i^{(t)}; \alpha, \Lambda, \eta)$ gives a lower bound to $\log p(\mathbf{x}_i, y_i | \alpha, \Lambda, \eta)$.

2. M-Step: An improved estimate of model parameters (α, Λ, η) are obtained by maximizing the aggregate lower bound:

$$(\alpha^{(t)}, \Lambda^{(t)}, \eta^{(t)}) = \underset{(\alpha, \Lambda, \eta)}{\arg\max} \sum_{i=1}^n L(\gamma_i^{(t)}, \phi_i^{(t)}, \xi_i^{(t)}; \alpha, \Lambda, \eta).$$

After t iterations, the objective function becomes $L(\gamma_i^{(t)}, \phi_i^{(t)}, \xi_i^{(t)}; \alpha^{(t)}, \Lambda^{(t)}, \eta^{(t)})$. In $(t+1)^{th}$ iterations, we have

$$\sum_{i=1}^{n} L(\gamma_i^{(t)}, \phi_i^{(t)}, \xi_i^{(t)}; \alpha^{(t)}, \Lambda^{(t)}, \eta^{(t)})$$

$$\leq \sum_{i=1}^{n} L(\gamma_i^{(t+1)}, \phi_i^{(t+1)}, \xi_i^{(t+1)}; \alpha^{(t)}, \Lambda^{(t)}, \eta^{(t)})$$

$$\leq \sum_{i=1}^{n} L(\gamma_i^{(t+1)}, \phi_i^{(t+1)}, \xi_i^{(t+1)}; \alpha^{(t+1)}, \Lambda^{(t+1)}, \eta^{(t+1)}) .$$

The first inequality holds because in the E-step, $(\gamma_i^{(t+1)}, \phi_i^{(t+1)}, \xi_i^{(t+1)})$ maximizes $L(\gamma_i, \phi_i, \xi_i; \alpha^{(t)}, \Lambda^{(t)}, \eta^{(t)})$. The second inequality holds because in the M-step, $(\alpha^{(t+1)}, \Lambda^{(t+1)}, \eta^{(t+1)})$ maximizes $L(\gamma_i^{(t+1)}, \phi_i^{(t+1)}, \xi_i^{(t+1)}; \alpha, \Lambda, \eta)$. Therefore, the objective function is non-decreasing until convergence.

16.6 Fast Variational Inference

The variational distribution in Section 16.5 exactly follows the idea proposed for latent Dirichlet allocation (LDA) (Blei et al., 2003), where every feature j of the data point $\mathbf{x}_i$ has a variational parameter ϕ_{ij} for the corresponding discrete distribution. This section introduces a different variational distribution with a smaller number of parameters, yielding a much faster variational inference algorithm. The fast variational inference is used for both DLDA and DMNB.

16.6.1 Variational Approximation

Given the lower bound to log-likelihood of each data point as (16.11), the variational distributions in (16.9) and (16.10) assign a separate discrete distribution ϕ_{ij} to each x_{ij} of the data point $\mathbf{x}_i$. The total number of ϕ_{ij} needed is hence $\sum_{i=1}^{n} m_i$, which is a huge number for high-dimensional data. Meanwhile, since in the E-step of the EM algorithm the optimization is performed over each variational parameter, a large number of variational parameters will lead to a large number of optimizations to perform, significantly slowing the algorithm down. To make the algorithm more efficient, a new family of variational distributions q_2 are introduced (Figure 16.2(b)). In particular, for DLDA,

$$q_2(\pi, \mathbf{z}|\phi, \gamma) = q_2(\pi|\gamma) \prod_{j=1}^{m_i} q_2(z_j|\phi) , \tag{16.41}$$

and for DMNB,

$$q_2(\pi, \mathbf{z}|\phi, \gamma) = q_2(\pi|\gamma) \prod_{\substack{j=1 \\ \exists x_j}}^{d} q_2(z_j|\phi) , \tag{16.42}$$

where γ and ϕ are k-dimensional variational parameters for Dirichlet and discrete distributions, respectively. Compared to $q_1(\pi, \mathbf{z}|\phi, \gamma)$ in (16.9) and (16.10), $q_2(\pi, \mathbf{z}|\phi, \gamma)$ only has one ϕ for each data point. The total number of ϕs needed hence decreases from $\sum_{i=1}^{n} m_i$ in (16.9) and (16.10), to n in (16.41) and (16.42). Accordingly, the number of optimizations over ϕ also decreases from $\sum_{i=1}^{n} m_i$ to n. Such a reduction implies a big saving on space and time, especially for high-dimensional data.

Given the variational distribution q_2, one could have a lower-bound function to the log-likelihood of each data point similar as (16.12):

$$L(\gamma_i, \phi_i, \xi_i; \alpha, \Lambda, \eta) = E_{q_2}[\log p(\pi_i|\alpha)] + E_{q_2}[\log p(\mathbf{z}_i|\pi_i)] + E_{q_2}[\log p(\mathbf{x}_i|\mathbf{z}_i, \Lambda)]$$
$$- E_{q_2}[\log q_2(\pi_i|\gamma_i)] - E_{q_2}[\log q_2(\mathbf{z}_i|\phi_i)] + E_{q_2}[\log p(y_i|\mathbf{z}_i, \eta)]. \quad (16.43)$$

For DLDA, the six terms in (16.43) are given as follows:

$$E_{q_2}[\log p(\pi_i|\alpha)] = \log \Gamma\left(\sum_{c=1}^{k} \alpha_c\right) - \sum_{c=1}^{k} \log \Gamma(\alpha_c)$$
$$+ \sum_{c=1}^{k} (\alpha_c - 1)\left(\Psi(\gamma_{ic}) - \Psi\left(\sum_{l=1}^{k} \gamma_{il}\right)\right), \quad (16.44)$$

$$E_{q_2}[\log p(\mathbf{z}_i|\pi_i)] = m_i \sum_{c=1}^{k} \phi_{ic}\left(\Psi(\gamma_{ic}) - \Psi\left(\sum_{l=1}^{k} \gamma_{il}\right)\right), \quad (16.45)$$

$$E_{q_2}[\log p(\mathbf{x}_i|\beta, \mathbf{z}_i)] = \sum_{j=1}^{m_i} \sum_{c=1}^{k} \sum_{v=1}^{V} \phi_{ic} x_{ij}^v \log \beta_{cv}, \quad (16.46)$$

$$E_{q_2}[\log q_2(\pi_i|\gamma_i)] = \log \Gamma\left(\sum_{c=1}^{k} \gamma_{ic}\right) - \sum_{c=1}^{k} \log \Gamma(\gamma_{ic})$$
$$+ \sum_{c=1}^{k} (\gamma_{ic} - 1)\left(\Psi(\gamma_{ic}) - \Psi\left(\sum_{l=1}^{k} \gamma_{il}\right)\right), \quad (16.47)$$

$$E_{q_2}[\log q_2(\mathbf{z}_i|\phi_i)] = m_i \sum_{c=1}^{k} \phi_{ic} \log \phi_{ic}, \quad (16.48)$$

$$E_{q_1}[\log p(y_i|\mathbf{z}_i, \eta)] \geq \frac{1}{m_i} \sum_{c=1}^{k} \phi_{ic} \sum_{j=1}^{m_i} \sum_{h=1}^{t-1} \left(\eta_{hc} y_{ih} - \frac{1}{\xi_i} \exp(\eta_{hc})\right)$$
$$+ 1 - \frac{1}{\xi_i} - \log(\xi_i). \quad (16.49)$$

For DMNB, the six terms in (16.43) are the same as in (16.44)–(16.49), except for (16.46), which is given by

$$E_{q_2}[\log p(\mathbf{x}_i|\mathbf{z}_i, \Theta)] = \sum_{\substack{j=1 \\ \exists x_{ij}}}^{d} \sum_{c=1}^{k} \phi_{ic} \log p_{\psi_j}(x_{ij}|\theta_{jc}) .$$

Maximizing $L(\gamma_i, \phi_i, \xi_i; \alpha, \Lambda, \eta)$ with respect to variational parameters yields the best lower bound. In particular, in DLDA,

$$\phi_{ic} \propto \exp\left(\Psi(\gamma_{ic}) - \Psi\left(\sum_{l=1}^{k} \gamma_{il}\right) + \frac{1}{m_i} \sum_{j=1}^{m_i} \sum_{v=1}^{V} x_{ij}^v \log \beta_{cv}\right), \quad (16.50)$$

$$\gamma_{ic} = \alpha_c + m_i \phi_{ic}, \quad (16.51)$$

$$\xi_i = 1 + \sum_{h=1}^{t-1} \sum_{c=1}^{k} \phi_{ic} \exp(\eta_{hc}) , \; [i]_1^n, \; [c]_1^k . \quad (16.52)$$

In DMNB,

$$\phi_{ic} \propto \exp\left(\Psi(\gamma_{ic}) - \Psi\left(\sum_{l=1}^{k}\gamma_{il}\right) + \frac{1}{m_i}\sum_{\substack{j=1\\\exists x_j}}^{d}\log p_{\Psi_j}(x_{ij}|\theta_{jc})\right.$$

$$\left. + \frac{1}{m_i}\sum_{h=1}^{t-1}(\eta_{hc}y_{ih} - \exp(\eta_{hc})/\xi_i)\right), \tag{16.53}$$

$$\gamma_{ic} = \alpha_c + m_i\phi_{ic}, \tag{16.54}$$

$$\xi_i = 1 + \sum_{h=1}^{t-1}\sum_{c=1}^{k}\phi_{ic}\exp(\eta_{hc}) \,, \ [i]_1^n, \ [c]_1^k \,. \tag{16.55}$$

Again, for a specific model of DMNB, such as DMNB-Gaussian, the updating equation for ϕ_{ic} could be obtained by replacing the corresponding distributions in place of $p_{\Psi_j}(x_{ij}|\theta_{jc})$ in (16.53). The form of the updates for γ_{ic} and ξ_i is independent of the exponential family being used.

16.6.2   Parameter Estimation

After obtaining the variational parameters, one can obtain a tractable lower bound of the log-likelihood as a function of the model parameters (α, Λ, η). The estimation for α is the same as in Section 16.5.2 using the Newton-Raphson algorithm with line search. The estimation for η is given by

$$\eta_{hc} = \log\frac{\sum_{i=1}^{n}y_{ih}\phi_{ic}}{\sum_{i=1}^{n}\phi_{ic}/\xi_i} \,, \ [c]_1^k \,, \ [h]_1^{t-1},$$

for both DLDA and DMNB.

For the estimation of Λ in DLDA, the update equation of β is given by

$$\beta_{cv} \propto \sum_{i=1}^{n}\left(\phi_{ic}\sum_{j=1}^{m_i}x_{ij}^v\right) \,, \ [c]_1^k \,, \ [v]_1^V \,. \tag{16.56}$$

For the estimation of Θ in DMNB, from a Bregman divergence perspective, assuming the expectation parameter for the jth feature of component c is τ_{jc}, the estimation for τ_{jc} is given by

$$\tau_{jc} = \frac{\sum_{i=1,\exists x_{ij}}^{n}\phi_{ic}s_{ij}}{\sum_{i=1,\exists x_{ij}}^{n}\phi_{ic}} \,, \ [j]_1^d \,, \ [c]_1^k \,, \tag{16.57}$$

where s_{ij} is the sufficient statistic and the natural parameter $\theta_{jc} = \nabla f_j(\tau_{jc})$ by conjugacy, given $f_j(\cdot)$ the conjugate of cumulant function ψ_j for each feature. For two special cases, DMNB-Gaussian and DMNB-Discrete, the closed-form parameter estimates are given below. Note that (16.57)–(16.60) are mild variants of (16.36)–(16.39), as ϕ_{ic} does not depend on feature j.

Fast DMNB-Gaussian: For Gaussians, the update equations for μ_{jc} and σ_{jc}^2 are given by

$$\mu_{jc} = \frac{\sum_{i=1,\exists x_{ij}}^{n}\phi_{ic}x_{ij}}{\sum_{i=1,\exists x_{ij}}^{n}\phi_{ic}}, \tag{16.58}$$

$$\sigma_{jc}^2 = \frac{\sum_{i=1,\exists x_{ij}}^{n}\phi_{ic}(x_{ij} - \mu_{jc})^2}{\sum_{i=1,\exists x_{ij}}^{n}\phi_{ic}} \,, \ [c]_1^k \,, \ [j]_1^d. \tag{16.59}$$

Fast DMNB-Discrete: For a discrete distribution p_{jc} over $r = 1,\ldots,r_j$ values for feature j, the update equation for $p_{jc}(r)$ is given by

$$p_{jc}(r) \propto \sum_{i=1}^{n} \phi_{ic}\mathbb{1}(x_{ij} = r) + \epsilon \,, \; [c]_1^k \,, \; [j]_1^d \,, \; [r]_1^{r_j}, \tag{16.60}$$

where $\mathbb{1}(x_{ij} = r)$ is the indicator of observing value r for feature j in observation $\mathbf{x}_i$.

Given the updates for variational and model parameters, a variational EM algorithm could be constructed as in Section 16.5.3.

16.7 Experimental Results

In this section, experimental results for discriminative mixed membership models are presented. For simplicity, DMM models with the regular variational inference algorithm following Blei et al. (2003) are referred to as "Standard DMM" (Std DMM), which includes Standard DLDA (Std DLDA) and Standard DMNB (Std DMNB). DMM models with the fast variational inference algorithm are referred to as "Fast DMM," which includes Fast DLDA and Fast DMNB. An overview of all DMM models is given in Figure 16.3. In this section, first, DMM models are compared to their unsupervised counterpart—mixed membership models. Second, Fast DMM models are compared to standard DMM models. Further, Fast DMM are compared to several state-of-the-art classification algorithms. Finally, several word lists of topics generated from Fast DLDA are presented. The experiments are performed using 10-fold cross-validation. In particular, the dataset is divided evenly into ten parts, one of which is picked as the test set, and the remaining nine parts are used as the training set. The process is repeated ten times, with each part used once as the test set. The mean and standard deviation of the results on test sets over 10 folds are presented.

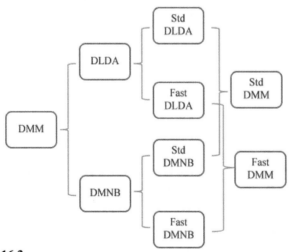

FIGURE 16.3
An overview of DMM models.

16.7.1 Datasets

Seven datasets from the UCI machine learning repository[1] are used for the experiments of DMNB. These datasets are represented as real-valued full matrices without missing entries. The numbers of data points, features, and classes in each dataset are listed in Table 16.1.

TABLE 16.1
The number of data points, features, and classes in each UCI dataset.

Dataset	Ecoli	Glass	Iono	Seg	Sona	Wdbc	Wine
Data points	336	214	351	2310	208	569	178
Features	7	9	32	19	60	30	13
Classes	8	6	2	7	2	2	3

Five text datasets are used for the experiments of DLDA. The details of the datasets are as follows:

1. **Nasa:** Nasa is a text dataset from the Aviation Safety Reporting System (ASRS) online database.[2] This database contains aviation safety reports submitted by pilots, controllers, and others. The dataset used is a subset of the whole database. It contains 4,226 documents about the anomalies originating from three sources: flight crew, maintenance, and passengers. The vocabulary size is 604.

2. **Classic3:** Classic3 (Dhillon et al., 2003) is a well-known text dataset. It contains 3,893 documents from three different classes including aeronautics, medicine, and information retrieval. The vocabulary size is 5,923.

3. **CMU Newsgroup:** The CMU Newsgroup is also a benchmark text dataset (Lang, 1995). The standard dataset of CMU Newsgroup contains 19,997 messages, collected from 20 different USENET newsgroups. Three subsets are used for the experiments: (1) **Diff** is a collection of 3000 messages from 3 different newsgroups with 1000 messages for each class: alt.atheism, rec.sport.baseball, and sci.space. The vocabulary size is 7,666. (2) **Sim** is a collection of 3000 messages from 3 somewhat similar newsgroups with 1000 messages for each class: talk.politics.guns, talk.politics.mideast, and talk.politics.misc. The vocabulary size is 10,083. (3) **Same** is a collection of 3000 messages from 3 very similar newsgroups with 1000 messages for each class: comp.graphics, comp.os.ms-windows, and comp.windows.x. The vocabulary size is 5,932.

16.7.2 DMM vs. MM

We first compare DMM models to corresponding MM models. In particular, we compare DLDA with LDA, and compare DMNB with MNB. Both the regular and fast variational inference are used for each model. In principle, MM models are not used for classification, but given the initialization we will introduce below, there is a one-to-one mapping between the component and the class; hence, we can measure the accuracy.

For initialization, the model parameters are initialized using all data points and their labels in the training set, in particular, the number of components k is set to be the number of classes t; the mean and standard deviation (for Gaussian case only) of the data points in each class are used to initialize Λ; and n_h/n are used to initialize each dimension of α, where n_h is the number of data points in class h and n is the total number of data points. η in DMM is set by cross validation. In particular, each η_h of $[h]_1^{t-1}$ in η takes value of ru_h, where u_h is a unit vector with the hth dimension being 1 and others being 0, and r takes values from 0 to 100 in steps of 10. The value of r which gives the best results on a validation set is used to set up η.

---

[1] See http://archive.ics.uci.edu/ml/.
[2] See http://akama.arc.nasa.gov/ASRSDBOnline/QueryWizard_Begin.aspx.

The results for DLDA and DMNB are presented in Tables 16.2 and 16.3, respectively. Comparing DMM models with the corresponding MM counterparts, one can see that while Std DMM models are not necessarily better than Std MM models, Fast DMM models are almost always better than Fast MM models. Overall, Fast DMM models achieve the highest accuracy among four algorithms. The higher accuracy of Fast DMM demonstrates the effects of logistic regression in accommodating label information for DMM models.

TABLE 16.2
Accuracy for LDA and DLDA ($k=t$). Fast DLDA has a higher accuracy on all datasets.

	Nasa	Classic3	Diff	Sim	Same
Std LDA	0.9140	0.6733	0.9677	0.8143	0.5633
	±0.0140	±0.0254	±0.0069	± 0.0161	±0.0243
Std DLDA	0.9220	0.6710	0.9600	0.8140	0.6267
	±0.0127	±0.0256	±0.0089	±0.0252	±0.0348
Fast LDA	0.9194	0.6748	0.9773	0.8553	0.7730
	±0.0148	± 0.0242	± 0.0110	±0.0197	±0.0205
Fast DLDA	**0.9237**	**0.6756**	**0.9800**	**0.8653**	**0.7900**
	±0.0163	**±0.0234**	**±0.0102**	**±0.0182**	**±0.0315**

TABLE 16.3
Accuracy for MNB and DMNB ($k=t$). Fast DMNB has a higher accuracy on most of the datasets.

	Ecoli	Glass	Iono	Seg	Sona	Wdbc	Wine
Std MNB	0.7895	**0.6190**	0.6829	0.6514	0.6300	0.9321	0.9606
	±0.0629	**±0.1052**	±0.0579	±0.0293	±0.0789	±0.0351	±0.0500
Std DMNB	0.7788	0.6048	0.7314	0.6398	0.6102	**0.9397**	0.9647
	±0.0554	±0.1231	±0.0895	±0.0397	±0.0822	**±0.0378**	±0.0411
Fast MNB	0.7950	0.5952	0.7486	0.6333	0.6100	0.9089	0.9470
	±0.0595	±0.0645	±0.0643	±0.0676	±0.0516	±0.0309	±0.0647
Fast DMNB	**0.8152**	0.5238	**0.8507**	**0.6701**	**0.6600**	0.9286	**0.9765**
	±0.0862	±0.1209	**±0.0891**	**±0.0487**	**±0.0876**	±0.0253	**±0.0304**

16.7.3   Fast DMM vs. Std DMM

From Tables 16.2 and 16.3, comparing Fast DMM with Std DMM, one can see that Fast DLDA has a higher accuracy than Std DLDA, and Fast DMNB generally also has a higher accuracy than Std DMNB, with only one exception.

One can also compare the running time between Std DMM and Fast DMM. The results for DLDA and DMNB are presented in Tables 16.4 and 16.5, respectively. In Table 16.5, although most of the datasets are small, Fast DMNB is already faster than Std DMNB. Fast DMM's advantage increases when it comes to the larger and higher-dimensional text data as in Table 16.4, where Fast DLDA is about 20 to 150 times faster than Std DLDA, showing Fast DMM models' significant superiority in terms of time efficiency. Therefore, Fast DMM models are generally more accurate and substantially faster than Std DMM models.

TABLE 16.4
Running time (seconds) of Std DLDA and Fast DLDA. Fast DLDA is computationally more efficient than Std DLDA.

	Nasa	Classic3	Diff	Sim	Same
Dimension	604	5923	7666	10083	5932
Std DLDA	549.17 ±5.74	2176.67 ±21.62	1752.78 ±22.36	2344.64 ±966.50	1981.46 ±289.24
Fast DLDA	**3.63** ±**0.21**	**114.34** ±**18.13**	**27.56** ±**0.61**	**36.10** ±**2.98**	**40.18** ±**5.83**
Speedup times	151	19	64	65	49

TABLE 16.5
Running time (seconds) of Std DMNB and Fast DMNB. Fast DMNB is computationally more efficient than Std DMNB.

	Ecoli	Glass	Iono	Seg	Sona	Wdbc	Wine
Dimension	7	9	32	19	60	30	13
Std DMNB	4.65 ±1.13	2.76 ±0.49	5.20 ±3.11	120.26 ±77.27	4.89 ±4.51	3.33 ±0.40	2.26 ±0.25
Fast DMNB	**3.97** ±**0.39**	**2.21** ±**0.21**	**0.82** ±**0.01**	**25.37** ±**6.32**	**1.03** ±**0.07**	**1.91** ±**0.11**	**1.10** ±**0.04**
Speedup times	1.17	1.25	6.34	4.74	4.75	1.74	2.05

We further investigate the cluster assignments of Fast DMM. The cluster membership of each data point could be considered as its probability belonging to different clusters. If one calculates the Shannon entropy of the cluster membership, a high entropy indicates a real mixed membership assignment, while a low entropy implies almost a sole membership. Figure 16.4 shows the histograms of cluster membership entropy for Std DLDA and Fast DLDA on Sim, and for Std DMNB and Fast DMNB on glass, where each bar denotes the number of data points falling into that range of entropy. While most data points from Std DMM have a large entropy over different ranges, the data points from Fast DMM mostly have a small entropy. The interesting observation indicates that fast variational inference actually generates somewhat sole membership while the regular variational inference generates real mixed membership. Such observation gives one possible explanation for Fast DMM's better classification performance than Std DMM: In a (single-label) classification scenario, each data point only belongs to one class; hence, the sole membership from Fast DMM would probably be more appropriate than the mixed membership.

One possible reason for the sole membership from fast variational inference is as follows: In the E-step, DMNB iterates through (16.31) and (16.32) to update ϕ and γ, while Fast MNB iterates through (16.53) and (16.54). The expression for γ in (16.54) contains the summation of ϕ_j over all features j. Since each ϕ_j may take different values, in the sense that each ϕ_j may peak at different components, the summation of ϕ_j may have several peaks on different components. Accordingly, γ will also have several peaks, leading to a mixed membership over those peaked components. In comparison, the expression for γ in (16.54) has a term of $m\phi$ instead, so no matter which component ϕ peaks at, the peak will be greatly enhanced in γ, and such enhancement in γ will further increase the "sole membership" nature of ϕ through the term $\exp\left(\Psi(\gamma_{ic}) - \Psi\left(\sum_{l=1}^{k} \gamma_{il}\right)\right)$ in (16.53). By

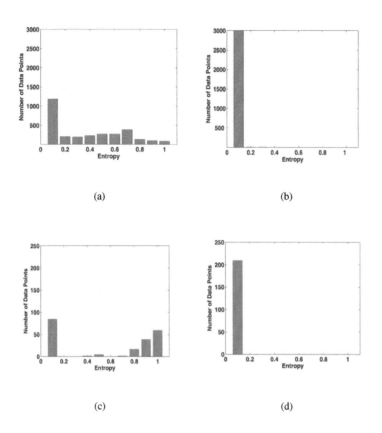

FIGURE 16.4
Histogram of cluster membership entropy on glass for Std DMNB and Fast DMNB. (a) is for Std DMNB and (b) is for Fast DMNB. Fast DMNB assigns most data points a "sole" membership, while Std DMNB assigns most data points a real mixed membership.

iterating through γ and ϕ, the accumulated enhancement finally leads to almost a sole membership on the peaked component.

16.7.4 Fast DMM vs. Other Classification Algorithms

Since Fast DMM models have better performance than Std DMM models, one can use Fast DMM to compare with other classification algorithms. In this chapter, Fast DMNB is compared with the support vector machine (SVM) (Chang and Lin, 2001), logistic regression (LR), and naive Bayes classifier (NBC) [2] on UCI data; and Fast DLDA is compared with SVM, NBC, LR, and a mixture of the von Mises-Fisher (vMF) model (Banerjee et al., 2005a) on text data. Since DMM is a combination of logistic regression and mixed membership models, it is also interesting to compare the results from DMM to the results from MM and logistic regression in two steps sequentially.

For Fast DMNB, the number of components k is set to be $(t, t+5, t+10)$, and for Fast DLDA, k is set to be $(t, t+15, t+30, t+50, t+100)$. The initialization of Λ is based on the mean and standard deviation (for Gaussian case only) of the training data in given classes plus some

---

[2]Note that naive Bayes used in this subsection is the classifier instead of the clustering algorithm.

perturbation if $k > t$. α is set to be $1/k$ on each dimension, and η is also from cross validation as in Section 16.7.2.

The results for Fast DLDA and DMNB are presented in Tables 16.6 and 16.7. The top parts of the tables are the results from the generative models, and the bottom parts are the results from discriminative classification algorithms. Bold is used for the best results among the generative models, and bold and italic are used for the best results among all algorithms. Three parts of information could be read from the tables:

1. Overall, on text datasets, Fast DLDA does better than all other algorithms, including SVM, on almost all datasets, which is a promising result, although more rigorous experiments are needed for further investigation; on UCI datasets, Fast DMNB also achieves higher accuracy than other algorithms on most of the datasets except SVM, which outperforms Fast DMNB five out of nine times.

2. The better performance of Fast DMM models compared to LR on original datasets indicates that the low-dimensional representation from Fast DMM helps the classification.

3. Interestingly, for Fast DMNB, the accuracy increases monotonically with k from t to $t + 10$ on most of the datasets. For Fast DLDA on text data, an increase of accuracy with a larger k is also observed, although the result goes up and down without a clear trend. One possible reason for the increasing accuracy is as follows: When k is too small, it is performing a drastic dimension

TABLE 16.6
Accuracy of Fast DLDA and other classification algorithms on text data. Fast DLDA has higher accuracy on most datasets.

	Nasa	Classic3	Diff	Sim	Same
Fast DLDA ($k=t$)	0.9237 ±0.0163	0.6756 ±0.0234	0.9800 ±0.0102	0.8653 ±0.0182	0.7900 ±0.0315
Fast DLDA ($k=t+15$)	0.9232 ±0.0144	0.6858 ±0.0216	0.9747 ±0.0121	0.8713 ±0.0264	0.8458 ±0.0214
Fast DLDA ($k=t+30$)	0.9301 ±0.0128	0.6838 ±0.0234	0.9817 ±0.0099	0.8707 ±0.0228	*0.8468* ±*0.0190*
Fast DLDA ($k=t+50$)	0.9237 ±0.0138	0.6854 ±0.0211	*0.9823* ±*0.0083*	0.8700 ±0.0230	0.8150 ±0.0184
Fast DLDA ($k=t+100$)	0.9261 ±0.0102	*0.6866* ±*0.0245*	0.9760 ±0.0108	*0.8718* ±*0.0182*	0.8347 ±0.0187
vMF	0.9216 ±0.0113	0.6509 ±0.0246	0.95301 ±0.0071	0.7447 ±0.0214	0.7600 ±0.0347
NBC	*0.9334* ±*0.0094*	0.6766 ±0.0230	0.9813 ±0.0069	0.8613 ±0.0216	0.8410 ±0.0262
LR	0.9209 ±0.0157	0.6396 ±0.0252	0.9553 ±0.0157	0.6750 ±0.1330	0.4823 ±0.1283
SVM	0.9192 ±0.0146	0.6854 ±0.0278	0.9563 ±0.0105	0.8357 ±0.0156	0.8120 ±0.203

reduction to represent each data point in a k-dimensional mixed membership representation, which may cause a huge loss of information, but the loss may decrease when k increases.

TABLE 16.7
Accuracy of Fast DMNB and other classification algorithms on UCI data. Fast DMNB has a higher accuracy, except for SVM.

	Ecoli	Glass	Iono	Seg	Sona	Wdbc	Wine
Fast DMNB ($k=t$)	0.8152 ±0.0862	0.5238 ±0.1209	0.8507 ±0.0891	0.6701 ±0.0487	0.6600 ±0.0876	0.9286 ±0.0253	0.9765 ±0.0304
Fast DMNB ($k=t+5$)	0.8392 ±0.0836	0.5248 ±0.0643	0.8543 ±0.0908	0.7632 ±0.0412	0.8100 ±0.0907	*0.9393* ±*0.0388*	*0.9882* ±*0.0284*
Fast DMNB ($k=t+10$)	*0.8485* ±*0.0515*	*0.5667* ±*0.1015*	*0.8943* ±*0.0786*	*0.7684* ±*0.0418*	*0.8200* ±*0.1509*	0.9375 ±0.0329	0.9765 ±0.0411
NBC	0.8363 ±0.0745	0.4333 ±0.1318	0.8114 ±0.0853	0.6850 ±0.0625	0.7268 ±0.0079	0.9339 ±0.0266	0.9705 ±0.0310
LR	0.8030 ±0.0610	0.5109 ±0.1234	0.8400 ±0.0276	0.8307 ±0.0358	0.7500 ±0.0816	0.9429 ±0.0250	0.7471 ±0.1469
SVM	0.8349 ±0.0670	0.4676 ±0.0875	*0.9171* ±*0.0594*	*0.9745* ±*0.0096*	0.7450 ±0.0896	*0.9536* ±*0.0173*	0.9765 ±0.0304

Fast DMM models do dimensionality reduction and classification in one step via a combination of Fast MM and logistic regression. In principle, one can use these two algorithms sequentially in two steps, i.e., first use Fast MM models to get a low-dimensional representation, and then apply logistic regression on the low-dimensional representation for classification. The results for these two strategies are presented in Figure 16.5. It is clear that Fast DMM models outperform the Fast MM+LR strategy. Therefore, by combining Fast MM and logistic regression together, Fast DMM achieves supervised dimensionality reduction to obtain a better low-dimensional representation than Fast MM, which helps classification.

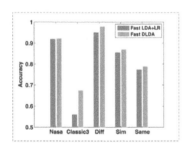 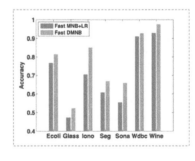

(a) (b)

FIGURE 16.5
Comparison between using Fast MM+LR and Fast DMM. (a) is for Fast DLDA on text data, and (b) is for Fast DMNB on UCI data. Fast DMM achieves higher accuracy, indicating the advantage of supervised dimension reduction.

16.7.5   Topics from Fast DLDA

As mentioned before, DMM models generate interpretable results. An example of several topic word lists on Nasa generated by Fast DLDA ($k = t + 30$) is given in Table 16.8. It is also an interesting result demonstrating the effect of allowing a larger number of components than the number of classes ($k > t$), that is, Fast DLDA may discover topics which are not explicitly specified in class labels, while maintaining the predefined number of classes. The first three topics in Table 16.8 correspond to three classes in Nasa, respectively, but Topic 4, which we call "passenger medical emergency," could be considered as a subcategory of the "passenger" class, and it is not specified in the labels. Neither NBC nor SVM is able to generate this type of results.

TABLE 16.8
Extracted Topics from Nasa dataset using Fast DLDA.

Topic 1	Topic 2	Topic 3	Topic 4
runway	maintenance	passenger	passenger
aircraft	aircraft	flight	flight
approach	flight	attendant	medical
tower	minimum equipment list	told	attendant
cleared	time	captain	emergency
landing	check	seat	aircraft
airport	engine	asked	doctor
turn	mechanical	back	landing
taxi	installed	attendants	attendants
traffic	part	aircraft	captain
final	inspection	lavatory	oxygen
controller	work	crew	paramedics

16.8   Conclusion

In this chapter, we have discussed discriminative mixed membership models as a combination of unsupervised mixed membership models and multi-label logistic regression. We introduced a fast variational inference algorithm which is substantially faster than the mean field approximation used in LDA (Blei et al., 2003) and leads to even better classification performance. An important property of DMM models is that they allow the number of components k to be different from the number of classes c. Interestingly, a larger k helps to discover the components not specified in labels and increases classification accuracy. In addition, DMM models are competitive with the state-of-the-art classification algorithms in terms of their accuracy, especially on text data, and are able to generate interpretable results.

References

Airoldi, E. M., Blei, D. M., Fienberg, S. E., and Xing, E. P. (2008). Mixed membership stochastic blockmodels. *Journal of Machine Learning Research* 9: 1981–2014.

Banerjee, A. (2007). An analysis of logistic models: Exponential family connections and online performance. In *Proceedings of the 7th SIAM International Conference on Data Mining (SDM '07)*. Minneapolis, MN, USA: SIAM.

Banerjee, A., Dhillon, I., Ghosh, J., and Merugu, S. (2004). An information theoretic analysis of maximum likelihood mixture estimation for exponential families. In *Proceedings of the 21ˢᵗ International Conference on Machine Learning (ICML '04)*. New York, NY, USA: ACM, 8.

Banerjee, A., Dhillon, I., Ghosh, J., and Sra, S. (2005a). Clustering on the unit hypersphere using von Mises-Fisher distributions. *Journal of Machine Learning Research* 6: 1345–1382.

Banerjee, A., Merugu, S., Dhillon, I., and Ghosh, J. (2005b). Clustering with Bregman divergences. *Journal of Machine Learning Research* 6: 1705–1749.

Barndorff-Nielsen, O. (1978). *Information and Exponential Families in Statistical Theory*. Chichester: John Wiley & Sons, Ltd.

Blei, D. M. and Jordan, M. I. (2006). Variational inference for Dirichlet process mixtures. *Bayesian Analysis* 1: 121–144.

Blei, D. M. and McAuliffe, J. (2007). Supervised topic models. In Platt, J. C., Koller, D., Singer, Y., and Roweis, S. (eds), *Advances in Neural Information Processing Systems 20*. Cambridge, MA: The MIT Press, 121–128.

Blei, D. M., Ng, A. Y., and Jordan, M. I. (2003). Latent Dirichlet allocation. *Journal of Machine Learning Research* 3: 993–1022.

Burges, C. (1998). A tutorial on support vector machines for pattern recognition. *Data Mining and Knowledge Discovery* 2: 121–167.

Chang, C. and Lin, C. -Y. (2001). LIBSVM: A library for support vector machines. Software available at www.csie.ntu.edu.tw/~cjlin/papers/libsvm.pdf.

DeGroot, M. (1970). *Optimal Statistical Decisions*. Hoboken, NJ: John Wiley & Sons.

Dhillon, I., Mallela, S., and Modha, D. (2003). Information-theoretic co-clustering. In *Proceedings of the 9ᵗʰ ACM SIGKDD International Conference on Knowledge Discovery and Data Mining (KDD '03)*. New York, NY, USA: ACM, 89–98.

Fei-Fei, L. and Perona, P. (2005). A Bayesian hierarchical model for learning natural scene categories. In *Proceedings of the 10ᵗʰ IEEE Conference on Computer Vision and Pattern Recognition (CVPR 2005)*. Los Alamitos, CA, USA: IEEE Computer Society, 524–531.

Flaherty, P., Giaever, G., Kumm, J., Jordan, M. I., and Arkin, A. (2005). A latent variable model for chemogenomic profiling. *Bioinformatics* 21: 3286–3293.

Geman, S. and Geman, D. (1984). Stochastic relaxation, Gibbs distributions, and the Bayesian restoration of images. *IEEE Transactions on Pattern Analysis and Machine Intelligence* 6: 721–741.

Griffiths, T. L. and Steyvers, M. (2004). Finding scientific topics. *Proceedings of the National Academy of Science (PNAS)* 101: 5228–5225.

Koutsourelakis, P. -S. and Eliassi-Rad, T. (2008). Finding mixed-memberships in social networks. In *AAAI Spring Symposium: Social Information Processing*. Palo Alto, CA, USA: AAAI, 48–53.

Lacoste-Julien, S., Sha, F., and Jordan, M. I. (2008). DiscLDA: Discriminative learning for dimensionality reduction and classification. In Koller, D., Schuurmans, D., Bengio, Y., and Bottou, L. (eds), *Advances in Neural Information Processing Systems 21*. Red Hook, NY: Curran Associates, Inc., 897–904.

Lang, K. (1995). News Weeder: Learning to filter netnews. In *Proceedings of the 12th International Conference on Machine Learning (ICML '95)*. San Francisco, CA, USA: Morgan Kaufmann Publishers Inc., 331–339.

McCullagh, P. and Nelder, J. A. (1989). *Generalized Linear Models*. Chapman & Hall/CRC, 2nd edition.

Mimno, D. and McCallum, A. (2008). Topic models conditioned on arbitrary features with Dirichlet-multinomial regression. In *Proceedings of the 24th Conference on Uncertainty in Artificial Intelligence (UAI 2008)*. Corvallis, OR, USA: AUAI Press, 411–418.

Minka, T. P. (2003a). A Comparison of Numerical Optimizers for Logistic Regression. Tech. report, Massachusetts Institute of Technology.

Minka, T. P. (2003b). Estimating a Dirichlet Distribution. Tech. report, Massachusetts Institute of Technology.

Neal, R. M. and Hinton, G. E. (1998). A view of the EM algorithm that justifies incremental, sparse, and other variants. In Jordan, M. I. (ed.), *Learning in Graphical Models*. Cambridge, MA: The MIT Press, 355–368.

Pampel, F. (2000). *Logistic Regression: A Primer*. Thousand Oaks, CA: Sage Publications.

Ramage, D., Manning, C. D., and Dumais, S. (2011). Partially labeled topic models for interpretable text mining. In *Proceedings of the 17th ACM SIGKDD Conference on Knowledge Discovery and Data Mining (KDD '11)*. New York, NY, USA: ACM, 457–465.

Redner, R. and Walker, H. (1984). Mixture densities, maximum likelihood and the EM algorithm. *SIAM Review* 26: 195–239.

Shan, H. and Banerjee, A. (2010). Mixed-membership naive Bayes models. *Data Mining and Knowledge Discovery* 23: 1–62.

Wainwright, M. and Jordan, M. I. (2008). Graphical models, exponential families, and variational inference. *Foundations and Trends in Machine Learning* 1: 1–305.

Wang, C. and Blei, D. M. (2011). Collaborative topic modeling for recommending scientific articles. In *Proceedings of the 17th ACM SIGKDD Conference on Knowledge Discovery and Data Mining (KDD '11)*. New York, NY, USA: ACM, 448–456.

Wang, C., Blei, D. M., and Fei-Fei, L. (2009). Simultaneous image classificationn and annotation. In *Proceedings of the 2009 IEEE Conference on Computer Vision and Pattern Recognition (CVPR 2009)*. Los Alamitos, CA, USA: IEEE Computer Society, 1903–1910.

Wang, H., Huang, M., and Zhu, X. (2008). A generative probabilistic model for multi-label classification. In *Proceedings of the 8th IEEE International Conference on Data Mining (ICDM 2008)*. Los Alamitos, CA, USA: IEEE Computer Society, 628–637.

17

Mixed Membership Matrix Factorization

Lester Mackey

Computer Science Division, University of California, Berkeley, CA 94720, USA

David Weiss

Computer and Information Science, University of Pennsylvania, Philadephia, PA 19104, USA

Michael I. Jordan

Computer Science Division and Department of Statistics, University of California, Berkeley, CA 94720, USA

CONTENTS

Discrete mixed membership modeling and continuous latent factor modeling (also known as matrix factorization) are two popular, complementary approaches to dyadic data analysis. In this chapter, we develop a fully Bayesian framework for integrating the two approaches into unified Mixed Membership Matrix Factorization (M^3F) models. We introduce two M^3F models, derive Gibbs sampling inference procedures, and validate our methods on the EachMovie, MovieLens, and Netflix Prize collaborative filtering datasets. We find that even when fitting fewer parameters, the M^3F models outperform state-of-the-art latent factor approaches on all benchmarks, yielding the greatest gains in accuracy on sparsely-rated, high-variance items.

17.1 Introduction

This chapter is concerned with unifying discrete mixed membership modeling and continuous latent factor modeling for probabilistic dyadic data prediction. The ideas contained herein are based on the work of Mackey et al. (2010). In the dyadic data prediction (DDP) problem (Hofmann et al., 1998), we observe labeled *dyads*, i.e., ordered pairs of objects, and form predictions for the labels of unseen dyads. For example, in the collaborative filtering setting we observe U users, M items, and a training set $\mathcal{T} = \{(u_n, j_n, r_n)\}_{n=1}^N$ with real-valued ratings r_n representing the preferences of certain users u_n for certain items j_n. The goal is then to predict unobserved ratings based on users' past preferences. Other concrete examples of DDP include link prediction in social network analysis, binding affinity prediction in bioinformatics, and click prediction in web searches.

Matrix factorization methods (Rennie and Srebro, 2005; DeCoste, 2006; Salakhutdinov and Mnih, 2007; 2008; Takács et al., 2009; Koren et al., 2009; Lawrence and Urtasun, 2009) represent the state of the art for dyadic data prediction tasks. These methods view a dyadic dataset as a sparsely observed ratings matrix, $R \in \mathbb{R}^{U \times M}$, and learn a constrained decomposition of that matrix as a product of two latent factor matrices: $R \approx A^\top B$ for $A \in \mathbb{R}^{D \times U}$, $B \in \mathbb{R}^{D \times M}$, and D small. While latent factor methods perform remarkably well on the DDP task, they fail to capture the heterogeneous nature of objects and their interactions. Such models, for instance, do not account for the fact that a user's ratings are influenced by instantaneous mood, that protein interactions are affected by transient functional contexts, or even that users with distinct behaviors may be sharing a single account or web browser.

The fundamental limitation of continuous latent factor methods is a result of the static way in which ratings are assumed to be produced: a user generates all of his item ratings using the same factor vector without regard for context. Discrete mixed membership models, such as latent Dirichlet allocation (Blei et al., 2003), were developed to address a similar limitation of mixture models. Whereas mixture models assume that each generated object is underlyingly a member of a single latent topic, mixed membership models represent objects as distributions over topics. Mixed membership dyadic data models such as the mixed membership stochastic blockmodel (Airoldi et al., 2008) for relational prediction and Bi-LDA (Porteous et al., 2008) for rating prediction introduce context dependence by allowing each object to select a new topic for each new interaction. However, the relatively poor predictive performance of Bi-LDA suggests that the blockmodel assumption—that objects only interact via their topics—is too restrictive.

In this chapter we develop a fully Bayesian framework for wedding the strong performance and expressiveness of continuous latent factor models with the context dependence and topic clustering of discrete mixed membership models. In Section 17.2, we provide additional background on matrix factorization and mixed membership modeling. We introduce our Mixed Membership Matrix Factorization (M^3F) framework in Section 17.3, and discuss procedures for inference and prediction under two specific M^3F models in Section 17.4. Section 17.5 describes experimental evaluation and analysis of our models on a variety of real-world collaborative filtering datasets . The results demonstrate that mixed membership matrix factorization methods outperform their context-blind counterparts and simultaneously reveal interesting clustering structure in the data. Finally, we present a conclusion in Section 17.6.

17.2 Background

17.2.1 Latent Factor Models

We begin by considering a prototypical latent factor model, Bayesian Probabilistic Matrix Factorization (BPMF) of Salakhutdinov and Mnih (2008) (see Figure 17.1). Like most factor models, BPMF associates with each user u an unknown factor vector $\mathbf{a}_u \in \mathbb{R}^D$ and with each item j an unknown factor vector $\mathbf{b}_j \in \mathbb{R}^D$. A user generates a rating for an item by adding Gaussian noise to the inner product, $r_{uj} = \mathbf{a}_u \cdot \mathbf{b}_j$. We refer to this inner product as the *static rating* for a user-item pair, because, as discussed in the introduction, the latent factor rating mechanism does not model the context in which a rating is given and does not allow a user to don different moods or "hats" in different dyadic interactions. Such contextual flexibility is desirable for capturing the context-sensitive nature of dyadic interactions, and, therefore, we turn our attention to mixed membership models.

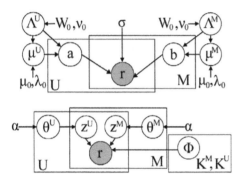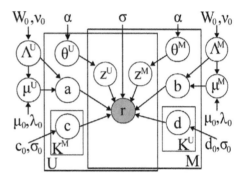

FIGURE 17.1
Graphical model representations of BPMF (top left), Bi-LDA (bottom left), and M³F-TIB (right).

17.2.2 Mixed Membership Models

Two recent examples of dyadic mixed membership (DMM) models are the mixed membership stochastic blockmodel (MMSB) (Airoldi et al., 2008) and Bi-LDA (Porteous et al., 2008) (see Figure 17.1). In DMM models, each user u and item j has its own discrete distribution over topics, represented by topic parameters θ_u^U and θ_j^M. When a user desires to rate an item, both the user and the item select interaction-specific topics according to their distributions; the selected topics then determine the distribution over ratings.

One drawback of DMM models is the reliance on purely group-wise interactions: one learns how a user group interacts with an item group but not how a user group interacts directly with a particular item. M³F models address this limitation in two ways—first, by modeling interactions between groups and specific users or items and second, by incorporating the user-item specific static rating of latent factor models.

17.3 Mixed Membership Matrix Factorization

In this section, we present a general mixed membership matrix factorization framework and two specific models that leverage the predictive power and static specificity of continuous latent factor models while allowing for the clustered context-sensitivity of mixed membership models. In each M^3F model, users and items are endowed both with latent factor vectors ($\mathbf{a}_u$ and $\mathbf{b}_j$) and with topic distribution parameters (θ_u^U and θ_j^M). To rate an item, a user first draws a topic z_{uj}^U from his distribution, representing, for example, his mood at the time of rating (in the mood for romance vs. comedy), and the item draws a topic z_{uj}^M from its distribution, representing, for example, the context under which it is being rated (in a theater on opening night vs. in a high school classroom). The selected user and item topics, $z_{uj}^U = i$ and $z_{uj}^M = k$, together with the identity of the user and item, u and j, jointly specify a rating bias, β_{uj}^{ik}, tailored to the user-item pair. Different M^3F models will differ principally in the precise form of this *contextual bias*. To generate a complete rating, the user-item-specific static rating $\mathbf{a}_u \cdot \mathbf{b}_j$ is added to the contextual bias β_{uj}^{ik}, along with some noise. Rather than learn point estimates under our M^3F models, we adopt a fully Bayesian methodology and place priors on all parameters of interest. Topic distribution parameters θ_u^U and θ_j^M are given independent exchangeable Dirichlet priors, and the latent factor vectors $\mathbf{a}_u$ and $\mathbf{b}_j$ are drawn independently from $\mathcal{N}\left(\mu^U, (\Lambda^U)^{-1}\right)$ and $\mathcal{N}\left(\mu^M, (\Lambda^M)^{-1}\right)$, respectively. As in Salakhutdinov and Mnih (2008), we place normal-Wishart priors on the hyperparameters (μ^U, Λ^U) and (μ^M, Λ^M). Suppose K^U is the number of user topics and K^M is the number of item topics. Then, given the contextual biases β_{uj}^{ik}, ratings are generated according to the following M^3F generative process:

$$\Lambda^U \sim \text{Wishart}(\mathbf{W}_0, \nu_0), \Lambda^M \sim \text{Wishart}(\mathbf{W}_0, \nu_0)$$

$$\mu^U \sim \mathcal{N}\left(\mu_0, (\lambda_0 \Lambda^U)^{-1}\right), \mu^M \sim \mathcal{N}\left(\mu_0, (\lambda_0 \Lambda^M)^{-1}\right).$$

For each $u \in \{1, \dots, U\}$:

$$\mathbf{a}_u \sim \mathcal{N}\left(\mu^U, (\Lambda^U)^{-1}\right)$$
$$\theta_u^U \sim \text{Dir}(\alpha/K^U).$$

For each $j \in \{1, \dots, M\}$:

$$\mathbf{b}_j \sim \mathcal{N}\left(\mu^M, (\Lambda^M)^{-1}\right)$$
$$\theta_j^M \sim \text{Dir}(\alpha/K^M).$$

For each rating r_{uj}:

$$z_{uj}^U \sim \text{Multi}(1, \theta_u^U), z_{uj}^M \sim \text{Multi}(1, \theta_j^M)$$
$$r_{uj} \mid z_{uj}^U = i, z_{uj}^M = k \sim \mathcal{N}\left(\beta_{uj}^{ik} + \mathbf{a}_u \cdot \mathbf{b}_j, \sigma^2\right).$$

For each of the following models discussed, we let Θ^U denote the collection of all user parameters (e.g., $\mathbf{a}, \theta^U, \Lambda^U, \mu^U$), Θ^M denote all item parameters, and Θ_0 denote all global parameters (e.g., $\mathbf{W}_0, \nu_0, \mu_0, \lambda_0, \alpha, \sigma_0^2, \sigma^2$). We now describe in more detail the specific forms of two M^3F models and their contextual biases.

17.3.1 The M^3F Topic-Indexed Bias Model

The M^3F Topic-Indexed Bias (TIB) model assumes that the contextual bias decomposes into a latent user bias and a latent item bias. The user bias is influenced by the interaction-specific topic selected by the item. Similarly, the item bias is influenced by the user's selected topic. We denote the latent rating bias of user u under item topic k as c_u^k and denote the bias for item j under user topic i as d_j^i. The contextual bias for a given user-item interaction is then found by summing the two latent biases and a fixed global bias, χ_0:[1]

$$\beta_{uj}^{ik} = \chi_0 + c_u^k + d_j^i.$$

Topic-indexed biases c_u^k and d_j^i are drawn independently from Gaussian priors with variance σ_0^2 and means c_0 and d_0, respectively. Figure 17.1 compares the graphical model representations of M^3F-TIB, BPMF, and Bi-LDA. Note that M^3F-TIB reduces to BPMF when K^U and K^M are both zero. Intuitively, the topic-indexed bias model captures the *"Napoleon Dynamite* effect," whereby certain movies provoke strongly differing reactions from otherwise similar users (Thompson, 2008). Each user-topic-indexed bias d_j^i represents one of K^U possible predispositions towards liking or disliking each item in the database, irrespective of the static latent factor parameterization. Thus, in the movie-recommendation problem, we expect the variance in user reactions to movies such as *Napoleon Dynamite* to be captured in part by a corresponding variance in the bias parameters d_j^i (see Section 17.5). Moreover, because the model is symmetric, each rating is also influenced by the item-topic-indexed bias c_u^k. This can be interpreted as the predisposition of each perceived item class towards being liked or disliked by each user in the database. Finally, because M^3F-TIB is a mixed membership model, each user and item can choose a different topic and hence a different bias for each rating (e.g., when multiple users share a single account).

17.3.2 The M^3F Topic-Indexed Factor Model

The M^3F Topic-Indexed Factor (TIF) model assumes that the joint contextual bias is an inner product of topic-indexed factor vectors, rather than the sum of topic-indexed biases as in the TIB model. Each item topic k maintains a latent factor vector $\mathbf{c}_u^k \in \mathbb{R}^{\tilde{D}}$ for each user, and each user topic i maintains a latent factor vector $\mathbf{d}_j^i \in \mathbb{R}^{\tilde{D}}$ for each item. Each user and each item additionally maintains a single static rating bias, ξ_u and χ_j, respectively. The joint contextual bias is formed by summing the user bias, the item bias, and the inner product between the topic-indexed factor vectors:

$$\beta_{uj}^{ik} = \xi_u + \chi_j + \mathbf{c}_u^k \cdot \mathbf{d}_j^i.$$

The topic-indexed factors $\mathbf{c}_u^k$ and $\mathbf{d}_j^i$ are drawn independently from $\mathcal{N}\left(\tilde{\mu}^U, (\tilde{\Lambda}^U)^{-1}\right)$ and $\mathcal{N}\left(\tilde{\mu}^M, (\tilde{\Lambda}^M)^{-1}\right)$ priors, and conjugate normal-Wishart priors are placed on the hyperparameters $(\tilde{\mu}^U, \tilde{\Lambda}^U)$ and $(\tilde{\mu}^M, \tilde{\Lambda}^M)$. The static user and item biases, ξ_u and χ_j, are drawn independently from Gaussian priors with variance σ_0^2 and means ξ_0 and χ_0, respectively.[2]

Intuitively, the topic-indexed factor model can be interpreted as an extended matrix factorization with both *global* and *local* low-dimensional representations. Each user u has a single global factor $\mathbf{a}_u$ but K^U local factors $\mathbf{c}_u^k$; similarly, each item j has both a global factor $\mathbf{b}_j$ and multiple local factors $\mathbf{d}_j^i$. A strength of latent factor methods is their ability to discover globally predictive intrinsic properties of users and items. The topic-indexed factor model extends this representation to allow for intrinsic properties that are predictive in some but perhaps not all contexts. For example, in the movie-recommendation setting, is *Lost In Translation* a dark comedy or a romance film? The answer may vary from user to user and thus may be captured by different vectors $\mathbf{d}_j^i$ for each user-indexed topic.

---

[1] The global bias, χ_0, is suppressed in the remainder of the paper for clarity.
[2] Static biases ξ and χ are suppressed in the remainder of the paper for clarity.

17.4 Inference and Prediction

The goal in dyadic data prediction is to predict unobserved ratings $\mathbf{r}^{(h)}$ given observed ratings $\mathbf{r}^{(v)}$. As in Salakhutdinov and Mnih (2007; 2008) and Takács et al. (2009), we adopt root mean squared error (RMSE)[3] as our primary error metric and note that the Bayes optimal prediction under RMSE loss is the posterior mean of the predictive distribution $p(\mathbf{r}^{(h)}|\mathbf{r}^{(v)}, \Theta_0)$.

In our M³F models, the predictive distribution over unobserved ratings is found by integrating out all topics and parameters. The posterior distribution $p(\mathbf{z}^U, \mathbf{z}^M, \Theta^U, \Theta^M|\mathbf{r}^{(v)}, \Theta_0)$ is thus our main inferential quantity of interest. Unfortunately, as in both LDA and BPMF, analytical computation of this posterior is infeasible due to complex coupling in the marginal distribution $p(\mathbf{r}^{(v)}|\Theta_0)$ (Blei et al., 2003; Salakhutdinov and Mnih, 2008).

17.4.1 Inference via Gibbs Sampling

In this work, we use a Gibbs sampling MCMC procedure (Geman and Geman, 1984) to draw samples of topic and parameter variables $\{(\mathbf{z}^{U(t)}, \mathbf{z}^{M(t)}, \Theta^{U(t)}, \Theta^{M(t)})\}_{t=1}^T$ from their joint posterior. Our use of conjugate priors ensures that each Gibbs conditional has a simple closed form.

Algorithm 1 displays the Gibbs sampling algorithm for the M³F-TIB model; the M³F-TIF Gibbs sampler is similar. The exact conditional distributions of both models are presented in the Appendix. Note that we choose to sample the topic parameters θ^U and θ^M rather than integrate them out as in a collapsed Gibbs sampler (see, e.g., Porteous et al., 2008). This decision allows us to sample the interaction-specific topic variables in parallel. Indeed, each loop in Algorithm 1 corresponds to a block of parameters that can be sampled in parallel. In practice, such parallel computation yields substantial savings in sampling time for large-scale dyadic datasets.

17.4.2 Prediction

Given posterior samples of parameters, we can approximate the true predictive distribution by the Monte Carlo expectation

$$\hat{p}(\mathbf{r}^{(h)}|\mathbf{r}^{(v)}, \Theta_0) = \frac{1}{T}\sum_{t=1}^T \sum_{\mathbf{z}^U, \mathbf{z}^M} p(\mathbf{z}^U, \mathbf{z}^M|\Theta^{U(t)}, \Theta^{M(t)})$$
$$p(\mathbf{r}^{(h)}|\mathbf{z}^U, \mathbf{z}^M, \Theta^{U(t)}, \Theta^{M(t)}, \Theta_0), \qquad (17.1)$$

where we have integrated over the unknown topic variables. Equation (17.1) yields the following posterior mean prediction for each user-item pair under the M³F-TIB model:

$$\frac{1}{T}\sum_{t=1}^T \left(\mathbf{a}_u^{(t)} \cdot \mathbf{b}_j^{(t)} + \sum_{k=1}^{K^M} c_u^{k(t)} \theta_{jk}^{M(t)} + \sum_{i=1}^{K^U} d_j^{i(t)} \theta_{ui}^{U(t)} \right).$$

Under the M³F-TIF model, posterior mean prediction takes the form

$$\frac{1}{T}\sum_{t=1}^T \left(\mathbf{a}_u^{(t)} \cdot \mathbf{b}_j^{(t)} + \sum_{i=1}^{K^U}\sum_{k=1}^{K^M} \theta_{ui}^{U(t)} \theta_{jk}^{M(t)} c_u^{k(t)} \cdot \mathbf{d}_j^{i(t)} \right).$$

---

[3]For work linking improved RMSE with better top-K recommendation rankings, see Koren (2008).

Input: $(\mathbf{a}^{(0)}, \mathbf{b}^{(0)}, \mathbf{c}^{(0)}, \mathbf{d}^{(0)}, \theta^{U(0)}, \theta^{M(0)}, \mathbf{z}^{M(0)})$
  for $t = 1$ **to** T **do**
    // Sample Hyperparameters
    for $(u, j) \in \mathcal{T}$ **do**
      $(\mu^U, \Lambda^U)^t \sim \mu^U, \Lambda^U \mid \mathbf{a}^{t-1}, \Theta_0$
      $(\mu^M, \Lambda^M)^t \sim \mu^M, \Lambda^M \mid \mathbf{b}^{t-1}, \Theta_0$
    end for
    // Sample Topics
    for $(u, j) \in \mathcal{T}$ **do**
      $z_{uj}^{U(t)} \sim z_{uj}^U \mid (z_{uj}^M, \theta_u^U, \mathbf{a}_u, \mathbf{b}_j, \mathbf{c}_u, \mathbf{d}_j)^{t-1}, \mathbf{r}^{(v)}, \Theta_0$
      $z_{uj}^{M(t)} \sim z_{uj}^M \mid (\theta_j^M, \mathbf{a}_u, \mathbf{b}_j, \mathbf{c}_u, \mathbf{d}_j)^{t-1}, z_{uj}^{U(t)}, \mathbf{r}^{(v)}, \Theta_0$
    end for
    // Sample User Parameters
    for $u = 1$ **to** U **do**
      $\theta_u^{U(t)} \sim \theta_u^U \mid \mathbf{z}^{U(t)}, \Theta_0$
      $\mathbf{a}_u^t \sim \mathbf{a}_u \mid (\Lambda^U, \mu^U, \mathbf{z}_u^U, \mathbf{z}^M)^t, (\mathbf{b}, \mathbf{c}_u, \mathbf{d})^{t-1}, \Theta_0$
      for $i = 1$ **to** K^M **do**
        $c_u^{i(t)} \sim c_u^i \mid (\mathbf{z}^U, \mathbf{z}^M, \mathbf{a}_u)^t, (\mathbf{b}, \mathbf{d})^{t-1}, \mathbf{r}^{(v)}, \Theta_0$
      end for
    end for
    // Sample Item Parameters
    for $j = 1$ **to** M **do**
      $\theta_j^{M(t)} \sim \theta_j^M \mid \mathbf{z}^{M(t)}, \Theta_0$
      $\mathbf{b}_j^t \sim \mathbf{b}_j \mid (\Lambda^U, \mu^U, \mathbf{z}_u^U, \mathbf{z}^M, \mathbf{a}, \mathbf{c}_u)^t, \mathbf{d}^{t-1}, \Theta_0$
      for $k = 1$ **to** K^U **do**
        $d_j^{k(t)} \sim d_j^k \mid (\mathbf{z}^U, \mathbf{z}^M, \mathbf{a}, \mathbf{b}_j, \mathbf{c})^t, \mathbf{r}^{(v)}, \Theta_0$
      end for
    end for
  end for

Algorithm 1: Gibbs Sampling for M³F-TIB.

17.5 Experimental Evaluation

We evaluate our models on several movie rating collaborative filtering datasets, including the Netflix Prize dataset,[4] the EachMovie dataset, and the 1M and 10M MovieLens datasets.[5] The Netflix Prize dataset contains 100 million ratings in $\{1, \ldots, 5\}$ distributed across 17,770 movies and 480,189 users. The EachMovie dataset contains 2.8 million ratings in $\{1, \ldots, 6\}$ distributed across 1,648 movies and 74,424 users. The 1M MovieLens dataset has 6,040 users, 3,952 movies, and 1 million ratings in $\{1, \ldots, 5\}$. The 10M MovieLens dataset has 10,681 movies, 71,567 users, and 10 million ratings on a .5 to 5 scale with half-star increments. In all experiments, we set W_0 equal to the identity matrix, ν_0 equal to the number of static matrix factors, μ_0 equal to the all-zeros vector, χ_0 equal to the mean rating in the dataset, and $(\lambda_0, \sigma^2, \sigma_0^2) = (10, .5, .1)$. For M³F-TIB experiments, we set $(c_0, d_0, \alpha) = (0, 0, 10000)$, and for M³F-TIF, we set $\tilde{W}_0$ equal to the identity matrix, $\tilde{\nu}_0$ equal to the number of topic-indexed factors, $\tilde{\mu}_0$ equal to the all-zeros vector, and $(\tilde{D}, \xi_0, \alpha, \tilde{\lambda}_0) = (2, 0, 10, 10000)$. Free parameters were selected by grid search on an EachMovie

---
[4]See http://www.netflixprize.com/.
[5]See http://www.grouplens.org/.

hold-out set, disjoint from the test sets used for evaluation. Throughout, reported error intervals are of plus or minus one standard error from the mean.

17.5.1　1M MovieLens and EachMovie Datasets

We first evaluated our models on the smaller datasets, 1M MovieLens and EachMovie. We conducted the "weak generalization" ratings prediction experiment of Marlin (2004), where, for each user in the training set, a single rating was withheld for the test set. All reported results were averaged over the same three random train-test splits used in Marlin (2003), Marlin (2004), Rennie and Srebro (2005), DeCoste (2006), Park and Pennock (2007), and Lawrence and Urtasun (2009). Our Gibbs samplers were initialized with draws from the prior and ran for 3000 samples for M^3F-TIB and 512 samples for M^3F-TIF. No samples were discarded for "burn-in."

Table 17.1 reports the predictive performance of our models for a variety of static factor dimensionalities (D) and topic counts (K^U, K^M). We compared all models against BPMF as a baseline by running the M^3F-TIB model with K^U and K^M set to zero. For comparison with previous results that report the normalized mean average error (NMAE) of Marlin (2004), we additionally ran M^3F-TIB with $(D, K^U, K^M) = (300, 2, 1)$ on EachMovie and achieved a weak RMSE of $(\mathbf{1.0878} \pm 0.0025)$ and a weak NMAE of $(\mathbf{0.4293} \pm 0.0013)$.

On both the EachMovie and the 1M MovieLens datasets, both M^3F models systematically outperformed the BPMF baseline for almost every setting of latent dimensionality and topic counts. For $D = 20$, increasing K^U to 2 provided a boost in accuracy for both M^3F models equivalent to doubling the number of BPMF static factor parameters $(D = 40)$. We also found that the M^3F-TIB model outperformed the more recent Gaussian process matrix factorization model of Lawrence and Urtasun (2009).

The results indicate that the mixed membership component of M^3F offers greater predictive power than simply increasing the dimensionality of a pure latent factor model. While the M^3F-TIF model sometimes failed to outperform the BPMF baseline due to overfitting, the M^3F-TIB model always outperformed BPMF regardless of the setting of K^U, K^M, or D. Note that the increase in

TABLE 17.1

1M MovieLens and EachMovie RMSE scores for varying static factor dimensionalities and topic counts for both M^3F models. All scores are averaged across 3 standardized cross-validation splits. Parentheses indicate topic counts (K^U, K^M). For M^3F-TIF, $\tilde{D} = 2$ throughout. L&U (2009) refers to Lawrence and Urtasun (2009). Best results for each D are boldened. Asterisks indicate significant improvement over BPMF under a one-tailed, paired t-test with level 0.05.

Method	1M MovieLens				EachMovie			
	D=10	D=20	D=30	D=40	D=10	D=20	D=30	D=40
BPMF	0.8695	0.8622	0.8621	0.8609	1.1229	1.1212	1.1203	1.1163
M^3F-TIB (1,1)	0.8671	0.8614	0.8616	0.8605	1.1205	1.1188	1.1183	1.1168
M^3F-TIF (1,2)	0.8664	0.8629	0.8622	0.8616	1.1351	1.1179	1.1095	1.1072
M^3F-TIF (2,1)	0.8674	0.8605	0.8605	0.8595	1.1366	1.1161	1.1088	1.1058
M^3F-TIF (2,2)	**0.8642**	**0.8584***	0.8584	0.8592	1.1211	1.1043	1.1035	1.1020
M^3F-TIB (1,2)	0.8669	0.8611	0.8604	0.8603	1.1217	1.1081	1.1016	1.0978
M^3F-TIB (2,1)	0.8649	0.8593	**0.8581***	**0.8577***	1.1186	1.1004	1.0952	1.0936
M^3F-TIB (2,2)	0.8658	0.8609	0.8605	0.8599	**1.1101***	**1.0961***	**1.0918***	**1.0905***
L&U (2009)	0.8801 (RBF)		0.8791 (Linear)		1.1111 (RBF)		1.0981 (Linear)	

the number of parameters from the BPMF model to the M^{3}F models is independent of D (M^{3}F-TIB requires $(U + M)(K^U + K^M)$ more parameters than BPMF with equal D), and therefore the ratio of the number of parameters of BPMF and M^{3}F approaches 1 if D increases while K^U, K^M, and $\tilde{D}$ are held fixed. Nonetheless, the modeling of joint contextual bias in the M^{3}F-TIB model continues to improve predictive performance even as D increases, suggesting that the M^{3}F-TIB model is capturing aspects of the data that are not captured by a pure latent factor model.

Finally, because the M^{3}F-TIB model offered superior performance to the M^{3}F-TIF model in most experiments, we focus on the M^{3}F-TIB model in the remainder of this section.

17.5.2 10M MovieLens Dataset

For the larger datasets, we initialized the Gibbs samplers with MAP estimates of a and b under simple Gaussian priors, which we trained with stochastic gradient descent. This is similar to the PMF initialization scheme of Salakhutdinov and Mnih (2008). All other parameters were initialized to their model means.

For the 10M MovieLens dataset, we averaged our results across the r_a and r_b train-test splits provided with the dataset after removing those test set ratings with no corresponding item in the training set. For comparison with the Gaussian process matrix factorization model of Lawrence and Urtasun (2009), we adopted a static factor dimensionality of $D = 10$. Our M^{3}F-TIB model with $(K^U, K^M) = (4, 1)$ achieved an RMSE of (**0.8447** $\pm$ 0.0095), representing a significant improvement ($p = 0.034$) over BPMF with RMSE (**0.8472** $\pm$ 0.0093) and a substantial increase in accuracy over the Gaussian process model with RMSE (**0.8740** $\pm$ 0.0197).

17.5.3 Netflix Prize Dataset

The unobserved ratings for the 100 million dyad Netflix Prize dataset are partitioned into two standard sets, known as the Quiz Set and the Test Set. Prior to September 2009, public evaluation was only available on the Quiz Set, and, as a result, most prior published "test set" results were evaluated on the Quiz Set. In Table 17.2, we compare the performance of BPMF and M^{3}F-TIB with $(K^U, K^M) = (4, 1)$ on the Quiz Set, the Test Set, and on their union (the Qualifying Set), across a wide range of static dimensionalities. We also report running times of our Matlab/MEX implementation on dual quad-core 2.67GHz Intel Xeon CPUs. We used the initialization scheme described in Section 17.5.2 and ran the Gibbs samplers for 500 iterations.

In addition to outperforming the BPMF baselines of comparable dimensionality, the M^{3}F-TIB models routinely proved to be more accurate than higher-dimensional BPMF models with longer running times and many more learned parameters. This major advantage of M^{3}F modeling is highlighted in Figure 17.2, which plots error as a function of the number of parameters modeled per user or item $(D + K^U + K^M)$.

To determine for which users and movies our models were providing the most improvement over BPMF, we divided the Qualifying Set into bins based on the number of ratings associated with each user and movie in the database. Figure 17.3 displays the improvements of BPMF/60, M^{3}F-TIB/40, and M^{3}F-TIB/60 over BPMF/40 as a function of the number of user or movie ratings. Consistent with our expectations, we found that adopting an M^{3}F model yielded improved accuracy for movies of small rating counts, with the greatest improvement over BPMF occurring for those high-variance movies with relatively few ratings. Moreover, the improvements realized by either M^{3}F-TIB model uniformly dominated the improvements realized by BPMF/60 across movie rating counts. At the same time, we found that the improvements of the M^{3}F-TIB models were skewed toward users with larger rating counts.

TABLE 17.2
Netflix Prize results for BPMF and M^{3}F-TIB with $(K^U, K^M) = (4, 1)$. Hidden ratings are partitioned into Quiz and Test sets; the Qualifying set is their union. Best results in each block are boldened. Reported times are average running times per sample.

Method	Test	Quiz	Qual	Time
BPMF/15	0.9125	0.9117	0.9121	27.8s
TIB/15	**0.9093**	**0.9086**	**0.9090**	46.3s
BPMF/30	0.9049	0.9044	0.9047	38.6s
TIB/30	**0.9018**	**0.9012**	**0.9015**	56.9s
BPMF/40	0.9029	0.9026	0.9027	48.3s
TIB/40	**0.8992**	**0.8988**	**0.8990**	70.5s
BPMF/60	0.9004	0.9001	0.9002	94.3s
TIB/60	**0.8965**	**0.8960**	**0.8962**	97.0s
BPMF/120	0.8958	0.8953	0.8956	273.7s
TIB/120	**0.8937**	**0.8931**	**0.8934**	285.2s
BPMF/240	0.8939	0.8936	0.8938	1152.0s
TIB/240	**0.8931**	**0.8927**	**0.8929**	1158.2s

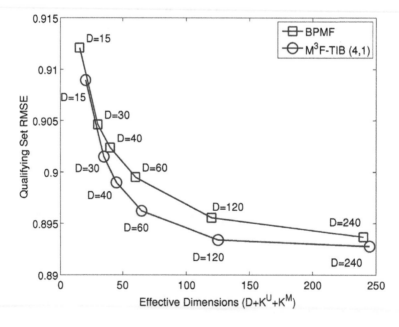

FIGURE 17.2
RMSE performance of BPMF and M^{3}F-TIB with $(K^U, K^M) = (4, 1)$ on the Netflix Prize Qualifying set as a function of the number of parameters modeled per user or item.

M^3F & The *Napoleon Dynamite* Effect

In our introduction to the M^3F-TIB model we discussed the joint contextual bias as a potential solution to the problem of making predictions for movies that have high variance. To investigate whether or not M^3F-TIB achieved progress towards this goal, we analyzed the correlation between the improvement in RMSE over the BPMF baseline and the variance of ratings for the 1000 most popular movies in the database. While the improvements for BPMF/60 were not significantly correlated with movie variance ($\rho = -0.016$), the improvements of the M^3F-TIB models were strongly correlated with $\rho = 0.117 (p < 0.001)$ and $\rho = 0.15$ ($p < 10^{-7}$) for the $(40, 4, 1)$ and $(60, 4, 1)$ models, respectively. These results indicate that a strength of the M^3F-TIB model lies in the ability of the topic-indexed biases to model variance in user biases toward specific items.

To further illuminate this property of the model, we computed the posterior expectation of the movie bias parameters, $\mathbb{E}(\mathbf{d}_j | \mathbf{r}^{(v)})$, for the 200 most popular movies in the database. For these movies, the variance of $\mathbb{E}(d_j^i | \mathbf{r}^{(v)})$ across topics and the variance of the ratings of these movies were very strongly correlated ($\rho = 0.682, p < 10^{-10}$). The five movies with the highest and lowest variance in $\mathbb{E}(d_j^i | \mathbf{r}^{(v)})$ across topics are shown in Table 17.3. The results are easily interpretable, with high-variance movies such as *Napoleon Dynamite* dominating the high-variance positions and universally acclaimed blockbusters dominating the low-variance positions.

TABLE 17.3
Top 200 movies from the Netflix Prize dataset with the highest and lowest cross-topic variance in $\mathbb{E}(d_j^i | \mathbf{r}^{(v)})$. Reported intervals are of the mean value of $\mathbb{E}(d_j^i | \mathbf{r}^{(v)})$, plus or minus one standard deviation.

Movie Title	$\mathbb{E}(d_j^i \| \mathbf{r}^{(v)})$
Napoleon Dynamite	-0.11 ± 0.93
Fahrenheit 9/11	-0.06 ± 0.90
Chicago	-0.12 ± 0.78
The Village	-0.14 ± 0.71
Lost in Translation	-0.02 ± 0.70
LotR: The Fellowship of the Ring	0.15 ± 0.00
LotR: The Two Towers	0.18 ± 0.00
LotR: The Return of the King	0.24 ± 0.00
Star Wars: Episode V	0.35 ± 0.00
Raiders of the Lost Ark	0.29 ± 0.00

17.6 Conclusion

In this chapter, we developed a fully Bayesian dyadic data prediction framework for integrating the complementary approaches of discrete mixed membership modeling and continuous latent factor modeling. We introduced two mixed membership matrix factorization models, presented MCMC inference procedures, and evaluated our methods on the EachMovie, MovieLens, and Netflix Prize datasets. On each dataset, we found that M^3F-TIB significantly outperformed BPMF and other state-of-the-art baselines, even when fitting fewer parameters. We further discovered that the greatest

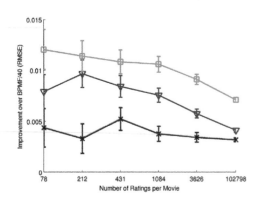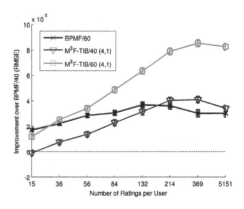

FIGURE 17.3
RMSE improvements over BPMF/40 on the Netflix Prize as a function of movie or user rating count. Left: Improvement as a function of movie rating count. Each x-axis label represents the average rating count of 1/6 of the movie base. Right: Improvement over BPMF as a function of user rating count. Each bin represents 1/8 of the user base.

performance improvements occurred for the high-variance, sparsely-rated items, for which accurate DDP is typically the hardest.

Appendix: Gibbs Sampling Conditionals for M³F Models

The M³F-TIB Model

In this section, we specify the conditional distributions used by the Gibbs sampler for the M³F-TIB model.

Normal-Wishart Parameters

$$\Lambda^U | rest \backslash \{\mu^U\} \sim \text{Wishart}((\mathbf{W}_0^{-1} + \sum_{u=1}^{U} (\mathbf{a}_u - \bar{\mathbf{a}})(\mathbf{a}_u - \bar{\mathbf{a}})^t + \frac{\lambda_0 U}{\lambda_0 + U}(\mu_0 - \bar{\mathbf{a}})(\mu_0 - \bar{\mathbf{a}})^t)^{-1}, \nu_0 + U), \text{ where } \bar{\mathbf{a}} = \frac{1}{U}\sum_{u=1}^{U} \mathbf{a}_u.$$

$$\Lambda^M | rest \backslash \{\mu^M\} \sim \text{Wishart}((\mathbf{W}_0^{-1} + \sum_{j=1}^{M} (\mathbf{b}_j - \bar{\mathbf{b}})(\mathbf{b}_j - \bar{\mathbf{b}})^t + \frac{\lambda_0 M}{\lambda_0 + M}(\mu_0 - \bar{\mathbf{b}})(\mu_0 - \bar{\mathbf{b}})^t)^{-1}, \nu_0 + M), \text{ where } \bar{\mathbf{b}} = \frac{1}{M}\sum_{j=1}^{M} \mathbf{b}_j.$$

$$\mu^U | rest \sim \mathcal{N}\left(\frac{\lambda_0\mu_0 + \sum_{u=1}^{U} \mathbf{a}_u}{\lambda_0 + U}, (\Lambda^U(\lambda_0 + U))^{-1}\right).$$

$$\mu^M | rest \sim \mathcal{N}\left(\frac{\lambda_0\mu_0 + \sum_{j=1}^{M} \mathbf{b}_j}{\lambda_0 + M}, (\Lambda^M(\lambda_0 + M))^{-1}\right).$$

Bias Parameters

For each u and $i \in \{1, \ldots, K^M\}$,

$$c_u^i | rest \sim \mathcal{N} \left(\frac{\frac{c_0}{\sigma_0^2} + \sum_{j \in V_u} \frac{1}{\sigma^2} z_{uji}^M (r_{uj} - \chi_0 - d_j^{z_{uj}^U} - \mathbf{a}_u \cdot \mathbf{b}_j)}{\frac{1}{\sigma_0^2} + \sum_{j \in V_u} \frac{1}{\sigma^2} z_{uji}^M}, \frac{1}{\frac{1}{\sigma_0^2} + \sum_{j \in V_u} \frac{1}{\sigma^2} z_{uji}^M} \right).$$

For each j and $i \in \{1, \ldots, K^U\}$,

$$d_j^i | rest \sim \mathcal{N} \left(\frac{\frac{d_0}{\sigma_0^2} + \sum_{u:j \in V_u} \frac{1}{\sigma^2} z_{uji}^U (r_{uj} - \chi_0 - c_u^{z_{uj}^M} - \mathbf{a}_u \cdot \mathbf{b}_j)}{\frac{1}{\sigma_0^2} + \sum_{u:j \in V_u} \frac{1}{\sigma^2} z_{uji}^U}, \frac{1}{\frac{1}{\sigma_0^2} + \sum_{u:j \in V_u} \frac{1}{\sigma^2} z_{uji}^U} \right).$$

Static Factors

For each u,

$$\mathbf{a}_u | rest \sim \mathcal{N} \left((\Lambda_u^{U*})^{-1} (\Lambda^U \mu^U + \sum_{j \in V_u} \frac{1}{\sigma^2} \mathbf{b}_j (r_{uj} - \chi_0 - c_u^{z_{uj}^M} - d_j^{z_{uj}^U})), (\Lambda_u^{U*})^{-1} \right),$$

where $\Lambda_u^{U*} = (\Lambda^U + \sum_{j \in V_u} \frac{1}{\sigma^2} \mathbf{b}_j (\mathbf{b}_j)^t)$.

For each j,

$$\mathbf{b}_j | rest \sim \mathcal{N} \left((\Lambda_j^{M*})^{-1} (\Lambda^M \mu^M + \sum_{u:j \in V_u} \frac{1}{\sigma^2} \mathbf{a}_u (r_{uj} - \chi_0 - c_u^{z_{uj}^M} - d_j^{z_{uj}^U})), (\Lambda_j^{M*})^{-1} \right),$$

where $\Lambda_j^{M*} = (\Lambda^M + \sum_{u:j \in V_u} \frac{1}{\sigma^2} \mathbf{a}_u (\mathbf{a}_u)^t)$.

Dirichlet Parameters

For each u, $\theta_u^U | rest \sim Dir(\alpha/K^U + \sum_{j \in V_u} z_{uj}^U)$.

For each j, $\theta_j^M | rest \sim Dir(\alpha/K^M + \sum_{u:j \in V_u} z_{uj}^M)$.

Topic Variables

For each u and $j \in V_u$, $z_{uj}^U | rest \sim \text{Multi}(1, \theta_{uj}^{U*})$, where

$$\theta_{uji}^{U*} \propto \theta_{ui}^U \exp \left(-\frac{(r_{uj} - \chi_0 - c_u^{z_{uj}^M} - d_j^i - \mathbf{a}_u \cdot \mathbf{b}_j)^2}{2\sigma^2} \right).$$

For each j and $u : j \in V_u$, $z_{uj}^M | rest \sim \text{Multi}(1, \theta_{uj}^{M*})$, where

$$\theta_{uji}^{M*} \propto \theta_{ji}^M \exp \left(-\frac{(r_{uj} - \chi_0 - c_u^i - d_j^{z_{uj}^U} - \mathbf{a}_u \cdot \mathbf{b}_j)^2}{2\sigma^2} \right).$$

The M^{3}F-TIF Model

In this section, we specify the conditional distributions used by the Gibbs sampler for the M^{3}F-TIF model.

Normal-Wishart Parameters

$$\Lambda^U | rest \backslash \{\mu^U\} \sim \text{Wishart}((\mathbf{W}_0^{-1} + \sum_{u=1}^{U} (\mathbf{a}_u - \bar{\mathbf{a}})(\mathbf{a}_u - \bar{\mathbf{a}})^t + \frac{\lambda_0 U}{\lambda_0 + U}(\mu_0 - \bar{\mathbf{a}})(\mu_0 - \bar{\mathbf{a}})^t)^{-1}, \nu_0 + U), \text{ where } \bar{\mathbf{a}} = \frac{1}{U} \sum_{u=1}^{U} \mathbf{a}_u.$$

$$\Lambda^M | rest \backslash \{\mu^M\} \sim \text{Wishart}((\mathbf{W}_0^{-1} + \sum_{j=1}^{M} (\mathbf{b}_j - \bar{\mathbf{b}})(\mathbf{b}_j - \bar{\mathbf{b}})^t + \frac{\lambda_0 M}{\lambda_0 + M}(\mu_0 - \bar{\mathbf{b}})(\mu_0 - \bar{\mathbf{b}})^t)^{-1}, \nu_0 + M), \text{ where } \bar{\mathbf{b}} = \frac{1}{M} \sum_{j=1}^{M} \mathbf{b}_j.$$

$$\mu^U | rest \sim \mathcal{N}\left(\frac{\lambda_0 \mu_0 + \sum_{u=1}^{U} \mathbf{a}_u}{\lambda_0 + U}, (\Lambda^U(\lambda_0 + U))^{-1}\right).$$

$$\mu^M | rest \sim \mathcal{N}\left(\frac{\lambda_0 \mu_0 + \sum_{j=1}^{M} \mathbf{b}_j}{\lambda_0 + M}, (\Lambda^M(\lambda_0 + M))^{-1}\right).$$

$$\tilde{\Lambda}^U | rest \backslash \{\tilde{\mu}^U\} \sim \text{Wishart}((\tilde{\mathbf{W}}_0^{-1} + \sum_{u=1}^{U} \sum_{i=1}^{K^M} (\mathbf{c}_u^i - \bar{\mathbf{c}})(\mathbf{c}_u^i - \bar{\mathbf{c}})^t + \frac{\tilde{\lambda}_0 U K^M}{\tilde{\lambda}_0 + U K^M}(\tilde{\mu}_0 - \bar{\mathbf{c}})(\tilde{\mu}_0 - \bar{\mathbf{c}})^t)^{-1}, \tilde{\nu}_0 + U K^M), \text{ where } \bar{\mathbf{c}} = \frac{1}{U K^M} \sum_{u=1}^{U} \sum_{i=1}^{K^M} \mathbf{c}_u^i.$$

$$\tilde{\Lambda}^M | rest \backslash \{\tilde{\mu}^M\} \sim \text{Wishart}((\tilde{\mathbf{W}}_0^{-1} + \sum_{j=1}^{M} \sum_{i=1}^{K^U} (\mathbf{d}_j^i - \bar{\mathbf{d}})(\mathbf{d}_j^i - \bar{\mathbf{d}})^t + \frac{\tilde{\lambda}_0 M K^U}{\tilde{\lambda}_0 + M K^U}(\tilde{\mu}_0 - \bar{\mathbf{d}})(\tilde{\mu}_0 - \bar{\mathbf{d}})^t)^{-1}, \tilde{\nu}_0 + M K^U), \text{ where } \bar{\mathbf{d}} = \frac{1}{M K^U} \sum_{j=1}^{M} \sum_{i=1}^{K^U} \mathbf{d}_j^i.$$

$$\tilde{\mu}^U | rest \sim \mathcal{N}\left(\frac{\tilde{\lambda}_0 \tilde{\mu}_0 + \sum_{u=1}^{U} \sum_{i=1}^{K^M} \mathbf{c}_u^i}{\tilde{\lambda}_0 + U K^M}, (\tilde{\Lambda}^U(\tilde{\lambda}_0 + U K^M))^{-1}\right).$$

$$\tilde{\mu}^M | rest \sim \mathcal{N}\left(\frac{\tilde{\lambda}_0 \tilde{\mu}_0 + \sum_{j=1}^{M} \sum_{i=1}^{K^U} \mathbf{d}_j^i}{\tilde{\lambda}_0 + M K^U}, (\tilde{\Lambda}^M(\tilde{\lambda}_0 + M K^U))^{-1}\right).$$

Bias Parameters

For each u,

$$\xi_u | rest \sim \mathcal{N}\left(\frac{\frac{\xi_0}{\sigma_0^2} + \sum_{j \in V_u} \frac{1}{\sigma^2}(r_{uj} - \chi_j - \mathbf{a}_u \cdot \mathbf{b}_j - \mathbf{c}_u^{z_{uj}^M} \cdot \mathbf{d}_j^{z_{uj}^U})}{\frac{1}{\sigma_0^2} + \sum_{j \in V_u} \frac{1}{\sigma^2}}, \frac{1}{\frac{1}{\sigma_0^2} + \sum_{j \in V_u} \frac{1}{\sigma^2}}\right).$$

For each j,

$$\chi_j | rest \sim \mathcal{N}\left(\frac{\frac{\chi_0}{\sigma_0^2} + \sum_{u:j \in V_u} \frac{1}{\sigma^2}(r_{uj} - \xi_u - \mathbf{a}_u \cdot \mathbf{b}_j - \mathbf{c}_u^{z_{uj}^M} \cdot \mathbf{d}_j^{z_{uj}^U})}{\frac{1}{\sigma_0^2} + \sum_{u:j \in V_u} \frac{1}{\sigma^2}}, \frac{1}{\frac{1}{\sigma_0^2} + \sum_{u:j \in V_u} \frac{1}{\sigma^2}}\right).$$

Static Factors

For each u,

$$\mathbf{a}_u | rest \sim \mathcal{N}\left((\Lambda_u^{U*})^{-1}(\Lambda^U \mu^U + \sum_{j \in V_u} \frac{1}{\sigma^2} \mathbf{b}_j(r_{uj} - \xi_u - \chi_j - \mathbf{c}_u^{z_{uj}^M} \cdot \mathbf{d}_j^{z_{uj}^U})), (\Lambda_u^{U*})^{-1}\right),$$

where $\Lambda_u^{U*} = (\Lambda^U + \sum_{j \in V_u} \frac{1}{\sigma^2} \mathbf{b}_j(\mathbf{b}_j)^t).$

For each j,

$$\mathbf{b}_j | rest \sim \mathcal{N}\left((\Lambda_j^{M*})^{-1}(\Lambda^M \mu^M + \sum_{u:j \in V_u} \frac{1}{\sigma^2}\mathbf{a}_u(r_{uj} - \xi_u - \chi_j - \mathbf{c}_u^{z_{uj}^M} \cdot \mathbf{d}_j^{z_{uj}^U})), (\Lambda_j^{M*})^{-1}\right),$$

where $\Lambda_j^{M*} = (\Lambda^M + \sum_{u:j \in V_u} \frac{1}{\sigma^2}\mathbf{a}_u(\mathbf{a}_u)^t)$.

Topic-indexed Factors

For each u and each $i \in 1, \ldots, K^M$,

$$\mathbf{c}_u^i | rest \sim \mathcal{N}\left((\tilde{\Lambda}_{ui}^{U*})^{-1}(\tilde{\Lambda}^U \tilde{\mu}^U + \sum_{j \in V_u} \frac{1}{\sigma^2}z_{uji}^M \mathbf{d}_j^{z_{uj}^U}(r_{uj} - \xi_u - \chi_j - \mathbf{a}_u \cdot \mathbf{b}_j)), (\tilde{\Lambda}_{ui}^{U*})^{-1}\right),$$

where $\tilde{\Lambda}_{ui}^{U*} = (\tilde{\Lambda}^U + \sum_{j \in V_u} \frac{1}{\sigma^2}z_{uji}^M \mathbf{d}_j^{z_{uj}^U}(\mathbf{d}_j^{z_{uj}^U})^t)$.

For each j and each $i \in 1, \ldots, K^U$,

$$\mathbf{d}_j^i | rest \sim \mathcal{N}\left((\tilde{\Lambda}_{ji}^{M*})^{-1}(\tilde{\Lambda}^M \tilde{\mu}^M + \sum_{u:j \in V_u} \frac{1}{\sigma^2}z_{uji}^U \mathbf{c}_u^{z_{uj}^M}(r_{uj} - \xi_u - \chi_j - \mathbf{a}_u \cdot \mathbf{b}_j)), (\tilde{\Lambda}_{ji}^{M*})^{-1}\right),$$

where $\tilde{\Lambda}_{ji}^{M*} = (\tilde{\Lambda}^M + \sum_{u:j \in V_u} \frac{1}{\sigma^2}z_{uji}^U \mathbf{c}_u^{z_{uj}^M}(\mathbf{c}_u^{z_{uj}^M})^t)$.

Dirichlet Parameters

For each u, $\theta_u^U | rest \sim Dir(\alpha/K^U + \sum_{j \in V_u} z_{uj}^U)$.

For each j, $\theta_j^M | rest \sim Dir(\alpha/K^M + \sum_{u:j \in V_u} z_{uj}^M)$.

Topic Variables

For each u and $j \in V_u$, $z_{uj}^U | rest \sim \text{Multi}(1, \theta_{uj}^{U*})$, where

$$\theta_{uji}^{U*} \propto \theta_{ui}^U \exp\left(-\frac{(r_{uj} - \xi_u - \chi_j - \mathbf{a}_u \cdot \mathbf{b}_j - \mathbf{c}_u^{z_{uj}^M} \cdot \mathbf{d}_j^i)^2}{2\sigma^2}\right).$$

For each j and $u : j \in V_u$, $z_{uj}^M | rest \sim \text{Multi}(1, \theta_{uj}^{M*})$, where

$$\theta_{uji}^{M*} \propto \theta_{ji}^M \exp\left(-\frac{(r_{uj} - \xi_u - \chi_j - \mathbf{a}_u \cdot \mathbf{b}_j - \mathbf{c}_u^i \cdot \mathbf{d}_j^{z_{uj}^U})^2}{2\sigma^2}\right).$$

Acknowledgments

This work was supported by the following Science Foundation Ireland (SFI) grants: Research Frontiers Programme (09/RFP/MTH2367), Strategic Research Cluster (08/SRC/I1407) and Research Centre (SFI/12/RC/2289).

References

Airoldi, E. M., Blei, D. M., Fienberg, S. E., and Xing, E. P. (2008). Mixed membership stochastic blockmodels. *Journal of Machine Learning Research* 9: 1981–2014.

Blei, D. M., Ng, A. Y., and Jordan, M. I. (2003). Latent Dirichlet allocation. *Journal of Machine Learning Research* 3: 993–1022.

DeCoste, D. (2006). Collaborative prediction using ensembles of maximum margin matrix factorizations. In *Proceedings of the 23rd International Conference on Machine Learning (ICML '06)*. New York, NY, USA: ACM, 249–256.

Geman, S. and Geman, D. (1984). Stochastic relaxation, Gibbs distributions, and the Bayesian restoration of images. *IEEE Pattern Analysis and Machine Intelligence* 6: 721–741.

Hofmann, T., Puzicha, J., and Jordan, M. I. (1998). Learning from dyadic data. In Kearns, M. J., Solla, S. A., and Cohn, D. A. (eds), *Advances in Neural Information Processing Systems 11*. Cambridge, MA: The MIT Press, 466–472.

Koren, Y. (2008). Factorization meets the neighborhood: A multifaceted collaborative filtering model. In *Proceedings of the 14th ACM SIGKDD International Conference on Knowledge Discovery and Data Mining (KDD '08)*. New York, NY, USA: ACM, 426–434.

Koren, Y., Bell, R. M., and Volinsky, C. (2009). Matrix factorization techniques for recommender systems. *IEEE Computer* 42: 30–37.

Lawrence, N. D. and Urtasun, R. (2009). Non-linear matrix factorization with Gaussian processes. In *Proceedings of the 26th Annual International Conference on Machine Learning (ICML '09)*. Omnipress, 601–608.

Mackey, L., Weiss, D., and Jordan, M. I. (2010). Mixed membership matrix factorization. In Fürnkranz, J. and Joachims, T. (eds), *Proceedings of the 27th International Conference on Machine Learning (ICML '10)*. Omnipress, 711–718.

Marlin, B. M. (2004). Collaborative Filtering: A Machine Learning Perspective. Master's thesis, University of Toronto, Toronto, Ontario, Canada.

Marlin, B. M. (2003). Modeling user rating profiles for collaborative filtering. In Thrun, S., Saul, L. K., and Schölkopf, B. (eds), *Advances in Neural Information Processing Systems 16*. Cambridge, MA: The MIT Press, 627–634.

Park, S. -T. and Pennock, D. M. (2007). Applying collaborative filtering techniques to movie search for better ranking and browsing. In *Proceedings of the 13th ACM SIGKDD International Conference on Knowledge Discovery and Data Mining (KDD '07)*. New York, NY, USA: ACM, 550–559.

Porteous, I., Bart, E., and Welling, M. (2008). Multi-HDP: A non parametric Bayesian model for tensor factorization. In *Proceedings of the 23rd National Conference on Artificial Intelligence - Volume 3 (AAAI '08)*. Palo Alto, CA, USA: AAAI, 1487–1490.

Rennie, J. D. M. and Srebro, N. (2005). Fast maximum margin matrix factorization for collaborative prediction. In *Proceedings of the 22nd International Conference on Machine Learning (ICML '05)*. New York, NY, USA: ACM, 713–719.

Salakhutdinov, R. and Mnih, A. (2007). Probabilistic matrix factorization. In Platt, J. C., Koller, D., Singer, Y., and Roweis, S. T. (eds), *Advances in Neural Information Processing Systems 20*. Red Hook, NY: Curran Associates, Inc., 1257–1264.

Salakhutdinov, R. and Mnih, A. (2008). Bayesian probabilistic matrix factorization using Markov chain Monte Carlo. In *Proceedings of the 25th International Conference on Machine Learning (ICML '08)*. New York, NY, USA: ACM, 880–887.

Takács, G., Pilászy, I., Németh, B., and Tikk, D. (2009). Scalable collaborative filtering approaches for large recommender systems. *Journal of Machine Learning Research* 10: 623–656.

Thompson, C. (2008). If you liked this, you're sure to love that. *New York Times Magazine*. November 21.

18

Discriminative Training of Mixed Membership Models

Jun Zhu

Department of Computer Science and Technology, State Key Laboratory of Intelligent Technology and Systems; Tsinghua National Laboratory for Information Science and Technology, Tsinghua University, Beijing 100084, China

Eric P. Xing

School of Computer Science, Carnegie Mellon University, Pittsburgh, PA 15213, USA

CONTENTS

Mixed membership models have shown great promise in analyzing genetics, text documents, and social network data. Unlike most existing likelihood-based approaches to learning mixed membership models, we present a discriminative training method based on the maximum margin principle to utilize supervising side information such as ratings or labels associated with documents to discover more predictive low-dimensional representations of the data. By using the linear expectation operator, we can derive efficient variational methods for posterior inference and parameter estimation. Empirical studies on the 20 Newsgroup dataset are provided. Our experimental results demonstrate qualitatively and quantitatively that the max-margin-based mixed membership model (topic model in particular for modeling text): 1) discovers sparse and highly discriminative topical representations; 2) achieves state-of-the-art prediction performance; and 3) is more efficient than existing supervised topic models.

18.1 Introduction

Mixed membership models are hierarchical extensions of finite mixture models where each data point exhibits multiple components. They have been successfully applied to analyze genetics (Pritchard et al., 2000), social networks (Airoldi et al., 2008), and text documents. For text analysis, probabilistic latent aspect models such as latent Dirichlet allocation (LDA) (Blei et al., 2003) have recently gained much popularity for stratifying a large collection of documents by projecting every document into a low-dimensional space spanned by a set of bases that capture the semantic aspects, also known as *topics*, of the collection. LDA posits that each document is an admixture of latent topics, of which each topic is represented as a unigram distribution over a given vocabulary. The document-specific admixture proportion vector θ is modeled as a latent Dirichlet random variable, and can be regarded as a low-dimensional representation of the document in a topical space. This low-dimensional representation can be used for downstream tasks such as classification, clustering, or merely as a tool for structurally visualizing the otherwise unstructured document collection.

LDA is typically built on a discrete bag-of-words representation of input contents, which can be texts (Blei et al., 2003), images (Fei-Fei and Perona, 2005), or multi-type data (Blei and Jordan, 2003). However, in many practical applications, we can easily obtain useful side information besides the document or image contents. For example, when online users post their reviews for products or restaurants, they usually associate each review with a rating score or a thumbs-up/thumbs-down opinion; web sites or pages in the public Yahoo! Directory[1] can have their categorical labels; and images in the LabelMe (Russell et al., 2008) database are organized by a visual ontology and additionally each image is associated with a set of annotation tags. Furthermore, there is an increasing trend towards using online crowdsourcing services (such as Amazon Mechanical Turk[2]) to collect large collections of labeled data with a reasonably low price. Such side information often provides useful high-level or direct summarization of the content, but it is not directly utilized in the original LDA to influence topic inference. One would expect that incorporating such information into latent aspect modeling could guide a topic model towards discovering secondary (or non-dominant) but semantically more salient statistical patterns (Chechik and Tishby, 2002) that may be more interesting or relevant to the user's goal, such as making predictions on unlabeled data.

To explore this potential, developing new topic models that appropriately capture side information mentioned above has recently gained increasing attention. Representative attempts include the supervised topic model (sLDA) (Blei and McAuliffe, 2007), which captures real-valued document ratings as a regression response; multi-class sLDA (Wang et al., 2009), which directly captures discrete labels of documents as a classification response; and discriminative LDA (DiscLDA) (Lacoste-Julien et al., 2008), which also performs classification, but with a mechanism different from that of sLDA. All these models focus on the document-level side information such as document categories or review rating scores to supervise model learning. More variants of supervised topic models can be found in a number of applied domains, such as the aspect rating model (Titov and McDonald, 2008) for predicting ratings for each aspect of a hotel. In computer vision, various supervised topic models have been designed for understanding complex scene images (Sudderth et al., 2005; Fei-Fei and Perona, 2005).

It is worth pointing out that among existing supervised topic models for incorporating side information, there are two classes of approaches, namely, *downstream supervised topic models* (DSTM) and *upstream supervised topic models* (USTM). In a DSTM, the response variable is predicted based on the latent representation of the document, whereas in a USTM the response variable is being conditioned to generate the latent representation of the document. Examples of USTM

---

[1] See http://dir.yahoo.com/.
[2] See https://www.mturk.com/.

include DiscLDA and the scene understanding models (Sudderth et al., 2005; Fei-Fei and Perona, 2005), whereas sLDA is an example of DSTM. Another distinction between existing supervised topic models is the training criterion, or more precisely, the choice of objective function in the optimization-based learning. The sLDA models are trained by maximizing the *joint* likelihood of the content data (e.g., text or image) and the responses (e.g., labeling or rating), whereas DiscLDA models are trained by maximizing the *conditional* likelihood of the responses given contents.

In this chapter, we present maximum entropy discrimination latent Dirichlet allocation (MedLDA), a supervised topic model leveraging the maximum margin principle for making more effective use of side information during estimation of latent topical representations. Unlike existing supervised topic models mentioned above, MedLDA employs an arguably more discriminative max-margin learning technique within a probabilistic framework; and unlike the commonly adopted two-stage heuristic which first estimates a latent topic vector for each document using a topic model and then feeds them to another downstream prediction model, MedLDA integrates the mechanism behind max-margin prediction models (e.g., SVMs) with the mechanism behind hierarchical Bayesian topic models (e.g., LDA) under a unified constrained optimization framework. It employs a composite objective motivated by a tradeoff between two components—the negative log-likelihood of an underlying topic model which measures the goodness-of-fit for document contents, and a measure of prediction error on training data. It then seeks a regularized posterior distribution of the predictive function in a feasible space defined by a set of *expected* max-margin constraints generalized from the SVM-style margin constraints. Our proposed approach builds on earlier developments in maximum entropy discrimination (MED) (Jaakkola et al., 1999; Jebara, 2001) and partially observed maximum entropy discrimination Markov network (PoMEN) (Zhu et al., 2008). In MedLDA, because of the influence of both the likelihood function over content data and max-margin constraints induced by the side information, the discovery of latent topics is therefore coupled with the max-margin estimation of model parameters. This interplay can yield latent topical representations that are more discriminative and more suitable for supervised prediction tasks, as we demonstrate in the experimental section. We also present an efficient variational approach for inference under MedLDA, with a running time comparable to that of an unsupervised LDA and lower than other likelihood-based supervised LDAs. This advantage stems from the fact that MedLDA can directly optimize a margin-based loss instead of a likelihood-based one, and thereby avoids dealing with the normalization factor resultant from a full probabilistic generative formulation, which generally makes learning harder.

Finally, although we have focused on topic models, we emphasize that the methodology we develop is quite general and can be applied to perform max-margin learning for various mixed membership models, including the relational model (Airoldi et al., 2008). Moreover, the ideas can be extended to nonparametric Bayesian models (Zhu et al., 2011a; Zhu, 2012; Xu et al., 2012).

The rest of this chapter is structured as follows. Section 18.2 introduces the preliminaries that are needed to present MedLDA. Section 18.3 presents the MedLDA model for classification, together with an efficient algorithm. Section 18.4 presents empirical studies of MedLDA. Finally, Section 18.5 concludes this chapter with future research directions discussed.

18.2 Preliminaries

We begin with a brief overview of the fundamentals of mixed membership models, support vector machines, and maximum entropy discrimination (Jaakkola et al., 1999), which constitute the major building blocks of the proposed MedLDA.

18.2.1 Hierarchical Bayesian Mixed Membership Models

A general formulation of mixed membership models was presented in Erosheva et al. (2004), which characterizes these models in terms of assumptions at four levels: *population*, *subject*, *latent variable*, and *sampling scheme*. Population level assumptions describe the general structure of the population that is common to all subjects. Subject level assumptions specify the distribution of observed responses given individual membership scores. Latent variable level assumptions are about whether the membership scores are fixed or random. Finally, the last level of assumptions specify the number of distinct observed characteristics (attributes) and the number of replications for each characteristic.

(1) **Population Level**. Assume that there are K components or basis subpopulations in the populations of interest. For each subpopulation k, we denote by $f(x_{dn}|\beta_{kn})$ the probability distribution of the nth response variable for the dth subject, where β_k is an M-dimensional vector of parameters. Within a subpopulation, the observed responses are assumed to be independent across subjects and characteristics.

(2) **Subject Level**. For each subject d, a membership vector $\theta_d = (\theta_{d1}, \ldots, \theta_{dK})$ represents the degrees of the subject's membership to the various subpopulations. The distribution of the observed response x_{dn} for each subject given the membership scores θ_d is then $p(x_{dn}|\theta_d) = \sum_k \theta_{dk} f(x_{dn}|\beta_{kn})$. Conditional on the mixed membership scores, the response variables x_{dn} are independent of each other, and also independent across subjects.

(3) **Latent Variable Level**. With respect to the membership scores, one could assume they are either fixed unknown constants or random realizations from some underlying distribution. For Bayesian mixed membership models, which are our focus, the latter strategy is adopted, that is, assume that θ_d are realizations of latent variables from some distribution D_α, parameterized by a vector α. The probability of observing x_{dn} is then $p(x_{dn}|\alpha, \beta) = \int \left(\sum_k \theta_{dk} f(x_{dn}|\beta_{kn})\right) D_\alpha(d\theta)$.

(4) **Sampling Scheme Level**. Suppose R independent replications of M distinct characteristics are observed for the dth subject. The conditional probability of observing $\mathbf{x}_d = \{x_{d1}^r, \ldots, x_{dM}^r\}_{r=1}^R$ given the parameters is then

$$p(\mathbf{x}_d|\alpha, \beta) = \int \left(\prod_{n=1}^M \prod_{r=1}^R \sum_{k=1}^K \theta_{dk} f(x_{dn}^r|\beta_{kn})\right) D_\alpha(d\theta). \tag{18.1}$$

Hierarchical Bayesian mixed membership models have been widely used in analyzing various forms of data, including discrete text documents (Blei et al., 2003), population genetics (Pritchard et al., 2000), social networks (Airoldi et al., 2008), and disability survey data (Erosheva, 2003). Below, we will study the mixed membership models for discrete text documents (i.e., topic models) as a test bed for mixed membership modeling ideas. But we emphasize that the methodology we will develop is applicable to a broad range of hierarchical Bayesian models.

18.2.2 Hierarchical Bayesian Topic Models

Latent Dirichlet allocation (LDA) (Blei et al., 2003) is a Bayesian mixed membership model for modeling discrete text documents. In LDA, the components or subpopulations are *topics*, of which each topic is a multinomial distribution over the M words in a given vocabulary, i.e., $\beta_k \in \mathcal{P}$, where $\mathcal{P}$ is the space of probability distributions with an appropriate dimension which will be omitted when the context is clear; and the membership scores θ_d for document d is a mixing proportion vector over the K topics. We denote the vector of words appearing in document d as $\mathbf{w}_d = (w_{d1}, \ldots, w_{dN_d})$. For the same word that appears for multiple times, there are multiple place holders in $\mathbf{w}_d$. Thus, $\mathbf{w}_d$

can be seen as a replication of appearing words. Let $\boldsymbol{\beta} = [\boldsymbol{\beta}_1; \ldots; \boldsymbol{\beta}_K]$ denote the $K \times M$ matrix of topic parameters. Under LDA, the likelihood of a document corresponds to the following generative process:

1. For document d, draw a topic mixing proportion vector $\boldsymbol{\theta}_d$: $\boldsymbol{\theta}_d | \boldsymbol{\alpha} \sim \text{Dir}(\boldsymbol{\alpha})$;
2. For the nth word in document d, where $1 \leq n \leq N_d$,
   (a) Draw a topic assignment z_{dn} according to $\boldsymbol{\theta}_d$: $z_{dn} | \boldsymbol{\theta}_d \sim \text{Mult}(\boldsymbol{\theta}_d)$;
   (b) Draw the word w_{dn} according to z_{dn}: $w_{dn} | z_{dn}, \boldsymbol{\beta} \sim \text{Mult}(\boldsymbol{\beta}_{z_{dn}})$,

where z_{dn} is a K-dimensional indicator vector (i.e., only one element is 1; all others are 0), an instance of the topic assignment random variable Z_{dn}, and $\text{Dir}(\boldsymbol{\alpha})$ is a K-dimensional Dirichlet distribution, parameterized by $\boldsymbol{\alpha}$. With a little abuse of notations, we have used $\boldsymbol{\beta}_{z_{dn}}$ to denote the topic that is selected by the non-zero element of z_{dn}.

Let $\mathbf{z}_d = \{z_{dn}\}_{n=1}^{N_d}$ denote the set of topic assignments for all the words in document d. For a corpus $\mathcal{D}$ that contains D documents, we let $\boldsymbol{\Theta} = \{\boldsymbol{\theta}_d\}_{d=1}^{D}$, $\mathbf{Z} = \{\mathbf{z}_d\}_{d=1}^{D}$, and $\mathbf{W} = \{\mathbf{w}_d\}_{d=1}^{D}$. According to the above generative process, an *unsupervised* LDA defines the joint distribution

$$p(\boldsymbol{\Theta}, \mathbf{Z}, \mathbf{W} | \boldsymbol{\alpha}, \boldsymbol{\beta}) = \prod_{d=1}^{D} p(\boldsymbol{\theta}_d | \boldsymbol{\alpha}) \left(\prod_{n=1}^{N} p(z_{dn} | \boldsymbol{\theta}_d) p(w_{dn} | z_{dn}, \boldsymbol{\beta}) \right). \tag{18.2}$$

For LDA, the learning task is to estimate the unknown parameters $(\boldsymbol{\alpha}, \boldsymbol{\beta})$. Maximum likelihood estimation (MLE) is usually applied, which solves the problem

$$\max_{\boldsymbol{\alpha}, \boldsymbol{\beta}} \ \log p(\mathbf{W} | \boldsymbol{\alpha}, \boldsymbol{\beta}), \ \text{s.t} : \boldsymbol{\beta}_k \in \mathcal{P}. \tag{18.3}$$

Once an LDA model is given (i.e., after learning), we can apply it to perform exploratory analysis for discovering underlying patterns. This task is done by deriving the posterior distribution using Bayes' rule, that is,

$$p(\boldsymbol{\Theta}, \mathbf{Z} | \mathbf{W}, \boldsymbol{\alpha}, \boldsymbol{\beta}) = \frac{p(\boldsymbol{\Theta}, \mathbf{Z}, \mathbf{W} | \boldsymbol{\alpha}, \boldsymbol{\beta})}{p(\mathbf{W} | \boldsymbol{\alpha}, \boldsymbol{\beta})}. \tag{18.4}$$

Computationally, however, the likelihood $p(\mathbf{W} | \boldsymbol{\alpha}, \boldsymbol{\beta})$ is intractable to compute exactly. Therefore, approximate inference algorithms based on variational (Blei et al., 2003) or Markov chain Monte Carlo (MCMC) methods (Griffiths and Steyvers, 2004) have been widely used for parameter estimation and posterior inference under LDA.

Note that we have restricted ourselves to treat $\boldsymbol{\beta}$ as an unknown parameter, as done in Blei and McAuliffe (2007); Wang et al. (2009). Extension to a Bayesian treatment of $\boldsymbol{\beta}$ (i.e., by putting a prior over $\boldsymbol{\beta}$ and inferring its posterior) can be easily done in LDA as shown in the literature (Blei et al., 2003), where posterior inference is to find $p(\boldsymbol{\Theta}, \mathbf{Z}, \boldsymbol{\beta} | \mathbf{W}, \boldsymbol{\alpha})$ by using Bayes' rule. As we shall see, MedLDA can also be easily extended to the full Bayesian setting under a general framework of regularized Bayesian inference.

The LDA described above does not utilize side information for learning topics and inferring topic vectors $\boldsymbol{\theta}$, which could limit their power for predictive tasks. To address this limitation, supervised topic models (sLDA) (Blei and McAuliffe, 2007) introduce a response variable Y to LDA for each document, as shown in Figure 18.1. For regression, where $y \in \mathbb{R}$, the generative process of sLDA is similar to LDA, but with an additional step—*Draw a response variable:* $y | \mathbf{z}_d, \boldsymbol{\eta}, \delta^2 \sim \mathcal{N}(\boldsymbol{\eta}^\top \bar{\mathbf{z}}_d, \delta^2)$ *for each document* d—where $\bar{\mathbf{z}}_d = \frac{1}{N} \sum_n z_{dn}$ is the average topic assignment over all the words in document d; $\boldsymbol{\eta}$ is the regression weight vector; and δ^2 is a noise variance parameter. Then, the joint distribution of sLDA is

$$p(\boldsymbol{\Theta}, \mathbf{Z}, \mathbf{y}, \mathbf{W} | \boldsymbol{\alpha}, \boldsymbol{\beta}, \boldsymbol{\eta}, \delta^2) = p(\boldsymbol{\Theta}, \mathbf{Z}, \mathbf{W} | \boldsymbol{\alpha}, \boldsymbol{\beta}) p(\mathbf{y} | \mathbf{Z}, \boldsymbol{\eta}, \delta^2), \tag{18.5}$$

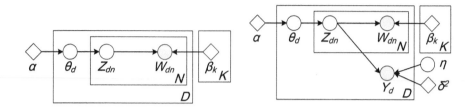

FIGURE 18.1
Graphical illustration of LDA (left) (Blei et al., 2003); and supervised LDA (right) (Blei and McAuliffe, 2007).

where $\mathbf{y} = \{y_d\}_{d=1}^{D}$ is the set of labels and $p(\mathbf{y}|\mathbf{Z}, \boldsymbol{\eta}, \delta^2) = \prod_d p(y_d|\boldsymbol{\eta}^\top \bar{\mathbf{z}}_d, \delta^2)$ due to the model's conditional independence assumption. In this case, the likelihood is $p(\mathbf{y}, \mathbf{W}|\boldsymbol{\alpha}, \boldsymbol{\beta}, \boldsymbol{\eta}, \delta^2)$ and that task of posterior inference is to find the posterior distribution $p(\boldsymbol{\Theta}, \mathbf{Z}|\mathbf{W}, \mathbf{y}, \boldsymbol{\alpha}, \boldsymbol{\beta}, \boldsymbol{\eta}, \delta^2)$ by using Bayes' rule. Again, due to the intractability of the likelihood, variational methods were used to do approximate inference and MLE.

By changing the likelihood model of Y, sLDA can deal with various types of responses, such as discrete ones for classification (Wang et al., 2009) using the multi-class logistic regression

$$p(y|\mathbf{z}_d, \boldsymbol{\eta}) = \frac{\exp(\boldsymbol{\eta}_y^\top \bar{\mathbf{z}}_d)}{\sum_{y'} \exp(\boldsymbol{\eta}_{y'}^\top \bar{\mathbf{z}}_d)}, \tag{18.6}$$

where $\boldsymbol{\eta}_y$ is the vector of parameters associated with class y. However, posterior inference in an sLDA classification model can be more challenging than that in the sLDA regression model. This is because the non-Gaussian probability distribution in Equation (18.6) is highly nonlinear in $\boldsymbol{\eta}$ and $\mathbf{z}$, and its normalization factor can make the topic assignments of different words in the same document strongly coupled. If we perform fully Bayesian inference, the likelihood is non-conjugate with the commonly used priors, e.g., a Gaussian prior over $\boldsymbol{\eta}$, and this imposes further challenges on posterior inference. Variational methods were successfully used to approximate the normalization factor (Wang et al., 2009) in an EM algorithm, but they can be computationally expensive as we shall demonstrate in the experimental section.[3]

DiscLDA (Lacoste-Julien et al., 2008) is another supervised topic model for classification. DiscLDA is an upstream model, and the unknown parameter is the transformation matrix used to generate the document latent representations conditioned on class labels. This transformation matrix is learned by maximizing the conditional marginal likelihood of the text given class labels.

This progress notwithstanding, most current developments of supervised topic models have been built on a likelihood-driven probabilistic inference paradigm. In contrast, the max-margin-based techniques widely used in learning discriminative models (Vapnik, 1998; Taskar et al., 2003) have been rarely exploited to learn supervised topic models. Our work in Zhu et al. (2012) presents the first formulation of max-margin supervised topic models,[4] followed by various work on image

---
[3]For fully Bayesian sLDA, a Gibbs sampling algorithm was developed in Zhu et al. (2013) by exploring data augmentation techniques.

[4]A preliminary version was first published in 2009 (Zhu et al., 2009).

annotation (Yang et al., 2010), classification (Wang and Mori, 2011), and entity relationship extraction (Li et al., 2011). In this chapter, we present a novel formulation of MedLDA under the general framework of regularized Bayesian inference. Below, we briefly review the max-margin principle using the example of support vector machines.

18.2.3 Support Vector Machines

Depending on the nature of the response variable, the max-margin principle can be exploited in both classification and regression. Below we use document classification as an example to recapitulate the ideas behind SVMs, which we will shortly leverage to build our max-margin topic models.

Let $\mathcal{D} = \{(\mathbf{x}_1, y_1), \cdots, (\mathbf{x}_D, y_D)\}$ be a training set, where $\mathbf{x} \in \mathcal{X}$ are inputs such as document-feature vectors, and y are categorical response values taking values from a finite set $\mathcal{Y} = \{1, \cdots, L\}$. We consider the general multi-class classification where L is greater than 2. The goal of SVMs is to find a discriminant function $h(y, \mathbf{x}; \boldsymbol{\eta}) \in \mathcal{F}$ that could make accurate predictions with the argmax rule $\hat{y} = \arg \max_y h(y, \mathbf{x}; \boldsymbol{\eta})$. One common choice of the function family $\mathcal{F}$ is linear functions, that is, $h(y, \mathbf{x}; \boldsymbol{\eta}) = \boldsymbol{\eta}_y^\top \mathbf{f}(\mathbf{x})$, where $\mathbf{f} = (f_1, \cdots, f_I)^\top$ is a vector of feature functions $f_i : \mathcal{X} \to \mathbb{R}$, and $\boldsymbol{\eta}_y$ is the corresponding weight vector associated with class y. Formally, the linear SVM finds an optimal linear function by solving the following constrained optimization problem (Crammer and Singer, 2001):[5]

$$
\min_{\boldsymbol{\eta}, \boldsymbol{\xi}} \quad \frac{1}{2} \|\boldsymbol{\eta}\|_2^2 + C \sum_{d=1}^D \xi_d \tag{18.7}
$$
$$
\text{s.t.} : \quad h(y_d, \mathbf{x}_d; \boldsymbol{\eta}) - h(y, \mathbf{x}_d; \boldsymbol{\eta}) \geq \ell_d(y) - \xi_d, \forall d, \forall y,
$$

where $\boldsymbol{\eta} = [\boldsymbol{\eta}_1^\top, \cdots, \boldsymbol{\eta}_L^\top]^\top$ is the concatenation of all subvectors; $\boldsymbol{\xi}$ are non-negative slack variables that tolerate some errors in the training data; C is a positive regularization constant; and $\ell_d(y)$ is a non-negative function that measures the cost of predicting y if the ground truth is y_d. It is typically assumed that $\ell_d(y_d) = 0$, i.e., no cost for correct predictions. The quadratic programming (QP) problem can be solved in a Lagrangian dual formulation. Samples with non-zero Lagrange multipliers are called support vectors.

18.2.4 Maximum Entropy Discrimination

The standard SVM formulation does not consider uncertainties of unknown variables, and it is thus far difficult to see how to incorporate the max-margin principle into Bayesian mixed membership models or topic models in particular. One significantly further step towards uniting the principles behind Bayesian generative modeling and max-margin learning is the maximum entropy discrimination (MED) formalism (Jebara, 2001), which learns a distribution of all possible classification models that belong to a particular parametric family, subject to a set of margin-based constraints. For instance, the MED classification model learns a distribution $q(\boldsymbol{\eta})$ through solving the following optimization problem:

$$
\min_{q(\boldsymbol{\eta}) \in \mathcal{P}, \boldsymbol{\xi}} \quad \mathrm{KL}(q(\boldsymbol{\eta}) \| p_0(\boldsymbol{\eta})) + C \sum_{d=1}^D \xi_d \tag{18.8}
$$
$$
\text{s.t.} : \quad \mathbb{E}_q[h(y_d, \mathbf{x}_d; \boldsymbol{\eta})] - \mathbb{E}_q[h(y, \mathbf{x}_d; \boldsymbol{\eta})] \geq \ell_d(y) - \xi_d, \forall d, \forall y,
$$

where $p_0(\boldsymbol{\eta})$ is a prior distribution over the parameters, and $\mathrm{KL}(p \| q) \triangleq \mathbb{E}_p[\log(p/q)]$ is the Kullback-Leibler (KL) divergence.

---

[5]The formulation implies that $\xi_d \geq 0$, since all possible predictions including y_d are included in the constraints.

As studied in Jebara (2001), this MED problem leads to an entropic-regularized posterior distribution of the SVM coefficients, $q(\boldsymbol{\eta})$; and the resultant predictor $\hat{y} = \arg\max_{y} \mathbb{E}_{q(\boldsymbol{\eta})}[h(y, \mathbf{x}; \boldsymbol{\eta})]$ enjoys several nice properties and subsumes the standard SVM as special cases when the prior $p_0(\boldsymbol{\eta})$ is standard normal. Moreover, as shown in Zhu and Xing (2009) and Zhu et al. (2011b), with different choices of the prior over $\boldsymbol{\eta}$, such as a sparsity-inducing Laplace or a nonparametric Dirichlet process, the resultant $q(\boldsymbol{\eta})$ can exhibit a wide variety of characteristics and are suitable for diverse utilities such as feature selection or learning complex non-linear discriminating functions. Finally, the recent developments of the maximum entropy discrimination Markov network (MaxEnDNet) (Zhu and Xing, 2009) and partially observed MaxEnDNet (PoMEN) (Zhu et al., 2008) have extended the basic MED to the much broader scenarios of learning structured prediction functions with or without latent variables.

In applying the MED idea to learn a supervised topic model, a major difficulty is the presence of heterogeneous latent variables in the topic models, such as the topic vector $\boldsymbol{\theta}$ and topic indicator Z. In the sequel, we present a novel formalism called maximum entropy discrimination LDA (MedLDA) that extends the basic MED to make this possible, and at the same time discovers latent discriminating topics present in the study corpus based on available discriminant side information.

18.3 MedLDA: Max-Margin Supervised Topic Models

Now we present a new class of supervised topic models that explicitly employ labeling information in the context of document classification.[6] To make our methodology general, we formalize MedLDA under the framework of regularized Bayesian inference (Zhu et al., 2011a), which can in principle be applied to any Bayesian mixed membership models with a slight change of adding some posterior constraints to consider the supervising side information.

18.3.1 Bayesian Inference as a Learning Model

As shown in Equation (18.4), Bayesian inference can be seen as an information processing rule that projects the prior p_0 and empirical data to a posterior distribution via the Bayes' rule. Under this classic interpretation, a natural way to consider supervising information is to extend the likelihood model to incorporate it, as adopted in sLDA models.

A fresh interpretation of Bayesian inference was given by Zellner (1988), which provides a novel and more natural interpretation of MedLDA, as we shall see. Specifically, the posterior distribution by Bayes' rule is in fact the solution of an optimization problem. For instance, the posterior $p(\Theta, \mathbf{Z}|\mathbf{W}, \boldsymbol{\alpha}, \boldsymbol{\beta})$ of LDA is equivalent to the optimum solution of

$$\min_{q(\Theta, \mathbf{Z}) \in \mathcal{P}} \mathrm{KL}(q(\Theta, \mathbf{Z}) \| p_0(\Theta, \mathbf{Z}|\boldsymbol{\alpha}, \boldsymbol{\beta})) - \mathbb{E}_q[\log p(\mathbf{W}|\Theta, \mathbf{Z}, \boldsymbol{\beta})]. \qquad (18.9)$$

We will use $\mathcal{L}_0(q(\Theta, \mathbf{Z}), \boldsymbol{\alpha}, \boldsymbol{\beta})$ to denote the objective function. In fact, we can show that the optimum objective value is the negative log-likelihood $-\log p(\mathbf{W}|\boldsymbol{\alpha}, \boldsymbol{\beta})$. Therefore, the MLE problem can be equivalently written in the variational form

$$\min_{\boldsymbol{\alpha}, \boldsymbol{\beta}} \left(\min_{q(\Theta, \mathbf{Z}) \in \mathcal{P}} \mathcal{L}_0(q(\Theta, \mathbf{Z}), \boldsymbol{\alpha}, \boldsymbol{\beta}) \right) = \min_{\boldsymbol{\alpha}, \boldsymbol{\beta}, q(\Theta, \mathbf{Z}) \in \mathcal{P}} \mathcal{L}_0(q(\Theta, \mathbf{Z}), \boldsymbol{\alpha}, \boldsymbol{\beta}), \qquad (18.10)$$

which is the same as the objective of the EM algorithm (Blei et al., 2003) if no mean field assumptions are made. For the case where β is random, we have the same equality as above but with β

---

[6]For regression, MedLDA can be developed as in Zhu et al. (2009).

moved from the set of unknown parameters into the distributions. For the fully Bayesian models (either treating α as random too or leaving it pre-specified), we can solve an optimization problem similar as above to infer the posterior distribution.

18.3.2 Regularized Bayesian Inference

For the standard Bayesian inference, the posterior distribution is determined by a prior distribution and a likelihood model through the Bayes' rule. Either the prior or the likelihood model *indirectly* influences the behavior of the posterior distribution. However, under the above optimization formulation of Bayes' rule, we can have an additional channel of bringing in additional side information to *directly* regularize the properties of the desired posterior distributions. Let $\mathcal{M}$ be a model containing all the variables (e.g., Θ and $\mathbf{Z}$ for LDA) whose posterior distributions we are trying to infer. Let $\mathcal{D}$ be the data (e.g., $\mathbf{W}$) whose likelihood model is defined, and let τ be hyperparameters. One formal implementation of this idea is the *regularized Bayesian inference* as introduced in Zhu et al. (2011a), which solves the constrained optimization problem

$$\min_{q(\mathcal{M}),\boldsymbol{\xi}} \mathrm{KL}(q(\mathcal{M})\|p_0(\mathcal{M}|\boldsymbol{\tau})) - \mathbb{E}_q[\log p(\mathcal{D}|\mathcal{M},\boldsymbol{\tau})] + U(\boldsymbol{\xi}) \tag{18.11}$$
$$\text{s.t.} : q(\mathcal{M}) \in \mathcal{P}_{\text{post}}(\boldsymbol{\xi}),$$

where $\mathcal{P}_{\text{post}}(\boldsymbol{\xi})$ is a subspace of distributions that satisfy a set of constraints. We assume $\mathcal{P}_{\text{post}}(\boldsymbol{\xi})$ is non-empty for all $\boldsymbol{\xi}$. The auxiliary parameters $\boldsymbol{\xi}$ are usually nonnegative and interpreted as slack variables. $U(\boldsymbol{\xi})$ is a convex function, which usually corresponds to a surrogate loss (e.g., hinge loss) of a prediction rule, as we shall see. Under the above formulation, Zhu et al. (2011a) presented the infinite latent SVM models for classification and multi-task learning. Below, we present MedLDA as another instantiation of regularized Bayesian models.

18.3.3 MedLDA: A Regularized Bayesian Model

Let $\mathcal{D} = \{(\mathbf{w}_d, y_d)\}_{d=1}^D$ be a given fully-labeled training set, where the response variable Y takes values from the finite set $\mathcal{Y}$. MedLDA consists of two parts. The first part is an LDA likelihood model for describing input documents. We choose to use an unsupervised LDA, which defines a likelihood model for $\mathbf{W}$. The second part is a mechanism to consider supervising signal. Since our goal is to discover latent representations $\mathbf{Z}$ that are good for classification, one natural solution is to connect $\mathbf{Z}$ directly to our ultimate goal. MedLDA obtains such a goal by building a classification model on $\mathbf{Z}$. One good candidate of the classification model is the max-margin method which avoids defining a normalized likelihood model.

Formally, let $\boldsymbol{\eta}$ denote the parameters of the classification model. As in MED, we treat $\boldsymbol{\eta}$ as random variables and want to infer the joint posterior distribution $q(\boldsymbol{\eta}, \Theta, \mathbf{Z}|\mathcal{D}, \boldsymbol{\alpha}, \boldsymbol{\beta})$, or $q(\boldsymbol{\eta}, \Theta, \mathbf{Z})$ for short. The classification model is defined as follows. If the latent topic representation $\mathbf{z}$ is given, MedLDA defines the linear discriminant function as

$$F(y, \boldsymbol{\eta}, \mathbf{z}; \mathbf{w}) = \boldsymbol{\eta}^\top \mathbf{f}(y, \bar{\mathbf{z}}), \tag{18.12}$$

where $\mathbf{f}(y, \bar{\mathbf{z}})$ is an LK-dimensional vector whose elements from $(y-1)K$ to yK are $\bar{\mathbf{z}}$ and all others are zero; and $\boldsymbol{\eta}$ is an LK-dimensional vector concatenating L class-specific sub-vectors. In order to predict on input data, MedLDA defines the *effective discriminant function* using the expectation operator

$$F(y; \mathbf{w}) = \mathbb{E}_{q(\boldsymbol{\eta}, \mathbf{z})}[F(y, \boldsymbol{\eta}, \mathbf{z}; \mathbf{w})], \tag{18.13}$$

which is a linear functional of q.

With the above definitions, a natural prediction rule for a given posterior distribution q is

$$\hat{y} = \arg\max_{y \in \mathcal{Y}} F(y; \mathbf{w}). \tag{18.14}$$

Then, we would like to "regularize" the properties of the latent topic representations to make them suitable for a classification task. Here, we adopt the framework of regularized Bayesian inference and impose the following max-margin constraints on the posterior distributions:

$$F(y_d; \mathbf{w}_d) - F(y; \mathbf{w}_d) \geq \ell_d(y), \ \forall y \in \mathcal{Y}, \ \forall d. \tag{18.15}$$

That is, we want to find a "posterior distribution" that can predict correctly on all the training data using the prediction rule (18.14). However, in many cases, these hard constraints would be too strict. In order to learn a robust classifier for the datasets which are not separable, a natural generalization is to impose the soft max-margin constraints

$$F(y_d; \mathbf{w}_d) - F(y; \mathbf{w}_d) \geq \ell_d(y) - \xi_d, \ \forall y \in \mathcal{Y}, \ \forall d, \tag{18.16}$$

where $\boldsymbol{\xi} = \{\xi_d\}$ are non-negative slack variables. Let

$$\mathcal{L}_1(q(\boldsymbol{\eta}, \boldsymbol{\Theta}, \mathbf{Z}), \boldsymbol{\alpha}, \boldsymbol{\beta}) = \mathrm{KL}(q(\boldsymbol{\eta}, \boldsymbol{\Theta}, \mathbf{Z}) \| p_0(\boldsymbol{\eta}, \boldsymbol{\Theta}, \mathbf{Z} | \boldsymbol{\alpha}, \boldsymbol{\beta})) - \mathbb{E}_q[\log p(\mathbf{W} | \mathbf{Z}, \boldsymbol{\beta})].$$

We define the soft-margin MedLDA model as solving

$$\min_{q(\boldsymbol{\eta}, \boldsymbol{\Theta}, \mathbf{Z}) \in \mathcal{P}, \boldsymbol{\alpha}, \boldsymbol{\beta}, \boldsymbol{\xi}} \mathcal{L}_1(q(\boldsymbol{\eta}, \boldsymbol{\Theta}, \mathbf{Z}), \boldsymbol{\alpha}, \boldsymbol{\beta}) + \frac{C}{D} \sum_{d=1}^{D} \xi_d \tag{18.17}$$

$$\text{s.t.} : \quad \mathbb{E}_q[\boldsymbol{\eta}^\top \Delta \mathbf{f}(y, \bar{\mathbf{z}}_d)] \geq \ell_d(y) - \xi_d, \ \xi_d \geq 0, \forall d, \forall y,$$

where the prior is $p_0(\boldsymbol{\eta}, \boldsymbol{\Theta}, \mathbf{Z} | \boldsymbol{\alpha}, \boldsymbol{\beta}) = p_0(\boldsymbol{\eta}) p_0(\boldsymbol{\Theta}, \mathbf{Z} | \boldsymbol{\alpha}, \boldsymbol{\beta})$, and $\Delta \mathbf{f}(y, \bar{\mathbf{z}}_d) = \mathbf{f}(y_d, \bar{\mathbf{z}}_d) - \mathbf{f}(y, \bar{\mathbf{z}}_d)$. By removing slack variables, problem (18.17) can be equivalently written as

$$\min_{q(\boldsymbol{\eta}, \boldsymbol{\Theta}, \mathbf{Z}) \in \mathcal{P}, \boldsymbol{\alpha}, \boldsymbol{\beta}} \mathcal{L}_1(q(\boldsymbol{\eta}, \boldsymbol{\Theta}, \mathbf{Z}), \boldsymbol{\alpha}, \boldsymbol{\beta}) + C\mathcal{R}(q(\boldsymbol{\eta}, \boldsymbol{\Theta}, \mathbf{Z})), \tag{18.18}$$

where

$$\mathcal{R} = \frac{1}{D} \sum_d \arg\max_y \left(\ell_d(y) - \mathbb{E}_q[\boldsymbol{\eta}^\top \Delta \mathbf{f}(y, \bar{\mathbf{z}}_d)] \right)$$

is the hinge loss, an upper bound of the prediction error on training data.

Based on the equality in Equation (18.10), we can see the rationale underlying MedLDA, which is that we want to find latent topical representations $q(\boldsymbol{\Theta}, \mathbf{Z})$ and a model parameter distribution $q(\boldsymbol{\eta})$ which on one hand tends to predict as accurate as possible on training data, while on the other hand tends to explain the data well. The two parts are closely coupled by the expected margin constraints.

Although in theory we can use either sLDA (Wang et al., 2009) or LDA as a building block of MedLDA to discover latent topical representations, as we have discussed in Section 18.2.2, inference under sLDA could be harder and slower because the probability model of discrete Y in Equation (18.6) is nonlinear over η and Z, both of which are latent variables in our case, and its normalization factor strongly couples the topic assignments of different words in the same document. Therefore, we choose to use LDA that only models the likelihood of document contents $\mathbf{W}$ but not document label Y as the underlying topic model to discover latent representations Z. Even with this likelihood model, document labels can still influence topic learning and inference because they induce margin constraints pertinent to the topical distributions. As we shall see, the resultant MedLDA classification model can be efficiently learned by utilizing existing high-performance SVM solvers. Moreover, since the goal of max-margin learning is to directly minimize a hinge loss (i.e., an upper bound of the empirical loss), we do not need a normalized distribution model for response variables Y.

Note that we have taken a full expectation to define $F(y; \mathbf{w})$ instead of taking the mode as done in latent SVMs (Felzenszwalb et al., 2010; Yu and Joachims, 2009), because expectation is a nice linear functional of the distributions under which it is taken, whereas taking the mode involves the highly nonlinear $argmax$ function for discrete Z, which could lead to a harder inference task. Furthermore, due to the same reason to avoid dealing with a highly nonlinear discriminant function, we did not adopt the method in Jebara (2001) either, which uses log-likelihood ratios to define the discriminant function when considering latent variables in MED. Specifically, in our case, the max-margin constraints would be

$$\forall d, \, \forall y, \; \log \frac{p(y_d | \mathbf{w}_d, \boldsymbol{\alpha}, \boldsymbol{\beta})}{p(y | \mathbf{w}_d, \boldsymbol{\alpha}, \boldsymbol{\beta})} \geq \ell_d(y) - \xi_d, \tag{18.19}$$

which are highly nonlinear due to the complex form of the marginal likelihood $p(y | \mathbf{w}_d, \boldsymbol{\alpha}, \boldsymbol{\beta})$. Our linear expectation operator is an effective tool to deal with latent variables in the context of maximum margin learning. In fact, besides the present work, we have successfully applied this operator to other challenging settings of learning latent variable structured prediction models with nontrivial dependence structures among output variables (Zhu et al., 2008) and learning nonparametric Bayesian models (Zhu et al., 2011b;a).

18.3.4 Optimization Algorithm for MedLDA

Although we have used the simple linear expectation operator to define max-margin constraints, the problem of MedLDA is still intractable to directly solve due to the intractability of $\mathcal{L}_1$. Below, we present a coordinate descent algorithm with a further constraint on the feasible distribution $q(\boldsymbol{\eta}, \boldsymbol{\Theta}, \mathbf{Z})$. Specifically, we impose the fully factorized mean field constraint that

$$q(\boldsymbol{\eta}, \boldsymbol{\Theta}, \mathbf{Z}) = q(\boldsymbol{\eta}) \prod_{d=1}^{D} q(\boldsymbol{\theta}_d | \boldsymbol{\gamma}_d) \prod_{n=1}^{N} q(z_{dn} | \boldsymbol{\phi}_{dn}), \tag{18.20}$$

where $\boldsymbol{\gamma}_d$ is a K-dimensional vector of Dirichlet parameters and each $\boldsymbol{\phi}_{dn}$ parameterizes a multinomial distribution over K topics. With this constraint, we have

$$F(y; \mathbf{w}_d) = \mathbb{E}_q[\boldsymbol{\eta}]^\top \mathbf{f}(y, \bar{\boldsymbol{\phi}}_d),$$

where $\bar{\boldsymbol{\phi}}_d = \mathbb{E}_q[\bar{\mathbf{z}}_d] = 1/N \sum_n \boldsymbol{\phi}_{dn}$; and the objective can be effectively evaluated since

$$\mathcal{L}_1(q(\boldsymbol{\eta}, \boldsymbol{\Theta}, \mathbf{Z}), \boldsymbol{\alpha}, \boldsymbol{\beta}) = \mathrm{KL}(q(\boldsymbol{\eta}) \| p_0(\boldsymbol{\eta})) + \mathcal{L}_0(q(\boldsymbol{\Theta}, \mathbf{Z}), \boldsymbol{\alpha}, \boldsymbol{\beta}), \tag{18.21}$$

where $\mathcal{L}_0$ can be computed as in Blei et al. (2003). By considering the unconstrained formulation (18.18), our algorithm alternates between the following steps:

1. **Solve for** $q(\boldsymbol{\eta})$: When $q(\boldsymbol{\Theta}, \mathbf{Z})$ and $(\boldsymbol{\alpha}, \boldsymbol{\beta})$ are fixed, the subproblem (in an equivalent constrained form) is to solve

$$\min_{q(\boldsymbol{\eta}) \in \mathcal{P}, \boldsymbol{\xi}} \; \mathrm{KL}(q(\boldsymbol{\eta}) \| p_0(\boldsymbol{\eta})) + \frac{C}{D} \sum_{d=1}^{D} \xi_d \tag{18.22}$$

$$\text{s.t.} : \mathbb{E}_q[\boldsymbol{\eta}]^\top \Delta \mathbf{f}(y, \bar{\boldsymbol{\phi}}_d) \geq \ell_d(y) - \xi_d, \forall d, \forall y.$$

By using Lagrangian methods, we have the optimum solution

$$q(\boldsymbol{\eta}) = \frac{1}{\Psi} p_0(\boldsymbol{\eta}) \exp\left(\boldsymbol{\eta}^\top (\sum_d \sum_y \mu_d^y \Delta \mathbf{f}(y, \bar{\boldsymbol{\phi}}_d)) \right), \tag{18.23}$$

where the Lagrange multipliers $\boldsymbol{\mu}$ are the solution of the dual problem:

$$\max_{\boldsymbol{\mu}} \quad -\log \Psi + \sum_d \sum_y \mu_d^y \Delta \ell_d(y) \tag{18.24}$$

$$\text{s.t.} : \quad \sum_y \mu_d^y \in \left[0, \frac{C}{D}\right], \forall d.$$

We can choose different priors in MedLDA for various regularization effects. Here, we consider the normal prior. For the standard normal prior $p_0(\boldsymbol{\eta}) = \mathcal{N}(0, I)$, we can get: $q(\boldsymbol{\eta})$ is a normal with a shifted mean, i.e., $q(\boldsymbol{\eta}) = \mathcal{N}(\boldsymbol{\lambda}, I)$, where $\boldsymbol{\lambda} = \sum_d \sum_y \mu_d^y \Delta \mathbf{f}(y, \bar{\boldsymbol{\phi}}_d)$, and the dual problem is

$$\max_{\boldsymbol{\mu}} \quad -\frac{1}{2} \| \sum_d \sum_y \mu_d^y \Delta \mathbf{f}(y, \bar{\boldsymbol{\phi}}_d) \|_2^2 + \sum_d \sum_y \mu_d^y \Delta \ell_d(y) \tag{18.25}$$

$$\text{s.t.} : \quad \sum_y \mu_d^y \in \left[0, \frac{C}{D}\right], \forall d.$$

The primal form of problem (18.25) is a multi-class SVM (Crammer and Singer, 2001):

$$\min_{\boldsymbol{\lambda}, \boldsymbol{\xi}} \quad \frac{1}{2} \|\boldsymbol{\lambda}\|_2^2 + \frac{C}{D} \sum_{d=1}^{D} \xi_d \tag{18.26}$$

$$\text{s.t.} : \quad \boldsymbol{\lambda}^\top \mathbb{E}[\Delta \mathbf{f}_d(y)] \geq \Delta \ell_d(y) - \xi_d, \ \forall d, \ \forall y.$$

We denote the optimum solution by $q^*(\boldsymbol{\eta})$ and its mean by $\boldsymbol{\lambda}^*$.

2. **Solve for ϕ and γ**: By keeping $q(\boldsymbol{\eta})$ at its previous optimum solution and fixing $(\boldsymbol{\alpha}, \boldsymbol{\beta})$, we have the subproblem as solving

$$\min_{\boldsymbol{\phi}, \boldsymbol{\gamma}} \quad \mathcal{L}_0(q(\boldsymbol{\Theta}, \mathbf{Z}), \boldsymbol{\alpha}, \boldsymbol{\beta}) + \frac{C}{D} \sum_{d=1}^{D} \max_{y \in \mathcal{Y}} \left(\ell_d(y) - (\boldsymbol{\lambda}^*)^\top \Delta \mathbf{f}(y, \bar{\boldsymbol{\phi}}_d)\right). \tag{18.27}$$

Since q is fully factorized, we can perform the optimization on each document separately. We observe that the constraints in MedLDA are not dependent on γ and $q(\boldsymbol{\eta})$ is also not directly connected with γ. Thus, optimizing $\mathcal{L}$ with respect to γ_d leads to the same update rule as in LDA:

$$\gamma_d \leftarrow \boldsymbol{\alpha} + \sum_{n=1}^{N} \phi_{dn}. \tag{18.28}$$

For ϕ, the constraints do affect its solution. Although in theory we can solve this subproblem using Lagrangian dual methods, it would be hard to derive the dual objective function (if possible at all). Here, we choose to update ϕ using sub-gradient methods. Specifically, let $g(\boldsymbol{\phi}, \boldsymbol{\gamma})$ be the objective function of problem (18.27). The sub-gradient is

$$\frac{\partial g(\boldsymbol{\phi}, \boldsymbol{\gamma})}{\partial \phi_{dn}} = \frac{\partial \mathcal{L}_0}{\partial \phi_{dn}} + \frac{C}{ND} (\boldsymbol{\lambda}_{\bar{y}_d}^* - \boldsymbol{\lambda}_{y_d}^*), \tag{18.29}$$

where $\bar{y}_d = \arg\max_y (\ell_d(y) + (\boldsymbol{\lambda}^*)^\top \mathbf{f}(y, \bar{\boldsymbol{\phi}}_d))$ is the loss-augmented prediction. By setting the sub-gradient equal to zero, we can get

$$\phi_{dn} \propto \exp\left(\mathbb{E}[\log \boldsymbol{\theta}_d | \gamma_d] + \log p(w_{dn} | \boldsymbol{\beta}) + \frac{C}{ND} (\boldsymbol{\lambda}_{y_d}^* - \boldsymbol{\lambda}_{\bar{y}_d}^*)\right). \tag{18.30}$$

We can see that the first two terms in Equation (18.30) are the same as in unsupervised LDA (Blei et al., 2003), and the last term is due to the max-margin formulation of MedLDA and reflects our intuition that the discovered latent topical representation is influenced by the margin constraints. Specifically, for those examples that are misclassified (i.e., $\bar{y}_d \neq y_d$), the last term will not be zero, and it acts as a regularization term that biases the model towards discovering latent representations that tend to make more accurate prediction on these difficult examples. Moreover, this term is fixed for words in the document and thus will directly affect the latent representation of the document (i.e., γ_d) and therefore leads to a discriminative latent representation. As we shall see in Section 18.4, such an estimate is more suitable for the classification task: for instance, MedLDA needs many fewer support vectors than the max-margin classifiers that are built on raw text or the topical representations discovered by LDA.

3. **Solve for α and β**: The last substep is to solve for (α, β) with $q(\eta)$ and $q(\Theta, \mathbf{Z})$ fixed. This subproblem is the same as the problem of estimating (α, β) in LDA, since the constraints do not directly act on (α, β). Therefore, we have the same update rules:

$$\beta_{kw} \propto \sum_d \sum_n \mathbb{I}(w_{dn} = w)\phi_{dn}^k, \tag{18.31}$$

where $\mathbb{I}(\cdot)$ is an indicator function that equals 1 if the condition holds, 0 otherwise. For α, the same gradient descent algorithm as in Blei et al. (2003) can be applied to find a numerical solution.

The above formulation of MedLDA has a slack variable associated with each document. This is known as the *n-slack* formulation (Joachims et al., 2009). Another equivalent formulation, which can be more efficiently solved, is the so called *1-slack* formulation. The 1-slack MedLDA can be written as follows:

$$\min_{q(\eta, \Theta, \mathbf{Z}), \alpha, \beta, \xi} \quad \mathcal{L}_1(q(\eta, \Theta, \mathbf{Z}), \alpha, \beta) + C\xi \tag{18.32}$$

$$\text{s.t.:} \quad \frac{1}{D}\sum_d \mathbb{E}_q[\eta^\top \Delta\mathbf{f}_d(\bar{y}_d)] \geq \frac{1}{D}\sum_d \Delta\ell_d(\bar{y}_d) - \xi, \forall(\bar{y}_1, \cdots, \bar{y}_D).$$

By using the above alternating minimization algorithm and the cutting plane algorithm for solving the 1-slack as well as n-slack multi-class SVMs (Joachims et al., 2009), which is implemented in the SVM^{struct} package,[7] we can solve the 1-slack or n-slack MedLDA model efficiently, as we shall see in Section 18.4.3. SVM^{struct} provides the solutions of the primal parameters λ as well as the dual parameters μ, which are needed to do inference.

18.4 Experiments

In this section, we provide qualitative as well as quantitative evaluation of MedLDA on topic estimation and document classification. For MedLDA and other topic models (except DiscLDA, whose implementation details are explained in Footnote 12), we optimize the K-dimensional Dirichlet parameters α using the Newton-Raphson method (Blei et al., 2003). For initialization, we set ϕ to be uniform and each topic β_k to be a uniform distribution plus a very small random noise; we set

---

[7]See http://svmlight.joachims.org/svm_multiclass.html.

the posterior mean of η to be zero. We have released our implementation for public use.[8] In all the experimental results, we also report the standard deviation for a topic model with five randomly initialized runs.

18.4.1 Topic Estimation

We begin with an empirical assessment of topic estimation by MedLDA on the 20 Newsgroups dataset with a standard list of stopwords[9] removed. The dataset contains about 20,000 postings in 20 related categories. We compare this with unsupervised LDA.[10] We fit the dataset to a 110-topic MedLDA model, which exploits the supervised category information, and a 110-topic unsupervised LDA, which ignores category information.

Figure 18.2 shows the 2D embedding of the inferred topic proportions θ by MedLDA and LDA using the t-SNE stochastic neighborhood embedding method (van der Maaten and Hinton, 2008), where each dot represents a document and each color-shape pair represents a category. Visually, the max-margin based MedLDA produces a good separation of the documents in different categories, while LDA does not produce a well-separated embedding, and documents in different categories tend to mix together. This is consistent with our expectation that MedLDA could produce a strong connection between latent topics and categories by doing supervised learning, while LDA ignores supervision and thus builds a weaker connection. Intuitively, a well-separated representation is more discriminative for document categorization. This is further empirically supported in Section 18.4.2. Note that a similar embedding was presented in Lacoste-Julien et al. (2008), where the transformation matrix in their model is pre-designed. The results of MedLDA in Figure 18.2 are *automatically* learned.

It is also interesting to examine the discovered topics and their relevance to class labels. In Figure 18.3a we show the top topics in four example categories as discovered by both MedLDA and LDA. Here, the semantic meaning of each topic is represented by the first ten high probability words.

To visually illustrate the discriminative power of the latent representations, i.e., the topic proportion vector θ of documents, we illustrate and compare the per-class distribution over topics for each model at the right side of Figure 18.3a. This distribution is computed by averaging the expected topic vector of the documents in each class. We can see that MedLDA yields sharper, sparser, and fast decaying per-class distributions over topics. For the documents in different categories, we can see that their per-class average distributions over topics are very different, which suggests that the topical representations by MedLDA have a good discrimination power. Also, the sharper and sparser representations by MedLDA can result in a simpler max-margin classifier (e.g., with fewer support vectors), as we shall see in Section 18.4.2. All these observations suggest that the topical representations discovered by MedLDA have a better discriminative power and are more suitable for prediction tasks (see Section 18.4.2 for prediction performance). This behavior of MedLDA is in fact due to the regularization effect enforced over ϕ as shown in Equation (18.30). On the other hand, the fully unsupervised LDA seems to discover topics that model the fine details of documents with no regard for their discrimination power (i.e., it discovers different variations of the same topic which results in a flat per-class distribution over topics). For instance, in the class *comp.graphics*, MedLDA mainly models documents using two salient, discriminative topics (T69 and T11), whereas LDA results in a much flatter distribution. Moreover, in the cases where LDA and MedLDA discover comparably the same set of topics in a given class (like *politics.mideast* and *misc.forsale*), MedLDA results in a sharper low-dimensional representation.

---

[8] See http://www.ml-thu.net/$\sim$jun/software.shtml.

[9] See http://mallet.cs.umass.edu/.

[10] We implemented LDA based on the public variational inference code by David Blei, using the same data structures as MedLDA for fair comparison.

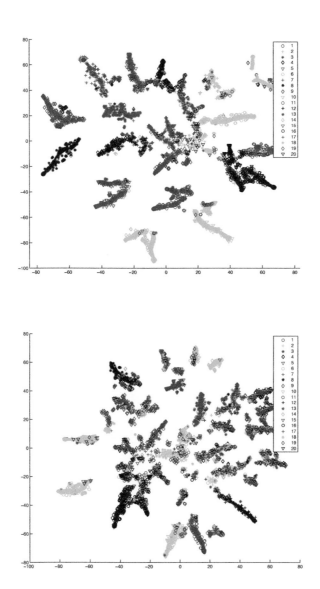

FIGURE 18.2
t-SNE 2D embedding of the topical representation by MedLDA (above) and unsupervised LDA (below). The mapping between each index and category name can be found in:
http://people.csail.mit.edu/jrennie/20Newsgroups/.

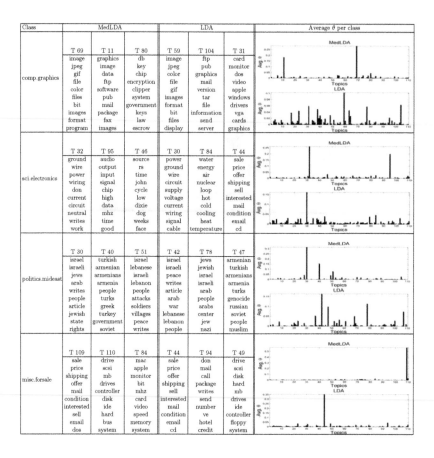

(a)

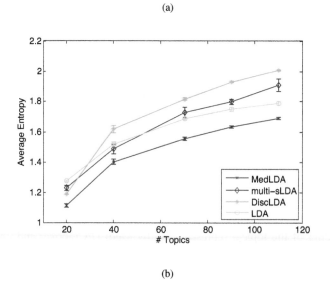

(b)

FIGURE 18.3

Top topics under each class as discovered by the MedLDA and LDA models (a). The average entropy of θ over documents on 20 Newsgroups data (b).

A quantitative measure for the sparsity or sharpness of the distributions over topics is the entropy. We compute the entropy of the inferred topic proportion for each document and take the average over the corpus. Here, we compare MedLDA with LDA, sLDA for multi-class classification (multi-sLDA) (Wang et al., 2009),[11] and DiscLDA (Lacoste-Julien et al., 2008).[12] For DiscLDA, as in Lacoste-Julien et al. (2008), we fix the transformation matrix and set it to be diagonally sparse. We use the standard training/testing split[13] to fit the models on training data and infer the topic distributions on testing documents. Figure 18.3b shows the average entropy of different models on testing documents when different topic numbers are chosen. For DiscLDA, we set the class-specific topic number $K_0 = 1, 2, 3, 4, 5$ and correspondingly $K = 22, 44, 66, 88, 110$. We can see that MedLDA yields the smallest entropy, which indicates that the probability mass is concentrated on quite a few topics, consistent with the observations in Figure 18.3a. In contrast, for LDA the probability mass is more uniformly distributed on many topics (again consistent with Figure 18.3a), which results in a higher entropy. For DiscLDA, although the transformation matrix is designed to be diagonally sparse, the distributions over the class-specific topics and shared topics are flat. Therefore, the entropy is also high. Using automatically learned transition matrices might improve the sparsity of DiscLDA.

18.4.2 Prediction Accuracy

We perform binary and multi-class classification on the 20 Newsgroup dataset. To obtain a baseline, we first fit all the data to an LDA model, and then use the latent representation of the training[14] documents as features to build a binary or multi-class SVM classifier. We denote this baseline as *LDA+SVM*.

Binary Classification

As in Lacoste-Julien et al. (2008), the binary classification is to distinguish postings of the newsgroup *alt.atheism* and the postings of the group *talk.religion.misc*. The training set contains 856 documents with a split of 480/376 over the two categories, and the test set contains 569 documents with a split of 318/251 over the two categories. Therefore, the *naive baseline* that predicts the most frequent category for all test documents has accuracy 0.672.

We compare the binary MedLDA with sLDA, DiscLDA, LDA+SVM, and the standard binary SVM built on raw text features. For supervised LDA, we use both the regression model (sLDA) (Blei and McAuliffe, 2007) and classification model (multi-sLDA) (Wang et al., 2009). For the sLDA regression model, we fit it using the binary representation (0/1) of the classes, and use a threshold 0.5 to make prediction. For MedLDA, to see whether a second-stage max-margin classifier can improve the performance, we also build a method of *MedLDA+SVM* similar to LDA+SVM. For DiscLDA, we fix the transition matrix. Automatically learning the transition matrix can yield slightly better results, as reported in Lacoste-Julien (2009). For all the above methods that utilize the class label information, they are fit *ONLY* on the training data.

---

[11]We thank the authors for providing their implementation, on which we made necessary slight modifications, e.g., improving the time efficiency and optimizing α.

[12]DiscLDA is a conditional model that uses class-specific topics and shared topics. Since the code is not publicly available, we implemented an in-house version by following the same strategy as in Lacoste-Julien et al. (2008) and share K_1 topics across classes and allocate K_0 topics to each class, where $K_1 = 2K_0$, and we varied $K_0 = \{1, 2, \cdots\}$. We should note here that Lacoste-Julien et al. (2008) and Lacoste-Julien (2009) gave an optimization algorithm for learning the topic structure (i.e., a transformation matrix), however, since the code is not available, we resorted to one of the fixed splitting strategies mentioned in the paper. Moreover, for the multi-class case, the authors only reported results using the same fixed splitting strategy we mentioned above. For the number of iterations for training and inference, we followed Lacoste-Julien (2009). Moreover, following Lacoste-Julien (2009) and personal communication with the first author, we used symmetric Dirichlet priors on β and θ, and set the Dirichlet parameters to 0.01 and $0.1/(K_0 + K_1)$, respectively.

[13]See http://people.csail.mit.edu/jrennie/20Newsgroups/.

[14]We use the training/testing split in: http://people.csail.mit.edu/jrennie/20Newsgroups/.

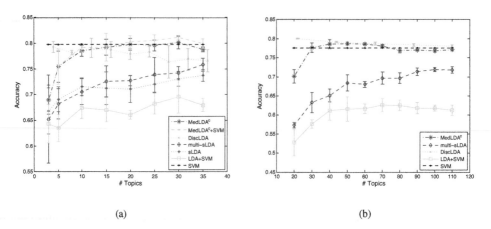

(a)   (b)

FIGURE 18.4
Classification accuracy of different models for: (a) binary and (b) multi-class classification on the 20 Newsgroup data.

We use the SVM-light (Joachims, 1999), which provides both primal and dual parameters, to build SVM classifiers and to estimate the posterior mean of η in MedLDA. The parameter C is chosen via 5-fold cross-validation during training from $\{k^2 : k = 1, \cdots, 8\}$. For each model, we run the experiments five times and take the average as the final results. The prediction accuracy of different models with respect to the number of topics is shown in Figure 18.4(a). For DiscLDA, we follow Lacoste-Julien et al. (2008) and set $K = 2K_0 + K_1$, where K_0 is the number of class-specific topics, K_1 is the number of shared topics, and $K_1 = 2K_0$. Here, we set $K_0 = 1, \cdots, 8, 10$.

We can see that the max-margin MedLDA outperforms the likelihood-based downstream models, including multi-sLDA, sLDA, and LDA+SVM. The best performances of the two discriminative models, MedLDA and DiscLDA, are comparable. However, MedLDA is easier to learn and faster in testing, as we shall see in Section 18.4.3. Moreover, the different approximate inference algorithms used in MedLDA (i.e., variational approximation) and DiscLDA (i.e., Monte Carlo sampling methods) can also make the performance different. We tried the collapsed variational inference (Teh et al., 2006) for MedLDA and it can give slightly better results. However, the collapsed variational method is computationally more expensive. Finally, since MedLDA already integrates the max-margin principle into its training, our conjecture is that the combination of MedLDA and SVM does not further improve the performance much on this task. We believe that the slight differences between MedLDA and MedLDA+SVM are due to the tuning of regularization parameters. For efficiency, we do not change the regularization constant C during training MedLDA. The performance of MedLDA would be improved if we selected a good C in different iterations because the data representation is changing.

Multi-Class Classification

We perform multi-class classification on 20 Newsgroups with all the 20 categories. The dataset has a balanced distribution over the categories. For the test set, which contains 7,505 documents in total, the smallest category has 251 documents and the largest category has 399 documents. For the training set, which contains 11,269 documents, the smallest and the largest categories contain 376 and 599 documents, respectively. Therefore, the naive baseline that predicts the most frequent category for all the test documents has the classification accuracy 0.0532.

We compare MedLDA with LDA+SVM, multi-sLDA, DiscLDA, and the standard multi-class SVM built on raw text. We use the SVM^{struct} package with a cost function as $\Delta \ell_d(y) \triangleq \ell \mathbb{I}(y \neq y_d)$ to solve the sub-step of learning $q(\eta)$ and build the SVM classifiers for LDA+SVM. The parameter

ℓ is selected with 5-fold cross-validation. The average results, as well as standard deviations over 5 randomly initialized runs, are shown in Figure 18.4(b). For DiscLDA, we use the same equation as in Lacoste-Julien et al. (2008) to set the number of topics and set $K_0 = 1, \cdots, 5$. We can see that supervised topic models discover more predictive representations for classification, and the discriminative max-margin MedLDA and DiscLDA perform comparably, slightly better than the standard multi-class SVM (about 1.3 ± 0.3 percent improvement in accuracy). However, as we have stated and will show in Section 18.4.3, MedLDA is simpler to implement and faster in testing than DiscLDA. As we shall see shortly, MedLDA needs much fewer support vectors than standard SVM.

Figure 18.5(a) shows the classification accuracy on the 20 Newsgroups dataset for MedLDA with 70 topics. We show the results with ℓ manually set to $1, 4, 8, 12, \cdots, 32$. We can see that although the common $0/1$-cost works well for MedLDA, we can get better accuracy by using a larger cost to penalize wrong predictions. The performance is quite stable when ℓ is set to be larger than 8. The reason why ℓ affects the performance is that ℓ as well as C control: 1) the scale of the posterior mean of η and the Lagrangian multipliers μ, whose dot-product regularizes the topic mixing proportions in Equation (18.30); and 2) the goodness-of-fit of the MED large-margin classifier on the data. For practical reasons, we only try a small subset of candidate C values in parameter search, which can also influence the difference on performance in Figure 18.5(a). Performing very careful parameter search on C could possibly shrink the difference. Finally, for a small ℓ (e.g., 1 for the $0/1$-cost), we usually need a large C in order to obtain good performance. But, our empirical experience with SVM^{struct} shows that the multi-class SVM with a larger C (and smaller ℓ) is typically more expensive to train than the SVM with a larger ℓ (and smaller C). That is one reason why we choose to use a large ℓ.

Figure 18.5(b) shows the number of support vectors for MedLDA, LDA+SVM, and the multi-class SVM built on raw text features, which are high-dimensional ($\sim$60,000 dimensions for the 20 Newsgroup data) and sparse. Here we consider the traditional n-slack formulation of multi-class SVM and n-slack MedLDA using the SVM^{struct} package, where a support vector corresponds to a document-label pair. For MedLDA and LDA+SVM, we set $K = 70$. For MedLDA, we report both the number of support vectors at the final iteration and the average number of support vectors over all iterations. We can see that both MedLDA and LDA+SVM generally need many fewer support vectors than the standard SVM on raw text. The major reason is that both MedLDA and LDA+SVM use a much lower-dimensional and more compact representation for each document. Moreover, MedLDA needs (about 4 times) fewer support vectors than LDA+SVM. This could be because

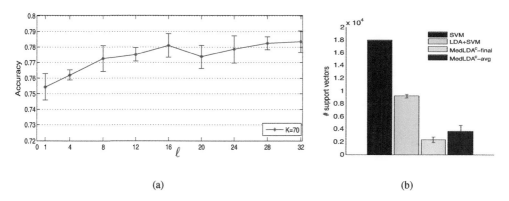

(a)   (b)

FIGURE 18.5
Sensitivity to the cost parameter ℓ for the MedLDA (a); and the number of support vectors for n-slack multi-class SVM, LDA+SVM, and n-slack MedLDA (b). For MedLDA, we show both the number of support vectors at the final iteration and the average number during training.

MedLDA makes use of both text contents and the supervising class labels in the training data, and its estimated topics tend to be more discriminative when being used to infer the latent topical representations of documents, i.e., using these latent representations by MedLDA, the documents in different categories are more likely to be well-separated, and therefore the max-margin classifier is simpler (i.e., needs fewer support vectors). This observation is consistent with what we have observed on the per-class distributions over topics in Figure 18.3a. Finally, we observe that about 32% of the support vectors in MedLDA are also the support vectors in multi-class SVM on the raw features.

18.4.3 Time Efficiency

Now, we report empirical results on time efficiency in training and testing. All the following results are achieved on a standard desktop with a 2.66GHz Intel processor. We implement all the models in C++ language.

Training Time

Figure 18.6 shows the average training time together with standard deviations on both binary and multi-class classification tasks with 5 randomly initialized runs. Here, we do not compare with DiscLDA because learning the transition matrix is not fully implemented in Lacoste-Julien (2009), but we will compare the testing time with it. From the results, we can see that for binary classification, MedLDA is more efficient than multi-class sLDA and is comparable with LDA+SVM. The slowness of multi-class sLDA is because the normalization factor in the distribution model of y strongly couples the topic assignments of different words in the same document. Therefore, the posterior inference is slower than that of LDA and MedLDA, which uses LDA as the underlying topic model. For the sLDA regression model, it takes even more training time due to the mismatch between its normal assumption and the non-Gaussian binary response variables, which prolongs the E-step.

For multi-class classification, the training time of MedLDA is mainly dependent on solving a multi-class SVM problem. Here, we implemented both 1-slack and n-slack versions of multi-class

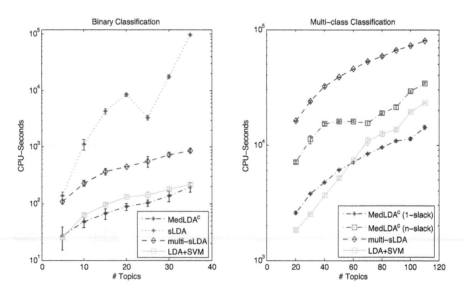

FIGURE 18.6
Training time (CPU seconds in log-scale) of different models for both binary (left) and multi-class classification (right).

SVM (Joachims et al., 2009) for solving the sub-problem of estimating $q(\eta)$ and Lagrangian multipliers in MedLDA. As we can see from Figure 18.6, the MedLDA with 1-slack SVM as the subsolver can be very efficient, comparable to unsupervised LDA+SVM. The MedLDA with n-slack SVM solvers is about three times slower. Similar to the binary case, for the multi-class supervised sLDA (Wang et al., 2009), because of the normalization factor in the category probability model (i.e., a softmax function), the posterior inference on different topic assignment variables (in the same document) is strongly correlated. Therefore, the inference is about ten times slower than that on LDA and MedLDA, which takes LDA as the underlying topic model.

We also show the time spent on inference and the ratio it takes over the total training time for different models in Figure 18.7(a). We can clearly see that the difference between 1-slack MedLDA and n-slack MedLDA is on the learning of SVMs. Both methods have similar inference time. We can also see that for LDA+SVM and multi-sLDA, more than 95% of the training time is spent on inference, which is very expensive for multi-sLDA. Note that LDA+SVM takes a longer inference time than MedLDA because we use more data (both training and testing) to learn unsupervised LDA.

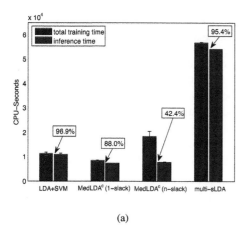 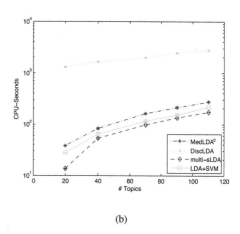

(a)   (b)

FIGURE 18.7
The inference time and total training time for learning different models, as well as the ratio of inference time over total training time (a). For MedLDA, we consider both the 1-slack and n-slack formulations; for LDA+SVM, the SVM classifier is the fast 1-slack formulation; and (b) Testing time of different models with respect to the number of topics for multi-class classification.

Testing Time

Figure 18.7(b) shows the average testing time with standard deviation on the 20 Newsgroup testing data with five randomly initialized runs. We can see that MedLDA, multi-class sLDA, and unsupervised LDA are comparable in testing time, faster than that of DiscLDA. This is because all three models of MedLDA, multi-class sLDA, and LDA are *downstream* models (see the Introduction for definition). In testing, they do exactly the same tasks, i.e., inferring the overall latent topical representation and doing prediction with a linear model. Therefore, they have comparable testing time. However, DiscLDA is an *upstream* model, for which the inference to find the category-dependent latent topic representations is done multiple times. Therefore, in principle, the testing time of an upstream topic model is about $|\mathcal{C}|$ times slower than that of its downstream counterpart model, where $\mathcal{C}$ is the finite set of categories. The results in Figure 18.7(b) show that DiscLDA is roughly twenty times slower than other downstream models. Of course, the different inference algorithms can also make the testing time different.

18.5   Conclusions and Discussions

We have presented maximum entropy discrimination LDA (MedLDA), a supervised topic model that uses the discriminative max-margin principle to estimate model parameters such as topic distributions underlying a corpus, and infer latent topical vectors of documents. MedLDA integrates the max-margin principle into the process of topic learning and inference via optimizing one single objective function with a set of *expected* margin constraints. The objective function is a tradeoff between the goodness-of-fit of an underlying topic model and the prediction accuracy of the resultant topic vectors in a max-margin classifier. We provide empirical evidence which appears to demonstrate that this integration could yield predictive topical representations that are suitable for prediction tasks, such as classification. Our results demonstrate that MedLDA is an attractive supervised topic model, which can achieve state-of-the-art performance for topic discovery and prediction accuracy while needing fewer support vectors than competing max-margin methods that are built on raw text or the topical representations discovered by unsupervised LDA.

The results of prediction accuracy on the 20 Newsgroups dataset show that MedLDA works slightly better than the SVM classifiers built on raw input features. These slight improvements tend to raise the question, "When and why should we choose MedLDA?" We have two possible answers:

1. MedLDA is a topic model. Besides predicting on unseen data, MedLDA can discover semantic patterns underlying complex data. In contrast, SVM models are more like black box machines which take raw input features and find good decision boundaries or regression curves, but that are incapable of discovering or considering hidden structures of complex data.[15] As an extension of SVM, MedLDA performs both exploratory analysis (i.e., topic discovery) and predictive tasks (e.g., classification) simultaneously. So, the first selection rule is that if we want to disclose some underlying patterns besides doing prediction, MedLDA should be preferred to SVM.

2. Even if our goal is prediction performance, MedLDA should also be considered as a competitive alternative. As shown in the synthetic experiments (Zhu et al., 2012) as well as the follow-up work (Yang et al., 2010; Wang and Mori, 2011; Li et al., 2011), depending on the data and problems, max-margin supervised topic models can outperform SVM models, or at least they are comparable if no gains are obtained. One reason that leads to our current results on 20 Newsgroups is that the fully factorized mean field assumption could be too restricted and lead to inaccurate estimates. In fact, we have tried more sophisticated inference methods such as collapsed variational inference (Teh et al., 2006) and collapsed Gibbs sampling,[16] both of which could lead to superior prediction performance.

Finally, MedLDA presents one of the first successful attempts, in the context of Bayesian mixed membership models (or topic models in particular), towards pushing forward the interface between max-margin learning and Bayesian generative modeling. As further demonstrated in others' work (Yang et al., 2010; Wang and Mori, 2011; Li et al., 2011) as well as our recent work on regularized Bayesian inference (Chen et al., 2012; Zhu et al., 2011a;b; Zhu, 2012; Xu et al., 2012), the max-margin principle could be a fruitful addition to "regularize" the desired posterior distributions of Bayesian models for performing better prediction in a broad range of scenarios, such as image annotation/classification, multi-task learning, social link prediction, low-rank matrix factorization, etc. Of course, the flexibility on performing max-margin learning brings in new challenges. For example, the learning and inference problems of such models need to deal with some non-smooth loss

---

[15]Some strategies like sparse feature selection can be incorporated to make an SVM more interpretable in the original feature space, but this is beyond the scope of this discussion.

[16]Sampling methods for MedLDA can be developed by using Lagrangian dual methods. Details are reported in Jiang et al. (2012).

functions (e.g., the hinge loss in MedLDA), for which developing efficient algorithms for large-scale applications is a challenging research problem. Moreover, although we have good theoretical understandings of the generalization ability of max-margin methods without latent variables (e.g., SVMs), it is a challenging problem to provide theoretical guarantees for the generalization performance of max-margin models with latent variables.

References

Airoldi, E. M., Blei, D. M., Fienberg, S. E., and Xing, E. P. (2008). Mixed membership stochastic blockmodels. *Journal of Machine Learning Research* : 1981–2014.

Blei, D. M. and Jordan, M. I. (2003). Modeling annotated data. In *Proceedings of the 26th Annual International ACM SIGIR Conference on Research and Development in Information Retrieval (SIGIR '03).* New York, NY, USA: ACM, 127–134.

Blei, D. M. and McAuliffe, J. (2007). Supervised topic models. In Platt, J. C., Koller, D., Singer, Y., and Roweis, S. (eds), *Advances in Neural Information Processing Systems 20.* Cambridge, MA: The MIT Press, 121–128.

Blei, D. M., Ng, A. Y., and Jordan, M. I. (2003). Latent Dirichlet allocation. *Journal of Machine Learning Research* : 993–1022.

Chechik, G. and Tishby, N. (2002). Extracting relevant structures with side information. In Becker, S., Thrun, S., and Obermayer, K. (eds), *Advances in Neural Information Processing Systems 15.* Cambridge, MA: The MIT Press, 857–864.

Chen, N., Zhu, J., Sun, F., and Xing, E. P. (2012). Large-margin predictive latent subspace learning for multiview data analysis. *IEEE Transactions on Pattern Analysis and Machine Intelligence (TPAMI)* 34: 2365–2378.

Crammer, K. and Singer, Y. (2001). On the algorithmic implementation of multiclass kernel-based vector machines. *Journal of Machine Learning Research* : 265–292.

Erosheva, E. A. (2003). Bayesian estimation of the Grade of Membership model. In Bernardo, J. M., Bayarri, M. J., Berger, J. O., Dawid, A. P., Heckerman, D., Smith, A. F. M., and West, M. (eds), *Bayesian Statistics 7.* New York, NY: Oxford University Press, 501–510.

Erosheva, E. A., Fienberg, S. E., and Lafferty, J. D. (2004). Mixed-membership models of scientific publications. *Proceedings of National Academy of Sciences* 101 : 5220–5227.

Fei-Fei, L. and Perona, P. (2005). A Bayesian hierarchical model for learning natural scene categories. In *Proceedings of the 10th IEEE Computer Vision and Pattern Recognition (CVPR 2005).* San Diego, CA, USA: IEEE Computer Society, 524–531.

Felzenszwalb, P., Girshick, R., McAllester, D., and Ramanan, D. (2010). Object detection with discriminatively trained part based models. *IEEE Transactions on Pattern Analysis and Machine Intelligence* 32: 1627 – 1645.

Griffiths, T. L. and Steyvers, M. (2004). Finding scientific topics. *Proceedings of the National Academy of Sciences* 101 Suppl 1: 5228–5235.

Jaakkola, T., Meilă, M., and Jebara, T. (1999). Maximum entropy discrimination. In Solla, S. A., Leen, T. K., and Müller, K. -R. (eds), *Advances in Neural Information Processing Systems 12*. Cambridge, MA: The MIT Press, 470–476.

Jebara, T. (2001). Discriminative, Generative and Imitative Learning. Ph.D. thesis, Media Laboratory, Massachusetts Institute of Technology, Cambridge, Massachusetts, USA.

Jiang, Q., Zhu, J., Sun, M., and Xing, E. P. (2012). Monte Carlo methods for maximum margin supervised topic models. In Bartlett, P., Pereira, F. C. N., Burges, C. J. C., Bottou, L., and Weinberger, K. Q. (eds), *Advances in Neural Information Processing Systems 25*. Red Hook, NY: Curran Associates, Inc., 1601–1609.

Joachims, T. (1999). Making large-scale SVM learning practical. In Schölkopf, B., Burges, C. J. C., and Smola, A. J. (eds), *Advances in Kernel Methods–Support Vector Learning*. Cambridge, MA: The MIT Press.

Joachims, T., Finley, T., and Yu, C. -N. (2009). Cutting-plane training of structural SVMs. *Machine Learning Journal* 77 : 27–59.

Lacoste-Julien, S. (2009). Discriminative Machine Learning with Structure. Ph.D. thesis, EECS Department, University of California, Berkeley, California, USA.

Lacoste-Julien, S., Sha, F., and Jordan, M. I. (2008). DiscLDA: Discriminative learning for dimensionality reduction and classification. In Koller, D., Schuurmans, D., Bengio, Y., and Bottou, L. (eds), *Advances in Neural Information Processing Systems 21*. Red Hook, NY: Curran Assoiates, Inc., 897–904.

Li, D., Somasundaran, S., and Chakraborty, A. (2011). A combination of topic models with max-margin learning for relation detection. In *Proceedings of TextGraphs-6: Workshop on Graph-based Methods for Natural Language Processing (ACL-HLT 2011)*. The Association for Computational Linguistics, 1–9.

Pritchard, J. K., Stephens, M., Rosenberg, N. A., and Donnelly, P. (2000). Association mapping in structured populations. *American Journal of Human Genetics* 67: 170–181.

Russell, B. C., Torralba, A., Murphy, K. P., and Freeman, W. T. (2008). LabelMe: A database and web-based tool for image annotation. *International Journal of Computer Vision* 77: 157–173.

Sudderth, E. B., Torralba, A., Freeman, W., and Willsky, A. S. (2005). Learning hierarchical models of scenes, objects, and parts. In *Proceedings of the 10th IEEE International Conference on Computer Vision (ICCV 2005), Vol. 2*. Los Alamitos, CA, USA: IEEE Computer Society, 1331–1338.

Taskar, B., Guestrin, C., and Koller, D. (2003). Max-margin Markov networks. In Thrun, S., Saul, L. K., and Schölkopf, B. (eds), *Advances in Neural Information Processing Systems 16*. Cambridge, MA: The MIT Press, 25–32.

Teh, Y. W., Newman, D., and Welling, M. (2006). A collapsed variational Bayesian inference algorithm for latent Dirichlet allocation. In Schölkopf, B., Platt, J., and Hofmann, T. (eds), *Advances in Neural Information Processing Systems 19*. Red Hook, NY: Curran Associates, Inc., 1343–1350.

Titov, I. and McDonald, R. (2008). A joint model of text and aspect ratings for sentiment summarization. In *Proceedings of the Annual Meeting of the Association for Computational Linguistics (ACL-08)*. Columbus, OH, USA: Association for Computational Linguistics, 308–316.

van der Maaten, L. and Hinton, G. (2008). Visualizing data using t-SNE. *Journal of Machine Learning Research*9 : 2579–2605.

Vapnik, V. (1998). *Statistical Learning Theory*. New York, NY: John Wiley & Sons.

Wang, C., Blei, D. M., and Fei-Fei, L. (2009). Simultaneous image classificationn and annotation. In *Proceedings of the 2009 IEEE Conference on Computer Vision and Pattern Recognition (CVPR 2009)*. Los Alamitos, CA, USA: IEEE Computer Society, 1903–1910.

Wang, Y. and Mori, G. (2011). Max-margin latent Dirichlet allocation for image classification and annotation. In *Proceedings of the 22nd British Machine Vision Conference (BMVC 2011)*. BMVA Press.

Xu, M., Zhu, J., and Zhang, B. (2012). Bayesian nonparametric maximum margin matrix factorization for collaborative prediction. In Bartlett, P., Pereira, F. C. N., Burges, C. J. C., Bottou, L., and Weinberger, K. Q. (eds), *Advances in Neural Information Processing Systems 25*. Red Hook, NY: Curran Associates, Inc., 64–72.

Yang, S., Bian, J., and Zha, H. (2010). Hybrid generative/discriminative learning for automatic image annotation. In Grünwald, P. and Spirtes, P. (eds), *Proceedings of the 26th Conference on Uncertainty in Artificial Intelligence (UAI 2010)*. Corvallis, OR, USA: AUAI Press, 683–690.

Yu, C. -N. and Joachims, T. (2009). Learning structural SVMs with latent variables. In Bottou, L. and Littman, L. (eds), *Proceedings of the 26th International Conference on Machine Learning (ICML '09)*. Omnipress, 1169–1176.

Zellner, A. (1988). Optimal information processing and Bayes's theorem. *American Statistician* 42: 278–280.

Zhu, J. (2012). Max-margin nonparametric latent feature models for link prediction. In *Proceedings of the 29th International Conference on Machine Learning (ICML '12)*. Omnipress, 719–726.

Zhu, J., Ahmed, A., and Xing, E. P. (2009). MedLDA: Maximum margin supervised topic models for regression and classification. In *Proceedings of the 26th International Conference on Machine Learning (ICML '09)*. Omnipress, 1257–1264.

Zhu, J., Ahmed, A., and Xing, E. P. (2012). MedLDA: Maximum margin supervised topic models. *Journal of Machine Learning Research (JMLR)* 13 : 2237–2278.

Zhu, J., Chen, N., and Xing, E. P. (2011a). Infinite latent SVM for classification and multi-task learning. In Shawe-Taylor, J., Zemel, R. S., Bartlett, P., Pereira, F., and Weinberger, K. Q. (eds), *Advances in Neural Information Processing Systems 24*. Red Hook, NY: Curran Associates, Inc., 1620–1628.

Zhu, J., Chen, N., and Xing, E. P. (2011b). Infinite SVM: A Dirichlet process mixture of large-margin kernel machines. In *Proceedings of the 28th International Conference on Machine Learning (ICML '11)*. Omnipress, 617–624.

Zhu, J. and Xing, E. P. (2009). Maximum entropy discrimination Markov networks. *Journal of Machine Learning Research* 10 : 2531–2569.

Zhu, J., Xing, E. P., and Zhang, B. (2008). Partially observed maximum entropy discrimination Markov networks. In Koller, D., Schuurmans, D., Bengio, Y., and Bottou, L. (eds), *Advances in Neural Information Processing Systems 21*. Red Hook, NY: Curran Associates, Inc., 1924–1931.

Zhu, J., Zheng, X., and Zhang, B. (2013). Improved Bayesian logistic supervised topic models with data augmentation. In *Proceedings of the 51st Annual Meeting of the Association for Computational Linguistics (ACL 2013)*.

Part V

Special Methodology for Sequence and Rank Data

19

Population Stratification with Mixed Membership Models

Suyash Shringarpure

School of Medicine, Stanford University, Stanford, CA 94305, USA

Eric P. Xing

Machine Learning Department, Carnegie Mellon University, Pittsburgh, PA 15213, USA

CONTENTS

Population stratification (or population structure) is the presence of genotypic differences among different groups of individuals. Genotype-based clustering of individuals is an important way of summarizing the genetic similarities and differences between groups of individuals. It enables us to formulate and test hypotheses regarding evolutionary history of populations. Identifying population stratification is also important in genetic association studies and other population genetic analyses.

We present *blockmStruct*, a mixed membership model for identifying population stratification from single nucleotide polymorphism (SNP) data. Our model incorporates mutations in SNP haplotype blocks in the ancestry inference. We demonstrate using simulation data that *mStruct* recovers ancestry more accurately than similar methods without mutation models. We analyze SNP data from 597 individuals from The Human Genome Diversity Project and the HapMap project and show that the recovered population structure recapitulates geographic patterns of diversity.

19.1   Introduction

Due to the continuing improvements in sequencing technologies and their falling costs, a number of genomic datasets such as HapMap (Gibbs, 2003) and the Human Genome Diversity Project (HGDP) (Cavalli-Sforza, 2005) are now available for study. These datasets contain individuals from various ethno-linguistic groups and regions across the world. An important task, therefore, is to characterize the genetic variation present in a given sample of individuals. Genotype clustering is one way of identifying the population stratification in a given sample of individuals. It provides a summarization of individuals based on genetic similarity and differences that can be interpreted and visualized easily. We can use the resulting summary to propose and test hypotheses about the evolutionary history of populations. Detecting population stratification present in a sample of individuals is also essential for reducing false positives in genetic association studies.

A number of evolutionary processes such as mutation, recombination, selection, admixture, migrations, expansions, and bottlenecks affect the ancestry and genomes of a group of individuals. These processes result in genomes which have contributions from more than one ancestral population. Different parts of an individual's genome can be inherited from ancestors of different populations. The *Structure* model by Pritchard et al. (2000) was one of the early attempts to address the problem of clustering individuals while allowing partial membership in multiple ancestral populations. It used a mixed membership model to determine the fractional contributions from multiple ancestral populations to an individual genome. Various extensions to the underlying model that account for other evolutionary processes such as mutation (Shringarpure and Xing, 2009) and recombination (Falush et al., 2003) have also been proposed. Figure 19.1 shows the representation of ancestry vectors for a set of individuals from the HAPMAP dataset.

We consider here a mixed membership model that takes into account mutations during inheritance from ancestral populations to modern populations. Our model is a modification of the *mStruct* model proposed in Shringarpure and Xing (2009). It considers single nucleotide polymorphisms (SNPs) to occur in unlinked haplotype blocks and hypothesizes that mutations occur within the SNP blocks during inheritance. We validate the model on simulated data and show how modeling mutations in haplotype blocks can allow us to accurately recover ancestry when mutations are present. We show results on data from 597 individuals in the Human Genome Diversity Project (HGDP) at 10,000 SNPs on chromosome 1.

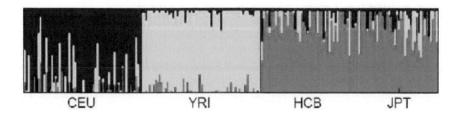

FIGURE 19.1
Analysis of individuals from the HAPMAP dataset assuming $K = 3$ ancestral populations. The mixed membership vector underlying each individual is represented as a thin vertical line of unit length and multiple colors, with the height of each color reflecting the fraction of the individual's genome originating from a certain ancestral population denoted by that color and formally represented by APs.

19.2 Related Work

A number of approaches have been proposed for detecting population stratification from genomic data. One class of methods uses low-dimensional projections and eigenanalysis to cluster individuals (Patterson et al., 2006). These methods do not assume any specific evolutionary model for the genomic data. Their advantages are their efficiency and the ability to describe the statistical significance of the stratification produced.

Another class of methods for population stratification assumes an explicit evolutionary model for the genomic data. These model-based stratification methods, starting with the *Structure* model by Pritchard et al. (2000), have become very popular due to their interpretability. The *Structure* method uses a hierarchical Bayesian model to capture the effect of admixture on modern genomes. The underlying framework for these methods is the mixed membership model of Erosheva et al. (2004). Such a model postulates that a genome, or the ensemble of genetic markers of an individual, is made up of independently and identically distributed (*iid*) samples (Pritchard et al., 2000) from multiple population-specific multinomial distributions (known as *allele frequency profiles*, or AP) of marker alleles. The mixed membership model represents each ancestral population by a specific AP which defines a unique vector of allele frequencies of each marker in each ancestral population. The fraction of contributions from each AP in a modern individual genome is represented as an *admixing vector* (also known as an *ancestral proportion vector* or *structure vector*) in a *structural map* over the population sample.

The model parameters are inferred using Markov Chain Monte Carlo (MCMC) sampling. The drawback of the *Structure* method is that it is slow compared to eigenanalysis and cannot be used efficiently for large datasets. Recently, some methods such as ADMIXTURE (Alexander et al., 2009) and Frappe (Tang et al., 2005) have proposed computational improvements to *Structure* using faster optimization methods for learning the *Structure* model parameters.

Various extensions to *Structure* have been proposed to account for evolutionary processes such as mutation (Shringarpure and Xing, 2009) and recombination (Falush et al., 2003). Shringarpure and Xing (2009) extend the *Structure* model by allowing allele mutations in the assumed evolutionary model. Their results on microsatellite data from the 52 populations in the HGDP show that modeling allele mutations affects the ancestry inference and the inferred ancestry proportions (mixed memberships) for the individuals. They also present results on the accumulated mutation among the individuals relative to the inferred ancestral populations. However, the *mStruct* model fails to extract any mutation information from the HGDP SNP data, possibly due to a simplistic mutation model for SNP data.

In the following sections, we present a modification of the *mStruct* model, which we call *blockmStruct*, for analysis of SNP data. Our model assumes that a modern genome is composed of SNP halotype blocks which are not linked to each other and that mutations occur within a haplotype block. We will first introduce the *Structure* model, present the *mStruct* model as an improvement to *Structure*, and then develop *blockmStruct* as a modification that can analyze SNP data and account for linkage disequilibrium.

19.3 The *Structure* Model

The *Structure* model by Pritchard et al. (2000) represents the earliest uses of mixed membership models in the context of modeling genetic data. It assumes that genomes of modern individuals are composed of a mixture of ancestral populations. The details of the *Structure* model can be understood by examining the choice of representation of individuals, ancestral populations, and the underlying generative process.

19.3.1 Representation of Individuals

For the following discussion, we will assume that the genomic data for each of the N individuals is a set of I loci and at the ith loci; the alleles observed are represented as $\{a_{i,1}, \cdots, a_{i,L_i}\}$ (therefore L_i denotes the number of alleles observed at locus i). We consider the case of diploid human data, i.e., each chromosome has two copies. Therefore, the eth copy ($e \in \{1,2\}$) of the ith locus in the nth individual can be represented as $x_{i,n_e} \in \{a_{i,1}, \cdots, a_{i,L_i}\}$. This representation can be used for all polymorphic markers, for instance, microsatellites (repeats of a 6–8 base pair DNA unit, represented as integer counts) and SNPs (single nucleotide polymorphisms, represented as 0/1).

The *Structure* model assumes that all the loci in an individual's genome are independent of each other. The genome for the nth individual is therefore given by the set of alleles $\{x_{1,n_1}, x_{1,n_2}, \cdots, x_{I,n_1}, x_{I,n_2}\}$.

19.3.2 Representation of Ancestral Populations

An intuitive representation for characterizing the allelic diversity observed at a polymorphic locus is in terms of their allele frequencies. These (multinomial) allele frequency distributions are called *allele frequency profiles* (or APs) (Falush et al., 2003). The *Structure* model represents ancestral populations as a collection of allele frequency profiles, one per locus. We can represent an ancestral population k by a unique set of population-specific multinomial distributions $\lambda^k = \{\lambda_i^k, i = 1, \cdots, I\}$, where $\lambda_i^k = [\lambda_{i,1}^k, \cdots, \lambda_{i,L_i}^k]$ is the vector of multinomial parameters, also known as an AP (Falush et al., 2003), of the allele distribution at locus i in ancestral population k; L_i denotes the total number of observed marker alleles at locus i; and I denotes the total number of marker loci. This representation, known as population-specific allele-frequency profiles, is used by the program *Structure*.

Under an AP, the probability of an allele x at locus i given its ancestral population of origin k is given by

$$P(x_i|\lambda^k) = \sum_{l=1}^{L_i} \mathcal{I}[x_i = a_l^i]\lambda_{i,l}^k, \tag{19.1}$$

where $\mathcal{I}[.]$ is the indicator function which takes value 1 when the included condition is true and 0 otherwise.

In the case of SNPs, the multinomial distribution can be reduced to a Bernoulli distribution since there are only two alleles. Then we can represent λ_k^i as the parameter of the Bernoulli distribution for the kth ancestral population at the ith SNP.

19.3.3 Generative Process

In a general mixed membership model, the nth individual is represented by a mixed membership vector (or ancestry vector) $\lambda_n = \{\lambda_{n,1}, \cdots, \lambda_{n,K}\}$ that represents the individual-specific fractional contributions of the K ancestral populations to the genome. For every individual, the alleles at all loci may be inherited from founders in different ancestral populations, each represented by a unique distribution of founding alleles and the way they can be inherited. Formally, this scenario can be captured in the following generative process:

1. For each individual n, draw the mixed membership vector (or ancestry vector) $\lambda_n \sim P(\cdot|\alpha)$, where $P(\cdot|\alpha)$ is a pre-chosen structure prior.

2. For each marker allele $x_{i,n_e} \in \mathbf{x}_n$

    (a) Draw the latent *ancestral-population-origin* indicator $z_{i,n_e} \sim \text{Multinomial}(\cdot|\lambda_n)$.

    (b) Draw the allele $x_{i,n_e}|z_{i,n_e} = k \sim P_k(\cdot|\lambda_i^k)$.

In *Structure*, the ancestral populations are represented by a set of population-specific APs. Thus the distribution $P_k(\cdot|\lambda^k)$ from which an observed allele can be sampled is a multinomial distribution defined by the frequencies of all observed alleles in the ancestral population, i.e., $x_{i,n_e}|z_{i,n_e} = k \sim \text{Multinomial}(\cdot|\beta_i^k)$. Figure 19.2 shows the graphical model representation of *Structure*.

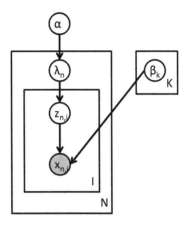

FIGURE 19.2
Graphical model representation of *Structure*. For convenience, we have ignored the diploid nature of the observation. The shaded node indicates the variables we observe.

This model has been generalized to allow linked loci and correlated allele frequencies (Falush et al., 2003). It has been successfully applied to human genetic data in Rosenberg et al. (2002). Figure 19.3 shows the ancestry vectors representation for 1,048 individuals from 53 groups in the Human Genome Diversity Project (HGDP) (Rosenberg et al., 2002). From the figure, we can see that ancestry vectors for individuals within a continent are more similar to each other than ancestry vectors for individuals in different continents.

19.4 The *mStruct* Model

The *Structure* model provides a method for inferring the ancestry of individuals as contributions from multiple (hypothetical) ancestral populations represented as APs. But a serious pitfall of using such a model is that there is no mutation model for individual alleles with respect to the common prototypes, i.e., every unique allele measurement at a particular locus is assumed to correspond to a unique ancestral allele, rather than allowing the possibility of it just being derived from some common ancestral allele at that locus as a result of a mutation (Excoffier and Hamilton, 2003). The *mStruct* model was proposed as an extension of the *Structure* model to account for the possibility of allele mutations (Shringarpure and Xing, 2009). We will present the *mStruct* model in terms of its differences with respect to the *Structure* model, which is the representation of ancestral populations and the generative process.

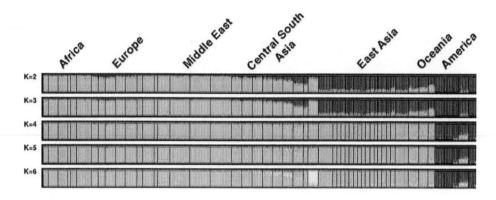

FIGURE 19.3
Ancestry vectors for 1,048 individuals from the Human Genome Diversity Project with K ranging from 2 to 6.

19.4.1 Representation of Ancestral Populations

An AP does not enable us to model the possibility of mutations, i.e., there is no way of representing a situation where two observed alleles might have been derived from a single ancestral allele by two different mutations. This possibility can be represented by a genetically more realistic statistical model known as the *population-specific mixture of ancestral alleles* (MAA). For each locus i, an MAA for ancestral population k is a set $\theta_i^k \equiv \{\mu_i^k, \delta_i^k, \beta_i^k\}$ consisting of three components: 1) a set of *ancestral* (or founder) alleles $\mu_i^k \equiv \{\mu_{i,1}^k, \ldots, \mu_{i,L_i'}^k\}$, which can differ from their descendent alleles in the modern population; 2) a mutation parameter δ_i^k associated with the locus, which can be further generalized to be allele-specific if necessary; and 3) an AP β_i^k which now represents the frequencies of the *ancestral* alleles. Here L_i' denotes the total number of *ancestral* alleles at loci i, which is different from L_i in the previous subsection, which denotes the total number of *observed* alleles at loci i. By explictly associating a mutation model with an ancestral population, we can now capture mutation events as described above. It is important to note that the mutation parameter δ is not the mutation rate commonly referred to in the literature. As we shall see later, it is a measure of the variability of a locus which can be described approximately as the combined effect of the per-generation mutation rate and the age of the population.

An MAA is more expressive than an AP, because the incorporation of a mutation model helps to capture details about the population structure which an AP cannot; and the MAA reduces to the AP when the mutation rates (and hence the mutation parameters) become zero and the founders are identical to their descendents. MAA is also arguably more realistic because it allows mutation rates (and mutation parameters) to be different for different founder alleles, even within the same ancestral population, as is commonly the case with many genetic markers. For example, the mutation rates for microsatellite alleles are believed to be dependent on their length (number of repeats). As we shall show shortly, with an MAA, one can examine the mutation parameters corresponding to each ancestral population via Bayesian inference from genotype data; this might enable us to infer the age of alleles and also estimate population divergence times subject to a calibration constant.

Under an MAA specific to an ancestral population k, the correspondence between a marker allele X_{i,n_e} and a founder $\mu_{i,l}^k \in \mu_i^k$ is not directly observable. For each allele founder $\mu_{i,l}^k$, we associate with it an inheritance model $p(\cdot|\mu_{i,l}^k, \delta_{i,l}^k)$ from which descendants can be sampled. Then, given specifications of the ancestral population from which X_{i,n_e} is derived, which is denoted by hidden indicator variable Z_{i,n_e}, the conditional distribution of X_{i,n_e} under MAA follows a mixture of population-specific inheritance models:

$$P(x_{i,n_e} = a_{i,l'} \mid z_{i,n_e} = k) = \sum_{l=1}^{L} \lambda_{i,l}^{k} \times p(x_{i,n_e} \mid \mu_{i,l}^{k}, \delta_{i,l}^{k}). \qquad (19.2)$$

Comparing to the counterpart of this function under AP: $P(x_{i,n_e} = a_{i,l'} \mid z_{i,n_e} = k) = \lambda_{i,l'}^{k}$, we can see that the latter cannot explicitly model allele diversities in terms of molecular evolution from the founders.

19.4.2 Generative Process

Recall that in an MAA for each locus we define a finite set of founders with prototypical alleles $\mu_i^k \equiv \{\mu_{i,1}^k, \ldots, \mu_{i,L_i}^k\}$ that can be different from the alleles observed in a modern population; each founder is associated with a unique frequency $\beta_{i,l}^k$, and a unique (if desired) mutation model from the prototype allele parameterized by rate $\delta_{i,l}^k$. Under this representation, the distribution $P_k(\cdot \mid \beta_i^k)$ from which an observed allele can be sampled becomes a mixture of inheritance models each defined on a specific founder; and the ensuing sampling module that can be plugged into the general admixture scheme outlined earlier (to replace step 2) becomes a two-step generative process:

1. Draw the latent founder indicator $c_{\cdot i, n_e} \mid z_{\cdot i, n_e} = k \sim \text{Multinomial}(\cdot \mid \beta_i^k)$;

2. Draw the allele $x_{\cdot i, n_e} \mid c_{\cdot i, n_e} = l, z_{\cdot i, n_e} = k \sim P_m(\cdot \mid \mu_{i,l}^k, \delta_{i,l}^k)$,

where $P_m()$ is a mutation model that can be flexibly defined based on whether the genetic markers are microsatellites or single nucleotide polymorphisms.

For simplicity of presentation, in the model described above, we assume that the set of founder alleles (but not their frequencies) at a particular locus is the same for all ancestral populations (i.e., $\mu_i^k \equiv \mu_i$). We shall also assume that the mutation parameters for each population at any locus are independent of the alleles at that locus (i.e., $\delta_{i,l}^k \equiv \delta_i^k$). Also, our model assumes Hardy-Weinberg equilibrium within populations and that loci are not linked to each other.

Figure 19.4 shows the graphical model representation of *mStruct*. Comparing it to Figure 19.2, we can see that *mStruct* includes an extra step in the generative process that allows for mutations.

Microsatellite Mutation Model

Microsatellites are a class of tandem-repeat loci that involve a DNA unit that is 1–4 basepair in length. Microsatellite DNA has significantly higher mutation rates as compared to other DNA, with mutation rates as high as 10^{-3} or 10^{-4} (Kelly et al., 1991; Henderson and Petes, 1992). The large amount of variations present in microsatellite DNA make it ideal for differentiating founder patterns between closely related populations. Microsatellite loci have been used before in DNA fingerprinting (Queller et al., 1993), linkage analysis (Dietrich et al., 1992), and in the reconstruction of human phylogeny (Bowcock et al., 1994). By applying theoretical models of microsatellite evolution to data, questions such as time of divergence of two populations can be attempted to be addressed (Pisani et al., 2004; Zhivotovsky et al., 2004).

The choice of a suitable microsatellite mutation model is important, for both computational and interpretation purposes. Below we discuss the mutation model that we use and the biological interpretation of the parameters of the mutation model. We begin with a stepwise mutation model for microsatellites widely used in forensic analysis (Valdes et al., 1993; Lin et al., 2006). This model defines a conditional distribution of a progeny allele b given its progenitor allele a,

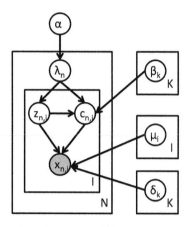

FIGURE 19.4
Graphical model representation of *mStruct*. For convenience, we have ignored the diploid nature of
the observation. The shaded node indicates the variables we observe.

both of which take continuous values:

$$p(b|a) = \frac{1}{2}\xi(1 - \delta)\delta^{|b-a|-1},\tag{19.3}$$

where ξ is the mutation rate (probability of any mutation), and δ is the factor by which mutation decreases as distance between the two alleles increases. Although this mutation distribution is
not stationary (i.e., it does not ensure allele frequencies to be constant over the generations), it is
commonly used in forensic inference due to its simplicity. To some degree δ can be regarded as a
parameter that controls the probability of unit-distance mutation, as can be seen from the following
identity: $p(b+1|a)/p(b|a) = \delta$.

In practice, the alleles for almost all microsatellites are represented by discrete counts. The
two-parameter stepwise mutation model described above complicates the inference procedure. We
propose a discrete microsatellite mutation model that is a simplification of Equation (19.3), but cap-
tures its main idea. We posit that: $P(b|a) \propto \delta^{|b-a|}$. Since $b \in [1, \infty)$, the normalization constant of
this distribution is:

$$\begin{aligned}
\sum_{b=1}^{\infty} P(b|a) &= \sum_{b=1}^{a} \delta^{a-b} + \sum_{b=a+1}^{\infty} \delta^{b-a}\\
&= \frac{1 - \delta^a}{1 - \delta} + \frac{\delta}{1 - \delta}\\
&= \frac{1 + \delta - \delta^a}{1 - \delta},
\end{aligned}$$

which gives the mutation model as

$$P(b|a) = \frac{1 - \delta}{1 - \delta^a + \delta}\delta^{|b-a|}.\tag{19.4}$$

We can interpret δ as a variance parameter, the factor by which probability drops as a fuction of
the distance between the mutated version b of the allele a. Figure 19.5 shows the discrete pdf for
various values of δ.

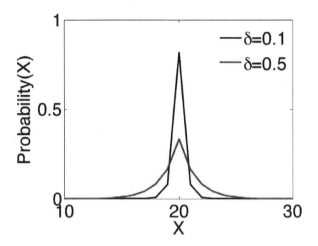

FIGURE 19.5
Discrete pdf for two values of mutation parameter.

Determination of Founder Set at Each Locus

According to our model assumptions, there can be a different number of founder alleles at each locus. This number is typically smaller than the number of alleles observed at each marker since the founder alleles are "ancestral." To estimate the appropriate number and allele states of founders, we fit finite mixtures (of fixed size, corresponding to the desired number of ancestral alleles) of microsatellite mutation models over all the measurements at a particular marker for all individuals. We use the Bayesian information criterion (BIC) (Schwarz, 1978) to determine the best number and states of founder alleles to use at each locus, since information criteria tend to favor smaller numbers of founder alleles which fit the observed data well.

For each locus, we fit many different finite-sized mixtures of mutation distributions, with the size varying from 1 to the number of observed alleles at the locus. For each mixture size, the likelihood is optimized and a BIC value is computed. The number of founder alleles is chosen to be the size of the mixture that has the best (minimum) BIC value. We can do this as a pre-processing step before the actual inference or estimation procedures. This is possible since we assumed that the set of founder alleles at each locus was the same for all populations.

19.4.3 Result on HGDP Data

Analyzing the HGDP data described earlier allows us to gain more insights about the structure of human populations. Figure 19.6 shows the results of population structure analysis of the HGDP data using *mStruct* for $K = 4$ (chosen using BIC) compared with results using *Structure* for the same value of K.

From the figure, we can see that while both methods achieve similar clusters of individuals by continent, the *mStruct* ancestral proportions indicate a significant level of similiarity even between individuals from different continents.

Analysis of Mutations

mStruct also models the mutations from alleles in ancestral populations to the observed alleles in modern populations. This information can be used to reconstruct the estimated mutation

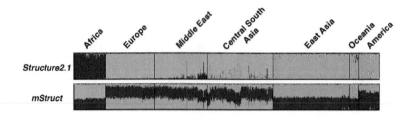

FIGURE 19.6
Population structure using *Structure* and *mStruct* for $K = 4$.

accumulated within an individual. Using the latitude and longitude labels associated with the locations for each individual, we can construct a function that maps the geographical coordinates (latitude and longitude) to an estimate of the accumulated mutation at that location. Figure 19.7 shows the contours of this function overlaid on the world map.

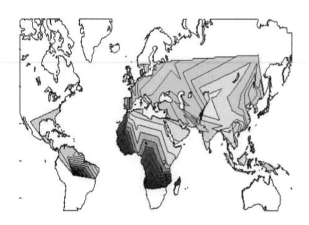

FIGURE 19.7
Contours of accumulated mutations in modern populations using *mStruct* for $K = 4$. Darker (more red) colors indicate higher accumulated mutations and lighter (more blue) colors indicate lower accumulated mutations.

The accumulated mutation at a location, which is plotted in Figure 19.7, depends on two factors which are effective simultaneously—the mutation rate per base per generation and the number of generations between the ancestral and modern populations at a particular location. If we assume that mutation rates do not vary significantly by geography, then the accumulated mutation at a location is a proxy for the number of the generations between the ancestral and modern populations at a location, i.e, how long ago the location was first inhabited. In this aspect, Figure 19.7 agrees with the "Out of Africa" models that are commonly agreed to explain human migrations across the world (Hammer et al., 1998).

19.5 The *blockmStruct* Model

The MAA model for populations described earlier is effective in modeling the mutations in microsatellite marker lengths. Experiments with SNP data show that such a model is inadequate to model mutations in SNPs, which have only two allelic states. Due to its higher density across the genome, SNP data is also often found to show much larger linkage among adjacent loci than microsatellites.

We modify the "population-specific mixture of ancestral alleles" representation to propose a "population-specific mixture of ancestral haplotype blocks" (MAH, or mixture of ancestral haplotypes). For a locus i and ancestral population k, we assume that there are three components: (1) a set of ancestral (or founder) haplotype blocks $\mu_i^k \equiv \{\mu_{i,1}^k, \ldots, \mu_{i,L_i'}^k\}$, which can differ from their descendant haplotype blocks in the modern population; (2) a mutation parameter δ_i^k associated with the locus, which can be further generalized to be allele-specific if necessary; and (3) an AP β_i^k, which now represents the frequencies of the ancestral haplotype blocks. Here L_i' denotes the total number of ancestral haplotype blocks present at locus i, which is different from L_i in the previous section, which denotes the total number of observed haplotype blocks at loci i. By explicitly associating a mutation model with an ancestral population, we can now capture mutation events. It is important to note that the mutation parameter δ is not the mutation rate commonly referred to in the literature. As we shall see later, it is a measure of the variability of a locus that can be described approximately as the combined effect of the per-generation mutation rate and the age of the population.

19.5.1 Representation of Modern Genomes

The *blockmStruct* model assumes that each chromosome of an individual's genome is composed of J unlinked haplotype blocks. The genome of the nth individual is therefore given by the set $\{y_{1,n_0}, y_{1,n_1}, \cdots, y_{J,n_0}, y_{J,n_1}\}$. The length of the jth haplotype block, given by l_j, is the number of SNPs included in the jth haplotype block, leading to the identity that $\sum_{j=1}^{J} l_j = I$. The jth haplotype block for the nth individual, $y_{j,n_e} = \{x_{j_1,n_e}, \cdots, x_{j_1+l_j-1,n_e}\}$ where $j_1, j_1 + l_j - 1 \in \{1, \cdots, I\}$, are indices denoting the left-most and right-most boundaries of the jth block, respectively. Figure 19.8 shows the *blockmStruct* representation of haplotype blocks.

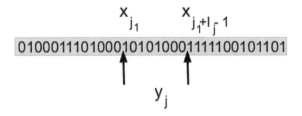

FIGURE 19.8
Representation of haplotype blocks in *blockmStruct*. For notational convenience, we drop the e subscript denoting ploidy and the n subscript denoting an individual in the diagram. Haplotype block y_j is composed of l_j SNPs x_{j_1} to $x_{j_1+l_j-1}$. In this example, $y_j = 10101000$ and $l_j = 8$.

The lengths and boundaries of the haplotype blocks are assumed to be the same for all individuals and can be chosen according to different strategies. In our experiments, we assume that all haplotype blocks have fixed length b. The following discusses the possible strategies of choosing haplotype blocks and their advantages and disadvantages.

Strategies for Choosing Haplotype Blocks

Haplotype blocks can be chosen according to various criteria. A few commonly used criteria are:

- Fixed number of SNPs per block. This is the simplest strategy for choosing haplotype blocks and is the most efficient to compute.

- Length in KB or MB of the haplotype block. This requires knowledge of the positions of the SNPs in the genome and can produce blocks of variable length. A useful heuristic is to use the knowledge of the range of linkage disequilibrium to pick a single block length.

- Choosing boundaries when the linkage disequilibrium (or correlation) between adjacent SNPs drop below a pre-specified threshold. This allows us to create blocks consistent with the earlier assumption of unlinked haplotype blocks.

- By using a haplotype inference program. A number of programs are available for inferring haplotype blocks, and using one of them is likely to produce the most accurate haplotype blocks. However, this inference is often too computationally expensive to be efficient.

19.5.2 The Generative Process

We propose to represent each ancestral population by a set of population-specific MAHs. This results in a generative process similar to the one defined for *mStruct*:

- For each individual n, draw the mixed membership vector (or ancestry vector): $\lambda_n \sim P(\cdot|\alpha)$, where $P(\cdot|\alpha)$ is a pre-chosen structure prior.

- For each marker allele $x_{i,n_e} \in \mathbf{x}_n$

    - 2.1: Draw the latent *ancestral-population-origin* indicator $z_{i,n_e} \sim \text{Multinomial}(\cdot|\lambda_n)$;
    - 2.2a: Draw the latent founder indicator $c_{i,n_e}|z_{i,n_e} = k \sim \text{Multinomial}(\cdot|\beta_i^k)$;
    - 2.2b: Draw the haplotype block $y_{i,n_e}|c_{i,n_e} = l, z_{i,n_e} = k \sim P_m(\cdot|\mu_{i,l}^k, \delta_{i,l}^k)$,

where $P_m()$ is a mutation model that we will define below to capture mutations within haplotype blocks.

Figure 19.9 shows the graphical model representation of *blockmStruct*. We can see by comparing it to Figure 19.4 that *blockmStruct* differs from *mStruct* in its representation of modern individual genomes. This difference also affects the ancestral alleles inferred by the methods. *mStruct* and *blockmStruct* also differ in the mutation models they use.

Mutation Model

We assume a mutation model for the haplotype that assumes that within a haplotype block, mutations occur independently of each other with a fixed probability. Let $d(y, \mu)$ be the number of SNPs which are different in the two haplotypes y and μ (also referred to as the Manhattan distance between the two haplotype blocks). The parameter $\delta \in [0, 1]$ is the probability of a single mutation in the haplotype block. If the size of both blocks y and μ is assumed to be b, it is easy to show that

$$P(y|\mu, \delta) = \delta^{d(y,\mu)}(1 - \delta)^{b-d(y,\mu)}. \tag{19.5}$$

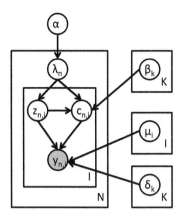

FIGURE 19.9
Graphical model representation of *blockmStruct*. For convenience, we have ignored the diploid nature of the observation. The shaded node indicates the variables we observe.

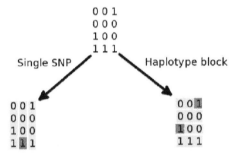

FIGURE 19.10
A demonstration of how using different mutation models can lead to different inferences of ancestry. The dark shaded squares indicate sites of mutation. On the left, the shaded columns indicate use of single SNP mutation models. On the right, the shaded rows indicate use of mutation models over SNP blocks.

This model is an intuitive representation of the possibility of SNPs switching allelic state. The assumption of independence among mutations within a haplotype block is similar to a mutation model for single SNPs.

Figure 19.10 uses toy data to demonstrate the effects of different mutation models on ancestry inference. In a mutation model over single SNPs, mutations are inferred to have occurred at a single location. In the haplotype mutation model, inference leads to two potential sites of mutation. Thus, the additional constraints imposed by the existence of ancestral haplotype blocks can lead to the model capturing more mutational information.

Another advantage of the block mutation model combined with the MAH representation of populations is that it implicitly models linkage between adjacent loci. By constraining the ancestral alleles to be haplotype blocks and not individual SNPs, we can account for linkage between alleles at physically adjacent loci.

19.5.3   Inference and Parameter Estimation

For notational convenience, we will ignore the diploid nature of observations in the analysis that follows. With the understanding that the analysis is carried out for an arbitrary nth individual, we will drop the subscript n. We overload the indicator arrays z_i and c_i to also use them as scalar index variables, as well as scalars with a value equal to the index at which the array forms have 1s. In other words: $z_i \in \{1, \ldots, K\}$ or $z_i = [z_{i,1}, \ldots, z_{i,K}]$, where $z_{i,k} = \mathcal{I}[z_i = k]$, and $\mathcal{I}[\cdot]$ denotes an indicator function equal to 1 when the predicate argument is true and 0 otherwise. A similar overloading is also assumed for the c_i variables. We use $f(y_i|\mu_{i,c_i}, \delta_{i,z_i})$ to denote our mutation model for haplotypes.

The joint probability distribution of the the data and the relevant variables under the *blockm-Struct* model can then be written as:

$$P(y, z, c, \lambda | \alpha, \beta, \mu, \delta)$$
$$= p(\lambda|\alpha) \prod_{i=1}^{I} P(z_i|\lambda) \, P\left(c_i | z_i, \beta_i^{k=1:K}\right) P\left(y_i | c_i, z_i, \mu_i, \delta_i^{k=1:K}\right).$$

The marginal likelihood of the data can be computed by summing/integrating out the latent variables:

$$P(y|\alpha, \beta, \mu, \delta) = \frac{\Gamma\left(\sum_{k=1}^{K} \alpha_k\right)}{\prod_{k=1}^{K} \Gamma(\alpha_k)} \int \left(\prod_{k=1}^{K} \lambda_k^{\alpha_k - 1}\right) \cdots$$
$$\times \prod_{i=1}^{I} \sum_{k=1}^{K} \left(\prod_{k=1}^{K} \lambda_k^{z_{i,k}}\right) \sum_{i=1}^{I} \prod_{k=1}^{K} \prod_{l=1}^{L_i} \left(\beta_{i,l}^{k}\right)^{c_{i,l} z_{i,k}} \cdots$$
$$\times P\left(y_i | \mu_{i,l}, \delta_i^{k}\right)^{c_{i,l} z_{i,k}} d\lambda.$$

However, a closed-form solution to this summation/integration is not possible, and indeed exact inference on hidden variables such as the mixed membership vector λ and estimation of model parameters such as the mutation rates δ under *blockmStruct* is intractable. We use a variational inference algorithm as described in Shringarpure and Xing (2009).

Variational Inference

We use a mean field approximation for performing inference on the model. This approximation method estimates an intractable joint posterior $p()$ of all the hidden variables in the model by a product of marginal distributions $q() = \prod q_i()$, each over only a single hidden variable. The optimal parameterization of $q_i()$ for each variable is obtained by minimizing the Kullback-Leibler divergence between the variational approximation q and the true joint posterior p. Using results from the generalized mean field theory (Xing et al., 2003), we can write the variational distributions of the latent variables in *blockmStruct* as follows:

$$q(\lambda) \propto \prod_{k=1}^{K} \lambda_k^{\alpha_k - 1 + \sum_{i=1}^{I} \langle z_{i,k} \rangle},$$

$$q(c_i) \propto \prod_{l=1}^{L} \left(\prod_{k=1}^{K} \left(\beta_{i,l}^{k} f(y_i|\mu_{i,l}, \delta_i^{k})\right)^{\langle z_{i,k} \rangle}\right)^{c_{i,l}},$$

$$q(z_i) \propto \prod_{k=1}^{K} \left(e^{\langle \log(\lambda_k) \rangle} \left(\prod_{l=1}^{L} \beta_{i,l}^{k} f(y_i|\mu_{i,l}, \delta_i^{k})^{\langle c_{i,l} \rangle}\right)\right)^{z_{i,k}}.$$

In the distributions above, the '$\langle \cdot \rangle$' are used to indicate the expected values of the enclosed random variables. A close inspection of the above formulas reveals that these variational distributions have the form $q(\lambda) \sim \text{Dirichlet}(\gamma_{.1}, \ldots, \gamma_{.K})$, $q(z_i) \sim \text{Multinomial}(\rho_{.i,1}, \ldots, \rho_{.i,K})$, and $q(c_i) \sim \text{Multinomial}(\xi_{.i,1}, \ldots, \xi_{.i,L})$, respectively, of which the parameters γ_k, $\rho_{i,k}$, and $\xi_{i,l}$ are given by the following equations:

$$\gamma_k = \alpha_k + \sum_{i=1}^{I} \langle z_{i,k} \rangle,$$

$$\rho_{i,k} = \frac{e^{\langle \log(\lambda_k) \rangle} \left(\prod_{l=1}^{L} \beta_{i,l}^k f(x_i | \mu_{i,l}, \delta_i^k)^{\langle c_{i,l} \rangle} \right)}{\sum_{k=1}^{K} \left(e^{\langle \log(\lambda_k) \rangle} \left(\prod_{l=1}^{L} \beta_{i,l}^k f(x_i | \mu_{i,l}, \delta_i^k)^{\langle c_{i,l} \rangle} \right) \right)},$$

$$\xi_{i,l} = \frac{\prod_{k=1}^{K} \left(\beta_{i,l}^k f(x_i | \mu_{i,l}, \delta_i^k) \right)^{\langle z_{i,k} \rangle}}{\sum_{k=1}^{K} \left(\prod_{k=1}^{K} \left(\beta_{i,l}^k f(x_i | \mu_{i,l}, \delta_i^k) \right)^{\langle z_{i,k} \rangle} \right)},$$

and they have the properties: $\langle \log(\lambda_k) \rangle = \psi(\gamma_k) - \psi(\sum_k \gamma_k)$, $\langle z_{i,k} \rangle = \rho_{i,k}$, and $\langle c_{i,l} \rangle = \xi_{i,l}$, which suggest that they can be computed via fixed point iterations. (The digamma function $\psi()$ used above is the first derivative of the logarithm of the gamma function $\Gamma()$.) It can be shown that this iteration will converge to a local optimum, similar to what happens in an EM algorithm. Empirically, a near global optimal can be obtained by multiple random restarts of the fixed point iteration. Upon convergence, we can easily compute an estimate of the ancestry vector λ for each individual from $q(\lambda)$.

Parameter Estimation

The parameters of our model are the centroids μ, the mutation parameters δ, the ancestral allele frequency distributions β, and the Dirichlet hyperparameter that is the prior on ancestral populations— α. For the hyperparameter estimation, we perform empirical Bayes estimation using the variational expectation maximization (variational EM) algorithm described in Blei et al. (2003). The variational inference described in Section 19.5.3 provides us with a tractable lower bound on the log-likelihood as a function of the current values of the hyperparameters. We can thus maximize it with respect to the hyperparameters. If we alternately carry out variational inference with fixed hyperparameters, followed by a maximization of the lower bound with respect to the hyperparameters for fixed values of the variational parameters, we can get an empirical Bayes estimate of the hyperparameters. The derivation, details of which we will not show here, leads to the following iterative algorithm:

1. (*E-step*) For each individual, find the optimizing values of the variational parameters $(\gamma_n, \rho_n, \xi_n; n \in 1, \ldots, N)$ using the variational updates described above.

2. (*M-step*) Maximize the resulting variational lower bound on the likelihood with respect to the model parameters, namely $\alpha, \beta, \mu, \delta$.

The two steps are repeated until the lower bound on the log-likelihood converges. The estimation of hyperparameters for *blockmStruct* is identical to that of *mStruct*, with the alleles x replaced by the haplotype blocks y. We therefore refer the reader to Shringarpure and Xing (2009) for mathematical details of the parameter estimation.

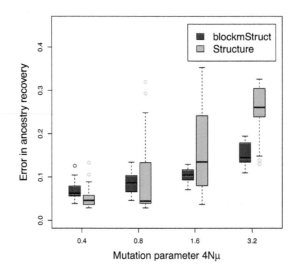

FIGURE 19.11
Effect of varying mutation rate on accuracy of individual ancestry recovery.

19.6   Experiments

We first validate *blockmStruct* using simulation data to examine the accuracy of ancestry recovery with varying mutation rates. We then use the *blockmStruct* model to analyze data from the HGDP.

19.6.1   Simulation Experiments

We will demonstrate the accuracy of the *blockmStruct* model through the task of recovering individual ancestry using simulated data. We use the coalescent software *ms* (Hudson, 1990) with recombination to generate data for our simulation. Most coalescent simulation software, including *ms*, assumes an infinite site model of mutation, disallowing recurrent mutations. This would generate simulation data that would violate the assumptions of the *blockmStruct* model. We therefore use the coalescent software to generate genealogy trees at 500 loci. At each locus, we assume that the unit of inheritance is a block of 5 SNPs. Mutations are placed on the branches of the genealogy trees according to a poisson distribution with probability proportional to the branch lengths and applied to the blocks. We simulate a two-population admixture with 200 individuals in the resulting population. The recombination probability between adjacent bases is set to 10^{-8} per generation and the effective population size N is set to 10^4 to approximate parameter values for human populations. We assume that the mutation parameter $4N\mu$ is variable and has values of $\{0.4, 0.8, 1.6, 3.2\}$, corresponding to parameter values for human populations. To examine the effect of modeling mutation, we compare our results to *Structure* using the haplotype blocks as alleles. We compare the performance of *Structure* and *blockmStruct* in terms of their error in recovering the ancestry proportions λ.

The results are shown in Figure 19.11. We find that when the mutation rate $4N\mu$ is low, *Structure* performs as good as, or even better than, *blockmStruct*. As the mutation rate rises, the error in ancestry recovery rises for both methods. For *Structure*, this is expected since it has no model for mutations. For *blockmStruct*, this effect is a result of the mismatch between the mutation model used to simulate data and the mutation model assumed by *blockmStruct*. However, since *blockmStruct* has a mutation model, its ancestry recovery error increases much slower than that for *Structure*. This demonstrates the utility of modeling mutations in haplotype blocks.

19.6.2 Analysis of HGDP+HapMap Data

We analyzed a dataset containing high-density SNP genotyping data for 597 individuals from the HGDP and the HapMap project. For computational reasons, we used 10,000 SNPs on chromosome 1 for our ancestry analysis. The number of ancestral populations K was varied from 2 to 6. We used SNP blocks of size 15, so that the correlation between the left (or right) endpoints of two consecutive windows was less than 0.25 in 90% of the windows.

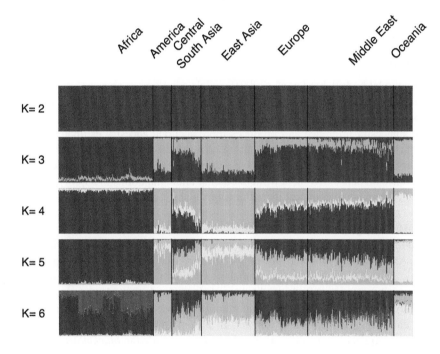

FIGURE 19.12
Ancestry vectors for 597 individuals from the Human Genome Diversity Project with K ranging from 2 to 6, inferred using *blockmStruct*.

From Figure 19.12, we can see that the ancestry vectors produced by *blockmStruct* cluster individuals by their continental divisions. For $K = 2$, the individuals are divided into an African and a non-African cluster. For $K = 3$, the non-African cluster separates into a cluster containing Europe, the Middle East, and Central-South Asia, and a second cluster containing the Americas, East Asia, and Oceania. At $K = 4$, the Oceanian populations separate into a cluster of their own. At $K = 5$, the population component corresponding to the Americas and East Asia splits into two components. One of the two components corresponds to American populations. The other component appears to varying degrees in the Asian, European, and Middle Eastern populations. At $K = 6$, the ancestral population component for the African population splits into two components which display varying degrees of admixture for different African groups. For larger values of K, the new population components add little interpretative value.

The population structure uncovered by *blockmStruct* shows significant similarities and differences with the population structure inferred by *Structure* and *mStruct* on the same data. Population structure analysis of the HGDP data by *Structure* (Figure 19.3) shows that ancestral populations correspond to geographical divisions, similar to the inference by *blockmStruct* for small values of K. For larger values of K, *Structure* infers population clusters that correspond completely to one (or more) regional groups. *blockmStruct*, on the other hand, infers population components for higher values of K that contribute partially to multiple regional populations. This behavior is similar to the results for *mStruct* seen in Figure 19.6. However, the ancestry proportions inferred by *blockmStruct* do not exhibit the same degree of ancestry sharing as seen in Figure 19.6.

We find that for all values of K, the model infers mutation parameters that are significantly larger than zero. However, in all cases, the method fails to uncover meaningful mutational structure. We discuss this behavior further in the Section 19.7.

19.7 Discussion

The *Structure* model by Pritchard et al. (2000) models admixing of ancestral populations but does not model allele mutations. The *mStruct* model by Shringarpure and Xing (2009) extends *Structure* by modeling allele mutations for microsatellite markers and demonstrates that modeling allele mutations affects ancestry inference and produces more accurate ancestry estimates when mutations are present. However, *mStruct* fails to account for mutations in SNP data and produces results identical to that of *Structure* on SNP data. We have developed *blockmStruct*, a model for performing ancestry inference on dense SNP data while modeling allele mutations in haplotype blocks.

We validated our model using simulated genotype data to conclude that the method can recover ancestry more accurately than previous methods even with high mutation rates in SNPs. Our analysis of the HGDP+HapMap data indicates that the mutation model in *blockmStruct* affects the results of ancestry inference and produces population structure that shares similarities with both *Structure* and *mStruct*. As in the *Structure* analysis, the population structure corresponds broadly to continental geographic divisions. Like the *mStruct* analysis, the inferred ancestral population components show partial membership in multiple modern populations, producing a higher degree of population sharing. However, the results also show important differences compared to results from analyses of the same data using *Structure* and *mStruct*. In *Structure*, most individuals are assigned membership almost completely to a single ancestral population while the *blockmStruct* analysis assigns admixed ancestry to a number of individuals. The degree of admixture assigned to individuals by *blockmStruct* is not as high as that inferred by *mStruct*. This suggests that the choice of representing modern individuals as haplotype blocks allows us to capture relationships between individuals more accurately using SNP data than either *Structure* or *mStruct* permit.

Another advantage of the haplotype block representation is that it offers us a way of modeling linkage between adjacent loci. *Structure* and *mStruct* both assume no linkage between loci. While both methods are robust to some degree of linkage disequilibrium, such an assumption is not appropriate for high-density SNP data. As demonstrated earlier, a haplotype blockmodel indirectly accounts for linkage disequilibrium through the ancestral haplotype blocks. This is in alternative to the explicit modeling of linkage using hidden markov models, for instance in Falush et al. (2003), which requires more complex inference procedures. *blockmStruct* offers a computationally efficient way of modeling linkage and mutations simultaneously in population structure. In our analyses, we assumed a fixed-length model for haplotype blocks by examining the range of linkage in the observed SNPs. More accurate modeling of haplotype blocks, using variable-length blocks based on linkage decay or a haplotype inference method, can enable more accurate modeling of linkage at the cost of increased computation. A computationally efficient choice of haplotype blocks that can account for linkage more accurately remains a question for further study.

The ability to analyze accumulated mutation is an important advantage that *mStruct* offers over *Structure* when analyzing microsatellite data. With SNP data, *mStruct* fails to capture any mutational information, resulting in most mutation rates being inferred to be close to zero, resulting in *mStruct* and *Structure* producing identical population structure with SNP data. The mutation model of *blockmStruct* based on haplotype blocks is an extension of the single-SNP mutation model of *mStruct*. However, even though it infers non-zero values for most mutation parameters, it fails to recover any meaningful spatial structure in the accumulated mutation. This is likely to be a result of the mismatch between the assumed mutation model and the true mutation model over haplotype blocks. Alternative models of haplotype mutation may improve the recovery of mutational information in *blockmStruct*.

References

Alexander, D. H., Novembre, J., and Lange, K. (2009). Fast model-based estimation of ancestry in unrelated individuals. *Genome Research* 19: 1655–1664.

Blei, D. M., Ng, A. Y., and Jordan, M. I. (2003). Latent Dirichlet allocation. *Journal of Machine Learning Research* 3: 993–1022.

Bowcock, A. M., Ruiz-Linares, A., Tomfohrde, J., Minch, E., Kidd, J. R., and Cavalli-Sforza, L. L. (1994). High resolution of human evolutionary trees with polymorphic microsatellites. *Nature* 368: 455–457.

Cavalli-Sforza, L. L. (2005). The Human Genome Diversity Project: Past, present and future. *Nature Reviews Genetics* 6: 333–340.

Dietrich, W., Katz, H., Lincoln, S. E., Shin, H. S., Friedman, J., Dracopoli, N. L., and Lander, E. S. (1992). A genetic map of the mouse suitable for typing intraspecific crosses. *Genetics* 131: 423–447.

Erosheva, E. A., Fienberg, S. E., and Lafferty, J. D. (2004). Mixed-membership models of scientific publications. *Proceedings of the National Academy of Sciences* 101: 5220–5227.

Excoffier, L. and Hamilton, G. (2003). Comment on genetic structure of human populations. *Science* 300: 1877.

Falush, D., Stephens, M., and Pritchard, J. K. (2003). Inference of population structure using multilocus genotype data: Linked loci and correlated allele frequencies. *Genetics* 164: 1567–1587.

Gibbs, R. A. (2003). The International HapMap Project. *Nature* 426: 789–796.

Hammer, M. F., Karafet, T., Rasanayagam, A., Wood, E. T., Altheide, T. K., Jenkins, T., Griffiths, R. C., Templeton, A. R., and Zegura, S. L. (1998). Out of Africa and back again: Nested cladistic analysis of human Y chromosome variation. *Molecular Biology and Evolution* 15: 427–441.

Henderson, S. T. and Petes, T. D. (1992). Instability of simple sequence DNA in *Saccharomyces cerevisiae*. *Molecular and Cellular Biology* 12: 2749–2757.

Hudson, R. R. (1990). Gene genealogies and the coalescent process. *Oxford Surveys in Evolutionary Biology* 7: 1–44.

Kelly, R., Gibbs, M., Collick, A., and Jeffreys, A. J. (1991). Spontaneous mutation at the hypervariable mouse minisatellite locus Ms6-hm: Flanking DNA sequence and analysis of germline and early somatic mutation events. *Proceedings: Biological Sciences* 245: 235–245.

Lin, T., Myers, E. W., and Xing, E. P. (2006). Interpreting anonymous DNA samples from mass disasters–Probabilistic forensic inference using genetic markers. *Bioinformatics* 22: e298.

Patterson, N., Price, A. L., and Reich, D. (2006). Population structure and eigenanalysis. *PLoS Genetics* 2: e190.

Pisani, D., Poling, L. L., Lyons-Weiler, M., and Hedges, S. B. (2004). The colonization of land by animals: Molecular phylogeny and divergence times among arthropods. *BMC Biology* 2: 1.

Pritchard, J. K., Stephens, M., and Donnelly, P. (2000). Inference of population structure from multilocus genotype data. *Genetics* 155: 945–959.

Queller, D. C., Strassmann, J. E., and Hughes, C. R. (1993). Microsatellites and kinship. *Trends in Ecology & Evolution* 8: 285–288.

Rosenberg, N. A., Pritchard, J. K., Weber, J. L., Cann, H. M., Kidd, K. K., Zhivotovsky, L. A., and Feldman, M. W. (2002). Genetic structure of human populations. *Science* 298: 2381–2385.

Schwarz, G. (1978). Estimating the dimension of a model. *Annals of Statistics* 6: 461–464.

Shringarpure, S. and Xing, E. P. (2009). mStruct: Inference of population structure in light of both genetic admixing and allele mutations. *Genetics* 182: 575–593.

Tang, H., Peng, J., Wang, P., and Risch, N. J. (2005). Estimation of individual admixture: Analytical and study design considerations. *Genetic Epidemiology* 28: 289–301.

Valdes, A. M., Slatkin, M., and Freimer, N. B. (1993). Allele frequencies at microsatellite loci: The stepwise mutation model revisited. *Genetics* 133: 737–749.

Xing, E. P., Jordan, M. I., and Russell, S. (2003). A generalized mean field algorithm for variational inference in exponential families. In *Proceedings of the 19th Conference on Uncertainty in Artificial Intelligence (UAI 2003)*. San Francisco, CA, USA: Morgan Kaufmann Publishers Inc., 583–591.

Zhivotovsky, L. A., Underhill, P. A., Cinnioglu, C., Kayser, M., Morar, B., Kivisild, T., Scozzari, R., Cruciani, F., Destro-Bisol, G., Spedini, G., et al. (2004). The effective mutation rate at Y chromosome short tandem repeats, with application to human population-divergence time. *American Journal of Human Genetics* 74: 50–61.

20

Mixed Membership Models for Time Series

Emily B. Fox

Department of Statistics, University of Washington, Seattle, WA 98195, USA

Michael I. Jordan

Computer Science Division and Department of Statistics, University of California, Berkeley, CA 94720, USA

CONTENTS

In this chapter we discuss some of the consequences of the mixed membership perspective on time series analysis. In its most abstract form, a mixed membership model aims to associate an individual *entity* with some set of *attributes* based on a collection of observed *data*. For example, a person (*entity*) can be associated with various defining characteristics (*attributes*) based on observed pairwise interactions with other people (*data*). Likewise, one can describe a document (*entity*) as comprised of a set of topics (*attributes*) based on the observed words in the document (*data*). Although much of the literature on mixed membership models considers the setting in which exchangeable collections of data are associated with each member of a set of entities, it is equally natural to consider problems in which an entire time series is viewed as an entity and the goal is to characterize the time series in terms of a set of underlying dynamic attributes or *dynamic regimes*. Indeed, this perspective is already present in the classical hidden Markov model (Rabiner, 1989) and switching state-space model (Kim, 1994), where the dynamic regimes are referred to as "states," and the collection of states realized in a sample path of the underlying process can be viewed as a mixed membership characterization of the observed time series. Our goal here is to review some of the richer modeling possibilities for time series that are provided by recent developments in the mixed membership framework.

Much of our discussion centers around the fact that while in classical time series analysis it is commonplace to focus on a single time series, in mixed membership modeling it is rare to focus on a single entity (e.g., a single document); rather, the goal is to model the way in which multiple entities are related according to the overlap in their pattern of mixed membership. Thus we take a nontraditional perspective on time series in which the focus is on collections of time series. Each individual time series may be characterized as proceeding through a sequence of states, and the focus is on relationships in the choice of states among the different time series.

As an example that we review later in this chapter, consider a multivariate time series that arises when position and velocity sensors are placed on the limbs and joints of a person who is going through an exercise routine. In the specific dataset that we discuss, the time series can be segmented into types of exercise (e.g., jumping jacks, touch-the-toes, and twists). Each person may select a subset from a library of possible exercise types for their individual routine. The goal is to discover these exercise types (i.e., the "behaviors" or "dynamic regimes") and to identify which person engages in which behavior, and when. Discovering and characterizing "jumping jacks" in one person's routine should be useful in identifying that behavior in another person's routine. In essence, we would like to implement a combinatorial form of shrinkage involving subsets of behaviors selected from an overall library of behaviors.

Another example arises in genetics, where mixed membership models are referred to as "admixture models" (Pritchard et al., 2000). Here the goal is to model each individual genome as a mosaic of marker frequencies associated with different ancestral genomes. If we wish to capture the dependence of nearby markers along the genome, then the overall problem is that of capturing relationships among the selection of ancestral states along a collection of one-dimensional spatial series.

One approach to problems of this kind involves a relatively straightforward adaptation of hidden Markov models or other switching state-space models into a Bayesian hierarchical model: transition and emission (or state-space) parameters are chosen from a global prior distribution and each individual time series either uses these global parameters directly or perturbs them further. This approach in essence involves using a single global library of states, with individual time series differing according to their particular random sequence of states. This approach is akin to the traditional Dirichlet-multinomial framework that is used in many mixed membership models. An alternative is to make use of a beta-Bernoulli framework in which each individual time series is modeled by first selecting a subset of states from a global library and then drawing state sequences from a model defined on that particular subset of states. We will overview both of these approaches in the remainder of the chapter.

While much of our discussion is agnostic to the distinction between parametric and nonparametric models, our overall focus is on the nonparametric case. This is because the model choice issues that arise in the multiple time series setting can be daunting, and the nonparametric framework provides at least some initial control over these issues. In particular, in a classical state-space setting we would need to select the number of states for each individual time series, and do so in a manner that captures partial overlap in the selected subsets of states among the time series. The nonparametric approach deals with these issues as part of the model specification rather than as a separate model choice procedure.

The remainder of the chapter is organized as follows. In Section 20.1.1, we review a set of time series models that form the building blocks for our mixed membership models. The mixed membership analogy for time series models is aided by relating to a canonical mixed membership model: latent Dirichlet allocation (LDA), reviewed in Section 20.1.2. Bayesian nonparametric variants of LDA are outlined in Section 20.1.3. Building on this background, in Section 20.2 we turn our focus to mixed membership in time series. We first present Bayesian parametric and nonparametric models for single time series in Section 20.2.1 and then for collections of time series in Section 20.2.3.

Section 20.3 contains a brief survey of related Bayesian and Bayesian nonparametric time series models.

20.1 Background

In this section we provide a brief introduction to some basic terminology from time series analysis. We also overview some of the relevant background from mixed membership modeling, both parametric and nonparametric.

20.1.1 State-Space Models

The autoregressive (AR) process is a classical model for time series analysis that we will use as a building block. An AR model assumes that each observation is a function of some fixed number of previous observations plus an uncorrelated *innovation*. Specifically, a linear, time-invariant AR model has the following form:

$$y_t = \sum_{i=1}^{r} a_i y_{t-i} + \epsilon_t, \tag{20.1}$$

where y_t represents a sequence of equally spaced observations, ϵ_t the uncorrelated innovations, and a_i the time-invariant autoregressive parameters. Often one assumes normally distributed innovations $\epsilon_t \sim \mathcal{N}(0, \sigma^2)$, further implying that the innovations are *independent*.

A more general formulation is that of *linear state-space models*, sometimes referred to as *dynamic linear models*. This formulation, which is closely related to autoregressive moving average processes, assumes that there exists an underlying state vector $x_t \in \mathbb{R}^n$ such that the past and future of the dynamical process $y_t \in \mathbb{R}^d$ are conditionally independent. A linear time-invariant state-space model is given by

$$x_t = A x_{t-1} + e_t \qquad y_t = C x_t + w_t, \tag{20.2}$$

where e_t and w_t are independent, zero-mean Gaussian noise processes with covariances Σ and R, respectively. Here, we assume a *vector-valued* process. One could likewise consider a vector-valued AR process, as we do in Section 20.2.1.

There are several ways to move beyond linear state-space models. One approach is to consider smooth nonlinear functions in place of the matrix multiplication in linear models. Another approach, which is our focus here, is to consider *regime-switching* models based on a latent sequence of discrete states $\{z_t\}$. In particular, we consider *Markov switching processes*, where the state sequence is modeled as Markovian. If the entire state is a discrete random variable, and the observations $\{y_t\}$ are modeled as being conditionally independent given the discrete state, then we are in the realm of *hidden Markov models* (HMMs) (Rabiner, 1989). Details of the HMM formulation are expounded upon in Section 20.2.1.

It is also useful to consider hybrid models in which the state contains both discrete and continuous components. We will discuss an important example of this formulation—the autoregressive HMM—in Section 20.2.1. Such models can be viewed as a collection of AR models, one for each discrete state. We will find it useful to refer to the discrete states as "dynamic regimes" or "behaviors" in the setting of such models. Conditional on the value of a discrete state, the model does not merely produce independent observations, but exhibits autoregressive behavior.

20.1.2 Latent Dirichlet Allocation

In this section, we briefly overview the latent Dirichlet allocation (LDA) model (Blei et al., 2003) as a a canonical example of a mixed membership model. We use the language of "documents," "topics," and "words." In contrast to hard-assignment predecessors that assumed each document was associated with a single topic category, LDA aims to model each document as a mixture of topics. Throughout this chapter, when describing a mixed membership model, we seek to define some observed quantity as an *entity* that is allowed to be associated with, or have *membership* characterized by, multiple *attributes*. For LDA, the entity is a *document* and the attributes are a set of possible *topics*. Typically, in a mixed membership model, each entity represents a set of observations, and a key question is what structure is imposed on these observations. For LDA, each document is a collection of observed *words* and the model makes a simplifying *exchangeability* assumption in which the ordering of words is ignored.

Specifically, LDA associates each document d with a latent distribution over the possible topics, $\pi^{(d)}$, and each topic k is associated with a distribution over words in the vocabulary, θ_k. Each word $w_i^{(d)}$ is then generated by first selecting a topic from the document-specific topic distribution and then selecting a word from the topic-specific word distribution.

Formally, the standard LDA model with K topics, D documents, and N_d words per document d is given as

$$
\begin{aligned}
\theta_k &\sim \text{Dir}(\eta_1, \ldots, \eta_V) & k &= 1, \ldots, K \\
\pi^{(d)} &\sim \text{Dir}(\beta_1, \ldots, \beta_K) & d &= 1, \ldots D \\
z_i^{(d)} \mid \pi^{(d)} &\sim \pi^{(d)} & d &= 1, \ldots D, \; i = 1, \ldots, N_d \\
w_i^{(d)} \mid \{\theta_k\}, z_i^{(d)} &\sim \theta_{z_i^{(d)}} & d &= 1, \ldots D, \; i = 1, \ldots, N_d.
\end{aligned}
\tag{20.3}
$$

Here $z_i^{(d)}$ is a topic indicator variable associated with observed word $w_i^{(d)}$, indicating which topic k generated this ith word in document d. In expectation, for each document d we have $E[\pi_k^{(d)} \mid \beta] = \beta_k$. That is, the expected topic proportions for each document are identical a priori.

20.1.3 Bayesian Nonparametric Mixed Membership Models

The LDA model of Equation (20.3) assumes a finite number of topics K. Bayesian nonparametric methods allow for extensions to models with an unbounded number of topics. That is, in the mixed membership analogy, each entity can be associated with a potentially countably infinite number of attributes. We review two such approaches: one based on the hierarchical Dirichlet process (Teh et al., 2006) and the other based on the beta process (Hjort, 1990; Thibaux and Jordan, 2007). In the latter case, the association of entities with attributes is directly modeled as *sparse*.

Hierarchical Dirichlet Process Topic Models

To allow for a countably infinite collection of topics, in place of finite-dimensional topic-distributions $\pi^{(d)} = [\pi_1^{(d)}, \ldots, \pi_K^{(d)}]$ as specified in Equation (20.3), one wants to define distributions whose support lies on a countable set, $\pi^{(d)} = [\pi_1^{(d)}, \pi_2^{(d)}, \ldots]$.

The *Dirichlet process* (DP), denoted by $\text{DP}(\alpha H)$, provides a distribution over countably infinite discrete probability measures

$$
G = \sum_{k=1}^{\infty} \pi_k \delta_{\theta_k} \qquad \theta_k \sim H
\tag{20.4}
$$

defined on a parameter space Θ with base measure H. The mixture weights are sampled via a

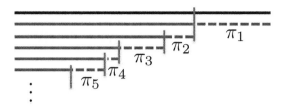

FIGURE 20.1
Pictorial representation of the stick-breaking construction of the Dirichlet process.

stick-breaking construction (Sethuraman, 1994):

$$\pi_k = \nu_k \prod_{\ell=1}^{k-1} (1 - \nu_\ell) \qquad \nu_k \sim \text{Beta}(1, \alpha). \tag{20.5}$$

This can be viewed as dividing a unit-length stick into lengths given by the weights π_k: the kth weight is a random proportion ν_k of the remaining stick after the first $(k-1)$ weights have been chosen. We denote this distribution by $\pi \sim \text{GEM}(\alpha)$. See Figure 20.1 for a pictorial representation of this process.

Drawing indicators $z_i \sim \pi$, one can integrate the underlying random stick-breaking measure π to examine the predictive distribution of z_i conditioned on a set of indicators $z_1, \ldots, z_{i-1}$ and the DP concentration parameter α. The resulting sequence of partitions is described via the *Chinese restaurant process* (CRP) (Pitman, 2002), which provides insight into the clustering properties induced by the DP.

For the LDA model, recall that each θ_k is a draw from a Dirichlet distribution (here denoted generically by H) and defines a distribution over the vocabulary for topic k. To define a model for multiple documents, one might consider independently sampling $G^{(d)} \sim \text{DP}(\alpha H)$ for each document d, where each of these random measures is of the form $G^{(d)} = \sum_{k=1}^{\infty} \pi_k^{(d)} \delta_{\theta_k^{(d)}}$. Unfortunately, the topic-specific word distribution for document d, $\theta_k^{(d)}$, is necessarily different from that of document d', $\theta_k^{(d')}$, since each are independent draws from the base measure H. This is clearly not a desirable model—in a mixed membership model we want the parameter that describes each attribute (*topic*) to be shared between entities (*documents*).

One method of sharing parameters θ_k between documents while allowing for document-specific topic weights $\pi^{(d)}$ is to employ the *hierarchical Dirichlet process* (HDP) (Teh et al., 2006). The HDP defines a shared set of parameters by drawing θ_k independently from H. The weights are then specified as

$$\beta \sim \text{GEM}(\gamma) \qquad \pi^{(d)} \mid \beta \sim \text{DP}(\alpha\beta). \tag{20.6}$$

Coupling this prior to the likelihood used in the LDA model, we obtain a model that we refer to as *HDP-LDA*. See Figure 20.2(a) for a graphical model representation, and Figure 20.3 for an illustration of the coupling of document-specific topic distributions via the global stick-breaking distribution β. Letting $G^{(d)} = \sum_{k=1}^{\infty} \pi_k^{(d)} \delta_{\theta_k}$ and $G^{(0)} = \sum_{k=1}^{\infty} \beta_k \delta_{\theta_k}$, one can show that the specification of Equation (20.6) is equivalent to defining a hierarchy of Dirichlet processes (Teh et al., 2006):

$$G^{(0)} \sim \text{DP}(\gamma H) \qquad G^{(d)} \mid G^{(0)} \sim \text{DP}\left(\alpha G^{(0)}\right). \tag{20.7}$$

Thus the name *hierarchical* Dirichlet process. Note that there are many possible alternative formulations one could have considered to generate different countably infinite weights $\pi^{(d)}$ with shared

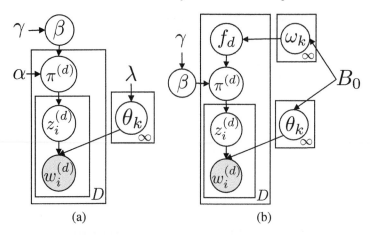

(a)   (b)

FIGURE 20.2

Graphical model of the (a) HDP-based and (b) beta-process-based topic model. The HDP-LDA model specifies a global topic distribution $\beta \sim \text{GEM}(\gamma)$ and draws document-specific topic distributions as $\pi^{(d)} \mid \beta \sim \text{DP}(\alpha\beta)$. Each word $w_i^{(d)}$ in document d is generated by first drawing a topic-indicator $z_i^{(d)} \mid \pi^{(d)} \sim \pi^{(d)}$ and then drawing from the topic-specific word distribution: $w_i^{(d)} \mid \{\theta_k\}, z_i^{(d)} \sim \theta_{z_i^{(d)}}$. The standard LDA model arises as a special case when β is fixed to a finite measure $\beta = [\beta_1, \dots, \beta_K]$. The beta process model specifies a collection of *sparse* topic distributions. Here, the beta process measure $B \sim \text{BP}(1, B_0)$ is represented by its masses ω_k and locations θ_k, as in Equation (20.8). The features are then conditionally independent draws $f_{dk} \mid \omega_k \sim \text{Bernoulli}(\omega_k)$, and are used to define document-specific topic distributions $\pi_j^{(d)} \mid f_d, \beta \sim \text{Dir}(\beta \otimes f_d)$. Given the topic distributions, the generative process for the topic-indicators $z_i^{(d)}$ and words $w_i^{(d)}$ is just as in the HDP-LDA model.

atoms θ_k. The HDP is a particularly simple instantiation of such a model that has appealing theoretical and computational properties due to its interpretation as a hierarchy of Dirichlet processes.

Via the construction of Equation (20.6), we have that $E[\pi_k^{(d)} \mid \beta] = \beta_k$. That is, all of the document-specific topic distributions are centered around the same stick-breaking weights β.

Beta-Bernoulli Process Topic Models

The HDP-LDA model defines countably infinite topic distributions $\pi^{(d)}$ in which every topic k has positive mass $\pi_k^{(d)} > 0$ (see Figure 20.3). This implies that each entity (*document*) is associated with infinitely many attributes (*topics*). In practice, however, for any finite length document d, only a finite subset of the topics will be present. The HDP-LDA model implicitly provides such *attribute counts* through the assignment of words $w_i^{(d)}$ to topics via the indicator variables $z_i^{(d)}$.

As an alternative representation that more directly captures the inherent sparsity of association between documents and topics, one can consider *feature-based* Bayesian nonparametric variants of LDA via the *beta-Bernoulli process*, such as in the *focused topic model* of Williamson et al. (2010). (A precursor to this model was presented in the time series context by Fox et al. (2010), and is discussed in Section 20.2.3.) In such models, each document is endowed with an infinite-dimensional binary feature vector that indicates which topics are associated with the given document. In contrast to HDP-LDA, this formulation directly allows each document to be represented as a *sparse* mixture of topics. That is, there are only a few topics that have positive probability of appearing in any document.

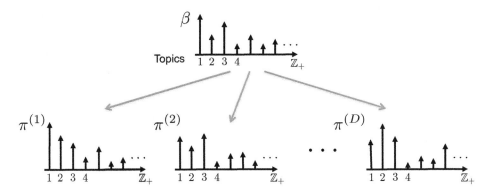

FIGURE 20.3

Illustration of the coupling of the document-specific topic distributions $\pi^{(d)}$ via the global stick-breaking distribution β. Each topic distribution has countably infinite support and, in expectation, $E[\pi^{(d)} \mid \beta] = \beta_k$.

Informally, one can think of the *beta process* (BP) (Hjort, 1990; Thibaux and Jordan, 2007) as defining an infinite set of coin-flipping probabilities and a Bernoulli process realization as corresponding to the outcome from an infinite coin-flipping sequence based on the beta-process-determined coin-tossing probabilities. The set of resulting *heads* indicate the set of selected *features*, and implicitly defines an infinite-dimensional feature vector. The properties of the beta process induce sparsity in the feature space by encouraging sharing of features among the Bernoulli process realizations.

More formally, let $\boldsymbol{f}_d = [f_{d1}, f_{d2}, \ldots]$ be an infinite-dimensional feature vector associated with document d, where $f_{dk} = 1$ if and only if document d is associated with topic k. The beta process, denoted $\text{BP}(c, B_0)$, provides a distribution on measures

$$B = \sum_{k=1}^{\infty} \omega_k \delta_{\theta_k}, \tag{20.8}$$

with $\omega_k \in (0, 1)$. We interpret ω_k as the feature-inclusion probability for feature k (e.g., the kth topic in an LDA model). This kth feature is associated with parameter θ_k.

The collection of points $\{\theta_k, \omega_k\}$ are a draw from a non-homogeneous Poisson process with rate $\nu(d\omega, d\theta) = c\omega^{-1}(1 - \omega)^{c-1}d\omega B_0(d\theta)$ defined on the product space $\Theta \otimes [0, 1]$. Here, $c > 0$ and B_0 is a base measure with total mass $B_0(\Theta) = \alpha$. Since the rate measure η has infinite mass, the draw from the Poisson process yields an infinite collection of points, as in Equation (20.8). For an example realization and its associated cumulative distribution, see Figure 20.4. One can also interpret the beta process as the limit of a finite model with K features:

$$B_K = \sum_{k=1}^{K} \omega_k \delta_{\theta_k} \qquad \omega_k \sim \text{Beta}\left(\frac{c\alpha}{K}, c(1 - \frac{\alpha}{K})\right) \qquad \theta_k \sim \alpha^{-1}B_0. \tag{20.9}$$

In the limit as $K \to \infty$, $B_K \to B$ and one can define stick-breaking constructions analogous to those in the Dirichlet process (Paisley et al., 2010; 2011). For each feature k, we independently sample

$$f_{dk} \mid \omega_k \sim \text{Bernoulli}(\omega_k). \tag{20.10}$$

That is, with probability ω_k, topic k is associated with document d. One can visualize this process

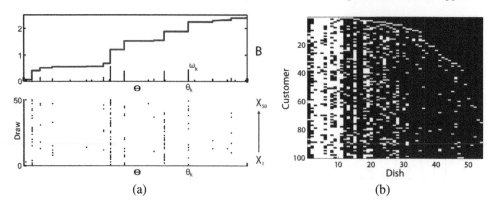

(a)   (b)

FIGURE 20.4

(a) *Top*: A draw B from a beta process is shown by the discrete masses, with the corresponding cumulative distribution shown above. Bottom : 50 draws X_i from a Bernoulli process using the beta process realization. Each dot corresponds to a coin flip at that atom in B that came up heads. (b) An image of a feature matrix associated with a realization from an Indian buffet process with $\alpha = 10$. Each row corresponds to a different customer, and each column to a different dish. White indicates a chosen feature.

as walking along the atoms of the discrete beta process measure B and, at each atom θ_k, flipping a coin with probability of heads given by ω_k. More formally, setting $X_d = \sum_{k=1}^{\infty} f_{dk}\delta_{\theta_k}$, this process is equivalent to sampling X_d from a *Bernoulli process* with base measure B: $X_d \mid B \sim \mathrm{BeP}(B)$. Example realizations are shown in Figure 20.4(a).

The characteristics of this beta-Bernoulli process define desirable traits for a Bayesian nonparametric featural model: we have a countably infinite collection of coin-tossing probabilities (one for each of our infinite number of features) defined by the beta process, but only a sparse, finite subset are active in any Bernoulli process realization. In particular, one can show that B has finite expected mass implying that there are only a finite number of successes in the infinite coin-flipping sequence that define X_d. Likewise, the sparse set of features active in X_d are likely to be similar to those of $X_{d'}$ (an independent draw from $\mathrm{BeP}(B)$), though variability is clearly possible. Finally, the beta process is conjugate to the Bernoulli process (Kim, 1999), which implies that one can analytically marginalize the latent random beta process measure B and examine the predictive distribution of f_d given $f_1, \ldots, f_{d-1}$ and the concentration parameter α. As established by Thibaux and Jordan (2007), the marginal distribution on the $\{f_d\}$ obtained from the beta-Bernoulli process is the *Indian buffet process* (IBP) of Griffiths and Ghahramani (2005), just as the marginalization of the Dirichlet-multinomial process yields the Chinese restaurant process. The IBP can be useful in developing posterior inference algorithms and a significant portion of the literature is written in terms of the IBP representation.

Returning to the LDA model, one can obtain the focused topic model of Williamson et al. (2010) within the beta-Bernoulli process framework as follows:

$$
\begin{aligned}
B &\sim \mathrm{BP}(1, B_0) \\
X_d \mid B &\sim \mathrm{BeP}(B) \quad d = 1, \ldots D \\
\pi^{(d)} \mid f_d, \beta &\sim \mathrm{Dir}(\beta \otimes f_d) \quad d = 1, \ldots D,
\end{aligned}
\tag{20.11}
$$

where Williamson et al. (2010) treat β as random according to $\beta_k \sim \mathrm{Gamma}(\gamma, 1)$. Here, f_d is the feature vector associated with X_d and $\mathrm{Dir}(\beta \otimes f_d)$ represents a Dirichlet distribution defined solely over the components indicated by f_d, with hyperparameters the corresponding subset of β. This implies that $\pi^{(d)}$ is a distribution with positive mass only on the *sparse* set of selected topics. See

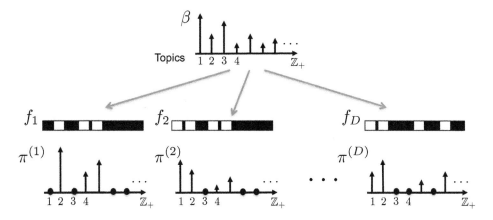

FIGURE 20.5
Illustration of generating the *sparse* document-specific topic distributions $\pi^{(d)}$ via the beta process specification. Each document's binary feature vector f_d limits the support of the topic distribution to the sparse set of selected topics. The non-zero components are Dirichlet distributed with hyperparmeters given by the corresponding subset of β. See Equation (20.11).

Figure 20.5. Given $\pi^{(d)}$, the $z_i^{(d)}$ and $w_i^{(d)}$ are generated just as in Equation (20.3). As before, we take $\theta_k \sim \text{Dir}(\eta_1, \ldots, \eta_V)$. The graphical model is depicted in Figure 20.2(b).

20.2 Mixed Membership in Time Series

Building on the background provided in Section 20.1, we can now explore how ideas of mixed membership models can be used in the time series setting. Our particular focus is on time series that can be well described using *regime-switching* models. For example, stock returns might be modeled as switches between regimes of volatility or an EEG recording between spiking patterns dependent on seizure type. For the exercise routines scenario, people switch between a set of actions such as jumping jacks, side twists, and so on. In this section, we present a set of regime-switching models for describing such datasets, and show how one can interpret the models as providing a form of mixed membership for time series.

To form the mixed membership interpretation, we build off of the canonical example of LDA from Section 20.1.2. Recall that for LDA, the entity of interest is a *document* and the set of attributes are the possible *topics*. Each document is then modeled as having membership in multiple topics (i.e., *mixed membership*). For time series analysis, the equivalent analogy is that the entity is the *time series* $\{y_t : t = 1, \ldots, T\}$, which we denote compactly by $y_{1:T}$. Just as a document is a collection of observed *words*, a time series is a sequence of observed *data points* of various forms depending upon the application domain. We take the attributes of a time series to be the collection of *dynamic regimes* (e.g., jumping jacks, arm circles, etc.). Our mixed membership time series model associates a single time series with a collection of dynamic regimes. However, unlike in text analysis, it is unreasonable to assume a *bag-of-words* representation for time series since the ordering of the data points is fundamental to the description of each dynamic regime.

The central defining characteristics of a mixed membership time series model are (i) the model used to describe each dynamic regime, and (ii) the model used to describe the switches between regimes. In Section 20.2.1 and in Section 20.2.2 we choose one switching model and explore multiple choices for the dynamic regime model. Another interesting question explored in Section 20.2.3 is how to jointly model multiple time series. This question is in direct analogy to the ideas behind the analysis of a *corpus* of documents in LDA.

20.2.1 Markov Switching Processes as a Mixed Membership Model

A flexible yet simple regime-switching model for describing a single time series with such patterned behaviors is the class of *Markov switching processes*. These processes assume that the time series can be described via Markov transitions between a set of latent dynamic regimes which are individually modeled via temporally independent or linear dynamical systems. Examples include the hidden Markov model (HMM), switching vector autoregressive (VAR) process, and switching linear dynamical system (SLDS).[1] These models have proven useful in such diverse fields as speech recognition, econometrics, neuroscience, remote target tracking, and human motion capture.

Hidden Markov Models

The hidden Markov model, or HMM, is a class of doubly stochastic processes based on an underlying, discrete-valued state sequence that is modeled as Markovian (Rabiner, 1989). Conditioned on this state sequence, the model assumes that the observations, which may be discrete or continuous valued, are independent. Specifically, let z_t denote the state, or *dynamic regime*, of the Markov chain at time t and let π_j denote the state-specific *transition distribution* for state j. Then, the Markovian structure on the state sequence dictates that

$$z_t \mid z_{t-1} \sim \pi_{z_{t-1}}. \tag{20.12}$$

Given the state z_t, the observation y_t is a conditionally independent emission

$$y_t \mid \{\theta_j\}, z_t \sim F(\theta_{z_t}) \tag{20.13}$$

for an indexed family of distributions $F(\cdot)$. Here, θ_j are the *emission parameters* for state j.

A Bayesian specification of the HMM might further assume

$$\pi_j \sim \text{Dir}(\beta_1, \dots, \beta_K) \qquad \theta_j \sim H \tag{20.14}$$

independently for each HMM state $j = 1, \dots, K$.

The HMM represents a simple example of a mixed membership model for time series: a given time series (*entity*) is modeled as having been generated from a collection of dynamic regimes (*attributes*), each with different mixture weights. The key component of the HMM, which differs from standard mixture models such as in LDA, is the fact that there is a Markovian structure to the assignment of data points to mixture components (i.e., dynamic regimes). In particular, the probability that observation y_t is generated from the dynamic regime associated with state j (via an assignment $z_t = j$) is dependent upon the previous state z_{t-1}. As such, the mixing proportions for the time series are defined by the transition matrix P with rows π_j. This is in contrast to the LDA model in which the mixing proportions for a given document are simply captured by a single vector of weights.

Switching VAR Processes

The modeling assumption of the HMM that observations are conditionally independent given the latent state sequence is often insufficient in capturing the temporal dependencies present in many datasets. Instead, one can assume that the observations have conditionally *linear* dynamics. The latent HMM state then models switches between a set of such linear models in order to capture more

---

[1]These processes are sometimes referred to as *Markov jump-linear systems* (MJLS) within the control theory community.

complex dynamical phenomena. We restrict our attention in this chapter to switching vector autoregressive (VAR) processes, or *autoregressive HMMs* (AR-HMMs), which are broadly applicable in many domains while maintaining a number of simplifying properties that make them a practical choice computationally.

We define an AR-HMM, with switches between order-r vector autoregressive processes,[2] as

$$\boldsymbol{y}_t = \sum_{i=1}^{r} A_{i,z_t} \boldsymbol{y}_{t-i} + \boldsymbol{e}_t(z_t), \qquad (20.15)$$

where z_t represents the HMM latent state at time t, and is defined as in Equation (20.12). The state-specific additive noise term is distributed as $\boldsymbol{e}_t(z_t) \sim \mathcal{N}(0, \Sigma_{z_t})$. We refer to $A_k = \{A_{1,k}, \ldots, A_{r,k}\}$ as the set of *lag matrices*. Note that the standard HMM with Gaussian emissions arises as a special case of this model when $A_k = 0$ for all k.

20.2.2 Hierarchical Dirichlet Process HMMs

In the HMM formulation described so far, we have assumed that there are K possible different dynamical regimes. This begs the question: what if this is not known, and what if we would like to allow for new dynamic regimes to be added as more data are observed? In such scenarios, an attractive approach is to appeal to Bayesian nonparametrics. Just as the hierarchical Dirichlet process (HDP) of Section 20.1.3 allowed for a collection of countably infinite topic distributions to be defined over the same set of topic parameters, one can employ the HDP to define an HMM with a set of countably infinite transition distributions defined over the same set of HMM emission parameters.

In particular, the HDP-HMM of Teh et al. (2006) defines

$$\beta \sim \text{GEM}(\gamma) \qquad \pi_j \mid \beta \sim \text{DP}(\alpha\beta) \qquad \theta_j \sim H. \qquad (20.16)$$

The evolution of the latent state z_t and observations y_t are just as in Equations (20.12) and (20.13). Informally, the Dirichlet process part of the HDP allows for this unbounded state-space and encourages the use of only a spare subset of these HMM states. The hierarchical layering of Dirichlet processes ties together the state-specific transition distribution (via β), and through this process, creates a *shared* sparse state-space.

The induced predictive distribution for the HDP-HMM state z_t, marginalizing the transition distributions π_j, is known as the *infinite HMM* urn model (Beal et al., 2002). In particular, the HDP-HMM of Teh et al. (2006) provides an interpretation of this urn model in terms of an underlying collection of linked random probability measures. However, the HDP-HMM omits the self-transition bias of the infinite HMM and instead assumes that each transition distribution π_j is *identical* in expectation ($E[\pi_{jk} \mid \beta] = \beta_k$), implying that there is no differentiation between self-transitions and moves between different states. When modeling data with state persistence, as is common in most real-world datasets, the flexible nature of the HDP-HMM prior places significant mass on state sequences with unrealistically fast dynamics.

To better capture state persistence, the *sticky* HDP-HMM of Fox et al. (2008; 2011b) restores the self-transition parameter of the infinite HMM of Beal et al. (2002) and specifies

$$\beta \sim \text{GEM}(\gamma) \qquad \pi_j \mid \beta \sim \text{DP}(\alpha\beta + \kappa\delta_j) \qquad \theta_j \sim H, \qquad (20.17)$$

where $(\alpha\beta + \kappa\delta_j)$ indicates that an amount $\kappa > 0$ is added to the jth component of $\alpha\beta$. In expectation,

$$E[\pi_{jk} \mid \beta, \kappa] = \frac{\alpha\beta_k + \kappa\delta(j,k)}{\alpha + \kappa}. \qquad (20.18)$$

---

[2]We denote an order-r VAR process by VAR(r).

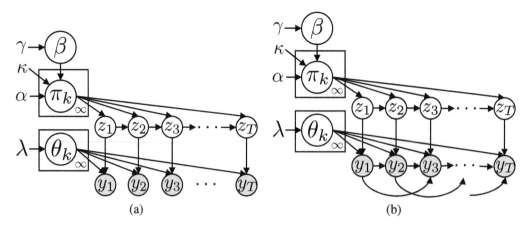

(a)   (b)

FIGURE 20.6

Graphical model of (a) the sticky HDP-HMM and (b) an HDP-based AR-HMM. In both cases, the state evolves as $z_{t+1} \mid \{\pi_k\}, z_t \sim \pi_{z_t}$, where $\pi_k \mid \beta \sim \mathrm{DP}(\alpha\beta + \kappa\delta_k)$ and $\beta \sim \mathrm{GEM}(\gamma)$. For the sticky HDP-HMM, the observations are generated as $y_t \mid \{\theta_k\}, z_t \sim F(\theta_{z_t})$ whereas the HDP-AR-HMM assumes conditionally VAR dynamics as in Equation (20.15), specifically in this case with order $r = 2$.

Here, $\delta(j, k)$ is the discrete Kronecker delta. From Equation (20.18), we see that the expected transition distribution has weights which are a convex combination of the global weights defined by β and state-specific weight defined by the sticky parameter κ. When $\kappa = 0$, the original HDP-HMM of Teh et al. (2006) is recovered. The graphical model for the sticky HDP-HMM is displayed in Figure 20.6(a).

One can also consider sticky HDP-HMMs with Dirichlet process mixture of Gaussian emissions (Fox et al., 2011b). Recently, HMMs with Dirichlet process emissions were also considered in Yau et al. (2011), along with efficient sampling algorithms for computations. Building on the sticky HDP-HMM framework, one can similarly consider HDP-based variants of the switching VAR process and switching linear dynamical system, such as represented in Figure 20.6(b); see Fox et al. (2011a) for further details. For the *HDP-AR-HMM*, Fox et al. (2011a) consider methods that allow for switching between VAR processes of unknown and potentially variable order.

20.2.3 A Collection of Time Series

In the mixed membership time series models considered thus far, we have assumed that we are interested in the dynamics of a single (potentially multivariate) time series. However, as in LDA where one assumes a *corpus* of documents, in a growing number of fields the focus is on making inferences based on a *collection* of related time series. One might monitor multiple financial indices, or collect EEG data from a given patient at multiple non-contiguous epochs. Recalling the exercise routines example, one might have a dataset consisting of multiple time series obtained from multiple individuals, each of whom performs some subset of exercise types. In this scenario, we would like to take advantage of the overlap between individuals, such that if a "jumping jack" behavior is discovered in the time series for one individual then it can be used in modeling the data for other individuals. More generally, one would like to discover and model the dynamic regimes that are shared among several related time series. The benefits of such joint modeling are twofold: we may more robustly estimate representative dynamic models in the presence of limited data, and we may also uncover interesting relationships among the time series.

Recall the basic finite HMM of Section 20.2.1 in which the transition matrix P defined the dynamic regime mixing proportions for a given time series. To develop a mixed membership model for a *collection* of time series, we again build on the LDA example. For LDA, the document-specific mixing proportions over topics are specified by $\pi^{(d)}$. Analogously, for each time series $y_{1:T_d}^{(d)}$, we denote the time-series specific transition matrix as $P^{(d)}$ with rows $\pi_j^{(d)}$. That is, for time series d, $\pi_j^{(d)}$ denotes the transition distribution from state j to each of the K possible next states. Just as LDA couples the document-specific topic distributions $\pi^{(d)}$ under a common Dirichlet prior, we can couple the *rows* of the transition matrix as

$$\pi_j^{(d)} \sim \text{Dir}(\beta_1, \dots, \beta_K). \tag{20.19}$$

A similar idea holds for extending the HDP-HMM to collections of time series. In particular, we can specify

$$\beta \sim \text{GEM}(\gamma) \qquad \pi_j^{(d)} \mid \beta \sim \text{DP}(\alpha\beta). \tag{20.20}$$

Analogously to LDA, both the finite and infinite HMM specifications above imply that the expected transition distributions are identical between time series ($E[\pi_j^{(d)} \mid \beta] = E[\pi_j^{(d')} \mid \beta]$). Here, however, the expected transition distributions are also identical between rows of the transition matrix.

To allow for state-specific variability in the expected transition distribution, one could similarly couple sticky HDP-HMMs, or consider a finite variant of the model via the weak-limit approximation (see Fox et al. (2011b) for details on finite truncations). Alternatively, one could independently center each row of the time-series-specific transition matrix around a state-specific distribution. For the finite model,

$$\pi_j^{(d)} \mid \beta_j \sim \text{Dir}(\beta_{j1}, \dots, \beta_{jK}). \tag{20.21}$$

For the infinite model, such a specification is more straightforwardly presented in terms of the Dirichlet random measures. Let $G_j^{(d)} = \sum \pi_j^{(d)} \delta_{\theta_k}$, with $\pi_j^{(d)}$ the time-series-specific transition distribution and θ_k the set of HMM emission parameters. Over the collection of D time series, we center $G_j^{(1)}, \dots, G_j^{(D)}$ around a common *state-j-specific transition measure* $G_j^{(0)}$. Then, each of the infinite collection of state-specific transition measures $G_1^{(0)}, G_2^{(0)}, \dots$ are centered around a global measure G_0. Specifically,

$$G_0 \sim \text{DP}(\gamma H) \qquad G_j^{(0)} \mid G_0 \sim \text{DP}(\eta G_0) \qquad G_j^{(d)} \mid G_j^{(0)} \sim \text{DP}\left(\alpha G_j^{(0)}\right). \tag{20.22}$$

Such a hierarchy allows for more variability between the transition distributions than the specification of Equation (20.20) by only directly coupling state-specific distributions between time series. The sharing of information between *states* occurs at a higher level in the latent hierarchy (i.e., one less directly coupled to observations).

Although they are straightforward extensions of existing models, the models presented in this section have not been discussed in the literature to the best of our knowledge. Instead, typical models for coupling multiple time series, each modeled via an HMM, rely on assuming exact sharing of the same transition matrix. (In the LDA framework, that would be equivalent to a model in which every document d shared the same topic weights, $\pi^{(d)} = \pi_0$.) With such a formulation, each time series (*entity*) has the exact same mixed membership with the global collection of dynamic regimes (*attributes*).

Alternatively, models have been proposed in which each time series d is hard-assigned to one of some M distinct HMMs, where each HMM is comprised of a unique set of states and corresponding transition distributions and emission parameters. For example, Qi et al. (2007) and Lennox et al. (2010) examine a Dirichlet process mixture of HMMs, allowing M to be unbounded. Based on a fixed assignment of time series to some subset of the global collection of HMMs, this model reduces to M' examples of exact sharing of HMM parameters, where M' is the number of unique HMMs assigned. That is, there are M' clusters of time series with the exact same mixed membership among a set of attributes (i.e., dynamic regimes) that are distinct between the clusters.

By defining a global collection of dynamic regimes and time-series-specific transition distributions, the formulations proposed above instead allow for commonalities between parameterizations while maintaining time-series-specific variations in the mixed membership. These ideas more closely mirror the LDA mixed membership story for a corpus of documents.

The Beta-Bernoulli Process HMM

Analogously to HDP-LDA, the HDP-based models for a collection of (or a single) time series assume that each time series has membership with an infinite collection of dynamic regimes. This is due to the fact that each transition distribution $\pi_j^{(d)}$ has positive mass on the countably infinite collection of dynamic regimes. In practice, just as a finite-length document is comprised of a finite set of instantiated topics, a finite-length time series is described by a limited set of dynamic regimes. This limited set might be related yet distinct from the set of dynamic regimes present in another time series. For example, in the case of the exercise routines, perhaps one observed individual performs jumping jacks, side twists, and arm circles, whereas another individual performs jumping jacks, arm circles, squats, and toe touches. In a similar fashion to the feature-based approach of the focused topic model described in Section 20.1.3, one can employ the beta-Bernoulli process to directly capture a *sparse* set of associations between time series and dynamic regimes.

The beta process framework provides a more abstract and flexible representation of Bayesian nonparametric mixed membership in a collection of time series. Globally, the collection of time series are still described by a shared library of infinitely many possible dynamic regimes. Individually, however, a given time series is modeled as exhibiting some sparse subset of these dynamic regimes.

More formally, Fox et al. (2010) propose the following specification: each time series d is endowed with an infinite-dimensional feature vector $\boldsymbol{f}_d = [f_{d1}, f_{d2}, \ldots]$, with $f_{dj} = 1$ indicating the inclusion of dynamic regime j in the membership of time series d. The feature vectors for the collection of D time series are coupled under a common beta process measure $B \sim \mathrm{BP}(c, B_0)$. In this scenario, one can think of B as defining coin-flipping probabilities for the global collection of dynamic regimes. Each feature vector $\boldsymbol{f}_d$ is implicitly modeled by a Bernoulli process draw $X_d \mid B \sim \mathrm{BeP}(B)$ with $X_d = \sum_k f_{dk} \delta_{\theta_k}$. That is, the beta-process-determined coins are flipped for each dynamic regime and the set of resulting heads indicate the set of selected features (i.e., via $f_{dk} = 1$).

The beta process specification allows flexibility in the number of total and time-series-specific dynamic regimes, and encourages time series to share similar subsets of the infinite set of possible dynamic regimes. Intuitively, the shared sparsity in the feature space arises from the fact that the total sum of coin-tossing probabilities is finite and only certain dynamic regimes have large probabilities. Thus, certain dynamic regimes are more prevalent among the time series, though the resulting set of dynamic regimes clearly need not be identical. For example, the lower subfigure in Figure 20.4(a) illustrates a collection of feature vectors drawn from this process.

To limit each time series to solely switch between its set of selected dynamic regimes, the feature vectors are used to form *feature-constrained transition distributions*:

$$\pi_j^{(d)} \mid \boldsymbol{f}_d \sim \mathrm{Dir}([\gamma, \ldots, \gamma, \gamma + \kappa, \gamma, \ldots] \otimes \boldsymbol{f}_d). \tag{20.23}$$

Again, we use $\mathrm{Dir}([\gamma, \ldots, \gamma, \gamma + \kappa, \gamma, \ldots] \otimes \boldsymbol{f}_d)$ to denote a Dirichlet distribution defined over the finite set of dimensions specified by $\boldsymbol{f}_d$ with hyperparameters given by the corresponding subset of $[\gamma, \ldots, \gamma, \gamma + \kappa, \gamma, \ldots]$. Here, the κ hyperparameter places extra expected mass on the component of $\pi_j^{(d)}$ corresponding to a self-transition $\pi_{jj}^{(d)}$, analogously to the sticky hyperparameter of the sticky

HDP-HMM (Fox et al., 2011b). This construction implies that $\pi_j^{(d)}$ has only a finite number of non-zero entries $\pi_{jk}^{(d)}$. As an example, if

$$\boldsymbol{f}_d = \begin{bmatrix} 1 & 0 & 0 & 1 & 1 & 0 & 1 & 0 & 0 & 0 \cdots \end{bmatrix},$$

then

$$\pi_j^{(d)} = \begin{bmatrix} \pi_{j1}^{(d)} & 0 & 0 & \pi_{j4}^{(d)} & \pi_{j5}^{(d)} & 0 & \pi_{j7}^{(d)} & 0 & 0 & 0 \cdots \end{bmatrix}$$

with $\begin{bmatrix} \pi_{j1}^{(d)} & \pi_{j4}^{(d)} & \pi_{j5}^{(d)} & \pi_{j7}^{(d)} \end{bmatrix}$ distributed according to a four-dimensional Dirichlet distribution. Pictorially, the generative process of the feature-constrained transition distributions is similar to that illustrated in Figure 20.5.

Although the methodology described thus far applies equally well to HMMs and other Markov switching processes, Fox et al. (2010) focus on the AR-HMM of Equation (20.15). Specifically, let $\mathbf{y}_t^{(d)}$ represent the observed value of the dth time series at time t, and let $z_t^{(d)}$ denote the latent dynamical regime. Assuming an order-r AR-HMM, we have

$$
\begin{aligned}
z_t^{(d)} \mid \{\pi_j^{(d)}\}, z_{t-1}^{(d)} &\sim \pi_{z_{t-1}^{(d)}}^{(d)} \\
\mathbf{y}_t^{(d)} &= \sum_{j=1}^{r} A_{j,z_t^{(d)}} \mathbf{y}_{t-j}^{(d)} + \mathbf{e}_t^{(d)}(z_t^{(d)}),
\end{aligned}
\tag{20.24}
$$

where $\mathbf{e}_t^{(d)}(k) \sim \mathcal{N}(0, \Sigma_k)$. Recall that each of the $\theta_k = \{A_k, \Sigma_k\}$ defines a different VAR(r) dynamic regime and the feature-constrained transition distributions $\pi^{(d)}$ restrict time series d to transition among dynamic regimes (indexed at time t by $z_t^{(d)}$) for which it has membership, as indicated by its feature vector $\boldsymbol{f}_d$.

Conditioned on the set of D feature vectors $\boldsymbol{f}_d$ coupled via the beta-Bernoulli process hierarchy, the model reduces to a collection of D switching VAR processes, each defined on the finite state-space formed by the set of selected dynamic regimes for that time series. Importantly, the beta-process-based featural model couples the dynamic regimes exhibited by different time series. Since the library of possible dynamic parameters is shared by all time series, posterior inference of each parameter set θ_k relies on pooling data among the time series that have $f_{dk} = 1$. It is through this pooling of data that one may achieve more robust parameter estimates than from considering each time series individually.

The resulting model is termed the *BP-AR-HMM*, with a graphical model representation presented in Figure 20.7. The overall model specification is summarized as: [3]

$$
\begin{aligned}
B &\sim \mathrm{BP}(1, B_0) \\
X_d \mid B &\sim \mathrm{BeP}(B), \quad d = 1, \ldots, D \\
\pi_j^{(d)} \mid \boldsymbol{f}_d &\sim \mathrm{Dir}([\gamma, \ldots, \gamma, \gamma + \kappa, \gamma, \ldots] \otimes \boldsymbol{f}_d), \quad d = 1, \ldots, D, \; j = 1, 2, \ldots \\
z_t^{(d)} \mid \{\pi_j^{(d)}\}, z_{t-1}^{(d)} &\sim \pi_{z_{t-1}^{(d)}}^{(d)}, \quad d = 1, \ldots, D, \; t = 1, \ldots, T_d \\
\mathbf{y}_t^{(d)} &= \sum_{j=1}^{r} A_{j,z_t^{(d)}} \mathbf{y}_{t-j}^{(d)} + \mathbf{e}_t^{(d)}(z_t^{(d)}), \quad d = 1, \ldots, D, \; t = 1, \ldots, T_d.
\end{aligned}
\tag{20.25}
$$

---

[3]One could consider alternative specifications for $\beta = [\gamma, \ldots, \gamma, \gamma + K, \gamma]$ such as in the focused topic model of (20.11) where each element β_k is an independent random variable. Note that Fox et al. (2010) treat γ, K as random.

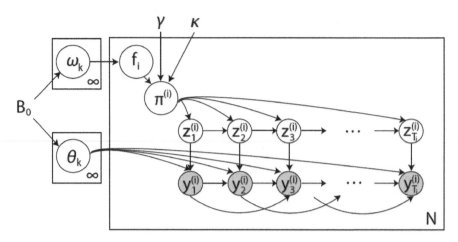

FIGURE 20.7

Graphical model of the BP-AR-HMM. The beta process distributed measure $B \mid B_0 \sim \text{BP}(1, B_0)$ is represented by its masses ω_k and locations θ_k, as in Equation (20.8). The features are then conditionally independent draws $f_{dk} \mid \omega_k \sim \text{Bernoulli}(\omega_k)$, and are used to define feature-constrained transition distributions $\pi_j^{(d)} \mid f_d \sim \text{Dir}([\gamma, \ldots, \gamma, \gamma + \kappa, \gamma, \ldots] \otimes f_d)$. The switching VAR dynamics are as in Equation (20.24).

Fox et al. (2010) apply the BP-AR-HMM to the analysis of multiple motion capture (MoCap) recordings of people performing various exercise routines, with the goal of jointly segmenting and identifying common dynamic behaviors among the recordings. In particular, the analysis examined six recordings taken from the CMU database (CMU, 2009), three from Subject 13 and three from Subject 14. Each of these routines used some combination of the following motion categories: running in place, jumping jacks, arm circles, side twists, knee raises, squats, punching, up and down, two variants of toe touches, arch over, and a reach-out stretch.

The resulting segmentation from the joint analysis is displayed in Figure 20.8. Each skeleton plot depicts the trajectory of a learned contiguous segment of more than two seconds, and boxes group segments categorized under the same behavior label in the posterior. The color of the box indicates the true behavior label. From this plot we can infer that although some true behaviors are split into two or more categories ("knee raises" [green] and "running in place" [yellow]),[4] the BP-AR-HMM is able to find common motions (e.g., six examples of "jumping jacks" [magenta]) while still allowing for various motion behaviors that appeared in only one movie (bottom left four skeleton plots.)

The key characteristic of the BP-AR-HMM that enables the clear identification of shared versus unique dynamic behaviors is the fact that the model takes a feature-based approach. The true feature matrix and BP-AR-HMM estimated matrix, averaged over a large collection of MCMC samples, are shown in Figure 20.9. Recall that each row represents an individual recording's feature vector f_d drawn from a Bernoulli process, and coupled under a common beta process prior. The columns indicate the possible dynamic behaviors (truncated to a finite number if no assignments were made thereafter.)

---

[4]The split behaviors shown in green and yellow correspond to the true motion categories of knee raises and running, respectively, and the splits can be attributed to the two subjects performing the same motion in a distinct manner.

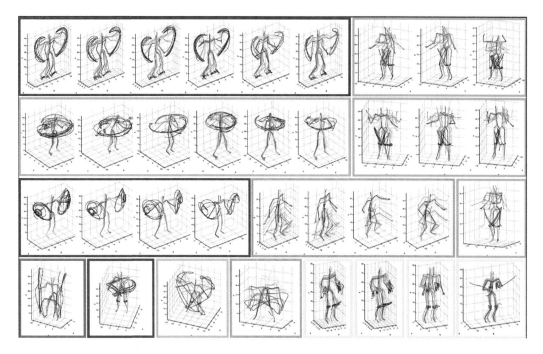

FIGURE 20.8

Each skeleton plot displays the trajectory of a learned contiguous segment of more than two seconds, bridging segments separated by fewer than 300 msec. The boxes group segments categorized under the same behavior label, with the color indicating the true behavior label (allowing for analysis of split behaviors). Skeleton rendering done by modifications to Neil Lawrence's Matlab MoCap toolbox (Lawrence, 2009).

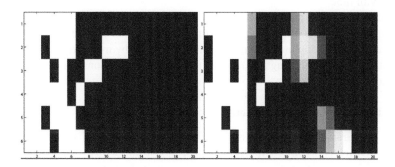

FIGURE 20.9

Feature matrices associated with the true MoCap sequences (left) and BP-AR-HMM estimated sequences over iterations 15,000 to 20,000 of an MCMC sampler (right). Each row is an individual recording and each column a possible dynamic behavior. The white squares indicate the set of selected dynamic behaviors.

20.3　Related Bayesian and Bayesian Nonparametric Time Series Models

In addition to the regime-switching models described in this chapter, there is large and growing literature on Bayesian parametric and nonparametric time series models, many of which also have interpretations as mixed membership models. We overview some of this literature in this section, aiming not to cover the entirety of related literature but simply to highlight three main themes: (i) *non-homogeneous* mixed membership models, and relatedly, *time-dependent processes*, (ii) other *HMM-based models*, and (iii) time-independent *mixtures of autoregressions*.

20.3.1　Non-Homogeneous Mixed Membership Models

Time-Varying Topic Models

The documents in a given corpus sometimes represent a collection spanning a wide range of time. It is likely that the prevalence and popularity of various topics, and words within a topic, change over this time period. For example, when analyzing scientific articles, the set of scientific questions being addressed naturally evolves. Likewise, within a given subfield, the terminology similarly develops—perhaps new words are created to describe newly discovered phenomena or other words go out of vogue.

To capture such changes, Blei and Lafferty (2006) proposed a *dynamic topic model*. This model takes the general framework of LDA, but specifies a Gaussian random walk on a set of topic-specific word parameters

$$\theta_{t,k} \mid \theta_{t-1,k} \sim \mathcal{N}(\theta_{t-1,k}, \sigma^2 I) \tag{20.26}$$

and document-specific topic parameters

$$\beta_t \mid \beta_{t-1} \sim \mathcal{N}(\beta_{t-1}, \delta^2 I). \tag{20.27}$$

The topic-specific word distribution arises via $\pi(\theta_{k,t,w}) = \frac{\exp(\theta_{k,t,w})}{\sum_w \exp(\theta_{k,t,w})}$. For the topic distribution, Blei and Lafferty (2006) specify $\eta \sim \mathcal{N}(\beta_t, a^2 I)$ and transform to $\pi(\eta)$. This formulation provides a *non-homogeneous* mixed membership model since the membership weights (i.e., topic weights) vary with time.

The formulation of Blei and Lafferty (2006) assumes discrete, evenly spaced corpora of documents. Often, however, documents are observed at uneven and potentially finely-sampled time points. Wang et al. (2008) explore a continuous time extension by modeling the evolution of $\theta_{t,k}$ as Brownian motion. As a simplifying assumption, the authors do not consider evolution of the global topic proportions β.

Time-Dependent Bayesian Nonparametric Processes

For Bayesian nonparametric time-varying topic modeling, Srebro and Roweis (2005) propose a time-dependent Dirichlet process. The Dirichlet process allows for an infinite set of possible topics, in a similar vein to the motivation in HDP-LDA. Importantly, however, this model does *not* assume a mixed membership formulation and instead takes each document to be hard-assigned to a single topic. The proposed time-dependent Dirichlet process models the changing popularity of various topics, but assumes that the topic-specific word distributions are static. That is, the Dirichlet process probability measures have time-varying weights, but static atoms.

More generally, there is a growing interest in time-dependent Bayesian nonparmetric processes. The dependent Dirichlet process was originally proposed by MacEachern (1998). A substantial focus has been on evolving the weights of the random discrete probability measures. Recently, Griffin and Steel (2011) examine a general class of autoregressive stick-breaking processes, and Mena et al. (2011) study stick-breaking processes for continuous-time modeling. Taddy (2010) considers an alternative autoregressive specification for Dirichlet process stick-breaking weights, with application to modeling the changing rate function in a dynamic spatial Poisson process.

20.3.2 Hidden-Markov-Based Bayesian Nonparametric Models

A number of other Bayesian nonparametric models have been proposed in the literature that take as their point of departure a latent Markov switching mechanism. Both the infinite factorial HMM (Van Gael et al., 2008) and the infinite hierarchical HMM (Heller et al., 2009) provide Bayesian nonparametric priors for infinite collections of latent Markov chains. The infinite factorial HMM provides a distribution on binary Markov chains via a Markov Indian buffet process. The implicitly defined time-varying infinite-dimensional binary feature vectors are employed in performing blind source separation (e.g., separating an audio recording into a time-varying set of overlapping speakers.) The infinite hierarchical HMM also employs an infinite collection of Markov chains, but the evolution of each depends upon the chain above. Instead of modeling binary Markov chains, the infinite hierarchical HMM examines finite multi-class state-spaces.

Another method that is based on a finite state-space is that of Taddy and Kottas (2009). The proposed model assumes that each HMM state defines an independent Dirichlet process regression. Extensions to non-homogenous Markov processes are considered based on external covariates that inform the latent state.

In Saeedi and Bouchard-Côté (2012), the authors propose a hierarchical gamma-exponential process for modeling recurrent continuous time processes. This framework provides a continuous-time analog to the discrete-time sticky HDP-HMM.

Instead of Markov-based regime-switching models that capture repeated returns to some (possibly infinite) set of dynamic regimes, one can consider changepoint methods in which each transition is to a new dynamic regime. Such methods often allow for very efficient computations. For example, Xuan and Murphy (2007) base such a model on the product partition model[5] framework to explore changepoints in the dependency structure of multivariate time series, harnessing the efficient dynamic programming techniques of Fearnhead (2006). More recently, Zantedeschi et al. (2011) explore a class of dynamic product partition models and online computations for predicting movements in the term structure of interest rates.

20.3.3 Bayesian Mixtures of Autoregressions

In this chapter, we explored two forms of switching autoregressive models: the HDP-AR-HMM and the BP-AR-HMM. Both models assume that the switches between autoregressive parameters follow a discrete-time Markov process. There is also substantial literature on nonlinear autoregressive modeling via mixtures of autoregressive processes, where the mixture components are independently selected over time. Lau and So (2008) consider a Dirichlet process mixture of autoregressions. That is, at each time step the observation is modeled as having been generated from one of an unbounded collection of autoregressive processes, with the mixing distribution given by a Dirichlet process. A variational approach to Dirichlet process mixtures of autoregressions with unknown orders has recently been explored in Morton et al. (2011). Wood et al. (2011) aim to capture the idea of *structural breaks* by segmenting a time series into contiguous blocks of L observations and assigning each *segment* to one of a finite mixture of autoregressive processes; implicitly, all L observations are associated with a given mixture component. Key to the formulation is the inclusion of time-varying mixture weights, leading to a nonstationary process, as in Section 20.3.1.

---

[5]A product partition model is a model in which the data are assumed independent across some set of unknown partitions (Hartigan, 1990; Barry and Hartigan, 1992). The Dirichlet process is a special case of a product partition model.

As an alternative formulation that captures Markovianity, but not directly in the latent mixture component, Müller et al. (1997) consider a model in which the probability of choosing a given autoregressive component is modeled via a kernel based on the previous set of observations (and potential covariates). The maximal set of K mixture components is fixed, with the associated autoregressive parameters taken to be draws from a Dirichlet process, implying that only $k \leq K$ will take distinct values.

20.4　Discussion

In this chapter, we have discussed a variety of time series models that have interpretations in the mixed membership framework. Mixed membership models are comprised of three key components: entities, attributes, and data. What differs between mixed membership models is the type of data associated with each entity, and how the entities are assigned membership with the set of possible attributes. Abstractly, in our case each time series is an entity that has membership with a collection of dynamic regimes, or attributes. The partial memberships are determined based on the temporally structured observations, or data, for the given time series. This structured data is in contrast to the typical focus of mixed membership models on exchangeable collections of data per entity (e.g., a bag-of-words representation of a document's text.)

Throughout the chapter, we have focused our attention on the class of *Markov switching processes*, and further restricted our exposition to Bayesian parametric and nonparametric treatments of such models. The latter allows for an unbounded set of attributes by modeling processes with Markov transitions between an infinite set of dynamic regimes. For the class of Markov switching processes, the mixed membership of a given time series is captured by the time-series-specific set of Markov transition distributions. Examples include the classical hidden Markov model (HMM), autoregressive HMM, and switching state-space model. In mixed membership modeling, one typically has a group of entities (e.g., a corpus of documents) and the goal is to allow each entity to have a unique set of partial memberships among a shared collection of attributes (e.g., topics). Through such modeling techniques, one can efficiently and flexibly share information between the data sources associated with the entities. Motivated by such goals, in this chapter we explored a nontraditional treatment of time series analysis by examining models for *collections* of time series. We proposed a Bayesian nonparametric model for multiple time series based on ideas analogous to Dirichlet-multinomial modeling of documents. We also reviewed a Bayesian nonparametric model based on a beta-Bernoulli framework that directly allows for sparse association of time series with dynamic regimes. Such a model enables decoupling the presence of a dynamic regime from its prevalence.

The discussion herein of time series analysis from a mixed membership perspective has been previously neglected, and leads to interesting ideas for further development of time series models.

References

Barry, D. and Hartigan, J. A. (1992). Product partition models for change point problems. *Annals of Statistics* 20: 260–279.

Beal, M. J., Ghahramani, Z., and Rasmussen, C. E. (2002). The infinite hidden Markov model. In

Dietterich, T. G., Becker, S., and Ghahramani, Z. (eds), *Advances in Neural Information Processing Systems 14*. Cambridge, MA: The MIT Press, 577–584.

Blei, D. M. and Lafferty, J. D. (2006). Dynamic topic models. In *Proceedings of the 23rd International Conference on Machine Learning (ICML '06)*. New York, NY, USA: ACM, 113–120.

Blei, D. M., Ng, A. Y., and Jordan, M. I. (2003). Latent Dirichlet allocation. *The Journal of Machine Learning Research* 3: 993–1022.

CMU (2009). Carnegie Mellon University graphics lab motion capture database. http://mocap.cs.cmu.edu.

Fearnhead, P. (2006). Exact and efficient Bayesian inference for multiple changepoint problems. *Statistics and Computing* 16: 203–213.

Fox, E. B., Sudderth, E. B., Jordan, M. I., and Willsky, A. S. (2008). An HDP-HMM for systems with state persistence. In *Proceedings of the 25th International Conference on Machine Learning (ICML '08)*. New York, NY, USA: ACM, 312–319.

Fox, E. B., Sudderth, E. B., Jordan, M. I., and Willsky, A. S. (2010). Sharing features among dynamical systems with beta processes. In Bengio, Y., Schuurmans, D., Lafferty, J., Williams, C. K. I., and Culotta, A. (eds), *Advances in Neural Information Processing Systems 22*. Red Hook, NY: Curran Associatees, Inc., 549–557.

Fox, E. B., Sudderth, E. B., Jordan, M. I., and Willsky, A. S. (2011a). Bayesian nonparametric inference of switching dynamic linear models. *IEEE Transactions on Signal Processing* 59: 1569–1585.

Fox, E. B., Sudderth, E. B., Jordan, M. I., and Willsky, A. S. (2011b). A sticky HDP-HMM with application to speaker diarization. *Annals of Applied Statistics* 5: 1020–1056.

Griffin, J. E. and Steel, M. F. J. (2011). Stick-breaking autoregressive processes. *Journal of Econometrics* 162: 383–396.

Griffiths, T. L. and Ghahramani, Z. (2005). Infinite Latent Feature Models and the Indian Buffet Process. Tech. report #2005-001, *Gatsby Computational Neuroscience Unit*.

Hartigan, J. A. (1990). Partition models. *Communications in Statistics–Theory and Methods* 19: 2745–2756.

Heller, K. A., Teh, Y. W., and Gorur, D. (2009). The infinite hierarchical hidden Markov model. In *Proceedings of the 12th International Conference on Artificial Intelligence and Statistics, (AISTATS 2009). Journal of Machine Learning Research – Proceedings Track* 5: 224–231.

Hjort, N. L. (1990). Nonparametric Bayes estimators based on beta processes in models for life history data. *Annals of Statistics* 18: 1259–1294.

Kim, C. -J. (1994). Dynamic linear models with Markov-switching. *Journal of Econometrics* 60: 1–22.

Kim, Y. (1999). Nonparametric Bayesian estimators for counting processes. *Annals of Statistics* 27: 562–588.

Lau, J. W. and So, M. K. P. (2008). Bayesian mixture of autoregressive models. *Computational Statistics & Data Analysis* 53: 38–60.

Lawrence, N. D. (2009). MATLAB motion capture toolbox. http://www.cs.man.ac.uk/~neill/mocap.

Lennox, K. P., Dahl, D. B., Vannucci, M., Day, R. and Tsai, J. W. (2010). A Dirichlet process mixture of hidden Markov models for protein structure prediction. *Annals of Applied Statistics* 4: 916–942.

MacEachern, S. N. (1998). Dependent nonparametric processes. In *Proceedings of the American Statistical Association: Section on Bayesian Statististical Science*. Alexandria, VA, USA: American Statistical Association, 50–55.

Mena, R. H., Ruggiero, M., and Walker, S. G. (2011). Geometric stick-breaking processes for continuous-time Bayesian nonparametric modeling. *Journal of Statistical Planning and Inference* 141: 3217–3230.

Morton, K. D., Torrione, P. A., and Collins, L. M. (2011). Variational Bayesian learning for mixture autoregressive models with uncertain-order. *IEEE Transactions on Signal Processing* 59: 2614–2627.

Müller, P., West, M., and MacEachern, S. N. (1997). Bayesian models for non-linear autoregressions. *Journal of Time Series Analysis* 18: 593–614.

Paisley, J., Blei, D. M., and Jordan, M. I. (2011). The Stick-Breaking Construction of the Beta Process as a Poisson Process. Tech. report available at http://arxiv.org/abs/1109.0343.

Paisley, J., Zaas, A., Woods, C. W., Ginsburg, G. S., and Carin, L. (2010). A stick-breaking construction of the beta process. In *Proceedings of the 27th International Conference on Machine Learning (ICML '10)*. Omnipress, 847–854.

Pitman, J. (2002). Combinatorial Stochastic Processes. Tech. report 621, Department of Statistics, U.C. Berkeley.

Pritchard, J. K., Stephens, M., Rosenberg, N. A., and Donnelly, P. (2000). Association mapping in structured populations. *The American Journal of Human Genetics* 67: 170–181.

Qi, Y., Paisley, J., and Carin, L. (2007). Music analysis using hidden Markov mixture models. *IEEE Transactions on Signal Processing* 55: 5209–5224.

Rabiner, L. R. (1989). A tutorial on hidden Markov models and selected applications in speech recognition. *Proceedings of the IEEE* 77: 257–286.

Saeedi, A. and Bouchard-Côté, A. (2012). Priors over recurrent continuous time processes. In Shawe-Taylor, J., Zemel, R. S., Bartlett, P., Pereira, F., and Weinberger, K. Q. (eds), *Advances in Neural Information Processing Systems 24*. Red Hook, NY: Curran Associates, Inc., 2052–2060.

Sethuraman, J. (1994). A constructive definition of Dirichlet priors. *Statistica Sinica* 4: 639–650.

Srebro, N. and Roweis, S. (2005). Time-Varying Topic Models Using Dependent Dirichlet Processes. Tech. report #2005-003, University of Toronto Machine Learning.

Taddy, M. A. and Kottas, A. (2009). Markov switching Dirichlet process mixture regression. *Bayesian Analysis* 4: 793–816.

Taddy, M. A. (2010). Autoregressive mixture models for dynamic spatial Poisson processes: Application to tracking intensity of violent crimes. *Journal of the American Statistical Association* 105: 1403–1417.

Teh, Y. W., Jordan, M. I., Beal, M. J., and Blei, D. M. (2006). Hierarchical Dirichlet processes. *Journal of the American Statistical Association* 101: 1566–1581.

Thibaux, R. and Jordan, M. I. (2007). Hierarchical beta processes and the Indian buffet process. In *Proceedings of the 11ᵗʰ International Conference on Artificial Intelligence and Statistics (AIS-TATS 2007). Journal of Machine Learning Research – Proceedings Track* 2: 564–571.

Van Gael, J., Saatci, Y., Teh, Y. W., and Ghahramani, Z. (2008). Beam sampling for the infinite hidden Markov model. In *Proceedings of the 25ᵗʰ International Conference on Machine Learning (ICML '08).* Vol. 307 of *ACM International Conference Proceedings Series.* New York, NY, USA: ACM. 1088–1095.

Wang, C., Blei, D. M., and Heckerman, D. (2008). Continuous time dynamic topic models. In *Proceedings of the 24ᵗʰ Conference on Uncertainty in Artificial Intelligence (UAI 2008).* Corvallis, OR, USA: AUAI Press, 579–586.

Williamson, S., Wang, C., Heller, K. A., and Blei, D. M. (2010). The IBP-compound Dirichlet process and its application to focused topic modeling. In *Proceedings of the 27ᵗʰ International Conference on Machine Learning (ICML '10).* Omnipress, 1151–1158.

Wood, S., Rosen, O., and Kohn, R. (2011). Bayesian mixtures of autoregressive models. *Journal of Computational and Graphical Statistics* 20: 174–195.

Xuan, X. and Murphy, K. (2007). Modeling changing dependency structure in multivariate time series. In Ghahramani, Z. (ed), *Proceedings of the 24ᵗʰ International Conference on Machine Learning (ICML '07).* Omnipress, 1055–1062.

Yau, C., Papaspiliopoulos, O., Roberts, G. O., and Holmes, C. C. (2011). Bayesian non-parametric hidden Markov models with applications in genomics. *Journal of the Royal Statistical Society, Series B* 73: 37–57.

Zantedeschi, D., Damien, P. L., and Polson, N. G. (2011). Predictive macro-finance with dynamic partition models. *Journal of the American Statistical Association* 106: 427–439.

21

Mixed Membership Models for Rank Data: Investigating Structure in Irish Voting Data

Isobel Claire Gormley

School of Mathematical Sciences & Complex and Adaptive Systems Laboratory, University College Dublin, Dublin 4, Ireland

Thomas Brendan Murphy

School of Mathematical Sciences & Complex and Adaptive Systems Laboratory, University College Dublin, Dublin 4, Ireland

CONTENTS

A mixed membership model is an individual-level mixture model where individuals have partial membership of the profiles that characterize a population. A mixed membership model for rank data is outlined and illustrated through the analysis of voting in the 2002 Irish general election. This particular election uses a voting system called Proportional Representation using a Single Transferable Vote (PR-STV), where voters rank some or all of the candidates in order of preference. The dataset considered consists of all votes in a constituency from the 2002 Irish general election. Interest lies in highlighting distinct voting profiles within the electorate and studying how voters affiliate themselves to these voting profiles. The mixed membership model for rank data is fitted to the voting data and is shown to give a concise and highly interpretable explanation of voting patterns in this election.

21.1 Introduction

Mixture models are a well-established tool for model-based clustering of data (McLachlan and Basford, 1988; Fraley and Raftery, 2002). Mixture models describe a population as a finite collection of homogeneous groups, each of which is characterized by a specific probability density. While based on a similar concept, mixed membership, or Grade of Membership (GoM), models allow every individual to have partial membership in each of the profiles that characterize the population. Thus, mixed membership models provide a method for model-based soft clustering of data. The mixed membership (or GoM) model for multivariate categorical data is developed in Erosheva (2002) and Blei et al. (2003), and this model has been used in a number of applications including Erosheva et al. (2004; 2007) and Airoldi et al. (2010), amongst others.

Rank data arise when a set of judges rank some (or all) of a set of objects. Rank data emerge in many areas of society; the final ordering of athletes in a race, league tables, the ranking of relevant results by internet search engines, and consumer preference data provide examples of such data. In this chapter, a mixed membership model for rank data that was originally developed in Gormley and Murphy (2009) is described and applied to the problem of finding structure in Irish voting data.

The Irish electoral system uses a voting system called Proportional Representation using a Single Transferable Vote (PR-STV). In this system, voters rank some or all of the candidates in order of preference. When drawing inferences from such data, the information contained in the different preference levels must be exploited by the use of appropriate modeling tools. An illustration of the mixed membership model for rank data methodology is provided through an examination of voting data from the 2002 Irish general election. Interest lies in highlighting voting profiles that occur within the electorate. The mixed membership model provides the scope to examine if and how voters exhibit mixed membership by sharing preference behavior described by more than one of these voting profiles.

A latent class representation of the mixed membership model for rank data is used for model fitting within the Bayesian paradigm. A Metropolis-within-Gibbs sampler is necessary to provide samples from the posterior distribution. Model selection is achieved using the deviance information criterion (DIC) and the adequacy of model fit is assessed using posterior predictive checks.

The chapter proceeds as follows: Section 21.2 outlines the Irish voting system and the details surrounding the 2002 Irish general election. We employ the Plackett-Luce model for rank data in this application as the rank data model; we discuss this model in Section 21.3.1. The specification of the mixed membership model for rank data follows in Section 21.3.2. Estimation of the mixed membership model for rank data is outlined in Section 21.4.1. Section 21.4.2 addresses the question of model choice. We present the application of the mixed membership model for rank data in the 2002 Irish general election data in Section 21.5. The chapter concludes in Section 21.6 with a discussion of the methodology.

21.2 The 2002 Irish General Election

Dáil Éireann is the main parliament in the Republic of Ireland; it has 166 members. Members (called Teachtaí Dála or TDs) are elected to the Dáil through a general election which must take place at least every five years. On May 17, 2002, a general election was held to elect the 29th Dáil; candidates ran in 42 constituencies. Each constituency elected either three, four, or five candidates, where the number of candidates to be elected is determined by the population of the constituency. The Ceann Comhairle is the position of Speaker of the House in Dáil Eireann. The Ceann Comhairle from the previous parliament is automatically re-elected in their constituency and thus the number of candidates elected through the general election in that constituency is reduced by one. The outgoing government consisted of a Fianna Fáil and Progressive Democrat coalition with Fianna Fáil having 77 seats and the Progressive Democrats having 4 seats. Thus, the outgoing government was a minority government who relied on a number of independent TDs for support. After the election, a coalition government involving Fianna Fáil and the Progressive Democrats was formed again, this time with a majority holding 81 and 8 seats, respectively. This was the first time that a government had been re-elected in an Irish general election in 30 years. Extensive descriptions of the 2002 election are provided by Kennedy (2002), Weeks (2002), Gallagher et al. (2003), and Marsh (2003).

In the 2002 general election, a trial was conducted in three constituencies where electronic voting was introduced: Dublin North, Dublin West, and Meath. The voting data from these three constituencies was made publicly available providing an unprecedented insight into the voting in Irish elections beyond what had previously been available in poll data. The data from the Dublin North constituency was analyzed because it contained a particularly diverse range of candidates and thus the data was expected to contain interesting voting behavior.

In 2002, the Dublin North constituency consisted of an electorate of 72,353 with four TDs to be elected from this constituency. A total of 43,942 people voted and twelve candidates ran for election: Fianna Fáil, the largest political party at the time, ran three candidates; Fine Gael, the largest opposition party, ran two candidates; the Labour, Green, and Sinn Féin parties ran one candidate each, and smaller parties like the Socialist, Christian Solidarity, and Independent Health Alliance parties also ran one candidate each. One independent candidate ran for election and the Progressive Democrats did not run any candidate in Dublin North. Four of the candidates were incumbent candidates from the 28th Dáil; however, Seán Ryan (Labour) was elected to the 28th Dáil through a by-election after the resignation of Ray Burke (Fianna Fáil) from his seat during the 28th Dáil.

The votes in the election were totaled through a series of counts where candidates are eliminated, their votes are distributed, and surplus votes are transferred between candidates. A detailed introduction to the PR-STV voting system in an Irish context is given in Sinnott (1999) and a good overall comparison of different voting systems is given by Farrell (2001) and Gallagher and Mitchell (2005).

Details of the counting and transfer of votes in the Dublin North constituency are shown in Table 21.1. The total valid poll consisted of 43,942 votes, so the number of votes required to guarantee election (called the droop quota) was 8,789. In the first count, the number of first preferences for each candidate was counted. If no candidate exceeded the droop quota, then the lowest candidates were eliminated and their votes were distributed using the next available preferences on their ballots; that is, a vote was transferred to the next preferred candidate on the ballot who had not been eliminated or already elected; if no such candidate existed then the vote was considered to be nontransferable. If a candidate was elected by exceeding the droop quota, then their surplus votes (the amount by which they exceed the droop quota) were distributed using the next available preferences on these surplus votes; the surplus votes to be transferred were sampled from the set of votes that brought the candidate over the droop quota. The procedure of eliminating low candidates and distributing surpluses continued until either four candidates exceeded the droop quota or only four candidates remained.

For example, Trevor Sargent was the first candidate to be elected; he was elected in round 6 of the count because he exceeded the droop quota on the basis of the 7,294 first preference votes and 2,491 votes that he received through transfers in rounds 1 to 5 of the count. Because he received 997 votes in excess of the droop quota, these excess votes were transferred in round 7; the 997 votes that were distributed were sampled from the 1,667 that he received in round 5, because these votes brought his total over the droop quota.

By the end of the vote count in Dublin North, two candidates reached the droop quota and two

TABLE 21.1
The counting and transfer of votes in the Dublin North constituency in the 2002 Irish general election. The incumbent candidates are marked with an asterisk. The point at which each candidate was elected is marked in bold.

Candidate (Abbreviation)	Party	1	2	3	4	5	6	7	8
Trevor Sargent* (Sa)	Green	7294	7380	7678	7818	8118	**9785**	8789	8789
			+86	*+298*	*+140*	*+300*	*+1667*	*-996*	
Seán Ryan* (Ry)	Labour	6359	6407	6535	6665	6847	8578	**9128**	9128
			+48	*+128*	*+130*	*+182*	*+1731*	*+550*	
Jim Glennon (Gl)	Fianna Fáil	5892	5945	6028	6152	6294	6511	6598	**8640**
			+53	*+83*	*+124*	*+142*	*+217*	*+85*	*+2044*
G V Wright* (Wr)	Fianna Fáil	5658	5707	5739	5777	5868	6139	6249	**8617**
			+49	*+32*	*+38*	*+91*	*+271*	*+110*	*+2368*
Clare Daly (Dy)	Socialist	5501	5551	5730	5796	6244	6590	6772	7523
			+53	*+179*	*+66*	*+448*	*+346*	*+182*	*+751*
Michael Kennedy (Ke)	Fianna Fáil	5253	5309	5368	5422	5532	5732	5801	
			+56	*+59*	*+54*	*+110*	*+200*	*+69*	*-5801*
Nora Owen* (Ow)	Fine Gael	4012	4030	4132	4720	4763			
			+18	*+102*	*+588*	*+43*	*-4763*		
Mick Davis (Dv)	Sinn Féin	1350	1382	1424	1440				
			+32	*+42*	*+16*	*-1440*			
Cathal Boland (Bo)	Fine Gael	1177	1189	1216					
			+12	*+27*	*-1216*				
Ciarán Goulding (Go)	Independents Health Alliance	914	1009						
			+95	*-1009*					
Eamon Quinn (Qu)	Independent	285							
			-285						
David Walshe (Wa)	Christian Solidarity Party	247							
			-247						
Non Transferable			33	92	152	276	607	607	1245
			+33	*+59*	*+60*	*+124*	*+331*		*+638*
Total		43,942							

were elected without reaching the quota. The four candidates elected were also the four candidates with the highest number of first preferences, but this does not necessarily happen.

21.3 Model Specification

The Dublin North general election voting data possess some unique properties which require careful statistical modeling. A mixed membership model can easily accommodate the differing preferences that voters may have for the candidates. Although a finite mixture model may be used for the same purpose (e.g., Gormley and Murphy, 2008a) the finite mixture model needs a large number of mixture components to account for the voting behavior exhibited in the electorate; conversely the mixed membership model can account for different behavior using a relatively small number of profiles. In order to account for the ranked nature of the preference voting data, the Plackett-Luce model for rank data is used.

21.3.1 The Plackett-Luce Model for Rank Data

Under the PR-STV electoral system, a voter ranks some or all of the candidates in order of preference. In order to appropriately model such data, a model for rank data is required. A large number of models for rank data have already been developed (Bradley and Terry, 1952; Mallows, 1957; Plackett, 1975), and these are reviewed in Marden (1995). In this study the Plackett-Luce model (Plackett, 1975) is utilized to model the rank nature of the data.

The Plackett-Luce model is parameterized by a 'support' parameter

$$\underline{p} = (p_1, p_2, \dots, p_N),$$

where N denotes the total number of electoral candidates. Note that $0 \leq p_j \leq 1$ and $\sum_{j=1}^{N} p_j = 1$. The parameter p_j has the interpretation of being the probability of candidate j being ranked first by a voter. The model assumes that the probability of candidate j being given a lower than first preference is proportional to their support parameter p_j, but conditional on a smaller number of candidates being available for selection at lower preferences. Hence, at preference levels lower than the first, the probabilities are re-normalized to provide valid probability values. Further, it can be shown that the Plackett-Luce model has a random utility choice model interpretation (Chapman and Staelin, 1982).

Let voter i record the vote $\underline{x}_i = \{c(i,1), c(i,2), \dots, c(i,n_i)\}$, where n_i is the number of preferences expressed by voter i. The Plackett-Luce model states that the probability of vote $\underline{x}_i$ is given as

$$
\begin{aligned}
\mathbf{P}\{\underline{x}_i | \underline{p}\} &= \prod_{t=1}^{n_i} \frac{p_{c(i,t)}}{p_{c(i,t)} + p_{c(i,t+1)} + \cdots + p_{c(i,N)}} \\
&= \prod_{t=1}^{n_i} \frac{p_{c(i,t)}}{\sum_{s=t}^{N} p_{c(i,s)}} = \prod_{t=1}^{n_i} q_{it},
\end{aligned}
\tag{21.1}
$$

where $c(i, n_i + 1), \dots, c(i, N)$ is any permutation of the unranked candidates. Note that the probability of the ranking is conditional on n_i, the number of preferences expressed, and it can easily be shown that (21.1) sums to 1 over all $n_i!$ possible permutations of the candidates ranked in the vote $\underline{x}_i$.

21.3.2 The Mixed Membership Model for Rank Data

Mixed membership models allow every individual in a population to have partial membership in each of the profiles that characterize the population; thus, a soft clustering of the population members is achievable. Herein we describe a mixed membership model for rank data as developed by Gormley and Murphy (2009).

Under the mixed membership model, each voter $i = 1, \dots, M$ has an associated *mixed membership parameter* $\underline{\pi}_i = (\pi_{i1}, \pi_{i2}, \dots, \pi_{iK})$ which is a direct parameter of the model. The mixed membership parameter $\underline{\pi}_i$ describes the degree of membership of individual i in each of the K profiles which characterize the electorate. Note that $0 \leq \pi_{ik} \leq 1$ and $\sum_{k=1}^{K} \pi_{ik} = 1$ for $i = 1, \dots, M$. Thus, if individual i is fully characterized by profile k, then $\pi_{ik} = 1$ and $\pi_{ij} = 0$ for $j \neq k$. Additionally, if individual i is characterized by profiles $\mathcal{K} \subset \{1, 2, \dots, K\}$, then $\pi_{ij} > 0$ for $j \in \mathcal{K}$ and $\pi_{ij} = 0$ for $j \notin \mathcal{K}$.

The mixed membership model for ranked data is formulated as follows: We assume that the probability of voter i ranking candidate j in position t on their ballot is a convex combination of the probability of the voter choosing candidate j in position t as described by each profile, where the weights in the convex combination are equal to the voter's mixed membership parameter. That is, the probability of voter i choosing candidate j at preference level t, conditional on voter i's mixed membership parameter π_i and the profile specific support parameters $\mathbf{p} = (\underline{p}_1, \underline{p}_2, \dots, \underline{p}_K)$, is given as

$$\mathbf{P}\{c(i,t) = j|\underline{\pi}_i, \mathbf{p}\} = \sum_{k=1}^{K} \pi_{ik} \left[\frac{p_{kj}}{\sum_{s=t}^{N} p_{kc(i,s)}} \right]. \tag{21.2}$$

Additionally, local independence is then assumed between each preference level t, given the mixed membership parameters. Thus, the conditional probability of ranking $\underline{x}_i$ given membership parameter π_i and support parameters $\mathbf{p}$ is

$$\mathbf{P}\{\underline{x}_i|\underline{\pi}_i, \mathbf{p}\} = \prod_{t=1}^{n_i} \left\{ \sum_{k=1}^{K} \pi_{ik} \left[\frac{p_{kc(i,t)}}{\sum_{s=t}^{N} p_{kc(i,s)}} \right] \right\},$$

and the likelihood function based on the data $\mathbf{x} = (\underline{x}_1, \underline{x}_2, \dots, \underline{x}_M)$ is therefore

$$\mathbf{P}\{\mathbf{x}|\boldsymbol{\pi}, \mathbf{p}\} = \prod_{i=1}^{M} \prod_{t=1}^{n_i} \left\{ \sum_{k=1}^{K} \pi_{ik} \left[\frac{p_{kc(i,t)}}{\sum_{s=t}^{N} p_{kc(i,s)}} \right] \right\}.$$

Note that under the mixed membership model, each voter has partial membership of each profile and mixing takes place at each preference level t rather than at the vote level as would be typical of a rank data mixture model (Stern, 1993; Murphy and Martin, 2003; Gormley and Murphy, 2006; Busse et al., 2007; Gormley and Murphy, 2008a;b). Modeling rank data in this manner provides a deeper insight into the structure within the electorate by allowing mixing to occur at a finer level. This is a desirable characteristic as it may be restrictive to assume a voter expresses all preferences in their vote as dictated by a single profile; it is likely that a voter may express some preferences in line with the support parameters of one profile, and other preferences in line with the support parameters of other profiles. This is clearer when we look at the latent class representation of the mixed membership model (Section 21.3.2).

A Latent Class Representation of the Mixed Membership Model

The mixed membership model for rank data can be expressed using a latent class representation in a manner similar to Erosheva (2006); this representation facilitates efficient inference for the model and it assists with model interpretation. The latent class representation of the mixed membership model for rank data involves augmenting the data for each voter i with categorical latent variables which record the profile that is used by voter i when recording preference level t. The discrete distribution for the latent classes has a functional form that depends on mixed membership parameters π_i for voter i.

For each voter i, we impute binary latent vectors $\underline{z}_{it} = (z_{it1}, \dots, z_{itK})$ for $t = 1, \dots, n_i$, where $\underline{z}_{it} \sim \text{Multinomial}(1, \underline{\pi}_i)$. The value of $\underline{z}_{it}$ records the voting profile that is used by voter i when recording preference level t.

It follows that under the mixed membership model the 'augmented' data likelihood function based on the data $\mathbf{x}$ and the binary latent variables $\mathbf{z}$ is therefore of the form

$$\mathbf{P}\{\mathbf{x}, \mathbf{z}|\boldsymbol{\pi}, \mathbf{p}\} = \prod_{i=1}^{M} \prod_{k=1}^{K} \prod_{t=1}^{n_i} \left\{ \pi_{ik} \left[\frac{p_{kc(i,t)}}{\sum_{s=t}^{N} p_{kc(i,s)}} \right] \right\}^{z_{itk}}. \tag{21.3}$$

Employing the latent class representation of the mixed membership model not only allows estimation of the characteristic parameters of each profile but also direct estimation of the mixed membership parameter for each voter, thus achieving a soft clustering of the voters. In addition, the mixed membership of each individual can be further probed to establish which profile is best appropriate for modeling voter i when they are making choice level t.

21.3.3   Prior and Posterior Distributions

A Bayesian approach is taken when estimating the mixed membership model for rank data and thus the specification of prior distributions for the parameters of the model is required. It is assumed that the mixed membership parameters follow a Dirichlet($\underline{\alpha}$) distribution and that the support parameters follow a Dirichlet($\underline{\beta}$) distribution, i.e.,

$$\underline{\pi}_i \sim \text{Dirichlet} \left\{ \underline{\alpha} = (\alpha_1, \alpha_2, \dots, \alpha_K) \right\}$$
$$\underline{p}_k \sim \text{Dirichlet} \left\{ \underline{\beta} = (\beta_1, \beta_2, \dots, \beta_N) \right\}.$$

The conjugacy of the Dirichlet distribution with the multinomial distribution means the use of a Dirichlet prior is naturally attractive. The use of a Dirichlet prior does, however, induce a negative correlation structure between parameters. The sensitivity of inferences drawn under the mixed membership model for rank data to this prior specification is considered in Gormley and Murphy (2009). For even moderate sized datasets it was found that the posterior inferences were not heavily influenced by the prior specification. In Gormley and Murphy (2009), the sensitivity of the choice of prior model and hyperparameters was considered. In practice, the prior parameters are fixed as $\underline{\alpha} = (0.5, \dots, 0.5)$ and $\underline{\beta} = (0.5, \dots, 0.5)$, which is the Jeffreys prior for the multinomial distribution (e.g., O'Hagan and Forster, 2004). These priors have positive mass near the corners of the parameter simplex and thus the posterior distributions of the parameters can have high probability in these regions. However, the choice of parameters also avoids the posterior concentrating exactly on the corners of the simplex.

In principle, the prior hyperparameters could be estimated as part of the inference procedure rather than fixed as done here, but this greatly increases the computational burden of model fitting and inference.

Given these prior distributions and the augmented data likelihood function (21.3) from the mixed membership model for rank data, the posterior distribution based on the data is:

$$\mathbf{P}\{\boldsymbol{\pi}, \mathbf{p}, \mathbf{z}|\mathbf{x}\} \propto \left[\prod_{i=1}^{M} \prod_{k=1}^{K} \prod_{t=1}^{n_i} \left\{ \pi_{ik} \left[\frac{p_{kc(i,t)}}{\sum_{s=t}^{N} p_{kc(i,s)}} \right] \right\}^{z_{itk}} \right] \left[\prod_{i=1}^{M} \prod_{k=1}^{K} \pi_{ik}^{\alpha_k - 1} \right] \left[\prod_{k=1}^{K} \prod_{j=1}^{N} p_{kj}^{\beta_j - 1} \right].$$

This posterior distribution differs from the posterior distribution in the case of the original mixed membership model (Erosheva, 2002; 2003) in the form of the likelihood function. In the original mixed membership model, discrete response variables are treated as independent given the mixed membership parameters. The likelihood function is therefore the product of independent Bernoulli distributions. In the mixed membership model for rank data, however, the dependence of choices within a rank response leads to a more complex likelihood function that is the product of terms that share parameter values.

21.4   Model Inference

21.4.1   Parameter Estimation

The mixed membership model for rank data can be efficiently fitted in a Bayesian framework. Due to the structure of the posterior distribution, Markov chain Monte Carlo (MCMC) methods are necessary to produce posterior samples of the model parameters. In particular, a Gibbs sampling step can be used in the algorithm if the full conditional distribution for a model parameter has a tractable form. For most of the model parameters in the mixed membership model for rank data this is indeed the case; however, in the case of the support parameters **p**, it is not.

The full conditional distributions of the latent variables z_{it} and the mixed membership parameters π_i are readily available. In particular,

$$z_{it} \sim \text{Multinomial}\left\{1, \left(\frac{\pi_{i1}q_{1it}}{\sum_{k'=1}^{K}\pi_{ik'}q_{k'it}}, \frac{\pi_{i2}q_{2it}}{\sum_{k'=1}^{K}\pi_{ik'}q_{k'it}}, \ldots, \frac{\pi_{iK}q_{Kit}}{\sum_{k'=1}^{K}\pi_{ik'}q_{k'it}}\right)\right\},$$

where q_{kit} is defined as in (21.1) for $k = 1, 2, \ldots, K$, $i = 1, \ldots, M$, $t = 1, \ldots, n_i$ and

$$\pi_i \sim \text{Dirichlet}\left(\alpha_1 + \sum_{t=1}^{n_i} z_{it1}, \ldots, \alpha_K + \sum_{t=1}^{n_i} z_{itK}\right) \qquad \text{for } i = 1, \ldots, M.$$

In the case of the support parameters, the full conditional distributions are

$$\mathbf{P}\{\underline{p}_k | \pi, \mathbf{x}, \mathbf{z}\} \propto \left[\prod_{i=1}^{M}\prod_{t=1}^{n_i}\left\{\frac{\pi_{ik}p_{kc(i,t)}}{\sum_{s=t}^{N}p_{kc(i,s)}}\right\}^{z_{itk}}\right]\left[\prod_{j=1}^{N}p_{kj}^{\beta_j - 1}\right]. \tag{21.4}$$

Due to the form of the likelihood function based on the rank data, the complete conditional distribution of the support parameters is not readily available for sampling and a Gibbs sampling step cannot be implemented. However, a Metropolis step can be used to sample the support parameters. Thus, a Metropolis-within-Gibbs sampler (Carlin and Louis, 2000) can be used to sample from the posterior for all model parameters.

In any Metropolis-based algorithm, the rate of convergence of the chain depends on the relationship between the proposal and target distributions. The use of a proposal distribution which closely mimics the shape and orientation of the target distribution provides an improved rate of convergence and good mixing.

We start to construct a proposal distribution by examining the logarithm of the full conditional of the support parameter $\underline{p}_k$ (21.4) which is of the form

$$\log\mathbf{P}\{\underline{p}_k | \pi, \mathbf{x}, \mathbf{z}\} \propto \sum_{i=1}^{M}\sum_{t=1}^{n_i} z_{itk}\left\{\log p_{kc(i,t)} - \log\sum_{s=t}^{N}p_{kc(i,s)}\right\} + \sum_{j=1}^{N}(\beta_j - 1)\log p_{kj}.$$

The function $-\log(\cdot)$ is a convex function and thus the term $-\log\sum_{s=t}^{N}p_{kc(i,s)}$ can be approximated (in fact lower-bounded) by a hyperplane that is tangent to the function at the currently sampled value of $\underline{p}_k$. The resulting function is the log of a gamma density and this can, in turn, be replaced by the log of a Gaussian density because the shape parameter is typically quite large. Thus, the proposal distribution for p_{kj} emerges as a Gaussian density with mean and variance dependent on the previously sampled values of the model parameters. As the Gaussian distribution extends beyond the $[0, 1]$ interval in which the support parameters lie, proposed values from this surrogate proposal must be suitably normalized.

When estimating parameters via MCMC algorithms, some special features of the mixed membership model for ranking data require attention. A fundamental issue in the fitting of any mixture-based model within a Bayesian framework is that of label switching. This arises because of the invariance of posterior distribution to permutations in the labeling of the profiles. The methods proposed for dealing with label switching, including Stephens (2000), Celeux et al. (2000), and Jasra et al. (2005) need to be considered to avoid this issue. The online relabeling algorithm of Stephens (2000) was found to be an effective method for handling this issue; this algorithm implements relabeling as the MCMC algorithm progresses rather than as a post-processing step.

Full details of the Metropolis-within-Gibbs algorithm for fitting this model are given in Gormley and Murphy (2009).

21.4.2 Model Selection

Another feature of the mixed membership model is the need to infer the model dimensionality, i.e., the number of voting profiles (K) needed to appropriately model the electorate. Within the Bayesian paradigm, the natural approach would be to base inference on the posterior distribution of K given the data $\mathbf{x}$, $\mathbf{P}\{K|\mathbf{x}\}$. However, this posterior can be highly dependent on the model definition and is typically computationally challenging to construct. A comprehensive overview and comparison of model selection criteria within the context of mixed membership models is provided in Joutard et al. (2008).

In this application of the mixed membership model for rank data, the deviance information criterion (DIC), introduced by Spiegelhalter et al. (2002), is used to choose an appropriate model. The DIC criterion penalizes the posterior mean deviance of a model by the "effective number of parameters." The effective number of parameters is derived to be the difference between the posterior mean of the deviance and the deviance at the posterior means of the parameters of interest. Explicitly for data $\mathbf{x}$ and parameters θ the DIC is

$$DIC = \overline{D(\theta)} + p_D,$$

where $D(\theta) = -2\log[\mathbf{P}(\mathbf{x}|\theta)] + 2\log[h(\mathbf{x})]$ is the Bayesian deviance and $h(\mathbf{x})$ is a function of the data only. The effective number of parameters is defined as $p_D = \overline{D(\theta)} - D(\bar{\theta})$. The criterion has an approximate decision theoretic justification. In any case, models with small DIC values are preferable to models with large DIC values. The choice of θ in the calculation of DIC is important, and we use $\theta = (\boldsymbol{\pi}, \mathbf{p})$ because these are the primary model parameters of interest.

21.5 Application to the 2002 Irish General Election

The mixed membership model for rank data was applied to the voting data from the Dublin North constituency in the 2002 Irish general election. This study aims to establish the existence of different voting profiles in the electorate and to establish how voters align themselves with these profiles. This investigation will thus provide an enhanced insight into the actual voting behaviors exhibited in this electorate.

The Metropolis-within-Gibbs sampler, as outlined in Section 21.4.1, was run over 50,000 iterations with a burn-in period of 10,000 iterations. The model was fitted with $K = 1, 2, \ldots, 7$ voting profiles in order to establish the appropriate number of profiles to adequately model the data.

For each value of K, the DIC value was computed (shown in Figure 21.1). The plot shows a sharply decreasing trend when K increases from 1 to 3, and the DIC values decrease slightly thereafter. Consequently, the fitted models for $K \geq 3$ were examined and it was determined that the $K = 3$ model was most appropriate because the models with $K > 3$ included extra extreme profiles that didn't differ greatly from those in the model with $K = 3$.

21.5.1 Support for the Candidates

The marginal posterior density of the support parameters for each candidate within the three voting profiles are illustrated in Figure 21.2; a violin plot (Hintze and Nelson, 1998; Adler, 2005) is used to show these marginal posterior densities. The violin plot combines a boxplot and a kernel density estimate; the length of the violin corresponds to the length of the box in a boxplot but the breadth of the violin shows a back-to-back plot of a kernel density estimate of the values. The marginal probabilities for the voting profiles are $(0.323, 0.324, 0.353)$, respectively.

The three voting profiles have distinct and intuitive interpretations within the context of the 2002

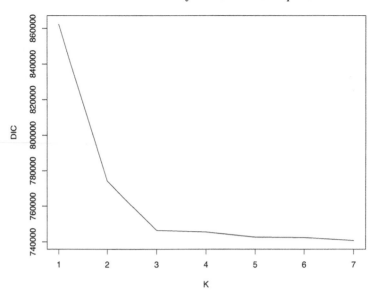

FIGURE 21.1
Values of the DIC for the mixed membership model for rank data fitted to the 2002 Dublin North constituency data over different values of the number of voting profiles K.

Irish general election. The four elected candidates have high support in at least one of the voting profiles and some other prominent candidates also have high support.

Voting Profile 1: Non-mainstream opposition and protest voters.
   Figure 21.2(a). The posterior mean support parameter estimates for the candidates in this voting profile suggest that a pure member of this voting profile should strongly support the non-mainstream opposition parties and single issue/protest candidates. Clare Daly (Socialist Party) has the largest support, and she would be characterized as a major candidate in the non-mainstream opposition in Ireland. Despite having such high support in this voting profile she failed to get elected. Trevor Sargent (Green Party) was leader of the Green Party at the time of the election and the 2002 election saw the party increase its number of seats in the Dáil from two to six seats thus moving them towards the mainstream opposition. Seán Ryan was a Labour party candidate; the Labour party has a diverse range of support within the Irish electorate so it could be considered to be a mainstream party, but it would also have to appeal to voters who don't support other mainstream parties. Interestingly, candidates that received very few first preference votes (e.g., Eamon Quinn and David Walshe) have appreciable support in this voting profile. The non-election of Claire Daly, despite having high support, can be explained by the fact that Trevor Sargent and Seán Ryan were only elected in the later counts (see Table 21.1), so Claire Daly didn't have the opportunity to receive transfers from voters who gave the other candidates higher preferences than her.

Voting Profile 2: Mainstream opposition voters.
   Figure 21.2(b). The support parameters for Trevor Sargent (Green Party), Seán Ryan (Labour), Nora Owen (Fine Gael), and Cathal Boland (Fine Gael) are all large relative to the other candidates. Fine Gael was the largest opposition party before the election and their support here suggests that this voting profile shows support for the mainstream opposition parties. Labour was the second largest opposition party and traditionally Labour and Fine Gael have formed coalition governments, so they share much support amongst the voters. The 2002 election saw

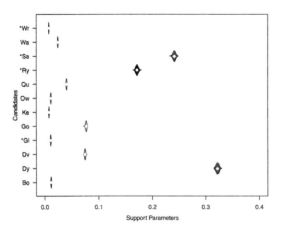

(a) Voting Profile 1: Non-mainstream opposition.

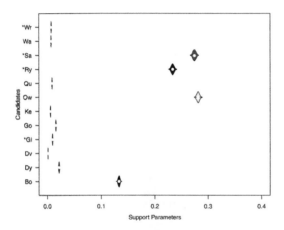

(b) Voting Profile 2: Mainstream opposition.

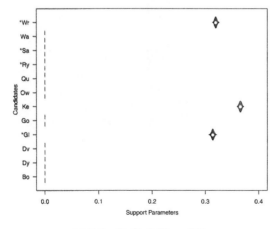

(c) Voting Profile 3: Fianna Fáil.

FIGURE 21.2
Violin plots of the posterior samples for the support parameters. The plot shows the marginal posterior density for each support parameter, for each of the twelve candidates and the three voting profiles. The abbreviation used for each candidate's name is given in Table 21.1. The elected candidates are marked with an asterisk.

the Green party move towards becoming a mainstream opposition party; this is reflected in this voting profile too. Prior to the election, there was some discussion in the print media about Fine Gael, Labour, and the Green party forming a coalition government if they gained enough seats, but this did not happen.

Voting Profile 3: Fianna Fáil voters.

Figure 21.2(c). The posterior mean support parameter estimates for the candidates in this profile reveal that only those voters with a high degree of profile membership should give strong support to the three Fianna Fáil candidates. All other candidates have very low support.

The division of the voters into three profiles provides a systematic method for decomposing the electorate into a small number of profiles. The relevance of the revealed profiles is supported by the exploratory analysis of these data in Laver (2004). Interestingly, the division of candidates amongst the profiles corresponds very closely to the hierarchical decomposition of the candidates and parties in Dublin North as found in Huang (2011) and Huang and Guestrin (2012).

21.5.2   Mixed Membership Parameters for the Electorate

The unique feature of the mixed membership model is that the partial memberships of the voting profiles for each voter are inferred directly when estimating the model. The entropy (Shannon, 1948) of each voter's mixed membership vector measures the degree to which they exhibit mixed membership across voting profiles. In fact, the exponential of the entropy can be seen as the effective number of profiles (Campbell, 1966; White et al., 2012) which are required to model voter i's preferences. Figure 21.3 shows a histogram of the exponentiated entropy values for the Dublin North voters. These show that there is significant evidence of mixed membership for the voters with many being effectively members of two or more of the profiles.

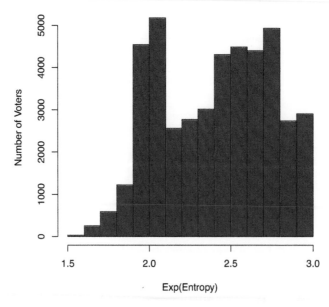

FIGURE 21.3
A histogram of the exponential of the entropy values for each voter's mixed membership parameter. The values shown give an "effective number of profiles" needed to model each voter.

The voter with the lowest effective number of profiles has a membership vector $\underline{\pi}_i =$

$(0.068, 0.885, 0.047)$ and they recorded the vote $\underline{x}_i$ =(Boland, Owen, Sargent, Ryan, Goulding, Quinn, Walsh, Daly, Glennon, Wright, Kennedy, Davis). Since their highest preference choices all have high support in Voting Profile 2, it is clear why they have particularly high membership to this profile and low membership to other profiles. The voter with the highest effective number of profiles has a membership vector $\underline{\pi}_i = (0.333, 0.336, 0.331)$ and they recorded the vote $\underline{x}_i$ =(Goulding, Daly, Ryan, Boland, Owen, Glennon, Wright, Kennedy). In this case, the voter's highest preference votes have high support in different profiles, so the mixed membership model suggests that all three profiles are needed to model their preferences.

We can further explore the mixed membership vectors by dividing the voters into groups, assigning each voter to the voting profile for which they have the highest membership score (i.e., their modal profile membership). We construct a kernel density estimate of the mixed membership parameter for each voting profile for each of the groups of voters (Figure 21.4). Clearly, a significant proportion of the voters who have the strongest affiliation to Voting Profiles 1 and 2 also have a strong affiliation to at least one other profile. In contrast, voters who have strongest affiliation to Voting Profile 3 tend to have very little affiliation to the other voting profiles. This suggests that Voting Profiles 1 and 2 are closer, thus voters exhibit more mixed membership between these two profiles. This makes intuitive sense within the context of the 2002 Irish general election as Voting Profile 3 represents the current government party, with Profiles 1 and 2 representing two different types of opposition.

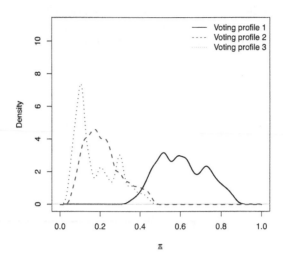

(a) Modal members of Profile 1.

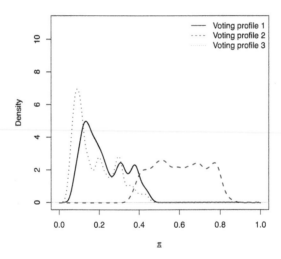

(b) Modal members of Profile 2.

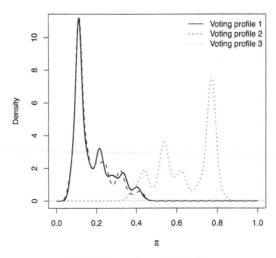

(c) Modal members of Profile 3.

FIGURE 21.4
Kernel density estimates of the membership parameters for those voters most likely to be characterized by each profile.

21.5.3 Posterior Predictive Model Checks

Posterior predictive simulation (Gilks et al., 1996) was employed to assess model fit. Subsequent to a burn-in period of 10,000 iterations, 40,000 samples thinned every 100th iteration were drawn from the posterior distribution $\mathbf{P}\{\boldsymbol{\pi}, \mathbf{p}, \mathbf{z}|\mathbf{x}\}$, giving $R = 400$ sets of parameters simulated from the posterior. A predictive election dataset $\mathbf{x}^r$ was then simulated from the mixed membership model for the rank data, given each of the $r = 1, \ldots, R$ draws of the parameters from the posterior distribution. Due to the discrete and structured nature of the data, it is difficult to fully assess model fit, so first order summaries were used. For the simulated votes, the number of first preference votes obtained by the twelve candidates was recorded. Figure 21.5 illustrates the number of first preferences received by each candidate in each simulated posterior predictive dataset, and in the Dublin North voting data.

The posited model appears to capture the main structure of the data, but there is some discrepancy between the observed and the simulated values. The discrepancy can be explained by the fact that the support parameters $\mathbf{p}$ are used to model the probability of candidate selection at all preference levels and thus the posterior estimates for these parameters depend on all preference levels rather than just first preferences. So, this may lead to a slight under or over estimation of the number of first preference selections for a candidate.

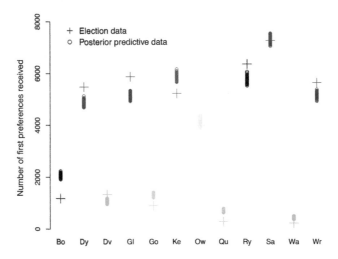

FIGURE 21.5
This plot shows the posterior predictive counts for each candidate in the Dublin North constituency. Each circle indicates the number of first preference votes received by the twelve candidates in each of 400 simulated posterior predictive datasets. The crosses indicate the number of first preferences received by each candidate in the actual voting data.

21.6 Conclusion

A mixed membership model for rank data has been described and applied to the analysis of a large election dataset. It has been shown that in the context of analyzing rank response data, the model provides scope to examine a population for the presence of preference profiles, to estimate the characteristics of these profiles, and to investigate the mixed membership of population members to the profiles on a case-by-case basis. The loss of information which may result from a hard clustering of the data is avoided by providing a soft clustering of the population. In particular, a hard clustering forces each voter to belong to one and only one cluster, so even if they are best characterized by a single cluster, any unusual aspects of their voting preferences are lost in the hard clustering. In contrast, the mixed membership model provides a parsimonious description of voting preferences because complex preference patterns can be captured using the mixed membership machinery.

The method provides an alternative modeling framework to the many mixture modeling approaches for rank data (Stern, 1993; Murphy and Martin, 2003; Gormley and Murphy, 2006; Busse et al., 2007; Gormley and Murphy, 2008a; Meilă and Chen, 2010). In particular, Gormley and Murphy (2008a) developed a finite mixture of Plackett-Luce models for modeling PR-STV data which provides a modeling framework. However, when studying large voting datasets with diverse candidates, a large number of mixture components are needed to appropriately model the data. In contrast, the mixed membership model can represent voting in such elections with many fewer profiles.

The model described herein can be fitted in a Bayesian paradigm using an efficient Markov chain Monte Carlo scheme. The method is able to explore the posterior efficiently because the proposal distributions developed for sampling the support parameters, which don't have a closed-form conditional posterior, are accurate approximations of the parameter conditional posterior distributions. Recently, Caron and Doucet (2012) developed a Gibbs sampling method for the Plackett-Luce model and this could be adapted to fit the mixed membership model outlined herein, thus improving the accuracy of model inference. An alternative method for fitting such models would be to use variational Bayesian (VB) methods or expectation propagation (EP); Weng and Lin (2011) developed an online VB algorithm and Guiver and Snelson (2009) developed an EP algorithm for a single Plackett-Luce model; there is potential to extend these methods to the mixed membership model herein.

The mixed membership model for rank data could be developed in several directions. In terms of the application in this chapter, further model accuracy could be attained by imposing a hierarchical framework—a hyperprior could be introduced for the Dirichlet parameters $\underline{\alpha}$ and $\underline{\beta}$ of the mixed membership and support parameter priors, respectively; such hierarchical priors are employed in Pritchard et al. (2000) and Erosheva (2003).

The issue of model choice for mixed membership models is still problematic (Joutard et al., 2008). The combination of the use of DIC (Spiegelhalter et al., 2002) and posterior predictive model checks (Gilks et al., 1996) provided a suitable method in this application, but there were different numbers of extreme profiles (K) that achieved similar fit. Thus, there remains the need for more automatic model choice methods.

Recently, a number of models have been developed that capture underlying group structure for rank data when concomitant information for the voters is also available (Gormley and Murphy, 2008b; Francis et al., 2010; Lee and Yu, 2010; 2012; Li et al., 2012). It would be worthwhile to extend the mixed membership modeling framework for rank data to include such concomitant information. Such a modeling extension would help explain the structure revealed by the mixed membership model for ranked data.

Appendix : Data Sources

The 2002 Dublin North constituency voting data was made available by the Dublin County Returning Officer. The data are available from the authors on request.

Acknowledgments

This work was supported by the following Science Foundation Ireland (SFI) grants: Research Frontiers Programme (09/RFP/MTH2367), Strategic Research Cluster (08/SRC/I1407) and Research Centre (SFI/12/RC/2289).

References

Adler, D. (2005). vioplot: Violin plot. R package, Version 0.2.

Airoldi, E. M., Erosheva, E. A., Fienberg, S. E., Joutard, C., Love, T., and Shringarpure, S. (2010). Re-conceptualizing the classification of PNAS articles. *Proceedings of the National Academy of Sciences* 107: 20899–20904.

Blei, D. M., Ng, A. Y., and Jordan, M. I. (2003). Latent Dirichlet allocation. *Journal of Machine Learning Research* 3: 993–1022.

Bradley, R. A. and Terry, M. E. (1952). Rank analysis of incomplete block designs: I. The method of paired comparisons. *Biometrika* 39: 324–345.

Busse, L. M., Orbanz, P., and Buhmann, J. M. (2007). Cluster analysis of heterogeneous rank data. In *Proceedings of the 24th International Conference on Machine Learning (ICML '07)*. New York, NY, USA: ACM, 113–120.

Campbell, L. L. (1966). Exponential entropy as a measure of extent of a distribution. *Probability Theory And Related Fields* 5: 217–225.

Carlin, B. P. and Louis, T. A. (2000). *Bayes and Empirical Bayes Methods for Data Analysis*. New York, NY: Chapman & Hall, 2nd edition.

Caron, F. and Doucet, A. (2012). Efficient Bayesian inference for generalized Bradley-Terry models. *Journal of Computational & Graphical Statistics* 21: 174–196.

Celeux, G., Hurn, M., and Robert, C. P. (2000). Computational and inferential difficulties with mixture posterior distributions. *Journal of the American Statistical Association* 95: 957–970.

Chapman, R. and Staelin, R. (1982). Exploiting rank ordered choice set data within the stochastic utility model. *Journal of Marketing Research* 19: 288–301.

Erosheva, E. A. (2002). Grade of Membership and Latent Structure Models with Application to Disability Survey Data. Ph.D. thesis, Department of Statistics, Carnegie Mellon University, Pittsburgh, Pennsylvania, USA.

Erosheva, E. A. (2003). Bayesian estimation of the Grade of Membership model. In Bernardo, J., Bayarri, M., Berger, J., Dawid, A., Heckerman, D., Smith, A., and West, M. (eds), *Bayesian Statistics, 7*. New York, NY: Oxford University Press, 501–510.

Erosheva, E. A. (2006). Latent Class Representation of the Grade of Membership Model. Tech. report 492, Department of Statistics, University of Washington.

Erosheva, E. A., Fienberg, S. E., and Joutard, C. (2007). Describing disability through individual-level mixture models for multivariate binary data. *Annals of Applied Statistics* 1: 502–537.

Erosheva, E. A., Fienberg, S. E., and Lafferty, J. D. (2004). Mixed-membership models of scientific publications. *Proceedings of the National Academy of Sciences* 101: 5220–5227.

Farrell, D. M. (2001). *Electoral Systems: A Comparative Introduction*. New York, NY: St. Martin's Press.

Fraley, C. and Raftery, A. E. (2002). Model-based clustering, discriminant analysis, and density estimation. *Journal of the American Statistical Association* 97: 611–631.

Francis, B., Dittrich, R., and Hatzinger, R. (2010). Modeling heterogeneity in ranked responses by nonparametric maximum likelihood: How do Europeans get their scientific knowledge? *Annals of Applied Statistics* 4: 2181–2202.

Gallagher, M., Marsh, M., and Mitchell, P. (2003). *How Ireland Voted 2002*. Basingtoke: Palgrave Macmillian.

Gallagher, M. and Mitchell, P. (2005). *The Politics of Electoral Systems*. Oxford, UK: Oxford University Press.

Gilks, W. R., Richardson, S., and Spiegelhalter, D. J. (eds) (1996). *Markov Chain Monte Carlo in Practice*. London, UK: Chapman & Hall.

Gormley, I. C. and Murphy, T. B. (2009). A Grade of Membership model for rank data. *Bayesian Analysis* 4: 265–296.

Gormley, I. C. and Murphy, T. B. (2006). Analysis of Irish third-level college applications data. *Journal of the Royal Statistical Society, Series A* 169: 361—379.

Gormley, I. C. and Murphy, T. B. (2008a). Exploring voting blocs within the Irish electorate: A mixture modeling approach. *Journal of the American Statistical Association* 103: 1014–1027.

Gormley, I. C. and Murphy, T. B. (2008b). A mixture of experts model for rank data with applications in election studies. *Annals of Applied Statistics* 2: 1452–1477.

Guiver, J. and Snelson, E. (2009). Bayesian inference for Plackett-Luce ranking models. In *Proceedings of the 26th Annual International Conference on Machine Learning (ICML '09)*. Omnipress, 377–384.

Hintze, J. L. and Nelson, R. D. (1998). Violin plots: A box plot-density trace synergism. *The American Statistician* 52: 181–184.

Huang, J. (2011). Probabilistic Reasoning and Learning on Permutations: Exploiting structural decompositions of the symmetric group. Ph.D. thesis, Carnegie Mellon University.

Huang, J. and Guestrin, C. (2012). Uncovering the riffled independence structure of rankings. *Electronic Journal of Statistics* 6: 199–230.

Jasra, A., Holmes, C. C., and Stephens, D. A. (2005). Markov chain Monte Carlo Methods and the label switching problem in Bayesian mixture modeling. *Statistical Science* 20: 50–67.

Joutard, C., Airoldi, E. M., Fienberg, S. E., and Love, T. (2008). Discovery of latent patterns with hierarchical Bayesian mixed-membership models and the issue of model choice. In Poncelet, P., Masseglia, F., and Teisseire, M. (eds), *Data Mining Patterns: New Methods and Applications*. Pennsylvania: IGI Global.

Kennedy, F. (2002). The 2002 general election in Ireland. *Irish Political Studies* 17: 95–106.

Laver, M. (2004). Analysing structure of party preference in electronic voting data. *Party Politics* 10: 521–541.

Lee, P. H. and Yu, P. L. H. (2010). Distance-based tree models for ranking data. *Computational Statistics & Data Analysis* 54: 1672–1682.

Lee, P. H. and Yu, P. L. H. (2012). Mixtures of weighted distance-based models for ranking data with applications in political studies. *Computational Statistics & Data Analysis* 56: 2486–2500.

Li, J., Gu, M., and Hu, T. (2012). General partially linear varying-coefficient transformation models for ranking data. *Journal of Applied Statistics* 39: 1475–1488.

Mallows, C. L. (1957). Non-null ranking models. I. *Biometrika* 44: 114–130.

Marden, J. I. (1995). *Analyzing and Modeling Rank Data*. London, UK: Chapman & Hall.

Marsh, M. (2003). The Irish general election of 2002: A new hegemony for Fianna Fail? *British Elections & Parties Review* 13: 17–28.

McLachlan, G. J. and Basford, K. E. (1988). *Mixture Models: Inference and Applications to Clustering*. New York, NY: Marcel Dekker Inc.

Meilă, M. and Chen, H. (2010). Dirichlet process mixtures of generalized Mallows models. In Grünwald, P. and Spirtes, P. (eds), *Proceedings of the 26th Conference on Uncertainty in Artificial Intelligence (UAI 2010)*. Corvallis, OR, USA: AUAI Press, 358–367.

Murphy, T. B. and Martin, D. (2003). Mixtures of distance-based models for ranking data. *Computational Statistics and Data Analysis* 41: 645–655.

O'Hagan, A. and Forster, J. (2004). *Kendall's Advanced Theory of Statistics: Volume 2B Bayesian Inference*. London, UK: Arnold, 2nd edition.

Plackett, R. L. (1975). The analysis of permutations. *Applied Statistics* 24: 193–202.

Pritchard, J. K., Stephens, M., and Donnelly, P. (2000). Inference of population structure using multilocus genotype data. *Genetics* 155: 945–959.

Shannon, C. E. (1948). A mathematical theory of communication. *Bell System Technical Journal* 27: 379–423.

Sinnott, R. (1999). The electoral system. In Coakley, J. and Gallagher, M. (eds), *Politics in the Republic of Ireland*. London, UK: Routledge & PSAI Press, 3rd edition, 99–126.

Spiegelhalter, D. J., Best, N. G., Carlin, B. P., and van der Linde, A. (2002). Bayesian measures of model complexity and fit. *Journal of the Royal Statistical Society, Series B* 64: 583–639.

Stephens, M. (2000). Dealing with label-switching in mixture models. *Journal of the Royal Statistical Society, Series B* 62: 795–810.

Stern, H. S. (1993). Probability models on rankings and the electoral process. In Fligner, M. A. and Verducci, J. S. (eds), *Probability Models and Statistical Analyses For Ranking Data*. New York, NY: Springer-Verlag, 173–195.

Weeks, L. (2002). The Irish parliamentary election, 2002. *Representation* 39: 215–225.

Weng, R. C. and Lin, C. -J. (2011). A Bayesian approximation method for online ranking. *Journal of Machine Learning Research* 12: 267–300.

White, A., Chan, J., Hayes, C., and Murphy, T. B. (2012). Mixed membership models for exploring user roles in online fora. In Ellison, N., Shanahan, J., and Tufekci, Z. (eds), *Proceedings of the 6th International AAAI Conference on Weblogs and Social Media (ICWSM 2012)*. Palo Alto, CA, USA: AAAI.

Part VI

Mixed Membership Models for Networks

22

Hierarchical Mixed Membership Stochastic Blockmodels for Multiple Networks and Experimental Interventions

Tracy M. Sweet

Department of Human Development and Quantitative Methodology, University of Maryland, College Park, MD 20742, USA

Andrew C. Thomas

Department of Statistics, Carnegie Mellon University, Pittsburgh, PA 15213, USA

Brian W. Junker

Department of Statistics, Carnegie Mellon University, Pittsburgh, PA 15213, USA

CONTENTS

Quantitative methods for analyzing social networks have primarily focused on either single network statistical models (e.g., Airoldi et al., 2008; Hoff et al., 2002; Wasserman and Pattison, 1996) or summarizing multiple networks with descriptive statistics (e.g., Penuel et al., 2013; Moolenaar et al., 2010; Frank et al., 2004). Many experimental interventions and observational studies however involve several if not many networks.

To model such samples of independent networks, we use the Hierarchical Network Models framework (Sweet et al., 2013, HNM) to introduce hierarchical mixed membership stochastic blockmodels (HMMSBM) which extend single-network mixed membership stochastic blockmodels (Airoldi et al., 2008, MMSBM) for use with multiple networks and network-level experimental data. We also introduce how covariates can be incorporated into these models.

The HMMSBM is quite flexible in that it can be used on both intervention and observational data. Models can be specified to estimate a variety of treatment effects related to subgroup membership and well as covariates and additional hierarchical parameters. Using simulated data, we present several empirical examples involving network ensemble data to illustrate model fit feasibility and parameter recovery.

22.1   Introduction

A social network represents the relationships among a group of individuals or entities and is commonly illustrated by a graph. The nodes or vertices represent individuals or actors and the edges between them the ties or relationships between two individuals. These edges may be directed, suggesting a sender and receiver of the interaction, or undirected, suggesting reciprocity between the two nodes; a network depicting collaboration is likely to be an undirected graph whereas a network depicting advice-seeking would be a directed graph. Figure 22.1 shows advice-seeking ties among teachers regarding two different subjects. Network ties are part of a larger class of observations termed *relational data*, since these data reflect pairwise relationships, such as the presence and direction of pairwise ties. Since relationships are pervasive, it is unsurprising that relational data methodology has applications in a wide variety of fields, including biology (Airoldi et al., 2005), international relations (Hoff and Ward, 2004), education (Weinbaum et al., 2008), sociology (Goodreau et al., 2009), and organizational theory (Krackhardt and Handcock, 2007).

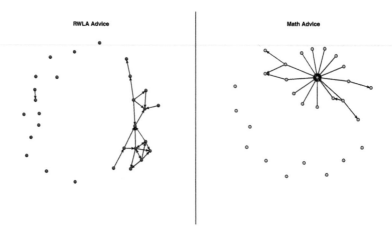

FIGURE 22.1
Two social networks, depicting asymmetric advice-seeking behavior among two groups of teachers, from Pitts and Spillane (2009). Vertices, or nodes, represent individual teachers. Arrows, or directed edges, point from advice seeking teachers to advice providing teachers.

Two prominent quantitative methods for analyzing social networks are descriptive network statistics and statistical modeling. Descriptive network statistics are useful for exploring, summarizing, and identifying certain features of networks, which are then used as covariates in other statistical models. Common statistics include density, the total number of ties; degree, the number of ties for any one node; betweenness, the extent that a node connects other nodes; and other observed structural elements such as triangles. Kolaczyk (2009) provides a comprehensive list. Descriptive statistics are inherently aggregate, so using them to represent a network or to compare networks is problematic. For example, Figure 22.1 show two networks with similar density (22 ties among 27 nodes and 19 ties among 28 nodes); however, the structure of the networks is quite different.

Alternatively, a statistical social network model formalizes the probability of observing the entire network and its various structural features. Current methods generally fall into one of three categories, exponential random graph models (ERGM), latent space models (LSM), and mixed membership stochastic blockmodels (MMSB); see Goldenberg et al. (2009) for a comprehensive review. An exponential random graph model(Wasserman and Pattison, 1996) represents the probability of observing a particular network as a function of network statistics. The latent space model (Hoff et al., 2002) assumes each node occupies a position in a latent social space. The probability of a tie between two individuals is modeled as a function of the pairwise distance in this space. Stochastic blockmodels cluster nodes to one of a fixed number of finite groups, and the probability of a tie between two nodes is determined by the group membership of each node. The mixed membership stochastic blockmodel (Airoldi et al., 2008) allows nodes to belong to multiple groups so that group membership may vary by node interaction.

Most modeling methodology for social networks focuses on modeling a single network, but in many applications more than one network may be of interest. The study of multiple networks can be divided into three classes: studying multiple types of ties among nodes of one network (e.g., friendship ties and collaboration ties), studying one network over time, and studying a single measure on multiple isolated networks. There has been a fair amount of work done for the first two cases. Fienberg et al. (1985) showed how loglinear models can be used to model multiple measures on a single network and Pattison and Wasserman (1999) extended this work for the logit forms of p^* models. Longitudinal methods to model a single network over time have been extensively studied. The three categories of models each have known longitudinal extensions: Hanneke et al. (2010) introduced temporal ERGMs which are based on a discrete Markov process; Westveld and Hoff (2011) embedded an auto-regressive structure in LSMs; and Xing et al. (2010) added a state-space model to the MMSBM.

Modeling a sample of isolated networks has only recently attracted sustained attention. Motivated by social networks of teachers in education research, Sweet et al. (2013) introduced hierarchical network models (HNM), a class of models for modeling ensembles of networks. The purpose of this paper is to use the HNM framework to formally introduce hierarchical mixed membership stochastic blockmodels (HMMSBM) which extend the MMSBM for use with relational data from multiple isolated networks.

In the next section, we formally define the MMSBM for a single network, present a covariate version of a MMSBM, and introduce an MCMC algorithm for estimation. In Section 22.3, we present the HNM framework and formally define the HMMSBM. Extending our MCMC algorithm for a single network, we present an algorithm for fitting the HMMSBM that we illustrate with two examples. We conduct a simple simulation study for sensitivity analysis and conclude with some remarks regarding estimation and utility of these models.

22.2 Modeling a Single Network

A single social network Y among n individuals can be represented by an adjacency matrix of dimension $n \times n$,

$$Y = \begin{bmatrix} Y_{11} & Y_{12} & \cdots & Y_{1n} \\ \vdots & \vdots & \ddots & \vdots \\ Y_{n1} & Y_{n2} & \cdots & Y_{nn} \end{bmatrix}, \tag{22.1}$$

where Y_{ij} is the value of the tie from i to j. These ties might be binary, indicating the presence or absence of a tie, or an integer or real number, indicating the frequency of interaction or strength of a tie. For the purposes of this paper, we restrict ourselves to binary ties.

In many contexts, individuals in the network belong to certain subgroups. In a school faculty network, for example, teachers belong to departments. However, these group memberships are often not directly observed and can only be inferred through the network structure. Figure 22.2 (left) shows an adjacency matrix for networks generated from a stochastic blockmodel in which individuals belong to one of four groups. A black square indicates the presence of a tie between two individuals. Ties within groups are much more likely than ties across groups. Blockmodels are most appropriate for relational data with this structure and a variety of blockmodels have been studied (see Anderson and Wasserman, 1992).

Stochastic blockmodels assign each individual membership to a block or group, and assignment may either be observed or latent. Tie probabilities are then determined through group membership; usually within-group tie probabilities are modeled to be much larger than between-group tie probabilities, resulting in the block structure shown in Figure 22.2.

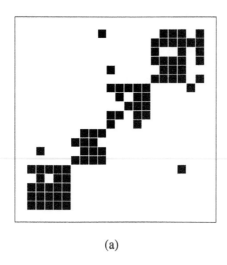   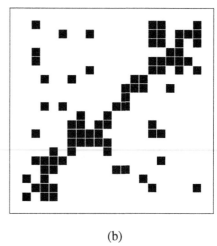

(a)   (b)

FIGURE 22.2

A network generated from a stochastic blockmodel where group membership is not mixed (a). Each node is assigned a group membership which determines the probability of ties. A network generated from a MMSBM (b). Node membership may vary with each pairwise interaction. Note, the $n \times n$ sociomatrix displays a black box for each tie and white otherwise.

Mixed membership stochastic blockmodels (MMSBM) instead allow block membership to be defined for each interaction with a new partner. Rather than assuming individual i is a member of block k for all interactions, the block membership is determined anew for each interaction. Individual i might belong to block k when interacting with individual j but belong to block k' when interacting with individual j'.

We define the MMSBM as a hierarchical Bayesian model (Airoldi et al., 2008),

$$Y_{ij} \sim Bernoulli(S_{ij}{}^T B R_{ji})$$
$$S_{ij} \sim Multinomial(1, \theta_i)$$
$$R_{ji} \sim Multinomial(1, \theta_j)$$
$$\theta_i \sim Dirichlet(\lambda)$$

$$B_{\ell m} \sim Beta(a_{\ell m}, b_{\ell m}) \,, \tag{22.2}$$

where the group membership probability vector for individual i is θ_i, and specific group memberships are determined through a multinomial distribution. S_{ij} is the group membership indicator vector of i when initiating interaction with j, and R_{ji} is the group membership indicator vector of j when acting in response to i. Notice the stochastic nature of S_{ij} and R_{ji}; each is sampled for every interaction from i to j, allowing individual group memberships to vary. The value of Y_{ij} is determined based on a block dependent probability matrix B, where $B_{\ell m}$ is the probability of a tie from an individual in group ℓ to an individual in group m.

The hyperparameter λ may be fixed and known or estimated as a parameter. The dimension of λ identifies the number of groups (g) and the value of λ determines the shape of the Dirichlet distribution on the g-simplex. The hyperparameters (a_ℓ, b_m) are generally elicited so that within-group tie probabilities are higher than across-group tie probabilties.

22.2.1 Single Network Model Estimation

We developed a Markov chain Monte Carlo (MCMC; Gelman et al., 2004) algorithm to fit the MMSBM. The joint likelihood of the model can be written as the following product:

$$P(Y|S, R, \theta, \lambda, B)P(S|\theta)P(R|\theta)P(\theta|\lambda)P(B)P(\lambda)$$
$$= \prod_{i \neq j} P(Y_{ij}|S_{ij}, R_{ji}, \theta_i, \theta_j, \lambda, B) \prod_{i \neq j} P(S_{ij}|\theta_i)P(R_{ji}|\theta_j) \prod_i P(\theta_i|\lambda) \prod_{\ell,m} P(B)P(\lambda) \,. \tag{22.3}$$

The complete conditionals for θ, R, S, B can be written in a closed form, so we use Gibbs updates for each. Full conditional posterior probability distributions are listed below. Define $\ldots$ to represent all other parameters and data in the model, and let $\ell^\star$ represent the group indicated by R_{ji} and $m^\star$ represent the group indicated by S_{ij}.

$$P(\theta_i|\ldots) \propto Dirichlet \left(\lambda + \sum_j S_{ij} + \sum_j R_{ij}\right)$$
$$P(S_{ij}|\ldots) \propto Multinomial (p)$$
$$p_k = \theta_{ik} B_{k\ell^\star}{}^{Y_{ij}} (1 - B_{k\ell^\star})^{(1-Y_{ij})}$$
$$P(R_{ji}|\ldots) \propto Multinomial (q)$$
$$q_k = \theta_{ik} B_{m^\star k}{}^{Y_{ij}} (1 - B_{m^\star k})^{(1-Y_{ij})}$$
$$P(B_{\ell m}|\ldots) \propto Beta \left(a_\ell + \sum_{(ij)^\star} Y_{ij}, b_m + \sum_{(ij)^\star} Y_{ij} \,, \right. \tag{22.4}$$

where $(ij)^\star$ is an (ℓ, m)-specific subset of $i = 1, .., n$ and $j = 1, .., n$ such that $S_{ij} = \ell$ and $R_{ji} = m$. In addition, we incorporate a sparsity parameter ρ (Airoldi et al., 2008). The absence of ties can be attributed to either rarity of interaction across groups or lack of interest in making across-group ties. For example, teachers in departments in schools may have few collaborative ties outside of their department because they interact less often with teachers outside their department but also because they would rather interact with those who teach the same subjects. The sparsity parameter helps to account for sparsity in the adjacency matrix due to lack of interaction. The probability of ties from group ℓ to m is therefore modeled as $\rho B_{\ell m}$.

If λ is estimated, we use a common parameterization and let $\lambda = \gamma \xi$ where $\gamma = \sum_{k=1}^g \lambda$ and $\sum_{k=1}^g \xi = 1$ (Erosheva, 2003). We can think of γ as a measure of how extreme the Dirichlet distribution is, i.e., small values of γ imply greater mass in the corners of the g−simplex. Since ξ sums to 1, it is an indirect measure of the probability of belonging to each group. Equal values of ξ suggest equal sized groups. As defined, γ and ξ are independent and we update each using Metropolis steps.

To update γ, we use a gamma proposal distribution with shape parameter ν_γ and rate parameter selected so that the proposal distribution has a mean at the current value of γ. The value of ν_γ is then tuned to ensure an appropriate acceptance rate. Then the proposed value of γ^{s+1} is accepted with probability $min\{1, R\}$ where $R = \frac{P(\gamma^{(s+1)}|...)}{P(\gamma^{(s)}|...)} \frac{P(\gamma^{(s)}|\gamma^{(s+1)})}{P(\gamma^{(s+1)}|\gamma^{(s)})}$.

To update ξ, we use a uniform Dirichlet proposal distribution centered at the current value of ξ. Thus, $\xi^{(s+1)} \sim Dirichlet(\nu_\xi g \xi^{(s)})$ where ν_ξ is the appropriate tuning parameter. The proposed value of $\xi^{(s+1)}$ is accepted with probability $min\{1, R\}$ where now, $R = \frac{P(\xi^{(s+1)}|...)}{P(\xi^{(s)}|...)} \frac{P(\xi^{(s)}|\xi^{(s+1)})}{P(\xi^{(s+1)}|\xi^{(s)})}$.

22.2.2 Empirical Example

To illustrate fitting a single network MMSBM, we use the Monk data of Sampson (1968). While staying with a group of monks as a graduate student, Sampson recorded relational data among the monks at different time periods during his year-long stay. Toward the end of his stay, there was a political crisis which resulted in several monks being expelled and several others leaving.

We use relational data from three time periods prior to the crisis. For each time period, we have nominations for the three monks they like best. These data have been aggregated into a single adjacency matrix, where $Y_{ij} = 1$ if monk i nominated j as one of his top three choices during any of the three time periods. Y_{ii} is undefined.

Based on past work suggesting three subgroups of Monks (Breiger et al., 1975), we fit the following MMSBM:

$$Y_{ij} \sim Bernoulli(S_{ij}^T B R_{ji})$$
$$S_{ij} \sim Multinomial(1, \theta_i)$$
$$R_{ji} \sim Multinomial(1, \theta_j)$$
$$\theta_i \sim Dirichlet(\gamma)\xi$$
$$B_{\ell\ell} \sim Beta(3, 1)$$
$$B_{\ell m} \sim Beta(1, 10), \ell \neq m$$
$$\gamma \sim Gamma(1, 5)$$
$$\xi \sim Dirichlet(1, 1, 1), \tag{22.5}$$

where $1 - \rho = 1 - \frac{\sum_{ij} Y_{ij}}{N(N-1)}$, and $N = 18$, the number of monks.

We sample MCMC chains of length 15,000, keeping the last 10,000, and retaining 1 out of every 25 steps for a posterior approximation of 401 samples. To assess our fit, we compare the original sociomatrix shown on the left in Figure 22.3 to our fitted model. Using posterior means for each parameter, we illustrate the probability of a tie between two monks by color, with low probabilities in shades of blue and high probabilities in red, orange, and yellow ((c), Figure 22.3).

22.2.3 Incorporating Covariates into a MMSBM

While the MMSBM captures block structure, network ties may also form based on other individual similarities independent of or unrelated to the existing block structure. While teachers in schools may belong to departments, they may also belong to groups based on unobserved characteristics. But some ties might also form based on proximity in the school building, teaching the same group of students, or attending new teacher seminars together, independently of the overarching grouping mechanism.

We present a simple extension for the MMSBM to include covariates as

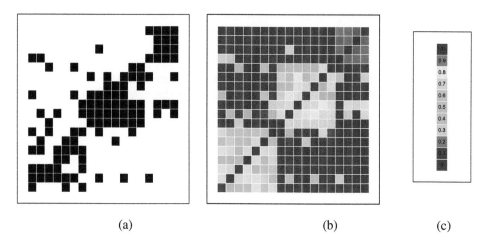

(a) (b) (c)

FIGURE 22.3
The original sociomatrix (a) versus the probability of a tie as determined by our model using posterior means (b). In general, estimated tie probabilities mirror the true tie structure. The legend shows increments of 0.1, with all values except 0 being the upper endpoint of the continuous class of colors (c).

$$Y_{ij} \sim Bernoulli(p_{ij})$$
$$p_{ij} = \frac{\exp\{\text{logit}\,(S_{ij}{}^T \text{logit}\,(B)R_{ji}) + \alpha X_{ij}\}}{1 + \exp\{\text{logit}\,(S_{ij}{}^T B R_{ji}) + \alpha X_{ij}\}}$$
$$S_{ij} \sim Multinomial(1, \theta_i)$$
$$R_{ji} \sim Multinomial(1, \theta_j)$$
$$\theta_i \sim Dirichlet(\lambda)$$
$$B_{\ell m} \sim Beta(a_{\ell m}, b_{\ell m}) \,, \tag{22.6}$$

where X_{ij} is a covariate and α is the coefficient for that covariate.

Model (22.6) can be fit using a MCMC algorithm similar to (22.4), with a more complicated sampling distribution. We use the same Gibbs update for θ, and the same Metropolis updates for γ and ξ, as presented in our standard MMSBM (22.4). We use the following Gibbs updates for S_{ij} and R_{ji},

$$P(S_{ij}|\dots) \propto Multinomial\,(p)$$
$$p_k = \theta_{ik} \frac{\exp\{\text{logit}\,(B)_{k\ell^\star} + \alpha X_{ij}\}^{Y_{ij}}}{1 + \exp\{\text{logit}\,(B)_{k\ell^\star} + \alpha X_{ij}\}}$$
$$P(R_{ji}|\dots) \propto Multinomial\,(q)$$
$$q_k = \theta_{ik} \frac{\exp\{\text{logit}\,(B)_{m^\star k} + \alpha X_{ij}\}^{Y_{ij}}}{1 + \exp\{\text{logit}\,(B)_{m^\star k} + \alpha X_{ij}\}} \,, \tag{22.7}$$

where again ℓ is the group indicated by R_{ji} and m is the group indicated by S_{ij}. We reparameterize B and use logit (B) throughout our MCMC algorithm. The entries in B no longer have a direct sampling and we instead use Metropolis–Hastings updates.

To take advantage of random walk updates we reparameterize B as logit (B), and having an unbounded support for our proposal distributions allows us to update the diagonal and off-diagonal elements using the same proposal distribution. Note that $S_{ij}{}^T \text{logit}(B) R_{ji}$ and $\text{logit}(S_{ij}{}^T B R_{ji})$ are equivalent.

Thus, to update an entry of $\text{logit}(B)$, $\text{logit}(B)_{\ell m}^s$, we propose a new entry, $\text{logit}(B)_{\ell m}^{s+1}$, using a normal random walk with mean $\text{logit}(B)_{\ell m}^s$ where variance is determined by a tuning parameter to ensure appropriate acceptance rates. The probably of accepting this new entry is $min\{1, R\}$, where $R = \frac{P(\text{logit}(B)_{\ell m}^{s+1}|...)}{P(\text{logit}(B)_{\ell m}^s|...)}$. We illustrate this algorithm in Section 22.3.4.

22.3   Modeling an Ensemble of Networks

22.3.1   The Hierarchical Network Framework

Consider a collection of K networks $\mathbb{Y} = (Y_1, \dots, Y_k)$ where $Y_k = (Y_{11k}, \dots, Y_{n_k n_k k})$. The hierarchical network framework for this collection $\mathbb{Y}$ is given as

$$P(Y|X, \Theta) = \prod_{k=1}^{K} P(Y_k|X_k, \Theta_k = (\theta_{1k}, .., \theta_{pk}))$$
$$(\Theta_1, \dots, \Theta_K) \sim F(\Theta_1, \dots, \Theta_K|W_1, \dots, W_K, \psi), \qquad (22.8)$$

where $P(Y_k|X_k, \Theta_k = (\theta_{1k}, .., \theta_{pk}))$ is a probability model for network k with covariates X_k.

Notice that this model structure specifies that networks may be independent of each other depending on choice of W, but need not be. Additional hierarchical structure can be specified by including additional parameters ψ. Notice also that we purposely omit any within-network dependence assumptions. Thus, this framework allows for a variety of dependence assumptions both across and within networks but is also flexible in that any social network model can be used. For example, Sweet et al. (2013) uses this framework to introduce hierarchical latent space models, a latent space modeling approach for multiple isolated networks.

22.3.2   The Hierarchical Mixed Membership Stochastic Blockmodel

Let Y_{ijk} be a binary tie from node i to node j in network k. The hierarchical mixed membership stochastic blockmodel is specified as

$$P(Y|S, R, B, \theta, \gamma)$$
$$= \prod_{k=1}^{K} \prod_{i \neq j} P(Y_{ijk}|S_{ijk}, R_{jik}, B_k, \theta_k, \gamma_k) P(S_{ijk}|\theta_{ik}) P(R_{jik}|\theta_{jk}) \prod_i P(\theta_{ik}|\lambda_k), \qquad (22.9)$$

where S_{ijk} is the group membership indicator vector for person i when sending a tie to person j in network k, and R_{jik} is the group membership indicator vector for j when receiving a tie from i in network k; B_k is the network specific group-group tie probability matrix, and θ_{ik} is the group membership probability vector for node i in network k.

This is easily presented as a hierarchical Bayesian model:

$$Y_{ijk} \sim Bernoulli(S_{ijk}{}^T B_k R_{jik})$$
$$S_{jik} \sim Multinomial(\theta_{ik}, 1)$$
$$R_{jik} \sim Multinomial(\theta_{jk}, 1)$$
$$\theta_{ik} \sim Dirichlet(\lambda_k)$$

$$B_{\ell mk} \sim Beta(a_{\ell mk}, b_{\ell mk}) \,. \tag{22.10}$$

We impose our hierarchical structure by requiring that the parameters come from some common distribution, and in fact, this framework becomes particularly interesting in cases where parameters are shared across networks. We present several examples in the next section.

Examples of HMMSBMs

The hierarchical structure of the HMMSBM naturally lends itself to pooling information across networks and we present several extensions of (22.10).

A simple extension is an HMMSBM for experimental data in which the treatment is hypothesized to affect a single parameter. The networks in the treatment condition would be generated from the same model and the control condition networks would be generated from a different model. For example, suppose we examine teacher collaboration networks in high schools. Typically we would expect to see teachers collaborating within their own departments and these departments operating mostly in isolation. But we could imagine an intervention whose aim is to increase collaboration across departments. In contrast, teachers in treatment schools are more likely to have across department ties than teachers in control schools. Such a model is given as

$$
\begin{aligned}
Y_{ijk} &\sim Bernoulli(S_{ijk}{}^T B_k R_{jik}) \\
S_{ijk} &\sim Multinomial(\theta_{ik}, 1) \\
R_{jik} &\sim Multinomial(\theta_{jk}, 1) \\
\theta_{ik} &\sim Dirichlet(\lambda_k), \text{ where } \lambda_k = \lambda_0 + T_k(\mathbf{1} - \lambda_0)(1 - \alpha) \\
B_{\ell mk} &\sim Beta(a_{\ell mk}, b_{\ell mk}) \\
\alpha &\sim Uniform(0, 1) \,,
\end{aligned}
\tag{22.11}
$$

where T_k is the indicator for being in the treatment group, $\mathbf{1}$ is the vector $(1, .., 1)$ with length g, and g is the number of groups. The treatment effect α is a proportion of how similar the group membership profiles are to the control group as compared to a uniform distribution on the simplex.

Rather than constraining each network to have a constant network level parameter, e.g., λ_0, we might instead model network parameters generated from a single distribution, introducing an additional level to the hierarchy. Suppose we are interested in how variable the membership probabilities vectors are across networks, for example we expect teacher collaboration networks to vary depending on the organizational structure in the schools. Then we could estimate the distributional hyperparameters that generate these membership probabilities (θ).

An example of such a model is

$$
\begin{aligned}
Y_{ijk} &\sim Bernoulli(S_{ijk}{}^T B_k R_{jik}) \\
S_{ijk} &\sim Multinomial(\theta_{ik}, 1) \\
R_{jik} &\sim Multinomial(\theta_{jk}, 1) \\
\theta_{ik} &\sim Dirichlet(\gamma_k \xi_k) \\
B_{\ell mk} &\sim Beta(a_{\ell mk}, b_{\ell mk}) \\
\gamma_k &\sim Gamma(\tau, \beta) \\
\xi_k &\sim Dirichlet(c) \\
\tau &\sim Gamma(a_\tau, b_\tau) \\
\beta &\sim Gamma(a_\beta, b_\beta) \,.
\end{aligned}
\tag{22.12}
$$

Thus we allow γ_k to vary by network and then estimate an overall mean and variance as determined by (τ, β).

Finally, we introduce a covariate MMSBM in Section 22.2.3 , which we can easily extend for multiple networks. Consider again our networks of teacher collaboration. A tie-level variable indicating whether two teachers serve on the same committee may be a covariate of interest, such as $X_{ijk} = 1$ if teacher i and j in school k serve on the same committee, and we may want to estimate this effect across all networks. A simple model in which the covariate effect is the same across networks is given as

$$Y_{ijk} \sim Bernoulli(p_{ijk})$$

$$p_{ijk} = \frac{\exp\{\text{logit}\,(S_{ijk}{}^T \text{logit}\,(B_k) R_{jik}) + \alpha X_{ijk}\}}{1 + \exp\{\text{logit}\,(S_{ijk}{}^T B_k R_{jik})\}}$$

$$S_{ijk} \sim Multinomial(\theta_{ik}, 1)$$

$$R_{jik} \sim Multinomial(\theta_{jk}, 1)$$

$$\theta_{ik} \sim Dirichlet(\gamma_k \xi_k)$$

$$B_{\ell m k} \sim Beta(a_{\ell m k}, b_{\ell m k}) . \tag{22.13}$$

These are merely a few models from the myriad of possibilities. Network-level experiments can affect other parameters in the model; indeed we can include additional hierarchical structure when modeling experimental data. Moreover, observational data may not need the full structure specified above and covariates can be incorporated in other ways as well.

22.3.3   Model Estimation

We use an MCMC algorithm for fitting HMMSMs that is similar to the one used for fitting the single network MMSBM. We first present MCMC steps for fitting the model given in (22.10), and then we discuss how these steps need to be augmented for models (22.11)–(22.13).

For each network k, we use Gibbs updates for θ_k, S_k, R_k, B_k. The complete conditionals for our Gibbs updates are given as:

$$P(\theta_{ik} | \dots) \propto Dirichlet \left(\gamma_k \xi_k + \sum_j S_{ijk} + \sum_j R_{ijk}\right)$$

$$P(S_{ijk} | \dots) \propto Multinomial\,(p)$$

$$p_h = \theta_{ikh} B_{h\ell^\star}{}^{Y_{ijk}} (1 - B_{h\ell^\star})^{(1-Y_{ijk})}$$

$$P(R_{ji} | \dots) \propto Multinomial\,(q)$$

$$q_h = \theta_{ikh} B_{m^\star h}{}^{Y_{ijk}} (1 - B_{m^\star h})^{(1-Y_{ijk})}$$

$$B_{\ell m k} \propto Beta \left(a_{\ell k} + \sum_{(ijk)^\star} Y_{ijk}, b_{mk} + \sum_{(ijk)^\star} Y_{ijk}\right),$$

$$\tag{22.14}$$

where $\ell^\star{}_k$ is the group membership indicated by R_{jik} and $m^\star{}_k$ is the group membership indicated by S_{ijk}. Again, let $(ijk)^\star$ be a specific subset of $i = 1, .., n_k$ and $j = 1, .., n_k$ such that $S_{ijk} = \ell$ and $R_{jik} = m$. Again, we incorporate a sparsity parameter ρ to account for the absence of ties due to lack of interaction.

For the intervention and covariate examples (22.11) and (22.13), respectively, the additional parameter α uses Metropolis or Metropolis-Hastings updates. For example, if we use a random walk method for proposing new values of α, we accept α^{s+1} with probability $min\{1, R\}$ where $R = \frac{P(\alpha^{(s+1)}|...)}{P(\alpha^{(s)}|...)}$. Note that R is a function of all K networks.

For models with additional levels of hierarchy we can update additional parameters using Metropolis within Gibbs steps. For example, in (22.12), β is updated using Gibbs steps,

$$\beta \propto Gamma\left(K\tau + a_\beta, \sum_{k=1}^{K} \gamma_k + b_\beta\right),$$

and we use Metropolis updates for γ_k, ξ_k, and τ. We update each γ_k using an analogous Metropolis step as for the single network. We use a network-specific tuning parameter $\nu_{\gamma,k}$ which is also the shape parameter and rate parameter of $\frac{\nu_{\gamma,k}}{\gamma_k^{(s)}}$, ensuring the proposal distribution has mean at the current value of γ_k. Each ξ_k is updated in the same way using a Dirichlet proposal distribution.

To update τ we use a Gamma proposal distribution not unlike those used for γ_k with shape parameter as the tuning parameter ν_τ and rate parameter such that the proposal distribution has mean of the current value of τ. Then the proposed value of τ^{s+1} is accepted with probability $min\{1, R\}$ where $R = \frac{P(\tau^{(s+1)}|...)}{P(\tau^{(s)}|...)} \frac{P(\tau^{(s)}|\tau^{(s+1)})}{P(\tau^{(s+1)}|\tau^{(s)})}$.

22.3.4 Empirical Examples

We present two examples to illustrate fitting HMMSBMs and use two simulated datasets, with and without a covariate.

In the first example we demonstrate fitting an HMMSBM similar to the example given in (22.12) where each network has a network-specific Dirichlet hyperparameter, λ_k, used to generate the membership probability vectors. Our goal is to assess parameter recovery on three levels: the hyperparameters of the distribution that generates λ_k, the λ_k themselves, and the lower-level parameters, R, S, and B that determine the probability of a tie.

We simulate data from 20 networks, each with 20 nodes and 4 groups using the following model to generate our first set of data:

$$Y_{ijk} \sim Bernoulli(S_{ijk}{}^T B_k R_{jik})$$
$$S_{ijk} \sim Multinomial(\theta_{ik}, 1)$$
$$R_{jik} \sim Multinomial(\theta_{jk}, 1)$$
$$\theta_{ik} \sim Dirichlet(\gamma_k \xi)$$
$$\gamma_k \sim Gamma(10, 50),\tag{22.15}$$

where $\xi = (0.25, 0.25, 0.25, 0.25)$. The group-group tie probability matrix is defined as

$$B = \begin{bmatrix} 0.9 & 0.05 & 0.05 & 0.05 \\ 0.05 & 0.8 & 0.05 & 0.05 \\ 0.05 & 0.05 & 0.7 & 0.05 \\ 0.05 & 0.05 & 0.05 & 0.6 \end{bmatrix}.$$

We constrain B to be the same for each network and select the hyperparameters with which to generate γ_k to ensure small enough values for block structure with low variability. Figure 22.4 shows adjacency matrices for these 20 networks.

We fit the following HMMSBM to these data using the MCMC algorithm described in Section 22.3.3. We let $\xi = (0.25, 0.25, 0.25, 0.25)$ and use a sparsity parameter equal to $\frac{\sum_{ijk} Y_{ijk}}{KN(N-1)}$ for all K networks. The model is given as

$$Y_{ijk} \sim Bernoulli(S_{ijk}{}^T B_k R_{jik})$$

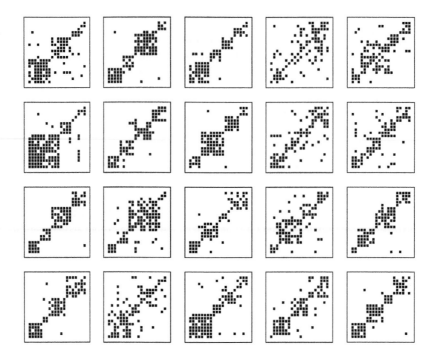

FIGURE 22.4
Networks with 20 nodes generated from a HMMSBM with group membership probabilities from a network-specific Dirichlet parameter $\gamma_k \sim Gamma(50, 10)$.

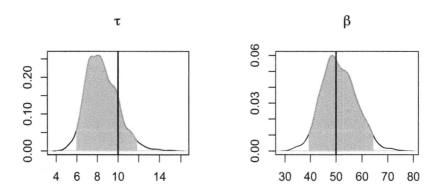

FIGURE 22.5
Posterior density for τ and β, the hyperparameters for the distribution of γ_k for each network. The vertical lines mark the value used to simulate the data, and the 95% equal-tailed credible intervals are indicated with gray. Densities show good recovery of the true value of each parameter.

$$S_{ijk} \sim Multinomial(\theta_{ik}, 1)$$
$$R_{jik} \sim Multinomial(\theta_{jk}, 1)$$
$$\theta_{ik} \sim Dirichlet(\gamma_k \xi_k)$$
$$B_{\ell\ell k} \sim Beta(3, 1)$$
$$B_{\ell m k} \sim Beta(1, 10), \ell \neq m$$
$$\lambda_k \sim Gamma(\tau, \beta)$$
$$\tau \sim Gamma(50, 1)$$
$$\beta \sim Gamma(10, 1).\tag{22.16}$$

We fit the model using our MCMC algorithm and run chains of length 30,000. We remove the first 5000 steps and retain every 25th iteration for a posterior sample of size 1001. The posterior samples for τ and β are illustrated as densities in Figure 22.5. The vertical lines show the true value for each parameter and the gray region indicates the 95% credible interval, suggesting accurate parameter recovery for τ and β. Similar plots for γ_k (see Figure 22.6) for each network k depict the variability in both the true value of γ_k as well as the accuracy of recovery.

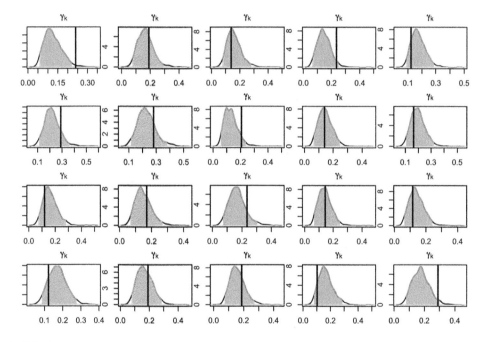

FIGURE 22.6

Posterior densities for γ_k where $k = 1, .., 20$. The 95% equal-tailed credible intervals contain the true value of γ_k for all but one of the simulated networks, suggesting good recovery.

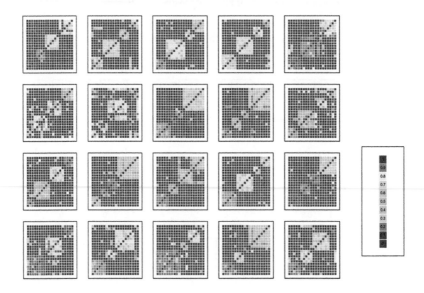

FIGURE 22.7
Tie probability matrix as estimated by posterior means. Visual comparisons to Figure 22.4 suggest accurate estimation of tie probability.

We use predicted probability tie matrices to assess recovery of lower-level parameters (Figure 22.7). Ties with high probability are shown as shades of red, orange, and yellow, and ties with low probability are shown as shades of blue and purple. Visual comparisons to Figure 22.4 reveal that the estimated tie probabilities align with the simulated data; between-pairwise ties are reflected as having higher probability in the fitted model than non-ties. We do note, however, that ties that exist across groups (those shown outside of the block structure) tend to have smaller estimated probabilities than ties within groups.

The second simulation serves two purposes: to illustrate fitting a MMSBM with covariates and to provide a second example of fitting HMMSBMs. We generate data for 10 networks, each with 15 nodes and 3 groups. We use a single edge-level indicator covariate X_{ijk}, such that $X_{ijk} = 1$ implies that individual i in network k and individual j in network k have the same characteristic and $X_{ijk} = 0$ otherwise. In the context of teacher relationships in school k, for example, X_{ijk} might represent teaching the same grade, serving on the same committee, having classrooms in the same wing of the building, etc. For these data, we randomly assigned each node to one of 5 groups, and $X_{ijk} = 1$ if nodes belong to the same group. The formal model used to generate these data is:

$$
\begin{aligned}
Y_{ijk} &\sim Bernoulli(p_{ijk}) \\
p_{ijk} &= \frac{\exp\{\text{logit}\,(S_{ijk}{}^T \text{logit}\,(B_k) R_{jik}) + 4X_{ijk}\}}{1 + \exp\{\text{logit}\,(S_{ijk}{}^T B_k R_{jik})\}} \\
S_{ijk} &\sim Multinomial(\theta_{ik}, 1) \\
R_{jik} &\sim Multinomial(\theta_{jk}, 1) \\
\theta_{ik} &\sim Dirichlet(\gamma_k \xi_k) \\
B_{\ell\ell k} &\sim Beta(12, 4) \\
B_{\ell m k} &\sim Beta(3, 30) \ \ell \neq m \\
\gamma_k &\sim Gamma(10, 60) \,,
\end{aligned}
\tag{22.17}
$$

where $\xi = (\frac{1}{3}, \frac{1}{3}, \frac{1}{3})$. Priors for γ_k were selected to ensure small enough values for block structure with low variability. We use different priors for the diagonal entries of B_k than the off-diagonal entries to model higher within-group tie probabilities. Hyperparameters of these priors were chosen to yield high and low probabilities for the diagonal and off-diagonal entries, respectively, without extreme values of almost 0 or 1.

The adjacency matrices for each of the 10 simulated networks are shown in Figure 22.8. We expect to see more variability in the block structure in these networks as compared to the first simulation study for two reasons. Foremost, we have included a covariate with a strong effect so that there are now many more across-group ties. In addition, we have varied the group-group tie probability matrix B_k by allowing these entries to both differ across networks and be generated (instead of deliberately chosen). As a result, the block structure that we do see varies across networks as the values of the diagonal entries of B_k vary.

We fit the following model on these simulated data:

$$
\begin{aligned}
Y_{ijk} &\sim Bernoulli(p_{ijk}) \\
p_{ijk} &= \frac{\exp\{\text{logit}\,(S_{ijk}{}^T \text{logit}\,(B_k) R_{jik}) + \alpha X_{ijk}\}}{1 + \exp\{\text{logit}\,(S_{ijk}{}^T B_k R_{jik})\}} \\
S_{ijk} &\sim Multinomial(\theta_{ik}, 1) \\
R_{jik} &\sim Multinomial(\theta_{jk}, 1) \\
\theta_{ik} &\sim Dirichlet(\gamma_k \xi_k) \\
B_{\ell\ell k} &\sim Beta(12, 4)
\end{aligned}
\tag{22.18}
$$

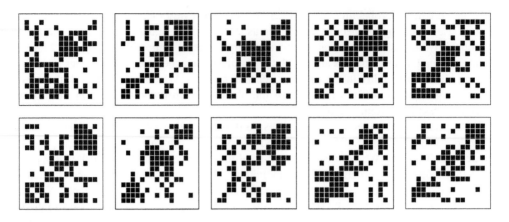

FIGURE 22.8
Networks generated from a single covariate HMMSBM with group membership probabilities from a network-specific Dirichlet parameter $\gamma_k \sim Gamma(60, 10)$. Despite a strong block structure specified by small values of γ_k, the networks have many across-group ties due to the high value of the regression coefficient of X_{ijk}.

$$B_{\ell mk} \sim Beta(3, 30) \ \ell \neq m$$
$$\gamma_k \sim Gamma(\tau, \beta)$$
$$\tau \sim Gamma(1, 0.1)$$
$$\beta \sim Gamma(6, 0.1)$$
$$\alpha \sim Normal(0, 100) \ ,$$

where $\xi = (\frac{1}{3}, \frac{1}{3}, \frac{1}{3})$.

We run MCMC chains of length 30,000, remove the first 5000 iterations, and keep every 25th step. With a posterior sample size of 1001, we assess parameter recovery. We begin with our high-level parameters. Figure 22.9 and Figure 22.10 show the posterior densities for α and τ and β, respectively. The true value of each parameter is indicated by a vertical line and 95% credible interval regions are shown in gray.

The estimation of α is accurate, but the estimates for τ and β are much less precise. The posterior distribution for τ is centered at a higher value than the value of $\tau = 10$ used to generate the data. The distribution for β is centered at a value slightly lower than the true value $\beta = 60$. Similarly, the distributions for each γ_k are skewed toward higher values. As shown in Figure 22.11, only 3 of the 10 posterior samples for γ_k contain the true value in their 95% credible interval. We suspect the lack of block structure contributes to these biases even though the covariate was the primary influence for across-group ties in the data generation process.

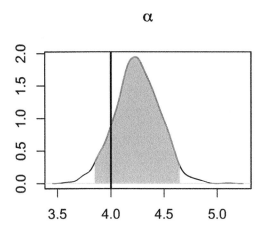

FIGURE 22.9

Posterior density for α, the regression coefficient in Equation (22.13). The true value of α is 4 and is displayed as the vertical line. The 95% equal-tailed credible interval is implied by the gray region and suggests good recovery.

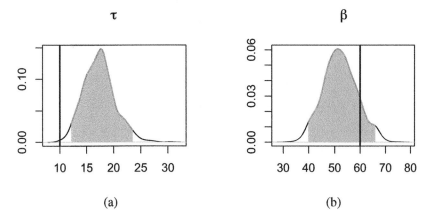

FIGURE 22.10

Posterior density for τ (a) and β (b) with the true values shown as vertical lines. The 95% equal-tailed credible intervals are implied by the gray region. Much of the posterior distribution for τ falls to the right of the true value used to generate the data.

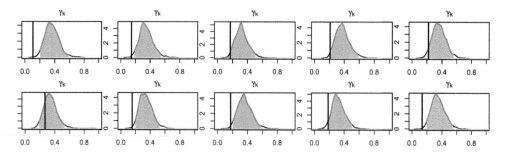

FIGURE 22.11

Posterior densities for γ_k where $k = 1, ..., 10$. The 95% equal-tailed credible intervals contain the true value of γ_k for only one of the networks and overestimating the value of γ_k.

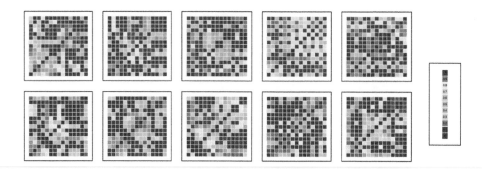

FIGURE 22.12

Tie probability matrix as estimated by posterior means. Ties with high probability are shown as shades of red, orange, and yellow and ties with low probability are shown as shades of blue and purple. Pairwise probabilities align well with the original adjacency matrices (Figure 22.8).

We plot the pairwise probability of a tie in each network in Figure 22.12. Due to the lack of block structure, visually comparing the predicted tie probabilities to the original dataset may seem inconclusive, but in fact predicted probabilities align well with the data.

22.3.5 HMMSBM Extension: Sensitivity Analysis

Given the small number of networks used in our simulations, we are interested in the extent to which our prior specification dominates our model fit. Recall from (22.15), we generated data with $\tau = 10$ and $\beta = 50$ and in the model fit illustrated in Section 22.3.4, we used the following prior distributions:

$$\tau \sim Gamma(10, 1)$$
$$\beta \sim Gamma(50, 1) \, .$$

We repeat model estimation twice using a less strong prior and a weak prior, such that both are centered at the true values. The moderate and weak priors used are given as

$$\tau \sim Gamma(1, 0.1)$$
$$\beta \sim Gamma(5, 0.1),$$

$$\tau \sim Gamma(0.1, 0.01)$$
$$\beta \sim Gamma(0.5, 0.01) \, .$$

We first compare the posterior distributions for τ and β. While the posterior distribution for τ and β contain the true values for each fit, the variance of the posterior sample increases as the variance of prior distribution increases (Figure 22.13). We do note that the scale of the increase is less than the 10-fold increase of the prior distribution variance (50, 500, and 500 for β and 10, 100, and 100 for τ). The posterior mean for τ varies little as the prior changes: 8.5, 8.2, and 10.5 under a strong, less strong, and weak prior, respectively. The posterior mean for β is much less accurate when the weak prior is used. The respective means are 51.1, 51.8, and 120.3.

To assess the prior distributions, we compare the 95% credible regions posterior distribution for γ_k, $k = 1, .., 20$ with the true values (Figure 22.14). We notice the following patterns: if a 95% credible interval γ_k does not cover the true value when the prior distribution is strong, it fails to cover the true value when the prior is moderate or weak. There is little difference in parameter recovery between the strong prior and the less strong prior. The weak prior fit recovers few of the γ_k well, and is strongly biased toward smaller values of γ_k.

Finally, we are interested in how these differences translate to tie probabilities. Figure 22.15 shows the adjacency matrix and posterior mean of the pairwise probability of a tie determined for each model fit. We also include a measure of variability, the width of the 95% credible interval for each pairwise tie probability. For brevity, we only show the first five networks from the data. The posterior pairwise tie probability varies little across each fit and what is even more surprising is that the 95% credible interval widths also vary little.

Based on this simple sensitivity analysis, we offer several conclusions. The prior specification of the high-level parameters τ and β have moderate influence of mid-level parameters γ_k and very little influence on low-level parameters R, S, and B, even with poor recovery of mid-level parameters. Furthermore, using a prior with much larger variance does not necessarily increase the variability in the low-level parameter estimates.

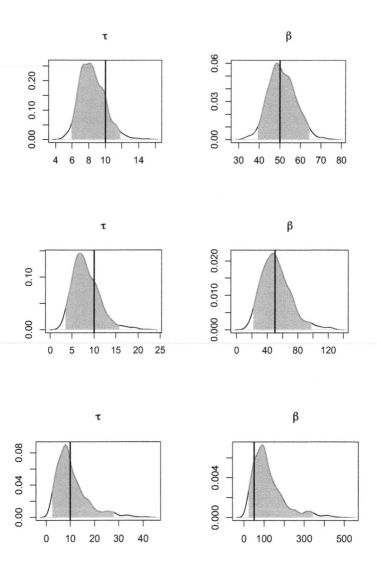

FIGURE 22.13
A comparison of posterior distributions for τ and β given three different prior gamma distributions. Hyperparameters are (10,1), (1,0.1), (0.1, 0.01) and (50,1), (5,0.1), (0.5, 0.01) for τ and β, respectively, and plots are shown top to bottom.

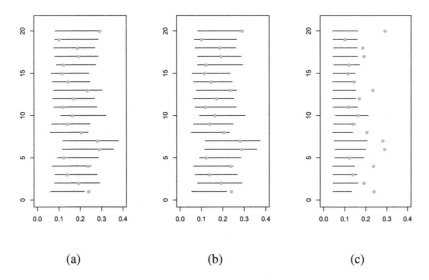

(a)   (b)   (c)

FIGURE 22.14

For each choice of prior specification, strong (a), less strong (b), and weak (c), 95% credible intervals for γ_k are shown in black and the true value of γ_k is shown in green. Parameter recovery is good when the strong or less strong priors are used but is poor when the weak prior is used.

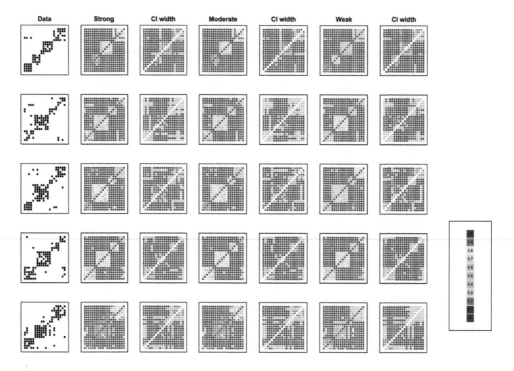

FIGURE 22.15
The adjacency matrix (a) can be compared to the posterior tie probabilities for three model fits that vary by prior distribution specification for τ and β. Priors for each parameter have the same mean but increase in variance by a factor of 10. The width of the 95% credible interval serves as a measure of variability.

22.4 Discussion

We have presented the hierarchical network models (HNM) framework for modeling ensembles of networks and introduced the hierarchical mixed membership stochastic blockmodel (HMMSBM) as an example of an HNM for networks with subgroup structure. This fills a substantial methodological void: while both single and mixed membership stochastic blockmodels have been used to incorporate grouping structure into models for relational data, very little prior work has focused on jointly modeling an ensemble of networks, and none of that work has focused on blockmodels. In addition we have presented a method for incorporating tie covariates into these models, addressing another void in the literature.

We presented several examples of HMMSBMs to demonstrate both the generality and wide utility of these models. We used two simulated datasets, one with a covariate and one without a covariate, to illustrate model fitting using our MCMC algorithm. Posterior tie probabilities from our fits align well with simulated true ties and non-ties, and in most cases parameters were recovered well. High-level parameters, those furthest away from the data, were recovered with less consistency in the simulation study involving tie covariates. Finally, we investigated the effects of prior specification and found that, as expected, high-level parameters were most affected by choice of prior but that priors had little influence on predicted tie probabilities.

With respect to the class of HMMSBMs and model fitting, our work reveals several areas for future work. Ties perhaps can form independently of subgroup structure due to common attributes; including covariates to account for this should produce preferable models. An important area for future research is understanding how the covariate effects and block effects interact with each other. Finally, high-level parameter estimates seem to depend strongly on hyperpriors, suggesting that estimation of these parameters is not yet data-dominated. Understanding how this situation improves as more networks (and perhaps larger networks) are added to the ensemble is also clearly important. On the other hand, it appears that priors have little effect on the low-level tie probabilities.

We have illustrated a proof of concept for HMMSBMs and the HNM framework in general. HMMSBMs are appropriate models for ensembles of networks with block structure and can be fit using relatively simple methods. The HNM framework is larger than HMMSBMs alone since most single network statistical models can be extended to model an ensemble of networks. Sweet et al. (2013) introduced hierarchical latent space models as a class of HNM models, and the authors are currently working on extending work done by Zijlstra, van Duijn and Snijders (2006) and Templin, Ho, Anderson and Wasserman (2003) for hierarchical exponential random graph models, and relating it to the general HNM framework.

Acknowledgments

This work was supported by the Program for Interdisciplinary Education Research Grant R305B040063 and by Hierarchical Network Models for Education Research Grant R305D120004, supported by the Institute for Education Sciences, Department of Education.

References

Airoldi, E. M., Blei, D. M., Fienberg, S. E., and Xing, E. P. (2008). Mixed membership stochastic blockmodels. *Journal of Machine Learning Research* 9: 1981–2014.

Airoldi, E. M., Blei, D. M., Xing, E. P., and Fienberg, S. E. (2005). A latent mixed membership model for relational data. In *Proceedings of the 3rd International Workshop on Link Discovery (LINKKDD '05)*. New York, NY, USA: ACM, 82–89.

Anderson, C. and Wasserman, S. (1992). Building stochastic blockmodels*. *Social Networks* 14: 137–161.

Breiger, R., Boorman, S., and Arabie, P. (1975). An algorithm for clustering relational data with applications to social network analysis and comparison with multidimensional scaling. *Journal of Mathematical Psychology* 12: 328–383.

Erosheva, E. A. (2003). Bayesian estimation of the Grade of Membership model. In Bernardo, J. M., Bayarri, M. J., Berger, J. O., Dawid, A. P., Heckerman, D., Smith, A. F. M., and West, M., (eds), *Bayesian Statistics 7*, 501–510. New York, NY: Oxford University Press.

Fienberg, S. E., Meyer, M., and Wasserman, S. (1985). Statistical analysis of multiple sociometric relations. *Journal of the American Statistical Association* 80: 51–67.

Frank, K. A., Zhao, Y., and Borman, K. (2004). Social capital and the diffusion of innovations within organizations: Application to the implementation of computer technology in schools. *Sociology of Education* 77: 148–171.

Gelman, A., Carlin, J. B., Stern, H. S., and Rubin, D. B. (2004). *Bayesian Data Analysis*. Chapman & Hall/CRC, 2nd edition.

Goldenberg, A., Zheng, A. X., Fienberg, S. E., and Airoldi, E. M. (2009). A survey of statistical network models. *Foundations and Trends in Machine Learning* 2: 129–133.

Goodreau, S., Kitts, J. A., and Morris, M. (2009). Birds of a feather, or friend of a friend? Using exponential random graph models to investigate adolescent social networks. *Demography* 46: 103–125.

Hanneke, S., Fu, W., and Xing, E. P. (2010). Discrete temporal models of social networks. *Electronic Journal of Statistics* 4: 585–605.

Hoff, P. D., Raftery, A. E., and Handcock, M. S. (2002). Latent space approaches to social network analysis. *Journal of the American Statistical Association* 97: 1090–1098.

Hoff, P. D. and Ward, M. D. (2004). Modeling dependencies in international relations networks. *Political Analysis* 12: 160–175.

Kolaczyk, E. (2009). In Airoldi, E. M., Blei, D. M., Fienberg, S. E., Goldenberg, A., Xing, E. P., and Zheng, A. X. (eds), *Statistical Analysis of Network Data: Methods and Models*. New York, NY: Springer.

Krackhardt, D. and Handcock, M. S. (2007). Heider vs Simmel: Emergent features in dynamic structures. *Statistical Network Analysis: Models, Issues, and New Directions* : 14–27.

Moolenaar, N. M., Daly, A. J., and Sleegers, P. J. C. (2010). Occupying the principal position: Examining relationships between transformational leadership, social network position, and schools' innovative climate. *Educational Administration Quarterly* 46: 623–670.

Pattison, P. and Wasserman, S. (1999). Logit models and logistic regressions for social networks: II. Multivariate relations. *British Journal of Mathematical and Statistical Psychology* 52: 169–193.

Penuel, W. R., Frank, K. A., Sun, M., Kim, C., and Singleton, C. (2013). The organization as a filter of institutional diffusion. *Teachers College Record* 115.
http://www.tcrecord.org/Content.asp?ContentID=16742.

Pitts, V. and Spillane, J. (2009). Using social network methods to study school leadership. *International Journal of Research & Method in Education* 32: 185–207.

Sampson, S. (1968). A Novitiate in a Period of Change: An Experimental and Case Study of Social Relationships. Ph.D. thesis, Cornell University, Ithaca, New York, USA.

Sweet, T. M., Thomas, A. C., and Junker, B. W. (2013). Hierarchical network models for education interventions. *Journal of Educational and Behavioral Statistics* 38: 295–318. doi:10.3102/1076998612458702.

Templin, J., Ho, M. -H., Anderson, C., and Wasserman, S. (2003). Mixed effects p* model for multiple social networks. In *Proceedings of the American Statistical Association: Section on Bayesian Statistical Science*. Alexandria, VA, USA: American Statistical Association, 4198–4024.

Wasserman, S. and Pattison, P. (1996). Logit models and logistic regressions for social networks: I. An introduction to Markov graphs and p*. *Psychometrika* 61: 401–425.

Weinbaum, E., Cole, R., Weiss, M., and Supovitz, J. (2008). Going with the flow: Communication and reform in high schools. In Supovitz, J. and Weinbaum, E. (eds), *The Implementation Gap: Understanding Reform in High Schools*. New York, NY: Teachers College Press, 68–102.

Westveld, A. and Hoff, P. D. (2011). A mixed effects model for longitudinal relational and network data, with applications to international trade and conflict. *Annals of Applied Statistics* 5: 843–872.

Xing, E. P., Fu, W., and Song, L. (2010). A state-space mixed membership blockmodel for dynamic network tomography. *Annals of Applied Statistics* 4: 535–566.

Zijlstra, B., van Duijn, M., and Snijders, T. A. B. (2006). The multilevel p_2 model. *Methodology: European Journal of Research Methods for the Behavioral and Social Sciences* 2: 42–47.

23

Analyzing Time-Evolving Networks using an Evolving Cluster Mixed Membership Blockmodel

Qirong Ho

Machine Learning Department, Carnegie Mellon University, Pittsburgh, PA 15213, USA

Eric P. Xing

School of Computer Science, Carnegie Mellon University, Pittsburgh, PA 15213, USA

CONTENTS

Time-evolving networks are a natural representation for dynamic social and biological interactions. While latent space models are gaining popularity in network modeling and analysis, previous works mostly ignore networks with temporal behavior and multi-modal actor roles. Furthermore, prior knowledge, such as division and grouping of social actors or biological specificity of molecular functions, has not been systematically exploited in network modeling. In this chapter, we develop a network model featuring a state-space mixture prior that tracks complex actor latent role changes through time. We provide a fast variational inference algorithm for learning our model, and validate it with simulations and held-out likelihood comparisons on real-world, time-evolving networks. Finally, we demonstrate our model's utility as a network analysis tool, by applying it to United States Congress voting data.

23.1 Introduction

Social and biological systems can often be represented as a series of temporal networks over actors, and these networks may undergo systematic rewiring or experience large topological changes over time. The dynamics of these *time-evolving networks* pose many interesting questions. For instance, what are the latent roles played by these networked actors? How will these roles dictate the way two actors interact? Furthermore, how do actors play *multiple* roles (multi-functionality) in different social and biological contexts, and how does an actor's set of roles *evolve* over time? By knowing which actors play what roles as well as the relationships between different roles, we can gain insight as to how social or biological communities form in networks. For example, we can elucidate how actors with diverse role compositions group together, and how these groupings change over time.

In particular, we want network actors to be capable of *multiple* roles, because assuming a single role per actor may simply be too restrictive. As an example, consider a social network composed of working adults. We can imagine the participants play at least two roles: one when at work (say, being a manager or a worker), and one when at home (perhaps a parent, or possibly unmarried). These two classes of roles are orthogonal to each other, thus one cannot account for all network behaviors with just one class.

The time-evolving aspect of the network is equally important—we do not expect each actor's roles to remain static over time, but anticipate that they will change, giving rise to rewiring in the network. Returning to the previous example, we might imagine a newly-pregnant mother increasing her "parent" role, or a promoted employee shifting from worker to manager. In fact, *multiple* roles could change at once—a working father caught in an accident would be less active both as a worker and as a parent, for instance.

A final, crucial assumption is that the *relationships* between roles remain constant over time, like how a manager always delegates work to a subordinate, or how parents are always involved in raising children. This static relationship between roles provides a reference point for actor role mixtures to evolve over time; it is difficult to interpret actor role changes if the roles themselves are also changing! In fact, allowing both actor roles and role relationships to change arbitrarily makes for an ill-posed problem; it becomes unclear if a given network change should be explained in terms of actor roles or role relationships, or even a combination of both.

In this chapter, we present a mixed membership solution to understanding time-evolving networks, which we call a dynamic mixture of mixed membership stochastic blockmodels (dM^3SB). This model employs the regular mixed membership stochastic blockmodel (MMSB) as the basic building block, but augments it with a multi-modal mixture prior that captures each actor's role-mixture trajectory in a statistically flexible manner. Essentially, we conjoin the MMSB with a set of state-space models, one over each mixture component, and each state-space trajectory corresponds to the average evolution of the role mixtures of a *group* of actors.

Compared to MMSB, this evolving mixture prior presents additional challenges to parameter learning and latent variable inference. We overcome these difficulties by developing a variational EM algorithm inspired by ideas from Ghahramani and Hinton (2000) and dMMSB (an earlier version of dM^3SB) (Xing et al., 2010), which allow for efficient approximate inference and parameter learning. In the following sections, we first develop the dM^3SB model and variational EM algorithm, after which we present validation experiments on both synthetic and real data. Finally, we conclude with a demonstration of dM^3SB towards analyzing voting data from the United States Congress.

23.2 Related Work

There is increasing interest in employing latent space models for network analysis[1] (Hoff et al., 2002; Handcock et al., 2007; Heaukulani and Ghahramani, 2013; Soufiani and Airoldi, 2012), of which dM³SB is one kind. However, most of these models assume static networks and a single, fixed role for each actor. Hence, they cannot model evolution of multiple actor roles over time, making them unsuitable for analyzing complex temporal networks.

With respect to addressing these issues, Airoldi et al. (2008) provided a foundation with MMSB which permits actors to have role mixtures instead of single roles. Later, Xing et al. (2010) developed a dynamic extension of MMSB, called dMMSB, which addresses temporal evolution of actor role mixtures. The dMMSB places a time-evolving, unimodal prior on all network actors; specifically, it employs a time-evolving logistic normal distribution similar to a state-space model.

Although an important first step towards dynamic network analysis, dMMSB offers very weak modeling power—because it employs a unimodal logistic normal for the role distribution of all actors, it is only applicable to networks where the role mixtures of all actors follow similar, unimodal dynamics. A direct solution might be to introduce a separate dynamic process for each actor, but not only is this computationally impractical for large networks with many actors, it is also statistically unsatisfactory from a Bayesian standpoint as the actors no longer share any common pattern and coupling, leaving the model prone to over-fitting and unable to support activity and anomaly detection.

This challenge naturally leads us to explore "evolving clusters" of actors. By modeling dynamic processes on clusters, rather than on individuals or on the whole network, we can increase inferential power while retaining a common, yet expressive, multi-modal mixture model prior over each actor. Such a prior allows dM³SB to accommodate the non-stationary and heterogeneous behaviors of actors.

23.3 Problem Formulation

We consider a sequence of interaction networks or graphs, denoted by $\{\mathcal{G}^{(t)}\}_{t=1}^{T}$, where each $\mathcal{G}^{(t)} \equiv \{\mathcal{V}, \mathcal{E}^{(t)}\}$ represents the network observed at time t. We assume the set of actors $\mathcal{V} = \{1, \ldots, N\}$ is constant. Furthermore, we permit $\mathcal{E}^{(t)} \equiv \{e_{ij}^{(t)}\}_{i,j=1}^{N,N}$, the set of interactions between actors, to evolve with time. We ignore self edges $e_{ii}^{(t)}$.

Our goal is to infer the time-evolving actor role mixtures that give rise to this network sequence. An actor's role mixture is essentially a probability distribution over network roles. For example, a person in a social network could be 0.5 manager and 0.5 parent, meaning that half of his interactions (and non-interactions) can be explained in terms of manager role behavior, while the other half can be explained in terms of parenting behavior. The precise definition of an actor role mixture will be made clear later.

We approach this problem by extending the mixed membership stochastic blockmodel (MMSB) (Airoldi et al., 2008), a static network model that treats each actor as having a mixture of network roles. The key modification is the addition of a time-evolving (i.e., dynamic) prior on top of the MMSB, which allows it to account for temporally-evolving network dynamics. This prior is a *mixture* of time-evolving logistic normal distributions, which is *multi-modal, time-evolving*, and

---

[1]Also, see the chapter entitled "Mixed Membership Blockmodels for Dynamic Networks with Feedback" (Cho et al., 2014).

captures *correlations* between roles. In particular, it is similar to the factorial hidden Markov model, for which variational inference techniques have been developed (Ghahramani and Hinton, 2000). With this prior, the resulting MMSB model is able to fit complex, time-evolving data densities that the static, unimodal, uncorrelated Dirichlet prior used in MMSB cannot.

23.4   Time-Evolving Network Models

Rather than directly introduce the full dM³SB model, we shall start by introducing the regular MMSB, and gradually extend it to become dM³SB. We hope that this presentation will not only be easier to understand, but will also make the connection between MMSB and dM³SB more clear.

23.4.1   The Mixed Membership Stochastic Blockmodel (MMSB)

We begin by describing the mixed membership stochastic blockmodel (Airoldi et al., 2008), which serves as the foundation for our model. The MMSB is a static network model, meaning that we only consider one network $\mathcal{E} \equiv \{e_{ij}\}_{i,j=1}^{N,N}$. Furthermore, it assumes each actor $v_i \in \mathcal{V}$ possesses a latent mixture of K roles, which determine observed network interactions. This role mixture formalizes the notion of actor multi-functionality, and we denote it by a normalized $K \times 1$ vector π_i, referred to as a *mixed membership* or MM vector. We assume these vectors are drawn from some prior $p(\pi)$.

Given MM vectors π_i, π_j for actors i and j, the network edge e_{ij} is stochastically generated as follows: first, actor i (the *donor*) picks one role $z_{\to ij} \sim p(z|\pi_i)$ to *interact* with actor j. Next, actor j (the *receiver*) also picks one role $z_{\leftarrow ij} \sim p(z|\pi_j)$ to *receive the interaction* from i. Both $z_{\to ij}, z_{\leftarrow ij}$ are $K \times 1$ unit indicator vectors. Finally, the chosen roles of i, j determine the network interaction $e_{ij} \sim p(e|z_{\to ij}, z_{\leftarrow ij})$, where $e_{ij} \in \{0, 1\}$. The specific distributions over $z_{\to ij}, z_{\leftarrow ij}, e_{ij}$ are:

- $z_{\to ij} \sim \text{Multinomial}(\pi_i)$. Actor i's donor role indicator.

- $z_{\leftarrow ij} \sim \text{Multinomial}(\pi_j)$. Actor j's receiver role indicator.

- $e_{ij} \sim \text{Bernoulli}(z_{\to ij}^\top B z_{\leftarrow ij})$. Interaction outcome from actor i to j,

where B is a $K \times K$ *role compatibility matrix*. Intuitively, the bilinear form $z_{\to ij}^\top B z_{\leftarrow ij}$ selects a single element of B; the indicators $z_{\to ij}, z_{\leftarrow ij}$ behave like indices into B.

This generative model has two noteworthy features. First, observed relations $\mathcal{E}$ result from actor latent roles interacting. In the case of social networks, the latent roles are naturally interpretable as social functions, like manager, worker, parent, or single adult. Note that actor i's latent membership indicators $z_{\to i\cdot}, z_{\leftarrow\cdot i}$ are *unique to each interaction*; he/she may assume different roles for interacting with each actor.

Second, the role compatibility matrix B completely determines the affinity between latent roles. For example, a diagonally-dominant B signifies that actors of the same role are more likely to interact. Conversely, off-diagonal entries in B suggest interactions between actors of different roles. The MMSB's expressive power lies in its ability to control the interaction strength between any pair of roles, by specifying the corresponding entries of B.

An Example

We now provide a simple example to explain how MMSBs generate interactions. Say we have two social network actors i, j, with MM vectors:

- $\pi_i = [\text{parent} = 0.3, \text{worker} = 0.7]$,

- $\pi_j = [\text{child} = 1]$.

Let us assume that i is the biological father of j, and that the presence or absence of the directed edge e_{ij} signifies whether i has given orders to j. Finally, suppose that the role compatibility matrix has the following entries:

- $B_{\text{parent,child}} = 0.5$,

- $B_{\text{worker,child}} = 0.01$,

where we ignore the other entries of B as they are irrelevant to this discussion. Intuitively, this B reflects how people acting as parents are likely to order their children to do things, whereas people acting as (office) workers are unlikely to interact with children at all. Then, the probability of $e_{ij} = 1$ is computed as:

$$
\begin{aligned}
&p(e_{ij} = 1 \mid \pi_i, \pi_j, B) \\
&= \sum_{z_{\rightarrow ij}, z_{\leftarrow ij}} p(e_{ij} = 1 \mid z_{\rightarrow ij}, z_{\leftarrow ij}, B)\, p(z_{\rightarrow ij} \mid \pi_i)\, \mathbb{P}(z_{\leftarrow ij} \mid \pi_j) \\
&= p(e_{ij} = 1 \mid z_{\rightarrow ij} = \text{parent}, z_{\leftarrow ij} = \text{child}, B) \\
&\quad \times p(z_{\rightarrow ij} = \text{parent} \mid \pi_i)\, p(z_{\leftarrow ij} = \text{child} \mid \pi_j) \\
&\quad + p(e_{ij} = 1 \mid z_{\rightarrow ij} = \text{worker}, z_{\leftarrow ij} = \text{child}, B) \\
&\quad \times p(z_{\rightarrow ij} = \text{worker} \mid \pi_i)\, p(z_{\leftarrow ij} = \text{child} \mid \pi_j) \\
&= (0.5)(0.3)(1) + (0.01)(0.7)(1) \\
&= 0.15 + 0.007 \\
&= 0.157.
\end{aligned}
$$

We see that most of the interaction probability comes from the parent $\rightarrow$ child relationship, rather than the worker $\rightarrow$ child relationship.

23.4.2 Mixture of MMSBs (M³SB)

The actor MM prior $p(\pi)$ significantly affects MMSB's expressive power. In the previous section, we say that MMSB uses a Dirichlet prior, which is conjugate to the multinomial role indicator distribution $p(z \mid \pi)$. The advantage of this conjugacy is that one can derive a clean variational inference algorithm (Airoldi et al., 2008). However, a Dirichlet prior over roles is fairly restrictive in a statistical sense: it is not multi-modal and cannot capture correlations between roles.

To overcome these shortcomings, we shall extend the MMSB by making $p(\pi)$ a *logistic normal mixture prior*, which is both multi-modal (due to the mixture) and permits correlations (due to the normal distribution). This adds the following generative process over the MM vectors π:

- $c_i \sim \text{Multinomial}(\delta)$. Mixture component indicator.

- $\gamma_i \sim \text{Normal}(\mu_{c_i}, \Sigma_{c_i})$. Unnormalized MM vector.

- $\pi_i = \text{Logistic}(\gamma_i)$. Logistic-transformed MM vector, where $[\text{Logistic}(\gamma)]_k = \frac{\exp\{\gamma_k\}}{\sum_{l=1}^K \exp\{\gamma_l\}}$.

Combining this generative process over π with the MMSB model gives rise to what we call a mixture of MMSBs (M³SB). Here, c_i is a $C \times 1$ cluster selection indicator for π_i, where C is the number of mixture components. Thus, π_i is drawn from a logistic normal distribution with mean and covariance selected by c_i, while c_i itself is drawn from a prior multinomial distribution δ.

The M³SB accounts for role correlations using its logistic normal distribution, and has the flexibility to fit complex data densities by virtue of its multi-modal mixture prior. In the next section, we shall exploit these properties to design a time-varying network model that tracks the role mixture trajectories of *clusters* of actors. This is in contrast to the dMMSB model of Xing et al. (2010), which tracks a single, average trajectory.

23.4.3　Dynamic M^{3}SB (dM^{3}SB)

In a time-evolving network, we assume that the actor MM vectors $\pi^{(t)}$ and their prior $p^{(t)}(\pi)$ change with time, and the goal is to infer their dynamic trajectories. Inferring the *dynamic* actor MM vectors allows us to detect large-scale temporal network trends, particularly groups of actors whose MM vectors π shift from one set of roles to another. For example, if a company suddenly goes out of business, then its employees will also shift from the "worker" role to the "unemployed" role.

In order to model time-evolution in the network, we place a state-space model on *every* logistic normal distribution in the mixture prior $p(\pi)$, similar to a Kalman filter. Let N denote the number of actors and T the number of time points in the evolving network. Also, let K denote the number of MMSB latent roles and C the number of mixture components. We begin with an outline of our full generative process; see Figure 23.1 for a graphical model representation.

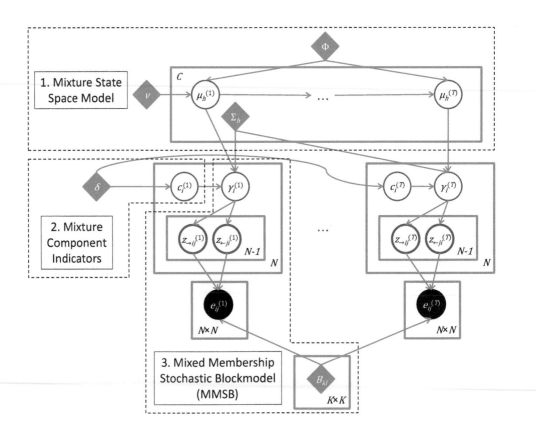

FIGURE 23.1
Graphical model representation of dM^{3}SB.

1. **Mixture State-Space Model for MM Vectors**

   - $\mu_h^{(1)} \sim \text{Normal}(\nu, \Phi)$ for $h = 1 \ldots C$. Mixture means for the MM prior at $t = 1$.
   - $\mu_h^{(t)} \sim \text{Normal}(\mu^{(t-1)}, \Phi)$ for $h = 1 \ldots C, t = 1 \ldots T$. Mixture means for $t > 1$.

2. **Mixture Component Indicators**

   - $\{c_i^{(t)}\}_{i=1}^N \sim \text{Multinomial}(\delta)$ for $t = 1 \ldots T$. Mixture indicator for each MM vector.

3. **Mixed Membership Stochastic Blockmodel**

   - $\{\gamma_i^{(t)}\}_{i=1}^N \sim \text{Normal}(\mu_{c_i^{(t)}}^{(t)}, \Sigma_{c_i^{(t)}})$ for $t = 1 \ldots T$. Unnormalized MM vectors *according to the mixture indicated by* $c_i^{(t)}$.
   - $\pi_i^{(t)} = \text{Logistic}(\gamma_i^{(t)})$, $[\text{Logistic}(\gamma)]_k = \frac{\exp\{\gamma_k\}}{\sum_{l=1}^K \exp\{\gamma_l\}}$. Logistic transform $\gamma_i^{(t)}$ into MM vector $\pi_i^{(t)}$.
   - For every actor pair $(i, j \neq i)$ and every time point $t = 1 \ldots T$:
     - $z_{\rightarrow ij}^{(t)} \sim \text{Multinomial}(\pi_i^{(t)})$. Actor i's donor role indicator.
     - $z_{\leftarrow ij}^{(t)} \sim \text{Multinomial}(\pi_j^{(t)})$. Actor j's receiver role indicator.
     - $e_{ij}^{(t)} \sim \text{Bernoulli}(z_{\rightarrow ij}^{(t)\top} B z_{\leftarrow ij}^{(t)})$. Interaction outcome from actor i to j.

We refer to this model as the dynamic mixture of MMSBs (dM³SB for short). The general idea is to apply the state-space model (SSM) used in object tracking to the MMSB model. Specifically, the MMSB becomes the emission model to the SSM; a distinct MMSB model is "emitted" at each time point (Figure 23.1). Furthermore, the SSM contains C distinct trajectories μ_h, each modeling the mean trajectory for a subset of MM vectors $\pi_i^{(t)}$. The SSM has two parameters ν, Φ, representing the prior mean and variance of the C trajectories. Each trajectory evolves according to a linear transition model $\mu_h^{(t)} = A\mu_h^{(t-1)} + w_h^{(t)}$, where A is a transition matrix and $w_h^{(t)} \sim \text{Normal}(0, \Phi)$ is Gaussian transition noise. We assume A to be the identity matrix, which corresponds to random walk dynamics; generalization to arbitrary A is straightforward.

Each MM vector $\pi_i^{(t)}$ is then drawn from one of the C trajectories $\mu_h^{(t)}$. The choice of trajectory for $\pi_i^{(t)}$ is given by the indicator vector $c_i^{(t)}$, which is drawn from some prior. For simplicity, we have used a single multinomial prior with parameter δ for all $c_i^{(t)}$. Observe that $c_i^{(t)}$ *can change over time*, allowing actors to switch clusters if that would fit the data better. Given $c_i^{(t)}$, the MM vector $\pi_i^{(t)}$ is drawn according to $\mathcal{LN}(\mu_{c_i^{(t)}}^{(t)}, \Sigma_{c_i^{(t)}})$, where the variances $\Sigma_1, \ldots, \Sigma_C$ are model parameters. $\mathcal{LN}$ denotes a logistic normal distribution, the result of applying a logistic transformation to a normal distribution.

Once $\{\pi_i^{(t)}\}_{i=1}^N$ have been drawn for some t, the remaining variables $z_{\rightarrow ij}^{(t)}, z_{\leftarrow ij}^{(t)}, e_{ij}^{(t)}$ follow the MMSB exactly. We assume the role compatibility B to be a model parameter, although we note that more sophisticated assumptions can be found in the literature, such as a state-space model prior (Xing et al., 2010).

23.5 dM³SB Inference and Learning

As with other mixed membership models, neither exact latent variable inference nor parameter learning are computationally tractable in dM³SB. The mixture prior on $\pi_i^{(t)}$, a factorial hidden

Markov model, presents the biggest difficulty—it is analytically un-integrable, its likelihood is subject to many local maxima, and it requires exponential time for exact inference. Moreover, its logistic normal distribution does not admit closed-form integration with the multinomial distribution of $z|\pi$. Finally, the space of possible discrete role indicators z is exponentially large in the number of actors N and time points T.

We address all these difficulties with a variational EM procedure (Ghahramani and Beal, 2001) based on the generalized mean field (GMF) algorithm (Xing et al., 2003), and using techniques from Ghahramani and Hinton (2000) and dMMSB (Xing et al., 2010). Our algorithm simultaneously performs inference and learning for dM^{3}SB in a computationally-effective fashion.

Throughout this section, we shall present just the final dM^{3}SB update equations. For more thorough derivations, the interested reader is referred to the Appendix.

Briefly, variational inference attempts to approximate the true posterior distribution with a simpler factored distribution on which inference is computationally more tractable. Let $\Theta = \{\nu, \Phi, \{\Sigma_h\}_{h=1}^C, \delta, B\}$ denote all model parameters. We approximate the joint posterior $p(\{z^{(t)}, \gamma^{(t)}, c^{(t)}, \{\mu_h^{(t)}\}_{h=1}^C\}_{t=1}^T \mid \{\mathcal{E}^{(t)}\}_{t=1}^T; \Theta)$ by a variational distribution over factored marginals,

$$q = q_\mu \left(\{\mu_h^{(t)}\}_{t,h}^{T,C} \right) \prod_{t,i=1}^{T,N} \left[q_\gamma(\gamma_i^{(t)}) q_c(c_i^{(t)}) \prod_{j=1}^N q_z(z_{\to ij}^{(t)}, z_{\leftarrow ij}^{(t)}) \right].$$

The variational factors q_z, q_γ, and q_c are the marginal distributions over the MMSB latent variables z, γ, and mixture indicators c, respectively. The last variational factor q_μ is the marginal distribution over the mixture of C SSMs over time. The idea is to approximate latent variable inference under p (intractable) with feasible inference under q. In particular, Ghahramani and Hinton (2000) have demonstrated that it is feasible to have one marginal q_μ over all μs.

The GMF algorithm maximizes a lower bound on the marginal distribution $p(\{\mathcal{E}^{(t)}\}_{t=1}^T; \Theta)$ over arbitrary choices of $q_z, q_\gamma, q_c, q_\mu$. We use the GMF solutions to the variational distributions q as the E-step of our variational EM algorithm, and derive the M-step through direct maximization of our variational lower bound with respect to Θ. Under GMF, the optimal solution to a marginal $q(\mathbf{X})$ for some latent variable set $\mathbf{X}$ is $p(\mathbf{X}|\mathbf{Y}, \mathbb{E}_q[\phi(\mathcal{MB}_\mathbf{X})])$, the distribution of $\mathbf{X}$ conditioned on the observed variables $\mathbf{Y}$ and the *expected exponential family sufficient statistics* (under variational distribution q) of $\mathbf{X}$'s Markov blanket variables (Xing et al., 2003). Hence, our E-step iteratively computes $q(\mathbf{X}) := p(\mathbf{X}|\{\mathcal{E}^{(t)}\}_{t=1}^T, \mathbb{E}_q[\phi(\mathcal{MB}_\mathbf{X})])$ for $\mathbf{X} = \{u_h^{(t)}\}_{t,h}^{T,C}, \gamma_i^{(t)}, c_i^{(t)}$ and $\{z_{\to ij}^{(t)}, z_{\leftarrow ij}^{(t)}\}$. For brevity, we present only the final E-step equations; exact derivations can be found in the Appendix.

E-step for q_z:

From here, we drop time indices t whenever appropriate. q_z is a categorical distribution over K^2 elements,

$$q_z(z_{\to ij} = k, z_{\leftarrow ij} = l) \sim \text{Multinomial}(\omega_{(ij)}), \tag{23.1}$$
$$\omega_{(ij)kl} \propto (B_{kl})^{e_{ij}} (1 - B_{kl})^{1-e_{ij}} \exp(\langle \gamma_{ik} \rangle + \langle \gamma_{jl} \rangle),$$

where $\omega_{(ij)}$ is a normalized $K^2 \times 1$ vector indexed by (k,l).[2] The notation $\langle X \rangle$ denotes the expectation of X under q; for example, the expectations of z under q_z are $\langle z_{(\to ij)k} \rangle := \sum_l \omega_{(ij)kl}$ and $\langle z_{(\leftarrow ij)l} \rangle := \sum_k \omega_{(ij)kl}$.

---

[2] k, l correspond to roles indicated by $z_{i \to j}, z_{i \leftarrow j}$.

E-step for q_γ:

q_γ does not have a closed form, because the logistic-normal distribution of γ is not conjugate to the multinomial distribution of z. We apply a Laplace approximation to q_γ, making it normally distributed (Xing et al., 2010; Ahmed and Xing, 2007). Define $\Psi(a, b, C) := \exp\{-\frac{1}{2}(a-b)^\top C^{-1}(a-b)\}$. The approximation to q_γ is

$$q_\gamma(\gamma_i) \propto \Psi(\gamma_i, \tau_i, \Lambda_i), \text{ where} \tag{23.2}$$

$$\Lambda_i = \left((2N-2)H_i + \sum_{h=1}^{C} \Sigma_h^{-1} \langle c_{ih} \rangle \right)^{-1},$$

$$\tau_i = u + \Lambda_i \{ \sum_{j \neq i}^{N} (\langle z_{\rightarrow ij} \rangle + \langle z_{\leftarrow ji} \rangle) - (2N-2)(g_i + H_i(u - \hat{\gamma}_i)) \},$$

$$u = \left(\sum_{h=1}^{C} \Sigma_h^{-1} \langle c_{ih} \rangle \right)^{-1} \left(\sum_{h=1}^{C} \Sigma_h^{-1} \langle c_{ih} \rangle \langle \mu_h \rangle \right),$$

$\hat{\gamma}_i$ is a Taylor expansion point, and g_i and H_i are the gradient and Hessian of the vector-valued function $\log(\sum_{l=1}^{K} \exp \gamma_i)$ evaluated at $\gamma_i = \hat{\gamma}_i$. We set $\hat{\gamma}_i$ to $\langle \gamma_i \rangle$ from the previous E-step iteration, keeping the expansion point close to the current expectation of γ_i.

E-step for q_c:

q_c is discrete over C elements,

$$q_c(c_i = h) \propto \delta_h |\Sigma_h|^{-1/2} \exp \left\{ -\frac{1}{2} \text{tr} \left[\Sigma_h^{-1} \left(\langle \gamma_i \gamma_i^\top \rangle \right. \right. \right.$$
$$\left. \left. \left. - \langle \mu_h \rangle \langle \gamma_i \rangle^\top - \langle \gamma_i \rangle \langle \mu_h \rangle^\top + \langle \mu_h \mu_h^\top \rangle \right) \right] \right\}.$$

Note the dependency on second order moments $\langle \gamma_i \gamma_i^\top \rangle$ and $\langle \mu_h \mu_h^\top \rangle$. Since q_γ, q_μ are Gaussian, these moments are simple to compute.

E-step for q_μ:

The GMF solution to q_μ factors across clusters h:

$$q_\mu \left(\{\mu_h^{(t)}\}_{t,h}^{T,C} \right) := \prod_{h=1}^{C} q_{\mu,h} \left(\{\mu_h^{(t)}\}_t^T \right), \text{ where} \tag{23.3}$$

$$q_{\mu,h} \left(\{\mu_h^{(t)}\}_t^T \right) \propto$$

$$\Psi(\mu_h^{(1)}, \nu, \Phi) \text{Ob}(1, h) \prod_{t=1}^{T} \Psi(\mu_h^{(t)}, \mu_h^{(t-1)}, \Phi) \text{Ob}(t, h),$$

$$\text{Ob}(t, h) := \Psi \left(\frac{\sum_{i=1}^{N} \langle c_{ih}^{(t)} \rangle \langle \gamma_i^{(t)} \rangle}{\sum_{i=1}^{N} \langle c_{ih}^{(t)} \rangle}, \mu_h^{(t)}, \frac{\Sigma_h}{\sum_{i=1}^{N} \langle c_{ih}^{(t)} \rangle} \right).$$

Notice that factor $q_{\mu,h}(\{\mu_h^{(t)}\}_t^T)$ resembles a state-space model for cluster h, with "observation probability" at time t proportional to $\text{Ob}(h, t)$. Hence the mean and covariance of each μ can be efficiently computed using the Kalman smoothing algorithm.

Input: Temporal sequence of networks $\{\mathcal{G}^{(t)}\}_{t=1}^{T}$.
Output: Variational distributions $q_z, q_\gamma, q_c, q_\mu$ and model parameters $B, \delta, \nu, \Phi, \{\Sigma_h\}_{h=1}^{C}$.
Initialize parameters $B, \delta, \nu, \Phi, \{\Sigma_h\}_{h=1}^{C}$.
Sample initial values for $\mu^{(t)}, \gamma^{(t)}, c^{(t)}$.
repeat
  repeat
    Update $q_z(z_{i\rightarrow j}^{(t)}, z_{i\leftarrow j}^{(t)})$ for all i, j, t.
    Update B.
    Update $q_\gamma(\gamma_i^{(t)})$ for all i, t.
  until convergence
  Update $q_\mu(\{\mu_h^{(t)}\}_{t,h=1}^{T,C})$.
  Update ν, Φ.
  Update $q_c(c_i^{(t)})$ for all i, t.
  Update $\delta, \{\Sigma_h\}_{h=1}^{C}$.
until convergence

Algorithm 1: Variational EM for dM^3SB.

23.5.1 Parameter Estimation (M-step)

Given GMF solutions to each q from our E-step, we take our variational lower bound on the log marginal likelihood, and maximize it jointly with respect to all parameters Θ (for details, refer to the Appendix). Let $\mathbb{S}(A) := A + A^\top$. The parameter solutions are:

$$\hat{\beta}_{kl} := \frac{\sum_{t,i,j\neq i}^{T,N,N} \omega_{(ij)kl}^{(t)} e_{ij}^{(t)}}{\sum_{t,i,j\neq i}^{T,N,N} \omega_{(ij)kl}^{(t)}}, \quad \hat{\nu} := \sum_{h}^{C} \frac{\langle \mu_h^{(1)} \rangle}{C}, \quad \hat{\delta} := \sum_{t,i}^{T,N} \frac{\langle c_i^{(t)} \rangle}{TN}$$

$$\hat{\Phi} := \frac{1}{TC} \left[\sum_{h=1}^{C} \langle \mu_h^{(1)} \mu_h^{(1)\top} \rangle - \mathbb{S}\left(\langle \mu_h^{(1)} \rangle \hat{\nu}^\top \right) + \hat{\nu}\hat{\nu}^\top \right.$$
$$\left. + \sum_{t=2}^{T} \langle \mu_h^{(t)} \mu_h^{(t)\top} \rangle - \mathbb{S}\left(\langle \mu_h^{(t)} \mu_h^{(t-1)\top} \rangle \right) + \langle \mu_h^{(t-1)} \mu_h^{(t-1)\top} \rangle \right]$$

$$\hat{\Sigma}_h := \frac{\sum_{t,i}^{T,N} \langle c_{ih}^{(t)} \rangle [\langle \gamma_i^{(t)} \gamma_i^{(t)\top} \rangle - \mathbb{S}(\langle \gamma_i^{(t)} \rangle \langle \mu_h^{(t)} \rangle^\top) + \langle \mu_h^{(t)} \mu_h^{(t)\top} \rangle]}{\sum_{t,i}^{T,N} \langle c_{ih}^{(t)} \rangle}.$$

Our full inference and learning algorithm is summarized in Algorithm 1. This algorithm interleaves the E-step and M-step equations, yielding a coordinate ascent algorithm in the space of variational and model parameters. The algorithm is guaranteed to converge to a local optimum in our variational lower bound, and we use multiple random restarts to approach the global optimum. Similar to the MMSB variational EM algorithm (Airoldi et al., 2008), we update q_z, q_γ, and B more frequently for improved convergence. Note that each random restart can be run on a separate computational thread, making dM^3SB easily *parallelizable* and therefore highly *scalable*.

23.5.2 Variational Inference

23.5.3 Suitability of the Variational Approximation

Given that our true model is multi-modal, our variational approximation will only be useful if it also fits multi-modal data. Historically, *naive* mean field approximations, such as used in latent space models such as MMSB (Airoldi et al., 2008) and the latent Dirichlet allocation (Blei et al., 2003), approximate all latent variables with unimodal variational distributions. These unimodal distributions are unlikely to fit multi-modal densities well; instead, we employ a *structured* mean field approximation that approximates all μs with a single, multi-modal switching state-space distribution $q_\mu()$, essentially a collection of C Kalman filters. This ensures that the multi-modal structure of the prior on the MM vectors $\gamma_i^{(t)}$ is not lost. Moreover, although each $q_\gamma(\gamma_i^{(t)})$ for a given i, t is a unimodal Gaussian, it can be fitted to any mode in $q_\mu()$, independently of $q_\gamma(\gamma_i^{(t)})$ for other i, t. This flexibility ensures the variational posterior over all $\gamma_i^{(t)}$s remains multi-modal.

23.6   Validation

To validate dM³SB, we need to show that it fits multi-modal, correlated, time-varying data better than alternative models. For this purpose, we shall compare dM³SB to its unimodal predecessor dMMSB (Xing et al., 2010), and show that it improves over the latter in multiple respects, on both synthetic and real-world data. Later, we shall conduct a case study on a real-world dataset to demonstrate dM³SB's capabilities.

In the experiments that follow, we ran our algorithm for 50 outer loop iterations per random restart, with 5 iterations per inner loop. We also fixed $\Phi = \mathbb{I}_K$ and $\delta = 1/C$ instead of running their M-steps, as we found this yields more stable results. For the remaining parameters, we used their M-steps with the following initializations: $B_{kl} \sim \text{Uniform}(0, 1)$, $\Sigma_h = \mathbb{I}_K$. As for ν, we initialized $\langle \mu_h^{(1)} \rangle \sim \text{Uniform}([-1, 1]^K)$ for all h and set ν to their average. The remaining variational parameters were initialized via the generative process.

23.6.1   Synthetic Data

Previously, Xing et al. (2010) compared the performance of the dMMSB time-varying model against a naive sequence of disjoint MMSBs, one per network time point. In particular, when the roles are correlated, the logistic-normal prior provides a better fit to the data than the Dirichlet prior. Moreover, for time-varying networks, dMMSB provides a better fit than disjoint MMSBs on every time point.

We now demonstrate that dM³SB's multi-modal prior is an even better fit to time-varying network data than dMMSB's unimodal prior. In this experiment, we shall compare dM³SBs to dMMSBs in terms of model fit (measured by the log marginal likelihood) and actor MM recovery. We generate data with $N = 200$ actors and $T = 5$ time points, and assume a $K = 3$ role compatibility matrix $B = (B_1, B_2, B_3)^\top$, with rows $B_1 = (1, .25, 0)$, $B_2 = (0, 1, .25)$, and $B_3 = (0, 0, 1)$. The actors are divided into four groups of 50, with the first three groups having true MM vectors $(.9, .05, .05)$, $(.05, .9, .05)$ and $(.05, .05, .9)$, respectively, for all time points. The last group has MM vectors that move over time, according to the sequence $\pi^{(1)} = (.6, .3, .1)$, $\pi^{(2)} = (.3, .6, .1)$, $\pi^{(3)} = (.1, .8, .1)$, $\pi^{(4)} = (.1, .6, .3)$, $\pi^{(5)} = (.1, .3, .6)$. The generated B, MM vectors π, and networks $\mathcal{E}^{(t)}$ are visualized in Figure 23.2.

Thus far, we have not addressed model selection—specifically, selection of the number of roles K and the number of mixture components (clusters) C. To do so, we performed a gridsearch over $K \in \{2, 3, 4, 5, 6\}$ and $C \in \{1, 2, 3, 4, 5\}$ on the full network, using 200 random restarts per (K, C) combination. For all combinations, we observed convergence well within our limit of 50 outer iterations. Furthermore, completing all 200 restarts for each K, C took between 8 hours ($K = 2, C = 1$) and 28 hours ($K = 6, C = 5$) on a *single* processor. Since the random restarts can be run in

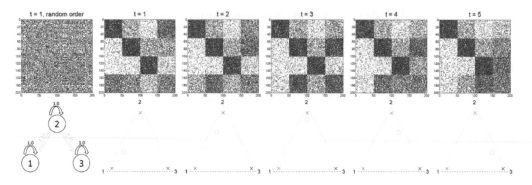

FIGURE 23.2
Synthetic data ground truth visualization. Top Row: Adjacency matrix visualizations, beginning on the left with $t = 1$ using random actor ordering, followed by $t = 1, \ldots, 5$ with actors grouped according to the ground truth. Bottom left: The role compatibility matrix B, shown as a graph. Circles represent roles, and numbered arrows represent interaction probabilities. **Bottom row:** True actor MM plots in the 3-role simplex for each t. Blue, green and red crosses denote the static MMs of the first 3 actor groups, and the cyan circle denotes the moving MM of the last actor group.

parallel, with sufficient computing power one could easily scale dM^{3}SB to much larger time-varying networks with thousands of actors and tens of time points.

For each (K, C) from the gridsearch, we selected its best random restart using the variational lower bound with a Bayesian information criterious (BIC) penalty. The best restart BIC scores are plotted in Figure 23.3; note that dMMSB corresponds to the special case $C = 1$. The optimal BIC score selects the correct number of roles $K = 3$ and clusters $C = 4$, making it a good substitute for held-out model selection.

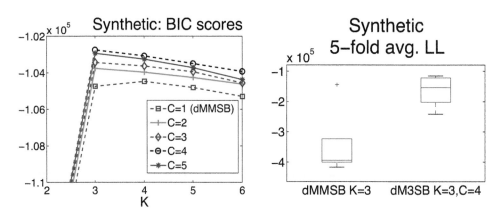

FIGURE 23.3
Synthetic data: BIC scores and 5-fold heldout log-likelihoods for dM^{3}SB and dMMSB.

Next, using the BIC-optimal (K, C), we ran dM^{3}SB on a 5-fold heldout experiment. In each fold, we randomly partitioned the dataset's actors into two equal sets, and used the two corresponding subnetworks as training and test data. In each training fold, we selected the best model parameters Θ from 100 random restarts using the variational lower bound. We then estimated the log marginal likelihood for these parameters on the corresponding test fold, using Monte Carlo integration with 2000 samples. This process was repeated for all 5 folds to get an average log marginal likelihood for dM^{3}SB . For comparison, we conducted the same heldout experiment for a dMMSB set to K from the optimal (K, C) pair. The average log marginal likelihood for both methods is shown in Figure 23.3, and we see that dM^{3}SB's greater heldout likelihood makes it a better statistical fit to this synthetic dataset than dMMSB.

Finally, we compared dM^{3}SB to dMMSB in role estimation (B) and actor role recovery $(\pi_i^{(t)})$, using their best restarts on the correct (K, C) (or just K for dMMSB). Table 23.1 shows, for both methods versus the ground truth, the average ℓ_2 error in $\pi_i^{(t)}$—specifically, we compared the ground truth to $\pi_i^{(t)}$'s posterior mean from either method—as well as the total variation in B. dM^{3}SB's average ℓ_2 error in $\pi_i^{(t)}$ is significantly lower than dMMSB's, at the cost of a higher total variation in B. However, dM^{3}SB's total variation of 0.1083 implies an average difference of only 0.012 in each of the nine entries of B, which is already quite accurate. The fact that dM^{3}SB accurately recovers $\pi_i^{(t)}$ confirms that its posterior over all $\pi_i^{(t)}$ is multi-modal, which validates our variational approximation.

TABLE 23.1
Synthetic data: Estimation accuracy of dM^{3}SB ($K = 3, C = 4$) and dMMSB ($K = 3$).

dM^{3}SB role matrix B, Total Variation	0.1083
dMMSB role matrix B, Total Variation	0.0135
dM^{3}SB MMs $\pi_i^{(t)}$, mean ℓ_2 difference	0.0266
dMMSB MMs $\pi_i^{(t)}$, mean ℓ_2 difference	0.0477

We also note that dM^{3}SB's mean cluster trajectories $\langle \mu_h^{(t)} \rangle$ accurately estimated the four groups' mean MM vectors with a maximum ℓ_2 error of 0.0761 for any group h and time t, except at $t = 5$, where dM^{3}SB exchanged group 3's trajectory with that of (moving) group 4. In conclusion, we have seen that dM^{3}SB provides a better fit to this synthetic dataset than dMMSB, thanks to the former's multi-modal prior.

23.6.2 Real Data

We now assess the model fitness of both dM^{3}SB and dMMSB on two real-world datasets: a 151 actor subset of the Enron email communications dataset (Shetty and Adibi, 2004) over the 12 months of 2001, and a 100 actor subset of the United States Congress voting data over the 8 quarters of 2005 and 2006 (described in the next section). As with the synthetic data, we shall use heldout log-likelihood to measure how well each model fits the data.

For both datasets, we first selected the optimal values of (K, C) via BIC score gridsearch with dM^{3}SB over $K \in \{3, 4, 5, 6\}, C \in \{2, 3, 4, 5\}$. Our previous synthetic experiment has demonstrated that model gridsearch using BIC produces good results. The optimal values were $K = 4, C = 2$ for the Senator dataset, and $K = 3, C = 4$ for the Enron dataset (Figure 23.4).

Using each dataset's optimal (K, C), we next ran dM^{3}SB on the 5-fold heldout experiment discussed in the previous section, obtaining average log marginal likelihoods. For comparison, we conducted the same heldout experiments for dMMSB set to K from the optimal (K, C) pair.

Plots of the heldout log marginal likelihoods for dM^{3}SB and dMMSB can be found in Figure 23.4. On the Senator dataset, dM^{3}SB has the higher log marginal likelihood, implying that it is a better statistical fit than dMMSB. For the Enron dataset, both methods have the same likelihood, showing that using dM^{3}SB with more mixture components at least incurs no statistical cost over

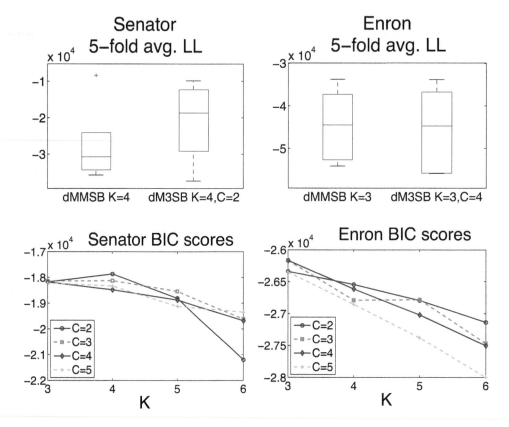

FIGURE 23.4
Senator/Enron data: BIC scores and 5-fold heldout log-likelihoods for dM^{3}SB and dMMSB.

dMMSB. These results demonstrate that dM^{3}SB's multi-modal prior is a better fit to some real-world, time-varying networks, compared to dMMSB unimodal prior.

23.7   Case Study: U. S. Congress Voting Data

We finish our discussion with an application of dM^{3}SB to the United States 109th Congress voting records. Here, we will show that dM^{3}SB not only recovers MM vectors and a role compatibility matrix that matches our intuitive expectations of the data, but that the MM vectors are useful for identifying outliers and other unusual phenomena.

The 109th Congress involved 100 senators and 542 bills spread over the dates January 1, 2005 through December 31, 2006. The original voting data[3] is provided in the form of yes/no votes for each senator and each bill. In order to create a time-varying network suitable for dM^{3}SB, we applied the method of Kolar et al. (2008) to recreate their network result.

The generated time-varying network contains 100 actors (senators), and 8 time points corresponding to 3-month epochs starting on January 1, 2005 and ending on December 31, 2006. The

---

[3] Available at http://www.senate.gov.

network is an undirected graph, where an edge between two senators indicates that their votes were mostly similar during that particular epoch. Conversely, a missing edge indicates that their votes were mostly different. Our intention is to discover how the political allegiances of different senators shifted from 2005 to 2006.

For our analysis, we used the optimal dM^{3}SB restart from the BIC gridsearch described in the previous held-out experiment. Recall that this optimal restart uses $K = 4$ roles and $C = 2$ clusters. The learned MM vectors π_i, compatibility matrix B, and most probable cluster assignments are summarized in Figure 23.5. The results are intuitive: Democratic party members have a high proportion of Role 1, while Republican party members have a high proportion of Role 2. Both Roles 1 and 2 interact exclusively with themselves, reflecting the tendency of both political parties to vote with their comrades and against the other party. The remaining two roles exhibit no interactions; senators with high proportions of these roles are unaligned and unlikely to vote with either political party. Observe that the two clusters perfectly capture party affiliations—Republican senators are almost always in cluster 1, while Democratic senators are almost always in cluster 2.

While it is reassuring to see results that reflect a contemporary understanding of U.S. politics, a more useful application of dM^{3}SB's mixed membership analysis is in identifying outliers. For instance, consider the Democrat Senator Ben Nelson (#75): from $t = 1$ through 7, his votes were unaligned with either Democrats or Republicans, though his votes were gradually shifting towards Republican. At $t = 8$ (the end of 2006), his voting becomes strongly Republican (Role 2), and he shifts from the Democrat cluster (1) to the Republican one (2). Sen. Nelson's trajectory through the role simplex is plotted in Figure 23.6. Incidentally, Sen. Nelson was re-elected as the Senator from Nebraska in late 2006, winning a considerable percentage of his state's Republican vote.

Next, observe how the senator from New Jersey, #28, started off unaligned from $t = 1$ to 4 but ended up Democratic from $t = 5$ to 8; his role trajectory is also plotted in Figure 23.6. There is an interesting reason for this: the seat for New Jersey was occupied by two senators during the Congress, Senator Jon Corzine in the first session ($t = 1$ to 4), and Senator Bob Menendez in the second session ($t = 5$ to 8). Sen. Corzine was known to have far-left views, reflected in #28's lack of both Republican *and* Democratic roles during his term (the Democrat role captures mainstream rather than extremist voting behavior). Once Sen. Menendez took over, #28's behavior fell in line with most Democrats.

Other notable outliers include Senator James Jeffords (#54), the sole Independent senator who votes like a Democrat, and three Republican senators with Democratic leanings: Senator Lincoln Chafee #19, Senator Susan Collins #25, and Senator Olympia Snowe #89. These senators exhibit MM vectors that deviate significantly from their party average, which make them obvious outliers under even a simple K-means cluster. Through examining these outliers, dM^{3}SB allows us to perform anomaly detection and analysis.

In summary, dM^{3}SB provides a latent space view of the 109th Congress voting network, which reveals both expected aggregate trends (voting along bipartisian lines) as well as unexpected anomalies (senators who differ from their party norm). We anticipate that dM^{3}SB can also be applied to understanding time-evolving biological networks, just as Xing et al. (2010) applied the earlier dMMSB model to such data in 2010.

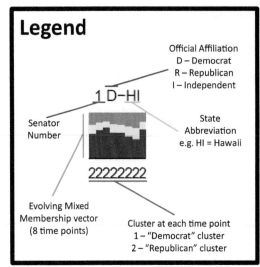

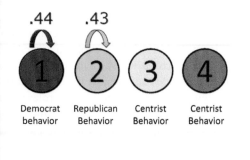

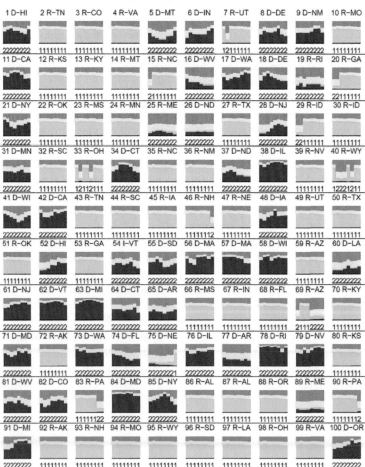

FIGURE 23.5

Congress voting network: Mixed membership vectors (colored bars) and most probable cluster assignments (numbers under bars) for all 100 senators, displayed as an 8-time-point series from left-to-right. The annotation beside a senator's number refers to that senator's political party (D for Democrat, R for Republican, I for Independent) and state (as a two-letter abbreviation). Refer to the legend for specific details. The learned role compatibility matrix is displayed at the top right.

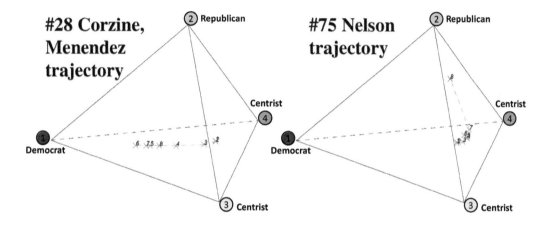

FIGURE 23.6

Congress voting network 3-simplex visualizations. Colors (green, blue) denote cluster membership. Left: MM vector time-trajectory for Senator #28 (D-NJ)—Jon Corzine during time points 1–4 and Bob Menendez during time points 5–8. Right: MM vector time-trajectory for Senator Ben Nelson (#75, D-NE).

23.8 Conclusion

dM^{3}SB is a probabilistic model for latent role analysis in time-varying networks, with an efficient variational EM algorithm for approximate inference and learning. This model is distinguished by its explict modeling of actor multi-functionalities (role MMs), as well as its multi-modal, time-evolving, logistic normal mixture prior over these multi-functionalities, which allows dM^{3}SB to fix complex latent role densities. We also note that dM^{3}SB's variational inference algorithm is trivial to run in parallel, since each random restart can be run on a separate computational thread.

Notably, dM^{3}SB is an evolution of the dMMSB (Xing et al., 2010) and MMSB (Airoldi et al., 2008) models, and shares much in common with them. Validation experiments show that dM^{3}SB's multi-modal prior outperforms the unimodal prior of dMMSB on both synthetic and real data, which underscores the importance of using statistically flexible priors. The most important uses of dM^{3}SB are exploration of actor latent roles and anomaly detection, which were demonstrated in a case study on the 109th U.S. Congress voting data.

Appendix

Derivation of the Variational EM Algorithm

This appendix provides detailed derivations of the dM^{3}SB variational EM algorithm. Recall that our goal is to find the posterior distribution of the latent variables μ, c, γ, z given the observed sequence network $E^{(1)}, \ldots, E^{(T)}$, under the maximum likelihood model parameters B, δ, ν, Φ, and Σ.

Finding the posterior (inference) or solving for the maximum likelihood parameters (learning) are both intractable under our original model. Hence we resort to a variational EM algorithm, which locally optimizes the model parameters with respect to a lower bound on the true marginal log-likelihood, while simultaneously finding a variational distribution that approximates the latent variable posterior. The marginal log-likelihood lower bound being optimized is

$$
\begin{aligned}
\log p\left(E \mid \Theta\right) &= \log \int_X p\left(E, X \mid \Theta\right) dX \\
&= \log \int_X q\left(X\right) \frac{p\left(E, X \mid \Theta\right)}{q\left(X\right)} dX \\
&\geq \int_X q\left(X\right) \log \frac{p\left(E, X \mid \Theta\right)}{q\left(X\right)} dX \quad \text{(Jensen's inequality)} \\
&= \mathbb{E}_q\left[\log p\left(E, X \mid \Theta\right) - \log q\left(X\right)\right] =: \mathcal{L}\left(q, \Theta\right),
\end{aligned}
$$

where X denotes the latent variables $\{\mu, c, \gamma, z\}$, Θ denotes the model parameters $\{B, \delta, \nu, \Phi, \Sigma\}$, and q is the variational distribution. This lower bound is iteratively maximized with respect to q's parameters (E-step) and the model parameters Θ(M-step).

In principle, the lower bound $\mathcal{L}\left(q, \Theta\right)$ holds for any distribution q; ideally q should closely approximate the true posterior $p(X \mid E, \Theta)$. In the next section, we define a factored form for q and derive its optimal solution.

Variational Distribution q

We assume a factorized form for q:

$$
q = q_\mu \left(\mu_1^{(1)}, \ldots, \mu_C^{(T)} \right) \prod_{t,i=1}^{T,N} \left[q_\gamma \left(\gamma_i^{(t)} \right) q_c \left(c_i^{(t)} \right) \prod_{\substack{j \neq i}}^{N} q_z \left(z_{i \to j}^{(t)}, z_{i \leftarrow j}^{(t)} \right) \right].
$$

We now make use of the generalized mean field (GMF) theory (Xing et al. 2003) to determine each factor's form. GMF theory optimizes a lower bound on the marginal distribution $p(E \mid \Theta)$ over arbitrary choices of q_μ, q_γ, q_c, and q_z. In particular, the optimal solution to q_X is $p(X \mid E, \mathbb{E}_q [\phi(\mathrm{MB}_X)])$, the distribution of the latent variable set X conditioned on the observed variables E and the *expected exponential family sufficient statistics* (under q) of X's Markov blanket variables. More precisely, q_X has the same functional form as $p(X \mid E, \mathrm{MB}_X)$, but where a variational parameter $\mathcal{V}$ replaces $\phi(Y)$ for each $Y \in \mathrm{MB}_X$, with optimal solution $\mathcal{V} := \mathbb{E}_q [\phi(Y)]$. In general, if $Y \in \mathrm{MB}_X$, then we use $\langle \phi(Y) \rangle$ to denote the variational parameter corresponding to Y.

We begin by deriving optimal solutions to q_μ, q_γ, q_c, and q_z in terms of the the variational parameters $\langle \phi(Y) \rangle$. After we have derived all factors, we present closed-form solutions to $\langle \phi(Y) \rangle$. These solutions form a set of fixed-point equations which, when iterated, converge to a local optimum in the space of variational parameters (thus completing the E-step).

Distribution of q_z

q_z is a discrete distribution since the zs are indicator vectors. We begin by deriving the distribution of the zs conditioned on their Markov blanket:

$$
p \left(z_{i \to j}^{(t)}, z_{i \leftarrow j}^{(t)} \mid \mathrm{MB}_{z_{i \to j}^{(t)}, z_{i \leftarrow j}^{(t)}} \right)
$$

$$
\propto p \left(E_{ij}^{(t)} \mid z_{i \to j}^{(t)}, z_{i \leftarrow j}^{(t)} \right) p \left(z_{i \to j}^{(t)} \mid \gamma_i^{(t)} \right) p \left(z_{i \leftarrow j}^{(t)} \mid \gamma_j^{(t)} \right)
$$

$$
= \left(\left(z_{i \to j}^{(t)} \right)^\top B z_{i \leftarrow j}^{(t)} \right)^{E_{ij}^{(t)}} \left(1 - \left(z_{i \to j}^{(t)} \right)^\top B z_{i \leftarrow j}^{(t)} \right)^{1 - E_{ij}^{(t)}}
$$

$$
\prod_{k=1}^{K} \left(\frac{\exp \gamma_{i,k}^{(t)}}{\sum_{l=1}^{K} \exp \gamma_{i,l}^{(t)}} \right)^{z_{i \to j,k}^{(t)}} \left(\frac{\exp \gamma_{j,k}^{(t)}}{\sum_{l=1}^{K} \exp \gamma_{j,l}^{(t)}} \right)^{z_{i \leftarrow j,k}^{(t)}}
$$

$$
\propto \exp \left\{ E_{ij}^{(t)} \log \left(\left(z_{i \to j}^{(t)} \right)^\top B z_{i \leftarrow j}^{(t)} \right) + \left(1 - E_{ij}^{(t)} \right) \log \left(1 - \left(z_{i \to j}^{(t)} \right)^\top B z_{i \leftarrow j}^{(t)} \right) \right.
$$

$$
\left. + \left(z_{i \to j}^{(t)} \right)^\top \gamma_i^{(t)} + \left(z_{i \leftarrow j}^{(t)} \right)^\top \gamma_j^{(t)} \right\}.
$$

The variables $\gamma_i^{(t)}, \gamma_j^{(t)}$ belong to other variational factors, and their exponential family sufficient statistics are just $\gamma_i^{(t)}$ and $\gamma_j^{(t)}$ themselves. Hence

$$
q_z \left(z_{i \to j}^{(t)}, z_{i \leftarrow j}^{(t)} \right)
$$

$$
:\propto \exp \left\{ E_{ij}^{(t)} \log \left(\left(z_{i \to j}^{(t)} \right)^\top B z_{i \leftarrow j}^{(t)} \right) + \left(1 - E_{ij}^{(t)} \right) \log \left(1 - \left(z_{i \to j}^{(t)} \right)^\top B z_{i \leftarrow j}^{(t)} \right) \right.
$$

$$+ \left(z_{i \to j}^{(t)}\right)^\top \left\langle \gamma_i^{(t)} \right\rangle + \left(z_{i \leftarrow j}^{(t)}\right)^\top \left\langle \gamma_j^{(t)} \right\rangle \bigg\},$$

with variational parameters $\left\langle \gamma_i^{(t)} \right\rangle$ and $\left\langle \gamma_j^{(t)} \right\rangle$. We can also express q_z in terms of indices k, l:

$$q_z \left(z_{i \to j}^{(t)} = k, z_{i \leftarrow j}^{(t)} = l\right)$$
$$:\propto \exp \left\{ E_{ij}^{(t)} \log B_{k,l} + \left(1 - E_{ij}^{(t)}\right) \log \left(1 - B_{k,l}\right) + \left\langle \gamma_{i,k}^{(t)} \right\rangle + \left\langle \gamma_{j,l}^{(t)} \right\rangle \right\}.$$

Distribution of q_γ

q_γ is a continuous distribution. The distribution of $\gamma_i^{(t)}$ conditioned on its Markov blanket is

$$p \left(\gamma_i^{(t)} \mid \mathrm{MB}_{\gamma_i^{(t)}}\right)$$

$$\propto p \left(\gamma_i^{(t)} \mid c_i^{(t)}, \mu_1^{(t)}, \dots, \mu_C^{(t)}\right) \prod_{j \neq i}^{N} p \left(z_{i \to j}^{(t)} \mid \gamma_i^{(t)}\right) p \left(z_{j \leftarrow i}^{(t)} \mid \gamma_i^{(t)}\right)$$

$$\propto \exp \left\{ \sum_{h=1}^{C} -\frac{1}{2} c_{i,h}^{(t)} \left(\gamma_i^{(t)} - \mu_h^{(t)}\right)^\top \Sigma_h^{-1} \left(\gamma_i^{(t)} - \mu_h^{(t)}\right) \right\}$$

$$\prod_{j \neq i}^{N} \prod_{k=1}^{K} \left(\frac{\exp \gamma_{i,k}^{(t)}}{\sum_{l=1}^{K} \exp \gamma_{i,l}^{(t)}}\right)^{z_{i \to j,k}^{(t)}} \left(\frac{\exp \gamma_{i,k}^{(t)}}{\sum_{l=1}^{K} \exp \gamma_{i,l}^{(t)}}\right)^{z_{j \leftarrow i,k}^{(t)}}$$

$$= \exp \left\{ \sum_{h=1}^{C} -\frac{1}{2} c_{i,h}^{(t)} \left(\gamma_i^{(t)} - \mu_h^{(t)}\right)^\top \Sigma_h^{-1} \left(\gamma_i^{(t)} - \mu_h^{(t)}\right) \right.$$

$$\left. + \sum_{j \neq i}^{N} \sum_{k=1}^{K} \left(z_{i \to j,k}^{(t)} \gamma_{i,k}^{(t)} + z_{j \leftarrow i,k}^{(t)} \gamma_{i,k}^{(t)}\right) - (2N - 2) \log \sum_{l=1}^{K} \exp \gamma_{i,l}^{(t)} \right\}$$

$$\propto \exp \left\{ \sum_{h=1}^{C} -\frac{1}{2} c_{i,h}^{(t)} \left[\left(\gamma_i^{(t)}\right)^\top \Sigma_h^{-1} \gamma_i^{(t)} - \left(\gamma_i^{(t)}\right)^\top \Sigma_h^{-1} \mu_h^{(t)} - \left(\mu_h^{(t)}\right)^\top \Sigma_h^{-1} \gamma_i^{(t)}\right] \right.$$

$$\left. + \left(\sum_{j \neq i}^{N} z_{i \to j}^{(t)} + z_{j \leftarrow i}^{(t)}\right)^\top \gamma_i^{(t)} - (2N - 2) \log \sum_{l=1}^{K} \exp \gamma_{i,l}^{(t)} \right\}.$$

The variables $c_i^{(t)}, \mu_1^{(t)}, \dots, \mu_C^{(t)}, z_{i \to 1}^{(t)}, \dots, z_{i \to N}^{(t)}, z_{1 \leftarrow i}^{(t)}, \dots, z_{N \leftarrow i}^{(t)}$ belong to other variational factors. The sufficient statistics for variables z are just $z_{i \to j}^{(t)}$ and $z_{j \leftarrow i}^{(t)}$ themselves. For variables c and μ, their sufficient statistics are $c_{i,h}^{(t)}$ and $c_{i,h}^{(t)} \left(\mu_h^{(t)}\right)^\top$. However, since c is marginally independent of μ under q, we can take their expectations independently, hence the variational parameters are just $\left\langle c_{i,h}^{(t)} \right\rangle$ and $\left\langle \mu_h^{(t)} \right\rangle$. Hence

$$q_\gamma \left(\gamma_i^{(t)}\right) :\propto \exp \left\{ \sum_{h=1}^{C} -\frac{1}{2} \left\langle c_{i,h}^{(t)} \right\rangle \left[\left(\gamma_i^{(t)}\right)^\top \Sigma_h^{-1} \gamma_i^{(t)} - \left(\gamma_i^{(t)}\right)^\top \Sigma_h^{-1} \left\langle \mu_h^{(t)} \right\rangle - \left\langle \mu_h^{(t)} \right\rangle^\top \Sigma_h^{-1} \gamma_i^{(t)}\right] \right.$$

$$\left. + \left(\sum_{j \neq i}^{N} \left\langle z_{i \to j}^{(t)} \right\rangle + \left\langle z_{j \leftarrow i}^{(t)} \right\rangle\right)^\top \gamma_i^{(t)} - (2N - 2) \log \sum_{l=1}^{K} \exp \gamma_{i,l}^{(t)} \right\},$$

with variational parameters $\left\langle c_i^{(t)} \right\rangle, \left\langle \mu_h^{(t)} \right\rangle, \left\langle z_{i \to j}^{(t)} \right\rangle, \left\langle z_{j \leftarrow i}^{(t)} \right\rangle$.

Laplace Approximation to q_γ

The term $\mathcal{Z}_\gamma \left(\gamma_i^{(t)} \right) := \log \sum_{l=1}^K \exp \gamma_{i,l}^{(t)}$ makes the exponent analytically un-integrable, which prevents us from computing the normalizer for $q_\gamma \left(\gamma_i^{(t)} \right)$. Thus, we approximate $\mathcal{Z}_\gamma \left(\gamma_i^{(t)} \right)$ with its second-order Taylor expansion around a chosen point $\hat{\gamma}_i^{(t)}$:

$$
\mathcal{Z}_\gamma \left(\gamma_i^{(t)} \right) \approx \mathcal{Z}_\gamma \left(\hat{\gamma}_i^{(t)} \right) + \left(g_i^{(t)} \right)^\top \left(\gamma_i^{(t)} - \hat{\gamma}_i^{(t)} \right)
$$
$$
+ \frac{1}{2} \left(\gamma_i^{(t)} - \hat{\gamma}_i^{(t)} \right)^\top H_i^{(t)} \left(\gamma_i^{(t)} - \hat{\gamma}_i^{(t)} \right) \tag{23.4}
$$
$$
g_{i,k}^{(t)} := \frac{\exp \hat{\gamma}_{i,k}^{(t)}}{\sum_{k'=1}^K \exp \hat{\gamma}_{i,k'}^{(t)}}
$$
$$
H_{i,kl}^{(t)} := \frac{\mathbb{I}[k=l] \exp \hat{\gamma}_{i,k}^{(t)}}{\sum_{k'=1}^K \exp \hat{\gamma}_{i,k'}^{(t)}} - \frac{\exp \hat{\gamma}_{i,k}^{(t)} \exp \hat{\gamma}_{i,l}^{(t)}}{\left(\sum_{k'=1}^K \exp \hat{\gamma}_{i,k'}^{(t)} \right)^2}.
$$

Note that $H_i^{(t)} = \text{diag}\left(g_i^{(t)} \right) - g_i^{(t)} \left(g_i^{(t)} \right)^\top$. Because the variational EM algorithm is iterative, we set $\hat{\gamma}_i^{(t)}$ to $\tilde{\gamma}_i^{(t)} := \mathbb{E}_q \left[\gamma_i^{(t)} \right]$ from the previous iteration, which should keep the point of expansion close to $\mathbb{E}_q \left[\gamma_i^{(t)} \right]$ for the current iteration. The point of this Taylor expansion is to approximate q_γ with a normal distribution; consider the exponent of q_γ,

$$
-\left\{ \sum_{h=1}^C \frac{\left\langle c_{i,h}^{(t)} \right\rangle}{2} \left[\left(\gamma_i^{(t)} \right)^\top \Sigma_h^{-1} \gamma_i^{(t)} - \left(\gamma_i^{(t)} \right)^\top \Sigma_h^{-1} \left\langle \mu_h^{(t)} \right\rangle - \left\langle \mu_h^{(t)} \right\rangle^\top \Sigma_h^{-1} \gamma_i^{(t)} \right] \right\}
$$
$$
+ \left(\sum_{j \neq i}^N \left\langle z_{i \to j}^{(t)} \right\rangle + \left\langle z_{j \leftarrow i}^{(t)} \right\rangle \right)^\top \gamma_i^{(t)} - (2N-2) \mathcal{Z}_\gamma \left(\gamma_i^{(t)} \right)
$$
$$
= \text{const}^{(1)} - \frac{1}{2} \left(\gamma_i^{(t)} - u \right)^\top S \left(\gamma_i^{(t)} - u \right)
$$
$$
+ \left(\sum_{j \neq i}^N \left\langle z_{i \to j}^{(t)} \right\rangle + \left\langle z_{j \leftarrow i}^{(t)} \right\rangle \right)^\top \gamma_i^{(t)} - (2N-2) \mathcal{Z}_\gamma \left(\gamma_i^{(t)} \right),
$$

where $\text{const}^{(i)}$ denotes a constant independent of $\gamma_i^{(t)}$, $S := \sum_{h=1}^C \Sigma_h^{-1} \left\langle c_{i,h}^{(t)} \right\rangle$ and $u := S^{-1} \left(\sum_{h=1}^C \Sigma_h^{-1} \left\langle c_{i,h}^{(t)} \right\rangle \left\langle \mu_h^{(t)} \right\rangle \right)$. Applying the Taylor expansion in Equation (23.4) gives

$$
\approx \text{const}^{(1)} - \frac{1}{2} \left(\gamma_i^{(t)} - u \right)^\top S \left(\gamma_i^{(t)} - u \right) + \left(\sum_{j \neq i}^N \left\langle z_{i \to j}^{(t)} \right\rangle + \left\langle z_{j \leftarrow i}^{(t)} \right\rangle \right)^\top \gamma_i^{(t)}
$$
$$
- (2N-2) \left[\mathcal{Z}_\gamma \left(\hat{\gamma}_i^{(t)} \right) + \left(g_i^{(t)} \right)^\top \left(\gamma_i^{(t)} - \hat{\gamma}_i^{(t)} \right) + \frac{1}{2} \left(\gamma_i^{(t)} - \hat{\gamma}_i^{(t)} \right)^\top H_i^{(t)} \left(\gamma_i^{(t)} - \hat{\gamma}_i^{(t)} \right) \right]
$$
$$
= \text{const}^{(2)} - \frac{1}{2} \left(\gamma_i^{(t)} - u \right)^\top S \left(\gamma_i^{(t)} - u \right) + \left(\sum_{j \neq i}^N \left\langle z_{i \to j}^{(t)} \right\rangle + \left\langle z_{j \leftarrow i}^{(t)} \right\rangle \right)^\top \gamma_i^{(t)}
$$

$$- (2N - 2) \left[\left(g_i^{(t)} \right)^\top \gamma_i^{(t)} + \frac{1}{2} \left(\gamma_i^{(t)} \right)^\top H_i^{(t)} \gamma_i^{(t)} - \left(\hat{\gamma}_i^{(t)} \right)^\top H_i^{(t)} \gamma_i^{(t)} \right]$$

$$= \mathrm{const}^{(2)} - \frac{1}{2} \left(\gamma_i^{(t)} - u \right)^\top S \left(\gamma_i^{(t)} - u \right)$$

$$+ \left[\left(\sum_{j \neq i}^N \left\langle z_{i \to j}^{(t)} \right\rangle + \left\langle z_{j \leftarrow i}^{(t)} \right\rangle \right)^\top - (2N - 2) \left(\left(g_i^{(t)} \right)^\top - \left(\hat{\gamma}_i^{(t)} \right)^\top H_i^{(t)} \right) \right] \gamma_i^{(t)}$$

$$- (N - 1) \left(\gamma_i^{(t)} \right)^\top H_i^{(t)} \gamma_i^{(t)}.$$

Define $A := \left(\sum_{j \neq i}^N \left\langle z_{i \to j}^{(t)} \right\rangle + \left\langle z_{j \leftarrow i}^{(t)} \right\rangle \right)^\top - (2N - 2) \left(\left(g_i^{(t)} \right)^\top - \left(\hat{\gamma}_i^{(t)} \right)^\top H_i^{(t)} \right)$ and $B := - (N - 1) H_i^{(t)}$, so that we obtain

$$= \mathrm{const}^{(2)} - \frac{1}{2} \left(\gamma_i^{(t)} - u \right)^\top S \left(\gamma_i^{(t)} - u \right) + A \gamma_i^{(t)} + \left(\gamma_i^{(t)} \right)^\top B \gamma_i^{(t)}$$

$$= \mathrm{const}^{(2)} - \frac{1}{2} \left(\gamma_i^{(t)} - u \right)^\top S \left(\gamma_i^{(t)} - u \right) + A \left(\gamma_i^{(t)} - u + u \right)$$

$$+ \left(\gamma_i^{(t)} - u + u \right)^\top B \left(\gamma_i^{(t)} - u + u \right)$$

$$= \mathrm{const}^{(3)} - \frac{1}{2} \left(\gamma_i^{(t)} - u \right)^\top (S - 2B) \left(\gamma_i^{(t)} - u \right) + \left(A + 2u^\top B \right) \left(\gamma_i^{(t)} - u \right).$$

Finally, define $D := A + 2u^\top B$ and $E := S - 2B$, resulting in

$$= \mathrm{const}^{(3)} - \frac{1}{2} \left(\gamma_i^{(t)} - u \right)^\top E \left(\gamma_i^{(t)} - u \right) + D \left(\gamma_i^{(t)} - u \right)$$

$$= \mathrm{const}^{(4)} - \frac{1}{2} \left(\gamma_i^{(t)} - u \right)^\top E \left(\gamma_i^{(t)} - u \right) + \left(E^{-1} D^\top \right)^\top E \left(\gamma_i^{(t)} - u \right)$$

$$- \frac{1}{2} \left(E^{-1} D^\top \right)^\top E \left(E^{-1} D^\top \right)$$

$$= \mathrm{const}^{(4)} - \frac{1}{2} \left(\gamma_i^{(t)} - u - E^{-1} D^\top \right)^\top E \left(\gamma_i^{(t)} - u - E^{-1} D^\top \right).$$

Hence $q_\gamma \left(\gamma_i^{(t)} \right)$ is approximately Normal $\left(\tau_i^{(t)}, \Lambda_i^{(t)} \right)$ with variance and mean

$$\Lambda_i^{(t)} := E^{-1}$$

$$= \left(\left[\sum_{h=1}^C \Sigma_h^{-1} \left\langle c_{i,h}^{(t)} \right\rangle \right] + (2N - 2) H_i \right)^{-1}$$

$$\tau_i^{(t)} := u + E^{-1} D^\top$$

$$= u + \Lambda_i^{(t)} \left\{ \left[\sum_{j \neq i}^N \left\langle z_{i \to j}^{(t)} \right\rangle + \left\langle z_{j \leftarrow i}^{(t)} \right\rangle \right] - (2N - 2) \left[g_i^{(t)} + H_i^{(t)} \left(u - \hat{\gamma}_i^{(t)} \right) \right] \right\}$$

$$u := \left(\sum_{h=1}^C \Sigma_h^{-1} \left\langle c_{i,h}^{(t)} \right\rangle \right)^{-1} \left(\sum_{h=1}^C \Sigma_h^{-1} \left\langle c_{i,h}^{(t)} \right\rangle \left\langle \mu_h^{(t)} \right\rangle \right).$$

Distribution of q_c

q_c is a discrete distribution. The distribution of $c_i^{(t)}$ conditioned on its Markov blanket is

$$p\left(c_i^{(t)} \mid \text{MB}_{c_i^{(t)}}\right)$$

$$\propto p\left(\gamma_i^{(t)} \mid c_i^{(t)}, \mu_1^{(t)}, \ldots, \mu_C^{(t)}\right) p\left(c_i^{(t)}\right)$$

$$\propto \left(\prod_{h=1}^{C} \left[|\Sigma_h|^{-1/2}\right]^{c_{i,h}^{(t)}}\right) \exp\left\{\sum_{h=1}^{C} -\frac{1}{2} c_{i,h}^{(t)} \left(\gamma_i^{(t)} - \mu_h^{(t)}\right)^{\top} \Sigma_h^{-1} \left(\gamma_i^{(t)} - \mu_h^{(t)}\right)\right\} \left(\prod_{h=1}^{C} \delta_h^{c_{i,h}^{(t)}}\right)$$

$$= \exp\left\{\sum_{h=1}^{C} -\frac{1}{2} c_{i,h}^{(t)} \left(\gamma_i^{(t)} - \mu_h^{(t)}\right)^{\top} \Sigma_h^{-1} \left(\gamma_i^{(t)} - \mu_h^{(t)}\right) + \sum_{h=1}^{C} c_{i,h}^{(t)} \log \frac{\delta_h}{|\Sigma_h|^{1/2}}\right\}$$

$$= \exp\left\{\sum_{h=1}^{C} -\frac{1}{2} c_{i,h}^{(t)} \left[\left(\gamma_i^{(t)}\right)^{\top} \Sigma_h^{-1} \gamma_i^{(t)} - \left(\gamma_i^{(t)}\right)^{\top} \Sigma_h^{-1} \mu_h^{(t)} - \left(\mu_h^{(t)}\right)^{\top} \Sigma_h^{-1} \gamma_i^{(t)} + \left(\mu_h^{(t)}\right)^{\top} \Sigma_h^{-1} \mu_h^{(t)}\right]\right.$$

$$\left. + \sum_{h=1}^{C} c_{i,h}^{(t)} \log \frac{\delta_h}{|\Sigma_h|^{1/2}}\right\}$$

$$= \exp\left\{\sum_{h=1}^{C} -\frac{1}{2} c_{i,h}^{(t)} \text{tr}\left[\Sigma_h^{-1} \left(\gamma_i^{(t)} \left(\gamma_i^{(t)}\right)^{\top} - \mu_h^{(t)} \left(\gamma_i^{(t)}\right)^{\top} - \gamma_i^{(t)} \left(\mu_h^{(t)}\right)^{\top} + \mu_h^{(t)} \left(\mu_h^{(t)}\right)^{\top}\right)\right]\right.$$

$$\left. + \sum_{h=1}^{C} c_{i,h}^{(t)} \log \frac{\delta_h}{|\Sigma_h|^{1/2}}\right\}.$$

The variables $\gamma_1^{(t)}, \ldots, \gamma_N^{(t)}, \mu_1^{(t)}, \ldots, \mu_C^{(t)}$ belong to other variational factors. The sufficient statistics of γ and μ are $\gamma_i^{(t)} \left(\gamma_i^{(t)}\right)^{\top}$, $\mu_h^{(t)} \left(\gamma_i^{(t)}\right)^{\top}$, and $\mu_h^{(t)} \left(\mu_h^{(t)}\right)^{\top}$, but since γ and μ are marginally independent under q, we can take their expectations separately. Hence

$$q_c\left(c_i^{(t)}\right)$$

$$:\propto \exp\left\{\sum_{h=1}^{C} -\frac{1}{2} c_{i,h}^{(t)} \text{tr}\left[\Sigma_h^{-1} \left(\left\langle \gamma_i^{(t)} \left(\gamma_i^{(t)}\right)^{\top}\right\rangle - \left\langle \mu_h^{(t)}\right\rangle \left\langle \gamma_i^{(t)}\right\rangle^{\top}\right.\right.\right.$$

$$\left.\left.\left. - \left\langle \gamma_i^{(t)}\right\rangle \left\langle \mu_h^{(t)}\right\rangle^{\top} + \left\langle \mu_h^{(t)} \left(\mu_h^{(t)}\right)^{\top}\right\rangle\right)\right] + \sum_{h=1}^{C} c_{i,h}^{(t)} \log \frac{\delta_h}{|\Sigma_h|^{1/2}}\right\},$$

with variational parameters $\left\langle \mu_h^{(t)} \left(\mu_h^{(t)}\right)^{\top}\right\rangle, \left\langle \gamma_i^{(t)} \left(\gamma_i^{(t)}\right)^{\top}\right\rangle, \left\langle \mu_h^{(t)}\right\rangle, \left\langle \gamma_i^{(t)}\right\rangle$. We can also express q_c in terms of indices h:

$$q_c\left(c_i^{(t)} = h\right)$$

$$:\propto \frac{\delta_h}{|\Sigma_h|^{1/2}} \exp\left\{-\frac{1}{2} \text{tr}\left[\Sigma_h^{-1} \left(\left\langle \gamma_i^{(t)} \left(\gamma_i^{(t)}\right)^{\top}\right\rangle - \left\langle \mu_h^{(t)}\right\rangle \left\langle \gamma_i^{(t)}\right\rangle^{\top}\right.\right.\right.$$

$$\left.\left.\left. - \left\langle \gamma_i^{(t)}\right\rangle \left\langle \mu_h^{(t)}\right\rangle^{\top} + \left\langle \mu_h^{(t)} \left(\mu_h^{(t)}\right)^{\top}\right\rangle\right)\right]\right\}.$$

Distribution of q_μ

q_μ is a continuous distribution. The distribution of $\mu_1^{(1)}, \ldots, \mu_C^{(T)}$ conditioned on its Markov blanket is

$$p\left(\mu_1^{(1)}, \ldots, \mu_C^{(T)} \mid \mathrm{MB}_{\mu_1^{(1)}, \ldots, \mu_C^{(T)}}\right)$$

$$\propto \left[\prod_{t=1}^{T} \prod_{i=1}^{N} p\left(\gamma_i^{(t)} \mid c_i^{(t)}, \mu_1^{(t)}, \ldots, \mu_C^{(t)}\right)\right] \left[\prod_{h=1}^{C} p\left(\mu_h^{(1)}\right) \prod_{t=2}^{T} p\left(\mu_h^{(t)} \mid \mu_h^{(t-1)}\right)\right]$$

$$\propto \exp\left\{\sum_{t=1}^{T} \sum_{i=1}^{N} \sum_{h=1}^{C} -\frac{1}{2} c_{i,h}^{(t)} \left(\gamma_i^{(t)} - \mu_h^{(t)}\right)^{\top} \Sigma_h^{-1} \left(\gamma_i^{(t)} - \mu_h^{(t)}\right)\right.$$

$$\left. + \sum_{h=1}^{C} \left[-\frac{1}{2}\left(\mu_h^{(1)} - \nu\right)^{\top} \Phi^{-1}\left(\mu_h^{(1)} - \nu\right) + \sum_{t=2}^{T} -\frac{1}{2}\left(\mu_h^{(t)} - \mu_h^{(t-1)}\right)^{\top} \Phi^{-1}\left(\mu_h^{(t)} - \mu_h^{(t-1)}\right)\right]\right\}$$

$$\propto \exp\left\{\sum_{t=1}^{T} \sum_{i=1}^{N} \sum_{h=1}^{C} -\frac{1}{2} c_{i,h}^{(t)} \left[-\left(\gamma_i^{(t)}\right)^{\top} \Sigma_h^{-1} \mu_h^{(t)} - \left(\mu_h^{(t)}\right)^{\top} \Sigma_h^{-1} \gamma_i^{(t)} + \left(\mu_h^{(t)}\right)^{\top} \Sigma_h^{-1} \mu_h^{(t)}\right]\right.$$

$$\left. + \sum_{h=1}^{C} \left[-\frac{1}{2}\left(\mu_h^{(1)} - \nu\right)^{\top} \Phi^{-1}\left(\mu_h^{(1)} - \nu\right) + \sum_{t=2}^{T} -\frac{1}{2}\left(\mu_h^{(t)} - \mu_h^{(t-1)}\right)^{\top} \Phi^{-1}\left(\mu_h^{(t)} - \mu_h^{(t-1)}\right)\right]\right\}.$$

The variables $\gamma_1^{(1)}, \ldots, \gamma_N^{(T)}, c_1^{(1)}, \ldots, c_N^{(T)}$ belong to other variational factors. The sufficient statistic of γ and c is $c_{i,h}^{(t)} \left(\gamma_u^{(t)}\right)^{\top}$, but since γ and c are marginally independent under q, we can take their expectations separately. Hence

$$q_\mu\left(\mu_1^{(1)}, \ldots, \mu_C^{(T)}\right)$$

$$:\propto \exp\left\{\sum_{t=1}^{T} \sum_{i=1}^{N} \sum_{h=1}^{C} -\frac{1}{2} \left\langle c_{i,h}^{(t)} \right\rangle \left[-\left\langle\gamma_i^{(t)}\right\rangle^{\top} \Sigma_h^{-1} \mu_h^{(t)} - \left(\mu_h^{(t)}\right)^{\top} \Sigma_h^{-1} \left\langle\gamma_i^{(t)}\right\rangle + \left(\mu_h^{(t)}\right)^{\top} \Sigma_h^{-1} \mu_h^{(t)}\right]\right.$$

$$\left. + \sum_{h=1}^{C} \left[-\frac{1}{2}\left(\mu_h^{(1)} - \nu\right)^{\top} \Phi^{-1}\left(\mu_h^{(1)} - \nu\right) + \sum_{t=2}^{T} -\frac{1}{2}\left(\mu_h^{(t)} - \mu_h^{(t-1)}\right)^{\top} \Phi^{-1}\left(\mu_h^{(t)} - \mu_h^{(t-1)}\right)\right]\right\}$$

$$\propto \prod_{h=1}^{C} \exp\left\{\sum_{t=1}^{T} \sum_{i=1}^{N} -\frac{1}{2} \left\langle c_{i,h}^{(t)} \right\rangle \left[-\left\langle\gamma_i^{(t)}\right\rangle^{\top} \Sigma_h^{-1} \mu_h^{(t)} - \left(\mu_h^{(t)}\right)^{\top} \Sigma_h^{-1} \left\langle\gamma_i^{(t)}\right\rangle + \left(\mu_h^{(t)}\right)^{\top} \Sigma_h^{-1} \mu_h^{(t)}\right]\right.$$

$$\left. -\frac{1}{2}\left(\mu_h^{(1)} - \nu\right)^{\top} \Phi^{-1}\left(\mu_h^{(1)} - \nu\right) + \sum_{t=2}^{T} -\frac{1}{2}\left(\mu_h^{(t)} - \mu_h^{(t-1)}\right)^{\top} \Phi^{-1}\left(\mu_h^{(t)} - \mu_h^{(t-1)}\right)\right\},$$

with variational parameters $\left\langle\gamma_i^{(t)}\right\rangle, \left\langle c_i^{(t)}\right\rangle$.

Kalman Smoother for q_μ

We can apply the Kalman smoother to compute the mean and covariance of each $\mu_h^{(t)}$ under q_μ. Let $\Psi(a, b, C) := \exp\left\{-\frac{1}{2}(a-b)^{\top} C^{-1}(a-b)\right\}$, then with some manipulation we obtain

$$q_\mu\left(\mu_1^{(1)}, \ldots, \mu_C^{(T)}\right) \propto \prod_{h=1}^{C} \left[\Psi\left(\mu_h^{(1)}, \nu, \Phi\right) \prod_{i=1}^{N} \Psi\left(\left\langle\gamma_i^{(1)}\right\rangle, \mu_h^{(1)}, \Sigma_h\right)^{\left\langle c_{i,h}^{(1)}\right\rangle}\right]$$

$$\left[\prod_{t=2}^{T} \Psi\left(\mu_h^{(t)}, \mu_h^{(t-1)}, \Phi\right) \prod_{i=1}^{N} \Psi\left(\left\langle\gamma_i^{(t)}\right\rangle, \mu_h^{(t)}, \Sigma_h\right)^{\left\langle c_{i,h}^{(t)}\right\rangle}\right]$$

$$\propto \prod_{h=1}^{C} \left[\Psi\left(\mu_h^{(1)}, \nu, \Phi\right) \Psi\left(\frac{\sum_{i=1}^{N}\left\langle c_{i,h}^{(1)}\right\rangle\left\langle \gamma_i^{(1)}\right\rangle}{\sum_{i=1}^{N}\left\langle c_{i,h}^{(1)}\right\rangle}, \mu_h^{(1)}, \frac{\Sigma_h}{\sum_{i=1}^{N}\left\langle c_{i,h}^{(1)}\right\rangle}\right) \right]$$

$$\left[\prod_{t=2}^{T} \Psi\left(\mu_h^{(t)}, \mu_h^{(t-1)}, \Phi\right) \Psi\left(\frac{\sum_{i=1}^{N}\left\langle c_{i,h}^{(t)}\right\rangle\left\langle \gamma_i^{(t)}\right\rangle}{\sum_{i=1}^{N}\left\langle c_{i,h}^{(t)}\right\rangle}, \mu_h^{(t)}, \frac{\Sigma_h}{\sum_{i=1}^{N}\left\langle c_{i,h}^{(t)}\right\rangle}\right) \right].$$

Notice that q_μ factorizes across cluster indices h:

$$q_\mu\left(\mu_1^{(1)}, \ldots, \mu_C^{(T)}\right) = \prod_{h=1}^{C} q_{\mu_h}\left(\mu_h^{(1)}, \ldots, \mu_h^{(T)}\right)$$

$$q_{\mu_h}\left(\mu_h^{(1)}, \ldots, \mu_h^{(T)}\right) :\propto \Psi\left(\mu_h^{(1)}, \nu, \Phi\right) \Psi\left(\frac{\sum_{i=1}^{N}\left\langle c_{i,h}^{(1)}\right\rangle\left\langle \gamma_i^{(1)}\right\rangle}{\sum_{i=1}^{N}\left\langle c_{i,h}^{(1)}\right\rangle}, \mu_h^{(1)}, \frac{\Sigma_h}{\sum_{i=1}^{N}\left\langle c_{i,h}^{(1)}\right\rangle}\right)$$

$$\left[\prod_{t=2}^{T} \Psi\left(\mu_h^{(t)}, \mu_h^{(t-1)}, \Phi\right) \Psi\left(\frac{\sum_{i=1}^{N}\left\langle c_{i,h}^{(t)}\right\rangle\left\langle \gamma_i^{(t)}\right\rangle}{\sum_{i=1}^{N}\left\langle c_{i,h}^{(t)}\right\rangle}, \mu_h^{(t)}, \frac{\Sigma_h}{\sum_{i=1}^{N}\left\langle c_{i,h}^{(t)}\right\rangle}\right) \right].$$

Observe that each factor $q_{\mu_h}\left(\mu_h^{(1)}, \ldots, \mu_h^{(T)}\right)$ is a linear system of the form

$$\mu_h^{(t+1)} = \mu_h^{(t)} + w_h^{(t)}$$
$$\alpha_h^{(t)} = \mu_h^{(t)} + v_h^{(t)},$$

where $\mu_h^{(t)}$ are latent variables and $\alpha_h^{(t)}$ are observed variables with value $\alpha_h^{(t)} = \frac{\sum_{i=1}^{N}\left\langle c_{i,h}^{(t)}\right\rangle\left\langle \gamma_i^{(t)}\right\rangle}{\sum_{i=1}^{N}\left\langle c_{i,h}^{(t)}\right\rangle}$.

Furthermore, $w_h^{(t)} \sim N(0, \Phi)$, $v_h^{(t)} \sim N\left(0, \Xi_h^{(t)}\right)$ with $\Xi_h^{(t)} = \frac{\Sigma_h}{\sum_{i=1}^{N}\left\langle c_{i,h}^{(t)}\right\rangle}$, and $\mu_h^{(1)} \sim N(\nu, \Phi)$.

Hence the distribution of each $\mu_h^{(t)}$ under q_μ is Gaussian, and its mean and covariance can be computed using the Kalman smoother equations

$$\hat{\mu}_h^{(t+1)|(t)} = \hat{\mu}_h^{(t)|(t)}$$
$$P_h^{(t+1)|(t)} = P_h^{(t)|(t)} + \Phi$$
$$K_h^{(t+1)} = P_h^{(t+1)|(t)} \left(P_h^{(t+1)|(t)} + \Xi_h^{(t+1)}\right)^{-1}$$
$$\hat{\mu}_h^{(t+1)|(t+1)} = \hat{\mu}_h^{(t+1)|(t)} + K_h^{(t+1)}\left(\alpha_h^{(t+1)} - \hat{\mu}_h^{(t+1)|(t)}\right)$$
$$P_h^{(t+1)|(t+1)} = \left(\mathbb{I} - K_h^{(t+1)}\right) P_h^{(t+1)|(t)}$$

and

$$L_h^{(t)} = P_h^{(t)|(t)} \left(P_h^{(t+1)|(t)}\right)^{-1}$$
$$\hat{\mu}_h^{(t)|(T)} = \hat{\mu}_h^{(t)|(t)} + L_h^{(t)}\left(\hat{\mu}_h^{(t+1)|(T)} - \hat{\mu}_h^{(t+1)|(t)}\right)$$
$$P_h^{(t)|(T)} = P_h^{(t)|(t)} + L_h^{(t)}\left(P_h^{(t+1)|(T)} - P_h^{(t+1)|(t)}\right)\left(L_h^{(t)}\right)^{\top}.$$

Thus, μ_h has mean $\hat{\mu}_h^{(t)|(T)}$ and covariance $P_h^{(t)|(T)}$ under q_μ.

E-Step: Solutions to Variational Parameters

In the E-step, we find locally optimal variational parameters for each factor of q. The solutions to the continuous parameters are

$$\left\langle \mu_h^{(t)} \right\rangle = \hat{\mu}_h^{(t)|(T)}$$

$$\left\langle \mu_h^{(t)} \left(\mu_h^{(t)} \right)^\top \right\rangle = \mathbb{E}_{q_\mu} \left[\mu_h^{(t)} \left(\mu_h^{(t)} \right)^\top \right]$$

$$= \mathbb{V}_{q_\mu} \left[\mu_h^{(t)} \right] + \mathbb{E}_{q_\mu} \left[\mu_h^{(t)} \right] \mathbb{E}_{q_\mu} \left[\mu_h^{(t)} \right]^\top$$

$$= P_h^{(t)|(T)} + \hat{\mu}_h^{(t)|(T)} \left(\hat{\mu}_h^{(t)|(T)} \right)^\top$$

$$\left\langle \gamma_i^{(t)} \right\rangle = \tau_i^{(t)}$$

$$\left\langle \gamma_i^{(t)} \left(\gamma_i^{(t)} \right)^\top \right\rangle = \mathbb{E}_{q_\gamma} \left[\gamma_i^{(t)} \left(\gamma_i^{(t)} \right)^\top \right]$$

$$= \mathbb{V}_{q_\gamma} \left[\gamma_i^{(t)} \right] + \mathbb{E}_{q_\gamma} \left[\gamma_i^{(t)} \right] \mathbb{E}_{q_\mu} \left[\gamma_i^{(t)} \right]^\top$$

$$= \Lambda_i^{(t)} + \tau_i^{(t)} \left(\tau_i^{(t)} \right)^\top,$$

while the solutions to the discrete parameters are

$$\left\langle c_{h,i}^{(t)} \right\rangle = q \left(c_i^{(t)} = h \right)$$

$$\left\langle z_{(i \to j),k}^{(t)} \right\rangle = \sum_{l=1}^{K} q_z \left(z_{i \to j}^{(t)} = k, z_{i \leftarrow j}^{(t)} = l \right)$$

$$\left\langle z_{(i \leftarrow j),l}^{(t)} \right\rangle = \sum_{k=1}^{K} q_z \left(z_{i \to j}^{(t)} = k, z_{i \leftarrow j}^{(t)} = l \right).$$

These solutions are used to update the variational parameters in each factor of q. Note that they form a set of fixed-point equations that converge to a local optimum in the space of variational parameters. Hence the E-step involves iterating these equations until some convergence threshold has been reached.

M-Step

In the M-step, we maximize $\mathcal{L}(q, \Theta)$ with respect to the model parameters $\Theta = \{B, \Sigma, \delta, \nu, \Phi\}$. Recall that

$$\mathcal{L}(q, \Theta) := \mathbb{E}_q \left[\log p(E, X \mid \Theta) - \log q(X) \right].$$

Note that the variational distribution q is not actually a function of the model parameters Θ; the model parameters that appear in the q's optimal solution come from the previous M-step, similar to regular EM. Hence it suffices to maximize

$$\mathcal{L}'(q, \Theta) := \mathbb{E}_q \left[\log p(E, X \mid \Theta) \right]$$

$$= \mathbb{E}_q \left[\log \left(\prod_{t,i=1}^{T,N} \prod_{\substack{j \neq i}}^{N} p \left(E_{i,j}^{(t)} \mid z_{i \to j}^{(t)}, z_{i \leftarrow j}^{(t)}; B \right) p \left(z_{i \to j}^{(t)} \mid \gamma_i^{(t)} \right) p \left(z_{i \leftarrow j}^{(t)} \mid \gamma_j^{(t)} \right) \right) \right.$$

$$\left. \left(\prod_{t,i=1}^{T,N} p \left(\gamma_i^{(t)} \mid c_i^{(t)}, \mu_1^{(t)}, \ldots, \mu_C^{(t)}; \Sigma_1, \ldots, \Sigma_C \right) p \left(c_i^{(t)}; \delta \right) \right) \right]$$

$$\left(\prod_{h=1}^{C} p\left(\mu_h^{(1)}; \nu, \Phi\right) \prod_{t=2}^{T} p\left(\mu_h^{(t)} \mid \mu_h^{(t-1)}; \Phi\right) \right) \Bigg]$$

$$= \mathbb{E}_q \left[\sum_{t,i=1}^{T,N} \sum_{j \neq i}^{N} \log p\left(E_{i,j}^{(t)} \mid z_{i \to j}^{(t)}, z_{i \leftarrow j}^{(t)}; B\right) \right]$$

$$+ \mathbb{E}_q \left[\sum_{t,i=1}^{T,N} \sum_{j \neq i}^{N} \log p\left(z_{i \to j}^{(t)} \mid \gamma_i^{(t)}\right) p\left(z_{i \leftarrow j}^{(t)} \mid \gamma_j^{(t)}\right) \right]$$

$$+ \mathbb{E}_q \left[\sum_{t,i=1}^{T,N} \log p\left(\gamma_i^{(t)} \mid c_i^{(t)}, \mu_1^{(t)}, \ldots, \mu_C^{(t)}; \Sigma_1, \ldots, \Sigma_C\right) \right]$$

$$+ \mathbb{E}_q \left[\sum_{t,i=1}^{T,N} \log p\left(c_i^{(t)}; \delta\right) \right]$$

$$+ \mathbb{E}_q \left[\sum_{h=1}^{C} \log p\left(\mu_h^{(1)}; \nu, \Phi\right) + \sum_{h=1}^{C} \sum_{t=2}^{T} \log p\left(\mu_h^{(t)} \mid \mu_h^{(t-1)}; \Phi\right) \right].$$

Maximizing B

Consider the B-dependent terms in $\mathcal{L}'(q, \Theta)$,

$$\mathbb{E}_q \left[\sum_{t,i=1}^{T,N} \sum_{j \neq i}^{N} \log p\left(E_{i,j}^{(t)} \mid z_{i \to j}^{(t)}, z_{i \leftarrow j}^{(t)}; B\right) \right]$$

$$= \sum_{t,i=1}^{T,N} \sum_{j \neq i}^{N} \mathbb{E}_q \left[\log p\left(E_{i,j}^{(t)} \mid z_{i \to j}^{(t)}, z_{i \leftarrow j}^{(t)}; B\right) \right]$$

$$= \sum_{t,i=1}^{T,N} \sum_{j \neq i}^{N} \sum_{z_{i \to j}^{(t)}} \sum_{z_{i \leftarrow j}^{(t)}} q_z\left(z_{i \to j}^{(t)}, z_{i \leftarrow j}^{(t)}\right) \log p\left(E_{i,j}^{(t)} \mid z_{i \to j}^{(t)}, z_{i \leftarrow j}^{(t)}; B\right)$$

(zs independent of other latent variables under q).

Since $z_{i \to j}^{(t)}, z_{i \leftarrow j}^{(t)}$ are indicator variables, we index their possible values with $k \in \{1, \ldots, K\}$ and $l \in \{1, \ldots, K\}$, respectively:

$$= \sum_{t,i=1}^{T,N} \sum_{j \neq i}^{N} \sum_{k,l=1}^{K,K} q_z\left(z_{i \to j}^{(t)} = k, z_{i \leftarrow j}^{(t)} = l\right) \log p\left(E_{i,j}^{(t)} \mid z_{i \to j}^{(t)} = k, z_{i \leftarrow j}^{(t)} = l; B\right)$$

$$= \sum_{t,i=1}^{T,N} \sum_{j \neq i}^{N} \sum_{k,l=1}^{K,K} q_z\left(z_{i \to j}^{(t)} = k, z_{i \leftarrow j}^{(t)} = l\right) \left(E_{i,j}^{(t)} \log B_{k,l} + \left(1 - E_{i,j}^{(t)}\right) \log\left(1 - B_{k,l}\right)\right).$$

$$(23.5)$$

Setting the first derivative wrt $B_{k,l}$ to zero yields the maximizer $\hat{B}_{k,l}$ for $\mathcal{L}'(q, \Theta)$:

$$0 = \frac{\partial}{\partial B_{k,l}} \sum_{t,i=1}^{T,N} \sum_{j \neq i}^{N} \sum_{k',l'=1}^{K,K} q_z\left(z_{i \to j}^{(t)} = k', z_{i \leftarrow j}^{(t)} = l'\right)$$

$$\left(E_{i,j}^{(t)} \log B_{k',l'} + \left(1 - E_{i,j}^{(t)} \right) \log \left(1 - B_{k',l'} \right) \right)$$

$$0 = \sum_{t,i=1}^{T,N} \sum_{j \neq i}^{N} q_z \left(z_{i \to j}^{(t)} = k, z_{i \leftarrow j}^{(t)} = l \right) \left(\frac{E_{i,j}^{(t)}}{B_{k,l}} - \frac{1 - E_{i,j}^{(t)}}{1 - B_{k,l}} \right)$$

$$0 = \sum_{t,i=1}^{T,N} \sum_{j \neq i}^{N} q_z \left(z_{i \to j}^{(t)} = k, z_{i \leftarrow j}^{(t)} = l \right) \left(E_{i,j}^{(t)} - B_{k,l} \right)$$

$$\hat{B}_{k,l} := B_{k,l} = \frac{\sum_{t,i=1}^{T,N} \sum_{j \neq i}^{N} q_z \left(z_{i \to j}^{(t)} = k, z_{i \leftarrow j}^{(t)} = l \right) E_{i,j}^{(t)}}{\sum_{t,i=1}^{T,N} \sum_{j \neq i}^{N} q_z \left(z_{i \to j}^{(t)} = k, z_{i \leftarrow j}^{(t)} = l \right)}.$$

Maximizing Σ

Consider the $\Sigma_1, \ldots, \Sigma_C$-dependent terms in $\mathcal{L}'(q, \Theta)$,

$$\mathbb{E}_q \left[\sum_{t,i=1}^{T,N} \log p \left(\gamma_i^{(t)} \mid c_i^{(t)}, \mu_1^{(t)}, \ldots, \mu_C^{(t)}; \Sigma_1, \ldots, \Sigma_C \right) \right]$$

$$= \sum_{t,i=1}^{T,N} \mathbb{E}_q \left[\log p \left(\gamma_i^{(t)} \mid c_i^{(t)}, \mu_1^{(t)}, \ldots, \mu_C^{(t)}; \Sigma_1, \ldots, \Sigma_C \right) \right]$$

$$= \sum_{t,i=1}^{T,N} \mathbb{E}_q \left[\log \prod_{h=1}^{C} \left((2\pi)^{-K/2} |\Sigma_h|^{-1/2} \exp \left\{ -\frac{1}{2} \left(\gamma_i^{(t)} - \mu_h^{(t)} \right)^\top \Sigma_h^{-1} \left(\gamma_i^{(t)} - \mu_h^{(t)} \right) \right\} \right)^{c_{i,h}^{(t)}} \right]$$

$$= \sum_{t,i=1}^{T,N} \sum_{h=1}^{C} \mathbb{E}_q \left[c_{i,h}^{(t)} \log \left((2\pi)^{-K/2} |\Sigma_h|^{-1/2} \right) - \frac{1}{2} c_{i,h}^{(t)} \left(\gamma_i^{(t)} - \mu_h^{(t)} \right)^\top \Sigma_h^{-1} \left(\gamma_i^{(t)} - \mu_h^{(t)} \right) \right]$$

$$= \sum_{t,i=1}^{T,N} \sum_{h=1}^{C} -\log \left((2\pi)^{K/2} |\Sigma_h|^{1/2} \right) \mathbb{E}_q \left[c_{i,h}^{(t)} \right]$$

$$- \frac{1}{2} \mathbb{E}_q \left[c_{i,h}^{(t)} \mathrm{tr} \left[\Sigma_h^{-1} \left(\gamma_i^{(t)} \left(\gamma_i^{(t)} \right)^\top - \mu_h^{(t)} \left(\gamma_i^{(t)} \right)^\top - \gamma_i^{(t)} \left(\mu_h^{(t)} \right)^\top + \mu_h^{(t)} \left(\mu_h^{(t)} \right)^\top \right) \right] \right].$$

Since c, μ, γ are independent of each other (and other latent variables) under q,

$$= \sum_{t,i=1}^{T,N} \sum_{h=1}^{C} -\log \left((2\pi)^{K/2} |\Sigma_h|^{1/2} \right) \left\langle c_{i,h}^{(t)} \right\rangle \tag{23.6}$$

$$- \frac{1}{2} \left\langle c_{i,h}^{(t)} \right\rangle \mathrm{tr} \left[\Sigma_h^{-1} \left(\left\langle \gamma_i^{(t)} \left(\gamma_i^{(t)} \right)^\top \right\rangle - \left\langle \mu_h^{(t)} \right\rangle \left\langle \gamma_i^{(t)} \right\rangle^\top \right. \right.$$

$$\left. \left. - \left\langle \gamma_i^{(t)} \right\rangle \left\langle \mu_h^{(t)} \right\rangle^\top + \left\langle \mu_h^{(t)} \left(\mu_h^{(t)} \right)^\top \right\rangle \right) \right],$$

where we have defined $\langle X \rangle := \mathbb{E}_q[X]$, and the solutions to $\langle X \rangle$ are identical to the E-step. Setting the first derivative wrt Σ_h to zero yields the maximizer $\hat{\Sigma}_h$ for $\mathcal{L}'(q, \Theta)$:

$$0 = \nabla_{\Sigma_h} \sum_{t,i=1}^{T,N} \sum_{h=1}^{C} -\log \left((2\pi)^{K/2} |\Sigma_h|^{1/2} \right) \left\langle c_{i,h}^{(t)} \right\rangle$$

$$-\frac{1}{2}\left\langle c_{i,h}^{(t)}\right\rangle \mathrm{tr}\left[\Sigma_h^{-1}\left(\left\langle \gamma_i^{(t)}\left(\gamma_i^{(t)}\right)^\top\right\rangle - \left\langle \mu_h^{(t)}\right\rangle\left\langle \gamma_i^{(t)}\right\rangle^\top - \left\langle \gamma_i^{(t)}\right\rangle\left\langle \mu_h^{(t)}\right\rangle^\top + \left\langle \mu_h^{(t)}\left(\mu_h^{(t)}\right)^\top\right\rangle\right)\right]$$

$$0 = \sum_{t,i=1}^{T,N} -\frac{1}{2}\left\langle c_{i,h}^{(t)}\right\rangle \Sigma_h^{-1}$$

$$+\frac{1}{2}\left\langle c_{i,h}^{(t)}\right\rangle \Sigma_h^{-1}\left(\left\langle \gamma_i^{(t)}\left(\gamma_i^{(t)}\right)^\top\right\rangle - \left\langle \gamma_i^{(t)}\right\rangle\left\langle \mu_h^{(t)}\right\rangle^\top - \left\langle \mu_h^{(t)}\right\rangle\left\langle \gamma_i^{(t)}\right\rangle^\top + \left\langle \mu_h^{(t)}\left(\mu_h^{(t)}\right)^\top\right\rangle\right)\Sigma_h^{-1}$$

$$0 = \sum_{t,i=1}^{T,N} -\left\langle c_{i,h}^{(t)}\right\rangle \Sigma_h$$

$$+\left\langle c_{i,h}^{(t)}\right\rangle\left(\left\langle \gamma_i^{(t)}\left(\gamma_i^{(t)}\right)^\top\right\rangle - \left\langle \gamma_i^{(t)}\right\rangle\left\langle \mu_h^{(t)}\right\rangle^\top - \left\langle \mu_h^{(t)}\right\rangle\left\langle \gamma_i^{(t)}\right\rangle^\top + \left\langle \mu_h^{(t)}\left(\mu_h^{(t)}\right)^\top\right\rangle\right)$$

$$\hat{\Sigma}_h := \Sigma_h = \frac{\sum_{t,i=1}^{T,N}\left\langle c_{i,h}^{(t)}\right\rangle\left(\left\langle \gamma_i^{(t)}\left(\gamma_i^{(t)}\right)^\top\right\rangle - \left\langle \gamma_i^{(t)}\right\rangle\left\langle \mu_h^{(t)}\right\rangle^\top - \left\langle \mu_h^{(t)}\right\rangle\left\langle \gamma_i^{(t)}\right\rangle^\top + \left\langle \mu_h^{(t)}\left(\mu_h^{(t)}\right)^\top\right\rangle\right)}{\sum_{t,i=1}^{T,N}\left\langle c_{i,h}^{(t)}\right\rangle}.$$

Maximizing δ

Consider the δ-dependent terms in $\mathcal{L}'(q,\Theta)$,

$$\mathbb{E}_q\left[\sum_{t,i=1}^{T,N}\log p\left(c_i^{(t)};\delta\right)\right]$$

$$= \sum_{t,i=1}^{T,N}\mathbb{E}_q\left[\log\prod_{h=1}^{C}\delta_h^{c_{i,h}^{(t)}}\right]$$

$$= \sum_{t,i=1}^{T,N}\sum_{h=1}^{C}\mathbb{E}_q\left[c_{i,h}^{(t)}\log\delta_h\right]$$

$$= \sum_{t,i=1}^{T,N}\sum_{h=1}^{C}\left\langle c_{i,h}^{(t)}\right\rangle\log\delta_h$$

$$= \left(\sum_{t,i=1}^{T,N}\left\langle c_i^{(t)}\right\rangle\right)^\top\log\delta, \tag{23.7}$$

where $\left\langle c_{i,h}^{(t)}\right\rangle := \mathbb{E}_q\left[c_{i,h}^{(t)}\right]$, and the solution to $\left\langle c_{i,h}^{(t)}\right\rangle$ is identical to the E-step. Taking the first derivative with respect to $\delta_1,\ldots,\delta_{C-1}$,

$$\frac{\partial}{\partial\delta_h}\sum_{t,i=1}^{T,N}\sum_{h'=1}^{C}\left\langle c_{i,h'}^{(t)}\right\rangle\log\delta_{h'} = \sum_{t,i=1}^{T,N}\frac{\partial}{\partial\delta_h}\left\langle c_{i,h}^{(t)}\right\rangle\log\delta_h + \frac{\partial}{\partial\delta_h}\left\langle c_{i,C}^{(t)}\right\rangle\log\left(1-\sum_{h'=1}^{C-1}\delta_{h'}\right)$$

$$= \sum_{t,i=1}^{T,N}\frac{\left\langle c_{i,h}^{(t)}\right\rangle}{\delta_h} - \frac{\left\langle c_{i,C}^{(t)}\right\rangle}{1-\sum_{h'=1}^{C-1}\delta_{h'}}.$$

By setting all the derivatives to zero and performing some manipulation, we obtain the maximizer $\hat{\delta}$ for $\mathcal{L}'(q,\Theta)$:

$$\hat{\delta} = \frac{\sum_{t,i=1}^{T,N}\left\langle c_i^{(t)}\right\rangle}{TN}.$$

Maximizing ν, Φ

Consider the ν, Φ-dependent terms in $\mathcal{L}'(q, \Theta)$,

$$\mathbb{E}_q\left[\sum_{h=1}^{C} \log p\left(\mu_h^{(1)}; \nu, \Phi\right) + \sum_{h=1}^{C}\sum_{t=2}^{T} \log p\left(\mu_h^{(t)} \mid \mu_h^{(t-1)}; \Phi\right)\right]$$

$$= \sum_{h=1}^{C} \mathbb{E}_q\left[\log p\left(\mu_h^{(1)}; \nu, \Phi\right)\right] + \sum_{h=1}^{C}\sum_{t=2}^{T} \mathbb{E}_q\left[\log p\left(\mu_h^{(t)} \mid \mu_h^{(t-1)}; \Phi\right)\right].$$

We begin by maximizing wrt ν, which only requires us to focus on the first term:

$$\sum_{h=1}^{C} \mathbb{E}_q\left[\log p\left(\mu_h^{(1)}; \nu, \Phi\right)\right]$$

$$= \sum_{h=1}^{C} \mathbb{E}_q\left[\log\left((2\pi)^{-K/2}|\Phi|^{-1/2}\exp\left\{-\frac{1}{2}\left(\mu_h^{(1)} - \nu\right)^{\top}\Phi^{-1}\left(\mu_h^{(1)} - \nu\right)\right\}\right)\right]$$

$$= \sum_{h=1}^{C} \mathbb{E}_q\left[\log\left((2\pi)^{-K/2}|\Phi|^{-1/2}\right)\right.$$
$$\left. -\frac{1}{2}\left(\left(\mu_h^{(1)}\right)^{\top}\Phi^{-1}\mu_h^{(1)} - \left(\mu_h^{(1)}\right)^{\top}\Phi^{-1}\nu - \nu^{\top}\Phi^{-1}\mu_h^{(1)} + \nu^{\top}\Phi^{-1}\nu\right)\right].$$

Dropping terms that do not depend on ν,

$$= \sum_{h=1}^{C} \mathbb{E}_q\left[-\frac{1}{2}\left(-\left(\mu_h^{(1)}\right)^{\top}\Phi^{-1}\nu - \nu^{\top}\Phi^{-1}\mu_h^{(1)} + \nu^{\top}\Phi^{-1}\nu\right)\right]$$

$$= \sum_{h=1}^{C} \frac{1}{2}\left\langle\mu_h^{(1)}\right\rangle^{\top}\Phi^{-1}\nu + \frac{1}{2}\nu^{\top}\Phi^{-1}\left\langle\mu_h^{(1)}\right\rangle - \frac{1}{2}\nu^{\top}\Phi^{-1}\nu,$$

where $\left\langle\mu_h^{(1)}\right\rangle := \mathbb{E}_q\left[\mu_h^{(1)}\right]$, and the solution to $\left\langle\mu_h^{(1)}\right\rangle$ is identical to the E-step. Setting the first derivative wrt ν to zero yields the maximizer $\hat{\nu}$ for $\mathcal{L}'(q, \Theta)$:

$$0 = \nabla_\nu \sum_{h=1}^{C} \frac{1}{2}\left\langle\mu_h^{(1)}\right\rangle^{\top}\Phi^{-1}\nu + \frac{1}{2}\nu^{\top}\Phi^{-1}\left\langle\mu_h^{(1)}\right\rangle - \frac{1}{2}\nu^{\top}\Phi^{-1}\nu$$

$$0 = \sum_{h=1}^{C} \Phi^{-1}\left\langle\mu_h^{(1)}\right\rangle - \Phi^{-1}\nu$$

$$\hat{\nu} := \nu = \frac{\sum_{h=1}^{C}\left\langle\mu_h^{(1)}\right\rangle}{C}.$$

We now substitute $\nu = \hat{\nu}$ and consider the Φ-dependent terms in $\mathcal{L}'(q, \Theta)$:

$$\mathbb{E}_q\left[\sum_{h=1}^{C} \log p\left(\mu_h^{(1)}; \hat{\nu}, \Phi\right) + \sum_{h=1}^{C}\sum_{t=2}^{T} \log p\left(\mu_h^{(t)} \mid \mu_h^{(t-1)}; \Phi\right)\right]$$

$$= \sum_{h=1}^{C} \mathbb{E}_q\left[\log p\left(\mu_h^{(1)}; \hat{\nu}, \Phi\right)\right] + \sum_{h=1}^{C}\sum_{t=2}^{T} \mathbb{E}_q\left[\log p\left(\mu_h^{(t)} \mid \mu_h^{(t-1)}; \Phi\right)\right]$$

$$
\begin{aligned}
= \sum_{h=1}^{C} &-\log\left((2\pi)^{K/2}|\Phi|^{1/2}\right) - \frac{1}{2}\mathbb{E}_q\left[\left(\mu_h^{(1)}-\hat{\nu}\right)^{\top}\Phi^{-1}\left(\mu_h^{(1)}-\hat{\nu}\right)\right] \\
+ \sum_{h=1}^{C}\sum_{t=2}^{T} &-\log\left((2\pi)^{K/2}|\Phi|^{1/2}\right) - \frac{1}{2}\mathbb{E}_q\left[\left(\mu_h^{(t)}-\mu_h^{(t-1)}\right)^{\top}\Phi^{-1}\left(\mu_h^{(t)}-\mu_h^{(t-1)}\right)\right] \\
= -TC\log&\left((2\pi)^{K/2}|\Phi|^{1/2}\right) \\
&- \sum_{h=1}^{C}\frac{1}{2}\mathrm{tr}\left[\Phi^{-1}\left(\left\langle\mu_h^{(1)}\left(\mu_h^{(1)}\right)^{\top}\right\rangle - \hat{\nu}\left\langle\mu_h^{(1)}\right\rangle^{\top} - \left\langle\mu_h^{(1)}\right\rangle\hat{\nu}^{\top} + \hat{\nu}\hat{\nu}^{\top}\right)\right] \\
-\sum_{h=1}^{C}\sum_{t=2}^{T}\frac{1}{2}\mathrm{tr}&\left[\Phi^{-1}\left(\left\langle\mu_h^{(t)}\left(\mu_h^{(t)}\right)^{\top}\right\rangle - \left\langle\mu_h^{(t-1)}\left(\mu_h^{(t)}\right)^{\top}\right\rangle\right.\right. \\
&\left.\left.- \left\langle\mu_h^{(t)}\left(\mu_h^{(t-1)}\right)^{\top}\right\rangle + \left\langle\mu_h^{(t-1)}\left(\mu_h^{(t-1)}\right)^{\top}\right\rangle\right)\right],
\end{aligned}
$$

where $\langle X\rangle := \mathbb{E}_q[X]$. The solutions to $\left\langle\mu_h^{(1)}\right\rangle$, $\left\langle\mu_h^{(t)}\left(\mu_h^{(t)}\right)^{\top}\right\rangle$, $\left\langle\mu_h^{(t-1)}\left(\mu_h^{(t-1)}\right)^{\top}\right\rangle$ are identical to the E-step. The remaining expectations are

$$
\left\langle\mu_h^{(t)}\left(\mu_h^{(t-1)}\right)^{\top}\right\rangle = \left\langle\mu_h^{(t-1)}\left(\mu_h^{(t)}\right)^{\top}\right\rangle^{\top} = P_h^{(t)|(T)}\left(L_h^{(t-1)}\right)^{\top} + \left\langle\mu_h^{(t)}\right\rangle\left\langle\mu_h^{(t-1)}\right\rangle^{\top},
$$

where P and L are defined in the section discussing the Kalman smoother. Setting the first derivative wrt Φ to zero yields the maximizer $\hat{\Phi}$ for $\mathcal{L}'(q,\Theta)$:

$$
\begin{aligned}
0 = \nabla_\Phi &- TC\log\left((2\pi)^{K/2}|\Phi|^{1/2}\right) - \sum_{h=1}^{C}\frac{1}{2}\mathrm{tr}\left[\Phi^{-1}\left(\left\langle\mu_h^{(1)}\left(\mu_h^{(1)}\right)^{\top}\right\rangle - \hat{\nu}\left\langle\mu_h^{(1)}\right\rangle^{\top} - \left\langle\mu_h^{(1)}\right\rangle\hat{\nu}^{\top} + \hat{\nu}\hat{\nu}^{\top}\right)\right] \\
&- \sum_{h=1}^{C}\sum_{t=2}^{T}\frac{1}{2}\mathrm{tr}\left[\Phi^{-1}\left(\left\langle\mu_h^{(t)}\left(\mu_h^{(t)}\right)^{\top}\right\rangle - \left\langle\mu_h^{(t-1)}\left(\mu_h^{(t)}\right)^{\top}\right\rangle\right.\right. \\
&\left.\left.\qquad - \left\langle\mu_h^{(t)}\left(\mu_h^{(t-1)}\right)^{\top}\right\rangle + \left\langle\mu_h^{(t-1)}\left(\mu_h^{(t-1)}\right)^{\top}\right\rangle\right)\right] \\
0 = -\frac{TC}{2}\Phi^{-1} &+ \sum_{h=1}^{C}\frac{1}{2}\Phi^{-1}\left(\left\langle\mu_h^{(1)}\left(\mu_h^{(1)}\right)^{\top}\right\rangle - \left\langle\mu_h^{(1)}\right\rangle\hat{\nu}^{\top} - \hat{\nu}\left\langle\mu_h^{(1)}\right\rangle^{\top} + \hat{\nu}\hat{\nu}^{\top}\right)\Phi^{-1} \\
&+ \sum_{h=1}^{C}\sum_{t=2}^{T}\frac{1}{2}\Phi^{-1}\left(\left\langle\mu_h^{(t)}\left(\mu_h^{(t)}\right)^{\top}\right\rangle - \left\langle\mu_h^{(t)}\left(\mu_h^{(t-1)}\right)^{\top}\right\rangle\right. \\
&\left.\qquad - \left\langle\mu_h^{(t-1)}\left(\mu_h^{(t)}\right)^{\top}\right\rangle + \left\langle\mu_h^{(t-1)}\left(\mu_h^{(t-1)}\right)^{\top}\right\rangle\right)\Phi^{-1} \\
0 = -TC\Phi &+ \sum_{h=1}^{C}\left\langle\mu_h^{(1)}\left(\mu_h^{(1)}\right)^{\top}\right\rangle - \left\langle\mu_h^{(1)}\right\rangle\hat{\nu}^{\top} - \hat{\nu}\left\langle\mu_h^{(1)}\right\rangle^{\top} + \hat{\nu}\hat{\nu}^{\top} \\
&+ \sum_{h=1}^{C}\sum_{t=2}^{T}\left\langle\mu_h^{(t)}\left(\mu_h^{(t)}\right)^{\top}\right\rangle - \left\langle\mu_h^{(t)}\left(\mu_h^{(t-1)}\right)^{\top}\right\rangle - \left\langle\mu_h^{(t-1)}\left(\mu_h^{(t)}\right)^{\top}\right\rangle + \left\langle\mu_h^{(t-1)}\left(\mu_h^{(t-1)}\right)^{\top}\right\rangle
\end{aligned}
$$

$$
\hat{\Phi} := \Phi = \frac{\sum_{h=1}^{C}\left\langle\mu_h^{(1)}\left(\mu_h^{(1)}\right)^{\top}\right\rangle - \left\langle\mu_h^{(1)}\right\rangle\hat{\nu}^{\top} - \hat{\nu}\left\langle\mu_h^{(1)}\right\rangle^{\top} + \hat{\nu}\hat{\nu}^{\top}}{TC}
$$
$$
+ \frac{\sum_{h=1}^{C}\sum_{t=2}^{T}\left\langle\mu_h^{(t)}\left(\mu_h^{(t)}\right)^{\top}\right\rangle - \left\langle\mu_h^{(t)}\left(\mu_h^{(t-1)}\right)^{\top}\right\rangle - \left\langle\mu_h^{(t-1)}\left(\mu_h^{(t)}\right)^{\top}\right\rangle + \left\langle\mu_h^{(t-1)}\left(\mu_h^{(t-1)}\right)^{\top}\right\rangle}{TC}.
$$

Computing the Variational Lower Bound $\mathcal{L}(q, \Theta)$

The marginal likelihood lower bound $\mathcal{L}(q, \Theta)$ can be used to test for convergence in the variational EM algorithm. It also functions as a surrogate for the true marginal likelihood $p(E \mid \Theta)$; this is useful when taking random restarts, as it enables us to select the highest likelihood restart. Recall that

$$
\begin{aligned}
& \mathcal{L}(q, \Theta) \\
&= \mathbb{E}_q \left[\log p(E, X \mid \Theta) - \log q(X) \right] \\
&= \mathbb{E}_q \left[\sum_{t,i=1}^{T,N} \sum_{j \neq i}^{N} \log p\left(E_{i,j}^{(t)} \mid z_{i \rightarrow j}^{(t)}, z_{i \leftarrow j}^{(t)}; B \right) \right] + \mathbb{E}_q \left[\sum_{t,i=1}^{T,N} \sum_{j \neq i}^{N} \log p\left(z_{i \rightarrow j}^{(t)} \mid \gamma_i^{(t)} \right) p\left(z_{i \leftarrow j}^{(t)} \mid \gamma_j^{(t)} \right) \right] \\
&\quad + \mathbb{E}_q \left[\sum_{t,i=1}^{T,N} \log p\left(\gamma_i^{(t)} \mid c_i^{(t)}, \mu_1^{(t)}, \ldots, \mu_C^{(t)}; \Sigma_1, \ldots, \Sigma_C \right) \right] + \mathbb{E}_q \left[\sum_{t,i=1}^{T,N} \log p\left(c_i^{(t)}; \delta \right) \right] \\
&\quad + \mathbb{E}_q \left[\sum_{h=1}^{C} \log p\left(\mu_h^{(1)}; \nu, \Phi \right) + \sum_{h=1}^{C} \sum_{t=2}^{T} \log p\left(\mu_h^{(t)} \mid \mu_h^{(t-1)}; \Phi \right) \right] \\
&\quad - \mathbb{E}_q \left[\log q_\mu \left(\mu_1^{(1)}, \ldots, \mu_C^{(T)} \right) \right] - \mathbb{E}_q \left[\sum_{t,i=1}^{T,N} \log q_\gamma \left(\gamma_i^{(t)} \right) \right] \\
&\quad - \mathbb{E}_q \left[\sum_{t,i=1}^{T,N} \log q_c \left(c_i^{(t)} \right) \right] - \mathbb{E}_q \left[\sum_{t,i,j=1}^{T,N,N} \log q_z \left(z_{i \rightarrow j}^{(t)}, z_{i \leftarrow j}^{(t)} \right) \right].
\end{aligned}
$$

It turns out that we cannot compute $\mathcal{L}(q, \Theta)$ exactly because of term 2, but we can lower-bound the latter to produce a lower bound $\mathcal{L}_{lower}(q, \Theta)$ on $\mathcal{L}(q, \Theta)$.

Closed forms for terms 1,3,4, and 5 are in Equations (23.5, 23.6, 23.7, and 23.8), respectively. We now provide closed forms for terms 6,7,8, and 9, as well as the aforementioned lower bound for term 2.

Lower Bound for Term 2

$$
\begin{aligned}
& \mathbb{E}_q \left[\sum_{t,i=1}^{T,N} \sum_{j \neq i}^{N} \log p\left(z_{i \rightarrow j}^{(t)} \mid \gamma_i^{(t)} \right) p\left(z_{i \leftarrow j}^{(t)} \mid \gamma_j^{(t)} \right) \right] \\
&= \sum_{t,i=1}^{T,N} \sum_{j \neq i}^{N} \mathbb{E}_q \left[\log \prod_{k=1}^{K} \left(\frac{\exp \gamma_{i,k}^{(t)}}{\sum_{l=1}^{K} \exp \gamma_{i,l}^{(t)}} \right)^{z_{i \rightarrow j,k}^{(t)}} \left(\frac{\exp \gamma_{j,k}^{(t)}}{\sum_{l=1}^{K} \exp \gamma_{j,l}^{(t)}} \right)^{z_{i \leftarrow j,k}^{(t)}} \right] \\
&= \sum_{t,i=1}^{T,N} \sum_{j \neq i}^{N} \sum_{k=1}^{K} \mathbb{E}_q \left[z_{i \rightarrow j,k}^{(t)} \gamma_{i,k}^{(t)} - z_{i \rightarrow j,k}^{(t)} \log \sum_{l=1}^{K} \exp \gamma_{i,l}^{(t)} + z_{i \leftarrow j,k}^{(t)} \gamma_{j,k}^{(t)} - z_{i \leftarrow j,k}^{(t)} \log \sum_{l=1}^{K} \exp \gamma_{j,l}^{(t)} \right].
\end{aligned}
$$

Since z, γ are independent of each other under q,

$$
\begin{aligned}
&= \sum_{t,i=1}^{T,N} \sum_{j \neq i}^{N} \sum_{k=1}^{K} \left\langle z_{i \rightarrow j,k}^{(t)} \right\rangle \left\langle \gamma_{i,k}^{(t)} \right\rangle - \left\langle z_{i \rightarrow j,k}^{(t)} \right\rangle \mathbb{E}_q \left[\log \sum_{l=1}^{K} \exp \gamma_{i,l}^{(t)} \right] \\
&\quad + \left\langle z_{i \leftarrow j,k}^{(t)} \right\rangle \left\langle \gamma_{j,k}^{(t)} \right\rangle - \left\langle z_{i \leftarrow j,k}^{(t)} \right\rangle \mathbb{E}_q \left[\log \sum_{l=1}^{K} \exp \gamma_{j,l}^{(t)} \right].
\end{aligned}
$$

Applying Jensen's inequality to the log-sum-exp terms,

$$
\geq \sum_{t,i=1}^{T,N} \sum_{j \neq i}^{N} \sum_{k=1}^{K} \left\langle z_{i \to j,k}^{(t)} \right\rangle \left\langle \gamma_{i,k}^{(t)} \right\rangle - \left\langle z_{i \to j,k}^{(t)} \right\rangle \log \mathbb{E}_q \left[\sum_{l=1}^{K} \exp \gamma_{i,l}^{(t)} \right]
$$

$$
+ \left\langle z_{i \leftarrow j,k}^{(t)} \right\rangle \left\langle \gamma_{j,k}^{(t)} \right\rangle - \left\langle z_{i \leftarrow j,k}^{(t)} \right\rangle \log \mathbb{E}_q \left[\sum_{l=1}^{K} \exp \gamma_{j,l}^{(t)} \right]
$$

$$
= \sum_{t,i=1}^{T,N} \sum_{j \neq i}^{N} \sum_{k=1}^{K} \left\langle z_{i \to j,k}^{(t)} \right\rangle \left\langle \gamma_{i,k}^{(t)} \right\rangle - \left\langle z_{i \to j,k}^{(t)} \right\rangle \log \left(\sum_{l=1}^{K} \left\langle \exp \gamma_{i,l}^{(t)} \right\rangle \right)
$$

$$
+ \left\langle z_{i \leftarrow j,k}^{(t)} \right\rangle \left\langle \gamma_{j,k}^{(t)} \right\rangle - \left\langle z_{i \leftarrow j,k}^{(t)} \right\rangle \log \left(\sum_{l=1}^{K} \left\langle \exp \gamma_{j,l}^{(t)} \right\rangle \right)
$$

$$
= \sum_{t,i=1}^{T,N} \sum_{j \neq i}^{N} \sum_{k=1}^{K} \left\langle z_{i \to j,k}^{(t)} \right\rangle \left(\left\langle \gamma_{i,k}^{(t)} \right\rangle - \log \sum_{l=1}^{K} \left\langle \exp \gamma_{i,l}^{(t)} \right\rangle \right)
$$

$$
+ \left\langle z_{i \leftarrow j,k}^{(t)} \right\rangle \left(\left\langle \gamma_{j,k}^{(t)} \right\rangle - \log \sum_{l=1}^{K} \left\langle \exp \gamma_{j,l}^{(t)} \right\rangle \right)
$$

$$
= \sum_{t,i=1}^{T,N} \sum_{k=1}^{K} \left(\left\langle \gamma_{i,k}^{(t)} \right\rangle - \log \sum_{l=1}^{K} \left\langle \exp \gamma_{i,l}^{(t)} \right\rangle \right) \left(\sum_{j \neq i}^{N} \left\langle z_{i \to j,k}^{(t)} \right\rangle + \left\langle z_{j \leftarrow i,k}^{(t)} \right\rangle \right)
$$

$$
= \sum_{t,i=1}^{T,N} \left(\left\langle \gamma_{i}^{(t)} \right\rangle - \log \sum_{l=1}^{K} \left\langle \exp \gamma_{i,l}^{(t)} \right\rangle \right)^{\top} \left(\sum_{j \neq i}^{N} \left\langle z_{i \to j}^{(t)} \right\rangle + \left\langle z_{j \leftarrow i}^{(t)} \right\rangle \right),
$$

where $\langle X \rangle := \mathbb{E}_q [X]$. The solutions to $\left\langle z_{i \to j,k}^{(t)} \right\rangle, \left\langle z_{i \leftarrow j,k}^{(t)} \right\rangle, \left\langle \gamma_{i,k}^{(t)} \right\rangle$ are in the E-step. As for $\left\langle \exp \gamma_{i,l}^{(t)} \right\rangle$, observe that for a univariate Gaussian random variable X with mean μ and variance σ^2,

$$
\mathbb{E} \left[\exp X \right] = \int_x \frac{1}{\sqrt{2\pi}\sigma} \exp \left\{ -\frac{(x - \mu)^2}{2\sigma^2} \right\} \exp \left\{ x \right\} dx
$$

$$
= \int_x \frac{1}{\sqrt{2\pi}\sigma} \exp \left\{ -\frac{x^2 - 2x \left(\mu + \sigma^2 \right) + \mu^2}{2\sigma^2} \right\} dx
$$

$$
= \int_x \frac{1}{\sqrt{2\pi}\sigma} \exp \left\{ -\frac{x^2 - 2x \left(\mu + \sigma^2 \right) + \left(\mu^2 + 2\mu\sigma^2 + \sigma^4 \right)}{2\sigma^2} \right\} \exp \left\{ \mu + \frac{\sigma^2}{2} \right\} dx
$$

$$
= \exp \left\{ \mu + \frac{\sigma^2}{2} \right\} \int_x \frac{1}{\sqrt{2\pi}\sigma} \exp \left\{ -\frac{\left(x - \left(\mu + \sigma^2 \right) \right)^2}{2\sigma^2} \right\} dx
$$

$$
= \exp \left\{ \mu + \frac{\sigma^2}{2} \right\}.
$$

Hence,

$$
\left\langle \exp \gamma_{i,l}^{(t)} \right\rangle = \exp \left\{ \left\langle \gamma_{i,l}^{(t)} \right\rangle + \frac{1}{2} \Lambda_{i,ll}^{(t)} \right\},
$$

where Λ_i is defined in the previous section discussing the Laplace approximation.

Term 6

Define

$$\mathcal{N}(a, b, C) := (2\pi)^{-\dim(C)/2} |C|^{-1/2} \exp\left\{-\frac{1}{2}(a-b)^\top C^{-1}(a-b)\right\}.$$

Thus,

$$-\mathbb{E}_q\left[\log q_\mu\left(\mu_1^{(1)}, \ldots, \mu_C^{(T)}\right)\right]$$

$$= -\mathbb{E}_q\left[\log \prod_{h=1}^C q_{\mu_h}\left(\mu_h^{(1)}, \ldots, \mu_h^{(T)}\right)\right]$$

$$= -\sum_{h=1}^C \mathbb{E}_q\left[\log \mathcal{N}\left(\mu_h^{(1)}, \nu, \Phi\right) \mathcal{N}\left(\alpha_h^{(1)}, \mu_h^{(1)}, \Xi_h^{(1)}\right)\right.$$

$$\left.\prod_{t=2}^T \mathcal{N}\left(\mu_h^{(t)}, \mu_h^{(t-1)}, \Phi\right) \mathcal{N}\left(\alpha_h^{(t)}, \mu_h^{(t)}, \Xi_h^{(t)}\right)\right]$$

$$= -\sum_{h=1}^C \mathbb{E}_q\left[\log \mathcal{N}\left(\mu_h^{(1)}, \nu, \Phi\right) + \log \mathcal{N}\left(\alpha_h^{(1)}, \mu_h^{(1)}, \Xi_h^{(1)}\right)\right.$$

$$\left. + \sum_{t=2}^T \log \mathcal{N}\left(\mu_h^{(t)}, \mu_h^{(t-1)}, \Phi\right) + \log \mathcal{N}\left(\alpha_h^{(t)}, \mu_h^{(t)}, \Xi_h^{(t)}\right)\right],$$

where $\alpha_h^{(t)}, \Xi_h^{(t)}$ are from the Kalman smoother. Also note our abuse of notation: ν, Φ refer to the values used to compute $\mu_h^{(t)|(T)}, P_h^{(t)|(T)}, L_h^{(t)}$ in the E-step (see Kalman smoother section), and not their current values (recall that q_μ is *not* a function of ν, Φ). Now define

$$\mathcal{Z}_N(C) := \log (2\pi)^{-\dim(C)/2} |C|^{-1/2}$$

$$\Psi(a, b, C) := \exp\left\{-\frac{1}{2}(a-b)^\top C^{-1}(a-b)\right\}$$

so we have

$$= -\left[CT\mathcal{Z}_N(\Phi) + \sum_{h=1}^C \sum_{t=1}^T +\mathcal{Z}_N\left(\Xi_h^{(t)}\right)\right]$$

$$- \sum_{h=1}^C \left[\mathbb{E}_q\left[\log \Psi\left(\mu_h^{(1)}, \nu, \Phi\right)\right] + \mathbb{E}_q\left[\log \Psi\left(\alpha_h^{(1)}, \mu_h^{(1)}, \Xi_h^{(1)}\right)\right]\right.$$

$$\left. + \sum_{t=2}^T \mathbb{E}_q\left[\log \Psi\left(\mu_h^{(t)}, \mu_h^{(t-1)}, \Phi\right)\right] + \mathbb{E}_q\left[\log \Psi\left(\alpha_h^{(t)}, \mu_h^{(t)}, \Xi_h^{(t)}\right)\right]\right],$$

where

$$\mathbb{E}_q\left[\log \Psi\left(\mu_h^{(1)}, \nu, \Phi\right)\right] = -\frac{1}{2}\mathrm{tr}\left[\Phi^{-1}\left(m_h^{(1)} - \nu\left(\hat{\mu}_h^{(1)|(T)}\right)^\top - \left(\hat{\mu}_h^{(1)|(T)}\right)\nu^\top + \nu\nu^\top\right)\right]$$

$$\mathbb{E}_q\left[\log \Psi\left(\mu_h^{(t)}, \mu_h^{(t-1)}, \Phi\right)\right] = -\frac{1}{2}\mathrm{tr}\left[\Phi^{-1}\left(m_h^{(t)} - \left(V_h^{(t,t-1)}\right)^\top - V_h^{(t,t-1)} + m_h^{(t-1)}\right)\right]$$

$$\forall t \in \{2, \ldots T\}$$

$$\mathbb{E}_q\left[\log\Psi\left(\alpha_h^{(t)},\mu_h^{(t)},\Xi_h^{(t)}\right)\right] = -\frac{1}{2}\text{tr}\left[\left(\Xi_h^{(t)}\right)^{-1}\left(\alpha_h^{(t)}\left(\alpha_h^{(t)}\right)^\top - \hat{\mu}_h^{(t)|(T)}\left(\alpha_h^{(t)}\right)^\top\right.\right.$$

$$\left.\left. -\alpha_h^{(t)}\left(\hat{\mu}_h^{(t)|(T)}\right)^\top + m_h^{(t)}\right)\right]\forall t\in\{1,\ldots,T\}$$

$$m_h^{(t)} := P_h^{(t)|(T)} + \hat{\mu}_h^{(t)|(T)}\left(\hat{\mu}_h^{(t)|(T)}\right)^\top$$

$$V_h^{(t,t-1)} := P_h^{(t)|(T)}\left(L_h^{(t-1)}\right)^\top + \hat{\mu}_h^{(t)|(T)}\left(\hat{\mu}_h^{(t-1)|(T)}\right)^\top,$$

and where $\mu_h^{(t)|(T)}, P_h^{(t)|(T)}, L_h^{(t)}$ are from the Kalman smoother section.

Term 7

Using definitions from the previous section,

$$-\mathbb{E}_q\left[\sum_{t,i=1}^{T,N}\log q_\gamma\left(\gamma_i^{(t)}\right)\right]$$

$$= -\sum_{t,i=1}^{T,N}\mathbb{E}_q\left[\log\mathcal{N}\left(\gamma_i^{(t)},\tau_i^{(t)},\Lambda_i^{(t)}\right)\right]$$

$$= -\sum_{t,i=1}^{T,N}\mathcal{Z}_N\left(\Lambda_i^{(t)}\right) - \sum_{t,i=1}^{T,N}\mathbb{E}_q\left[\log\Psi\left(\gamma_i^{(t)},\tau_i^{(t)},\Lambda_i^{(t)}\right)\right]$$

$$= -\sum_{t,i=1}^{T,N}\mathcal{Z}_N\left(\Lambda_i^{(t)}\right)$$

$$+ \sum_{t,i=1}^{T,N}\frac{1}{2}\text{tr}\left[\left(\Lambda_i^{(t)}\right)^{-1}\left(\mathbb{E}_q\left[\gamma_i^{(t)}\left(\gamma_i^{(t)}\right)^\top\right] - \tau_i^{(t)}\mathbb{E}_q\left[\gamma_i^{(t)}\right]^\top\right.\right.$$

$$\left.\left. -\mathbb{E}_q\left[\gamma_i^{(t)}\right]\left(\tau_i^{(t)}\right)^\top + \tau_i^{(t)}\left(\tau_i^{(t)}\right)^\top\right)\right]$$

$$= -\sum_{t,i=1}^{T,N}\mathcal{Z}_N\left(\Lambda_i^{(t)}\right) + \frac{TNK}{2},$$

where $\Lambda_i^{(t)}$ is from the Laplace approximation section.

Term 8

Term 8 is trivial to compute since q_c is discrete:

$$-\mathbb{E}_q\left[\sum_{t,i=1}^{T,N}\log q_c\left(c_i^{(t)}\right)\right] = -\sum_{t,i=1}^{T,N}\sum_{h=1}^{C}q_c\left(c_i^{(t)}=h\right)\log q_c\left(c_i^{(t)}=h\right).$$

Term 9

Term 9 is also trivial to compute since q_z is discrete:

$$-\mathbb{E}_q \left[\sum_{t,i=1}^{T,N} \sum_{j\neq i}^{N} \log q_z \left(z_{i\to j}^{(t)}, z_{i\leftarrow j}^{(t)} \right) \right]$$

$$= - \sum_{t,i=1}^{T,N} \sum_{j\neq i}^{N} \sum_{k,l=1}^{K,K} q_z \left(z_{i\to j}^{(t)} = k, z_{i\leftarrow j}^{(t)} = l \right) \log q_z \left(z_{i\to j}^{(t)} = k, z_{i\leftarrow j}^{(t)} = l \right).$$

References

Ahmed, A. and Xing, E. P. (2007). On tight approximate inference of logistic-normal admixture model. In *Proceedings of the 11th International Conference on Artificial Intelligence and Statistics (AISTATS 2007)*. *Journal of Machine Learning Research – Proceedings Track* 2: 16–26.

Airoldi, E. M., Blei, D. M., Fienberg, S. E., and Xing, E. P. (2008). Mixed membership stochastic blockmodels. *Journal of Machine Learning Research* 9: 1981–2014.

Blei, D. M., Ng, A. Y., and Jordan, M. I. (2003). Latent Dirichlet allocation. *Journal of Machine Learning Research* 3: 993–1022.

Cho, Y. -S., Ver Steeg, G., and Galstyan, A. (2014). Mixed membership blockmodels for dynamic networks with feedback. In Airoldi, E. M., Blei, D. M., Erosheva, E. A., and Fienberg, S. E. (eds), *Handbook of Mixed Membership Models and Its Applications*. Chapman & Hall/CRC.

Ghahramani, Z. and Beal, M. J. (2001). Propagation algorithms for variational Bayesian learning. In Leen, T. K., Dietterich, T. G., and Tresp, V. (eds) *Advances in Neural Information Processing Systems 13*. Cambridge, MA: The MIT Press, 507–513.

Ghahramani, Z. and Hinton, G. E. (2000). Variational learning for switching state-space models. *Neural Computation* 12: 831–864.

Handcock, M. S., Raftery, A. E., and Tantrum, J. M. (2007). Model-based clustering for social networks. *Journal of the Royal Statistical Society: Series A* 170: 1–22.

Heaukulani, C. and Ghahramani, Z. (2013). Dynamic probabilistic models for latent feature propagation in social networks. In *Proceedings of the 30th International Conference on Machine Learning (ICML '13)*. Omnipress, 275–283.

Hoff, P. D., Raftery, A. E., and Handcock, M. S. (2002). Latent space approaches to social network analysis. *Journal of the American Statistical Association* 97: 1090–1098.

Kolar, M., Song, L., Ahmed, A., and Xing, E. P. (2008). Estimating time-varying networks. *Annals of Applied Statistics*, to appear. http://arxiv.org/abs/0812.5087 [stat.ML].

Shetty, J. and Adibi, J. (2004). The Enron Email Dataset Database Schema and Brief Statistical Report. Tech. report, Information Sciences Institute, University of Southern California.

Soufiani, H. A. and Airoldi, E. M. (2012). Graphlet decomposition of a weighted network. In *Proceedings of the 15th International Conference on Artificial Intelligence and Statistics*. Vol. 22 of *Journal of Machine Learning Research: Workshop and Conference Proceedings*, 54–63.

Xing, E. P., Fu, W., and Song, L. (2010). A state-space mixed membership blockmodel for dynamic network tomography. *Annals of Applied Statistics* 4: 535–566.

Xing, E. P., Jordan, M. I., and Russell, S. (2003). A generalized mean field algorithm for variational inference in exponential families. In *Proceedings of the 19th Conference on Uncertainty in Artificial Intelligence (UAI 2003)*. San Francisco, CA, USA: Morgan Kaufmann Publishers Inc., 583–591.

24

Mixed Membership Blockmodels for Dynamic Networks with Feedback

Yoon-Sik Cho

Information Sciences Institute, University of Southern California, Marina del Rey, CA 90292, USA

Greg Ver Steeg

Information Sciences Institute, University of Southern California, Marina del Rey, CA 90292, USA

Aram Galstyan

Information Sciences Institute, University of Southern California, Marina del Rey, CA 90292, USA

CONTENTS

Real-world networks are inherently complex dynamical systems, where both node attributes and network topology change in time. These changes often affect each other, providing complex feedback mechanisms between node and link dynamics. Here we propose a dynamic mixed membership model of networks that explicitly take into account such feedback. In the proposed model, the probability of observing a link between two nodes depends on their current group membership vectors, while those membership vectors themselves evolve in the presence of a link between the nodes. Thus, the network is shaped by the interaction of stochastic processes describing the nodes, while the processes themselves are influenced by the changing network structure. We derive an efficient variational procedure for inference, and validate the model using both synthetic and real-world data.

24.1 Introduction

Networks are a useful paradigm for representing various social, biological, and technological systems. Modeling the structure and formation of networks is made more difficult when the nodes in the network and the topology of the network change over time. The growth of the internet and social media, in particular, has provided researchers with huge amounts of data that make such studies both feasible and highly desirable.

A standard approach to network modeling assumes a generative model for links based on node attributes. That is, the nodes or objects modeled are assumed to have some (possibly latent) attributes, e.g., group membership, and these latent properties determine the formation of links between nodes. A version of this approach which has achieved great success is the mixed membership stochastic blockmodel (MMSB) (Airoldi et al., 2008). MMSBs recognize that nodes often have multiple attributes (mixed membership) that may come into play when determining whether two nodes should be linked. Thus, MMSBs are a special case of a more general class of latent space models, which assume that nodes' attributes are described in some abstract space, and the formation of links between nodes depends on the distance between their attributes in that space (Hoff et al., 2002; Krioukov et al., 2009). In MMSBs each actor is characterized by a probability distribution over his attributes, so the corresponding latent space is a simplex (Airoldi, 2007; Blei and Fienberg, 2007).

A common limitation of these approaches is that the attributes of nodes are assumed to be unchanging over time. If the nodes represent people, for instance, we know that attributes like interests, location, or job may change over time and this may affect a person's connections to the network. In this case, it is necessary to model the dynamics of the nodes' hidden attributes as well. Despite recent progress in modeling time-varying networks (Fu et al., 2009; Ho et al., 2011; Kolar et al., 2010; Kolar and Xing, 2011; Xing et al., 2010), there are still some open problems. In particular, the existing models so far have neglected the possibility that the change in a node's attributes at one time step may depend on the network structure at previous time steps. The network structure, on there other hand, depends on the nodes' attributes, thus resulting in a feedback loop between node dynamics and network evolution.

A concrete example of this phenomena occurs in social networks. For instance, it is known that new friendship links are often formed as a result of selection effects like *homophily*: actors often befriend people with similar interests (Snijders et al., 2006). In turn, social actors introduce their friends to new ideas and interests in a process known as social influence or diffusion. Together, these dynamics cause both the nodes and the network structure to evolve simultaneously.

Our contribution is to combine a model of node dynamics that depends on network topology with an MMSB-inspired generative model for link formation that depends on changing node attributes. We use this model to describe the co-evolution of selection and influence for real-world dynamic network data. The rest of the chapter is structured as follows. We begin with a high-level description of dynamic networks and how we can adapt MMSBs to describe them, followed by a discussion of related work. In Section 24.2, we describe the details of our co-evolving mixed membership stochastic blockmodel (CMMSB), including a discussion of how to efficiently infer model parameters. In Section 24.3, we apply a CMMSB to a synthetic dataset and a real-world dataset consisting of the bill co-sponsorship network among U.S. senators. A discussion of results follows in Section 24.4. We provide detailed calculations in the Appendix.

24.1.1 Selection and Influence in Networks

Suppose we have N nodes and we observe a network structure among them at discrete time steps, $t = 0, 1, \ldots, T$. If there exists a directed link from node p to node q at time t, we say $Y_t(p, q) = 1$, otherwise 0. There are many examples of real-world data that fit this format including friendship ties in a social network and gene regulatory networks.

We suppose that the nodes themselves are described by some hidden attribute that changes over time, i.e., node p is described at time t by μ_p^t. For a social network, this vector could represent interests, group membership, or behavioral traits, while in a gene regulatory network this could indicate response to stages of a cell cycle. Then by *selection* we mean that the probability of a link between two nodes depends on their attribute vector:

$$\text{Prob}(Y_t(p, q) = 1) = g(\mu_p^t, \mu_q^t). \tag{24.1}$$

One of the most famous forms of selection is *homophily*, or *assortative mixing*, which states that nodes tend to interact with other nodes that have similar attributes. We stress, however, that different selection mechanisms are possible as well, i.e., disassortative mixing patterns such as buyer-seller relationship, etc.

The next step is to explicitly model the dynamics in the latent space. For instance, μ may drift over time, or perhaps it responds to either one-time or recurring external events. As discussed in the introduction, we are particularly interested in modeling the influence of a node's neighbors on his/her dynamics. Toward this end, we allow a feedback mechanism where an interaction at one time step affects the position of the node at the next time step. That is, we want to model dynamics of the form

$$\mu_p^{t+1} = f(\mu_p^t, \mu_{\mathcal{S}_p^t}^t), \tag{24.2}$$

where $\mathcal{S}_p^t$ denotes neighbors of node p at time t. For instance, to model positive social influence one should select a function f such that the distance between nodes contracts after the interaction. It is possible to have more general (e.g., *repulsive*) interactions as well, depending on the concrete scenario.

Together, Equations (24.1) and (24.2) provide a very high-level description of our approach. We would like to emphasize that while distance-based interactions (such as given by Equation (24.1)) are at the core of most prior work, introducing a feedback mechanism via the influence model as in Equation 24.2 is one of the main ideas distinguishing our approach from a previous attempt to formulate dynamic MMSBs in Fu et al. (2009).

Once we have specified a model for node dynamics, the task of fixing a model for link formation remains. Ideally, a generative model for link formation based on the node dynamics should capture our intuitions about real link formation while admitting some uncertainty and allowing efficient inference. For these reasons, we chose to adapt MMSBs, which we describe in the next section.

24.1.2 Mixed Membership Stochastic Blockmodels

In this paper we will use a latent space representation of the nodes based on MMSBs (Airoldi et al., 2008). In this section, we will purposely adhere to a high-level description of MMSBs and their dynamic extensions, whereas we will discuss a detailed implementation in Section 24.2. Starting with a static MMSB, we see that each node has a normalized mixed membership vector $\pi_p \in \mathbb{R}^K$, which describes the probability for node p to take one of K roles. The role that a node takes in a particular interaction is sampled according to the membership vector, and the probability of a link between p, q then depends on the roles they take and the role compatibility matrix, B. The generative process is as follows:

$$\pi_p \sim \text{Prior distribution}$$
$$\mathbf{z}_{p \to q} \sim \text{Multinomial}(\pi_p)$$
$$\mathbf{z}_{q \to p} \sim \text{Multinomial}(\pi_q)$$
$$Y(p, q) \sim \text{Bernoulli}(\mathbf{z}_{p \to q}^\top B \mathbf{z}_{p \leftarrow q}).$$

The most naive dynamic extension is to simply add a t index to all the variables in the previous expression. This amounts to learning T independent, static MMSBs and fails to take into account any of our knowledge of the underlying node dynamics. An extension considered in Fu et al. (2009) is to say that the prior distribution for the π^t should evolve over time. However, each mixed membership vector is still sampled from the same distribution at each time, so the effect is to model only aggregate dynamics.

In contrast, and as discussed in the previous section, we would prefer that the mixed membership vector of nodes evolved individually but under mutual influence. The particular form of influence we will study is

$$\mu_p^{t+1} = (1 - \beta_p)\mu_p^t + \beta_p \mu_{avg}^t + \text{noise term}, \qquad (24.3)$$

where $\mu_{avg}^t = \frac{1}{|S_p^t|} \sum_{q \in S_p^t} w_{p \leftarrow q} \mu_q^t$ is the weighted average of node q's neighbors' log-membership vectors. Thus, the membership vector of node q at time $t+1$ is a weighted average of his membership vector at time t as well as the membership vectors of the nodes he has interacted with at time t. This feature of our model has the desired effect of incorporating feedback between network structure and individual node dynamics. The relative importance of the neighbors is captured by the parameter $0 < \beta_p < 1$; larger β_p means that node p is more susceptible to influence from his neighbors.

Before proceeding further, we note that exact inference is not feasible even for static MMSBs, so adding dynamics to a model makes the inference problem much harder. Here we use a variational EM approach that allows us to do efficient approximate inference (Beal and Ghahramani, 2003; Xing et al., 2003).

24.1.3   Related Work

The problem of properly characterizing selection and influence has been a subject of extensive studies in sociology. For instance, Steglich et al. (2010) suggested a continuous time agent-based model of network co-evolution. In this model, each agent is characterized by a certain utility function that depends on the agent's individual attributes as well as his/her local neighborhood in the network. The agents evolve as continuous-time Markovian processes which, at randomly chosen time points, select an action to maximize their utility. Despite its intuitive appeal, a serious shortcoming of this model is that it cannot handle missing data well, thus most of the attributes have to be fully observable. This was addressed in Fan and Shelton (2009), where a continuous dynamic Bayesian approach was developed. Continuous-time models have certain advantages when the network observations are infrequent and well-separated in time. In situations where more fine-grained data is available, however, discrete-time models are more suitable (Hanneke et al., 2010).

The model represented here is based on MMSBs (Airoldi et al., 2008). MMSBs are an extension of stochastic blockmodels that have been studied extensively both in social sciences and in computer science (Holland et al., 1983; Goldenberg et al., 2010). In a stochastic blockmodel each node is assigned to a block (or a role), and the pattern of interactions between different nodes depends only on their block assignment. Many situations, however, are better described by multi-faceted interactions, where nodes can bear multiple latent roles that influence their relationships to others. MMSBs account for such "mixed" interactions by allowing each node to have a probability distribution over roles and by making the interactions role-dependent (Airoldi et al., 2008). A different approach to mixed membership community detection has been developed in physics (Ball et al., 2011; Ahn et al., 2010). In particular, Ahn et al. (2010) suggested a definition of communities in terms of links rather than nodes.

Previously, several works have considered a dynamic extension of the MMSB which we will henceforth refer to as dMMSB (Fu et al., 2009; Ho et al., 2011; Xing et al., 2010). In contrast to dMMSB, where the dynamics were imposed *externally*, our model assumes that the membership evolution is driven by the interactions between the nodes through a parametrized *influence* mechanism. At the same time, the patterns of those interactions themselves change due to the evolution of the node memberships. An advantage of the present model over dMMSB is that the latter models the aggregate dynamics, e.g., the *mean* of the logistic normal distribution from which the membership vectors are sampled. CMMSB, however, models each node's trajectory separately, thus providing better flexibility for describing system dynamics. Of course, more flexibility comes at a higher computational cost, as CMMSBs track the trajectories of all nodes individually. This additional cost, however, can be well justified in scenarios when the system as a whole is almost static (e.g., no shift in the mean membership vector), but different subsystems experience dynamic changes. One such scenario that deals with political polarization in the U.S. Senate is presented in our experimental results section.

24.2 Co-evolving Mixed Membership Blockmodel

Consider a set of N nodes, each of which can have K different roles, and let π_p^t be the mixed membership vector of node p at time t. Let Y_t be the network formed by those nodes at time t: $Y_t(p,q) = 1$ if the nodes p and q are connected at time t, and $Y_t(p,q) = 0$ otherwise. Further, let $Y_{0:T} = \{Y_0, Y_1, \ldots, Y_T\}$ be a time sequence of such networks. The generative process that induces this sequence is described below.

- For each node p at time $t = 0$, employ a logistic normal distribution[1] to sample an initial membership vector,

$$\pi_{p,k}^0 = \exp(\mu_{p,k}^0 - C(\mu_p^0)), \quad \mu_p^0 \sim \mathcal{N}(\alpha^0, A),$$

where $C(\mu) = \log(\sum_k \exp(\mu_k))$ is a normalization constant, and α^0, A are the prior mean and covariance matrix.

- For each node p at time $t > 0$, the mean of each normal distribution is updated due to *influence* from the neighbors at its previous step:

$$\alpha_p^t = (1 - \beta_p)\mu_p^{t-1} + \beta_p \mu_{\mathcal{S}_p^{t-1}},$$

where $\mu_{\mathcal{S}_p^t}$ is the average of the weighted membership vector μ-s of the nodes which node p is connected to at time t:

$$\mu_{\mathcal{S}_p^t} = \frac{1}{|\mathcal{S}_p^t|} \sum_{q \in \mathcal{S}_p^t} w_{p \leftarrow q}^t \mu_q.$$

β_p describes how easily the node p is influenced by its neighbors, while the weights, w, allow for different degrees of influence from different neighbors. The membership vector at time t is

$$\pi_{p,k}^t = \exp(\mu_{p,k}^t - C(\mu_p^t)), \quad \mu_p^t \sim \mathcal{N}(\alpha_p^t, \Sigma_\mu),$$

where the covariance Σ_μ accounts for noise in the evolution process.

- For each pair of nodes p, q at time t, sample role indicator vectors from multinomial distributions:

$$\mathbf{z}_{p \to q}^t \sim \text{Multinomial}(\pi_p^t), \mathbf{z}_{p \leftarrow q}^t \sim \text{Multinomial}(\pi_q^t).$$

Here $\mathbf{z}_{p \to q}$ is a unit indicator vector of dimension K, so that $z_{p \to q,k} = 1$ means node p undertakes role k while interacting with q.

---

[1]We found that the logistic normal form of the membership vector suggested in Fu et al. (2009) leads to more tractable equations compared to the Dirichlet distribution used for static MMSBs.

- Sample a link between p and q as a Bernoulli trial:

$$Y_t(p,q) \sim \text{Bernoulli}((1-\rho)\mathbf{z}_{p\to q}^{t\top}B^t\mathbf{z}_{p\gets q}^t),$$

where B is a $K \times K$ role compatibility matrix, so that B_{rs}^t describes the likelihood of interaction between two nodes in roles r and s at time t. When B^t is diagonal, the only possible interactions are among the nodes in the same role. Here ρ is a parameter that accounts for the sparsity of the network (Airoldi et al., 2008).

Thus, the coupling between dynamics of different nodes is introduced by allowing the role vector of a node to be influenced by the role vectors of its neighbors. To benefit from computational simplicity, we updated π by changing its associated μ. This update of μ is a linear combination of its current state and the values of its neighbors' current states. The influence is measured by a node-specific parameter β_p, and $w_{p\gets q}^t$, which need to be estimated from the data. β_p describes how easily the node p is influenced by its neighbors: $\beta_p = 0$ means it is not influenced at all, whereas $\beta_p = 1$ means the behavior is solely determined by the neighbors. On the other hand, $w_{p\gets q}^t$ reflects the weight of the specific influence that node q exerts on node p, so that larger values correspond to more influence.

24.2.1   Inference

Under the CEMMSB, the joint probability of the data $Y_{0:T}$ and the latent variables $\{\mu_{1:N}^t, \mathbf{z}_{p\to q}^t : p, q \in N, \mathbf{z}_{p\gets q}^t : p, q \in N\}$ can be written in the following factored form. To simplify the notation, we define $\mathbf{z}_{p,q}^t$ as a pair of $\mathbf{z}_{p\to q}^t$ and $\mathbf{z}_{p\gets q}^t$. Also, denote the sets of latent group indicators $\{\mathbf{z}_{p\to q}^t : p, q \in N\}$, and $\{\mathbf{z}_{p\gets q}^t : p, q \in N\}$ as $\mathbf{Z}_{\to}^t$, and $\mathbf{Z}_{\gets}^t$.

$$p(Y_{0:T}, \mu_{1:N}^{0:T}, \mathbf{Z}_{\to}^{0:T}, \mathbf{Z}_{\gets}^{0:T} | \alpha, A, B, \beta_p, w_{p\gets q}^t, \Sigma_\mu) = \tag{24.4}$$

$$\prod_t \prod_{p,q} P(Y_t(p,q) | \mathbf{z}_{p,q}^t, B^t) P(\mathbf{z}_{p,q}^t | \mu_p^t, \mu_q^t)$$

$$\times \prod_p P(\mu_p^0 | \alpha^0, A) \prod_{t\neq 0} P(\mu_p^t | \mu_p^{t-1}, \mu_{\mathcal{S}_p^{t-1}}, \Sigma_\mu, \beta_p).$$

In Equation (24.4), the term describing the dynamics of the membership vector is defined as follows:[2]

$$P(\mu_p^t | \mu_p^{t-1}, \mu_{\mathcal{S}_p^{t-1}}, \Sigma_\mu, Y_t, \beta_p) = f_G(\mu_p^t - f_b(\mu_p^{t-1}, \mu_{\mathcal{S}_p^{t-1}}), \Sigma_\mu) \tag{24.5}$$

$$f_G(\mathbf{x}, \Sigma_\mu) = \frac{1}{(2\pi)^{k/2}|\Sigma_\mu|^{1/2}} e^{-\frac{1}{2}x^T\Sigma_\mu^{-1}x}$$

$$f_b(\mu_p^{t-1}, \mu_{\mathcal{S}_p^{t-1}}) = (1-\beta_p)\mu_p^{t-1} + \beta_p\mu_{\mathcal{S}_p^{t-1}}.$$

As we already mentioned, performing exact inference with this model is not feasible. Thus, one needs to resort to approximate techniques. Here we use a variational EM approach (Beal and Ghahramani, 2003; Xing et al., 2003). The main idea behind variational methods is to posit a simpler distribution $q(X)$ over the latent variables with free (variational) parameters, and then fit those parameters so that the distribution is close to the true posterior in KL divergence.

$$D_{KL}(q||p) = \int_X q(X) \log \frac{q(X)}{p(X,Y)} dX. \tag{24.6}$$

Here we introduce the following factorized variational distribution:

---

[2]For simplicity, we will assume Σ_μ is a diagonal matrix.

$$q(\mu_{1:N}^{0:T}, Z_{\rightarrow}^{0:T}, Z_{\leftarrow}^{0:T} | \gamma_{1:N}^{0:T}, \Phi_{\rightarrow}^{0:T}, \Phi_{\leftarrow}^{0:T}) = \tag{24.7}$$
$$\prod_{p,t} q_1(\mu_p^t | \gamma_p^t, \Sigma_p^t) \prod_{p,q,t} (q_2(\mathbf{z}_{p \rightarrow q}^t | \phi_{p \rightarrow q}^t) q_2(\mathbf{z}_{p \leftarrow q}^t | \phi_{p \leftarrow q}^t)),$$

where q_1 is the normal distribution, and q_2 is the multinomial distribution, and $\gamma_p^t, \Sigma_p^t, \phi_{p \rightarrow q}^t, \phi_{p \leftarrow q}^t$ are the variational parameters. Intuitively, $\phi_{p \rightarrow q,g}^t$ is the probability of node p undertaking the role g in an interaction with node q at time t, and $\phi_{p \leftarrow q,h}^t$ is defined similarly.

For this choice of the variational distribution, we rewrite Equation (24.6) as follows:

$$D_{KL}(q||p) = \tag{24.8}$$
$$E_q[\log \prod_t \prod_p q_1(\mu_p^t | \gamma_p^t, \Sigma_p^t)] + E_q[\log \prod_t \prod_{p,q} q_2(\mathbf{z}_{p \rightarrow q}^t | \phi_{p \rightarrow q}^t)$$
$$+ E_q[\log \prod_t \prod_{p,q} q_2(\mathbf{z}_{p \leftarrow q}^t | \phi_{p \leftarrow q}^t) - E_q[\log \prod_t \prod_{p,q} P(Y_t(p,q) | \mathbf{z}_{p \rightarrow q}^t, \mathbf{z}_{p \leftarrow q}^t, B)]$$
$$- E_q[\log \prod_t \prod_{p,q} P(\mathbf{z}_{p \rightarrow q}^t | \mu_p^t) - E_q[\log \prod_t \prod_{p,q} P(\mathbf{z}_{p \leftarrow q}^t | \mu_q^t)$$
$$- E_q[\log \prod_{t \neq 0} \prod_p P(\mu_p^t | \mu_p^{t-1}, \mu_{\mathcal{S}_p^{t-1}}, \Sigma_\mu)] - E_q[\log \prod_p P(\mu_p^0 | \alpha^0, A)].$$

In the third line of the above equation, we need to compute the expected value of $\log[\sum_k \exp(\mu_k)]$ under the variational distribution, which is problematic. Toward this end, we introduce N additional variational parameters ζ, and replace the expectation of the log by its upper bound induced from the first-order Taylor expansion (Blei and Lafferty, 2007):

$$\log[\sum \exp(\mu_k)] \leq \log \zeta - 1 + \frac{1}{\zeta} \sum \exp(\mu_k). \tag{24.9}$$

The variational EM algorithm works by iterating between the E-step of calculating the expectation value using the variational distribution, and the M-step of updating the model (hyper)parameters so that the data likelihood is locally maximized. The pseudo-code is shown in Algorithm 1, and the details of the calculations are discussed below.

24.2.2 Variational E-step

In the variational E-step, we minimize the KL distance over the variational parameters. Taking the derivative of KL divergence with respect to each variational parameter and setting it to zero, we obtain a set of equations that can be solved via iterative or other numerical techniques. For instance, the variational parameters $(\phi_{p \rightarrow q}^t, \phi_{p \leftarrow q}^t)$ corresponding to a pair of nodes (p, q) at time t, can be found via the following iterative scheme:

$$\phi_{p \rightarrow q,g}^t \propto \exp(\gamma_{p,g}^t) \prod_h (B(g,h)^{Y_t(p,q)}(1 - B(g,h))^{1-Y_t(p,q)})^{\phi_{p \leftarrow q,h}^t}. \tag{24.10}$$

$$\phi_{p \leftarrow q,h}^t \propto \exp(\gamma_{q,h}^t) \prod_g (B(g,h)^{Y_t(p,q)}(1 - B(g,h))^{1-Y_t(p,q)})^{\phi_{p \rightarrow q,h}^t}. \tag{24.11}$$

In the above equations, $\phi_{p \rightarrow q,g}^t$ and $\phi_{p \leftarrow q,h}^t$ are normalized after each update. Note also that Equations (24.10) and (24.11) are coupled with each other as well as with the parameters $\gamma_{p,g}^t, \gamma_{q,h}^t$.

Sets of variational parameters, $\{\gamma\}^t$ and $\{\sigma\}^t$, are initialized at the beginning of variational EM.

Input: data $Y_t(p, q)$, size N, T, K
Initialize all $\{\gamma\}^t, \{\sigma\}^t$
Start with an initial guess for the model parameters.
repeat
  repeat
    for $t = 0$ **to** T **do**
      repeat
        Initialize $\phi^t_{p \to q}, \phi^t_{p \leftarrow q}$ to $\frac{1}{K}$ for all g, h
        repeat
          Update all $\{\phi\}^t$
        until convergence of $\{\phi\}^t$
        Find $\{\gamma\}^t, \{\sigma\}^t$
        Update all $\{\zeta\}^t$
      until convergence in time t
    end for
  until convergence across all time steps
  Update hyperparameters.
until convergence in hyperparameters

Algorithm 1: Variational EM.

For $\{\gamma\}^t$, we sample it from normal distribution $\mathcal{N}(\alpha^0, A)$, and for $\{\sigma\}^t$ we initialize it to the same value over all nodes across the whole time steps. Once the $\{\phi\}^t$ are converged to optimal points, we then update $\{\gamma\}^t$ and $\{\sigma\}^t$ using the update equations. Both of the variational parameters do not have closed forms of solution, and the details are given in the KL-Distance section of the Appendix. Here we simply note that their general form is:

$$\gamma^t_p = f_\gamma(\gamma^{t-1}_p, \gamma^{t+1}_p, \gamma^t_q, \phi^t_{p \to q}, \phi^t_{q \leftarrow p}, \zeta^t_p, \Sigma^t_p). \tag{24.12}$$

Thus, the parameter γ^t_p depends on its *immediate* past and future values, γ^{t-1}_p and γ^{t+1}_p, as well as the parameters of its neighbors.

For the variational parameters of a covariance matrix Σ^t_p, which is assumed to be a diagonal matrix with components $((\sigma^t_{p,1})^2, (\sigma^t_{p,2})^2, ...(\sigma^t_{p,k})^2)$, the general form of the optimal point is :

$$\sigma^t_{p,k} = f_\sigma(\gamma^t_{p,k}, \zeta^t_p). \tag{24.13}$$

Finally, for the variational parameters ζ we have

$$\zeta^t_p = \sum_i \exp(\gamma^t_{p,i} + \frac{(\sigma^t_{p,i})^2}{2}). \tag{24.14}$$

Note that the above equations can be solved via a simple iterative update as before. To expedite convergence, however, we combine the iterations with the Newton-Raphson method, where we solve for individual parameters while keeping the others fixed, and then repeat this process until all the parameters have converged.

24.2.3 Variational M-step

The M-step in the EM algorithm computes the parameters by maximizing the expected log-likelihood found in the E-step. The model parameters in our case are: B^t, the role compatibility matrix, the covariance matrix Σ_μ, β_p for each node, $w^t_{p \leftarrow q}$ for each pair, α, and A from the prior.

If we assume that the time variation of the block compatibility matrix is small compared to the evolution of the node attributes, we can neglect the time dependence in B and use its average across time, which yields:

$$\hat{B}(g,h) = \frac{\sum_{p,q,t} Y_t(p,q) \cdot \phi^t_{p\to q,g} \phi^t_{p\leftarrow q,h}}{\sum_{p,q,t} \phi^t_{p\to q,g} \phi^t_{p\leftarrow q,h}}. \tag{24.15}$$

Likewise, for the update of diagonal components of the noise covariance matrix Σ_μ,

$$(\hat{\eta}_k)^2 = \frac{1}{N(T-1)} E_q [\sum_{p,t} (\mu^t_{p,k} - (1-\beta)\mu^{t-1}_{p,k} - \beta\mu_{\mathcal{S}^{t-1}_p,k})^2]. \tag{24.16}$$

Similar equations are obtained for β_p and $w^t_{p\leftarrow q}$. The update equation of β_p and $w^t_{p\leftarrow q}$ is a function of γ and σ, which are related to the transition for specific node p.

$$\beta_p = \frac{\sum_{t>0}\sum_k (\gamma^{t-1^2}_{p,k} + \sigma^{t-1^2}_k - \gamma^t_{p,k}\gamma^{t-1}_{p,k} - \gamma^{t-1}_{p,k}\gamma_{\mathcal{S}^{t-1}_p,k} + \gamma^t_{p,k}\gamma_{\mathcal{S}^{t-1}_p,k})}{\sum_{t>0}\sum_k (\gamma^{t-1^2}_{p,k} + \sigma^{t-1^2}_k - 2\gamma^{t-1}_{p,k}\gamma_{\mathcal{S}^{t-1}_p,k}) + \sum_{t>0}\sum_k (\gamma^2_{\mathcal{S}^t_p,k} + \sigma^2_{\mathcal{S}^t_p,k})},$$

where $\gamma_{\mathcal{S}^t_p}$ and $\Sigma_{\mathcal{S}^t_p}$ are the mean and covariance of a set of nodes which node p is connected to at time t.

The priors of the model can be expressed in closed form as below:

$$\alpha^0 = \frac{1}{N}\sum_p \gamma^0_p. \tag{24.17}$$

$$a_k = \sqrt{\frac{1}{N}\sum((\gamma^0_{p,k})^2 + (\sigma^0_{p,k})^2 - 2\alpha^0_k\gamma^0_{p,k} + (\alpha^0_k)^2)}. \tag{24.18}$$

24.3 Results

24.3.1 Experiments on Synthetic Data

We tested our model by generating a sequence of networks according to the process described above, for 50 nodes, and $K = 3$ latent roles across $T = 8$ time steps. We used a covariance matrix of $A = 3I$, and mean α^0 having homogeneous values for the prior, so that initially nodes had a well-defined role (i.e., the membership vector would have peaked around a single role). More precisely, the majority of nodes had around 90% of membership probability mass centered at a specific role, and on average a third of those nodes had 90% on role k. For the role compatibility matrix, we gave high weight at the diagonal.

Starting from some initial parameter estimates, we performed variational EM and obtained re-estimated parameters which were very close to the original values (ground truth). With those learned parameters, we inferred the hidden trajectory of agents as given by their mixed membership vector for each time step. The results are shown in Figure 24.1, where, for three nodes, we plot the projection of trajectories onto the simplex. One can see that for all three nodes, the inferred trajectories are very close to the actual ones.

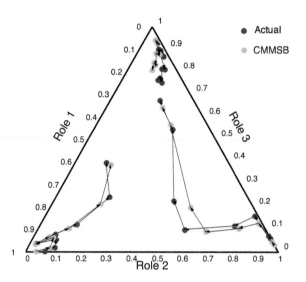

FIGURE 24.1
Actual and inferred mixed membership trajectories on a simplex.

24.3.2 Comparison with dMMSB

As a further verification of our results, we compare the performance of our inference method to the dynamic mixed membership stochastic blockmodel (dMMSB) (Fu et al., 2009). We use synthetic data generated in a manner similar to the previous section. This time, though, for simplicity we keep $K = 2$ and set all the βs to some constant for all the nodes: $\beta = 0.1$ in one trial and $\beta = 0.2$ in the other. In this case, we compare performance by evaluating the distance in L_2 norm between actual and inferred mixed membership vectors for each method. At each time step, we calculate the average over all nodes of the L_2 distance from the actual membership vector.

As we show in Figures 24.2(a) and 24.2(b), CMMSBs capture the dynamics better than the dMMSBs. This is due to the fact that our model tracks all of the nodes individually (internal dynamics), while dMMSBs regard the dynamism as an evolution of the environment (external dynamics). Here, we have only included results for relatively small and homogeneous dynamics. In fact, we noticed that our method tends to fare even better as we increase the degree of dynamics or the heterogeneity of dynamics across nodes (node-varying values of β). We believe heterogeneous dynamics are more prevalent in real systems, and so we expect our method to outperform dMMSB even more than is indicated by Figure 24.2(b).

24.3.3 U. S. Senate Co-Sponsorship Network

We have also performed some preliminary experiments for testing our model against real-world data. In particular, we used senate co-sponsorship networks from the 97th to the 104th Senate, by considering each senate as a separate time point in the dynamics. There were 43 senators who remained part of the senate during this period. For any pair of senators (p, q) in a given senate, we generated a directed link $p \rightarrow q$ if p co-sponsored at least three bills that q originally sponsored. The threshold of three bills was chosen to avoid having too dense of a network. With this data, we wanted to test (a) to what extent senators tend to follow others who share their political views (i.e., conservative vs. liberal) and (b) whether some senators change their political creed more easily than others.

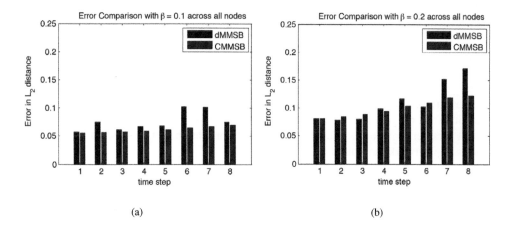

FIGURE 24.2
Inference error for dMMSB and CMMSB for synthetic data generated with $K = 2$ and $\beta = 0.1$ for all the nodes (a), and when $\beta = 0.2$ for all the nodes (b).

The number of roles $K = 2$ was chosen to reflect the mostly bi-polar nature of the U.S. Senate. The susceptibility of senator p to influence is measured by the corresponding parameter β_p, which is learned using the EM algorithm. High β means that a senator tends to change his/her role more easily. Likewise, the power of influence of senator q on senator p is measured by the parameter $w_{p \leftarrow q}^t$, where $w_{p \leftarrow q_1}^t > w_{p \leftarrow q_2}^t$ means senator q_1 is more influential on senator p than senator q_2. Here the direction of the arrow reflects the direction of the influence which is opposite to the direction of the link. To initialize the EM procedure, we assigned the same β and w to all the senators, and start with a matrix which is weighted at the diagonal for B.

Another method for validation is to compare the degree of influence. Our model handles and learns the degree of influence in the update equation. Sorting out influential senators is an area of active research. Recently, KNOWLEGIS has been ranking U.S. senators based on various criteria, including influence, since 2005. Since our data was extracted from the 97th Senate to the 104th Senate, direct comparison of the rankings was impossible. Another study (Maisel, 2010) ranked the 10 most influential senators in both parties who have been elected since 1955. We compared our top five influential senators, and were able to find three senators (Senator Robert Byrd, Senator Strom Thurmond, and Senator Bob Dole) on the list.

24.3.4 Interpreting Results

The role compatibility matrix learned from the variational EM has high values on the diagonal confirming our intuition that interaction is indeed more likely between senators that share the same role. Furthermore, the learned values of β showed that senators varied in their "susceptibility." In particular, Senator Arlen Spector was found to be the *most influenceable* one, while Sen. Dole was found to be one of the most *inert* ones. Note that while there are no direct ways of estimating the "dynamism" of senators, our results seem to agree with our intuition about both senators (e.g., Sen. Spector switched parties in 2009 while Sen. Dole became his party's candidate for President in 1996).

To get some independent verification, we compared our results to the yearly ratings that the ACU (American Conservative Union) and ADA (Americans for Democratic Action) assign to senators.[3]

---

[3] Accessible at http://www.conservative.org/, http://www.adaction.org/.

ACU/ADA rated every senator based on selected votes which they believe display a clear ideological distinction, so that high scores in the ACU mean that they are truly conservative, while lower scores in the ACU suggest they are liberal, and for the ADA vice versa. To compare the ratings with our predictions (given by the membership vector) we scaled the former to get scores in the range $[0, 1]$.

Figure 24.3 shows the relationship between these scores and our mixed membership vector score, confirming our interpretation of the two roles in our model as corresponding to liberal/conservative. Although these values cannot be used for quantitative agreement, we found that at least qualitatively, the inferred trajectories agree reasonably well with the ACU/ADA ratings. This agreement is rather remarkable since the ACU/ADA scores are based on selected votes rather than co-sponsorship network as in our data.

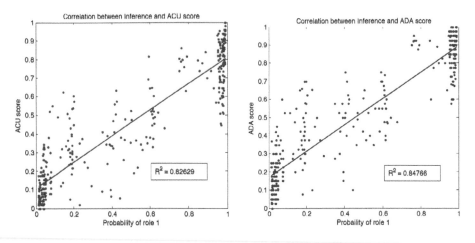

FIGURE 24.3
Correlation between ACU/ADA scores and inferred probabilities.

Of course, we are most interested in correctly identifying the dynamics for each senator. We compare our inferred trajectory of the most dynamic senator, and the inert senator to the scores of the ACU and ADA. In Figure 24.4 the scores of the ADA have been flipped, so that we can compare all of the scores in the same measurement. However, since ACU/ADA scores are rated for every senator each year, the dynamics of inference and the dynamics of ACU/ADA scores cannot be compared one to one. Not all senators showed high correlation of the trend like Sen. Specter and Sen. Dole.

24.3.5   Polarization Dynamics

The yearly ACU/ADA scores give a good comparison of the relative political position of senators scored in each year. However, they are not very appropriate for comparison between years, a point illustrated by the fact that the score is based on voting records for different bills in each year. Therefore, for validation of the dynamics we turn to another scoring system highly regarded by political scientists and used to observe historical trends, the DW-NOMINATE score. For the time period of our study, McCarty et al. (2006) shows that the political polarization of the senate was increasing. In particular, they show that the gap between the average DW-NOMINATE score of Republicans and Democrats is monotonically increasing, as we show in Figure 24.5. In fact, the polarization for the entire senate was stronger every year. This is due to the unbalanced seats in the entire senate. In other words, our data had 22 Republicans and 21 Democrats, while for the entire senate, majority outnumbered minority by around 10 seats. For comparison, for each time step we took the average of our inferred score for the 14 most and least conservative senators. As we show

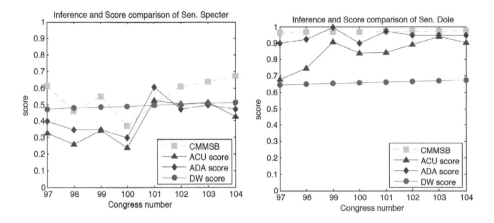

FIGURE 24.4

Comparison of inference results with ACU and ADA scores: Sen. Specter (left) and Sen. Dole (right).

in Figure 24.5, our inferred result agrees qualitatively with the results of McCarty et al. (2006), showing an increase in polarization for every senate in the studied time-window. Since the DW-NOMINATE score uses its own metric, and our polarization is measured by the difference between upper average and lower average probability, we should not expect to get quantitative agreement. We would like to highlight, however, that the direction of the trend is correctly predicted for each of the eight terms.

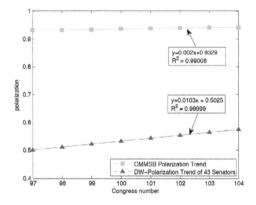

FIGURE 24.5

Polarization trends during the 97th–104th U. S. Congresses.

24.4 Discussion

We have presented the CMMSB for modeling inter-coupled node and link dynamics in networks. We used a variational EM approach for learning and inference with CMMSB, and were able to reproduce the hidden dynamics for synthetically generated data, both qualitatively and quantitatively. We also tested our model using the U.S. Senate bill co-sponsorship data, and obtained reasonable results in our experiments. In particular, CMMSBs were able to detect increasing polarization in the senate as reported by other sources that analyze individual voting records of the senators.

Our results with the U.S. Senate dataset suggest that our dynamical model can actually capture some nuances of individual dynamics. While we lack a ground truth for the true position of senators, third party analyses qualitatively support the findings of our model. Of course, many factors are not explicitly modeled in our approach, but we hope that by including individual dynamical terms we capture these effects implicitly. For instance, external events like upcoming re-election campaigns surely affect senator's actions. While the true chain of events may rely on these events, if all relevant external events are not or cannot be included in our model, then capturing dynamics through shifts in observed relationships is a good proxy.

The approach to modeling influence described in Section 24.2 is only one of several possibilities. Although we learned a static parameter β for each node, describing how easily influenced they are, we also pointed out the possibility of adding a weight that varies for each pair: that is, a node may be more influenced by one person than another. Additionally, someone's influence may change over time. Finally, we chose a simple linear influence mechanism. In principle, someone may be more influential along one axis than another. For instance, a node may be influenced by a friend's musical taste, but not by his politics.

As future work, we intend to test our model against different real-world data, such as communication networks or co-authorship networks of publications. We also plan to extend CMMSBs in several ways. A significant bottleneck of the current model is that it explicitly considers links between all the pairs of nodes, resulting in a quadratic complexity in the network size. Most real-world networks, however, are sparse, which is not accounted for in the current approach. Introducing sparsity into the model would greatly enhance its efficiency. We note that this is also a drawback for static MMSBs, but progress has already been made towards reducing this complexity (Mørup et al., 2011).

An additional drawback of of MMSBs (and stochastic blockmodels in general) is the inability to properly deal with degree heterogeneity. Indeed, MMSBs (or related latent space models) might assign nodes to the same group based merely on the frequency of their interactions with the other nodes. Possible remedies are found in the degree-correct blockmodel recently proposed in Karrer and Newman (2011) or in exponential random graph models that separately model node and group variability (Reichardt et al., 2011). The problem reveals a fundamental ambiguity about network modeling. A priori, we have no reason to believe that node connectivity is a less important dimension for clustering nodes than homophily for some hidden attribute. Our intuition leads us to expect otherwise for human networks, but this intuition must be explicitly modeled. In the co-sponsorship network studied here, most senators are well-connected and so the network structure is better explained by political views than node connectivity. However, large variability in node connectivity has been observed in many social networks where this effect will have to be explicitly modeled.

Acknowledgments

This research was supported in part by the National Science Foundation under grant No. 0916534, U.S. ARO MURI grant No. W911NF0610094, and US AFOSR MURI grant No. FA9550-10-1-0569.

Appendix

Alternative View of EM Algorithm

We start with the log-likelihood function where Y is the data, X is the set of latent variables, and Θ is the set of model parameters:

$$\log p(Y|\Theta) = \log \int p(Y, X|\Theta)dX \tag{24.19}$$

$$= \log \int q(X)\frac{p(Y, X|\Theta)}{q(X)}dX$$

$$\geq \int q(X) \log \frac{p(Y, X|\Theta)}{q(X)}dX \text{ (Jensen's Inequality)}.$$

We define the lower bound as free energy:

$$F(q, \Theta) = \int q(X) \log \frac{p(X|Y, \Theta)p(Y|\Theta)}{q(X)}dX \tag{24.20}$$

$$= \log P(Y|\Theta) - D_{KL}(q(X)||p(Y, X)).$$

The goal is to maximize the lower bound (free energy) by updating q and Θ. In E-step, we minimize the KL-distance of two distributions, and in M-step, we maximize the free energy under fixed q distribution obtained in E-step.

KL-Distance

Here we present the KL-distance between $q(X)$, and $p(Y, X)$:

$$D_{KL}(q||p) = \tag{24.21}$$

$$\sum_t \sum_p \left(-\frac{1}{2}E_q[(\mu_p^t - \gamma_p^t)^T(\Sigma_p^t)^{-1}(\mu_p^t - \gamma_p^t)] - \log(2\pi)^{k/2} - \log(|\Sigma_p^t|^{1/2}) \right)$$

$$+ \sum_t \sum_{p,q} \sum_g \phi_{p \to q,g}^t \log \phi_{p \to q,g}^t + \sum_t \sum_{p,q} \sum_h \phi_{p \leftarrow q,h}^t \log \phi_{p \leftarrow q,h}^t$$

$$- \sum_t \sum_{p,q} \sum_{g,h} \phi_{p \to q,g}^t \phi_{p \leftarrow q,h}^t f(Y_t(p, q), B(g, h))$$

$$- \sum_t \sum_{p,q} \sum_g \phi_{p \to q,g}^t \left(\gamma_{p,g}^t - \log \zeta_p^t + 1 - \frac{1}{\zeta_p^t} \sum_k \exp(\gamma_{p,k}^t + \frac{\sigma_{p,k}^{t\,2}}{2}) \right)$$

$$- \sum_t \sum_{p,q} \sum_h \phi_{p \leftarrow q,h}^t \left(\gamma_{q,h}^t - \log \zeta_q^t + 1 - \frac{1}{\zeta_q^t} \sum_k \exp(\gamma_{q,k}^t + \frac{\sigma_{q,k}^{t\,2}}{2}) \right)$$

$$- \sum_{t>0} \sum_p \left(-\frac{1}{2}E_q \left[(\mu_p^t - f_b(\mu_p^{t-1}, \mu_{\mathcal{S}_p^{t-1}}))^T \Sigma_\mu^{-1} \right. \right.$$

$$\left. \left. \cdot(\mu_p^t - f_b(\mu_p^{t-1}, \mu_{\mathcal{S}_p^{t-1}})) \right] - (k/2) \log(2\pi) - \log(|\Sigma_\mu|^{1/2}) \right)$$

$$- \sum_p \left(-\frac{1}{2}E_q[(\mu_p^0 - \alpha^0)^T A^{-1}(\mu_p^0 - \alpha^0)] - \log(2\pi)^{k/2} - \log(|A|^{1/2}) \right),$$

where $\exp(\gamma_{p,k} + \frac{\sigma_{p,k}^{t}{}^2}{2})$ comes from the moment-generating function of the normal distribution, $M_X(t) := E[e^{tX}]$, with t=1. The first line simplifies to const $- \sum_{t,p,k} \log \sigma_{p,k}^t$, where, once again, we have taken the covariance matrix to be diagonal.

Variational E-step

In the variational E-step, we minimize the KL distance over the variational parameters. Variational parameters $\{\gamma_p\}^t$ and $\{\sigma_{p,k}\}^t$ need to be solved analytically. We use the Newton-Raphson method as an optimization algorithm for tightening the bound with respect to those variational parameters.

First, we minimize the divergence with respect to γ_p^t. Since the other variational parameters Σ_p^t are assumed to be a diagonal matrix, we treat the multivariate normal distribution as a combination of independent normal distribution and update the mean and variance for each coordinate. We use the Newton-Raphson method for each coordinate where the derivative is :

$$dD_{KL}(q||p)/d\gamma_{p,k}^t = \tag{24.22}$$

$$-\sum_q \phi_{p\to q,k}^t + \sum_q \sum_g \phi_{p\to q,g}^t \frac{1}{\zeta_p^t} \exp(\gamma_{p,k}^t + \frac{(\sigma_{p,k}^t)^2}{2})$$

$$-\sum_q \phi_{q\leftarrow p,k}^t + \sum_q \sum_h \phi_{q\leftarrow p,h}^t \frac{1}{\zeta_p^t} \exp(\gamma_{p,k}^t + \frac{(\sigma_{p,k}^t)^2}{2})$$

$$+[\frac{\gamma_{p,k}^t}{|\eta_k|^2} + (1-\beta_p)^2 \frac{\gamma_{p,k}^t}{|\eta_k|^2} + \sum_{q\in\mathcal{S}_p^t} \beta_q^2 \frac{\gamma_{\mathcal{S}_q^t,k}^t/|\mathcal{S}_q^t|}{|\eta_k|^2}$$

$$-(1-\beta_p)\frac{\gamma_p^{t-1}}{|\eta_k|^2} - (1-\beta_p)\frac{\gamma_p^{t+1}}{|\eta_k|^2} + \beta_p(1-\beta_p)\frac{\gamma_{\mathcal{S}_p^t}^t}{|\eta_k|^2} - \beta_p\frac{\gamma_{\mathcal{S}_p^{t-1}}^{t-1}}{|\eta_k|^2}$$

$$+(1-\beta_p)\sum_{q\in\mathcal{S}_p^t} \beta_q \frac{\gamma_q^t/|\mathcal{S}_p^t|}{|\eta_k|^2} - \sum_{q\in\mathcal{S}_p^t} \beta_q \frac{\gamma_q^{t+1}/|\mathcal{S}_p^t|}{|\eta_k|^2}].$$

$\gamma_{\mathcal{S}_p^t,}^t \equiv \sum_{q\in\mathcal{S}_p^t} (w_{p\leftarrow q}^t)(\gamma_{q,k}^t)$ is the mean of set of neighbors of node p at time t, and $\sigma_{\mathcal{S}_p^t,k}^2 \equiv \sum_{q\in\mathcal{S}_p^t} (w_{p\leftarrow q}^t)^2 (\sigma_{q,k}^t)^2$ are the variance of set of neighbors of node p at time t. Mean and variance of neighbors can be easily computed since the components of neighbors are independent of each other and are Gaussian themselves. The derivative above is valid for $\gamma_{p,k}^t$ when $0 < t < T$; the form is slightly different when $t = 0$ or $t = T$.

Second, we minimize the divergence with respect to $((\sigma_{p,1}^t)^2, (\sigma_{p,2}^t)^2, ...(\sigma_{p,K}^t)^2)$ using the Newton-Raphson method. The derivative with respect to $\sigma_{p,k}^t$ is :

$$dD_{KL}(q||p)/d\sigma_{p,k}^t = \tag{24.23}$$

$$2(N-1)\frac{\sigma_{p,k}^t}{\zeta_p^t} \exp(\gamma_{p,k}^t + \frac{(\sigma_{p,k}^t)^2}{2}) - \frac{1}{\sigma_{p,k}^t}$$

$$+\frac{1 + (1-\beta_p)^2 + \sum_q Y_t(q,p)\beta_q^2 \frac{w_{q\leftarrow p}^t{}^2}{|\mathcal{S}_q^t|^2}}{\eta_k^2}\sigma_{p,k}^t,$$

where η_k^2 is the diagonal component of the covariance matrix Σ_μ. When $t = 0$ or $t = T$, the derivative slightly differs from the above equation.

Variational M-step

$$p(Y|\Theta) \geq \int_X q(X) \frac{\log p(Y, X|\Theta)}{q(X)}. \tag{24.24}$$

The M-step in the EM algorithm computes the hyperparameters by maximizing the lower bound under fixed q found in the E-step. The lower bound of the log-likelihood is from Jensen's inequality (Equation 24.24), and the expectation is taken with respect to a variational distribution. Hence the general form of the update equation at the kth step is as below:

$$\Theta_k = \arg \max_\Theta \int q_k(X) \log p(Y, X|\Theta) dX. \tag{24.25}$$

Since the final form of most model parameters are quite intuitive, we only derive Equation (24.17) in this section. To obtain the update equation of β_p, we start from differentiating the expected log-likelihood and setting it to zero:

$$0 = \sum_{t<T} \sum_k \left(-(1 - \beta_p)(\gamma_{p,k}^t{}^2 + \sigma_{p,k}^2) + \beta_p(\gamma_{\mathcal{S}_p^t,k}^2 + \sigma_{\mathcal{S}_p^t,k}^2) \right)$$

$$+ \sum_{t>0} \sum_k \left(\gamma_{p,k}^t \gamma_{p,k}^{t-1} - \gamma_{p,k}^t \gamma_{\mathcal{S}_p^{t-1},k} + (1 - 2\beta_p)\gamma_{p,k}^{t-1}\gamma_{\mathcal{S}_p^{t-1},k} \right).$$

Solving the equation above,

$$\beta_p = \frac{\sum_{t>0} \sum_k (\gamma_{p,k}^{t-1}{}^2 + \sigma_k^{t-1}{}^2 - \gamma_{p,k}^t \gamma_{p,k}^{t-1} - \gamma_{p,k}^{t-1}\gamma_{\mathcal{S}_p^{t-1},k} + \gamma_{p,k}^t \gamma_{\mathcal{S}_p^{t-1},k})}{\sum_{t>0} \sum_k (\gamma_{p,k}^{t-1}{}^2 + \sigma_k^{t-1}{}^2 - 2\gamma_{p,k}^{t-1}\gamma_{\mathcal{S}_p^{t-1},k}) + \sum_{t<T} \sum_k (\gamma_{\mathcal{S}_p^t,k}^2 + \sigma_{\mathcal{S}_p^t,k}^2)}. \tag{24.26}$$

For solving the optimal weight, we differentiate the lower bound with respect to $w_{p \leftarrow q_1}$ and set it to zero:

$$0 = \sum_k \left(\beta_p^2 \frac{w_{p \leftarrow q_1}}{|\mathcal{S}_p^t|^2}(\gamma_{q_1,k}^t{}^2 + \sigma_{q_1,k}^t{}^2) - \frac{\beta_p}{|\mathcal{S}_p^t|}\gamma_{q_1,k}^t \gamma_{p,k}^{t+1} \right.$$

$$\left. + \frac{\beta_p}{|\mathcal{S}_p^t|}(1 - \beta_p)\gamma_{q_1,k}^t \gamma_{p,k}^t + \frac{\beta_p^2}{|\mathcal{S}_p^t|^2}\left(\sum_{q \in \mathcal{S}_p^t, q \neq q_1} Y(p, q_1)w_{p \leftarrow q}\gamma_{q,k}^t \right) \right). \tag{24.27}$$

Finally, the update equation for weight becomes,

$$w_{p \leftarrow q_1} = \frac{\sum_k \frac{\beta_p}{|\mathcal{S}_p^t|}\gamma_{q_1,k}^t \gamma_{p,k}^{t+1} - \frac{\beta_p}{|\mathcal{S}_p^t|}(1 - \beta_p)\gamma_{q_1,k}^t \gamma_{p,k}^t - \frac{\beta_p^2}{|\mathcal{S}_p^t|^2}(\sum_{q \in \mathcal{S}_p^t, q \neq q_1} Y(p, q_1)w_{p \leftarrow q}\gamma_{q,k}^t)}{\sum_k \frac{\beta_p^2}{|\mathcal{S}_p^t|^2}(\gamma_{q_1,k}^t{}^2 + \sigma_{q_1,k}^t{}^2).} \tag{24.28}$$

References

Ahn, Y.-Y., Bagrow, J. P., and Lehmann, S. (2010). Link communities reveal multiscale complexity in networks. *Nature* 466: 761–764.

Airoldi, E. M. (2007). Discussion on the paper by Handcock, Raftery and Tantrum 'Model-based clustering for social networks.' *Journal of the Royal Statistical Society, Series A* 170: 330.

Airoldi, E. M., Blei, D. M., Fienberg, S. E., and Xing, E. P. (2008). Mixed membership stochastic blockmodels. *Journal of Machine Learning Research* 9: 1981–2014.

Ball, B., Karrer, B., and Newman, M. E. J. (2011). Efficient and principled method for detecting communities in networks. *Physical Review E* 84: 036103.

Beal, M. J. and Ghahramani, Z. (2003). The variational Bayesian EM algorithm for incomplete data: With application to scoring graphical model structures. *Bayesian Statistics* 7: 453–464.

Blei, D. M. and Fienberg, S. E. (2007). Discussion on the paper by Handcock, Raftery and Tantrum 'Model-based clustering for social networks.' *Journal of the Royal Statistical Society, Series A* 170: 332.

Blei, D. M. and Lafferty, J. D. (2007). A correlated topic model of Science. *Annals of Applied Statistics* 1: 17–35.

Fan, Y. and Shelton, C. R. (2009). Learning continuous-time social network dynamics. In *Proceedings of the 25th Conference on Uncertainty in Artificial Intelligence (UAI 2009)*. Corvallis, OR, USA: AUAI Press, 161–168.

Fu, W., Song, L., and Xing, E. P. (2009). Dynamic mixed membership blockmodel for evolving networks. In *Proceedings of the 26th International Conference on Machine Learning (ICML '09)*. Omnipress, 329–336.

Goldenberg, A., Zheng, A. X., Fienberg, S. E., and Airoldi, E. M. (2010). A survey of statistical network models. *Foundations and Trends in Machine Learning* 2: 129–233.

Hanneke, S., Fu, W., and Xing, E. P. (2010). Discrete temporal models of social networks. *Electronic Journal of Statistics* 4: 585–605.

Ho, Q., Song, L., and Xing, E. P. (2011). Evolving cluster mixed-membership blockmodel for time-evolving networks. In *Proceedings of the 14th International Conference on Artificial Intelligence and Statistics (AISTATS 2011)*. Palo Alto, CA, USA: AAAI.

Hoff, P. D., Raftery, A. E., and Handcock, M. S. (2002). Latent space approaches to social network analysis. *Journal of the American Statistical Association* : 1090–1098.

Holland, P. W., Laskey, K. B., and Leinhardt, S. (1983). Stochastic blockmodels: First steps. *Social Networks* 5: 109–137.

Karrer, B. and Newman, M. E. J. (2011). Stochastic blockmodels and community structure in networks. *Physical Review E* 83: 016107.

Kolar, M. and Xing, E. P. (2011). On time varying undirected graphs. In *Proceedings of the 14th International Conference on Artificial Intelligence and Statistics (AISTATS 2011). Journal of Machine Learning Research – Proceedings Track* 15 : 407–415.

Kolar, M., Song, L., Ahmed, A., and Xing, E. P. (2010). Estimating time-varying networks. *Annals of Applied Statistics* 4 : 94–123.

Krioukov, D., Papadopoulos, F., Vahdat, A., and Boguñá, M. (2009). Curvature and temperature of complex networks. *Physical Review E* 80: 035101.

Maisel, L. S. (2010). Rating United States Senators: The strength of Maine's delegation since 1955. *The New England Journal of Political Science* 4.

McCarty, N., Poole, K. T., and Rosenthal, H. (2006). *Polarized America: The Dance of Ideology and Unequal Riches*. Cambridge, MA: The MIT Press.

Mørup, M., Schmidt, M. N., and Hansen, L. K. (2011). Infinite Multiple Membership Relational Modeling for Complex Networks. Tech. report, http://arxiv.org/abs/1101.5097, comments: 8 pages, 4 figures.

Reichardt, J., Alamino, R., and Saad, D. (2011). The interplay between microscopic and mesoscopic structures in complex networks. *PLoS ONE* 6: e21282.

Snijders, T. A. B., Steglich, C. E. G., and Schweinberger, M. (2006). Modeling the co-evolution of networks and behavior. In van Montfort, K., Oud, H., and Satorra A. (eds), *Longitudinal Models in the Behavioral and Related Sciences*. Mahwah, NJ: Lawrence Erlbaum, 41–71.

Steglich, C. E. G., Snijders, T. A. B., and Pearson, M. (2010). Dynamic networks and behavior: Separating selection from influence. *Sociological Methodology* 40: 329–393.

Xing, E. P., Fu, W., and Song, L. (2010). A state-space mixed membership blockmodel for dynamic network tomography. *Annals of Applied Statistics* 4 : 535–566.

Xing, E. P., Jordan, M. I., and Russell, S. (2003). A generalized mean field algorithm for variational inference in exponential families. In *Proceedings of the 19th Conference on Uncertainty in Artificial Intelligence (UAI 2003)*. San Francisco, CA, USA: Morgan Kaufmann Publishers Inc., 583–591.

25

Overlapping Clustering Methods for Networks

Pierre Latouche

Laboratoire SAMM, Université Paris 1 Panthéon-Sorbonne, France

Etienne Birmelé

Laboratoire Statistique et Génome, Université d'Évry-val-d'Essonne, France
Laboratoire Biométrie et Biologie Evolutive, INRIA Rhône-Alpes, Lyon, France

Christophe Ambroise

Laboratoire Statistique et Génome, Université d'Évry-val-d'Essonne, France

CONTENTS

Networks allow the representation of interactions between objects. Their structures are often complex to explore and need some algorithmic and statistical tools for summarizing. One possible way to go about this is to cluster their vertices into groups having similar connectivity patterns.

This chapter aims to present an overview of clustering methods for network vertices. Common community structure searching algorithms are detailed. The well-known stochastic blockmodel (SBM) is then introduced and its generalization to overlapping mixed membership structure closes the chapter. Examples of application are also presented and the main hypothesis underlying the presented algorithms is discussed.

25.1   Introduction

Because networks are a straightforward formalism for representing interactions between objects of interest, they are used in many scientific fields. In biology, regulatory networks allow us to describe the regulation of gene expression through transcriptional factors (Milo et al., 2002), while metabolic networks focus on representing pathways of biochemical reactions (Lacroix et al., 2006). Besides, the binding procedures of proteins are often described as protein-protein interaction networks (Albert and Barabási, 2002; Barabási and Oltvai, 2004). In social sciences, networks are widely used to represent relational ties between actors (Snijders and Nowicki, 1997; Nowicki and Snijders, 2001; Palla et al., 2007). Other examples of networks are powergrids (Watts and Strogatz, 1998) and the World Wide Web (Zanghi et al., 2008).

As a network describes the presence or absence of links between objects, the notion of groups of nodes having a similar behavior naturally arises. In some cases, this notion of similarity is even the process from which the network originates. Common affinities will, for example, lead to edges in social networks, whereas gene duplication is the main growth process of protein-protein interaction networks.

The most widely assumed group structure is the partition, where each node belongs to only one group. When dealing with real-world applications, this assumption of empty intersections between groups is often too rigid. For instance, so-called *moonlighting proteins* are known to have several functions in the cells (Jeffery, 1999). Considering social networks, actors typically belong to several groups of interests (Palla et al., 2005). Thus, exploring structures which allow for more complex membership for each node is of great practical interest. One possibility consists in considering models where each individual is allowed to belong to all groups depending on a mixed membership coefficient, all membership coefficients summing to 1. This approach is considered, for example, in the latent Dirichlet allocation model (Blei et al., 2003), in the context of text mining, or in the mixed membership stochastic blockmodel (Airoldi et al., 2008) for the clustering of nodes in networks. An alternative is to consider that each node belongs to multiple groups, but that for each possible group the node either belongs to the group or it does not.

In this chapter, we propose to give an overview of the methods using the latter approach, i.e., retrieving group memberships of nodes based on their connectivity pattern, the memberships of each node being summarized in a $\{0, 1\}$-vector. The first section introduces the notion of networks and the characteristics of real networks one should have in mind when building models. The second section deals with the partitioning of nodes, i.e., methods assigning each vertex to exactly one group. The last section presents generalizations of those methods which allow for the overlapping groups of nodes.

25.2   Networks and Their Characteristics

25.2.1   Network Representations

A network is commonly represented by a graph $\mathcal{G} = (\mathcal{V}, \mathcal{E})$, where $\mathcal{V}$ is a set of N vertices and $\mathcal{E}$ is a set of edges between pairs of vertices. The graph is said to be directed (Figure 25.1) if the pairs (u, v) in $\mathcal{E}$ are ordered. Conversely, unordered pairs form an undirected graph (Figures 25.2 and 25.3). Note that the edges can be weighted by a function $w : \mathcal{E} \to \mathbb{F}$ for any set $\mathbb{F}$. However, we will concentrate only on binary graphs, i.e., $\mathbb{F} = \{0, 1\}$. The size of $\mathcal{G}$ is then given through the edge count $m = |\mathcal{E}|$. The graph is said to be dense if m is close to the maximal number M of edges, whereas a low value of m leads to a sparse graph. To characterize the density of $\mathcal{G}$, a criterion $\delta(\mathcal{G})$

is often used. It is defined as the ratio of the number m of existing edges over the number M of potential edges:

$$\delta(\mathcal{G}) = \frac{m}{M}.$$

For a directed graph, $M = N^2$ while $M = N(N+1)/2$ otherwise. If $\mathcal{G}$ does not contain any self loop, i.e., an edge from a vertex to itself, then $M = N(N-1)$ for a directed graph and $M = N(N-1)/2$ otherwise.

The neighborhood $N_{\mathcal{G}}(u)$ of vertex u is defined as the set of all the vertices connected to u. Its degree $d(u)$ is equal to its number of incident edges. Finally, a path from a vertex u to a vertex v is a sequence of edges in $\mathcal{E}$ starting at vertex $v_0 = u$ and ending at vertex $v_{k+1} = v$:

$$\{u, v_1\}, \{v_1, v_2\}, \ldots, \{v_k, v\}.$$

If there exists at least one path between every pair of vertices, then the graph is said to be connected. For instance, the graph in Figure 25.1 is connected contrary to the graphs in Figures 25.2 and 25.3, which have some isolated vertices.

A network can equivalently be represented by a so-called adjacency matrix $\mathbf{X}$, which describes the presence or absence of an edge in a graph. As mentioned already, we focus on binary graphs and therefore $\mathbf{X}$ is in $\{0, 1\}^{N \times N}$. Thus, if there exists an edge from vertex i to vertex j, then X_{ij} equals 1 and 0 otherwise. If the network is undirected, the matrix is symmetric, i.e., X_{ij} and X_{ji} are equals. Non-zero entries of the diagonal correspond to self-loops. Every property of a graph can be interpreted in terms of its adjacency matrix: the degree of a vertex is the sum of the row or the column corresponding to it, or the fact that two vertices (i, j) are in different connected compounds is equivalent to $(X^k)_{ij} = 0$ for all power $1 \leq k \leq N$.

25.2.2 Properties of Real Networks

Most real networks have been shown to share some properties (Albert et al., 1999; Broder et al., 2000; Dorogovtsev et al., 2000; Amaral et al., 2000; Strogatz, 2001) that we briefly recall in the following:

- **Sparsity:** The number of edges is linear in the number of vertices. In other terms, the mean degree remains bounded when N grows, implying that the density tends to 0.

- **Existence of a giant component:** Real networks are often disconnected. However, a majority of the vertices are contained in the same component, the other components being significantly smaller.

- **Degree heterogeneity:** A few vertices have a lot of connections while most of the vertices have very few links. The degrees of the vertices are sometimes characterized using a scale-free distribution (e.g., see Barabasi and Albert, 1999).

- **Small world:** The shortest path from one vertex to another is generally rather small, typically of size $O(\log N)$.

All the properties listed above can be verified through easy computable statistics which are the degrees and the paths of length at most N. As they are key properties in the interpretation of real network behaviors with respect to information diffusion (Pastor-Satorras and Vespignani, 2001) or attack tolerance (Albert et al., 2000), we would like our random graph models to produce networks with similar properties.

Most of the real networks exhibit another property, which is the one of interest in this chapter, namely an *underlying group structure*. This means that nodes can be spread into classes having similar connectivity patterns. In order to retrieve such structures, statistical and algorithmic tools have been developed.

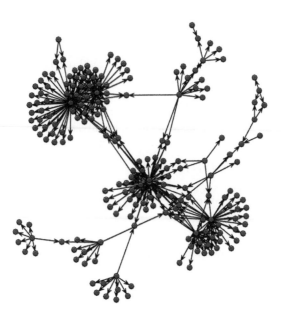

FIGURE 25.1
Subset of the yeast transcriptional regulatory network (Milo et al., 2002). Nodes of the directed network correspond to genes, and two genes are linked if one gene encodes a transcriptional factor that directly regulates the other gene.

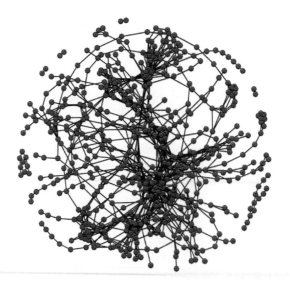

FIGURE 25.2
The metabolic network of bacteria *Escherichia coli* (Lacroix et al., 2006). Nodes of the undirected network correspond to biochemical reactions, and two reactions are connected if a compound produced by the first one is a part of the second one (or vice-versa).

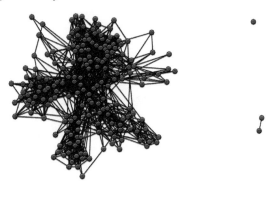

FIGURE 25.3
Subset of the French political blogosphere network. The data consists of a single day snapshot of political blogs automatically extracted on October 14th, 2006 and manually classified by the "Observatoire Présidentielle project" (Zanghi et al., 2008). Nodes correspond to hostnames and there is an edge between two nodes if there is a known hyperlink from one hostname to another (or vice-versa).

25.3 Graph Clustering

In this section, we concentrate on the classification of vertices depending on their connection profiles. There has been a wealth of literature on the topic which goes back to the earlier work of Moreno (1934). As shown in Newman and Leicht (2007), it appears that available methods can be grouped into three significant categories. First, some models look for community structure, also called assortative mixing (Newman, 2003; Danon et al., 2005), where vertices are partitioned into classes such that vertices of a class are mostly connected to vertices of the same class. Other models look for disassortative mixing in which vertices mostly connect to vertices of different classes. They are commonly used to analyze bipartite networks (Estrada and Rodriguez-Velazquez, 2005). Finally, a few procedures look for heterogeneous structure where vertices can have different types of connection profiles. In particular, they can be used to uncover both community structure and disassortative mixing.

In this section, we describe some of the most widely used graph clustering methods. Note that many model-free approaches exist (Fortunato, 2010). However, except for the algorithmic approach presented in Section 25.3.1, we concentrate in the following on methods which rely on statistical models only.

25.3.1 Community Structure

Most graph clustering methods aim at detecting community structure, also called assortative mixing, meaning the appearance of densely connected groups of vertices, with only sparser connections between groups (Figure 25.4). Most of them rely on the modularity score of Newman and Girvan (2004). However, we point out the recent work of Bickel and Chen (2009) who showed that these algorithms are (asymptotically) biased and that using modularity scores could lead to the discovery of an incorrect community structure, even for large graphs.

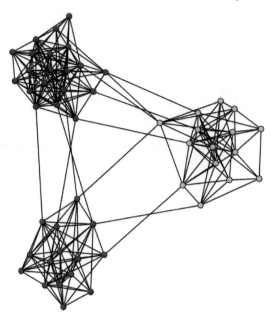

FIGURE 25.4
Example of an undirected affiliation network with 50 vertices. The network is made of three communities represented in red, blue, and green. Vertices connect mainly to vertices of the same community.

Modularity Score

Newman and Girvan (Girvan and Newman, 2002; Newman and Girvan, 2004) proposed several intuitive community detection algorithms which involve iterative removal of edges from the network to split it into communities. Edges to be removed are identified using one of a number of possible betweenness measures. All of them are based on the same idea: If two communities are joined by only a few *inter* community edges, then all paths from vertices in one community to vertices in the other must pass along one of those few edges. Therefore, given a suitable set of paths, we expect the number of paths that go along an edge to be the largest for *inter* community edges. First, Newman and Girvan introduced the edge betweenness, a generalization to edges of the vertex betweenness measure of Freeman (1977). The edge betweenness of an edge is defined as the number of shortest paths between all pairs of vertices in the network that run along that edge. Second, they considered the random walk betweenness. The expected number of times a random walk between a particular pair of vertices will pass down a particular edge is calculated. This expected value is then summed over all pairs of vertices to obtain the random walk betweenness of the edge. As shown in Newman and Girvan (2004), other scores can obviously be considered to obtain algorithms that may be more appropriate for some applications. However, it appears that the choice of measure does not highly influence the result of the algorithms. On the other hand, the recalculation step after each edge removal is crucial (see Algorithm 1).

All these algorithms produce a dendrogram (Figure 25.5) which represents an entirely nested

---

repeat
    Calculate betweenness scores for all edges   Remove the edge with the highest score
until *No edges remain*;

Algorithm 1: Example of a community structure detection algorithm with a betweenness score.

hierarchy of possible community divisions for the network. In order to select one of these divisions, Newman and Girvan (2004) proposed a modularity criterion. Consider a particular division with Q communities and denote e_{ql} as the fraction of all edges in the network that link vertices in community q to vertices in community l. Moreover, consider the fraction $a_q = \sum_{l=1}^{Q} e_{ql}$ of edges that connect to vertices of community q. The modularity criterion is then given by:

$$Q_{mod} = \sum_{q=1}^{Q} (e_{qq} - a_q^2). \tag{25.1}$$

The criterion is computed for all the divisions, and a division is chosen such that the modularity is maximized. Note that modularity can be generalized to both directed and valued graphs (Fortunato, 2010).

A limiting factor of these community detection algorithms is their poor scaling with the number m of edges and the number N of vertices in the network. For instance, calculating the shortest paths between a particular pair of vertices can be done in $O(m)$ (Ahuja et al., 1993; Cormen et al., 2001). Because they are $O(N^2)$ vertex pairs, the computational cost to compute all the edge betweenness scores is in $O(mN^2)$. This complexity was improved independently by Newman (2001) and Brandes (2001) finding all betweennesses in $O(mN)$. Since this calculation has to be repeated for the removal of each edge, the entire algorithm runs in worst-case time $O(m^2N)$. In other words, for dense networks, where m is in $O(N^2)$, it runs in $O(N^5)$ while it scales in $O(N^3)$ for sparse networks, where m is linear in N.

Rather than building the complete dendrogram (with edge removals) and then choosing the optimal division using the modularity criterion, Newman (2004) suggested to focus directly on the optimization of the modularity. Thus, he proposed an algorithm which falls in the general category of agglomerative hierarchical clustering methods (Everitt, 1974; Scott, 2000). Starting with a configuration in which each vertex is the sole member of one of N communities, the communities are iteratively joined together in pairs, choosing at each step the join that results in the greatest increase (or smallest decrease) in mod (25.1). Again, this leads to a dendrogram for which the best cut is chosen by looking for the maximal value of the modularity. The computational cost of the entire algorithm is in $O\left((m + N)N\right)$, or $O(N^3)$ for dense networks and $O(N^2)$ for sparse networks. It was shown to be capable of handling a collaboration network with 50,000 vertices in Newman (2004).

Latent Position Cluster Model

An alternative approach for community detection in networks is the latent position cluster model (LPCM) of Handcock et al. (2007). Consider a $N \times N$ binary adjacency matrix $\mathbf{X}$ such that X_{ij} equals 1 if there is an edge from vertex i to vertex j, and 0 otherwise. Moreover, let us define $\mathbf{Y}$ as covariate information where $\mathbf{Y}_{ij}$ denotes some observed characteristics about the pair (i, j) of vertices. This might represent, for instance, the traffic information of users from blog i to blog j in a blogosphere network (see Figure 25.3). Several characteristics can possibly be observed for each pair of vertices and therefore $\mathbf{Y}_{ij}$ can be vector valued. Note that a few other random graph models have been proposed in the literature to take covariates into account (see e.g., Zanghi et al., 2010; Mariadassou et al., 2010). They will not be considered in this chapter as we consider vertices clustered by the use of network topology only. Here, we describe LPCM in a general setting, as in Handcock et al. (2007), and emphasize that the algorithm can also be used if $\mathbf{Y}$ is not available simply by removing the terms in $\mathbf{Y}_{ij}$ in the following expressions.

LPCM assumes that the network does not contain any self loop while both directed and undirected relations can be analyzed. It is assumed that each vertex, usually called actor in social sciences, has an unobserved position in a d dimensional Euclidean latent space as in Hoff et al. (2002). Given the latent positions and the covariate information, the edges are assumed to be drawn from a Bernoulli distribution:

$$X_{ij} | \mathbf{Z}_i, \mathbf{Z}_j, \mathbf{Y}_{ij} \sim \text{Bern}\left(g(a_{\mathbf{Z}_i, \mathbf{Z}_j, \mathbf{Y}_{ij}})\right).$$

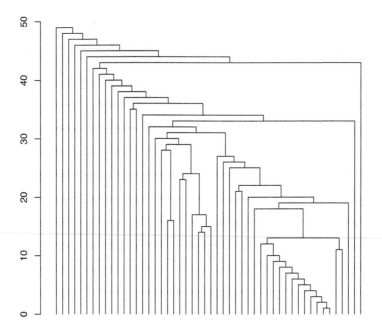

FIGURE 25.5
Dendrogram of a network with 50 vertices for the community detection algorithm with edge betweenness. It should be read from top to bottom. The algorithm starts with a single community which contains all the vertices. Edges with the highest edge betweenness are then removed iteratively splitting the network into several communities. After convergence, each vertex, represented by a leaf of the tree, is a sole member of one of the 50 communities.

The function $g(x) = (1 + e^{-x})^{-1}$ is the logistic sigmoid function. Moreover, $a_{\mathbf{Z}_i, \mathbf{Z}_j, \mathbf{Y}_{ij}}$ is given by:

$$a_{\mathbf{Z}_i, \mathbf{Z}_j, \mathbf{Y}_{ij}} = \mathbf{Y}_{ij}^{\mathsf{T}} \boldsymbol{\beta}_0 - \beta_1 |\mathbf{Z}_i - \mathbf{Z}_j|, \tag{25.2}$$

where $\boldsymbol{\beta}_0$ has the same dimensionality as $\mathbf{Y}_{ij}$ and β_1 is a scalar. Both $\boldsymbol{\beta}_0$ and β_1 are unknown parameters to be estimated. To represent clustering, the positions are assumed to be drawn from a finite mixture of Q multivariate normal distributions, each one representing a different class of vertices. Each multivariate distribution has its own mean vector as well as spherical covariance matrix:

$$\mathbf{Z}_i \sim \sum_{q=1}^{Q} \alpha_q \mathcal{N}(\boldsymbol{\mu}_q, \sigma_q^2 \mathbf{I}),$$

and $\boldsymbol{\alpha}$ denotes a vector of class proportions which satisfies $\alpha_q > 0, \forall q$ and $\sum_{q=1}^{Q} \alpha_q = 1$. Finally, according to LPCM, the latent positions $\mathbf{Z}_1, \ldots, \mathbf{Z}_N$ are i.i.d. and given this latent structure, all the edges are supposed to be independent. Consider now the second term on the right-hand side of (25.2). By construction, if β_1 is positive, we expect the L_1 distance $|\mathbf{Z}_i - \mathbf{Z}_j|$ to be smaller if vertices i and j are in the same class. In other words, the probability $g(a_{\mathbf{Z}_i, \mathbf{Z}_j, \mathbf{Y}_{ij}})$ of an edge between i and j is supposed to be higher for vertices sharing the same class. Note that this corresponds exactly to the definition of a community.

Handcock et al. (2007) proposed a two-stage maximum likelihood approach and a Bayesian algorithm, as well as a BIC criterion to estimate the number of latent classes. The two-stage maximum likelihood approach first maps the vertices in the latent space and then uses a mixture model to cluster the resulting positions. In practice, this procedure converges more quickly but loses some information by not estimating the positions and the cluster model at the same time. Conversely, the Bayesian algorithm (see Figure 25.6), based on Markov chain Monte Carlo, estimates both the latent positions and the mixture model parameters simultaneously. It gives better results but is time consuming. Both the maximum likelihood and the Bayesian approach are limited in the sense that they can handle networks with a few hundreds of vertices only.

25.3.2 Heterogeneous Structure

So far, we have seen some algorithms to uncover communities. However, some vertices may be grouped while exhibiting connection patterns differently from a dense group poorly linked to the rest of the network. In genetic regulatory networks, transcription factors co-regulating some biological process may, for example, not be linked to one another but act jointly on the regulated genes. Some other approaches which can look for heterogeneous structure in networks, where vertices can have different types of connection profiles, have therefore been developed.

Hofman and Wiggins' Model

Let us consider a binary adjacency matrix $\mathbf{X}$ representing a network $\mathcal{G}$. The model of Hofman and Wiggins (Hofman and Wiggins, 2008) associates to each vertex of the network a latent variable $\mathbf{Z}_i$ drawn from a multinomial distribution:

$$\mathbf{Z}_i \sim \text{Multinom}\left(1, \boldsymbol{\alpha} = (\alpha_1, \ldots, \alpha_Q)\right). \tag{25.3}$$

As in other standard mixture models, the vector $\mathbf{Z}_i$ has all its components set to zero except one such that Z_{iq} equals 1 if vertex i belongs to class q. Thus, $\sum_{q=1}^{Q} Z_{iq} = 1$, $\forall i$ and the vector $\boldsymbol{\alpha}$ satisfies $\alpha_q > 0$, $\forall q$ as well as $\sum_{q=1}^{Q} \alpha_q = 1$. The edges are then assumed to be drawn from a Bernoulli distribution:

$$X_{ij} \sim \text{Bern}(\lambda),$$

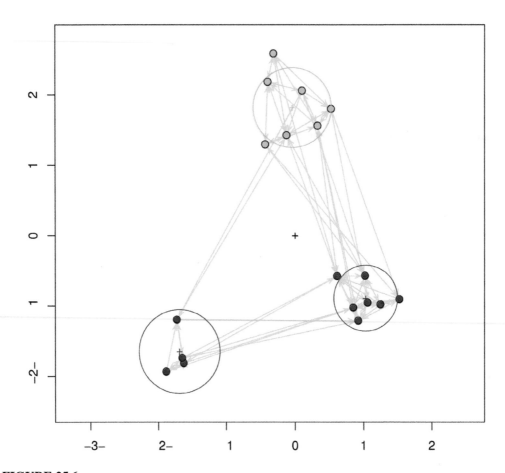

FIGURE 25.6
Directed network of social relations between 18 monks in an isolated American monastery (Sampson, 1969; White et al., 1976). Sampson collected sociometric information using interviews, experiments, and observations. This network focused on the relation of "liking." A monk is said to have a social relation of "like" to another monk if he ranked that monk in the top three monks for positive affection in any of the three interviews given. The positions of the vertices in the two data dimensional latent space have been calculated using the Bayesian approach for LPCM. The position of the three class centers found are indicated, as well as circles with radius equal to the square root of the class variances estimated.

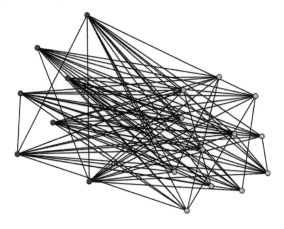

FIGURE 25.7
Example of an undirected network with 20 vertices. The connection probabilities between the two classes in red and green are higher than the *intra* class probabilities. Vertices connect mainly to vertices of a different class.

if vertices i and j are in the same class, i.e., $\mathbf{Z}_i = \mathbf{Z}_j$, and

$$X_{ij} \sim \mathrm{Bern}(\epsilon)$$

otherwise. Thus, the model is able to take into account both community structure ($\lambda > \epsilon$) (Figure 25.4) and disassortative mixing ($\lambda < \epsilon$) (Figure 25.7). As in the previous section, given the latent variables $\mathbf{Z}_1, \ldots, \mathbf{Z}_N$, all the edges are supposed to be independent. In order to estimate the posterior distribution $p(\mathbf{Z}, \boldsymbol{\alpha}, \lambda, \epsilon \,|\, \mathbf{X})$ over the latent variables and model parameters, Hofman and Wiggins (2008) used a variational Bayes expectation maximization (EM) algorithm with a factorized distribution.

Moreover, they proposed a model selection criterion to estimate the number of latent classes in networks. It relies on a variational approximation of the marginal log-likelihood $\log p(\mathbf{X})$ and has shown promising results.

Stochastic Blockmodels

Originally developed in social sciences, the stochastic blockmodel (SBM) is a probabilistic generalization (Fienberg and Wasserman, 1981; Holland et al., 1983) of the method described in White et al. (1976). Given a network, it assumes that each vertex belongs to a hidden class among Q classes and uses a matrix Π to describe the *intra* and *inter* connection probabilities (Frank and Harary, 1982). No assumption is made on the form of the connectivity matrix such that very different structures can be taken into account. In particular, SBM can characterize the presence of hubs which make networks locally dense (Daudin et al., 2008). Moreover, and to some extent, it generalizes many of the existing graph clustering techniques, as shown in Newman and Leicht (2007). For instance, the model of Hofman and Wiggins can be seen as a constrained SBM where the diagonal of Π is set to λ and all the other elements to ϵ.

Formally, SBM considers a latent variable $\mathbf{Z}_i$, drawn from a multinomial distribution (25.3), for each vertex in the network, as in Section 25.3.2. Thus, each vertex belongs to a single class, and that class is q if Z_{iq} equals 1. The edges are then assumed to be drawn from a Bernoulli distribution:

$$X_{ij} | Z_{iq} Z_{jl} = 1 \sim \mathrm{Bern}(\pi_{ql}),$$

where Π is a $Q \times Q$ matrix of connection probabilities. Again, given all the latent variables, the edges are supposed to be independent. Note that SBM was originally described in a more general setting (Nowicki and Snijders, 2001), allowing any discrete relational data. However, as explained in Section 25.2.1, we concentrate on binary edges only.

The identifiability of the parameters in SBM was studied by Allman et al. (2009; 2011), who showed that the model is generically identifiable up to a permutation of the classes. In other words, except in a set of parameters which has a null Lebesgue's measure, two parameters imply the same random graph model if and only if they differ only by the ordering of the classes.

Many methods have been proposed in the literature to jointly estimate SBM model parameters and cluster the vertices of the network. They all face the same difficulty. Indeed, contrary to many mixture models, the conditional distribution of all the latent variables $\mathbf{Z}$ and model parameters, given the observed data $\mathbf{X}$, can not be factorized due to conditional dependency. Therefore, optimization techniques such as the expectation maximization (EM) algorithm can not be used directly. In the case of SBM, Nowicki and Snijders (2001) proposed a Bayesian probabilistic approach. They introduced some prior Dirichlet distributions for the model parameters and used Gibbs sampling to approximate the posterior distribution over the model parameters and posterior predictive distribution. Their algorithm is implemented in the software BLOCKS, which is part of the package StoCNET (Boer et al., 2006). It gives accurate a posteriori estimates but can not handle networks with more than 200 vertices. Daudin et al. (2008) proposed a frequentist variational EM approach for SBM which can handle much larger networks and developed an integrated classification likelihood (ICL) criterion for the model selection. Latouche et al. (2011) adapted it in a Bayesian framework, yielding an algorithm which retrieves better small classes and does the model selection with a non-asymptotic criterion. Online strategies have also been developed (Zanghi et al., 2008), as well as extensions to deal with discrete or continuous edges (Mariadassou et al., 2010).

25.4 Overlapping Clustering

As mentioned previously, most graph clustering methods suffer from the restriction they impose by requiring that each vertex belongs to exactly one class. We present in this section some algorithmic and statistical adaptations of the existing clustering methods which tackle this issue. We focus here on the methods by assigning to each vertex a vector of $\{0, 1\}^Q$, where Q denotes the number of classes. In other words, each individual belongs completely to all groups it participates in. Methods using vectors of coefficients summing to 1 and giving the relative importance of each class in the individual behavior have also been developed (Blei et al., 2003; Airoldi et al., 2008).

25.4.1 Algorithmic Approaches

The issue of overlapping clustering has received growing attention in the last few years, starting with an algorithmic approach based on clique percolation developed by Palla et al. (2005) and implemented in the software CFinder (Palla et al., 2006). In this approach, a k-clique community is defined as the union of all k-cliques (complete sub-graphs of size k) that can be reached from each other through a series of adjacent k-cliques.[1] Given a network, the algorithm first locates all cliques and then identifies the communities using a clique-clique overlap matrix (Everett and Borgatti, 1998). By construction, the resulting communities can overlap. In order to select the optimal value of k, Palla et al. (2005) suggested a global criterion which looks for a community structure as highly connected as possible. Small values of k lead to a giant community which smears the details of a network by merging small communities. Conversely, when k increases, the communities tend

---

[1]Two k-cliques are adjacent if they share $k-1$ vertices.

to become smaller, more disintegrated, but also more cohesive. Therefore, they proposed a heuristic which consists of running their algorithm for various values of k and then selecting the lowest value such that no giant community appears.

Shen et al. (2009) adapted the classification method of Girvan and Newman to overlapping clusters in a method called EAGLE. To do so, they first built a bottom-up dendrogram starting with some well-chosen and possibly overlapping maximal cliques. At each step, a distance was computed for every pair of communities based on the proportion of edges linking those communities. The two nearest ones were then merged. The cut level of the dendrogram was chosen according to a generalization of the modularity to overlapping communities, namely:

$$Q_{ov} = \frac{1}{2m} \sum_q \sum_{ij} \frac{1}{O_i O_j} \left(X_{ij} - \frac{k_i k_j}{2m} \right) \delta(C_i, C_j),$$

where O_i is equal to the number of communities i belongs to. It can be shown that if all O_is are equal to 1, this expression is equal to the modularity defined in Equation (25.1). The contribution of each edge then decreases when its incident vertices belong to several communities.

However, those algorithmic procedures are limited to the detection of communities. Statistical tools are then needed to find overlapping heterogeneous structures.

25.4.2 Overlapping Stochastic Blockmodel

Let us now investigate the adaptation of the stochastic blockmodel to overlapping classes. The hidden structure can no longer be a mixture model, so the constraints $\sum_q Z_{iq} = 1$ and $\sum_q \alpha_q = 1$ present in SBM are relaxed. Thus, a new latent vector $\mathbf{Z}_i$ is introduced for each vertex i of the network. This vector is composed from Q independent Boolean variables $Z_{iq} \in \{0,1\}$ drawn from a multivariate Bernoulli distribution:

$$\mathbf{Z}_i \sim \prod_{q=1}^{Q} \mathrm{Bern}(Z_{iq}; \alpha_q) = \prod_{q=1}^{Q} \alpha_q^{Z_{iq}} (1 - \alpha_q)^{1-Z_{iq}}. \tag{25.4}$$

We point out that $\mathbf{Z}_i$ can also have all its components set to zero, which is a useful feature in practice as we shall see in Section 25.4.2. The edge probabilities are then given by:

$$X_{ij} | \mathbf{Z}_i, \mathbf{Z}_j \sim \mathrm{Bern}\left(X_{ij}; g(a_{\mathbf{Z}_i, \mathbf{Z}_j}) \right) = e^{X_{ij} a_{\mathbf{Z}_i, \mathbf{Z}_j}} g(-a_{\mathbf{Z}_i, \mathbf{Z}_j}),$$

where

$$a_{\mathbf{Z}_i, \mathbf{Z}_j} = \mathbf{Z}_i^\mathsf{T} \mathbf{W} \mathbf{Z}_j + \mathbf{Z}_i^\mathsf{T} \mathbf{U} + \mathbf{V}^\mathsf{T} \mathbf{Z}_j + W^*, \tag{25.5}$$

and $g(x) = (1 + e^{-x})^{-1}$ is the logistic sigmoid function. $\mathbf{W}$ is a $Q \times Q$ real matrix, whereas $\mathbf{U}$ and $\mathbf{V}$ are Q-dimensional real vectors. The first term in the right-hand side of (25.5) describes the interactions between the vertices i and j. If i belongs only to class q and j only to class l, then only one interaction term remains ($\mathbf{Z}_i^\mathsf{T} \mathbf{W} \mathbf{Z}_j = W_{ql}$). However, the model can take more complex interactions into account if one or both of these two vertices belong to multiple classes (Figure 25.8). Note that the second term in (25.5) does not depend on $\mathbf{Z}_j$. It models the overall capacity of vertex i to connect to other vertices. By symmetry, the third term represents the global tendency of vertex j to receive an edge. These two parameters $\mathbf{U}$ and $\mathbf{V}$ are related to the sender/receiver effects δ_i and γ_j in the latent cluster random effects model (LCREM) of Krivitsky et al. (2009). However, contrary to LCREM, $\delta_i = \mathbf{Z}_i^\mathsf{T} \mathbf{U}$ and $\gamma_j = \mathbf{V}^\mathsf{T} \mathbf{Z}_j$ depend on the classes. In other words, two different vertices sharing the same classes will have exactly the same sender/receiver effects, which is not the case in LCREM. Finally, we use the scalar W^* as a bias, to model sparsity.

If we associate to each latent variable $\mathbf{Z}_i$ a vector $\tilde{\mathbf{Z}}_i = \left(\mathbf{Z}_i, 1 \right)^\mathsf{T}$, then (25.5) can be written:

$$a_{\mathbf{Z}_i, \mathbf{Z}_j} = \tilde{\mathbf{Z}}_i^\mathsf{T} \tilde{\mathbf{W}} \tilde{\mathbf{Z}}_j, \tag{25.6}$$

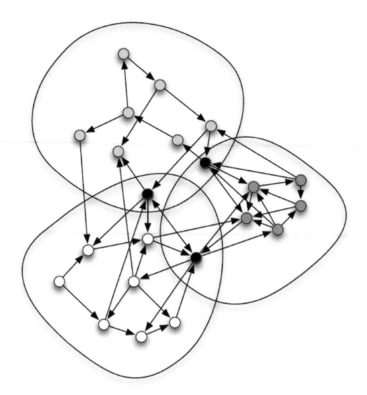

FIGURE 25.8
Example of a directed graph with three overlapping clusters.

where

$$\tilde{\mathbf{W}} = \begin{pmatrix} \mathbf{W} & \mathbf{U} \\ \mathbf{V}^\intercal & W^* \end{pmatrix}.$$

The $\tilde{Z}_{i(Q+1)}$s can be seen as random variables drawn from a Bernoulli distribution with probability $\alpha_{Q+1} = 1$. Thus, one way to think about the model is to consider that all the vertices in the graph belong to a $(Q + 1)$-th cluster which is overlapped by all the other clusters. In the following, we will use (25.6) to simplify the notations.

Finally, given the latent structure $\mathbf{Z} = \{\mathbf{Z}_1, \ldots, \mathbf{Z}_N\}$, all the edges are supposed to be independent. Thus, when considering directed graphs without self-loop, the overlapping stochastic blockmodel (OSBM) is defined through the following distributions:

$$p(\mathbf{Z} \mid \boldsymbol{\alpha}) = \prod_{i=1}^{N} \prod_{q=1}^{Q} \alpha_q^{Z_{iq}} (1 - \alpha_q)^{1 - Z_{iq}}, \tag{25.7}$$

and

$$p(\mathbf{X} \mid \mathbf{Z}, \tilde{\mathbf{W}}) = \prod_{\substack{i \neq j}}^{N} e^{X_{ij} a_{\mathbf{z}_i, \mathbf{z}_j}} g(-a_{\mathbf{z}_i, \mathbf{z}_j}).$$

The graphical model of OSBM is given in Figure 25.9.

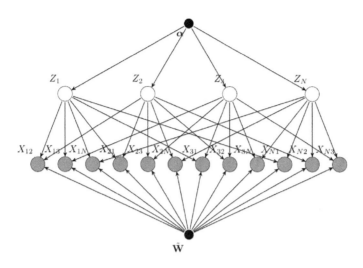

FIGURE 25.9
Directed acyclic graph representing the overlapping stochastic blockmodel. Nodes represent random variables, which are shaded when they are observed, and edges represent conditional dependencies.

Modeling Sparsity

As mentioned in 25.2, real networks are often sparse and it is crucial to distinguish the two sources of non-interaction. Sparsity might be the result of the rarity of interactions in general but it might also indicate that some class (*intra* or *inter*) connection probabilities are close to zero. For instance, social networks are often made of communities where vertices are mostly connected to vertices of the same community. This corresponds to classes with high *intra* connection probabilities and low *inter* connection probabilities. In (25.5), we notice that W^* appears in $a_{\mathbf{Z}_i, \mathbf{Z}_j}$ for every pair of vertices. Therefore, W^* is a convenient parameter to model the two sources of sparsity. Indeed, low values of W^* result from the rarity of interactions in general, whereas high values signify that sparsity comes from the classes (parameters in $\mathbf{W}$, $\mathbf{U}$, and $\mathbf{V}$).

Modeling Outliers

When applied to real networks, graph clustering methods often lead to giant classes of vertices having low output and input degrees (Daudin et al., 2008; Latouche et al., 2010). These classes are usually discarded and the analysis of networks focuses on more highly structured classes to extract useful information. The product of Bernoulli distributions (25.7) provides a natural way to encode these "outliers." Indeed, rather than using giant classes, OSBM uses the null component such that $\mathbf{Z}_i = \mathbf{0}$ if vertex i is an outlier and should not be classified in any class.

Identifiability

As in the case of the SBM, reordering the Q classes of the OSBM and doing the corresponding modification in α and $\tilde{\mathbf{W}}$ does not change the generative random graph model.

There is another family of operations which does not change the generative random graph model, which we call inversions. They correspond to fix a subset $S \subset 1, \ldots, Q$ and to exchange the labels 0 to 1 and vice-versa on the coordinates of the Z_is included in S. To give an intuition, let us consider the inversion with $S = 1$. If we denote by "cluster 1" the vertices whose Z_is have a 1 as the first coordinate, the initial graph sampling procedure consists of sampling the set "cluster 1" and then drawing the edges conditionally on that information. After the inversion, it samples the vertices which are not in "cluster 1" and draws the edges conditionally on that information, which is an equivalent procedure.

As shown in Latouche et al. (2011), the OSBM is generically identifiable up to permutations of the classes and inversions. In other words, except in a set of parameters which has a null Lebesgue's measure, two parameters imply the same random graph model if and only if the second can be obtained from the first by a permutation and an inversion.

Parameter Estimation

The log-likelihood of the observed dataset is defined through the marginalization $p(\mathbf{X} \mid \boldsymbol{\alpha}, \tilde{\mathbf{W}}) = \sum_{\mathbf{Z}} p(\mathbf{X}, \mathbf{Z} \mid \boldsymbol{\alpha}, \tilde{\mathbf{W}})$. This summation involves 2^{NQ} terms and quickly becomes intractable. To tackle this issue, the EM algorithm has been applied on many mixture models. However, the E-step requires the calculation of the posterior distribution $p(\mathbf{Z} \mid \mathbf{X}, \boldsymbol{\alpha}, \tilde{\mathbf{W}})$ which cannot be factorized in the case of networks (Daudin et al., 2008). In order to obtain a tractable procedure, some approximations based on global and local variational techniques have to be done.

The global variational technique consists of considering, for any distribution $q(\mathbf{Z})$, the decomposition

$$\log p(\mathbf{X} \mid \boldsymbol{\alpha}, \tilde{\mathbf{W}}) = \mathcal{L}_{ML}(q;\, \boldsymbol{\alpha}, \tilde{\mathbf{W}}) + \mathrm{KL}\left(q(\cdot) \,\|\, p(\cdot \mid \mathbf{X}, \boldsymbol{\alpha}, \tilde{\mathbf{W}})\right), \qquad (25.8)$$

where

$$\mathcal{L}_{ML}(q;\, \boldsymbol{\alpha}, \tilde{\mathbf{W}}) = \sum_{\mathbf{Z}} q(\mathbf{Z}) \log \left\{ \frac{p(\mathbf{X}, \mathbf{Z} \mid \boldsymbol{\alpha}, \tilde{\mathbf{W}})}{q(\mathbf{Z})} \right\}, \qquad (25.9)$$

and $\mathrm{KL}(\cdot \,\|\, \cdot)$ is the Kullback-Leibler divergence. The maximum $\log p(\mathbf{X} \mid \boldsymbol{\alpha}, \tilde{\mathbf{W}})$ of the lower bound $\mathcal{L}_{ML}$ (25.9) is reached when $q(\mathbf{Z}) = p(\mathbf{Z} \mid \mathbf{X}, \boldsymbol{\alpha}, \tilde{\mathbf{W}})$. Thus, if the posterior distribution $p(\mathbf{Z} \mid \mathbf{X}, \boldsymbol{\alpha}, \tilde{\mathbf{W}})$ was tractable, the optimizations of $\mathcal{L}_{ML}$ and $\log p(\mathbf{X} \mid \boldsymbol{\alpha}, \tilde{\mathbf{W}})$, with respect to $\boldsymbol{\alpha}$ and $\tilde{\mathbf{W}}$, would be equivalent. However, in the case of networks, $p(\mathbf{Z} \mid \mathbf{X}, \boldsymbol{\alpha}, \tilde{\mathbf{W}})$ cannot be calculated, and $\mathcal{L}_{ML}$ cannot be optimized over the entire space of $q(\mathbf{Z})$ distributions. Thus, the optimization is restricted to the class of distributions which satisfy:

$$q(\mathbf{Z}) = \prod_{i=1}^{N} q(\mathbf{Z}_i), \qquad (25.10)$$

with

$$q(\mathbf{Z}_i) = \prod_{q=1}^{Q} \mathrm{Bern}(Z_{iq};\, \tau_{iq}),$$

$$= \prod_{q=1}^{Q} \tau_{iq}^{Z_{iq}} (1 - \tau_{iq})^{1 - Z_{iq}}.$$

Each τ_{iq} is a variational parameter which corresponds to the posterior probability of node i belonging to class q.

This global variational approximation is sufficient to obtain a tractable problem in the case of SBM. Unfortunately, in the case of OSBM, a term $\mathbb{E}_{\mathbf{Z}_i, \mathbf{Z}_j}[\log g(-a_{\mathbf{Z}_i, \mathbf{Z}_j})]$ appears when writing down the complete formula of $\mathcal{L}_{ML}(q)$. Since the logistic sigmoid function is non linear, it cannot be computed analytically. Thus, we need a second level of approximation to optimize the lower bound of the observed dataset. It consists of again considering a lower bound and new parameters such that the bound is tight for the optimal values of the parameters.

More precisely, given a variational parameter ξ_{ij}, $\mathrm{E}_{\mathbf{Z}_i,\mathbf{z}_j}[\log g(-a_{\mathbf{Z}_i,\mathbf{z}_j})]$ satisfies:

$$\mathrm{E}_{\mathbf{Z}_i,\mathbf{z}_j}[\log g(-a_{ij})] \geq \log g(\xi_{ij}) - \frac{(\tilde{\tau}_i{}^\mathsf{T}\tilde{\mathbf{W}}\tilde{\tau}_j + \xi_{ij})}{2} - \lambda(\xi_{ij})\Big(\mathrm{E}_{\mathbf{Z}_i,\mathbf{z}_j}[(\tilde{\mathbf{Z}}_i{}^\mathsf{T}\tilde{\mathbf{W}}\tilde{\mathbf{Z}}_j)^2] - \xi_{ij}^2\Big). \quad (25.11)$$

Eventually, it leads to the two steps approximation:

$$\log p(\mathbf{X}\,|\,\boldsymbol{\alpha}, \tilde{\mathbf{W}}) \geq \mathcal{L}_{ML}(q;\,\boldsymbol{\alpha}, \tilde{\mathbf{W}}) \geq \mathcal{L}_{ML}(q;\,\boldsymbol{\alpha}, \tilde{\mathbf{W}}, \boldsymbol{\xi}). \quad (25.12)$$

The developed expression of $\mathcal{L}_{ML}(q;\,\boldsymbol{\alpha}, \tilde{\mathbf{W}}, \boldsymbol{\xi})$ is then tractable. It can be found in Latouche et al. (2011). The resulting variational EM algorithm (see Algorithm 2) alternatively computes the parameters ξ_{ij}, the posterior probabilities $\boldsymbol{\tau}_i$, and the parameters $\boldsymbol{\alpha}$ and $\tilde{\mathbf{W}}$ maximizing

$$\max_{\boldsymbol{\xi}} \mathcal{L}_{ML}(q;\,\boldsymbol{\alpha}, \tilde{\mathbf{W}}, \boldsymbol{\xi}).$$

Such a procedure is related to the work of Ghahramani and Jordan (1997) and their use of variational approximations to perform inference in factorial hidden Markov models.

```
// INITIALIZATION;
```
Initialize $\boldsymbol{\tau}$ with an ascendant hierarchical classification algorithm;
Sample $\tilde{\mathbf{W}}$ from a zero mean σ^2 spherical Gaussian distribution;

```
// OPTIMIZATION;
```
repeat
    `// ξ-transformation;`
    $\xi_{ij} \leftarrow \sqrt{\mathrm{Tr}\Big(\tilde{\mathbf{W}}^\mathsf{T}\tilde{\mathbf{E}}_i\tilde{\mathbf{W}}\Sigma_j\Big) + \tilde{\tau}_j{}^\mathsf{T}\tilde{\mathbf{W}}^\mathsf{T}\tilde{\mathbf{E}}_i\tilde{\mathbf{W}}\tilde{\tau}_j},\forall i \neq j$;
    `// M-step;`
    $\alpha_q \leftarrow \frac{\sum_{i=1}^N \tau_{iq}}{N},\forall q$;
    Optimize $\mathcal{L}_{ML}\Big(q;\boldsymbol{\alpha}, \tilde{\mathbf{W}}, \boldsymbol{\xi}\Big)$ with respect to $\tilde{\mathbf{W}}$, with a gradient-based optimization algorithm (e.g., quasi-Newton method of Broyden et al., 1970);
    `// E-step;`
    repeat
        for *i=1:N* **do**
            Optimize $\mathcal{L}_{ML}\Big(q;\boldsymbol{\alpha}, \tilde{\mathbf{W}}, \boldsymbol{\xi}\Big)$ with respect to $\boldsymbol{\tau}_i$, with a box constrained ($\tau_{iq} \in [0,1]$) gradient-based optimization algorithm (e.g., Byrd method, Byrd et al., 1995);
        end
    until $\boldsymbol{\tau}$ *converges*;
until $\mathcal{L}_{ML}\Big(q;\boldsymbol{\alpha}, \tilde{\mathbf{W}}, \boldsymbol{\xi}\Big)$ *converges*;

Algorithm 2: Overlapping stochastic blockmodel for directed graphs without self loop.

The computational cost of the algorithm is equal to $O(N^2Q^4)$. For comparison, the computational cost of the methods proposed by Daudin et al. (2008) and Latouche et al. (2010) for (non-overlapping) SBM is equal to $O(N^2Q^2)$. Analyzing a sparse network with 100 nodes takes about ten seconds on a dual core, and about a minute for dense networks.

25.5 Discussion

Clustering aims at summarizing the information of a dataset. When considering graphs, a widespread way of summarizing the set of vertices and edges consists of forming groups of vertices exhibiting similar connectivity patterns. In this chapter, we reviewed different models and algorithms dealing with this kind of clustering problem. The review went from simple to more complex approaches. Each presented approach assumed a particular structure. Although the overlapping structure generalizes simple community models or stochastic blockmodels, it does not mean that it should always be the default choice. Indeed the overlapping stochastic blockmodel has many parameters and may not be as stable as simpler models. Following the Occam's razor principle, preferring the simple model leads often to sounder solutions. This basic statistical remark leads us to state that model choice strategy is an important topic of research worth exploring for practical graph clustering application.

Choosing between models for clustering is usually performed using two kinds of strategies. The strategy tests different types of models and chooses the model maximizing a model choice criterion (Bayesian information criterion, ...) (Kemp and Tenenbaum, 2008). The second strategy explores the model space while estimating the parameters. Developing such strategies for choosing the number of clusters but also for choosing the type of model (SBM, OSBM, ...) would be of interest for the rapidly developing field of graph clustering.

References

Ahuja, R., Magnanti, T., and Orlin, J. (1993). *Network Flows: Theory, Algorithms, and Applications.* Upper Saddle River, NJ: Prentice Hall.

Airoldi, E. M., Blei, D. M., Fienberg, S. E., and Xing, E. P. (2008). Mixed membership stochastic blockmodels. *Journal of Machine Learning Research* 9: 1981–2014.

Albert, R. and Barabási, A. (2002). Statistical mechanics of complex networks. *Modern Physics* 74: 47–97.

Albert, R., Jeong, H., and Barabási, A. (1999). Diameter of the World-Wide Web. *Nature* 401: 130–131.

Albert, R., Jeong, H., and Barabási, A. (2000). Error and attack tolerance of complex networks. *Nature* 406: 378–382.

Allman, E. S., Matias, C., and Rhodes, J. A. (2009). Identifiability of parameters in latent structure models with many observed variables. *Annals of Statistics* 37: 3099–3132.

Allman, E. S., Matias, C., and Rhodes, J. A. (2011). Parameter identifiability in a class of random graph mixture models. *Journal of Statistical Planning and Inference* 141 : 1719–1736.

Amaral, L., Scala, A., Barthélémy, M., and Stanley, H. (2000). Classes of small-world networks. *Proceedings of the National Academy of Sciences* 97: 11149–11152.

Barabási, A. and Albert, R. (1999). Emergence of scaling in random networks. *Science* 286: 509–512.

Barabási, A. and Oltvai, Z. (2004). Network biology: Understanding the cell's functional organization. *Nature Reviews Genetics* 5: 101–113.

Bickel, P. J. and Chen, A. (2009). A non parametric view of network models and Newman-Girvan and other modularities. *Proceedings of the National Academy of Sciences* 106: 21068–21073.

Blei, D. M., Ng, A. Y., and Jordan, M. I. (2003). Latent Dirichlet allocation. *Journal of Machine Learning Research* 3: 993–1022.

Boer, P., Huisman, M., Snijders, T. A. B., Steglich, C. E. G., Wichers, L., and Zeggelink, E. (2006). StOCNET : An open software system for the advanced statistical analysis of social networks. Groningnen: ProGAMMA/ICS, Version 1.7.

Brandes, U. (2001). A faster algorithm for betweenness centrality. *Journal of Mathematical Sociology* 25: 163–177.

Broder, A., Kumar, R., Maghoul, F., Raghavan, P., Rajagopalan, S., Stata, R., Tomkins, A., and Wiener, J. (2000). Graph structure in the web. *Computer Networks* 33: 309–320.

Broyden, C., Fletcher, R., Goldfarb, D., and Shanno, D. (1970). BFGS method. *Journal of the Institute of Mathematics and Its Applications* 6: 76–90.

Byrd, R., Lu, P., Nocedal, J., and Zhu, C. (1995). A limited memory algorithm for bound constrained optimization. *Journal on Scientific and Statistical Computing* 16: 1190–1208.

Cormen, T., Leiserson, C., Rivest, R., and Stein, C. (2001). *Introduction to Algorithms*. Cambridge, MA: The MIT Press.

Danon, L., Diaz-Guilera, A., Duch, J., and Arenas, A. (2005). Comparing community structure identification. *Journal of Statistical Mechanics: Theory and Experiment* 9: P09010.

Daudin, J., Picard, F., and Robin, S. (2008). A mixture model for random graphs. *Statistics and Computing* 18: 1–36.

Dorogovtsev, S., Mendes, J., and Samukhin, A. (2000). Structure of growing networks with preferential linking. *Physical Review Letter* 85: 4633–4636.

Estrada, E. and Rodriguez-Velazquez, J. (2005). Spectral measures of bipartivity in complex networks. *Physical Review E* 72: 046105.

Everett, M. and Borgatti, S. (1998). Analyzing clique overlap. *Connections* 21: 49–61.

Everitt, B. (1974). *Cluster Analysis*. New York, NY: Wiley.

Fienberg, S. E. and Wasserman, S. (1981). Categorical data analysis of single sociometric relations. *Sociological Methodology* 12: 156–192.

Fortunato, S. (2010). Community detection in graphs. *Physics Reports* 3–5: 75–174.

Frank, O. and Harary, F. (1982). Cluster inference by using transitivity indices in empirical graphs. *Journal of the American Statistical Association* 77: 835–840.

Freeman, L. (1977). A set of measures of centrality based upon betweenness. *Sociometry* 40: 35–41.

Ghahramani, Z. and Jordan, M. I. (1997). Factorial hidden Markov models. *Machine Learning* 29: 245–273.

Girvan, M. and Newman, M. E. J. (2002). Community structure in social and biological networks. *Proceedings of the National Academy of Sciences* 99: 7821–7826.

Handcock, M. S., Raftery, A. E., and Tantrum, J. M. (2007). Model-based clustering for social networks. *Journal of the Royal Statistical Society* 170: 1–22.

Hoff, P. D., Raftery, A. E., and Handcock, M. S. (2002). Latent space approaches to social network analysis. *Journal of the Royal Statistical Society* 97: 1090–1098.

Hofman, J. and Wiggins, C. (2008). A Bayesian approach to network modularity. *Physical Review Letters* 100: 258701.

Holland, P. W., Laskey, K. B., and Leinhardt, S. (1983). Stochastic blockmodels: Some first steps. *Social Networks* 5: 109–137.

Jeffery, C. (1999). Moonlighting proteins. *Trends in Biochemical Sciences* 24: 8–11.

Kemp, C. and Tenenbaum, J. B. (2008). The discovery of structural form. *Proceedings of the National Academy of Sciences* 105: 10687–10692.

Krivitsky, P., Handcock, M. S., Raftery, A. E., and Hoff, P. D. (2009). Representing degree distributions, clustering, and homophily in social networks with latent cluster random effects models. *Social Networks* 31: 204–213.

Lacroix, V., Fernandes, C., and Sagot, M. -F. (2006). Motif search in graphs: Application to metabolic networks. *Transactions in Computational Biology and Bioinformatics* 3: 360–368.

Latouche, P., Birmelé, E., and Ambroise, C. (2010). Bayesian methods for graph clustering. In Fink, A., Lausen, B., Seidel, W., and Ultsch, A. (eds), *Advances in Data Analysis, Data Handling, and Business Intelligence*, Studies in Classification, Data Analysis, and Knowledge Organization. Berlin Heidelberg: Springer-Verlag. 229–239.

Latouche, P., Birmelé, E., and Ambroise, C. (2011). Overlapping stochastic block model with application to the French political blogosphere. *Annals of Applied Statistics* 5: 309–336.

Mariadassou, M., Robin, S., and Vacher, C. (2010). Uncovering latent structure in valued graphs: A variational approach. *Annals of Applied Statistics* 4: 715–742.

Milo, R., Shen-Orr, S., Itzkovitz, S., Kashtan, D., Chklovskii, D. B., and Alon, U. (2002). Network motifs: Simple building blocks of complex networks. *Science* 298: 824–827.

Moreno, J. (1934). *Who shall survive?: A New Approach to the Problem of Human Interrelations.* Washington D.C.: Nervous and Mental Disease Publishing.

Newman, M. E. J. (2001). Scientific collaboration networks: II. Shortest paths, weighted networks, and centrality. *Physical Review E* 64: 016132.

Newman, M. E. J. (2003). Mixing patterns in networks. *Physical Review E* 67: 026126.

Newman, M. E. J. (2004). Fast algorithm for detecting community structure in networks. *Physical Review Letter* 69.

Newman, M. E. J. and Girvan, M. (2004). Finding and evaluating community structure in networks. *Physical Review E* 69: 026113.

Newman, M. E. J. and Leicht, E. (2007). Mixture models and exploratory analysis in networks. *Proceedings of the National Academy of Sciences* 104: 9564–9569.

Nowicki, K. and Snijders, T. A. B. (2001). Estimation and prediction for stochastic blockstructures. *Journal of the American Statistical Association* 96: 1077–1087.

Palla, G., Barabási, A., and Vicsek, T. (2007). Quantifying social group evolution. *Nature* 446: 664–667.

Palla, G., Derenyi, I., Farkas, I., and Vicsek, T. (2005). Uncovering the overlapping community structure of complex networks in nature and society. *Nature* 435: 814–818.

Palla, G., Derenyi, I., Farkas, I., and Vicsek, T. (2006). CFinder, the community cluster finding program. Version 2.0.1.

Pastor-Satorras, R. and Vespignani, A. (2001). Epidemic spreading in scale-free networks. *Physical Review Letters* 86: 3200–3203.

Sampson, S. (1969). Crisis in a Cloister. Ph.D. thesis, Cornell University, Ithaca, New York, USA.

Scott, J. G. (2000). *Social Network Analysis: A Handbook*. London: Sage publications.

Shen, H., Cheng, X., Cai, K., and Hu, M. (2009). Detect overlapping and hierarchical structure in networks. *Physica A* : 1706–1712.

Snijders, T. A. B. and Nowicki, K. (1997). Estimation and prediction for stochastic block-structures for graphs with latent block structure. *Journal of Classification* 14: 75–100.

Strogatz, S. (2001). Exploring complex networks. *Nature* 410: 268–276.

Watts, D. and Strogatz, S. (1998). Collective dynamics of small-world networks. *Nature* 393: 440–442.

White, H., Boorman, S., and Breiger, R. (1976). Social structure from multiple networks. I. Block-models of roles and positions. *American Journal of Sociology* 81: 730–780.

Zanghi, H., Ambroise, C., and Miele, V. (2008). Fast online graph clustering via Erdös-Rényi mixture. *Pattern Recognition* 41: 3592–3599.

Zanghi, H., Volant, S., and Ambroise, C. (2010). Clustering based on random graph model embedding vertex features. *Pattern Recognition Letters* 31: 830–836.

Subject Index

Author Index

Printed and bound by CPI Group (UK) Ltd, Croydon, CR0 4YY

23/10/2024

01778257-0007